Complex Dynamics

Complex Dynamics

Families and Friends

Edited by
Dierk Schleicher

 CRC Press
Taylor & Francis Group
Boca Raton London New York

CRC Press is an imprint of the
Taylor & Francis Group, an **informa** business

AN A K PETERS BOOK

Editorial, Sales, and Customer Service Office

CRC Press
Taylor & Francis Group
6000 Broken Sound Parkway NW, Suite 300
Boca Raton, FL 33487-2742

First issued in paperback 2019

ISBN-13: 978-1-56881-450-6 (hbk)
ISBN-13: 978-0-367-38494-4 (pbk)

Library of Congress Cataloging-in-Publication Data

Complex dynamics : families and friends / edited by Dierk Schleicher.
   p. cm.
Includes bibliographical references.
ISBN 978-1-56881-450-6 (alk. paper)
  1. Holomorphic mappings. 2. Polynomials. 3. Dynamics. I. Schleicher,
Dierk, 1965-

QA331.C65346 2009
510–dc22

                                                    2008048528

In honor of John Hamal Hubbard

Contents

Foreword
Jean-Christophe Yoccoz

I first met John Hubbard in the spring of 1977 in Orsay, at the seminar dedicated that year to Thurston's fundamental work on Teichmüller spaces and surface diffeomorphisms. I was a second-year student at the Ecole Normale and still unsure about which kind of mathematics I wanted to do: I had attended an algebraic topology seminar the previous fall and would meet Michael Herman that same spring. On Jean Cerf's recommendation, I went to the seminar on Thurston's work, and it made a very strong and lasting impression on me.

The discussions between the participants — besides John, I remember best Adrien Douady, Albert Fathi, François Laudenbach, Valentin Poénaru, Larry Siebenmann, Mike Shub, and Dennis Sullivan — were extremely lively; still used to the dry reasoning of the "grandes écoles" entrance exams, I discovered another way to do mathematics, in which geometric intuition and vision play a central role.

In the preface to the book *The Mandelbrot Set, Theme and Variations* (edited by Tan Lei), John describes his first brush with holomorphic dynamics in the academic year 1976–77 in Orsay: "I was no computer whiz: at the time I was a complex analyst with a strong topological bent, with no knowledge of dynamical systems."

No computer whiz? John has been, since I started to meet him on a regular basis in the mid-80s, the mathematician in my fields of interest who made the most impressive use of computers, discovering new phenomena and suggesting the right statements and the way to prove them.

Analyst? In my book, an analyst is somebody who spends most of his time navigating through a sea of inequalities; I don't think that inequalities are John's preferred cup of tea. I see him above all as a geometer with a strong feeling for combinatorial structures: think of Branner-Hubbard tableaux or Hubbard trees or, more globally, of all of his work with Adrien Douady on the iteration of polynomials in the Orsay lecture notes.

Complex? Definitely! Genetically speaking, there are two subspecies in the holomorphic dynamics community: mathematicians who do *holomorphic* dynamics and mathematicians who do holomorphic *dynamics*. John definitely belongs to the first group, and I belong to the second. The same phenomenon occurs in another of our common fields of interest (and we are back to the subject of the Orsay/Thurston seminar), namely translation surfaces: holomorphic people like John see them as Riemann surfaces with a holomorphic 1-form. From my point of view, a translation surface is a simpler object than a Riemann surface, in the same way than a translation is a simpler object than a biholomorphism.

No knowledge of dynamical systems? This might have been true in 1977, but he has been doing his homework since! I still remember a seminar in Orsay in the late 80s or early 90s, when John was giving one of his first talks on holomorphic dynamics in higher dimensions, and I got very irritated as he claimed that it would be a shame to fail to understand the dynamics of the Hénon family for all complex parameters. What about Newhouse's phenomenon of infinitely many sinks? But such is his infectious enthusiasm for all kinds of mathematics that he is easily absolved.

Obviously, it is impossible to write about John without mentioning his collaboration with Adrien Douady. It is difficult to overestimate the influence that the Orsay Lecture notes and their paper on polynomial-like mappings have had on the whole subject. More than 20 years later, I still find the consequences of the straightening theorem for polynomial-like mappings absolutely baffling: for instance, take a quadratic polynomial with a Cremer-type fixed point at the origin; although one does not understand its dynamics on the Julia set in this case, one still knows from the straightening theorem that the dynamics will remain the same if you add a small cubic term. There was a very special kind of chemistry between Adrien and John that led to extraordinary results, and one cannot help but think of other celebrated pairs like Hardy and Littlewood.

My own debt to John is considerable. I have always found it hard to translate mathematics — amongst other things — from the mind to paper; on the other hand, he writes fluently and elegantly. This explains at least partially why, after waiting for several years for a text about the local connectivity results that I proved around 1990, he got impatient and wrote himself a text for Milnor's 60th birthday conference.

Another admirable feature of John's mathematical persona is the energy and time and generosity that he has spent in advising Ph.D. students and helping young mathematicians. Generically speaking, it seems that mathematicians are more concerned with that aspect of scientific activity than our senior colleagues in the experimental sciences. But even among mathematicians, John is really exceptional in this respect.[1]

John's mathematics has consistently been elegant, deep, and original. I hope that the papers in this volume are on the same level and that you will enjoy reading them.

---

[1] Editor's note: the editor of this volume, who was a Ph.D. student of John Hubbard, is in full agreement.

Introduction
Dierk Schleicher

Complex Dynamics: Families and Friends is a book on the mathematical theory of the dynamics of iterated holomorphic maps in one or several complex dimensions. In this theory, the iterated maps that one studies usually come in *families:* much of the pioneering research was done on the family of quadratic polynomials, often affectionately called "the quadratic family" and parametrized in the form $z \mapsto z^2 + c$ with a complex parameter c. A lot of further research has been focused on the exponential family, that is, the family of maps $z \mapsto \lambda e^z$ with a non-zero complex parameter λ. And whenever someone is working on different iterated holomorphic maps, the explanation often begins with the words "we study the following family of maps...".

Of course, some maps are quite prototypical, they are studied in different contexts, and their characteristic dynamical features often appear in a variety of different investigations (think of the "Douady rabbit"). It is no surprise that in many scientific presentations, the pioneers in the field discover some of the prototypes they had been studying for many years. One of these pioneers is John Hamal Hubbard. During such presentations, he would often exclaim: "this map is an *old friend* of mine!"

In this sense, this is a book about *families and friends* in complex dynamics. Many chapters are studies on selected families of iterated maps in complex dynamics and on particular "old friend" mappings.

The title of the book is also an allusion to another aspect. Complex dynamics is a broad field with many connections to other branches of mathematics and physics, including (hyperbolic) geometry, number theory, group theory, combinatorics, general dynamics, numerical analysis, renormalization theory, and many others. It is fair to say that these fields are friends of complex dynamics; there is a fruitful interplay between these fields, in the sense that good friends learn a lot from each other. Some of the contributions in this book relate more to these friends than to complex dynamics in a strict sense. A particularly good friend of complex dynamics is hyperbolic geometry — in view of Sullivan's dictionary, these are almost like twin brothers (the last chapter, for instance, by Antolín-Camarena, Maloney, and Roeder, discusses some aspects of hyperbolic geometry with relations to number theory).

Mathematical friendships are quite a dynamical object in their own right: especially the contributions in the last part, "Making New Friends", illustrate the fact that interesting research explores new areas and builds new relations. Complex dynamics acquires new friends. It also turns out that also within the field of complex dynamics, certain topics that were at some point seen as obstacles, perhaps even as enemies that needed to be overcome, could develop into good and interesting friends, the more so they better we know them. A very personal story to this effect is the chapter by Chéritat on the discovery of Julia sets of positive measure within the quadratic family.

This book brings together many aspects in the active and exciting field of complex dynamics. It includes some "classics", previously unpublished, that describe fundamental aspects of the field and that have given direction to the field over the years — see the chapters by Thurston, Shishikura, and Milnor. The other chapters contain aspects of current cutting-edge research, including some on mathematics in the making.

The organization of the book into four parts illustrates the logic and development of the field of complex dynamics. A lot of early work focused on the study of polynomial iteration, especially the prototypical family of quadratic polynomials. It remains an active research topic, the topological and geometric study of the Mandelbrot set being a key topic in the field. The corresponding chapters are collected in the first part, "Polynomial Dynamics from Combinatorics to Topology".

The opening chapter is a classical manuscript by William Thurston, "On the Geometry and Dynamics of Iterated Rational Maps", that has been circulating for more than two decades but never published. It consists of two parts. The first part is an introduction into the field of complex dynamics and helps interested readers get into the field and get used to its terminology. The second part of Thurston's chapter describes his classical theory of invariant laminations and especially his "quadratic minor lamination"; a theory that has inspired a lot of subsequent research and that is still being actively used today. It is closely related to the pioneering work of Douady and Hubbard in their "Orsay Notes". In particular, Thurston proved that quadratic polynomials don't have what's now called "Wandering Triangles" (three non-periodic dynamic rays landing at a common point), and he raised the question of whether an analogous result was true for polynomials of higher degrees. He writes, "The question whether this generalizes to higher degrees, and if not what is the nature of the counterexamples, seems to me the key unresolved issue about invariant laminations." There is also a new appendix to Thurston's manuscript, written by the editor of this volume, that describes the link between Thurston's theory of quadratic laminations and the classical theory of polynomial dynamics: quadratic laminations had been invented to described the topology of polynomial Julia sets and of the Mandelbrot set. (Thurston's original manuscript also contained a fragment of a third part; we do not publish it here in the hope that one day this fragment will be completed. The third part was concerned with a theorem of Thurston that is now sometimes called the "Fundamental Theorem of Complex Dynamics"; a proof of this theorem has meanwhile been published by Douady and Hubbard.)

Thurston's question on the existence of wandering triangles was answered more than 20 years later by Alexander Blokh and Lex Oversteegen: they proved quite recently that wandering triangles do exist (quite abundantly!) for polynomials of degree at least three. Chapter 2 in this book is their original publication of this result, "Wandering Gaps for Weakly Hyperbolic Polynomials".

Chapter 3 is a nod to Douady and Hubbard's "Orsay Notes": in "Combinatorics of Polynomial Iterations", Volodymyr Nekrashevych encodes the topological-combinatorial data associated with an iterated complex polynomial in a discrete group called the "iterated monodromy group". This algebraic approach has already led to important progress in the field, in particular to a solution to Hubbard's long-standing "twisted rabbit" problem.

In a series of seminal papers on the dynamics of the family of cubic polynomials from the 1990s, Bodil Branner and John Hubbard stated a conjecture on point components in disconnected polynomial Julia sets. This conjecture has recently been solved affirmatively by Kozlovski, Shen, and van Strien. Some of the new key ideas of this work are discussed by Tan Lei and Yin Yongcheng in Chapter 4 in the simpler context of unicritical polynomials (those polynomials having only a single finite critical point of possibly higher multiplicity): "The Unicritical Branner-Hubbard Conjecture".

One of the classical and fundamental conjectures in complex dynamics is the "local connectivity of the Mandelbrot set". This conjecture goes back to the early work of Adrien Douady and John Hubbard in their classical Orsay Notes; it is equally closely related to Thurston's Quadratic Minor Lamination described in Chapter 1. In Chapter 5, "A Priori Bounds For Some Infinitely Renormalizable Quadratics, III: Molecules", Jeremy Kahn and Mikhail Lyubich publish recent progress towards this conjecture, establishing new results on points where the Mandelbrot is locally connected.

Within the field of complex dynamics, polynomial iteration plays a fundamental role for a number of reasons. One is that the study is substantially simplified by the invariant basin of infinity (compare for instance Part II in Thurston's chapter). Another is related to universality and renormalization: results on iterated polynomials are prototypical, in a precise sense, for results in far greater generality. Among the pioneers for the mathematical theory behind renormalization and universality were Adrien Douady

Figure 0.1. The mathematical *family tree* of John Hubbard in five generations, each person being the former Ph.D. advisor of the next: Henri Cartan, Adrien Douady, John Hubbard, Dierk Schleicher, Alexandra Kaffl (far left: the wife of Henri Cartan). The picture was taken in June 2004 during the celebration of Henri Cartan's 100th birthday at the Ecole Normale Superieure in Paris (courtesy of François Tisseyre). (See Color Plate I.)

and John Hubbard with their theory of polynomial-like mappings. Further important aspects of one-dimensional complex dynamics can be found in the second part, "Beyond Polynomials: Rational and Transcendental Dynamics".

This part starts with another classic: "The Connectivity of the Julia Set and Fixed Points" by Mitsuhiro Shishikura (Chapter 6). This chapter goes back to a manuscript that has been circulating for some two decades and that has inspired the field significantly. Among the classical applications of this result is the fact that Newton maps of complex polynomials always have connected Julia sets; this is an important result in its own right.

Next comes a recent study of a particular family of rational maps with two critical points: "The Rabbit and Other Julia Sets Wrapped in Sierpiński Carpets" by Paul Blanchard, Robert Devaney, Antonio Garijo, Sebastian Marotta, and Elizabeth Russell. The title already indicates that

Figure 0.2. *Family and friends* of complex dynamics (picture taken at the conference in honor of John Hubbard on the occasion of his 60th birthday, June 2005 in Luminy/France). (See Color Plate I.)

several old friends, such as Douady's rabbit (from the family of quadratic polynomials), can be found in many other families of maps, for instance those studied in Chapter 7 and involving Sierpiński curves.

Chapter 8, "The Teichmüller Space of an Entire Function" by Núria Fagella and Christian Henriksen, brings together two important aspects of complex dynamics: transcendental iteration and Teichmüller theory. It has been known for a long time how useful Teichmüller theory can be for complex dynamics. There is by now quite a good deal of knowledge on this topic in the case of rational maps. The transcendental case is discussed in this chapter.

A current trend in complex dynamics is an increase in attention paid to several complex dimensions. This trend is reflected in the third part, "Two Complex Dimensions". Even in one-dimensional complex dynamics, two (or more) complex dimensions arise when maps come from a family of maps that depend on two or more complex parameters (such as the family of cubic polynomials); but of equal interest is the case when the dynamics takes place in several complex dimensions. The opening chapter in this part is again a classic: "Cubic Polynomial Maps with Periodic Critical Orbit, Part I" by John Milnor (Chapter 9). Up to affine conjugation, cubic polynomials depend on two complex parameters; another way of saying this is that a cubic polynomial has two independent critical points. This is a revised version of a manuscript that has been circulating for almost two decades. The study of complex two-dimensional parameter space is accomplished in terms of one-dimensional subspaces corresponding to maps with superattracting orbits of various periods. In the years since this manuscript was originally circulated, it has had a strong impact on a number of studies with related methods.

The subsequent Chapter 10, "Analytic Coordinates Recording Cubic Dynamics" by Carsten Petersen and Tan Lei, is another study of the space of cubic polynomials. The focus in Chapter 10 is on hyperbolic components (which are complex two-dimensional) and their boundary properties, a topic that is known to be notoriously difficult in the presence of two or more independent critical orbits.

Chapter 11, "Cubic Polynomials: A Measurable View of Parameter Space" by Romain Dujardin, is a study of the complex two-dimensional parameter space of cubic polynomials from the point of view of measure theory and complex analysis — nicely complementing the more topological classical studies by Branner and Hubbard.

Xavier Buff and Adam Epstein, in their work "Bifurcation Measure and Postcritically Finite Rational Maps" (Chapter 12), study the parameter space of rational maps (in one variable) of given degree d; this is a complex $(2d - 2)$-dimensional space. Like the previous chapter, this is an investigation from the point of view of measure theory: this measure is defined by the distribution of postcritically finite maps in parameter space.

Part three concludes with Chapter 13 on iteration theory in two dynamical variables: "Real Dynamics of a Family of Plane Birational Maps: Trapping Regions and Entropy Zero" by Eric Bedford and Jeffrey Diller. The restriction to two real variable brings the dynamics back into the plane, but it is really the real slice of a complex two-dimensional system.

The final part of the book, "Making New Friends", highlights the fact that mathematics is not a finished subject, but that mathematical progress happens because people make it happen.

In Chapter 14, "The Hunt for Julia Sets With Positive Measure", Arnaud Chéritat gives his very personal insight into the many-year quest for finding polynomial Julia sets of positive measure. The possibility of such Julia sets was sometimes seen as an obstacle towards proving density of hyperbolicity in the quadratic family, and thus as an enemy to combat. But the studies of Buff, Chéritat, and others helped us gain an understanding that ultimately turned Julia sets of positive measure into much more familiar objects. It is quite fitting that the final steps in the proof were made during the 2005 Luminy conference that Adrien Douady and I organized in honor of John Hubbard. We can almost hear him exclaim, upon seeing these examples, "these maps are *new friends* of mine!"

The next two chapters are on Thurston's theorem on complex dynamics. This theorem was originally announced in Part III of Thurston's manuscript on the geometry and dynamics of iterated rational maps (of which Parts I and II form Chapter 1 of this book). Even though Thurston's theorem is by now a classical result, it continues to inspire a lot of current work. In Chapter 15, "On Thurston's Pullback Map", Xavier Buff, Adam Epstein, Sarah Koch, and Kevin Pilgrim discuss the mapping properties of the map that is nowadays called "Thurston's σ-mapping"; they describe interesting features of the σ-map, including the recent surprising discovery of Curt McMullen that this map may be constant.

The σ-mapping acts on Teichmüller space. In Chapter 16, "On the Boundary Behavior of Thurston's Pullback Map", Nikita Selinger investigates the question of how this map can be extended to the boundary of Teichmüller space. This allows one to view Thurston's theorem in a different light, perhaps in a more geometric spirit (which may well be closer to what Thurston had originally envisioned).

The final chapter, Chapter 17, is on hyperbolic geometry, a close friend of complex dynamics. In "Computing Arithmetic Invariants for Hyperbolic Reflection Groups", Omar Antolín-Camarena, Gregory Maloney, and Roland Roeder investigate hyperbolic polyhedra generated by reflection groups. This is a study with interesting links to number theory, and with a strong experimental flavor: a good example of *making new friends*.

Editor's Remark. All chapters in this book are fully refereed; they are original research articles (except Chapter 16, which is a research announcement). An attempt was made to format them in a coherent way. However, some exceptions were made for those articles (by Thurston, Shishikura, and Milnor) that have been circulating for two decades or more: for these, we tried to keep the numbering as close to the originals as possible, sometimes at the expense of consistency within this book. Moreover, Thurston's chapter differs from the other chapters in a number of respects: it is the longest, it contains a part that is an introduction to complex dynamics, and it introduces a number of specific definitions. Therefore, in the interest of the reader, we have also treated this chapter in a special way by endowing it with a separate index.

Cover Art. The art on the front cover shows the parameter space of the family of Newton maps associated to the equation $\sin z = c$, where c is a complex parameter (courtesy of Matt Noonan). The corresponding Newton maps $N_c\colon z \mapsto z - (\sin z - c)/\cos z$ have (up to symmetries $z \mapsto z + 2\pi n$ for integers n) two "free" (non-fixed) critical points at $z = 0$ and $z = \pi$. Moreover, the Newton dynamics is invariant under the involution $z \mapsto \pi - z$ that relates the orbit of the two critical points, so all critical orbits are related by the dynamics. If N_c has an attracting periodic cycle other than the basins of the roots, then all free critical points converge to such a cycle. Such values of c in parameter space are colored black: they are organized by infinitely many homeomorphic copies of the Mandelbrot set (this follows from the fundamental theory of polynomial-like maps and renormalization by Adrien Douady and John Hubbard). Different colors indicate parameters c for which the critical point $z = 0$ converges to different solutions of $\sin z = c$. Different shades of the same color indicate different speeds of convergence. In the cover pictures, the complex c-plane is visualized as a Riemann sphere: all spheres show the same parameter space, rotated differently. Of course, the "starry" cover art comes from the imagination of an artist. This is not inappropriate for a book on dynamical systems, as celestial mechanics is a prime example of a dynamical system, one that has always inspired the study of dynamical systems.

The three pictures on the back illustrate some of John Hubbard's closest mathematical friends. The top two pictures are one-dimensional slices of parameter space for complex Hénon maps, with parameters chosen so that these maps are perturbations of quadratic polynomials. The top picture shows a perturbation of the Mandelbrot set, while the middle picture features the "fingering" phenomenon that also arises in various slices of cubic polynomials. These pictures were made with the help of the "SaddleDrop" program by Karl Papadantonakis. The third picture on the back shows a detail of the parameter space of Newton maps for cubic polynomials, illustrating the fact that the set of cubic polynomials with Newton maps having attracting cycles of periods greater than 1 is organized in the form of countably many homeomorphic copies of the Mandelbrot set.

Dedication. This book is dedicated to John Hamal Hubbard, the inspiring mathematician, teacher, colleague, co-founder of a mathematical school (a "research family"), and friend of many of us. Scope and style of the book were chosen following his inspirations. Most, if not all, of the contributors to the book were greatly inspired by his mathematical ideas and leadership. The selection of topics illustrates some (but by far not all) of his mathematical interests, including iteration theory in one and several complex variables, Teichmüller theory, and hyperbolic geometry. It is no coincidence that this book has a lot of color pictures; this pays tribute to the fact that John Hubbard was perhaps *the* pioneer of illustrating the theory with color pictures, and of getting the research inspired by computer experiments. For the same reason, it is quite appropriate that several of the articles have an exploratory character; see especially the last part, "Making New Friends".

The mathematical school that was jointly founded by Adrien Douady and John Hubbard has long grown into a large family with many friends. Many special semesters, such as those at the IHP in Paris in 2003 or at the Fields Institute in Toronto in 2006, almost had the character of family reunions. At both institutes, the staff — who run many such special semesters in many fields — remarked that the complex dynamics semesters had been especially lively and interactive. This has everything to do with the spirit of our mathematical family and the spirit of their founders.

In 2005, Adrien Douady and I organized a conference in Luminy in honor of John Hubbard's 60th birthday. During the organization of this conference, it was a wonderful experience to see how many friends John Hubbard has who were eager to contribute and to help, and who said they were pleased to pay back some of the support and friendship that Hamal (as his friends often call him) had offered them over the years. It was also during this Luminy conference that the plan to edit this book was launched.

Acknowledgments. Editing this book was a positive and inspiring experience, and it is a great pleasure to thank many people. First of all, this book owes everything to Adrien Douady. We had discussed the early plans of the book, and he helped me accept the challenge of engaging in this project. Working on it now that he is no longer around is different in a very sad way, and he is deeply missed by all of us. His influence and spirit shines through in much of the research in this field, and this book — the whole of it as well as the individual contributions — is certainly no exception.

The help of Misha Lyubich and Sebastian van Strien was crucial, especially in the early stages of planning the book; they helped me believe that the task could actually be accomplished. This wouldn't have happened without fresh ideas and constant support, in many ways, by Xavier Buff and Sarah Koch. Tan Lei shared her editorial experience with me more than once. Silvio Levy and John Smillie helped with numerous logistical aspects of Thurston's chapter. Several other contributors, in particular John Milnor, offered advice and encouragement along the way. Matt Noonan produced lots of beautiful pictures for the book cover, always ready to try yet another improvement after the previous ones already seemed almost perfect. Several of my colleagues, including Laurent Bartholdi, Walter Bergweiler, and Lasse Rempe, offered help and advice on various issues. Numerous anonymous referees helped improve the mathematics and the exposition of the individual chapters. Barbara Hubbard was always helpful when I needed advice or had questions, large or small. And Alice and Klaus Peters, as well as Ellen Ulyanova, offered useful advice on a number of topics. I would like to express my sincere gratitude to all of them. I apologize to those who helped over the years but whose names I forgot to mention here.

All of these people are members of the extended *complex family* of John Hubbard and of complex dynamics. But, of course, John Hubbard also has a *real family*: Barbara, Alex, Eleanor, Judith, and Diana; they have treated so many of us as their family members over all these years, and I would like to take this opportunity to thank them deeply for everything.

Anke Allner helped greatly and tirelessly with the LaTeX aspects of the book, assisting me in giving the various contributions a consistent look and in compiling many versions of all chapters until the formatting pleased both of us. Jan Cannizzo edited the English in those chapters that were written by non-native speakers, and also helped format the bibliographies in a consistent style. I am deeply thankful to both of them. I would also like to thank my colleagues and students for their patience when I was at times distracted from my other duties.

And, last but not least, I would like to thank my own family and friends, especially Anke and Diego, for their encouragement, their understanding, their support, and for so much more.

I. Polynomial Dynamics from Combinatorics to Topology

1   On the Geometry and Dynamics of Iterated Rational Maps

William P. Thurston

edited by Dierk Schleicher and Nikita Selinger

with an appendix by Dierk Schleicher

Introduction

Most of us are first introduced to functions such as $f(z) = z^2 - c$ as static objects. We would think of this function as describing a parabola. From the static point of view, it seems that we have learned about all there is to know about a quadratic function after we have finished a year of high school algebra.

There is another point of view which has gained considerable attention with the dawn of the age of easy access to computational power: the dynamical point of view. The dynamical aspect of functions appeals to people as a form of recreation, but it also has a good deal of theoretical and practical interest.

One common way people encounter functions as dynamical systems is in playing with a calculator. At a certain stage of boredom, a person is apt to begin hitting buttons in a physical pattern without any particular mathematical rationale, watching the display for its hypnotic effect. For instance, if you hit the sine button repeatedly, the display tends toward 0, although the convergence is quite slow if the units are radians. The effect of repeatedly hitting the cosine button if the units are radians is to converge to .739081133.

After exhausting single buttons, there is a whole world which can be explored by hitting buttons in repeated sequences.

The first few examples in this exercise tempt people to think that the behavior is always somewhat boring, and that no matter what sequence of buttons they hit, the display will converge to some fixed result. People who get as far as iterating a quadratic polynomial (which can be done with a few keystrokes) are in for a surprise: the phenomena are quite intricate. The

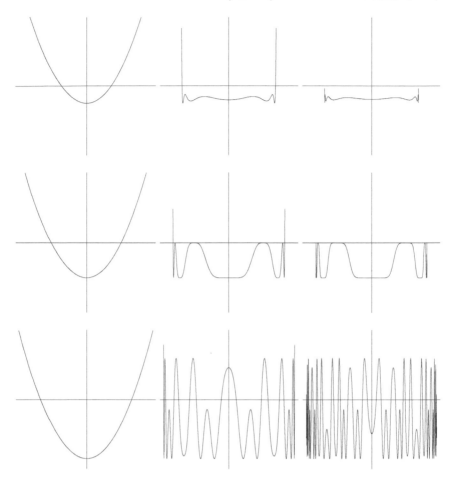

Figure 0.1. Graphs of assorted iterates of assorted quadratic polynomials.

display sometimes exhibits periodic behavior with unusual periods, and sometimes never settles down at all but goes through a seemingly random sequence of numbers.

For the polynomial $f(x) = x^2 - c$, the most interesting behavior is when c is in the range -0.25 to 2. When c is less than -0.25, every point on the real line shoots rapidly to $+\infty$; when c is greater than 2, almost every point shoots rapidly to $+\infty$, all except a Cantor set of points which remain bounded.

If you are unacquainted with the dynamical behavior of this function, it is worth getting out a calculator or computer to experiment. There are many simple programs which yield interesting results.

For $c = 0$, the origin is an attracting fixed point. As we shall establish later, many of the dynamic characteristics for any particular value of c are revealed by the the trajectory of 0, which is the critical point of this function. When c is increased, 0 is attracted to a fixed point whose location gradually decreases (moves to the left), and whose power of attraction gradually wanes. Finally, the fixed point loses all its attractive power, and at that moment gives birth to two nearby points which are interchanged by f. These two points constitute a periodic orbit of period two, which now attracts 0 as well as all nearby points. At first, the orbit attracts only weakly, but its attraction increases with c.

When c increases still further, the orbit of period two in turn begins to lose its attractive power, finally losing it altogether. At this moment, it gives birth to a new nearby periodic orbit of period four, whose attractive power waxes and then wanes. The process of period doubling continues, until a limiting case is reached where there is no attracting periodic point at all.

The interesting phenomena by no means stop at this limiting value for c, which is often called the Feigenbaum point. As c increases further, the behavior sometimes has a chaotic appearance, but occasionally there are windows (that is, intervals of values of c) for which most points tend toward a periodic orbit whose period is 3, or 4 or any other integer bigger than 2.

The general pattern of behavior as c varies is displayed in Figure 0.2, which can be easily generated (probably to a lower resolution) on a home computer.

This simply defined dynamical system has a great deal of intricate structure. It is complicated but many things about it are now well understood. The topic we will investigate is the complex version of functions such as $f(z) = z^2 - c$. It turns out that the dynamical analysis even of real polynomials benefits greatly by considering them in the wider context of complex polynomials, acting as maps of $\mathbb{C}$ to itself.

Figure 0.3 illustrates an example of dynamical behavior for complex quadratic polynomials: the complex plane is divided into points which remain bounded vs. points which eventually shoot off to infinity, for a certain choice of c.

One widely used example of a dynamical system is Newton's method. This is the technique for finding solutions to an equation $g(x) = 0$ by iterating the transformation

$$G(x) = x - \frac{g(x)}{g'(x)},$$

while hoping that it converges. When Newton's method is applied to the problem of finding the roots of a polynomial the transformation is a rational function. As we know, Newton's method converges rapidly to a

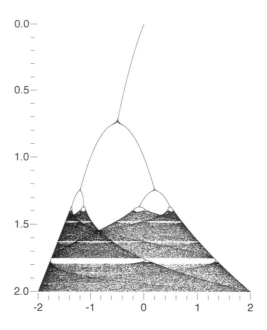

Figure 0.2. This figure illustrates what happens as the parameter c is varied for the map $f(x) = x^2 - c$. Each horizontal line in the figure describes the behavior for one value of c, with $c = 0$ at the top and $c = 2$ at the bottom. On each horizontal line, dots are plotted for the 100th to 200th images of 0 under iterations of f. Certain horizontal lines appear to intersect the figure in only finitely many points. These correspond to parameter values where there is an attracting periodic cycle. It turns out that there can be at most one such cycle. At other levels, the horizontal line intersects in a very large set.

root under good conditions: in particular, it works well when the initial guess is close to a simple root. Under other circumstances, convergence may happen slowly or not at all. For example, it sometimes happens that Newton's method has an attracting periodic orbit of period, say, three. In such a case, there are open sets of initial guesses leading to sequences of successive guesses which are doomed never to converge to a solution. In order to understand Newton's method as a global method, rather than just a local method which finds precise solutions once an approximate solution is known, it makes sense to understand a bit of the theory of iteration of rational functions (compare [HSS01, Rü08]).

A dynamical system is broadly defined as the action of a group or semigroup on a space: typically $\mathbb{R}$ (for an ordinary differential equation),

Figure 0.3. The quadratic polynomial $f(z) = z^2 - .5 - .6i$ has an attracting periodic orbit of period 5. As the polynomial is iterated, any point which is far from the origin shoots out rapidly to infinity, but there is a puddle around 0 which stays bounded forever. In this figure, the points which remain forever bounded under iterations of f are darkened.

$\mathbb{Z}$ (for an iterated homeomorphism), or the natural numbers $\mathbb{N}$ (for an iterated map which is not invertible). Dynamical systems are pervasive in mathematics and in the sciences — they describe the weather, the economy, geodesics on a Riemannian manifold... The only question a person might have about the value of the theory of dynamical systems is whether there is any theory which helps with a typical dynamical system. Is the behavior just very particular, so that the only reasonable method of looking at it is numerical integration or numerical iteration?

Rational polynomials are defined in a very simple way, as dynamical systems go, but their behavior is quite intricate — at first, they seem very unruly, not subject to any theory. Yet an amazing amount of structure and theory has been discovered for them. This gives hope that the theory of dynamical systems still has room to grow in value with typical difficult examples.

Another subject which has many parallels with the theory of iterated rational maps is the theory of Kleinian groups. Kleinian groups are discrete groups acting by Möbius transformations on the sphere. They are the same as fundamental groups of hyperbolic three-manifolds, not necessarily compact. Like rational maps, they also are one-complex-dimensional dynamical systems. Dennis Sullivan in particular has analyzed and exploited the relationships between the two subjects to find new theorems in both [S85a, S85b, MS98, S86]. This relationship was the main motivation for my interest in complex rational maps. Another motivation was my previous work with John Milnor on iterated maps of the unit interval [MT88], which is an independent but closely related theory. A third motivation for me has to do with the dynamics of pseudo-Anosov diffeomorphisms of surfaces. This theory also has parallels with the dynamics of rational maps. [T88, FLP79].

The present work is divided into three parts. Part I is a general discussion of the basic facts about iterated rational maps. We will define and analyze the basic properties of the Julia set, the equicontinuity set, periodic points and their classification. Our discussion will emphasize the Poincaré metric and the geometric method of analyzing complex functions.

Much of the material in Part I dates back to Fatou and Julia, who discovered many amazing things around 1918, although some of the material is of more recent origin. In particular, we will discuss sketchily the theory of quasiconformal deformations of rational maps as developed by Dennis Sullivan in analogy to the theory of quasiconformal deformations of Kleinian groups which Ahlfors, Bers and others had previously developed.

Part II focuses on the dynamics of polynomial mappings. For polynomial mappings, there is always a component of the domain of equicontinuity around infinity consisting of points that shoot off to infinity. The closure of this domain is the entire Julia set, so that it can serve as a handle by which to grab the entire dynamical picture. When the attracting domain of infinity is simply-connected, the Riemann mapping for it conveys a lot of information. Douady and Hubbard in particular have exploited this Riemann mapping to analyze quadratic mappings [DH84].

Our analysis in Part II is related to that of Douady and Hubbard, but we take a somewhat different point of view, involving laminations rather than trees as the main structure. This is done with the intention of bringing out parallels to the theory of surfaces and to the theory of Kleinian groups which are isomorphic to the fundamental group of a surface. In particular, in Part II we will give a simple definition of a topological space topologically embedded in the plane which we conjecture is homeomorphic to the frontier of the Mandelbrot set (for quadratic polynomials).

Part II leaves a feeling of incompleteness, in that the structures and techniques presented have clear directions for clarification and further development. I hope that this presentation will inspire other people to continue the development.

Part III (see editor's note below) is centered on the discussion and proof of a theorem which classifies a large class of rational mappings in terms of topological conditions. The theorem applies to those rational maps such that all critical points are either periodic points, or eventually land on periodic points. Using Sullivan's theory of quasiconformal deformations of rational maps, the classification can also be extended to rational maps such that all critical points are eventually attracted toward periodic orbits. It is often conjectured that such rational maps are dense in the space of rational maps.

The analysis for Part III uses orbifolds, Teichmüller spaces, and the Teichmüller metric. It is analogous to constructions for hyperbolic structures on three-dimensional manifolds. Even in the case of real polynomials considered as maps of an interval to itself, the classification theorem is new.

Much work also needs to be done to complete the theory of Part III. The existence theorem given there is expressed in terms of a topological condition involving infinitely many different simple closed curves. It should be possible to replace this condition with some finite combinatorial condition. We shall do this when the rational map is a polynomial (newer references include [BFH92, Po, HS94]), but the general case still seems a mystery.

Editor's Note. In his original manuscript, Thurston writes: "This is a partially completed manuscript, based on lectures I gave at an NSF-CBMS workshop in Duluth, Minnesota." Thurston's original manuscript had three parts. We include here parts I and II. Part III was titled "Topological classification of postcritically finite mappings"; it was only a very partial sketch. A proof of Thurston's main theorem for Part III has meanwhile been published by Douady and Hubbard [DH93]; Thurston writes: "The proofs I had planned to write had a more geometric flavor" (personal communication to the editor); we hope that he will some day find the time to write down his original proof. Finally, there is a new appendix that tries to link the topics of Parts I and II, i.e., the theories of polynomial dynamics and of invariant laminations, and that was inspired by a Section II.7 that Thurston planned to write but never did.

Parts I and II have undergone many editorial changes. Even though we tried to stay as close to the original as possible, we felt that a significant number of changes were called for, some of them more substantial than others. Since the original manuscript is quite classical and has been widely circulated, it might be useful to point out some of the major changes that have been incorporated into the text. In a number of cases, the statements of some results or definitions had to be changed: Theorem I.7.3 of the original manuscript is really a conjecture; in Section II.4, condition (G5) on gap invariance is changed; conditions (CL1) and (CL2) in Proposition II.6.4 are changed; in Proposition II.6.5, an additional condition was added; and one claim in Proposition II.6.12 was removed. We indicated these also by footnotes. A number of proofs were missing or incomplete and were added, especially for Proposition I.5.10 and Theorems II.6.13 and II.6.14.

For several reasons, a number of additional lemmas and corollaries were inserted: some of them were needed as supporting results for some missing proofs; for some it seemed that they had originally been planned but never written, and at a few places we felt that giving extra results would round off the picture or give additional insight. In order to maintain the original numbering of results, and to tell the added results apart, all results that were added by us have numbers ending in a lower-case letter (for instance, Definition II.6.10 is Thurston's, and the Lemma that was added afterwards is referred to as Lemma II.6.10a). All results in Section I.8 were added by us and have numbers like Theorem I.8.a.

We also added a few statements in results that had existed in the original, in particular in Theorem I.5.3 (several claims strengthened) and Proposition II.6.7 (Claim (d) added). In addition, we changed a number of the existing proofs, either because we felt they were incomplete or because they could be simplified, for example using some of the lemmas that were we added elsewhere. Among these, we would like to mention especially Proposition I.5.9, Proposition I.6.1, Theorem II.5.2 ("No Wandering Triangles"), and Proposition II.6.4.

At a number of places, it is quite visible that the original manuscript was written in the 1980s (for instance, when references to desktop computers of the time were made), and we kept it like that, even though elsewhere we felt free to add references to current literature.

We added all the pictures, as well as an index that Thurston had planned. We also added most of the references; we could not and did not make an attempt to do justice to the vast existing literature. We learned much of the theory from Adrien Douady, John Hubbard and John Milnor and their books, especially [DH84] and [M06], and this will shine through in many of the arguments that we inserted, even without explicit mention.

We changed some of the notation in Part I: Thurston champions meaningful terms like "equicontinuity set" and "non-linearizable irrationally indifferent periodic points"; we replaced these by the terms "Fatou set" and "Cremer points" because these are meanwhile standard.

Editing a classical manuscript like Thurston's is a challenging task, trying to find a balance between historical authenticity and improved exposition. Such a task is never finished and yet needs to be terminated; we hope that the readers will find some of our efforts helpful.

Acknowledgments. The editors would like to thank many colleagues and friends for interesting and useful discussions on laminations, and for constructive and useful comments; these include Laurent Bartholdi, Alexandra Kaffl, Jeremy Kahn, Jan Kiwi, Yauhen "Zhenya" Mikulich, Kevin Pilgrim, Mary Rees, Michael Stoll, and Vladlen Timorin, as well as Alexander Blokh, Clinton Curry, John Mayer and Lex Oversteegen from the laminations seminar in Birmingham/Alabama. Moreover, we would like to thank Silvio Levy and John Smillie for encouragement and "logistical" support, and Andrei Giurgiu for help with a number of the pictures.

Part I: The Basic Theory

I.1   Introduction to Part I

Here we will develop the foundations of the dynamical theory of rational maps. There are a number of basic definitions and theorems which help greatly to focus the picture. In particular, the definitions of the Julia set and the Fatou set reveal a dynamical dichotomy which is a tremendous aid to the formation of a mental image of a rational map.

First, in Section I.2, we will review some background from complex analysis, with a few sketchy proofs. We try to give geometric explanations rather than formulas, in the belief that formulas often serve to present a special reason for something, in the process hiding a more intuitive, general reason lurking not far away.

In Section I.3 we will define the Julia set, which is analogous to the limit set of a Kleinian group, and the equicontinuity set (also called *Fatou set*), which is analogous to the domain of discontinuity. We will prove basic properties of these sets. For instance, any neighborhood of any point in the Julia set eventually spreads out over all of the sphere, with the possible exception of small neighborhoods of one or two points (Theorem I.3.6). We will also show that periodic points are dense in the Julia set (Theorem I.3.11).

Section I.4 discusses a few examples to illustrate these concepts.

Often, rational maps can be analyzed by looking at their periodic points. In Section I.5 we give the definitions for repelling, attracting and indifferent periodic points, and more definitions which further categorize periodic points. Then we make an analysis of the local picture near a periodic point; the subtle cases are the ones involving indifferent periodic points. We show that there is a relation between critical points and periodic points which are attracting or indifferent, in Propositions I.5.6 and I.5.9.

In Section I.7, we begin to look at what happens as you vary the rational map. We show that in certain cases, the *hyperbolic* (or *Axiom A*) cases, the dynamics are stable over an open range of parameter values.

Finally, Section I.8 discusses how the theory of quasiconformal maps can be applied to the dynamics of rational maps, to give a conformal classification within certain topological conjugacy classes. This is particularly of interest in the hyperbolic case. Sullivan's no wandering domain Theorem (II.3.4) is also stated and discussed.

General background on the topics of Part I may be found in most textbooks on complex dynamics; see for instance Milnor [M06].

I.2   Preliminaries: Uniformization and the Poincaré Metric

In this section, we will summarize some of the fundamental background in complex analysis which will be useful for the theory of rational maps. There are many alternate sources for this material.

Definition I.2.1. A *Riemann surface* is a (real) two-manifold, equipped with a one-dimensional complex structure. In other words, it is a space which is locally modeled on $\mathbb{C}$, with holomorphic transition maps between local coordinate systems.

A Riemann surface is the same as a complex curve, and often just called a curve; it is not to be confused with a complex surface, which in reality has four dimensions. The use of the words "curve" and "surface" is a common place for confusion in conversations between people with different background.

We will be dealing a lot with a very special Riemann surface, the Riemann sphere. There are several ways to think of the Riemann sphere:

(a) The Riemann sphere is the one-point compactification of $\mathbb{C}$ (the complex numbers), with the coordinate system $1/z$ defining the complex structure in a neighborhood of the added point, which is named ∞.

(b) The Riemann sphere is an ordinary round sphere in Euclidean 3-space $\mathbf{E}^3$. Rotations of the sphere are holomorphic. The complex structure is determined by a covering by a set of coordinate maps to $\mathbb{C}$, defined by stereographic projection to the xy-plane (which is identified with $\mathbb{C}$), which may be preceded by an arbitrary rotation.

(c) The Riemann sphere is complex projective 1-space, that is $\mathbb{CP}^1 = (\mathbb{C}^2 - 0)/\mathbb{C}^*$, where $\mathbb{C}^*$ denotes the multiplicative group of non-zero complex numbers.

The relation of (a) with (c) is that the ratio $z = z_1/z_2$ associated with a point $(z_1, z_2) \in \mathbb{C}^2$ is invariant under scalar multiplication. This defines a holomorphic map of $\mathbb{CP}^1$ to the one-point compactification of $\mathbb{C}$. Projective transformations of $\mathbb{CP}^1$ are the transformations inherited from linear transformations of $\mathbb{C}^2$; in terms of the coordinate z, the projective transformation inherited from the linear transformation

$$\begin{bmatrix} z_1 \\ z_2 \end{bmatrix} \mapsto \begin{bmatrix} a & b \\ c & d \end{bmatrix} \begin{bmatrix} z_1 \\ z_2 \end{bmatrix}$$

is the fractional linear transformation or Möbius transformation

$$z \mapsto \frac{az + b}{cz + d}.$$

All holomorphic self-homeomorphisms of the Riemann sphere have the form of Möbius transformations.

Another very important Riemann surface is the open unit disk D^2 in $\mathbb{C}$ (which corresponds to the southern hemisphere on the Riemann sphere). The subgroup of Möbius transformations which preserve the unit disk preserves a certain Riemannian metric, the Poincaré metric

$$ds^2 = \frac{4(dx^2 + dy^2)}{(1 - x^2 - y^2)^2}.$$

This metric has constant negative curvature -1, so that it is identified with the hyperbolic plane (or Lobachevsky plane). Orientation-preserving hyperbolic isometries are the same as Möbius transformations which preserve the unit disk. In order to actually calculate with this group, it is usually convenient to transform by a Möbius transformation sending the unit disk to the half-plane $\{ (x, y) : y > 0 \}$; the group of Möbius transformations preserving the upper half-plane are exactly those which can be written with real coefficients.

The Poincaré metric has an elementary but extremely convenient property:

Proposition I.2.2. *A holomorphic map f of the unit disk into itself cannot expand the Poincaré metric. In other words, for any points w and z,*

$$d\big(f(w), f(z)\big) \leq d(w, z).$$

The inequality is strict unless f is a Möbius transformation.

Proof of I.2.2. In the special case that $w = f(w) = 0$, this is simply the Schwarz lemma. The general case can be reduced to the special case by pre- and postcomposing with Möbius transformations. $\boxed{\text{I.2.2}}$

Definition I.2.3. A family of mappings from one metric space to another is *equicontinuous* if for each x in the domain and for each ε there exists a δ such that for any map f in the family,

$$d(x, y) < \delta \implies d\big(f(x), f(y)\big) < \varepsilon.$$

This depends on the topology of the domain, not the particular metric, and if the range of the family is compact, it is independent of the metric in the range. If the domain is compact, the choice of $\delta(\varepsilon)$ need not depend on x.

When the range is not compact, there is often more than one useful metric — for instance, in the case of the open disk, both the Poincaré metric and the Euclidean metric induced from $\mathbb{C}$ are useful. It is a very significant observation that equicontinuity with respect to the Poincaré metric implies equicontinuity with respect to the Euclidean metric, since Poincaré distances are greater than a fixed constant times Euclidean distances.

From Proposition I.2.2 it follows that the set of holomorphic maps of the disk to itself is equicontinuous with respect to the Poincaré metric. Of course, it is equicontinuous with respect to the Euclidean metric as well.

The concept of equicontinuity is useful because of the Ascoli theorem:

Proposition I.2.4. *Let X and Y be metric spaces with countable bases, and let F be an equicontinuous family of maps from X to Y such that for some point $x_0 \in X$, the closure of the set of image points of x_0 under maps in F is compact. Then the closure of F in (say) the topology of pointwise convergence among the set of all continuous maps $X \to Y$ is compact.*

$$\boxed{\text{I.2.4}}$$

There is a beautiful theory of "uniformization" of Riemann surfaces, which strongly influences the way we see them — it basically says that any Riemann surface has a canonical geometric form, so that many sorts of questions about Riemann surfaces and holomorphic maps can be analyzed in terms of their topology, plus their geometry. The first step is the well-known *Riemann mapping theorem*:

Proposition I.2.5. *Let U be any connected, simply-connected open subset of the Riemann sphere. If the complement of U has more than one point, U is conformally equivalent to the open unit disk.*

Proof of I.2.5. The standard modern proof of this theorem is amazingly elementary: Choose some point $u \in U$, and consider the set of embeddings $g : U \to D^2$ which send u to the origin in the disk. Look for a holomorphic embedding which maximizes the derivative at u. The key point is that for any proper open subset V of the disk, there is a reembedding of V in the disk which strictly expands the Poincaré metric of D^2 (and which can be adjusted to send the origin to the origin). Such a map is easily constructed as a branch of the inverse of any map f from D^2 to itself which is proper, not one-to-one (in other words, a branched cover of the disk over itself) but does not have any critical value in V — traditionally, f is $z \mapsto z^2$, normalized to move its critical value out of V. Since f shrinks the Poincaré metric of D^2, its inverse must expand it. A map $g : U \to D^2$ with maximal derivative must exist because of the compactness property for holomorphic mappings which follows from Proposition I.2.4, together with the fact that pointwise limits of holomorphic functions are holomorphic; and it must be an isomorphism, by the argument above. $\boxed{\text{I.2.5}}$

Here is another case of the uniformization theorem which can be proven in much the same way. A lot of what we will do makes use of this statement, and does not require any uniformization theorem of greater generality.

Theorem I.2.6. *Let U be any open subset of the Riemann sphere whose complement contains at least three points. Then the universal covering is conformally equivalent to the unit disk.*

Proof of I.2.6. This time, instead of considering embeddings $U \to D^2$, consider the set E of holomorphic embeddings of pointed covering spaces of U into D^2. (*Pointed* means that a base point has been chosen). To make E into a manageable mathematical object, observe that there is a map of the universal cover $\tilde{U}$ of U into D^2 associated with any map in E. Therefore E is in one-to-one correspondence with the subset of the space of maps $\tilde{U} \to D^2$ which are covering spaces of their images. We will henceforth think of E as the latter set. Clearly, E is an equicontinuous family.

A limit of maps which are covering spaces of their images is also a covering space of its image, provided it is not a constant map. (In other words, E is a closed subset of the space of all non-constant maps $\tilde{U} \to D^2$.) The reason is that, first, the limit map cannot have any critical points of finite order — because any perturbation of it would still have critical points — and second, any identifications $f(x) = f(y)$ made by the limit map f must be approximated by identifications by the functions in the sequence, so any such identifications must come from some deck transformation of $\tilde{U}$ over U.

Now one needs to check that E is non-empty — this can be done by explicit construction, doing it first for the three-punctured sphere with the help of classical functions.

Given any element of E which is not surjective, it can be lifted to a branched cover of the disk (using a normalized square-root function as above, for instance). After normalizing so that the base-point goes to the origin, the new element of E has a greater first derivative at the origin.

Therefore, there is some element of E which takes the base point of E to the origin and maximizes the derivative among all elements of E which take the base point to the origin. It is surjective, and since the image is simply-connected, the map must be an isomorphism of the universal cover of U with the disk. $\boxed{\text{I.2.6}}$

Note that the universal covering of the Riemann sphere minus two points is conformally equivalent to the Riemann sphere minus one point, via the logarithm (or in the other direction, the exponential map).

Here is a stronger form of the theorem:

Theorem I.2.7.

(a) *A Riemannian metric on a surface plus an orientation determines a Riemann surface structure (complex structure).*

(b) *Every Riemann surface (complex curve) has universal cover conformally equivalent either to the disk, the plane, or the sphere.*

  (i) *If the Riemann surface has negative Euler number, its universal cover is equivalent to the disk.*

  (ii) *A closed Riemann surface of zero Euler number has universal cover $\mathbb{C}$. (Closed means that it is compact and without boundary.)*

  (iii) *A closed Riemann surface of positive Euler number has universal cover the sphere.*

  (iv) *Any non-compact surface of non-negative Euler number has some Riemann surface structures whose universal covers are the disk, and other structures with universal cover $\mathbb{C}$.* $\boxed{\text{I.2.7}}$

We will not attempt to explain the proof of this general version. Note, however, that it is sufficient to find some cover of the Riemann surface which embeds in the Riemann sphere, and apply the previous theorem, I.2.6.

Doubly connected Riemann surfaces sometimes have special importance; they can be classified completely.

Corollary I.2.7a. *Every doubly connected Riemann surface is conformally isomorphic to either the punctured plane $\mathbb{C} - \{0\}$, to the punctured disk $D^2 - \{0\}$, or to an annulus $A_r := \{z \in \mathbb{C} : 1/r < |z| < r\}$ for a unique $r > 1$.*

Proof of I.2.7a. This follows by considering the universal coverings of these surfaces and their groups of deck transformations: in the first case, the universal covering is $\mathbb{C}$, in the second case, it is the upper half plane, and in the third case it is the band $\{z \in \mathbb{C} \colon |\operatorname{Im} z| < (\log r)/2\pi\}$; these coverings are chosen so that in all three cases the group of deck transformations is translation by integers. $\boxed{\text{I.2.7a}}$

Every doubly connected Riemann surface has an associated *modulus*: if a surface is conformally isomorphic to A_r, then we define its modulus to be $\mu(A_r) := (\log r)/2\pi \in (0, \infty)$ in the other two cases, we define the modulus to be ∞.

The Poincaré metric transfers to a canonical metric of constant curvature -1 on any Riemann surface S whose universal cover is the disk, and this transferred metric is called the *Poincaré metric* for S. Because of the uniformization theorem, this metric has great significance — often the knowledge that this metric exists helps a lot in analyzing properties of Riemann surfaces for which we may have only indirect descriptions.

A surface equipped with a metric of constant curvature -1 is called a *hyperbolic surface*.

If the universal cover of a Riemann surface is $\mathbb{C}$, the surface does not have a canonical metric — since an affine transformation of $\mathbb{C}$ (which is a similarity of the Euclidean plane) may change the scale factor for the Euclidean metric — but the Euclidean metric is defined up to change of scale by a constant. You can think of the Euclidean metric as a limiting case of the Poincaré metric. In other words, you can represent the universal cover of the Riemann surface as an increasing union of proper simply-connected open sets. Each of these sets has an associated Poincaré metric, and the limit of these Poincaré metrics, rescaled so that they converge to something non-zero, is the Euclidean metric.

Proposition I.2.2 immediately extends to the more general case of maps between Riemann surfaces:

Proposition I.2.8.

(a) *A holomorphic map between Riemann surfaces whose universal covers are the disk can never increase the Poincaré metric. Unless the map is a covering map, it decreases the metric.*

(b) *The only holomorphic maps from a Riemann surface not covered by the disk to one which is covered by the disk are the constant maps.*

(c) *The only holomorphic maps from the Riemann sphere to anything else are constant.*

Proof of I.2.8. A continuous map between two spaces lifts to a map between their universal covers. Apply Proposition I.2.2 to deduce Part (a). Part (b) amounts to Liouville's theorem: it follows from the fact that there are maps of the disk to the plane and the Riemann sphere with arbitrarily high derivative at the origin. Composing with a non-constant map of the plane or sphere to the disk would give maps of arbitrarily high derivative from the disk to the disk, a contradiction. Part (c) follows from the fact that topologically, a diffeomorphism from the sphere to $\mathbb{R}^2$ has to reverse orientation at some points (since it must have degree zero) — this is impossible for a holomorphic map. $\boxed{\text{I.2.8}}$

Corollary I.2.9. *Any family of holomorphic maps from any Riemann surface to the Riemann sphere minus at least three points is equicontinuous.*
$\boxed{\text{I.2.9}}$

The following concept plays a key role in complex dynamics.

Definition I.2.9a (Normal Family). Let U be an open subset of the Riemann sphere. Then a family of holomorphic maps $f_i \colon U \to \hat{\mathbb{C}}$ is called *normal* if every sequence among the f_i has a subsequence that converges locally uniformly to a holomorphic map $f \colon U \to \hat{\mathbb{C}}$.

This is a local condition: if every $z \in U$ has the property that the family $\{f_i\}$ is normal in some neighborhood of z, then the family is normal on U. Moreover, it is not hard to see, using Ascoli's theorem (Proposition I.2.4), that for any rational map f and any open subset U of the Riemann sphere, the family of iterates $f^{\circ n}$ on U is equicontinuous if and only if it is normal. Corollary I.2.9 then takes the following form:

Theorem I.2.9b (Montel's Theorem). *If $U \subset \mathbb{C}$ is open and a, b, c are three different points in the Riemann sphere, then any family of holomorphic maps $f_i \colon U \to \hat{\mathbb{C}} - \{a, b, c\}$ is normal.*

Sometimes it is useful to extend this theorem as follows: if g_1, g_2, g_3 are three holomorphic functions on U with disjoint graphs (i.e., $g_i(z) \neq g_j(z)$ for all z if $i \neq j$), then any family of holomorphic maps f_i is normal if the graphs of all f_i and all g_j are disjoint (compare the proof of Theorem I.3.11).

Here are a couple of properties of holomorphic maps that will be useful:

Proposition I.2.10. *If f is any continuous function defined in an open set U and holomorphic except possibly on the intersection of a smooth arc with U, then f is holomorphic in all of U.*

Proof of I.2.10. This is a local proposition; for concreteness, and after rescaling, we may suppose that the open set is the unit disk D^2 whose boundary is cut twice by the arc, and that f is continuous on the closed disk. Let α be the intersection of the arc with the closed disk. Let g be the function defined by the Cauchy integral formula in terms of f, that is,

$$g(z) = \frac{1}{2\pi i} \int_{\partial D^2} \frac{f(w)}{w - z} \, dw$$

For any continuous function f defined on the circle, this formula defines a holomorphic function g. The only trouble is that g need not agree with f on the circle (unless f extends to a holomorphic function).

Let γ_1 and γ_2 be the two curves which go around the two halves of D^2 cut by α. Since γ_1 and γ_2 traverse α in opposite senses, we have

$$\int_{\partial D^2} \frac{f(w)}{w-z}\, dw = \int_{\gamma_1} \frac{f(w)}{w-z}\, dw + \int_{\gamma_2} \frac{f(w)}{w-z}\, dw.$$

For z in the part of the disk enclosed by γ_1, the integral around γ_1 agrees with f, while the integral around γ_2 is 0. Thus the holomorphic function g agrees with f on the half of the disk enclosed by γ_1, and in the other half as well, by symmetry. Consequently, f is holomorphic. $\boxed{\text{I.2.10}}$

There are, of course, much more general statements about removable singularities of holomorphic maps, but this will meet our needs.

Corollary I.2.11. *Let f be a holomorphic function on D^2. If there is any interval along ∂D^2 where f converges to a constant limiting value, then f is constant.*

Proof of I.2.11. If f converges to a constant value along any interval, one could tack on a little neighborhood on the outside of this interval and extend f to be constant there. By the preceding proposition, the extended f is holomorphic, hence constant everywhere. $\boxed{\text{I.2.11}}$

Definition I.2.12. A *rational map* is a holomorphic map of the Riemann sphere to itself.

Proposition I.2.13. *Any rational map $z \mapsto R(z)$ can be expressed as the ratio of two polynomials, $R(z) = P(z)/Q(z)$. Conversely, the ratio of two polynomials always extends to a holomorphic map of the Riemann sphere.*

Proof of I.2.13. The second half of the assertion is just a matter of expanding a rational function in terms of partial fractions, and interpreting the resulting expression in terms of the coordinate $w = 1/z$ near any pole.

The first half is also elementary. Given any holomorphic map R from the Riemann sphere to itself, look at the set of poles and zeros of R. Let $P(z)$ be a polynomial whose zeros match the finite zeros (including multiplicities) of R, and let $Q(z)$ be a polynomial whose zeros match the finite poles of R. The ratio $P(z)/Q(z)$ is now a rational function whose poles and zeros match those of R. (You can easily check that the multiplicities work out at ∞ as well as at the finite points). The quotient

$$\frac{R(z)}{P(z)/Q(z)}$$

is a rational map which does not take the values 0 or ∞, hence it is a constant; in this way, we obtain an expression for $R(z)$ as the ratio of two polynomials. $\boxed{\text{I.2.13}}$

The *degree* of a rational map can be thought of either algebraically or topologically. In the algebraic form, the degree of $P(z)/Q(z)$ is the maximum of the degree of P and the degree of Q, provided P and Q are relatively prime. In topological form, the degree is the topological degree of the map of the sphere to itself — and this can be computed as the number of preimages of any point on the sphere which is not a critical value, since any holomorphic map preserves local orientation, except at critical points where this does not make sense.

The two definitions are clearly equivalent, since the equation

$$P(z)/Q(z) = \text{constant}$$

has the right number of solutions.

A Riemann surface S has *finite type* if it can be expressed as a closed Riemann surface $\overline{S}$ minus a finite set of points. $\overline{S}$ is called the *completion* of S. The completion $\overline{S}$ is determined by S, both topologically and holomorphically: topologically, it is the end-compactification of S; the analytic structure is unique around any added point because a continuous map which is holomorphic in the complement of a point is holomorphic at the point as well. A point in $\overline{S}$ which is missing in S is called a *cusp* or *puncture* of S.

If S is hyperbolic, its Poincaré metric near a cusp can be estimated by comparing it to the Poincaré metric of a punctured disk embedded in the surface. The universal cover of the punctured disk is formed by taking the log; it becomes a half-plane, whose Poincaré metric we know. Working backward, it follows that a punctured disk has finite area in a neighborhood of the puncture, and hence that S has finite area in a neighborhood of the cusp. It follows that a Riemann surface of finite type has finite area. Conversely, every hyperbolic surface of finite area consists of the union of a compact part with a finite number of non-compact parts that are neighborhoods of isolated punctures. (See, for example, [T97] for a discussion of the structure of hyperbolic manifolds of finite volume.) Therefore, the Riemann surface structure associated with a hyperbolic surface has finite type if and only if the hyperbolic surface has finite area.

Elements of the fundamental group of a Riemann surface which are represented by loops going some number of times around a cusp are called *parabolic elements*. They play a special role, about halfway between the role of the other non-trivial elements — the *hyperbolic elements* — and the role of the trivial element of the group. Parabolic elements are distinguished from hyperbolic elements by the property that a parabolic element can be represented (up to free homotopy, i.e., homotopy which need not conserve base points) by an arbitrarily short loop in the Poincaré metric, while a hyperbolic element is represented by a unique shortest curve, a closed geodesic.

Proposition I.2.14. *Let R be a Riemann surface, $p \in R$ a point, and $f : (R - p) \to S$ a holomorphic map to a hyperbolic Riemann surface. Then either*

(a) *f extends to a holomorphic map defined on R, or*

(b) *S can be described as $S = T - q$, where T is another Riemann surface and q is a point, and f extends to a holomorphic map $\overline{f} : R \to T$ such that $f(p) = q$.*

This is a way of phrasing the "big Picard theorem" that a function f can omit at most two values near an essential singularity.

Proof of I.2.14. We need only consider the case that the original domain is $R - p = D^2 - 0$. A small loop around the origin is also small in the Poincaré metric, so its image in S must have a small diameter. There are two cases:

Case (a). The image of a small loop around the origin is null-homotopic in S. Then the map can be lifted to a map into the universal cover of S, that is, the disk. Consider the image of any cylinder between two concentric loops. The boundary of its image must be contained in the image of its boundary (since a neighborhood of any interior point maps to an open set); therefore there cannot be disjoint disks which contain the images of its two boundary components (since a cylinder is connected) — so the images of the boundary components are not only small, but they are close together. It follows that f extends to a continuous map $\overline{f}$ defined on R. Therefore, p is a removable singularity.

Case (b). The image of a small loop around the origin is not null-homotopic in S. Since it can be homotoped to be arbitrarily small, it follows that the element of the fundamental group of S must be parabolic, and the image of any representative of the loop which is sufficiently small must lie in a small neighborhood of a puncture. Complete S by adding the point for this puncture, and reduce to case (a). $\boxed{\text{I.2.14}}$

Proposition I.2.15. *Any Riemann surface of negative Euler number and finite type has only a finite number of non-constant holomorphic self-maps; each self-map is an isometry of finite order (under composition).*

Proof of I.2.15. The degree of a map $f : R \to S$ between Riemann surfaces of finite type is well-defined, since such a map extends to their completions.

Since the Poincaré metric can only be decreased, it follows that

$$\text{Area}(R) \geq \deg(f)\,\text{Area}(S)$$

Therefore, the degree of a self-map cannot be greater than one. If the degree is one, the area element must be preserved, so the map is an isometry. The group of isometries of a hyperbolic manifold of finite volume is finite. If the degree is zero, the map must be a constant. $\boxed{\text{I.2.15}}$

What about nonconstant maps between different Riemann surfaces of finite type? In the hyperbolic case, if there is any holomorphic map, the area of the image surface must be less than the first. By the Gauss-Bonnet theorem, the area of S is $2\pi\chi(S) \times K$, where K is the curvature and $\chi(S)$ is its Euler number. The interpretation of this is that the image surface must be simpler than the domain. Here is a more precise statement, called the *Riemann-Hurwitz formula*:

Proposition I.2.16. *Let $f : S \to T$ be a nonconstant holomorphic map between Riemann surfaces of finite type. Then*

$$\chi(S) = \deg(f)\chi(T) - c(f),$$

where $c(f)$ is the total multiplicity of the critical points of f. In particular, if S and T are both the sphere, the formula is

$$c(f) = 2(\deg(f) - 1)\,.$$

Proof of I.2.16. One can see this just by counting cells. Make a triangulation of $\overline{T}$ which includes as vertices all the special points — the points added at the punctures, and the critical values of f. The inverse images of all the cells give a triangulation of $\overline{S}$. To compute the Euler number, add up $(-1)^{\dim}$ over all the cells in the uncompleted surfaces S and T. Each regular cell in T counts for $\deg(f)$ cells in S, while the critical vertices count for $\deg(f) - m$, where m is the multiplicity of the critical value. That is the formula. The formula for the sphere is obtained by substituting $\chi(S^2) = 2$. $\boxed{\text{I.2.16}}$

I.3   The Fatou Set and the Julia Set

We will now begin the study of the dynamics of rational maps, with some classical material that was developed around 1918 independently by Fatou and Julia. A rational map of degree one is a Möbius transformation, and this case is not very interesting — we shall generally understand that our rational maps have degree greater than one, even when we forget to say it. Two rational maps f_1 and f_2 are *conjugate* if they are the same up to change of coordinates by a Möbius transformation, i.e., if there is a holomorphic one-to-one map g of the Riemann sphere such that $f_1 = g^{\circ -1} \circ f_2 \circ g$. (Since we are studying dynamics in $\mathbb{C}$, where arithmetic operations make sense, we use the notation $g^{\circ n}$ for the iterated composition $g \circ g \circ \cdots \circ g$ of g with itself n times, rather than g^n, which we reserve for the meaning $g \times g \times \cdots \times g$ of g multiplied with itself n times). We are really interested in the properties of rational maps up to conjugacy, since a conjugating map transfers any dynamical properties of one map to properties of the other.

Here is a fundamental concept:

Definition I.3.1. Let f be a rational map of the Riemann sphere to itself, and let $I(f) = \{\, f^{\circ i} : i \geq 0 \,\}$ be the sequence of its iterates. The *equicontinuity set* (or equivalently the *Fatou set*) E_f for f is the set of points z on the Riemann sphere for which there exists some neighborhood $N(z)$ so that the sequence $I(f)$ restricted to $N(z)$ is equicontinuous (or equivalently, where the sequence of iterates forms a normal family). The *Julia set* J_f is the complement of the Fatou set.

Definition I.3.2. A subset X of the Riemann sphere is *invariant* for f if the image of X by f is contained in X, that is, $f(X) \subset X$. The subset is *totally invariant* for f if

$$f^{\circ -1}(X) = X$$

This implies that $f(X) = X$.

Proposition I.3.3. *The following conditions are equivalent:*

(a) *X is totally invariant;*

(b) *X is invariant and $f^{\circ -1}(X) \subset X$;*

(c) *the complement of X is totally invariant;*

(d) *X and its complement are each invariant.*

Each of these conditions implies $f(X) = X$.

Proof of I.3.3. These properties all follow from the definitions (with a little concentration). $\boxed{\text{I.3.3}}$

Proposition I.3.4. *The Julia set and the Fatou set are totally invariant.*

Proof of I.3.4. It suffices to prove this for the Fatou set E_f. Let $z \in E_f$. The sequence of iterates of f at $f^{\circ -1}(z)$ is obtained by composing once more with f — clearly, this cannot destroy the property of equicontinuity (or normality), so $f^{\circ -1}(E_f) \subset E_f$. In the other direction, if $z \in E$ is a regular point for f, it is clear that $f(z) \in E$ as well, because the sequence of iterates of f in a small neighborhood of $f(z)$ is obtained from the sequence at z by composing with a branch of the inverse of f at $f(z)$. But even if z is a critical point of f, a similar argument works (although it uses the fact that f is holomorphic), because the inverse image of a small neighborhood of $f(z)$ is locally small — so epsilons which work for z transfer to epsilons which work for $f(z)$. $\boxed{\text{I.3.4}}$

Subsets of the Riemann sphere with fewer than three points are very different from closed sets with three or more points, since their complements do not have a Poincaré metric. We will give such a set a special name — an *elementary set* is a set with 0, 1 or 2 points. (This is in keeping with the terminology, in the theory of Kleinian groups, of an elementary group). Sometimes a rational map can have a non-empty totally invariant set which is elementary. For instance, if f is a polynomial, the point at infinity is totally invariant.

Proposition I.3.5. *Let f be a rational map of degree greater than one. Any finite set which is totally invariant is elementary. There is a unique maximal elementary totally invariant set E for f — it consists of 0, 1, or 2 points. E has a neighborhood basis of invariant (in fact, contracting) neighborhoods.*

E has two points if and only if f is conjugate to a map of the form $z \mapsto z^{\pm d}$, where d is the degree.

E has at least one point if and only if f is conjugate to a polynomial.

Proof of I.3.5. That there cannot be a finite totally invariant set with more than two elements follows immediately from Proposition I.2.15 (using the fact that $d > 1$). Since the union of totally invariant sets is totally invariant, there is a unique maximal elementary totally invariant set E. If E has one element, then this point can be moved to the point at infinity (by conjugacy with a Möbius transformation), and f becomes a polynomial. If E has two elements, then you can assume they are zero and ∞. Either f or $f \circ (z \mapsto z^{-1})$ fixes both zero and infinity. If f fixes zero and infinity, then it is a polynomial with a multiple zero of order d at zero, hence it is of the form $z \mapsto \lambda z^d$. Conjugating by the linear transformation $z \mapsto \lambda^{(1/d-1)}z$, we obtain the function $z \mapsto z^d$. (Notice the use of the condition $d > 1$. The number $\lambda + 1/\lambda$ is a conjugacy invariant in the case $d = 1$.) The other case also follows. The set E clearly has a neighborhood basis of contracting neighborhoods. $\boxed{\text{I.3.5}}$

We will call the maximal elementary totally invariant set by its acronym, the *metis* of f. Generically, of course, the metis is empty, but if we have a name for it anyway, we can avoid making lots of special exceptions when discussing rational maps.

The very definition of the Julia set says that f tends to expand, at least sporadically, near any point in the Julia set. Here is a crisper statement to the same effect.

Theorem I.3.6. *Let U be any open set which intersects the Julia set. Then the sequence $f^{\circ n}(U)$ is eventually increasing (with respect to inclusion), and it eventually envelopes the complement of the metis: more precisely, if V is any neighborhood of the metis, there is some n such that $f^{\circ m}(U) \cup V = S^2$ for all $m \geq n$.*

In the generic case when the metis is empty, this says that $f^{\circ n}(U)$ is the whole Riemann sphere for some n.

Proof of I.3.6. Let U be any open set which intersects the Julia set, and consider the sequence of images $f^{\circ n}(U)$. There cannot be an ε such that these images always omit three points whose distances (pairwise) are all at least ε, since the Poincaré metrics for the complements of uniformly separated triples are uniformly bounded below. (This is true for any particular triple, and the set of such triples is compact.)

We conclude that for every ε there is some n such that the complement of $f^{\circ n}(U)$ is contained in the ε-neighborhood of an elementary set E.

Let F be the set (not the union) of all elementary subsets $E \subset S^2$ such that for every ε, there is some n such that the complement of $f^{\circ n}(U)$ is contained in an ε-neighborhood of E. By the above, and since the sphere is compact, it follows that F is non-empty; we will show that F contains the metis of f. Define a transformation $c_f \colon A \mapsto S^2 - f(S^2 - A)$ on subsets of the sphere. This describes how f transforms complements of sets.

Lemma I.3.7. *If A is a set with k elements, then $c_f^{\circ k+2}(A)$ is contained in the metis of f.*

Proof of I.3.7. The transformation c_f always decreases the cardinality of a finite set A, except when $f^{\circ -1}\big(f(x)\big) = x$ for every $x \in A$, that is, when every element of A is a critical point of maximal multiplicity $d - 1$. Since the total multiplicity of critical points is $2(d - 1)$, there can be at most two critical points of maximal multiplicity. Even when all elements of A do have maximal multiplicity, the cardinality decreases after one step unless each such element maps to a critical point of maximal multiplicity. Thus the only way A can avoid decay is to be a subset of the metis. $\boxed{\text{I.3.7}}$

The set F defined above is clearly invariant under the transformation c_f. It follows from the lemma that F contains the metis (since F contains any superset of any of its elements). This says that for every neighborhood

V of the metis, there is some m so that $f^{\circ m}(U) \cup V = S^2$. Since the metis of f is attracting, there is some n such that this property is true for all $m > n$. $\boxed{\text{I.3.6}}$

Corollary I.3.8. *For any subset U of the Julia set which is relatively open, there is some n such that $f^{\circ n}(U) = J_f$. In other words, the Julia set is the closure of the backwards orbit of any point in the Julia set.* $\boxed{\text{I.3.8}}$

A non-empty subset of a topological space is called a *perfect set* if it is closed and contains no isolated point.

Corollary I.3.9. *The Julia set is perfect (for degree bigger than one), and in particular is non-empty.*

Proof of I.3.9. By the preceding corollary, this immediately reduces to showing that the Julia set has at least two points. If the whole Riemann sphere were the Fatou set, i.e., the equicontinuity set, then $f^{\circ n}$ could expand the area element by only a bounded amount, for all n. This cannot happen when $\deg(f) > 1$, for then $\deg(f^{\circ n})$ grows large. Thus, the Julia set contains at least one point. Therefore, the Julia set has infinitely many points (since it is invariant and not contained in the metis). It follows that the Julia set has no isolated points. $\boxed{\text{I.3.9}}$

Definition I.3.10. A *periodic point* of f is a point x such that $f^{\circ n}(x) = x$ for some n. The minimal $n > 0$ for which this is true is the *period* of x.

Another invariant set canonically associated with a rational map f is the set of all periodic points for f. This set is not totally invariant; however, it is closely related to a familiar totally invariant set:

Theorem I.3.11. *The Julia set is contained in the closure of the set of periodic points.*

Proof of I.3.11. Let x be any point in the Julia set. We can assume that $f^{\circ -1}(x)$ and $f^{\circ -2}(x)$ are disjoint from x, otherwise x is already a periodic point. Choose $x_0 = x$, x_1 and x_2 such that $f(x_{i+1}) = x_i$ ($i = 0, 1$). The points x_0, x_1 and x_2 are all in the Julia set. Consider an arbitrary neighborhood U of x; we will show that there is a periodic point for f in this neighborhood.

Let $U_0 = U$, and let U_{i+1} be the component of $f^{\circ -1}(U_i)$ containing x_{i+1}, for $i = 0, 1$. Consider the sequence $\{ f^{\circ n}|U_2 : n \geq 0 \}$, and their graphs in $U_2 \times S^2$. If any of these graphs intersect the graphs of the first three, it gives us a periodic point in U (since any periodic orbit intersecting U_2 or U_1 also travels through $U = U_0$.)

There is a holomorphic map of $U_2 \times S^2$ to itself, commuting with projection to U_2, and taking the graphs of id, f and $f^{\circ 2}$ to constant sections $U_2 \times 0$, $U_2 \times 1$, and $U_2 \times 2$. The graphs of the $f^{\circ n}$ transform to the graphs of some other sequence of functions. Since $\{f^{\circ n}\}$ is not equicontinuous, the new sequence is not equicontinuous either. Therefore, by Corollary I.2.9, some of the functions take the value 0, 1 or 2. Points which hit 0, 1, or 2 become periodic points when translated back to the original coordinate system.$\boxed{\text{I.3.11}}$

In fact, the Julia set equals the closure of the set of repelling periodic points: repelling periodic points are in the Julia set, and any rational map has only finitely many non-repelling periodic points (see Section I.5).

I.4   Some Simple Examples

It is time to look at a few examples.

Example I.4.1. An invariant set with two points.

For a polynomial f of the form $z \mapsto z^d$ the Julia set is the unit circle. Every point inside the unit circle tends toward the origin, and everything outside the unit circle tends toward infinity; in either case, the (spherical) derivative of f goes to zero and the point has a neighborhood whose radius gets squished to zero (in terms of the spherical metric). Note that the derivative of $f^{\circ n}$ grows quickly on the unit circle — it has norm d^n.

Example I.4.2. A nearly invariant set with two points.

Example I.4.1 has a built-in stability near its Julia set: any polynomial g of the form $z^d + R(z)$ also has a (topological) circle as its Julia set, when $R(z)$ is a small polynomial of lower degree. One way to prove this is to give a dynamical description of the topology of the circle. First we will describe the circle in terms of the dynamics of f, but the description will be robust enough that it carries over to g. Cover the circle by a finite number of neighborhoods $\{U_i\}$ that are open in $\mathbb{C}$ — say, $2d$ such sets — with the following properties:

(a) The inverse image $f^{\circ -1}(U_i)$ has d components, and the diameters of components of $f^{\circ -n}(U_i)$ go to zero with n. (For our particular example, the second half of this condition follows automatically from the first.)

(b) For each pair of indices i and j, at most one component of the closure of $f^{\circ -1}(U_i)$ intersects the closure of U_j.

(c) Each component of the closure of $f^{\circ -1}(U_i)$ is contained in some U_j.

These properties imply in particular that the set of components of $f^{\circ -n}(U_i)$ form a basis for the topology of the circle. A good way to get a collection $\{U_i\}$ is to divide the circle into $2d$ equal segments, starting at the point 1, and take small neighborhoods of these segments for our U_i.

What condition (b) means is that any component of $f^{\circ -1}(U_i)$ is determined by naming any U_j which it intersects. Therefore, a component of $f^{\circ -2}(U_i)$ is determined by naming a component of a $f^{\circ -1}(U_j)$ which it intersects. Continuing on in this way and applying (a), it follows that a point x on the circle is completely determined by naming a sequence $\{i_n\}$ such that $f^{\circ n}(x) \in U_{i_n}$. A neighborhood basis for the x is formed by sets of points which are in the same U_{i_n}'s for a finite number of i's.

What happens when we perturb f a little, to get g? If there are any non-generic intersections for property (b), where closures of sets intersect but their interiors do not, modify the sets a little bit to make them generic. Similarly, make intersections of the U_i with each other generic. Then the set of intersections which occur are stable, so we can suppose that g is close enough to f that it has exactly the same qualitative structure of intersections for the closures of the inverse images of the U_i and the closures of the U_j. We can also assume that there are no critical points in the U_i.

Unfortunately, the intersections of $g^{\circ -n}(U_j)$ with the U_i probably are different from those with f. This is where condition (c) helps. Say that a (finite or infinite) sequence of indices i_n is legitimate if $f(U_{i_n}) \supset U_{i_{n+1}}$ for each applicable n.

Lemma I.4.3. *For each legitimate sequence, the set*

$$\{\, x : g^{\circ n}(x) \subset U_{i_n} \text{ for all applicable } n \,\}$$

is non-empty.

Proof of I.4.3. For finite legitimate sequences, this is clear from the conditions on g. For infinite sequences, this set is the nested intersection of non-empty compact sets. $\boxed{\text{I.4.3}}$

Each infinite legitimate sequence corresponds to exactly one point on the circle. Define two legitimate sequences $\{i_n\}$ and $\{j_n\}$ to be adjacent if U_{i_n} intersects U_{j_n}, for each applicable n. There is exactly one legitimate sequence except for certain boundary cases, as in decimal expansions. Thus, a map h can be defined from the Julia set of f to some corresponding set for g. The sets of points admitting an infinite legitimate sequence beginning with a certain finite segment do not quite form a basis for the topology of the circle, any more than initial segments of decimal expansions define a basis for the topology of $\mathbb{R}$; but by taking unions of adjacent sets of this form, one obtains a basis. Therefore, h is continuous. Its image is closed and totally invariant by g, so the image $h(J_f)$ is the Julia set for g. Since h is one-to-one, it is a homeomorphism between the two Julia sets.

The Julia set for the polynomials of these two examples is also the boundary between the set of points whose orbits tend toward infinity and the set of orbits which stay bounded. This is a general phenomenon:

Proposition I.4.4. *The Julia set for a polynomial is the frontier of the set of points whose forward orbits are bounded.*

Proof of I.4.4. If a neighborhood consists of points whose orbits stay bounded, then the set of forward iterates of f on the neighborhood is clearly equicontinuous (by Proposition I.2.2). If a point tends toward infinity, it is also clearly a point of equicontinuity. On the other hand, if the orbit of a point is bounded but points arbitrarily close to it go to infinity, it is clearly not a point of equicontinuity. $\boxed{\text{I.4.4}}$

Example I.4.5. The Julia set can be a Cantor set.

Here is an example where the set of orbits which stay bounded has no interior. For concreteness, consider the map $f : z \mapsto z^2 + 4$. Let us analyze the set of points whose orbits are bounded. A point of modulus r goes to a point with modulus not less than $r^2 - 4$. If r is at least 3, then r increases, so the point tends toward infinity.

The disk of radius 3 around the origin maps to a disk of radius 9 centered about 4, as a two-fold branched cover. The set of points which do not escape the disk of radius 3 after one iterate is the inverse image of the disk of radius 3, which is two not-quite-round subdisks. The set of points which also do not escape the disk of radius 3 after two iterates is the union of two subdisks of each of these. To get the set of points which do not escape ever, continue this procedure, and form the infinite intersection.

Is the intersection a Cantor set, or something else? Note that for a point to remain in the disk of radius 3 after the first iterate, it must be outside the disk of radius 1, so the norm of the derivative of f is at least 2. Therefore, the diameters of the nested disks tend geometrically to zero, and the Julia set is indeed a Cantor set.

If you study examples of the form $z \mapsto z^2 + a$, letting a vary, you can find many different types of behavior. Indeed, it should be clear that something else must happen to allow a transition from the case when a is small (the Julia set is a circle) and the case that a is large (the Julia set is a Cantor set). In fact, many different things happen. It is worthwhile to experiment a little with a programmable calculator or a computer. Try, for example, the cases

$$z \mapsto z^2 + i$$

where the critical point 0 goes to i, $i - 1$, $-i$, $i - 1$, $-i$, ... and the case

$$z \mapsto z^2 + (-.1 + .6i)$$

where the critical point eventually spirals (at a fairly slow rate) toward a periodic orbit of period three. With a little experimentation, you can see that in the first case almost no orbits remain forever bounded, while in the second, there is a definite open set of points whose orbits remain bounded, and they all spiral toward the same periodic orbit of period three.

1.5   Periodic Points

There are important distinctions between different kinds of periodic points. If x is a periodic point of period p for f and U is a neighborhood of x, the composition $f^{\circ p}$ maps U to another neighborhood V of x. This locally defined map is the *return map* for x; it is well-defined except for the ambiguity of domain.

Definition I.5.1. The point x is called an *attracting* periodic point if there exists a neighborhood U of x such that the sequence of the iterates of the return map on this neighborhood tends uniformly toward the constant function $z \mapsto x$. It is *repelling* if there exists a neighborhood V of x such that the image of every neighborhood of x under high iterates of the return map contains V. If x is neither attracting nor repelling, it is *indifferent*.

Proposition I.5.2. *Let x be a periodic point, and let g be its return map.*

(a) x *is attracting* $\Longleftrightarrow$ $|g'| < 1$ $\Longleftrightarrow$ *there exists a neighborhood V of x with non-elementary complement mapped into a proper subset of itself by the return map;*

(b) x *is indifferent* $\Longleftrightarrow$ $|g'| = 1$;

(c) x *is repelling* $\Longleftrightarrow$ $|g'| > 1$ $\Longleftrightarrow$ *there exists a neighborhood V of x which is mapped diffeomorphically by g to a proper superset of itself.*

Proof of I.5.2. By Taylor's theorem, it is clear that if $|g'|$ is less than or greater than one, x is (respectively) attracting or repelling, and there is a neighborhood V as claimed in the statement. If there exists any neighborhood V of x which is mapped into a proper subset of itself by the return map (or by some iterate of the return map, in case that x is attracting in the sense of Definition I.5.1), the return map must decrease the Poincaré metric of V, so its derivative at x is less than one — provided, of course, that V is covered by the disk. Similarly, if any neighborhood is mapped to a strictly larger neighborhood, one can apply this reasoning to a locally defined inverse of g to conclude that the derivative must be strictly greater than one. When the derivative of g has modulus one, the only remaining possibility is that g is indifferent. $\boxed{\text{I.5.2}}$

In terms of derivatives, further distinctions can be made:

Definition I.5.3. The periodic point x is *super-attracting* of order k if the derivative of the return map vanishes to order k.

Definition I.5.4. An indifferent periodic point x is *rationally indifferent* (or *parabolic*) if the first derivative of its return map is a root of unity, and *irrationally indifferent* otherwise.

Proposition I.5.5. *Let x be a periodic point, and g its return map. If x is not indifferent, then g is determined up to conformal conjugacy, in any sufficiently small neighborhood, by its first derivative, or if this is zero, by the order of vanishing of its derivatives.*

Proof of I.5.5. It is not hard to prove this proposition by making formal manipulations of power series, then checking that the answers converge. However, it seems more satisfying to give topological-geometric arguments which explain why things have to work so.

Case (i): $0 < |g'(x)| < 1$. Let V be a simply connected neighborhood of x which the return map takes inside itself. Define a Riemann surface R by gluing together an infinite number of copies $V_{(0)}$, $V_{(1)}$, $V_{(2)}$, ... of V, identifying each copy $V_{(n+1)}$ to the previous copy $V_{(n)}$ by a copy of g.

The maps $g_{(i)}$ of the ith copy to itself commute with the identifications, so there is a well-defined map of R to itself, which we also call g, extending the original definition of g on $V_{(0)}$. This extension has a globally defined inverse. The Riemann surface R cannot be the disk (because of the canonical nature of the Poincaré metric), so it is $\mathbb{C}$.

In the coordinate system of $\mathbb{C}$, the map g must be affine: $g(z) = az + b$; by a change of coordinates so that the fixed point x is at the origin, it has the form $z \mapsto g'(x)z$.

This proof easily gives an actual formula for the linear coordinates. The uniformization for R is accomplished just by taking a sequence of embeddings of V in $\mathbb{C}$, so that the image $g^{\circ n}(V)$ (interpreted as the approximation for $V_{(0)}$) stays roughly constant in size, and passing to the limit. A formula for a linearizing map L is obtained by choosing a point $z_0 \neq x$ and setting

$$L(z) = \lim_{n \to \infty} \frac{g^{\circ n}(z)}{g^{\circ n}(z_0)}.$$

Case (ii): $|g'(x)| > 1$. Apply the argument of case (i) (which is strictly local in nature) to a local inverse of g.

Case (iii): $g'(x) = 0$. We will make an argument similar to case (i), but we need to take into account the fact that g is no longer a homeomorphism, but rather a branched covering.

Let V be a simply-connected neighborhood of x which is mapped into itself by g, and such that $g|V$ has a critical point only at x. Let k be the order of the first non-zero derivative of g at x, so that g has k-fold branching at x. Inductively define a sequence V_i by taking $V_0 = V$, and V_{i+1} to be the k-fold branched covering of V_i, branched over the point which maps to x. Let p_i be the covering projection $V_{i+1} \to V_i$. Thus, V_n is the k^n-fold cover of V, branched over x.

Since g is a k-fold branched covering, the map $g_0 = g$ of V_0 to itself lifts to a diffeomorphism $h_0 : V_0 \to V_1$. (This means that $p_0 \circ h_0 = g$.) The composition $g_1 = h_0 \circ p_0 : V_1 \to V_1$ is a k-fold branched cover of V_1 into itself. Now g_1 corresponds to g via the branched covering, i.e., $p_0 \circ g_1 = g \circ p_0$. Having inductively defined g_i, we define h_i and g_{i+1} in the same way.

Now we can form a Riemann surface R by gluing V_{i+1} to V_i by the embedding h_i. The maps g_i are compatible with the gluing, so they fit together to give a map g_∞ which is a k-fold branched covering of R over itself. By checking the two possibilities, one sees that R must be the disk.

There is only one k-fold branched cover of the disk branched only at the origin. One description of such cover is by the map $z \mapsto z^k$. Since the only isometries of the Poincaré metric of the disk which fix the origin are multiplication by constants, the covering map at hand must agree with this one, up to multiplication by a constant in the domain and range. The constant, of course, can be pushed backward or forward to the domain or range.

Observe that if we conjugate $z \mapsto z^k$ by the map $z \mapsto az$, we get the map $z \mapsto a^{k-1} z^k$. We can dispense with the constant (using the hypothesis that $k \neq 1$), and the proposition is proven.

Just as for the earlier cases, this proof translates into a formula for the conjugating map, although interpretation of the formula requires intelligence (or continuity) in deciding which k^nth roots to use. First, choose a local coordinate z so that the fixed point for the return map is at the origin and the kth derivative of the return map is 1. The formula for a local map h to the disk which conjugates g to $z \mapsto z^k$ is

$$h(z) = \lim_{n \to \infty} \left(g^{\circ k}(z) \right)^{1/k^n}. \qquad \boxed{\text{I.5.5}}$$

An attracting periodic point is always in the Fatou set, of course. Using the existence of the Poincaré metric, we can deduce considerably more about how it sits in the Fatou set.

Proposition I.5.6. *All the attracting periodic points are in different components of the Fatou set.*

Every point in a component of the Fatou set which contains an attracting periodic point p tends toward the periodic orbit of p in future time.

For each attracting periodic orbit, there is at least one critical point in the union of the components of the Fatou set which intersect the orbit.

Proof of I.5.6. If x is an attracting periodic point of period p, then $f^{\circ p}$ actually decreases the Poincaré metric of its component in E_f, since its derivative at x is less than 1 in modulus. There cannot be another periodic point in the same component, for the Poincaré distance between the two would have to decrease.

The map of one component of E_f to its image is a branched cover; it is a local isometry if it does not actually branch, i.e., if it does not contain any critical points. Hence the component of at least one of the periodic points in any attracting periodic orbit must contain a critical point. $\boxed{\text{I.5.6}}$

Corollary I.5.7. *There are at most $2(d-1)$ attracting periodic orbits for any rational map of degree d, and at most $d-1$ attracting periodic orbits in $\mathbb{C}$ for a polynomial of degree d.*

Proof of I.5.7. Since each attracting periodic orbit consumes at least one critical point, there can be no more attracting periodic orbits than there are critical points. $\boxed{\text{I.5.7}}$

In fact, a rational map of degree d can have at most $2(d-1)$ periodic orbits that are non-repelling. This statement is known as the *Fatou-Shishikura-inequality* [Sh87, Ep]. We will prove this result only in the case of polynomials, using a proof of Douady.

Corollary I.5.7a. *Every polynomial of degree d has at most $d-1$ non-repelling periodic orbits in $\mathbb{C}$.*

Proof of I.5.7a. A *polynomial-like map* is a proper holomorphic map $f\colon U \to V$, where U and V are simply connected bounded domains in $\mathbb{C}$ so that the closure of U is contained in V. In particular, if f is any polynomial, V is any sufficiently large disk and $U := f^{-1}(V)$, then the restriction $f\colon U \to V$ is a polynomial-like map.

It is easy to see that every polynomial-like map has a well-defined mapping degree $d \geq 1$, and that a polynomial-like map of degree d has exactly $d-1$ critical points, counting multiplicities. The proof of Corollary I.5.7 goes through to show that every polynomial-like map of degree d has at most $d-1$ attracting periodic orbits. The point is to show that given any polynomial-like map $f\colon U \to V$ and any finite number of indifferent periodic orbits, it is easy to perturb f and U slightly to a nearby polynomial-like map of the same degree in which each of finitely many chosen indifferent periodic orbits of f become attracting, while all attracting periodic orbits of f remain attracting. To see this, consider a set Z consisting of all points on all attracting cycles and finitely many indifferent cycles. Let $a \in Z$ be a point on an indifferent orbit. Construct, say, a polynomial h so that $h(z) = 0$ on all points of Z, while $h'(z) = 0$ for all points in $Z - \{a\}$ and $h'(a) = 1$. Now consider $f_\varepsilon := f + \varepsilon h$. Under f_ε, all points in Z are still periodic with equal period, and all multipliers of these orbits are unchanged, except that the multiplier on the orbit of a depends holomorphically on ε. If $|\varepsilon|$ is sufficiently small, then $f_\varepsilon \colon f_\varepsilon^{-1}(V) \to V$ is a polynomial-like map of degree d, and by an appropriate choice of ε one can make the orbit of a attracting. This can be repeated for any finite number of indifferent orbits, and this proves the claim. $\boxed{\text{I.5.7a}}$

Indifferent periodic points are considerably more subtle than attracting or repelling periodic points. What we really need to know is the qualitative description, especially for rationally indifferent periodic points. The local analysis of rationally indifferent periodic points with any given derivative shows that there are countably many different possible qualitative types, and for each qualitative type there is still an uncountable family of different local conjugacy classes.

We begin with the simplest possible case, when the derivative of the return map g is one. We can change coordinates so that the origin is the fixed point, and for some k the Taylor series for g has the form

$$z \mapsto z + a_k z^k + \cdots .$$

By a linear change of coordinates, we may assume that $a_k = 1$. To get a rough picture of the qualitative behavior of g, we can consider the special case that g is the time-one map for a flow (that is, it is obtained by following a vector field for one unit of time). An appropriate vector field is $V = z^k e$, where e is the unit vector field (often written $e = \partial/\partial z$) pointing in the x-direction in $\mathbb{C}$.

For $k = 2$ the vector field V is actually holomorphic on the whole Riemann sphere: in terms of the coordinate $w = 1/z$ it equals the constant vector field $-e$. The orbits of V on the Riemann sphere are all asymptotic to 0, as time goes to $\pm\infty$. Each orbit (except the orbit of the fixed point 0) is a circle minus one point; all the circles have a common tangent at the origin.

If $k > 2$, the vector field V can be obtained by pulling back the vector field for the case $k = 2$ by a $(k-1)$-fold covering map $h : z \mapsto (1/k-1)z^{k-1}$. In this case V blows up at ∞, so it is not globally integrable. The time-one map for its flow is nonetheless defined for some neighborhood of the origin — in fact, the only problem is along rays passing through $(k-1)$-st roots of unity. These rays are pushed outward along themselves, and a definite chunk goes out to the point ∞ in any positive time interval. The rays which pass through odd $2(k-1)$-st roots of unity get pulled inward, so that the image of an entire ray is a proper subset of itself after any positive time interval. All other orbits of the flow exist for all time, and they form curves which are asymptotic to 0 as time goes to $\pm\infty$. All of these typical orbits are tangent to a ray through an odd $2(k-1)$-st root of unity in positive time, and to a ray through an even $2(k-1)$-st root of unity in negative time. The whole picture is like a flower with $2(k-1)$ petals.

For a general return map g with first derivative one, the local qualitative picture is similar. When normalized so that the next non-zero term in the Taylor series is z^k, there are regions around the rays through odd $2(k-1)$-st roots of unity which tend toward 0 in positive time. In fact, for any reasonably small open disk U, the set F of points which tend toward 0 without ever leaving U has $k-1$ components, each of which is tangent on each side to a ray through a $(k-1)$-st root of unity. F also looks like a flower (at least for k big enough), but with only $(k-1)$ petals.

When the derivative of the return map g is a root of unity, but not one, then of course there is some iterate of g which has derivative one. Again, the set F consisting of points in a small disk which tend toward 0 without ever leaving the disk form a figure which is like a flower. Since F is invariant by g, the number of petals must be a multiple of the multiplicative order of the derivative of g.

Definition I.5.8. For any rationally indifferent periodic point, an *attracting flower* is the set of points in a sufficiently small disk around the periodic point whose forward orbits always remain in the disk under powers of the return map. A component of the attracting flower is an *attracting petal*.

Any holomorphic map f is invertible near each indifferent periodic point. The attracting petals of f^{-1} are called *repelling petals* of f; see Figure I.4.

We now have enough qualitative information to give a general discussion of the situation of rationally indifferent periodic points with respect to the Julia set and the Fatou set.

Proposition I.5.9. *A rationally indifferent periodic point is in the Julia set, but each of its attracting petals is in the Fatou set.*

Distinct attracting petals are in distinct components of the Fatou set; they are also distinct from any components of E_f which contain periodic orbits.

Every point in a component C of E_f containing an attracting petal tends toward the rationally indifferent periodic point in forward time. There is at least one critical point in $\bigcup_{i=1}^{pk} f^{\circ i}(C)$, where p is the period of the periodic point and k is the order of the first derivative of its return map.

Figure I.4. Attracting petals (bounded by solid curves) and repelling petals (dashed curves) of a rationally indifferent periodic point. The dynamics of the map f is indicated by small arrows.

Proof of I.5.9. It is obvious that any rationally indifferent periodic point x is in the Julia set, and that attracting petals are in E_f, but not so obvious that distinct attracting petals are in distinct components. The petals are somewhat artificial, depending on the choice of a small disk, and we do not have much control over what might happen to the material between petals once they leave this disk — why can't they be in E_f?

Consider any two points w and z in attracting flowers for rationally indifferent periodic points (possibly the same). Suppose that w and z are in the same component of E_f. Then there is a smooth path β joining w to z in E_f, with a certain Poincaré length L. Any image $f^{\circ n}(\beta)$ can only have shorter length.

Apply a large number N of iterates of f, where N is a multiple of the period of the periodic point x associated with w. Now $f^{\circ N}(w)$ is very close to x, so close that the disk of radius L around $f^{\circ N}(w)$ in the Poincaré metric of E_f is contained in the flower-disk for x. We see that z and w are associated with the same periodic point x.

It also follows that $f^{\circ N}(\beta)$ is contained in the attracting flower of x: indeed, for all powers m of the return map g, the image $g^{\circ m}\big(f^{\circ N}(\beta)\big)$ is contained in the flower-disk for x, by the choice of N. Consequently, its two endpoints are in the same attracting petal.

We will be able to deduce the existence of a critical point associated with each periodic cycle of attracting petals once we can prove that the Poincaré metrics of the components of E_f are not all mapped by isometries. This is not so obvious as it was for attracting periodic points — in fact, the local picture is topologically consistent with the hypothesis that the maps are all isometries.

Let x be a rationally indifferent periodic point, consider some attracting petal associated to x and let U be its Fatou component. Let g be the iterate of f that returns each petal to itself and suppose, by way of contradiction, that g has no critical points; it is then a covering map and hence a local isometry for the Poincaré metric of U. The map g is invertible in a neighborhood of x; we call this branch of g^{-1} the preferred branch.

Not all points in the petal are alike; some points have the property that the full orbit under g remains in the petal, while other points exit in backward time. More precisely, any attracting petal intersects its two adjacent repelling petals. Let z_0 and w_0 be two points in the same attracting petal and in different repelling petals so that their full orbits remain within the same attracting petal, and connect them by a Poincaré geodesic segment γ_0 in U. Set $z_n := g^{\circ n}(z_0)$ and $w_n := g^{\circ n}(w_0)$, for $n \geq 0$. Connect z_0 to z_1, as well as w_0 to w_1, by Poincaré geodesic segments in U. The preferred branch of g^{-1} sends z_1 to z_0 and w_1 to w_0 provided both z_0 and w_0 were chosen close enough to x. Using the preferred branch of g^{-1}, set $\gamma_n := g^{\circ n}(\gamma_0)$: if there is no critical point of g in U, then this is possible for all $n \in \mathbb{Z}$. By hypothesis, all γ_n have identical Poincaré lengths. Since w_n and z_n both converge to x as $n \to -\infty$ and thus to the boundary of U, it follows that the entire curves γ_n converge uniformly to x as $n \to -\infty$ (and the same is obviously true for $n \to +\infty$).

Let s_0 be a point in the same attracting petal as w_0 and z_0, but not in any repelling petal at x; instead, let s_0 be between the two adjacent repelling petals. Again, set $s_n := g^{\circ n}(s_0)$ for $n \geq 0$; then $s_n \to x$ as $n \to +\infty$ (the backwards orbit of s_n along the preferred branch of g^{-1} will leave a certain neighborhood of x, after which we have no control). Supposing that s_0 was chosen close enough to x, the preferred branch of g^{-1} sends s_{n+1} to s_n for $n \geq 0$. Connect s_0 to s_1 by a segment of a Poincaré geodesic, say α_0, and extend this by forward iteration to a curve connecting the forward orbit of s_0 and the point x in the limit. Denote this curve α.

Since the endpoints of γ_n are in different repelling petals of x, all γ_n (with n negative and sufficiently large) must intersect α. Under continued backwards iteration, any intersection point must converge to x (because such points are on γ_n), hence one can choose an n so that the backwards orbit of an intersection point of γ_n with α remains in any given ε-neighborhood of x. On the other hand, every point on α is eventually mapped by g^{-1} to α_0. This is a contradiction. $\boxed{\text{I.5.9}}$

The local picture near an irrationally indifferent periodic point is an entire story in itself. If the first derivative of the return map is written $e^{2\pi i \alpha}$, it turns out that the behavior depends on Diophantine properties of α. Siegel proved in the famous paper [Si42] that for a typical α which cannot be approximated extraordinarily well by rational numbers, the return map in a neighborhood of the periodic point is actually conjugate to a rigid rotation by the same angle! This implies that the periodic point is in the Fatou set, and that the return map is an isometry in a neighborhood of the periodic point and hence in the entire Fatou component.

Hence, there is an entire component of E_f where the return map is conjugate to an isometry which rotates the hyperbolic disk by $2\pi\alpha$. Such a component is known as a *Siegel disk*. Siegel's paper was the first revelation of a phenomenon which occurs in much more general dynamical situations, including such things as the solar system. This theory in dynamical systems has become known as the KAM theory, after Kolmogorov, Arnol'd and Moser, who made important contributions to it. Michel Herman has added much information to this theory in recent years, including in the case of iterated rational maps.

There is another case of irrationally indifferent periodic points that do not control Siegel disks, and are on the Julia set rather than in the Fatou set. An irrationally indifferent periodic point in the Julia set is called a *Cremer point*.

Proposition I.5.10. *Irrationally indifferent periodic points can lie in the Julia set; in other words, Cremer points exist.*

Proof of I.5.10. Let p_1 be a rational map for which 0 is a fixed point with $p_1'(0) = 1$. For $\lambda \in \mathbb{C} - \{0\}$, define a family of rational maps $p_\lambda := \lambda p_1$. The map p_λ has a fixed point at 0 with multiplier λ; we are interested in the case that $\lambda = e^{2\pi i\theta}$ with $\theta \in \mathbb{R}/\mathbb{Z}$. For $\varepsilon > 0$, let C_ε be the set of all θ so that either θ is rational, or p_λ has a periodic orbit (other than the fixed point 0) that is entirely contained in the open disk around 0 with radius ε. The set C_ε contains a neighborhood of each rational $\theta \in \mathbb{R}/\mathbb{Z}$ (when $\theta = p/q$ is rational, then $f^{\circ q}$ has the local form $z \mapsto z + O(z^2)$, so it has a double or multiple fixed point that splits up, when p_λ is slightly perturbed, into at least two fixed points near 0), so each C_ε is open and dense. Let $C := \bigcap_{\varepsilon>0} C_\varepsilon = \bigcap_{n\in\mathbb{N}} C_{1/n}$. This is a countable nested intersection of open and dense subsets of $\mathbb{R}/\mathbb{Z}$, so C is dense by Baire's Theorem. For $\theta \in C$, the point 0 is a limit point of periodic points. If we assume that all but finitely many periodic points are in the Julia set (which is a true fact that we have not shown), then we are done; here we give a direct proof without assuming this fact.

Suppose that 0 is in the Fatou set. Let U be the Fatou component containing 0 and let U_ε be the ε-neighborhood of 0 for the hyperbolic metric, for $\varepsilon > 0$ small enough so that U_ε is simply connected and contains no critical points. Since $p_\lambda : U \to U$ cannot expand the hyperbolic metric of U, it follows that $p_\lambda(U_\varepsilon) \subset U_\varepsilon$. If this is a strict inclusion, then by the Schwarz Lemma it follows that $|p_\lambda'(0)| < 1$, a contradiction. Hence $p_\lambda : U_\varepsilon \to U_\varepsilon$ is an isometry, so the restriction of p_λ to U_ε is conformally conjugate to a rigid rotation by the irrational angle θ. But then p_λ cannot have periodic points arbitrarily close to 0. $\boxed{\text{I.5.10}}$

I.6 The Fatou Set

In Section I.5 we have seen that non-repelling periodic points have associated components of the Fatou set, except in the case of certain of the irrationally indifferent periodic points. This raises the question: What does a general component of the Fatou set look like? Are all components associated with attracting or indifferent periodic points?

One fact to keep in mind is that often the existence of one component of the Fatou set implies the existence of many more. The Fatou set is saturated under both forward and inverse iteration. Even when a domain returns to itself after a finite number of forward iterations, it generally happens that it has an infinite number of inverse images, all but finitely many of which are preperiodic.

The problem of analyzing the general nature of components of the Fatou set was a troublesome one for many years, but it was solved recently by Sullivan in [S85a]. He proved in particular that every component of the Fatou set is eventually periodic, and further analyzed what these components could look like. We will discuss his theorem in more detail in Section I.8. It is the analogue of the Ahlfors' famous finite area theorem for Kleinian groups, and it is proved in a similar way, using the theory of quasiconformal maps and quasiconformal deformations. In this section, we will do what we can using the more elementary methods of topology and the geometry of the Poincaré metric.

There is one kind of component of the Fatou set which we have not yet described, called a Herman ring. A *Herman ring* is a component of the Fatou set which is topologically an annulus A that eventually returns to itself by a homeomorphism. An orientation-preserving isometry of the Poincaré metric of an annulus A can only be a rotation, or an involution which reverses the direction of the circle (conformal automorphisms of annuli $\{z \in \mathbb{C}: 1/r < |z| < r\}$ have the form $z \mapsto \lambda z$ or $z \mapsto \lambda/z$ with $|\lambda| = 1$). The return map cannot be periodic, or else the entire rational map would be periodic by analytic continuation. If the return map reversed the orientation of the circle, the second return map would be the identity. Therefore, the return map must be an irrational rotation of the annulus.

Michel Herman first constructed examples of Herman rings by using the KAM theory, applied to rational maps constructed with Blaschke products to leave a circle invariant and act on it by a homeomorphism. When the rotation number of the homeomorphism is a "good" irrational number, it has a neighborhood which is a Herman ring. Another construction, a sort of "connected sum" of Siegel disks, ironed out with the aid of quasiconformal deformation theory, was discovered later [Sh87].

Proposition I.6.1. *Let U be any component of the Fatou set for a rational map f. Then either U eventually arrives at a Herman ring, or for every loop α on U, some forward image $f^{\circ n}(\alpha)$ is null-homotopic.*

Corollary I.6.2. *A component of the Fatou set either is a disk, or it eventually lands on either a Herman ring or on a domain with a critical point.*

Proof of I.6.2. The fundamental group of a domain maps injectively to the fundamental group of its image, except when the domain contains a critical point. By the proposition, it follows that any domain which is not simply-connected eventually arrives at a Herman ring, or eventually must hit a critical point so that some fundamental group can die. $\boxed{\text{I.6.2}}$

This corollary at least tells us that the only conceivable wandering domains (domains which are not eventually periodic) are disks.

Proof of I.6.1. Let δ be a sufficiently small distance so that the image by f of a set of diameter less than δ cannot have complement with diameter less than δ. Our first claim is the following: if it ever happens that α and each of its forward images are contained in disks of diameter δ, then α must be contractible in E_f.

In fact, each iterated image of α cuts the Riemann sphere into some number of components, all but one of which have diameter less than δ. The image of any small component in the complement of $f^{\circ n}(\alpha)$ can not intersect the large complementary component of $f^{\circ(n+1)}(\alpha)$ (or it would have to contain it, in contradiction to the choice of δ). Since $f^{\circ(n+1)}(\alpha)$ has diameter at most δ, it follows that for each small complementary component W of α, all $f^{\circ n}(W)$ have diameter at most δ, and hence are contained in the Fatou set. This shows that α is contractible in E_f.

Let $P : \hat{\mathbb{C}} \to \mathbb{R}$ be the *scaling function* for the Poincaré metric, defined to be zero on the Julia set and equal to the ratio between the spherical metric and the Poincaré metric on E_f. The function P is continuous on $\hat{\mathbb{C}}$.

Up to homotopy, we may suppose that the loop α is smooth. Let L be its length in the Poincaré metric. Define a compact subset K of E_f to consist of those points where $P \geq \delta/L$.

If after some time iterates of α no longer intersect K, from then on the diameter of the iterates of α in the spherical metric is less than δ, so some forward iterate of α is contractible and we are done.

If α intersects K an infinite number of times, then in particular it eventually returns twice to the same component W of E_f (the components of the Fatou set form an open cover of K). If the return map for W is not a local isometry, then the Poincaré lengths of the images of α decrease by a definite proportion on each return to K, so eventually they have spherical diameter less than δ, and we are done.

The remaining possibility is that the return map for W acts as a local isometry of its Poincaré metric. The return map is therefore either a homeomorphism, or conceivably a covering map. Choose a base point $x \in \alpha$. Under iteration, it has a limit point in the Fatou set (at most distance L away from K in the Poincaré metric of the Fatou set); let p be such a limit point. For every neighborhood V of p there are thus $n, k \in \mathbb{N}$ so that $f^{\circ n}(x)$ and $f^{\circ(n+k)}(x)$ are in V, so $f^{\circ k}$ sends a point in V into V. The iterates of f thus have a subsequence that converges to a local isometry of W that fixes p and, after extracting another subsequence, has derivative 1 at p. But a covering that fixes some point with derivative 1 is the identity. It follows that the return map of W is a homeomorphism (if $f(x) = f(y)$ for $x \neq y$, then no sequence of iterates of f can converge to the identity).

If W has a simple closed hyperbolic geodesic γ, then any isometry of W sufficiently close to the identity must map γ to itself. If W has two simple closed hyperbolic geodesics that intersect, then any isometry of W sufficiently close to the identity must fix all intersection points, so it must fix both geodesics pointwise, and it thus is the identity. It follows that f must have finite order when restricted to W, hence globally, and this is impossible for a map of degree at least 2. If W has two disjoint simple closed geodesics, then let γ' be a shortest geodesic segment connecting these two geodesics; this is again rigid, so any isometry of W sufficiently close to the identity must fix γ' pointwise, and the same conclusion applies.

We can thus restrict to the case that the hyperbolic surface W does not have two different simple closed geodesics. Then W is either the disk, the disk with one or two punctures, or an annulus with finite modulus. But by Corollary I.3.9, a Fatou component cannot have isolated punctures. Therefore, if W is not simply connected, then it is an annulus with finite modulus and we have a Herman ring. $\boxed{\text{I.6.1}}$

A complete description of periodic Fatou components is as follows.

Theorem I.6.2a (Classification of Periodic Fatou Components). *Every periodic Fatou component has one of the following types:*

(a) *a component of the domain of attraction of an attracting periodic point; it may be simply or infinitely connected;*

(b) *a component of the domain of attraction of a rationally indifferent periodic point; it may be simply or infinitely connected;*

(c) *a Siegel disk; it is always simply connected;*

(d) *a Herman ring.*

Proof of I.6.2a. Let U be a periodic Fatou component, let g be the first return map of U, let c be the number of critical points of g (counting multiplicities), and let d be the degree of g. Corollary I.6.2 states that U is a disk or a Herman ring, or $c > 0$. Since the case of a Herman ring is clear, we will exclude it from now on.

Suppose we have $c > 0$. If U has finite connectivity, let $\chi(U)$ denote the Euler characteristic of U. The Riemann-Hurwitz formula (Proposition I.2.16) gives $d \cdot \chi(U) = \chi(U) + c$ (this formula was stated only for compact Riemann surfaces with finitely many punctures, but it is a topological result and thus also applies to surfaces with finite connectivity). We have

$$(d - 1)\chi(U) = c > 0\,,$$

hence $d \geq 2$ and $\chi(U) > 0$. Since the Julia set is never empty (Corollary I.3.9), we have $\chi(U) < 2$, so we must have $\chi(U) = 1$ and U is a disk. Therefore $c > 0$ implies that U is a disk or has infinite connectivity.

There are thus the following two possibilities (other than Herman rings): either U is a disk, or U is infinitely connected and $c > 0$.

Suppose there is a point $p \in U$ the orbit of which tends to the boundary of U. Let s be the Poincaré distance in U between p and $g(p)$. Then for all n, the Poincaré distance between $g^{\circ n}(p)$ and $g^{\circ(n+1)}(p)$ is at most s. Since $g^{\circ n}(p)$ tends to the boundary of U, it follows that the spherical distance between p and $g(p)$ tends to 0: hence any limit point of the orbit of p must be fixed by g. It is not hard to see that the set of these limit points must be connected, while g has only finitely many fixed points on ∂U. Therefore, the orbit of p converges to a unique point $q \in \partial U$; in fact, since any $p' \in U$ has finite Poincaré distance to p that can not expand during the iteration, it follows that all orbits of points in U converge to the same boundary point q. We need the following result the proof of which will be sketched below:

Lemma I.6.2b (The Snail Lemma). *Let U be a Fatou component and let g be the first return map of U. If all orbits in U tend to the same point $q \in \partial U$, then $g'(q) = 1$.*

It follows that U is a component of the domain of attraction of a rationally indifferent periodic point; the domain U may be simply or infinitely connected.

It is left to consider the case that every orbit has an accumulation point within U. If g is a local isometry of U, then U must be a disk (Corollary I.6.2). Checking automorphisms of the open unit disk D^2, it follows that the dynamics of g on U is conformally conjugate to a rigid rotation of D^2, i.e., to $z \mapsto e^{2\pi i \alpha} z$ for some $\alpha \in \mathbb{R}/\mathbb{Z}$. If α is rational, then g has finite order on U, so some iterate of g is the identity on U and hence everywhere. This is impossible if f and hence g have degree greater than 1. Therefore, U is a Siegel disk.

From now on, we may thus assume that g contracts the Poincaré metric of U, and every orbit has an accumulation point within U. Let $q \in U$ be an accumulation point of the orbit starting at some $p \in U$. Let V be the ε-neighborhood of q in U with respect to the Poincaré metric of U, for $\varepsilon > 0$ small enough so that V is simply connected. There is an $N > 0$ so that $g^{\circ N}$ contracts Poincaré distances in V by at least a factor 4, and there is an $n > N$ and a $z \in V$ so that $g^{\circ n}(z)$ is contained in the $\varepsilon/2$-neighborhood of q. This implies that $g^{\circ n}(V) \subset V$, hence $g^{\circ n}$ has an attracting fixed point in V, and this point must be q. Since all points in U have finite Poincaré distance from q, and the Poincaré distance is contracted, it follows that all points in U converge to q. Again, U may be simply or infinitely connected.

$\boxed{\text{I.6.2a}}$

Proof of I.6.2b (Sketch). The boundary point q must be fixed under g. It can not be an attracting point or the center of a Siegel disk, because these do not lie in the Julia set, and q cannot be repelling, because it is the limit of an orbit that is not eventually constant. Moreover, $g'(q)$ can not be a root of unity other than 1: in that case, g would permute different petals of the attracting flower, but these are in different components of the Fatou set by Proposition I.5.9 (of course, the multiplier of q under its first return map can be a root of unity; the map g is an iterate of the first return map of q). It remains to exclude the case that q is a Cremer point. We will not give a complete proof here, but only sketch the main idea (see [PM97, M06] for details).

Suppose that q is a Cremer point with multiplier $e^{2\pi i \alpha}$ and $\alpha \in (0, 1)$. Choose some point $p \in U$ and connect p to $g(p)$ by some curve $\gamma : [0, 1] \to U$ with $\gamma(0) = p$ and $\gamma(1) = g(p)$, and extend this to a curve $\gamma : [0, \infty) \to U$ by setting $\gamma(n + t) := g^{\circ n}(\gamma(t))$. We have $\gamma(t) \to q$ as $t \to \infty$, so this curve is eventually contained in a neighborhood of q in which g is univalent. By an appropriate choice of p and the initial segment $\gamma([0, 1])$, we may suppose that all of $\gamma([0, \infty))$ is contained in such a neighborhood, and $\gamma : [0, 1] \to U$ is injective. Since $\alpha \neq 0$, this curve must spiral infinitely often around q while converging to q. Connect a point $\gamma(t)$ with a point $\gamma(t + \tau)$ approximately "one turn around q later", so as to cut out a forward invariant simply connected neighborhood V of q (compare Figure I.5). The map g induces a holomorphic self-map of V with fixed point q and so that some orbit converges to q, so by the Schwarz lemma q must be attracting, a contradiction.

$\boxed{\text{I.6.2b}}$

By the "No Wandering Domains Theorem" of Sullivan (Theorem I.8.a), every Fatou component of a rational map is eventually periodic, so it eventually maps onto one of those components described in Theorem I.6.2a. Specifically for polynomials, one Fatou component is always the basin of attraction of the point at ∞; all other Fatou components are simply con-

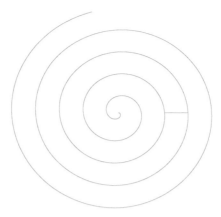

Figure I.5. Illustration of the proof of the Snail Lemma (Lemma I.6.2b): the curve γ spirals around q infinitely often while converging to q. A small cross-cut bounds a forward invariant neighborhood V of q, which by the Schwarz lemma implies that $|g'(q)| < 1$.

nected, by the maximum modulus theorem. In particular, all bounded Fatou components of a polynomial are components of attracting or rationally indifferent basins or Siegel disks: polynomials have no Herman rings.

I.7   The Parameter Spaces for Rational Maps

Until now we have concentrated on the theory of particular rational maps. It is also important to know what happens as the rational map varies.

First we should have in mind some idea of what the natural parameter space is for rational maps.

The parameter space can be described rather cleanly in the case of polynomial maps. A polynomial of degree d is determined by its $d + 1$ coefficients. However, from the dynamical point of view, two polynomials differ in only a trivial way if they are conjugate by an analytic homeomorphism, that is, by an affine transformation of $\mathbb{C}$. If $P(z) = a_d z^d + a_{d-1} z^{d-1} + \cdots + a_1 z + a_0$ is a polynomial of degree $d > 1$, then $P(z)$ can be conjugated by a unique translation $T(z) = z + b$ to make the coefficient a_{d-1} of z^{d-1} zero. Such a polynomial is called *centered*. Since $-a_{d-1}/a_d$ is the sum of the roots of a polynomial, this says that the origin is the center of mass of the d preimages of any point in $\mathbb{C}$ (counting multiplicities).

Once this normalization is made, we can consider the action of conjugating P by the linear transformation $L(z) = \lambda z$. The effect on the coefficient a_k of z^k is to multiply it by λ^{k-1}. With an appropriate choice of λ, the first coefficient a_d becomes 1. However, if λ is multiplied by any $(d-1)$-st root of unity, a_d also becomes 1. A polynomial is called *monic* if its leading coefficient a_d equals 1.

Therefore, the natural parameter space for polynomials of degree d is $\mathbb{C}^{d-1}$, modulo the action of $\mathbb{Z}_{d-1}$ on $\mathbb{C}^{d-1}$:

$$(a_0, a_1, a_2, \ldots, a_{d-2}) \mapsto (\zeta^{-1} a_0, a_1, \zeta a_2, \ldots, \zeta^{d-3} a_{d-2}),$$

where ζ ranges over the $(d-1)$-st roots of unity.

This action is isomorphic to the regular representation of $\mathbb{Z}_{d-1}$, that is, the action of the group on complex-valued functions on the group. The isomorphism suggests a more direct way to describe a polynomial mapping of degree d up to conjugacy. Explicitly, if f is a polynomial of degree d with leading coefficient 1 and next coefficient 0, then as invariants for f we can take the complex numbers

$$a_\zeta = \frac{f(\zeta) - \zeta}{\zeta}$$

which describe the shapes of the triangles formed by 0, ζ and $f(\zeta)$ where ζ ranges over all $(d-1)$-st roots of unity. Conjugation by a $(d-1)$-st root of unity leads to another map conjugate to f which satisfies the same conditions; the advantage is that the action of such a conjugation is just to permute the invariants a_ζ by rotating the subscripts.

The polynomial mapping f is easily reconstructed by an interpolation formula from the a_ζ:

$$f(z) = z^d + \frac{z^{d-1} - 1}{d-1} \sum_{\zeta^{d-1} = 1} \frac{a_\zeta}{z - \zeta}$$

when z is not a $(d-1)$-st root of unity.

In the case of quadratic polynomials, the group is trivial, so that the natural parameter space is $\mathbb{C}$. Any quadratic polynomial can be normalized up to conjugacy to have the form $z \mapsto z^2 + c$. Another normalization which is sometimes useful is $z \mapsto az + z^2$, which has a fixed point at the origin. However, this normalization is not quite canonical, since there are two fixed points.

Any cubic polynomial can be normalized to the form $z \mapsto z^3 + a_1 z + a_0$. The coefficient of the linear term, a_1, is uniquely determined, while the coefficient a_0 is determined only up to sign. The parameter space is $\mathbb{C}^2$, with coordinates (a_1, a_0^2).

The parameter spaces for polynomials of degree four or more are no longer homeomorphic to $\mathbb{C}^n$. For degree four, the space is the cone on the double suspension of the lens space $L(3, 1)$; the general case is similar.

The parameter spaces for rational maps are more complicated. The conditions on the coefficients of the polynomials P and Q for P/Q to be a rational function of degree d are not in general simple: one of P or Q must have a nonzero coefficient for the term of degree d, and the two polynomials must have no roots in common. The equivalence relation is given now by the action of the group of Möbius transformations, rather than just the complex affine group, and the action is more complicated. We will not worry about the exact description of the parameter space for rational maps of general degree. We note, though, that the parameter space has dimension $2(d-1)$.

For the case of degree 2 rational maps, here is what happens (compare also [M93]). Such a map has exactly two critical points: they cannot coalesce, since the Riemann sphere has no branched coverings with a single branch point. We may arrange these two points to be at 0 and ∞. If the rational map is written

$$R(z) = \frac{az^2 + bz + c}{Cz^2 + Bz + A},$$

the involution $z \mapsto z^{\circ -1}$ which exchanges 0 and ∞ acts by interchanging upper- and lowercase letters. The condition that 0 be a critical point is that the determinant of

$$\begin{bmatrix} b & c \\ B & A \end{bmatrix}$$

be zero. The condition that ∞ be a critical point is similar, and obtained by interchanging upper- and lowercase. The set of solutions such that in addition the numerator is not simply a scalar multiple of the denominator is more simply described by the conditions $b = B = 0$.

Thus, any degree-two rational map is conjugate to a map of the form

$$z \mapsto \frac{az^2 + c}{Cz^2 + A},$$

where the pair of points $(a/C, c/A)$ on the Riemann sphere (which are the critical values) must be distinct. This representation is not unique, however. We can further normalize by multiplying both numerator and denominator with any nonzero constant, and by conjugating with any linear map. Therefore, the quadruple (a, c, A, C) is equivalent to the quadruple $(\lambda\mu a, \lambda\mu^{-1}c, \lambda A, \lambda\mu^2 C)$, as well as to the quadruple obtained by switching capitalization.

If neither critical point maps to the other, then the rational map can be normalized to the form

$$z \mapsto \frac{z^2 + c}{Cz^2 + 1},$$

while if neither critical point maps to itself it can be normalized to the form

$$z \mapsto \frac{az^2 + 1}{z^2 + A}.$$

However, this second form is not unique: if ζ is a cube root of 1, then setting $\lambda = \mu = \zeta$, we get a new map which is conjugate to the original, with $a = \zeta^2 a$ and $A = \zeta A$.

Note that the special example $z \mapsto 1/z^2$ which exchanges the two critical points is conjugated to itself by this substitution. In the parameter space, this example represents a singular point, whose link is the lens space $L(3,2)$. The symmetry can be visualized using the observation that the three cube roots of 1 are the fixed points of $z \mapsto 1/z^2$. This map takes each of the sectors of the sphere bounded by meridians through cube roots of unity to the complementary 2/3 sector, while interchanging the two poles.

One simple invariant for a degree-two rational map is the cross-ratio of the two critical points and their critical values. In terms of the coefficients, this is cC/aA.

More significant than the description of the overall topology of the parameter space is the problem of describing mathematically the nature of the wide qualitative variations of rational maps as the parameters are varied.

The Fatou set for a rational map consists of points for which the behavior of nearby points is similar: nearby points have nearby orbits for all time. Such points are dynamically stable. There are analogous conditions one can ask for in the parameter space for rational maps. For all rational maps, there are points which are dynamically unstable — but it can happen that the pattern of this instability (i.e., the Julia set) is stable as the parameter varies.

Definition I.7.1. A rational map f is *structurally stable* if all nearby maps are topologically conjugate by a homeomorphism close to the identity. It is *J-stable* if for any nearby map g, there is a homeomorphism of J_f to J_g near the identity which conjugates $f|J_f$ to $g|J_g$. A rational map is called *hyperbolic* (or equivalently *axiom A*) if every critical point converges under the dynamics to an attracting periodic cycle (or eventually lands on one).

The *parameter limit set* (or equivalently the *bifurcation locus*) is the set of rational maps (up to conjugacy by Möbius transformations) that are *not* J-stable.

Note: this use of the word "hyperbolic" is standard, but it is unrelated to its use for "hyperbolic" Riemann surfaces, or for "hyperbolic" elements of the fundamental group of a Riemann surface.

We have already seen examples of maps f which are J-stable: the map $z \mapsto z^d$ is such an example. It is not structurally stable, because the superattracting fixed points at 0 and ∞ become ordinary attracting fixed points when f is perturbed. More generally,

Theorem I.7.2. *If f is a rational map such that the closure of the union of the forward orbits of the critical points are contained in the Fatou set, then f is J-stable.*

This condition is equivalent to the condition that f is hyperbolic.

Proof of I.7.2. Let P be the closure of the union of forward images of the critical points and $U := \mathbb{C} - P$. Let $V = f^{\circ -1}(U)$. By hypothesis, U and V contain the Julia set. Then f carries V to U as a covering map, since V is contained in U, which has no critical points.

Suppose that we are not in the special case when $\mathbb{C} - U$ has at most two points. Then f is acts as an isometry from the Poincaré metric of V to the Poincaré metric of U. Now $\mathbb{C} - U$ cannot be totally invariant, since it is not an elementary set and it is not the Julia set. Therefore, V is strictly smaller than U, so its Poincaré metric is strictly larger. In other words, f expands lengths of all tangent vectors to $\hat{\mathbb{C}}$ along J_f as measured with respect to the Poincaré metric of U.

In the special case that $\mathbb{C} - U$ consists of two points (it cannot have fewer, since there are at least two critical points), there is also a metric which is expanded by f: any Euclidean metric for the cylinder U.

The rest of the proof is a beautiful, standard argument in dynamical systems called "shadowing". We measure distances with respect to the expanding metric in a neighborhood of J_f. Let ε be small enough so that all distances less than ε are strictly increased, unless they are 0. In particular, if $D_\varepsilon(z)$ is the open ε-disk about a point $z \in J_f$, then $D_\varepsilon(z)$ is mapped injectively to its image by f.

Consider any perturbation f' of f such that the Riemannian metric is still expanded, and so that f' still maps the ε-disk about any point z in J_f injectively to a disk still containing the ε-disk about $f(z)$.

We claim that for each point $z \in J_f$, there is a unique point $h(z)$ whose f'-orbit stays within ε of the f-orbit of z for all forward time. In fact, $h(z)$ is very simply determined as the intersection of $D_\varepsilon(z)$ with $f'^{\circ -1}(D_\varepsilon(f(z)))$, with $f'^{\circ -2}(D_\varepsilon(f^{\circ 2}(z)))$, etc. The condition on f' implies that these disks are nested, so the intersection is not empty.

The fact that f' expands the Riemannian metric implies that the intersection consists of exactly one point $h(z)$. This also implies that $h(z) \in J_{f'}$.

Two distinct points of J_f cannot have orbits which always remain within ε of each other. Thus, at least if the perturbation f' is close enough to f, there can be no other point $z_1 \in J_f$ such that $h(z_1) = h(z)$. It is straightforward to check that h is continuous, so h gives the required conjugacy.

We must still show the equivalence of the two conditions stated in the theorem. It is easy to see that the first condition is implied by the second condition. Indeed, if all critical points tend toward or land on attracting periodic points, then the closure of the union of their forward orbits does not intersect the Julia set, so the first condition is obviously satisfied.

On the other hand, suppose the closure of the union of the forward orbits of the critical points does not intersect the Julia set. We will make use the fact that the Poincaré metric of U is expanded in a neighborhood of the Julia set to find attracting periodic orbits.

If c is any critical point, then c cannot travel through an infinite sequence of distinct components of E_f, otherwise the closure of its forward orbit would necessarily intersect J_f. Let W be a component of E_f which is visited more than once by c, and let g be the return map of f for W.

If we define W_ε to be W minus the open ε-neighborhood of J_f as measured with respect to the Poincaré metric of U, then we see that g maps W_ε into its interior. Impose the W_ε metric on the set $g(W_\varepsilon)$. It is a compact metric space which g maps into itself as a strict contraction. It follows that g has a unique attracting fixed point in W, so c tends toward an attracting periodic orbit of f. $\boxed{\text{I.7.2}}$

The proof of Theorem I.7.2 shows that a rational map is hyperbolic if and only if there exists a Riemannian metric in a neighborhood of the Julia set that is strictly expanded by the map, while the Poincaré metric on the Fatou set is strictly contracted. This corresponds to the general definition in dynamical systems that a map is hyperbolic if the space where the map acts (the "phase space") can be decomposed into invariant sets according to where expansion and contraction occurs.

Conjecture I.7.3.[1] *In the space of rational maps of any degree $d \geq 2$, or of polynomials of any degree $d \geq 2$, all J-stable rational maps f are hyperbolic.*

There are two standard equivalent definitions for the Julia set of a map: it can be defined either as the complement of the Fatou set, or as the set of accumulation points of periodic points. Analogously,

Theorem I.7.4. *Let f be any rational map which is in the parameter limit set. Then f can be approximated by rational maps with a critical point which eventually lands on a repelling periodic point.*

Also, f can be approximated by rational maps with indifferent periodic points whose return map has derivative 1.

Rather than giving a proof in the general case, we prove the following result for the space of quadratic polynomials. The Mandelbrot set **M** is the space of quadratic polynomials, parametrized as $z \mapsto z^2 + c$, for which the orbit

---

[1] In the original manuscript, this was stated, without proof, as Theorem I.7.3.

of the critical value c is bounded (we will see in Proposition II.3.1 that this is exactly the set of polynomials $z \mapsto z^2 + c$ for which the Julia set is connected).

Theorem I.7.4a. *Every quadratic polynomial $p_{c_0} : z \mapsto z^2 + c_0$ with $c_0 \in \partial\mathbf{M}$ can be approximated by quadratic polynomials p_c for which the finite critical value c eventually lands on a repelling periodic point. Also, c_0 can be approximated by quadratic polynomials for which the critical value is periodic, or by those that have an indifferent periodic point for which the derivative of the return map is a root of unity.*

Proof of I.7.4a. By definition, any neighborhood of p_{c_0} contains polynomials p_c for which the orbit of the finite critical value c is unbounded, and also polynomials for which it is bounded. The family of maps $P_n(c) := p_c^{\circ n}(c)$ is thus not normal in a neighborhood of c_0. It avoids the point ∞. If it also avoids the two fixed points of p_c, then it is normal by Montel's Theorem I.2.9b, so every neighborhood of c_0 contains parameters c in which the critical orbit is eventually fixed. This proves the first claim for all parameters $c_0 \in \partial\mathbf{M}$ for which both fixed points are distinct and repelling. Similarly, if there is a repelling periodic cycle of period 2, then a similar argument works. It is a simple explicit calculation that for every $c_0 \in \mathbb{C} - \{-0.75\}$, the polynomial $z \mapsto z^2 + c_0$ has two distinct repelling fixed points, or a repelling 2-cycle, or both. This proves the first claim.

For the second claim, we use a similar argument: if in a neighborhood of c_0, the maps $P_n(c)$ avoid the three points ∞, 0, and c, then they form a normal family and c_0 cannot be in $\partial\mathbf{M}$. But $P_n(c) \neq \infty$, and if $P_n(c) = 0$ or $P_n(c) = c$, then c is periodic.

Finally, consider a point $c_0 \in \partial\mathbf{M}$ and let $c_n \to c_0$ be a sequence of parameters for which the critical point is periodic. Let r_n be the radius of the largest disk around c_n that is contained in $\mathbf{M}$. Then clearly $r_n \to 0$, and it is easy to see that there is some c_n' with $|c_n' - c_n| \leq r_n$ so that $z \mapsto z^2 + c_n'$ has a periodic orbit for which the derivative of the return map maps absolute value 1, and there is some c_n'' with $|c_n'' - c_n| < 2r_n$ for which this derivative is a root of unity. $\boxed{\text{I.7.4a}}$

The fact that c_0 can be approximated by polynomials that have indifferent periodic points whose return map has derivative 1 can be shown using the methods from Part II: this follows from the fact that leaves with periodic endpoints are dense in QML; see the remark after Corollary II.6.13a.

For quadratic polynomials, $\partial\mathbf{M}$ is exactly the bifurcation locus of quadratic polynomials. In fact, every neighborhood of $z \mapsto z^2 + c_0$ contains polynomials with connected and with disconnected Julia sets, so c_0 is not J-stable. For the converse, see e.g., [Mc94, Theorem 4.2].

There is a filtration of the parameter space, according to how many of the critical points are on the Julia set. This filtration is analogous to the decomposition of the Riemann sphere into the Julia set and the Fatou set (but, unlike Fatou and Julia sets, this filtration is neither open nor closed).

I.8 Quasiconformal Deformations of Rational Maps

The theory of quasiconformal maps and conformal structures has proven to be a powerful tool in the theory of Kleinian groups, thanks to the work of Ahlfors, Bers and others.

Sullivan has shown how to extend the ideas which were useful for Kleinian groups to the theory of rational mappings. In this section, we shall sketch some of the results along these lines; see [S85a,S85b,MS98,S86] for more details.

We will need some background on quasiconformal homeomorphisms and the Measurable Riemann Mapping Theorem. We will only sketch their definition and main properties and refer to [A06] for details. There are many equivalent definitions of quasiconformal homeomorphisms. A geometric definition is as follows: a homeomorphism h between two Riemann surfaces is K-*quasiconformal* (for $K \in [1, \infty)$) if for every annulus A of finite modulus $\mu(A)$, the modulus of $h(A)$ satisfies

$$\frac{1}{K}\mu(A) \leq \mu(h(A)) \leq K\mu(A).$$

A homeomorphism between two Riemann surfaces is called *quasiconformal* if it is K-quasiconformal for some $K \geq 1$. Every quasiconformal homeomorphism is differentiable almost everywhere (but not every homeomorphism that is differentiable almost everywhere is quasiconformal, even if the derivatives are bounded). Every diffeomorphism of the Riemann sphere to itself is quasiconformal, but for our purposes non-smooth diffeomorphisms are important.

In Section I.2 we stated the uniformization Theorem I.2.6, which asserts that every Riemannian metric is conformally equivalent to a complete metric of constant curvature, unique up to a constant except when the underlying surface is a sphere. For the Riemann sphere, the metric is unique up to a constant factor and a Möbius transformation.

Morrey [Mo38] showed that this theorem applies to Riemannian metrics which are not necessarily smooth, but are defined by a measurable section of the bundle of quadratic forms which is positive definite, bounded and bounded away from non positive definite forms.

It is better to state the condition in terms of conformal structures, which are Riemannian metrics up to conformal changes. If we have one Riemannian metric g to use for reference, then any other Riemannian metric g' can be scaled until it has the same area form as g. If g' is now diagonalized with respect to g, it will have two eigenvalues $\lambda \geq 1$ and λ^{-1} along a pair of orthogonal axes. The conformal structure of g' is determined by the measurable function λ and a measurable line field defined where $\lambda > 1$ representing the axis for λ. The boundedness conditions say that λ is bounded. All of the above is up to sets of measure 0, of course. If g' is a conformal structure on the Riemann sphere and h is a quasiconformal homeomorphism from the Riemann sphere to itself, then g' can be transformed under h to a new conformal structure because h is differentiable almost everywhere.

We give the space of conformal structures a topology coming from the following metric: given two conformal structures g_1 and g_2, let K_0 be the infimum over all $K \geq 1$ so that there exists a K-quasiconformal homeomorphism that transforms g_1 to g_2; we then define the distance between g_1 and g_2 to be $\log K_0$.

The powerful Measurable Riemann Mapping Theorem [A06] is an extension of Morrey's result. One version of it says that for any measurable conformal structure g on the Riemann sphere, there is a quasiconformal homeomorphism h that transforms g to the standard structure on the Riemann sphere (in the sense that the pull-back under h of the standard structure is g). The map h is unique up to postcomposition with Möbius transformations, and it depends continuously on g (for the topology of uniform convergence of homeomorphisms on the Riemann sphere, properly normalized). In fact, the dependence of h is analytic in the following sense: if g depends analytic on some parameter t, and h is normalized so as to fix any three points on the sphere, then $h_t(z)$ is analytic with respect to t for every z in the sphere.

This gives rise to a powerful technique for constructing rational maps which are topologically conjugate to a given map f but not conjugate by Möbius transformations. All we have to do is find a measurable conformal structure which is invariant by f. If g' is any such structure, then by the Measurable Riemann Mapping Theorem there is some quasiconformal homeomorphism h of the Riemann sphere transforming g' to the standard conformal structure. The conjugate $f_{g'} = h \circ f \circ h^{\circ -1}$ is conformal with respect to the standard structure! Therefore, $f_{g'}$ is a rational map. For a typical choice of g', it is unlikely that $f_{g'}$ is conjugate to f by a Möbius transformation. On the other hand, sometimes $f_{g'}$ is Möbius-conjugate to f. In order for the method to be really useful, we need to be able to identify which pairs g_1 and g_2 give rise to equivalent rational maps f_{g_1} and f_{g_2}.

The *quasiconformal deformation space* of a rational map f is the space of all equivalence classes of rational maps (up to Möbius transformations, that is) which are conjugate to f by a quasiconformal homeomorphism.

The method of quasiconformal deformations of Julia sets was pioneered by Sullivan in his famous proof of the *No Wandering Domains Theorem* [S85a]:

Theorem I.8.a. *For every rational map, every component of the Fatou set is eventually periodic.*

We will need the following lemma in the proof.

Lemma I.8.b. *Let g_t be a family of conformal structures on the Riemann sphere that are invariant under a rational map f and that depend continuously on a parameter $t \in [0,1]$, and let h_t be a family of quasiconformal homeomorphisms that transform the structures g_t to the standard conformal structure on the Riemann sphere, and so that the h_t depend continuously on t. Suppose all rational maps $f_t := h_t \circ f \circ h_t^{-1}$ are equal. Then all h_t coincide on the Julia set of f.*

Proof of I.8.b. The maps h_t conjugate f to f_t, so they send periodic points of f to periodic points of f_t. These periodic points depend continuously on t (because h_t does). Since there are only finitely many periodic points of any given period and $f_t = f_0$, it follows that all h_t coincide on the set of periodic points. Since the Julia set is contained in the closure of the set of periodic points (Theorem I.3.11), it follows that the restrictions of all h_t to the Julia set coincide. $\boxed{\text{I.8.b}}$

Remark I.8.c. Given $d \geq 2$, we define a (real) $4d - 3$-dimensional manifold of diffeomorphisms from the circle $\mathbb{R}/\mathbb{Z}$ to itself, as follows. Consider $4d$ distinct points $a_1, \ldots, a_{4d-3}, b_0, b_1, b_2$ on $\mathbb{R}/\mathbb{Z}$. There is an $\varepsilon > 0$ so that for every $i = 1, \ldots, 4d - 3$ and $\varepsilon_i \in (-\varepsilon, \varepsilon)$, there is a smooth circle diffeomorphism that fixes all three points b_0, b_1, b_2 and that sends each a_i to $a_i + \varepsilon_i$. These diffeomorphisms are parametrized by an open cube in $\mathbb{R}^{4d-3}$. They are constructed so that they all have different cross-ratios among the points a_i and b_i, so no conformal automorphism of D^2 induces on ∂D^2 any of these circle diffeomorphisms (other than the one with all $\varepsilon_i = 0$).

It is well known and easy to see that each of these circle diffeomorphisms extends to a smooth diffeomorphism with bounded derivative, and hence to a quasiconformal homeomorphism, from D^2 to itself.

Proof of I.8.a. Let f be a rational map of some degree $d \geq 2$ and suppose that U is a wandering Fatou component of f. Replacing U by some $f^{\circ n}(U)$ if necessary, we may assume that neither U nor any Fatou component on its forward orbit contains a critical point. By Corollary I.6.2, we may also assume that U is simply connected. This implies that U maps univalently to all Fatou components on its entire forward orbit. We can then put any measurable conformal structure on U, transport it forward under f and backward under f^{-1} to a measurable invariant conformal structure on the Riemann sphere (univalence is used in order to assure that the push-forward of the conformal structure is well defined). The space of invariant conformal structures is an infinite-dimensional vector space. Each invariant conformal structure yields a rational map of the same degree as f. The key point is to assure that enough of these maps must be different; this would be a contradiction to the fact that the parameter space of rational maps of fixed degree d has complex dimension $2(d-1)$ (see Section I.7). (Note however, that there are wandering open sets in the Fatou set near any attracting periodic point that map forward univalently; in this case, the infinite dimensional space of invariant conformal structures only yields a finite dimensional space of rational maps.)

Suppose first that U has locally connected boundary, so any Riemann map $\phi \colon D^2 \to U$ extends continuously to the boundary. In Remark I.8.c, we constructed a real $4d-3$-dimensional space C of quasiconformal homeomorphisms of D^2 so that none of them, other than the identity, induces the same circle diffeomorphism on ∂D^2 as a conformal automorphism of D^2. We may choose the points $a_1, \ldots, a_{4d-3}$ and b_0, b_1, b_2 that define the circle diffeomorphisms in C in such a way that the extension of ϕ from $\overline{D^2}$ to $\overline{U}$ is injective on the set of these $4d$ points.

Use the Riemann map ϕ to define a $4d-3$-dimensional space of conformal structures on U, and extend them by f to invariant conformal structures on the entire Riemann sphere (even if these structures are constructed so that they are smooth on U, their extensions to the sphere will be discontinuous on the Julia set). Each of these defines a rational map of degree d, so we obtain a family of rational maps that is parametrized by $4d-3$ real coordinates, and so that the rational maps depend smoothly on these coordinates. Since the space of rational maps up to conformal conjugacy has complex dimension $2d-2$, it follows that there exists a smooth injective curve $\gamma \colon [0,1] \to C$ into the space of circle diffeomorphisms so that the corresponding rational maps are all conformally conjugate; denote these rational maps by f_t. Let h_t be the quasiconformal homeomorphisms that send the conformal structure induced by $\gamma(t)$ to the standard complex structure so that we have $f_t = h_t \circ f \circ h_t^{-1}$. Postcomposing all h_t with Möbius transformations, we may suppose that all f_t are the same, and we can apply Lemma I.8.b. It follows that all h_t coincide on the Julia set. Therefore, for all $t, t' \in [0,1]$, the composition $h_t^{-1} \circ h_{t'}$ is the identity on ∂U (and even on all of U because this is conformal on U). But the family h_t has been constructed in such a way that cross ratios on ∂U are different for h_t and $h_{t'}$, a contradiction.

If the boundary of U is not locally connected, then the Riemann map $\phi \colon D^2 \to U$ may not extend continuously to the boundaries, but it is well

known that it has radial limits in almost all directions [A73], and we use
points a_i and b_i on ∂D^2 for which radial limits exist. $\boxed{\text{I.8.a}}$

One of the fundamental conjectures in the field of complex dynamics is
the following conjecture:

Conjecture I.8.d. *In the space of rational maps for any degree $d \geq 2$, or
in the space of polynomials of degree $d \geq 2$, hyperbolic dynamics is dense.*

Note that this conjecture does not apply to arbitrary analytic spaces
of rational maps: for instance, in the 2-dimensional space of cubic polynomials, there is a 1-dimensional subspace consisting of maps that have an
irrationally indifferent fixed point of constant multiplier λ with $|\lambda| = 1$,
and these maps are all non-hyperbolic.

According to a theorem of Mañé, Sad, and Sullivan [MSS83] and Lyubich [L83], structurally stable maps are dense in every analytic space of
rational maps. Conjecture I.8.d can thus be restated as saying that in
the spaces of rational maps or polynomials of given degree $d \geq 2$, every
structurally stable component is hyperbolic (this is Conjecture I.7.3).

Hyperbolic rational maps are locally parametrized by the derivative at
the attracting periodic points, provided each critical point has its own attracting basin. More generally, they are parametrized by conformal structures on tori with finitely many punctures: each torus corresponds to the
quotient of a neighborhood of an attracting periodic point, where points on
the same orbit are identified and the punctures correspond to the critical
orbits.

Part II: Polynomials and the Riemann Mapping near Infinity

II.1 Julia Sets for Polynomial Mappings: Introduction

For a polynomial mapping, the boundary of the attracting basin of infinity equals the Julia set (see Proposition II.3.2). Therefore, the understanding of the combinatorics of polynomial mappings can be approached by backward extrapolation from infinity. This technique has been exploited with great success by Adrien Douady and John Hubbard, particularly for the quadratic case.

Polynomial mappings are analogous to function groups in the theory of Kleinian groups. Function groups are Kleinian groups which leave invariant at least one component of their domain of discontinuity. They are closely tied to the geometry and topology of hyperbolic surfaces.

In Part II, we will give at least the beginnings of a development of the theory of polynomial mappings using techniques parallel to some which have been used successfully in the theory of Kleinian groups. This is closely related to part of the work of Douady and Hubbard, but from a slightly different point of view.

II.2 The Limiting Behavior of the Riemann Map

The Riemann mapping theorem constructs a biholomorphic map from any simply-connected open subset $U \subset S^2$ to the disk, provided $S^2 - U$ has more than one point. What happens, though, near the frontier ∂U? In particular, suppose that $g : D^2 \to U$ is a one-to-one holomorphic mapping (in the inverse direction from the usual statement of the Riemann mapping theorem). Under what circumstances does g extend to a continuous map on all of $\overline{D^2}$? Here is the most relevant definition.

Definition II.2.1. A set X is *locally connected* if for every $x \in X$ and every neighborhood W of x there is a smaller neighborhood $V \subset W$ of x that is connected (i.e., if $A \subset W$ and $B \subset W$ are disjoint relatively closed subsets whose union is W, then one of them is disjoint from V). A set X is *locally path-connected* if for every $x \in X$ and for every neighborhood W of x there is a smaller neighborhood $V \subset W$ of x such that any two points in V are connected by a path in W. A *path* is a continuous map from the unit interval to the space.

Local path-connectedness easily implies local connectedness. For a metric space (in particular, a subset of S^2) which is, say, locally compact, it is not too hard to prove the converse — that if it is locally connected, it is also locally path-connected. To do this, first "join" any two nearby points by a sequence of points spaced no more than say 1/10th the distance apart — then join each adjacent pair of points in this sequence by a new sequence spaced no more than a 1/100th of the original distance apart, and not wandering too far from the pair they are joining — and continue on in this way, then pass to the limit. The basic property behind this argument is that any compact connected and locally connected set is uniformly locally connected: for every $\varepsilon > 0$ there is a $\delta > 0$ so that any two points x, y with distance less than δ are contained in a connected set of diameter less than ε.

Whenever the inverse Riemann mapping g extends continuously to the closed disk, its frontier ∂U must be locally path-connected because of the following result:

Lemma II.2.2. *The continuous image of a compact, locally path-connected space is locally path-connected.*

Proof of II.2.2. Let $f : X \to Y$ be a continuous surjection, where X is compact and locally path-connected. Let $y \in W$ be a point and neighborhood in Y. For each element $x \in f^{\circ -1}(y)$, there is an open neighborhood U_x any two of whose points are connected by a path in $f^{\circ -1}(W)$. Let U be the union of any collection of such open neighborhoods which cover $f^{\circ -1}(y)$. The image $V = f(U)$ is a neighborhood of y, since $f(X - U)$ is a compact set which does not contain y. Any point in V can be connected to y by a path in W. Therefore Y is locally path-connected. $\boxed{\text{II.2.2}}$

It turns out that local connectedness of $S^2 - U$ is, in fact, the precise condition for the inverse Riemann mapping to extend:

Theorem II.2.3. *The inverse Riemann mapping g extends to a continuous map $\overline{g}$ defined on the closed disk $\overline{D^2}$ if and only if the complement $S^2 - U$ is locally connected.*

This remarkable theorem is a consequence of Carathéodory's beautiful theory of prime ends. Rather than getting involved in this theory, we will give a direct proof.

Proof of II.2.3. We have essentially taken care of the easy direction already, that if the inverse Riemann mapping extends, then $S^2 - U$ is locally connected. The lemma shows that if g extends continuously to ∂D^2, the frontier ∂U is locally path-connected. It is also clear that if ∂U is locally path-connected, $S^2 - U$ is also locally path-connected: draw a straight line between any two of its points; if it doesn't intersect U, it is a path connecting the two points, and if it does, replace a segment of it by a path in the frontier of U. Local connectivity follows.

Sufficiency of the condition that $S^2 - U$ be locally connected is somewhat trickier. We will prove that g extends without constructing the extension ahead of time, by showing that for every ε and every point $x \in \partial D^2$ there is a neighborhood V of x in the closed disk such that the image of its intersection with the open disk has diameter less than ε. This is really a form of Cauchy's condition for convergence.

We first construct a neighborhood of any point $x \in \partial D^2$ bounded in the open disk D^2 by an arc whose image in U has an arbitrarily short length. This first step works for any simply-connected set U, even if its complement is not locally connected. (Note that in the following lemma and its corollary, as well as in many arguments elsewhere, we use the spherical metric of S^2 for curve lengths and areas.)

Lemma II.2.4. *Let $f : Q \mapsto \mathbb{C}$ be any holomorphic embedding of the unit square $Q = (0,1) \times (0,1)$ in the Riemann sphere. If A is the area of $f(Q)$, at least one of the arcs of the form $f(\{s\} \times (0,1))$ and at least one of the arcs of the form $f((0,1) \times \{t\})$ is rectifiable and has length less than $\sqrt{A}$.*

Proof of II.2.4. We have

$$A = \iint_Q |f'|^2 \, ds \, dt \geq \left(\iint_Q |f'| \, ds \, dt \right)^2$$

by the Schwarz inequality. The integral on the right is the average length of an arc, by Fubini's theorem. $\boxed{\text{II.2.4}}$

Corollary II.2.5. *For any point $x \in \partial D^2$ and any $\varepsilon > 0$, there is a neighborhood V of x in $\overline{D^2}$ whose intersection with D^2 is bounded by an arc α such that $g(\alpha)$ is rectifiable and has length less than ε.*

Proof of II.2.5. Map the unit square Q via the map $z \mapsto z^2$; then the images of two sides cover the interval $(-1,1)$. Then map the upper half plane by a Möbius transformation M to the disk so that 0 goes to x; then Q maps to the intersection of a neighborhood of x with the open disk. Restrict Q to a sub-square $Q_\delta := (0, \delta) \times (0, \delta)$ so that the area of the image of Q_δ is less than $\varepsilon^2/4$. Using Lemma II.2.4, we obtain $\delta_1, \delta_2 > 0$ so that the image of $(\{\delta_1\} \times (0, \delta_2]) \cup (0, \delta_1] \times \{\delta_2\})$ is a curve as desired. $\boxed{\text{II.2.5}}$

Local connectedness comes in precisely through the following.

Proposition II.2.6. *Let $U \subset S^2$ be any simply-connected open set whose complement is locally connected. For every ε there is a δ such that if α is any arc with interior in U and endpoints on ∂U, and if α has diameter less than δ, one or the other of the two components of $U - \alpha$ has diameter less than ε.*

Proof of II.2.6. First, recall that if a compact subset K of S^2 (such as $S^2 - U$ in our case) is locally connected, then it is uniformly locally connected: for every ε there is a $\delta < \varepsilon$ such that any δ-ball in K intersects only one component of the concentric ε-ball.

If α is an arc as in the proposition contained in a disk of radius δ, consider how the two components of $U - \alpha$ might intersect the concentric circle of radius ε. If both components were to intersect this circle, there would be an arc β in the intersection of U with the disk of radius ε, beginning and ending on the boundary, and cutting across α exactly once. The arc β would cut the disk into two halves, and the two endpoints of α would be in opposite halves: this contradicts the choice of δ.

The only possible conclusion is that one or the other of the two components of $U - \alpha$ does not intersect the circle of radius ε, and hence is contained in the disk of radius ε. $\boxed{\text{II.2.6}}$

$\boxed{\text{II.2.3}}$

II.3   Identifications in the Limit of the Riemann Mapping

Consider a polynomial map P of degree d which satisfies the condition that the forward orbits of the finite critical points are bounded. We may assume that P is *monic*, i.e., it has leading coefficient 1, after conjugating by a complex affine transformation.

Let E_∞ be the set of all points in the sphere that converge to ∞ under iteration of P. The point at infinity is a super-attracting fixed point of degree d for P. Therefore, by Proposition I.5.5, P is conjugate to the map $z \mapsto z^d$ in a neighborhood of infinity. We may assume that the conjugacy is close to the identity near infinity. With this proviso, the conjugating homeomorphism g is unique. (Note, however, that there is non-uniqueness in the picture coming from the fact that the map $z \mapsto \omega z$, where ω is a $(d-1)$-st root of unity, conjugates any polynomial of degree d with leading coefficient 1 to another such polynomial.)

Suppose that a conjugating homeomorphism g has been defined for some open disk D containing infinity with the property that $P(D) \subset D$. Every point inside D tends toward infinity with positive time. The inverse image $P^{-1}(D)$ is also a disk: it is obtained by taking a certain d-fold branched cover of the Riemann sphere, branched over the critical values of P. Since none of the finite critical points tends toward infinity, none of the critical values lies in D, so the branched cover is also a disk. There is a unique way to extend g analytically to this disk $P^{\circ -1}(D)$. We can continue indefinitely until we have defined g on the entire set E_∞. This proves the following.

Proposition II.3.1. *For any polynomial P of degree $d \geq 2$ such that no finite critical points tend to infinity in forward time, the set $E_\infty \subset S^2$ consisting of all points which are attracted to infinity is a simply connected set. The Riemann mapping g from E_∞ to the disk $\{|z| > 1\}$ conjugates P in this region to the map $z \mapsto z^d$. The conjugating Riemann mapping g is unique up to multiplication by a $(d-1)$-st root of unity.* $\boxed{\text{II.3.1}}$

There is a converse to this result: if E_∞ has a Riemann map to the open unit disk (or equivalently, if the Julia set of P is connected), then no finite critical point of P tends to infinity: the Riemann map conjugates the dynamics of P to a holomorphic self-map of the disk so that the critical point coming from ∞ has maximal multiplicity, and there cannot be another critical point in the disk.

Proposition II.3.2. *Every neighborhood of every point in the Julia set of a polynomial map intersects E_∞.*

Proof of II.3.2. This follows from the fact that the union of the forward images of any neighborhood of any point in the Julia set is the sphere minus at most two points (Proposition I.3.8). $\boxed{\text{II.3.2}}$

This elementary proposition immediately suggests two questions as holding clues to the theory of polynomial self-maps of $\mathbb{C}$: Under what circumstances does the inverse Riemann map $g^{\circ -1}$ extend to a continuous map of the disk to $\hat{\mathbb{C}}$? (By the results of Section II.2 this is equivalent to the question of whether or not the Julia set is locally connected.) If $g^{\circ -1}$ does extend to a continuous map on the disk, what identifications does g make on the boundary of the disk?

There are certainly big restrictions on what identifications can occur. First, there are restrictions which can be deduced from the separation properties of the sphere. These are conveniently described by the following proposition:

Proposition II.3.3. *Let $f : \overline{D^2} \to S^2$ be a continuous map of the closed disk to the sphere so that the restriction to the interior of the disk is an embedding. Let K_1 and K_2 be inverse images of two points in the image of the boundary of the disk. (In other words, they are equivalence classes in the identification.) Then the convex hull of K_1 is disjoint from the convex hull of K_2.*

Proof of II.3.3. Otherwise, there would be paths α_i beginning and ending on K_i which intersect once transversely in the disk. In the image, they would become closed curves which cross exactly once — a contradiction.

$$\boxed{\text{II.3.3}}$$

In view of the proposition, there is a convenient way to draw a picture of the set of identifications which occur on the boundary of the disk — as a collection of disjoint convex subsets of the disk which are the convex hulls of sets of points on the circle; see Figure II.1. Note that for the convex hulls one can equivalently use the Euclidean metric or the hyperbolic metric of D^2 (Euclidean geodesics in D^2 are the same as hyperbolic geodesics in the Klein model; usually the Poincaré disk model of hyperbolic geometry is used because this yields the clearest pictures). It is worth emphasizing that the convex hulls must be disjoint in the closed disk, not just in the open disk — this condition is not always easy to discern from a sketch. The collection of convex hulls must be closed, in the sense that if a pair of points is the limit of a sequence of pairs of points such that each pair of the sequence belongs to some convex hull in the collection, then the limiting pair must also belong to a convex hull in the collection (see Douady [D93] for more details). Any equivalence relation satisfying these conditions can be shown to arise as the set of identifications created by some continuous map of the disk into the sphere which is an embedding of the open disk.

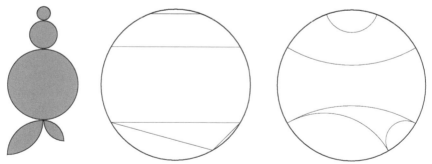

Figure II.1. Left: A compact connected set $K \subset \mathbb{C}$; center: the corresponding set of boundary identifications for the inverse Riemann map, using the Euclidean metric of D^2 for the convex hull; right: the same boundary identifications for the hyperbolic Poincaré metric of D^2 (visualizations in this metric generally look clearer).

It is convenient to think of the picture of the convex hulls in the closed unit disk $\overline{D}_1 \subset S^2$ as representing a set of identifications on the boundary of the complementary disk $D_2 = S^2 - \overline{D}_1$. Then the image of $\overline{D}_2$ is homeomorphic to the space obtained from $\overline{D}_2$ union all the convex hulls by collapsing each convex hull to a point. The map of $\overline{D}_2 \to S^2$ is surjective if and only if the union of the convex hulls is all of $\overline{D}_1$.

More restrictions stem from the fact that the identifications must be compatible with the dynamics.

(a) If a set K of points on the circle is to be identified to a single point, then its image under $z \mapsto z^d$ must also be identified to a single point.

(b) Suppose that the set K is an equivalence class under the identification. Let L be the set of dth roots of elements of K. Then L is divided into at most d equivalence classes $L_1 \ldots L_k$. Each L_i maps with a certain constant multiplicity m_i, i.e., it contains exactly m_i of the dth roots of each element of K.

The assertion about multiplicities needs a little explanation. Think of points on the circle as endpoints of rays in the disk. Points on the circle are identified when rays of the disk, mapped via the inverse Riemann mapping to the complement of the Julia set, have the same endpoints. Any point p in the Julia set has some punctured neighborhood U so that the map $P : P^{\circ -1}(U) \to U$ is a covering map, and the components of $P^{\circ -1}(U)$ are in one-to-one correspondence with the elements of $P^{\circ -1}(p)$. Then for any element $q \in P^{\circ -1}(p)$ and for any ray converging to p, there are exactly $\deg(V \to U)$ inverse images of the ray converging to q, where V is the component of $P^{\circ -1}(U)$ corresponding to q.

Every equivalence class therefore has a well-defined multiplicity under the map $z \mapsto z^d$: the number of points equivalent to any point in the class which have the same image as that point. When the multiplicity is greater than one, the equivalence class is a *critical* equivalence class.

When we form the convex hulls of all the equivalence classes, there may or may not be regions of the disk left over, not filled by any convex hull. If R is any such region, its boundary is made up of parts of the circle, and secants; no two secants can have an endpoint in common. The image of ∂R after the identification is homeomorphic to a circle. Since any circle on S^2 is the boundary of two disks, this implies that the Fatou set for P has a bounded component, bounded by this circle, corresponding to R.

The analysis will be greatly simplified by a theorem of Sullivan [S85a], which answered a longstanding problem (see Section I.8):

Theorem II.3.4. *There are no wandering domains in the Fatou set. Every component of the Fatou set for a rational map is eventually periodic, i.e., its forward images eventually start repeating in some cycle.* $\boxed{\text{II.3.4}}$

This means that any component of the complement of the union of the convex hulls of the equivalence classes is eventually periodic.

When any region returns to itself, there is an induced map on the quotient space of its boundary obtained by collapsing all secants to points. The return map of this quotient circle is a covering map.

II.4   Invariant Laminations

In the preceding section (II.3) we began the analysis of what the picture can look like when the Julia set of a polynomial is connected and locally connected. In this section, we will carry this line of investigation forward, but on a more abstract level. We will define the notion of an invariant lamination (which we secretly think of as determining a potential set of identifications for a polynomial map), and we will see how far we can go toward the goal of classifying all invariant laminations.

Definition II.4.1. A *geodesic lamination* is a set L of chords in the closed unit disk $\overline{D^2}$, called *leaves* of L, satisfying the following conditions:

(GL1) elements of L are disjoint, except possibly at their endpoints;

(GL2) the union of L is closed.

A *lamination* on the disk $\overline{D^2}$ is a collection of embedded arcs, disjoint except possibly at their endpoints, whose intersection with any sufficiently small neighborhood in the closed disk is topologically equivalent to the intersection of a geodesic lamination with some disk. A *gap* of a lamination L is the closure of a component of the complement of $\bigcup L$. Any gap for a geodesic lamination is the convex hull of its intersection with the boundary of the disk. If G is a gap, we denote this intersection by G_0. When G_0 is finite, then G is a polygon.

Definition II.4.2. A geodesic lamination L is *forward invariant* for the map

$$f : S^1 \to S^1, \qquad f(z) = z^d,$$

if it satisfies the following conditions:

(GL3) Forward invariance: if any leaf $\overline{pq}$ is in L, then either $f(p) = f(q)$, or $\overline{f(p)f(q)}$ is in L.

The lamination L is *invariant* if it is forward invariant and satisfies conditions (GL4) and (GL5):

(GL4) Backward invariance: if any leaf $\overline{pq}$ is in L, then there exists a collection of d disjoint leaves, each joining a preimage of p to a preimage of q.

(GL5) Gap invariance: for any gap G, the convex hull of the image of $G_0 = G \cap S^1$ is either

- a gap, or
- a leaf, or
- a single point.

In the case of a gap, the boundary of G must map locally mono-tonically (not necessarily strictly monotonically) to the boundary of the image gap with positive orientation, that is, the image of a point moving clockwise around G must move clockwise around the image of G (this is a covering map, except that the boundary map may be constant on some intervals; the boundary map from a gap to an image gap thus has a well-defined positive degree)[2].

The notion of an invariant lamination is tricky, because the map $z \mapsto z^d$ is not a homeomorphism. Condition (GL4) is not as strong as one might think: it is not specific about which of the preimages of p and q are joined by leaves, and there is really no good rule to decide this in general. It does not say anything about additional leaves beyond the minimum. Condition (GL5) helps control the possibilities, by guaranteeing that at least the image of a gap cannot cross leaves of the lamination. It sometimes happens, for instance, that there is a polygonal gap with $2k$ sides, whose vertices all map to the endpoints of a single leaf. If the sides of this polygon are isolated, then without condition (GL5), one could erase half of its sides (and all preimages of them) and still have a lamination satisfying the other conditions.

For any forward invariant geodesic lamination L, the map $z \mapsto z^d$ of the unit circle to itself extends to a map of the circle union $\bigcup L$ to itself. This extension is formed by mapping each leaf linearly to the leaf (or point) joining the images of its endpoints. It is not quite so clear how to extend the map over the gaps, but it can still be done, provided the lamination is invariant. One way to define an actual map is to decree that the center of mass of any gap maps to the center of mass of its image, and that each ray from the center of mass to a point on the boundary of a gap maps linearly to the appropriate ray in the image.

Definition II.4.3. Let L be an invariant lamination and G a gap of L. The *degree of G* is defined to be the degree of the map from G to its image gap, provided its image is a gap. If the image of G is a leaf l, the degree of G is half the number of leaves on ∂G mapping to l. If the image is a point, the degree is zero. If the degree of G is not one we call G a *critical gap*.

---

[2]This definition has slightly changed with respect to the original manuscript.

A leaf $l \in L$ is a *critical leaf* if the images of its two endpoints are the same (so the leaf maps to a point). (Note that any gap with degree zero is a polygon bounded by critical leaves.)

The idea of an invariant lamination has a generalization which is often useful. Suppose we have two laminations L and M on two copies of the disk, and suppose that f is a map from the boundary circle of one disk to the boundary circle of the other, which is topologically a covering map. Then L and M *correspond* by f if conditions (GL3), (GL4) and (GL5) above are satisfied for f. The notions of the degree of a gap, and the definition of a critical leaf, go over to this more general situation.

Proposition II.4.4. *For any two laminations L and M which correspond by a map f of degree d, the total number of critical leaves of L with respect to f, plus the sum for critical gaps of their degree minus one, equals $d - 1$.*

Proof of II.4.4. If there are any critical leaves, we can inductively reduce the proposition to cases with lower degree. Simply cut the circle for the domain lamination along any critical leaf l, then glue the endpoints of the two resulting arcs to form two circles. The lamination L produces laminations L_1 and L_2 on the two pieces. The original map f splits into two new maps f_1 and f_2 of total degree d which make the two new laminations correspond to M. Each critical leaf of the original lamination yields a critical leaf in one of the new laminations, except that if l is on a triangle made of three critical leaves, then the remaining two sides merge to give only one critical leaf — but there is also one less critical gap of degree 0. Any other kind of critical gap yields a critical gap of the same degree in one of the new laminations. By induction, this reduces the proposition to simpler cases.

If there are no critical leaves, the extension of f over the disk, as discussed above, is a branched covering. The branch points correspond to critical gaps, and the degrees of branching $b_1, \ldots, b_k$ are the degrees of the gaps. A simple Euler characteristic computation gives

$$d\chi(\overline{D^2}) - \sum_{b_i}(b_i - 1) = \chi(\overline{D^2}),$$

and the proposition follows because the Euler characteristic $\chi(\overline{D^2})$ is 1.

$$\boxed{\text{II.4.4}}$$

We now describe a basic and very useful construction for invariant geodesic laminations. The idea is to start with a lamination L which is not invariant — only forward invariant — but which has a complete set of critical leaves and gaps. We then work backwards, producing a sequence of laminations $L_0 = L, L_1, \ldots$ such that L_{i+1} corresponds to L_i by $f : z \mapsto z^d$. Finally, we take the closure L_∞ of the union of the L_i to obtain an invariant lamination. The original lamination might consist only of critical leaves. With a small amount of initial information, one constructs the lamination L_∞ which is essentially the only lamination having the given set of critical leaves. Here is the formal statement:

Proposition II.4.5. *Let L be a lamination without critical gaps which satisfies the following conditions:*

(a) *L is forward invariant by the map $f : z \mapsto z^d$, and*

(b) *there are $d - 1$ critical leaves — enough to satisfy the formula of the preceding proposition.*

Then there is an invariant geodesic lamination, the f-saturation $S(L)$ of L, which contains L.

Proof of II.4.5. We first construct the sequence $L_0 = L$, L_1, L_2, ... such that L_{i+1} corresponds by f to L_i. Consider a map $\overline{f}$ of $\overline{D^2}$ to itself which extends the map f over the entire disk, mapping regular leaves to leaves (so that Euclidean arc length scales proportionally along the leaf), mapping critical leaves to points on the boundary, and extending the map over gaps as a homeomorphism on the open gap if possible, otherwise collapsing to the image leaf (according to definition, images of non-critical gaps can be leaves but not points).

We can define a lamination M whose leaves are mainly the inverse images of leaves of L by $\overline{f}$, together with the critical leaves of L. This is closely related to the proof of the preceding proposition — we are thinking of the domain disk as being made up of d pieces (the complementary components of the $d - 1$ critical leaves), each of which maps one-to-one to its image, and just imposing a copy of L in each one. The only trouble with this definition is with leaves l of L which have an endpoint on the image (by $\overline{f}$) of a critical leaf of L. In this case, for each component of $f^{\circ -1}(l)$ which intersects a critical line, we put in two leaves, going to the two ends of the critical line or chain of contiguous critical lines (if the component of the preimage is a gap, use leaves on the boundary of the gap for this purpose); see Figure II.2. If both endpoints of l meet images of critical lines, we still put in at most two leaves. Now straighten out all the leaves of M without changing their endpoints, to make a geodesic lamination L_1. The lamination L_1 contains the original lamination L, by the condition on surjectivity of f.

We repeat this construction a countable number of times, to obtain our sequence $\{L_i\}$ of geodesic laminations, first straightening out the map $\overline{f}$ at each step so that it really maps L_{i+1} to L_i. We pass to the limit, forming the closure of the union of the L_i. Conditions (GL3), (GL4) and (GL5)

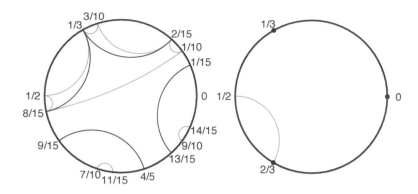

Figure II.2. Illustration of the proof of Proposition II.4.5 (for degree 5). If a leaf has a common endpoint with the image of an isolated critical leaf, then it pulls back to a triangle in each preimage component. The black leaves on the left are critical leaves; they all map to degenerate leaves at angles 0, $1/3$, $2/3$. The grey leaves on the left are all non-critical and map to the single grey leaf on the right.

are satisfied by the closure of the union because they are satisfied at each stage. $\boxed{\text{II.4.5}}$

As an example of this construction, we can take the case $d = 2$, with a geodesic lamination having a single leaf l which is any diameter of the circle. A diameter is automatically a critical leaf for $z \mapsto z^2$. There is a convenient notation for specifying a diameter: it is determined by the square of the two endpoints, which we represent as $e^{2\pi i \alpha}$ for some $\alpha \in [0, 1)$. We denote by $L(\alpha)$ the lamination constructed from the α-diameter.

For simplicity, consider the case that neither endpoint of the diameter is a periodic point under $z \mapsto z^2$ (this is equivalent to the condition that α is not a rational number with odd denominator). Then L_1 has 3 leaves, the two new leaves being on the two sides of the chosen diameter; L_2 has 7 leaves, one new leaf in each of the four components of the complement of the existing three; L_3 has 15 leaves, and so forth; see Figure II.3 for this and the following examples.

As a further example, the geodesic lamination with a single leaf going from the $1/3$ point of the circle to the $2/3$ point is forward invariant by the map $z \mapsto z^2$. Adjoin to this lamination any diameter which does not intersect this leaf — that is, a diameter with one of its endpoints between the $1/6$ point and the $1/3$ point. The f-saturation of this example contains the f-saturation of the corresponding example from the last paragraph (generated by a single diameter only). One would expect that, at least for a typical choice of diameter, the two laminations would be the same, since

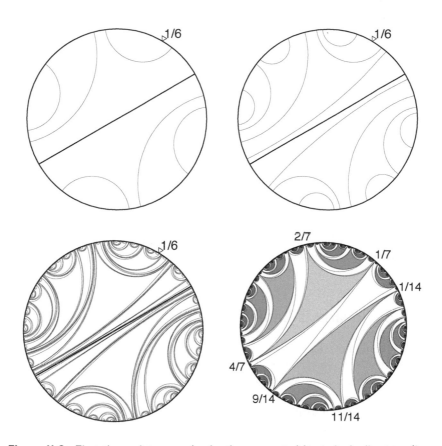

Figure II.3. First three pictures: a lamination generated by a single diameter (here with $\alpha = 1/6$); the image of the diameter (at angle α) is marked on the outside of the circle. Two initial stages in generating this lamination are shown in the top row (L_2 and L_3); the complete lamination is shown in the lower left. Lower Right: a lamination generated by the triangle with vertices $1/7$, $2/7$, $4/7$ and a diameter.

there is not usually much room left over to enlarge an invariant lamination. (We shall prove something to this effect below in Proposition II.6.7).

Similarly, there is a forward-invariant lamination with 3 leaves which make a triangle joining the $1/7$ point, the $2/7$ point, and the $4/7$ point. The map $z \mapsto z^2$ combinatorially rotates this triangle by $1/3$. Any diameter of the circle with one endpoint between the $1/14$ point and the $1/7$ point is compatible with this triangle (doesn't intersect it), so that, applying the proposition, one obtains a whole interval of laminations, each containing this triangle.

As an illustration of what can happen when we begin with a critical line which has a periodic endpoint, let us consider the simplest case — we begin with the lamination L having a single leaf, joining the 0-point to the 1/2-point. Then L_1 has four new leaves — joining the 1/4-point and the 3/4-point to the ends of the original diameter. L_2 has eight new leaves, with two new leaves in each quarter-circle, joining the midpoint of the quarter-circle to its two ends. The process continues in the same way. This lamination is not good for the purpose of defining a set of identifications of the circle which might occur for a polynomial map of degree 2, because it would force all dyadic points on the circle, and hence all points, to be identified to a single point. However, it is still significant as a prototype for other laminations having critical leaves with periodic endpoints.

The construction raises two questions: first, what are the possible choices for critical leaves and critical gaps in an invariant lamination, and second, to what extent is the invariant lamination determined by these critical leaves and gaps?

II.5   Invariant Laminations for the Quadratic Case

We shall now specialize to an analysis of invariant laminations for the degree-two case. Much of what we shall do seems to have natural generalizations to the cases of higher degree, and I hope that someone will undertake a careful analysis of these cases. However, there is a particular simplicity and clarity in the phenomena for degree two, and it seems worthwhile to concentrate on this.

It is convenient to define the *length* of a leaf l of a lamination to be the distance between its endpoints, measured *along the circle* in units of radians/2π. Thus, the maximum length of a leaf is $1/2$; a leaf of length $1/2$ is a critical leaf.

Consider the sequence of lengths L_i of the forward images l_i of a leaf l of a forward invariant lamination (for $z \mapsto z^2$). If L_i is less than $1/3$ we have $L_{i+1} > L_i$, while $L_i > 1/3$ implies $L_{i+1} < L_i$. Given a leaf l, let m be the (nonempty but possibly finite) sequence of indices m_i such that L_{m_i} is not less than any preceding length, and $L_{m_i} \geq 1/3$.

If l is any leaf, $-l$ is a line which does not cross any leaves of the lamination (as you see by looking at the images under $z \mapsto z^2$). Let $C(l)$ be the closed subset of the open disk D^2 containing the origin and bounded by $l \cup -l$.

Lemma II.5.1. *The leaf $l_{m_{i+1}}$ is the first leaf l_j, for $m_i + 1 \leq j \leq m_{i+1}$, to intersect the region $C(l_{m_i})$.*

Proof of II.5.1. First, note that $l_{m_{i+1}}$ must indeed be contained in $C(l_{m_i})$, since l_{m_i}, $-l_{m_i}$ and $l_{m_{i+1}}$ all have length at least $1/3$.

The length of the image of l_{m_i} is $1 - 2L_{m_i}$, while the two intervals in which $C(l_{m_i})$ touches the boundary of the disk are shorter, each having length $1/2 - L_{m_i}$. Any leaf which intersects $C(l_{m_i})$ must have length at most $1/2 - L_{m_i}$ or at least L_{m_i}. The leaf l_{m_i+1} is the shortest until $l_{m_{i+1}+1}$, so the lemma follows. $\boxed{\text{II.5.1}}$

Theorem II.5.2 (No Wandering Triangles). *Let L be an invariant lamination for $f : z \mapsto z^2$, and let G be a gap of L. Then either*

(a) *G eventually collapses to a leaf: it is a triangle which eventually maps to a triangle with a critical leaf as one of its sides, or a quadrilateral which eventually maps to a quadrilateral containing the origin, or*

(b) *G is eventually periodic: there is some integer $s \geq 0$ and some $p > 0$ such that $f^{\circ s+p}(G) = f^{\circ s}(G)$. The first iterate $f^{\circ s}$ which maps G to a periodic gap may either be*

   i) *an orientation preserving homeomorphism,*

   ii) *a degree two covering, or*

   iii) *it may collapse one edge of G to a point, and map G otherwise homeomorphically.*

The cases of eventually collapsing triangles and quadrilaterals are closely related. When there is a collapsing quadrilateral, it can be subdivided into two triangles (in two ways) by putting in a critical leaf; a new invariant lamination with collapsing triangles is obtained by adding preimages for the new critical leaf. Any collapsing triangle has a symmetric mate. As long as the critical leaf has no preimages which are other sides of the collapsing triangle, this construction can be reversed by removing the critical leaf and all its preimages.

Similarly, the cases of a gap which maps as a degree two covering to a periodic gap and a gap which maps to a periodic gap after one of its edges is collapsed are closely related. If G maps as a degree two covering to another gap which is eventually periodic, then a critical leaf may be added to G which joins any two diametrically opposite boundary points. By adjoining all the preimages of the new leaf, an invariant lamination is obtained with a gap which eventually maps to a periodic gap, after first collapsing a side. Conversely, if G maps by collapsing one side to an eventually periodic gap, then there is a symmetric gap on the other side of its critical leaf. As long as the critical leaf has no other preimages which are sides of G, removal of the critical leaf and all its predecessors reverses the construction.

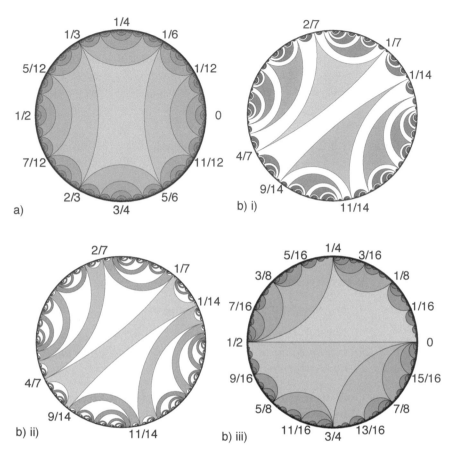

Figure II.4. Invariant laminations illustrating gaps in the four cases of Theorem II.5.2: a) a collapsing quadrilateral, b)i) a periodic polygonal gap that maps to itself homeomorphically, b)ii) a periodic gap that maps to itself as a degree two covering (a gap of Cantor double cover type), b)iii) a periodic gap for which the return map collapses one edge (a collapsing chain gap); see also Figure II.5 (the Siegel case). The gaps satisfying these conditions are shaded; they are drawn together with their backwards orbits.

Proof of II.5.2. First let us settle that if a gap eventually maps to a periodic gap, it maps in one of the three ways described. All gaps in the lamination map homeomorphically to their images, preserving the orientation, except for at most two: there might be one gap (containing the origin) which maps by a degree two covering, or else there might be two gaps which have a common side which is a critical leaf. If G eventually hits a gap of the first exceptional type, it can never hit an exceptional gap again unless the exceptional gap is periodic. If G eventually hits a gap G_1 of the second exceptional type, it might also eventually hit the symmetric exceptional gap G_2: but $f(G_1) = f(G_2)$ is then periodic, so that only one side of G is collapsed before it hits a periodic gap.

From now on, suppose that G is a gap that is disjoint from all its forward images. Let a, b, and c be distinct points on $\partial D^2 \cap \overline{G}$. Let G_i be the sequence of images of G, and a_i, b_i, and c_i the sequences of images of a, b and c; and let d_i be the minimum distance between any two of the three points a_i, b_i, c_i.

If $d_i = 0$ for some i, the gap G maps at some time to one which contains the origin (possibly on the boundary). Since we assume that G is not eventually periodic, this can not happen a second time by the argument above. If there are no more than four points on $\partial D^2 \cap \overline{G}$, we are done, since G is a quadrilateral or triangle which eventually collapses to a leaf. Otherwise, we can assume that a, b and c are chosen so that d_i is always greater than zero, and moreover that G is the triangle with vertices a, b, c (such a triangle is called a "wandering triangle").

The Euclidean area in $\overline{D^2}$ of G_i is bounded below only in terms of d_i, and the total area of $\overline{D^2}$ is finite. This implies $d_i \to 0$. There are thus infinitely many d_i that are smaller than all previous d_i, and arbitrarily close to 0. But such short leaves cannot be images of shorter leaves, so they must be the images of leaves of lengths converging to $1/2$.

Consider a G_m so that its longest side, say of length L_m, is longer than any side of any G_i with $i < m$, and also $L_m > 1/2 - d_1$. Let $k > m$ be minimal so that G_k has a side longer than L_m. Since G_m and its symmetric copy $-G_m$ bound a domain in $\overline{D^2}$ that contains G_k, it follows that $d_k \leq 1/2 - L_m < d_1$. By construction, all sides of all G_i with $i < k$ have lengths at most L_m. All sides of G_2 are then images of leaves of lengths in $[d_1, L_m]$. Hence they all have length at least $\min(2d_1, 1 - 2L_m) = 1 - 2L_m$. By induction, all sides of G_k have length at least $1 - 2L_m$, so in particular $d_k \geq 1 - 2L_m \geq 2d_k$. This is a contradiction. $\boxed{\text{II.5.2}}$

This theorem is invaluable for classifying laminations in the degree-two case. The question whether this generalizes to higher degrees, and if not what is the nature of the counterexamples, seems to me the key unresolved question about invariant laminations.[3]

---

[3]Much of the theory of this manuscript goes through for "unicritical polynomials" of degrees $d \geq 3$: these are polynomials with only a single critical point of maximal multiplicity; the corresponding laminations are those with d-fold rotational symmetry. For the non-existence of "Wandering Triangles" in this case, see [Le98] or [Sch1]. For general degree d laminations, there are no wandering $d + 1$-gons [Ki02]. However, the existence of wandering triangles in invariant laminations of degree 3 or greater has been shown by Blokh and Oversteegen [BO04, BO08]; upper bounds on how many such triangles there may be have been given by Blokh and Levin [BL02].

We are now in a position to make a qualitative analysis of gaps for degree two invariant laminations.

Theorem II.5.3. *A gap G of a degree two invariant lamination intersects ∂D^2 in either a finite set of points, a Cantor set, or a countable set of points.*

If G is not periodic under f, then either

(a) *it maps by some iterate of f to a periodic gap (in one of the three ways mentioned in the preceding theorem), or*

(b) *it is a triangle or quadrilateral which eventually collapses.*

If G is periodic, then the return map g may have either

(a) *degree one, in which case G may intersect the circle in*

    i) *(**Polygon case**) a finite number of points, in which case G is a finite-sided polygon on which g acts by a homeomorphism which permutes the sides transitively (as a rational rotation). Any two different eventually periodic polygons have disjoint sets of vertices, unless both are strictly preperiodic and share a common boundary leaf that eventually maps to a critical leaf.*

    ii) *(**Siegel case**) a Cantor set. In this case, g acts on ∂G as an irrational rotation blown up along the backward orbit of some point (see discussion below).*

    iii) *(**Collapsing chain case**) a countable set of points. For this to happen, there must be at least one point on the boundary of G which has finite order q under the return map. There is at most one side S of G which is fixed by the return map (and only if $q = 1$); except for S, all sides of G eventually collapse to points, and they are connected to each other in q one-sided-infinite chains. One endpoint of the collapsing leaf has period q under the return map of G.*

(b) *or degree two (**Cantor double cover case**), in which case G intersects the circle in a Cantor set, where g acts as the shift map on two symbols (see the discussion below). The gap G has a boundary leaf that is fixed by g; in particular, if G is the central gap, then the fixed boundary leaf is one of the two longest leaves of the lamination.*

Before proving the theorem, we will construct examples as mentioned in its statement.

The Polygon Case. Let $\alpha = p/q$ be rational in lowest terms. Subdivide the unit circle T^1 into q intervals of lengths $1/(2^q - 1)$, $2/(2^q - 1)$, $\ldots$, $2^{q-1}/(2^q - 1)$, ordered cyclically so each interval other than the longest one has an interval of twice its length exactly p positions further on the circle in counterclockwise order. Define a map $h \colon T^1 \to T^1$ so that it sends each interval other than the longest one linearly onto the interval of twice its length, and so that it maps the longest interval linearly with constant derivative 2, covering the shortest interval twice and the rest of T^1 once. This defines a continuous map $h \colon T^1 \to T^1$ with constant derivative 2 and hence degree 2, so it is conjugate to $z \mapsto z^2$. The q interval endpoints form an invariant q-gon with combinatorial rotation number p/q. (This construction can also be viewed as a modification of the construction in the Siegel case, when the orbit of x_0 under the rotation is periodic.)

This polygon is forward invariant, but it does not define an invariant lamination. Let $I := [\vartheta^-(\alpha), \vartheta^+(\alpha)]$ be the shortest boundary interval. Then one can add a critical leaf (a diameter) so that its image is an arbitrary point in I. This diameter defines an invariant lamination uniquely as before.

The Siegel Case. For any irrational angle α (as measured in our units, radians/2π), let $R_\alpha \colon S^1 \to S^1$ be the rotation of the circle by α. Another map $h_\alpha \colon S^1 \to S^1$ can be constructed as follows (compare also Figure II.5). Construct a probability measure ν on the circle by picking an arbitrary point x_0 (say 1), giving x_0 measure $1/2$, giving $R_\alpha^{\circ-1}(x_0)$ measure $1/4$, giving $R_\alpha^{\circ-2}(x_0)$ measure $1/8$, etc. It is easy to construct a continuous map $m \colon T^1 \to S^1$, where T^1 is a second copy of the circle such that m pushes the Lebesgue measure μ (normalized to have total mass one) of T^1 to the measure ν on S^1. (This process blows up the backward orbit of x_0.) Define the map $h_\alpha \colon T^1 \to T^1$ as follows: h_α sends each blown-up interval of length 2^{-k} ($k = 2, 3, 4, \ldots$) linearly with constant derivative 2 to the image interval, and the interval $m^{-1}(x_0)$ of length $1/2$ wraps around all of T^1, beginning and ending at $m^{-1}(R_\alpha(x_0))$. Since all blown-up intervals occupy the full length of T^1, this implies that h_α has constant derivative 2 everywhere, so it has degree two. The map m is not quite a semi-conjugacy of h_α to R_α, but it acts as a semi-conjugacy everywhere except on the longest blown-up interval (which has length $1/2$).

It follows that h_α is conjugate to the map $z \mapsto z^2$. There is a forward invariant lamination for h_α, consisting of all the lines which join endpoints of the intervals that have been obtained by blowing up points. The lamination has one critical leaf (a diameter). By the construction in Proposition II.4.5 there is an invariant lamination L_α containing this one. It has an invariant gap, which maps to itself by a blown-up rotation. This example is the general model: we will say that the return map for a gap is a blown-up

Figure II.5. A periodic gap in the Siegel case. The shaded gap G is fixed; it has the property that the dynamics acts on ∂G as an irrational rotation when the boundary leaves are collapsed to points.

rotation of angle R_α if it is topologically conjugate to the return map for this model. In this case, we say that the gap is of Siegel type with rotation number α.

The Polygon and Siegel cases yield invariant laminations having periodic gaps with arbitrary rational or irrational rotation numbers. The rotation number depends monotonically on the angle of the diameter in the lamination: this can be seen from the construction given here (in the construction in the Siegel case, the angle of the diameter equals the distance between the diameter and its degenerate image leaf, and this is determined by preimages of the point x_0 in the rigid rotation that are contained between x_0 and its image); we will give a different proof in Lemma II.6.10c.

The Collapsing Chain Case. Now we discuss the gaps which touch the boundary of the disk in a countable set of points; these are annoying special cases. See Figure II.6. The simplest example comes by joining points in the sequence 0, 1/2, 1/4, 1/8, 1/16, ... This forms an invariant gap, where the fixed leaf S_0 reduces to a point. By the technique of Proposition II.4.5 it can be extended to an invariant lamination. Note that for this invariant lamination, there is a chain of leaves of the lamination coming arbitrarily near any pair of points: this lamination could not arise from a set of identifications.

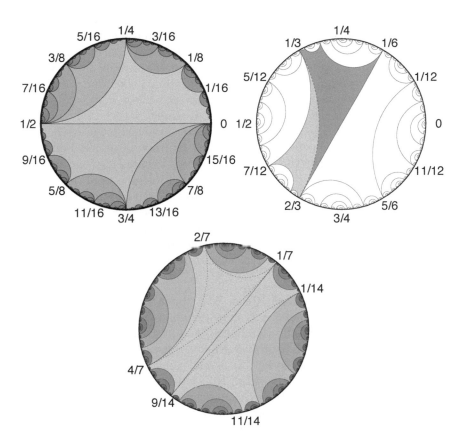

Figure II.6. Periodic gaps with collapsing chains. Top left: the simplest example of an invariant lamination with a fixed gap that intersects the circle in countably many points (at angles $0, 1/2, 1/4, 1/8, \dots$). Top right: an invariant lamination with a gap G of period 2 ; adjacent boundary points include $\frac{1}{3}$, $\frac{2}{3}$, $\frac{1}{6} = \frac{1}{3} - \frac{1}{6}$, $\frac{7}{24} = \frac{1}{3} - \frac{1}{24}$, $\frac{31}{96} = \frac{1}{3} - \frac{1}{96}, \dots$. The gap G and its image are indicated in two different shades of grey. Removing the initial leaf connecting $1/3$ and $2/3$ (dashed), we obtain a fixed gap with rotation number $1/2$. Bottom: a similar example with rotation number $1/3$, based on the polygon with vertices $1/7$, $2/7$, $4/7$. The central gap has two long boundary leaves, connecting the boundary points at angles $1/7$ and $4/7$, and at angles $1/14$ and $9/14$.

As an example where there is a side S fixed by the return map, form a gap G by joining points in the sequence

$$\frac{1}{3}, \frac{2}{3}, \frac{1}{6}, \frac{7}{24}, \frac{31}{96}, \dots$$

The gap G returns with period 2; the leaf $1/3 - 2/3$ flips under $z \mapsto z^2$, so it is fixed by the return map. The leaf $2/3 - 1/6$ is a critical leaf; the other leaves form a never-ending line marching forward to the slaughter.

If you remove the leaf $1/3 - 2/3$ from the above example, you obtain an example of a gap which has period of return 1, but where there are two chains of sides which are interchanged by the return map.

Other examples can be constructed similarly with arbitrary rational rotation numbers p/q, by modifying the invariant laminations constructed above which have polygonal gaps which return with period 1 and rotation number p/q. Next to such a polygon there is a central gap of Cantor double cover type, which returns with period q and degree two. Transplant the example $0 - 1/2 - 1/4 - 1/8 - \ldots$ into the central gap to obtain a gap which returns with period q, and rotation number 1, and for which there is one leaf (a side of the original polygon) fixed by the return map. Finally, remove each copy of the original polygon from the invariant lamination, to obtain an example with a gap which returns with period 1 and rotation number p/q.

The Cantor Double Cover Case. The standard tertiary Cantor set in the unit interval may be defined as the set of numbers which possess base-three expansions in which only the digits 0 and 2 occur. The *shift map* (or the one-sided Bernoulli shift on two symbols) discards the first digit, shifting each of the others to the left. It is analogous to the squaring map of the circle, since it maps the bottom half and the top half of the Cantor set linearly to the whole. To say that g acts on a set as the shift map is to say that there is a homeomorphism of the set to the standard Cantor set conjugating the map to the shift.

Proof of II.5.3. The classification in this theorem reduces to the statements about the periodic cases, by the preceding theorem.

Among all the sides of the gaps in the future orbit of G, there is a side P which is largest. We may as well assume that P is a side of G.

There is at most one other side of G whose length is at least $1/3$. We will call this other side Q, if it exists.

Every leaf l of the lamination eventually lands on a leaf of length at least $1/3$, by some forward iterate of f. If l is a side of G, the first time that a forward image of l has length greater than $1/3$ need not be when G returns to itself. However, by Lemma II.5.1, it is easy to see by an inductive argument that any sides of the other gaps in the orbit of G which have length at least $1/3$ map to the long sides of G upon their first return to G. (We only need to consider the case when Q exists; let s be its length. Then all boundary leaves of all gaps in the orbit of G, except P, have lengths at most s. All short boundary leaves of G have length at most $1/2 - s$. Any leaf of length in $[1/2 - s, s]$ has an image leaf of length at least $1 - 2s$, so the image cannot be a side of G other than P or Q. Thus a leaf of length at least $1/2 - s < 1/3$ can return to a short leaf of G only after mapping to P first.)

It follows that there are at most two orbits of sides of G under the return map g: each side eventually maps to P or to Q (if Q exists). We shall see that in most cases, there is only one orbit.

There is a variety of different situations, which we shall now analyze individually.

(a) First we take up the case that g has degree one. This means that G does not contain the origin in its interior.

1. Suppose that P is not a critical leaf. Then g is a homeomorphism on the boundary of G, so it acts as a bijection on the sides of G. The leaves P and Q (if Q exists) must eventually return to P or Q, so their future orbits are finite. It follows that G has a finite number of sides, because of the bijectivity and the fact that every side eventually hits P or Q. It follows that the action of g on the sides cannot be the trivial permutation. Therefore, if Q exists, g maps it to a different side, which by Lemma II.5.1 can only be P: hence, g permutes the sides transitively whether or not Q exists.

   The sides of G must touch, since the backward orbit of P and Q is dense in the boundary of G. Hence G is indeed a polygon, and g maps it as described; this is the *polygon case* in the theorem.

   If a periodic polygon G has a common vertex with another periodic polygon G', then by the transitive action each vertex of G must be a vertex of a polygon on the orbit of G', and each vertex of G' must be a vertex of a polygon on the orbit of G. This forces infinitely many further periodic polygons on the orbits of G and G', a contradiction. Now suppose that two periodic or preperiodic polygons G and G' share a vertex. If G and G' have different images, then we may replace G and G' by their image polygons. Repeat this argument as often as possible. Since we just proved that G and G' cannot both be periodic, after finitely many iterations we may assume that G and G' have the same image polygon; therefore, they are symmetric to each other with respect to rotation by a half turn. Let p be the common vertex of G and G'; the symmetry implies that G and G' have a common boundary leaf which is necessarily a diameter. We excluded this situation if G or G' are periodic, but it may happen if both are preperiodic.

2. If P is a critical leaf, the return map g maps the sides of G other than P bijectively to the sides of G. Therefore, G must have an infinite number of sides.

   2.1. If there are any periodic points for g on the boundary of G, they all have the same period, since g preserves the circular order. Not all points on the boundary of G are periodic points, since g is not injective. This will be the *collapsing chain case*.

      2.1.a) Suppose that the periodic points for g are fixed points. We claim that there is either exactly one fixed point or one fixed leaf on the boundary of G. Indeed, if there were more, we could obtain a new lamination by adjoining three leaves which join any three fixed points for g (if there are exactly two fixed points, they either are at the end of a common fixed leaf, or we add an extra leaf joining these two fixed points). By adding also all the preimages of the new leaves, we would obtain an invariant lamination with a gap without critical boundary leaf whose return map is the identity, violating the conclusion from Case 1.

      If Q does not exist, there is only one orbit of sides of G under g. But even if Q exists and g does not fix it, then it can never return to itself (since there are no periodic points other than fixed points), so it eventually maps to P, and there is only one orbit of sides of G under g.

      If x is an endpoint of a leaf and not a fixed point for g, then one of the two intervals on ∂G between x and $g(x)$ does not contain a fixed point. This interval must form exactly one leaf: otherwise there would be more than one orbit of leaves of G which are not fixed. Thus, the non-fixed leaves form infinite collapsing chains as described in the theorem. There must be fixed points at both ends of each one-sided infinite collapsing chain, hence there is only a single chain and one endpoint of the collapsing leaf P is fixed.

      2.1.b) If the period q of the periodic points of g is not 1, then we will construct a new lamination, thereby reducing to Case 2.1.a) when $q = 1$.

      Under this hypothesis, the lamination can be subdivided by adding leaves to form the boundary of the convex hull of all periodic points for g on G_0, and also adding all their preimages. If $q > 2$, then the convex hull of the periodic points is a new gap, which by Case 1 must have exactly q sides; if $q = 2$, there can only be two periodic points, so their convex hull is a leaf. The new lamination has a gap with a countable number of sides and period q times as great; its return map has fixed points, reducing the statement to Case 2.1.a): a collapsing chain of higher period. Note that the q collapsing chains are all on a single orbit under the action of f; moreover, the gap G can have no periodic leaf on its boundary (in the subdivided lamination, the new leaves are the periodic boundary leaves of the gaps of collapsing chain type).

2.2. This is the case that the longest leaf P of G is a critical leaf, and there are no periodic points on the boundary of G. This will be the *Siegel case*.

As before, we conclude that there is only one orbit of the sides of G under g, since in the absence of periodic points no side can ever return to itself: hence, all sides must eventually land on P and collapse.

No two sides of G can touch, for otherwise two distinct iterates of their common point would hit the image of P. This would violate the assumption that the boundary of G has no periodic points.

Since the endpoints of the leaves for any invariant lamination are dense on the circle, it follows that G touches the circle in a set obtained by removing a countable dense sequence of open intervals so that there are no isolated points; that is a Cantor set.

If we form the quotient space of the boundary of G by the equivalence relation which identifies each leaf to a point, we obtain a circle, and g induces a homeomorphism h on the circle. It remains to show that h is topologically conjugate to a rotation.

It is not true that all homeomorphisms of the circle which have no periodic points are topologically conjugate to rotations: if you start with a rotation and blow up an orbit in both forward and backward time, you can obtain a counterexample. This construction is due to Denjoy, who showed that such homeomorphisms can even be constructed as C^1 diffeomorphisms. What is true is that if a homeomorphism h of the circle has a dense orbit, then h is conjugate to an irrational rotation. This is the case here: since the boundary of G intersects the circle in a Cantor set, preimages of the critical leaf are dense in the boundary of G, and hence h has a dense backwards orbit.

We conclude that h is actually conjugate to a rotation as claimed in the theorem (the Siegel case).

(b) Now we consider the case that the return map $g : G \to G$ has degree two. This will be the *Cantor double cover case* in the theorem. We can suppose that G contains the origin.

In this case, Q exists and has the same length as P, and they have the same image after one application of f. By Lemma II.5.1, it follows that either $g(P) = P$, or $g(Q) = Q$. Up to labeling, we can assume we have the first case.

The return map g acts on the boundary of G as a two-fold covering. Since every side of G eventually lands on P, and sides are dense, it easily follows that the intersection of G with the circle is a Cantor set, and that the return map is topologically conjugate to the shift on two symbols. Let G be the central gap and G' its image gap. The two longest leaves of the lamination are on the boundary of G; let m be their common image leaf. Then m is the longest boundary leaf of G'. The first return map of G' fixes m: a leaf on the forward orbit of m can be shorter than m for the first time only if the leaf in the iteration before was longer than the preimage leaves of m; but there are no such leaves.

This exhausts all possible cases and proves the theorem. $\boxed{\text{II.5.3}}$

The following statement about eventually collapsing quadrilaterals will be useful later on.

Lemma II.5.3a. *In an invariant quadratic lamination with a collapsing quadrilateral, let s be the length of the shorter pair of sides of the central collapsing quadrilateral. Then any eventually collapsing quadrilateral, other than the central gap, has a pair of opposite sides of length at most $s/2$.*

Proof of II.5.3a. Let G be the central gap. If its short boundary sides have length s, then the long boundary sides have length $1/2 - s$, and all boundary sides of G map to a single leaf, say m, of length $2s$.

First observe that any leaf with length greater $s/2$ can, during its forward orbit, never land on a short side of G: it is too long to land directly on a short boundary side of G, and under iteration, it can never become longer than the longer side of G and thus never shorter than m. It follows that every eventually collapsing quadrilateral has a pair of opposite sides with length no more than $s/2$. $\boxed{\text{II.5.3a}}$

II.6   The Structure of the Set of Quadratic Invariant Laminations

In the preceding section, we gained some insight into the structure of particular quadratic invariant laminations. Now we will work toward an understanding of the relationships among different quadratic invariant laminations and of the structure of the set of quadratic invariant laminations.

We will need the following.

Proposition II.6.1. *If L is an invariant lamination other than the lamination whose leaves connect z to $\bar{z}$ for all $z \in \partial D^2$, then the gaps of L are dense in $\overline{D^2}$.*

Proof of II.6.1. Suppose that L is a lamination such that gaps are not dense. Then there is some one-parameter family l_t of distinct leaves of L. Let q_t and p_t be the moving endpoints of l_t.

There is an interval of time such that at least one of p_t and q_t is not constant. Therefore, we have an interval I where there is a continuous map τ defined such that for each x in the domain of τ, x is connected to $\tau(x)$. By forward invariance, the definition of τ can be extended to the whole of the circle, since I eventually engulfs everything, and τ is locally determined by the leaves of L. If τ was locally constant somewhere, by forward invariance it would have to be constant everywhere, a contradiction. We conclude that there are no gaps at all in L. It follows that τ is a homeomorphism of the circle of order two, and it conjugates $z \mapsto z^2$ on the boundary of the disk to itself.

Since 1 is the only fixed point, we have $\tau(1) = 1$ and thus $\tau(-1) = -1$. Since τ is not the identity, we have $\tau(i) = -i$, and it follows that $\tau(z) = \bar{z}$ for all points $z = e^{2\pi i k/2^n}$. By continuity, the claim follows. $\boxed{\text{II.6.1}}$

Here is an important concept:

Definition II.6.2. A leaf l of a forward-invariant lamination L (in the degree two case) is *principal* if l is at least as long as any of its forward images. For any invariant lamination L, let M be the greatest length of any leaf of L. Any leaf of length M is a principal leaf. Unless $M = 1/2$, L has two leaves of length M. The longest leaf or leaves we will call *major* leaves of L. The images of the major leaves of L coincide; this image may be either a point or leaf. This image we will call the *minor* leaf of L (by abuse of notation, since it might not be a leaf at all).

The minor leaf of an invariant lamination has the property that its length is the *infimum* of the *lim infs* of the sequence of lengths of the forward images of any leaf of L.

Lemma II.6.2a. *The endpoints of a non-degenerate minor leaf are either both periodic with the same period, or both preperiodic with the same preperiod and same period, or both not eventually periodic. In the latter case, both endpoints are on different orbits, and the minor leaf is approximated from both sides by leaves that have no common endpoints with the minor.*

Proof of II.6.2a. Consider a lamination with a non-degenerate minor m. This lamination has a central gap G. We use the results of Theorem II.5.3. If G has Cantor double cover type, then both endpoints of m are periodic with equal period. If G is a polygon with at least 6 sides, then G and its image gap must be strictly preperiodic, and both endpoints of m are on the same orbit, so they have the same period; since any two different gaps on the orbit of G have disjoint sets of vertices, both endpoints of m also have the same preperiod.

The final case is that G is a collapsing quadrilateral. Let M be a major leaf of G. We distinguish three cases: the leaf M could be, on the other side than G, adjacent to an eventually periodic polygon, to an eventually collapsing quadrilateral, or to no other gap.

If M is adjacent to an eventually periodic polygon, then both endpoints of M (and in fact, all vertices of the polygon) have equal period and preperiod. Now suppose that M is adjacent to an eventually collapsing quadrilateral, say G'. The image leaf of M is m, and since M is on the boundary of G', it eventually returns to a boundary leaf of G; thus m is a periodic leaf, and we need to show that both endpoints have equal period. If k is the number of iterations it takes G' to map to G, then m maps to M or its symmetric copy $-M$ after $k-1$ iterations (the other two boundary leaves of G are too short to be on the orbit of m), so either M or $-M$ is periodic. Possibly by replacing G' with the symmetric copy $-G'$, we may assume that M is periodic. After k iterations, G' maps to G in an orientation preserving way, so the common boundary leaf M of G and G' maps to itself, interchanging the two endpoints of M. This implies that both endpoints are on the same orbit, so they have equal period.

We are thus left with the case that M is a limit of other leaves. Let M' be a short boundary leaf of G. If M' is adjacent to an eventually periodic polygon, we conclude as above. If M' is adjacent to an eventually collapsing quadrilateral, then we may erase M' and its symmetric leaf $-M'$ together with their backwards orbits, and obtain a lamination with the same minor leaf m (for details, see the discussion after the proof of Proposition II.6.4). In this case, the central gap can only be of Cantor double cover type since m is periodic; this brings us into a case that was treated earlier. We may thus assume that M' is a limit of other leaves as well. We still need to show that the leaves approximating M and M' have no common endpoints with G.

If all the leaves approximating M (or those approximating M') have a common endpoint with G and thus with M (or with M'), then a vertex of G is an isolated vertex of infinitely many gaps accumulating at a side of G. The lengths of two adjacent sides of these gaps are close to the lengths of M (or M'), so by Lemma II.5.3a these gaps cannot be eventually collapsing quadrilaterals. These gaps must thus be eventually periodic polygons, but (in the absence of a critical leaf) none of these can have common vertices, a contradiction.

We thus conclude that both M and M' are limits of leaves with disjoint endpoints, so m is the limit of leaves from both sides. These leaves imply that once an endpoint of m visits a point that it had visited before on its orbit, the same must be true for the other endpoint. $\boxed{\text{II.6.2a}}$

Our goal is to describe the set of laminations in terms of their minor leaves. This goal is complicated by the fact that it is possible for two different laminations to have the same minor leaf: let's look at some examples (compare Figure II.7).

In the first place, any lamination L having a gap with a critical leaf as one of its sides, one of whose endpoints is fixed by the return map for the gap, can be subdivided by putting in leaves joining this fixed point to all of the vertices of the gap. After adjoining all the backwards images of the new leaves, one obtains an invariant lamination L_1 with the same major leaf and hence the same minor leaf as L. This turns a gap of collapsing chain type into many eventually collapsing triangles.

Another class of examples for the non-uniqueness of an invariant lamination with a given minor leaf is obtained from any lamination L with a gap G which is not a quadrilateral but which contains the origin (so that it maps to its image with degree two: either a gap of Cantor double cover type or a preperiodic polygon). It has two major leaves, P and Q. Adjoin two new leaves connecting the ends of P with the ends of Q, forming a collapsing quadrilateral. The quadrilateral separates G into two additional components (besides itself). Each of these components maps by a homeomorphism to the image of G. The lamination obtained by adding the two new leaves is hence forward-invariant.

There are two subcases. If G is a polygon, then the two additional components are eventually periodic polygons. The preimages of G can be subdivided in the same pattern, and a new invariant lamination L_Q is obtained which has the same minor leaf as L.

If the central gap G returns to G, then it has infinitely many sides; it is of Cantor double cover type. In this case, each of the two new components eventually return to all of G. Again, backward images for the extra two leaves can be adjoined, yielding an invariant lamination L_Q with a collapsing quadrilateral, having the same major leaves and minor leaf as

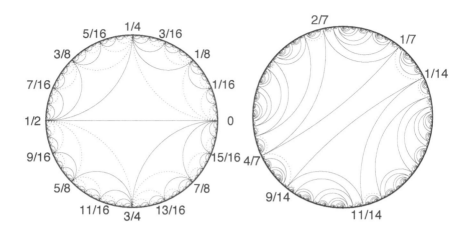

Figure II.7. Two examples of pairs of different invariant laminations with identical minor leaves. Dashed are additional sets of leaves that may or may not be part of the invariant lamination without changing the minor leaf. Left: adding extra leaves in the collapsing chain case (the minor leaf is degenerate at angle 0). Right: the Cantor double cover case: a gap containing the origin with infinitely many sides (the minor leaf connects the angles $1/7$ and $2/7$).

the original. This process can also be reversed; see the discussion after the proof of Proposition II.6.4.

In all these cases, at least one of the two laminations with a given minor leaf has a shortcoming: it has leaves which intersect (on the circle) but are not sides of a single gap. Such laminations cannot arise from the process of outlining the convex hulls of the collections of points which are identified as limiting values of a conformal (or even continuous) mapping from the disk to the complement of a closed connected set in the sphere.

Definition II.6.3. A lamination is *clean* if any two leaves with an endpoint in common are sides of a common gap. Otherwise, it is *unclean*.

Equivalently, a lamination is clean if no three leaves have a common endpoint.

Only the clean laminations have potential direct applicability to the theory of rational maps, but unclean laminations are indirectly useful in helping to explain the structure of the set of laminations. They can arise as the limiting case of a sequence of clean laminations. The reason for not building cleanness into the original definition of a lamination was so that the condition that a set of lines form a lamination be a readily verifiable closed condition. In making constructions, for instance, it is convenient to be able to form an increasing sequence of laminations and then pass to the closure of the union. One can worry afterwards whether the result is clean.

Clean laminations can be distinguished from unclean laminations by their critical gaps or critical leaves:

Proposition II.6.4. *A lamination L is clean if and only if the following two conditions are satisfied:*[4]

(CL1) *no side of any gap G collapses under f unless G is a periodic gap which returns by a blown up irrational rotation (the Siegel case), or a preperiodic preimage thereof which becomes periodic after the first iteration,*

(CL2) *no boundary leaf of an eventually collapsing quadrilateral is isolated.*

Moreover, every clean lamination satisfies the following condition:

(CL3) *no vertex of a collapsing quadrilateral is a periodic point.*

In particular, clean laminations do not have collapsing triangles or collapsing chain gaps.

Proof of II.6.4. If a lamination L has a gap G with a side that is a critical leaf, then there is another gap on the other side of the critical leaf. If G does not return by a blown-up irrational rotation, then by examining the various cases in Theorem II.5.3, we see that the critical leaf touches other leaves of each of the two gaps, so L is not clean.

More specifically, in Theorem II.5.3 the Cantor double cover case has a central gap that is rotationally symmetric, so there is no gap with a collapsing side. If there is a gap G with collapsing side in the polygon case, then there is a rotated gap $-G$ adjacent to G, and the critical leaf makes the lamination unclean. In the collapsing chain case, let G be a gap with collapsing side; again there is rotated gap $-G$ adjacent to G. The collapsing side of G has at least one adjacent side. Now let s by any boundary side of G in the collapsing chain, so that s maps after finitely many iterations to the collapsing side. Then s has two immediately adjacent boundary sides of G. The side s is the joint boundary of G and another gap G' that maps to $-G$ when s maps to the collapsing side, so G' has at least one extra boundary side with a common endpoint with s, and the lamination is unclean.

This leaves the case that G eventually maps to a periodic gap of Siegel type. Among the periodic gaps, there is one with a critical leaf on the boundary, so this must be G or $-G$, and the gaps G and $-G$ have a

---

[4]In this proposition, the statements of the first two conditions have changed with respect to the original version of the manuscript.

common periodic image; hence if G is not periodic, then its immediate image is. Thus, any clean lamination satisfies (CL1).

If there is a side of an eventually collapsing quadrilateral which is an isolated leaf, then the gap on its other side must be a polygon (if a quadratic lamination has a collapsing quadrilateral, this is necessarily the central gap, so there can be no collapsing leaves or gaps of Cantor double cover type), and L is not clean. Therefore, any clean lamination satisfies (CL2).

If there is a collapsing quadrilateral Q with a vertex v which is a periodic point, then some other quadrilateral having v as a vertex must map to Q when v returns. Therefore, L is not clean. This shows that a clean lamination also satisfies (CL3), so the proof in the forward direction is finished.

For the converse, suppose that the lamination L is not clean, so that there are two leaves l_1 and l_2 which touch at a common endpoint, but are not sides of a common gap. We will show that L fails (CL1) or (CL2).

Since l_1 and l_2 are not sides of a common gap, there is at least one other leaf l_3 between them, with the same endpoint p. There must also be at least two gaps G_1 and G_2 which touch the circle at p, because gaps are dense in the disk (Lemma II.6.1).

Neither of the G_i can be eventual degree two periodic regions (Cantor double cover case) or eventual irrational rotation regions (Siegel case), since G_1 and G_2 each have two sides that touch at p: the point p is truly a vertex (an isolated boundary point on the circle). If we suppose that (CL1) does not fail, then no G_i has a collapsing leaf, so by Theorem II.5.3, we may assume that the G_i are either eventually collapsing quadrilaterals or eventually periodic polygons. But if two eventually periodic polygons have a common vertex, then they have a common boundary leaf that eventually collapses (again Theorem II.5.3), and this contradicts (CL1). It follows that at least one of G_1 and G_2 is an eventually collapsing quadrilateral. If G_1 and G_2 have a common side, then (CL2) fails and we are done.

Otherwise, there must be at least one more gap separating G_1 and G_2 with a vertex at p. Continuing in this way, we find an infinite collection of gaps all touching at the same vertex. We may assume that at least most of these gaps are eventually collapsing quadrilaterals, otherwise we reduce to the previous case. Between any pair of such quadrilaterals, there are infinitely many further ones. But this implies that the two long boundary leaves of the central gap are approximated by eventually collapsing quadrilaterals, so these have two adjacent boundary sides with lengths close to the length of the major leaves. This contradicts Lemma II.5.3a.   II.6.4

We will examine a little more closely the situation for a lamination N which is unclean because of a collapsing quadrilateral with a pair of isolated sides.

If the shorter pair of sides of Q are isolated, then there is a clean lamination with the same minor obtained by removing the shorter sides of the collapsing quadrilateral and all their preimages. These shorter sides can never be in the forward image of the larger sides (because the larger sides map to the minor of the lamination, which is also the image of the shorter sides and thus longer than the latter), so the resulting lamination is certainly a non-empty forward invariant lamination L. Since the longer sides of the central quadrilateral already are enough to satisfy the backward-invariance property for the image of the quadrilateral, this lamination is also backward invariant. The new gap containing the old central quadrilateral maps with degree two to its image (it is either of Cantor double cover type or a preperiodic polygon); all other gaps of the new lamination map with degree one. This new lamination is clean (it satisfies (CL1) and (CL2)) and has the same minor as N (we saw this example earlier, in the discussion before Definition II.6.3 and in the proof of Lemma II.6.2a).

If the longer sides of the collapsing quadrilateral are isolated, but the shorter sides are not, then a new invariant lamination L can still be obtained by removing the isolated sides of all eventually collapsing quadrilaterals, but the new lamination has a different minor. In such a case, the gaps of N adjacent to the quadrilateral must be polygons with a finite number $n \geq 3$ sides (all other types of gaps need either a collapsing leaf, which N does not have, or a central gap other than a collapsing quadrilateral). The minor leaf of N can never return to any side of the central quadrilateral (the short sides are shorter than the minor, and the long sides are isolated which the minor is not). Therefore, the adjacent polygons are not eventually collapsing quadrilaterals. They are eventually periodic polygons which are not periodic. The lamination L has a central polygonal gap with $2n$ sides. Conversely, for any lamination L with a central polygonal gap having $2n > 4$ sides, a collapsing quadrilateral together with preimages can be inserted in n different ways to yield a new lamination. All but one of these ways yield a different minor.

Given a chord or boundary point m of the disk, when can it be the minor leaf of a lamination? Obvious conditions on m are the following[5]:

(a) all forward images of m have disjoint interiors,

(b) the length of a forward image of m is never less than the length of m,

(c) if m is a non-degenerate leaf, then m and all leaves on the forward orbit of m are disjoint from the interiors of the two preimage leaves of m of length at least $1/3$ (these would be the major leaves).

---

[5]The third condition has been added with respect to the original version of the manuscript. This condition excludes for m leaves such as the one connecting the angles 0 and $1/3$.

Suppose that m satisfies these three conditions. We will show that there exists an invariant lamination with minor m.

In particular, m cannot have length greater than $1/3$, otherwise its immediate image would be shorter. Therefore, the four preimages of the endpoints of m (or two when m is degenerate) can be joined in pairs in exactly one way by leaves M_1 and M_2 of length at least $1/3$. These will be the major leaves of the lamination. There is a forward invariant lamination $F(m)$ generated by M_1 and M_2.

There is also a fully invariant lamination $I(m)$ having m as a minor leaf. In the degenerate case, when m is a single point, we simply note that the lamination with a single (critical) leaf M is forward invariant and has enough critical leaves, so that the construction in Proposition II.4.5 applies to give us $I(m)$.

The construction in the case that m is a true leaf is similar, and goes as follows:

The leaf m cuts the disk into two pieces: a small piece D_0 and a large piece D_1. The two major leaves M_1 and M_2 cut the disk into three pieces, E_0 in the middle, and E_1 and E_2 on the two sides. Let I_0 consist of all the forward images of M_1 and M_2. (We do not take their closure at this stage, to avoid unnecessary problems with ambiguous liftings). Having constructed I_i, I_{i+1} is constructed inductively by adjoining preimages to leaves in I_i. Both E_1 and E_2 map to D_1; the map is a homeomorphism. We adjoin to E_1 and E_2 all leaves of I_i in D_1, transplanted by the inverse homeomorphism.

The region E_0, which maps to D_0, is also straightforward, as long as we keep in mind that we must not create any leaves separating M_1 from M_2. For any leaf l in D_0, there are exactly four points on the boundary of E_0 which map to its endpoints, and there is only one way to join them in pairs without creating any leaves longer than M_1 or M_2. Do it. (If m is a periodic minor, then there is some ambiguity for taking preimages of m; two of the possible preimages are M_1 and M_2 and must be in the lamination, and there are two further, shorter, possible preimages of m. For the sake of minimality, use only M_1 and M_2 as preimages of m. This implies that, whenever m is periodic, the central gap in $I(m)$ has Cantor double cover type.)

By inductive hypothesis, I_i is forward invariant. This forces I_{i+1} to contain I_i, and it follows that I_{i+1} is forward invariant.

Finally, we define $I(m)$ as the closure of the union of the I_i. $I(m)$ is fully invariant, and m is its minor leaf.

Here are three cogent questions concerning $I(m)$:

(L1) Under what conditions is $I(m)$ the unique lamination with m as minor leaf?

(L2) Under what conditions is $I(m)$ clean?

(L3) Under what conditions is $I(m)$ the smallest lamination having m as a minor leaf?

Property (L3) is the easiest to characterize.

Proposition II.6.5. *If m is a non-degenerate chord in $\overline{D^2}$ such that m and its forward images have disjoint interiors, no forward image of m is shorter than m, and m and its forward images are disjoint* [6] *from the interiors of the two preimage leaves of m of length at least $1/3$, then*

(a) *m is the minor leaf of an invariant lamination $I(m)$ (constructed above), which has the property that any other lamination having minor leaf m contains $I(m)$.*

(b) *m is the leaf of a forward invariant lamination $F(m)$, which has the property that any other forward invariant lamination containing m contains $F(m)$.*

Proof of II.6.5. At each step of the construction, the existence of previous leaves forced all the new leaves. $\boxed{\text{II.6.5}}$

Proposition II.6.6. *Let m be any point on the circle. Then $I(m)$ is the minimal lamination with m as minor leaf if and only if m is not periodic under $z \mapsto z^2$.*

Proof of II.6.6. The construction for $I(m)$ involves inductively adjoining preimages at each stage, beginning with a diameter M.

If m is periodic of period q, then there is a $q - 1$st preimage of M which has an endpoint at m, and the construction for the next stage involves taking two lifts for any such leaf. This creates two triangles with a common side M. Continuing in the construction, one obtains an infinite collection of triangles forming a triangulation of a region in the center of the disk. By the construction we have already seen, some of the edges can be removed to leave a lamination still containing M.

If, on the other hand, m is not periodic, then no ambiguities ever arise in the construction, so that any invariant lamination containing M must contain $I(m)$. $\boxed{\text{II.6.6}}$

As a result, if a clean invariant lamination has degenerate minor m, then m is non-periodic: if m is a periodic degenerate minor, the construction forces an unclean lamination.

---

[6]See Footnote 5.

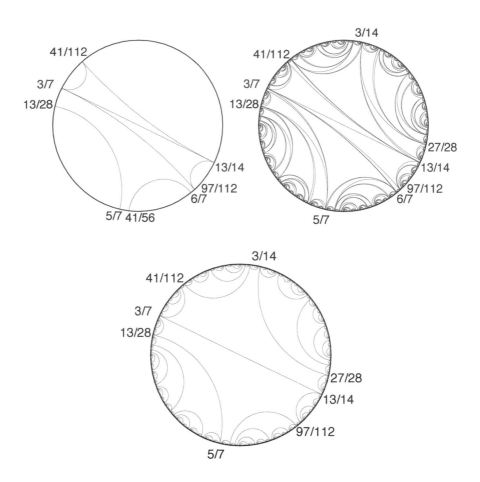

Figure II.8. Illustration of the proof of Proposition II.6.6, where m is a degenerate periodic minor of period q so that its immediate preimage is a diameter M (here m is at angle $6/7$ and $q = 3$). Top left: the $q − 1$-st preimage of M is a leaf ending in m, and its preimage are two triangles with longest side M (we already have three leaves with a common endpoint, so the lamination will be unclean). Top right: continued backwards images yield an invariant lamination $I(m)$ containing eventually collapsing triangles. Bottom: removing leaves yields another lamination with the same major M and the same minor m: it has a fixed gap with three collapsing chains of gaps. (Note that the lamination at the top right has a leaf connecting the angles $5/7$ and $6/7$: this is a non-degenerate minor leaf, of a different lamination, for which $m = 6/7$ is an endpoint. In a similar way, it can be shown that every periodic angle α has another angle β, automatically periodic with equal period, so that there exists a minor connecting the angles α and β.)

Remark II.6.6a. For every periodic angle α, there is a unique angle β of equal period, called the *conjugate angle of* α, so that the leaf joining α and β is the minor leaf of a clean invariant lamination.

The proof of this fact is analogous to the construction of the laminations $I(m)$ above: it starts with the diameter P connecting the two preimages of α and taking subsequent preimages, so that one endpoint of a preimage is always a periodic angle on the forward orbit of α, and no preimage leaf thus constructed intersects the diameter P. If n is the period of α, then every nth leaf on this orbit will have a common endpoint with P, and this leaves two choices: choose the one that yields the longer leaf; this assures that the sequence leaves of preimage leaves has lengths uniformly bounded below. Every nth leaf leaf has α as endpoint, and the other endpoints of these leaves converge to the conjugate angle β (compare Figure II.8). The details are not difficult but somewhat tedious. All these leaves will of course not be part of a clean lamination (there are infinitely many leaves with endpoints at angle α), but the leaf connecting α to β is the minor leaf of a clean invariant lamination that has a periodic gap of Cantor double cover type.

Here are some more observations concerning the three properties mentioned above (a point $p \neq 1$ on the circle is called dyadic if it eventually maps to 1 under $z \mapsto z^2$).

Proposition II.6.7.

(a) *Any invariant lamination which is clean is the minimal lamination with its given minor leaf, say m; this lamination equals $I(m)$.*

(b) *If the minor leaf is non-degenerate and its endpoints are not eventually periodic, then $I(m)$ satisfies all three properties: it is clean, minimal and unique.*

(c) *If the minor leaf is degenerate and not eventually periodic, then $I(m)$ is the unique invariant lamination with minor leaf m. It is unclean if and only if there exists an invariant lamination with a non-degenerate minor one of whose endpoints is m.*

(d) *Any minor m with a dyadic endpoint is degenerate, and $I(m)$ is clean, minimal and unique.*

Proof of II.6.7. Let L be any clean lamination. Let m be its minor leaf. If L' is any other lamination with minor leaf m, then L' cannot be contained in L: if it were, then L' would be clean as well. The additional leaves of L would have to fit in the gaps of L', and since L is clean, the only possible gaps of L' would be those which intersect the circle as a Cantor set: the Siegel and the Cantor double cover cases. But gaps of Siegel type are impossible because they have no invariant subdivisions at all.

The only remaining possibility is that L' has a central gap which returns with degree two (the Cantor double cover case); in this case, one of the major leaves is periodic. Then, since L has the same major leaves, it must also have a central gap. It cannot be a degree two gap, since such a gap is determined by the major leaves; it could only be an eventually periodic polygon or a collapsing quadrilateral. But a collapsing quadrilateral would make L unclean (Proposition II.6.4, (CL3)), and the other case is impossible: the two major leaves in the Cantor double cover case must remain major leaves in L, hence they still bound a common gap; but one of these boundary leaves is periodic and the other is not, which is impossible for an eventually periodic polygon.

To prove Part (a), suppose first that m is non-degenerate. By Proposition II.6.5, L contains $I(m)$ and hence equals $I(m)$. If m is degenerate and non-periodic, the same is true by Proposition II.6.6. Finally, if m is degenerate and periodic, then the immediate preimage of m is a diameter M which has a common endpoint with one of the leaves on its backwards orbit, so M is the common boundary of two gaps; this makes the lamination unclean (Proposition II.6.4).

As for Part (b), suppose that L is an invariant lamination with a non-degenerate minor leaf m whose endpoints are not eventually periodic. Then L contains $I(m)$. The central gap of L in this case is a collapsing quadrilateral, and all other gaps are either eventually collapsing quadrilaterals or eventually periodic polygons (all other types of gaps require either a collapsing leaf or a different central gap). None of the sides of the central quadrilateral can be sides of other gaps, because of the non-periodicity of the endpoints of m. Therefore, $I(m)$ is clean (Proposition II.6.4). The only gaps of $I(m)$ which can be subdivided to form another invariant lamination are eventually collapsing quadrilaterals, but this subdivision changes the minor leaf.

Part (c) is proven similarly, by examining the gaps. We already know that $I(m)$ is minimal (Proposition II.6.6), so we just need to show that it cannot be subdivided. If there are gaps adjacent to the major leaf M, then they can only be irrational rotation domains, preperiodic polygons, or collapsing triangles. Gaps which do not eventually touch M can only be polygons. All three kinds of gaps are indivisible. This proves uniqueness of $I(m)$. If $I(m)$ is unclean, then by Proposition II.6.4, there is a collapsing triangle or a polygon with a collapsing boundary leaf. These can be viewed as collapsing quadrilaterals, or symmetric polygons, with a diagonal drawn in. In the first case (of a pair of collapsing triangles), the diagonal can be removed together with its backward orbit, yielding another invariant lamination with a non-degenerate minor m' having m as endpoint. In the case of a polygon with $2n$ sides, the diagonal can be replaced by a pair of major leaves that have one endpoint each at the endpoints of the replaced diagonal, subdividing the polygon into one collapsing quadrilateral and two n-gons that both map homeomorphically onto the same image gap, and so that the new lamination has, as before, a non-degenerate minor m' with endpoint m.

For part (d), let p be a dyadic point on the circle. If p is the endpoint of a non-degenerate minor, then the other endpoint must be dyadic as well (Lemma II.6.2a); but then both endpoints eventually land at the same point 1, so the minor leaf eventually collapses to a point. This contradiction shows that any minor with dyadic endpoint is degenerate. If m is a degenerate dyadic leaf, then the corresponding major leaf M (a diameter) is non-isolated: arbitrarily short preimages of M accumulate at 1 and thus at m. Thus there is neither a central gap nor a gap with a collapsing side, so all gaps are eventually periodic polygons and thus indivisible. Therefore $I(m)$ is clean by Proposition II.6.4, minimal by Proposition II.6.6, and unique (as in Part (c)). $\boxed{\text{II.6.7}}$

Rather than to pursue all the intricacies of questions (L1), (L2) and (L3), it is more important for us to investigate the set of all minor leaves. It turns out that the collection of all possible minor leaves has a beautiful structure, a structure which should by now be familiar:

Theorem II.6.8. *The set QML of all chords in $\overline{D^2}$ which are minors of some invariant lamination for $z \mapsto z^2$ forms a lamination.*

Proof of II.6.8. Let m_1 and m_2 be any two minor leaves of laminations. We claim that m_1 and m_2 have disjoint interiors. Let $D(m_i)$ denote the closure of the small region of the disk cut off by m_i. The claim is trivial if the interiors of $D(m_1)$ and $D(m_2)$ do not intersect; so we may assume they do. Choose any point m in their intersection which is not eventually periodic, and let D be the diameter joining its two preimages. Then $D \bigcup F(m_i)$ is a forward-invariant lamination containing m_i, so it is contained in an invariant lamination L_i for $i = 1, 2$ (Proposition II.4.5). By the preceding result,

$$L_1 = L_2 = I(m)$$

so that, as leaves of a common lamination, m_1 and m_2 necessarily have disjoint interiors.

Since the condition to be a minor leaf of a lamination is closed (Proposition II.6.5), it follows that the set of all minors of all quadratic invariant laminations forms a lamination. $\boxed{\text{II.6.8}}$

Definition II.6.9. The lamination M of the disk by all minors of all invariant laminations is the *quadratic minor lamination*, abbreviated QML.

There is quite a bit of structure that can be found in the quadratic minor lamination.

Figure II.9. The quadratic minor lamination QML.

There is a convenient partial order which can be imposed on the set of leaves of this lamination:

Definition II.6.10. If m_1 and m_2 are minor leaves for two quadratic invariant laminations, we will say that

$$m_1 < m_2$$

whenever

$$D(m_1) \supset D(m_2) \,,$$

where $D(m)$ is

- the closed half-disk bounded by m, as above (so that $1 \notin \partial D(m)$), when m is non-degenerate,

- the entire disk if m is the special degenerate minor 1,

- and the single point m in the general degenerate case.

Thus 1 is the minimum in the partial order: its lamination $I(1)$ is the empty lamination. Every other degenerate minor is a maximal element. Intuitively, larger minors correspond (roughly) to larger laminations although they have smaller half-disks $D(m)$ (compare Lemma II.6.10a below).

The partial order $<$ has the property that if $m < n$ and $m' < n$ for leaves $m' \neq m$, then either $m < m'$ or $m' < m$ — that is to say, that the set of minors less than any given minor is linearly ordered.

Furthermore, there are greatest lower bounds for $<$: the intersection of all those half-disks of the form $D(m)$ which contain a given set of leaves S is a half-disk bounded by a leaf of QML which is the greatest lower bound. (The set of such half-disks is non-empty, in virtue of the fact that 1 is less than everything else.)

Every non-degenerate leaf l of QML is of course a leaf of some quadratic invariant lamination — in fact, it is a leaf for many such laminations, since it is contained in most laminations $I(m)$ whenever $m > l$. A formal statement is this:

Lemma II.6.10a.

(a) *If n is any non-degenerate leaf in QML and m is any non-degenerate leaf in $I(n)$ contained in $D(n)$, then m is also a leaf in QML and $m > n$ — unless m maps under iteration again into $D(n)$ so that its length doubles in each iteration step.*

(b) *If $m > n$ are two leaves in QML, then n also forms a leaf in $I(m)$ (here n is non-degenerate, but m may as well be degenerate).*

Proof of II.6.10a. (a) Under iteration, the length of m will continue to increase until it is at least $1/3$, and after that the image leaves cannot become shorter than n. Since all leaves on the forward orbit of m are contained in $I(n)$, the first two of the three hypotheses in Proposition II.6.5 are satisfied.

Let M_1 and M_2 be the two leaves of length at least $1/3$ that are immediate preimages of m (these would be the major leaves in $I(m)$; they are longer than the major leaves in $I(n)$ and thus not leaves in $I(n)$). In order to be able to apply Proposition II.6.5, we need to show that no leaf on the forward orbit of m intersects the interior of M_1 and M_2. Since no leaf in $I(n)$ can intersect the major leaves of $I(n)$, a leaf in $I(n)$ could intersect M_1 or M_2 only if its length is at most half the length of n.

If a leaf m' on the forward orbit of m has length greater than $1/3$ (but of course less than the major leaves of $I(n)$), then the image leaf is shorter than m', but at least as long as n; from then on, no leaf on the orbit of m' can intersect the interior of M_1 or M_2. Therefore, if a leaf on the orbit of m intersects M_1 or M_2, it can do so only during the initial iterations while the length of the leaf gets doubled; and the image of the intersecting leaf is contained in $D(n)$. If this does not happen, then by Proposition II.6.5 m is the minor of an invariant lamination and thus contained in QML; clearly $m > n$. This shows part a) of the claim.

(b) Suppose first that m is non-degenerate. We claim that the leaf n does not intersect a leaf in $I(m)$. Since $I(m)$ is constructed by the backwards saturation of the leaf m, any intersection of n with a leaf in $I(m)$ would imply that a forward iterate of n intersects m, in contradiction to the assumption $m > n$ (this argument would fail if n could intersect the major leaves in $I(m)$; but the two major leaves in $I(n)$ bound a central domain in the disk that is disjoint from leaves in the forward orbit of n but that contains the major leaves in $I(m)$).

In $I(m)$, there is a periodic central gap G that is of Cantor double cover type, or a collapsing quadrilateral, or a polygon with at least 6 sides. Each gap in $I(m)$ is either an eventually periodic polygon or eventually maps to G. If n is not a leaf in $I(m)$, then it must subdivide a gap in $I(m)$: but this is impossible in the case of an eventually periodic polygon and of a collapsing quadrilateral; in the Cantor double cover case, n would have to map eventually into G and then into the image of G, and this is impossible for the minor n.

If m is degenerate, then $I(m)$ is constructed as the backwards saturation of the diameter connecting the two preimages of m, and as above this diameter is contained in the central domain bounded by the two major leaves in $I(n)$; if n intersected a leaf in $I(m)$, a forward iterate of n would have to intersect the diameter, and this is not the case. Each gap in $I(m)$ is either an eventually periodic polygon, a gap of Siegel type, a collapsing chain gap, or an eventually collapsing triangle. None of these can be subdivided by n: this is clear in all cases except for gaps with collapsing chains. For the latter, recall that the boundary of a gap G of collapsing chain type consists of $q \geq 1$ chains of eventually collapsing leaves, plus at most one periodic leaf that is fixed under the return map of G. The vertices of G are preperiodic endpoints of eventually collapsing leaves, except the endpoints of the fixed boundary leaf (only if $q = 1$) and the q periodic limit points of the infinite collapsing chains (if $q \geq 2$). All preperiodic boundary vertices of G have different preperiods (because the eventually collapsing boundary leaves of G map to the collapsing diameter at different times). By Lemma II.6.2a, both endpoints of n are periodic, or both are preperiodic with equal preperiod. If n subdivides G, it must thus either equal the unique periodic boundary leaf of G (and hence n is contained in $I(m)$), or (if $q \geq 2$) it must connect two periodic boundary points of G; but these are permuted transitively preserving the cyclic order, so n would have to intersect one of its iterates, a contradiction.

II.6.10a

Proposition II.6.10b (The α Fixed Point). *Every invariant quadratic lamination with minor other than the degenerate minor at the point 1 contains either*

(a) *a polygon which is mapped to itself, by a rational rotation,*

(b) *a gap mapped to itself by an irrational rotation (the Siegel case), or*

(c) *a leaf which is mapped to itself by a 180° flip. (Such a leaf automatically is the chord joining 1/3 to 2/3.) This should really be thought of as a special case of* (a).

Furthermore, there is exactly one such gap, or leaf in the case of (c).

The shortest sides of these rotating gaps, and the invariant leaf in case (c), *are exactly the immediate successors of the minor 1 in the partial ordering on the leaves of QML.*

Proof of II.6.10b. To establish this, we use the Brouwer fixed point theorem. Consider any lamination L which has a minor other than 1. There is some continuous map f of the disk to itself which maps the leaves of L to leaves of L and gaps to gaps. If L contains a diameter or collapsing gap, this particular leaf or gap maps to a point. Either the major is a diameter other than the one connecting 1 to -1, or there is a pair of major leaves of length less than $1/2$ so that 1 is not on the boundary of the domain bounded by the pair of major leaves. In both cases, backwards iteration yields arbitrarily short leaves separating the boundary point 1 from the center of the disk. The local behavior of the map near 1 is to expand. Using the Brouwer fixed point theorem (in the disk from which a small neighborhood of 1 is removed, and precomposing f with a homotopy in the neighborhood of -1), it follows that f has a fixed point in the interior of the disk.

Such a fixed point may occur either in a gap of L or on a leaf of L. In the latter case, we have case (c). In the former case, we have case (a) or (b) (compare Theorem II.5.3: the only Cantor double cover gap of period 1 is all of the disk, and a collapsing chain gap of period 1 has a collapsing boundary leaf whose image is the minor at the point 1). In any case, the invariant leaf l or the invariant gap G, together with its symmetric copy (rotated by a half turn), subdivide the disk into finitely or countably many components so that the only fixed points within the disk can be on l or in G (the other components are disjoint from their images). This proves that there exists exactly one gap or leaf as claimed.

If case (a) or case (c) happens, then the minor for L is greater than (or equal to) the smallest leaf of the rotating polygon in the partial ordering (because the longest side of the polygon separates the polygon from the major leaf). By Proposition II.6.5, the smallest leaf of the rotating polygon is itself the minor leaf of some invariant quadratic lamination. Case (b) is really a limiting case of (a); it can happen only for the single lamination whose minor is the image of the collapsing leaf of the rotating gap. It follows that the minor for L is greater than (or equal to) one of these examples. Because of uniqueness of the rotating gap, these examples are incomparable with respect to the ordering, so they are the exactly the immediate successors of 1.

$$\boxed{\text{II.6.10b}}$$

An alternate way to compute the immediate successors of 1 would have been to compute them directly. The minors for the rotating gaps are not hard to compute, so it is not hard to see that they bound a region in the disk which contains no other possible minors. This region touches the unit circle on a set of measure 0 (see [GM93, Appendix C], [De94, p. 23], [BS94], [BKS]). Even though this approach is more concrete, the use of the fixed point theorem gives a better explanation of what is going on.

The following result is based on ideas from [BS94].

Lemma II.6.10c. *In Proposition II.6.10b, every rational rotation number $\rho \neq 1/2$ is generated by a fixed polygonal gap ($\rho = 1/2$ is generated by the leaf with endpoints 1/3 and 2/3), and every irrational rotation number is generated by a fixed gap of Siegel type, and the cyclic order of the corresponding minor leaves on the circle coincides with the cyclic order of the associated rotation numbers.*

Proof of II.6.10c. For $\alpha \in [0,1]$, define the a map $f_\alpha \colon \mathbb{R}/\mathbb{Z} \to \mathbb{R}/\mathbb{Z}$ via

$$f_\alpha(x) = \begin{cases} \alpha & \text{if} \quad 0 \leq x \leq \alpha/2 \text{ or } \alpha/2 + 1/2 \leq x \leq 1 \\ 2x \pmod 1 & \text{if} \quad \alpha/2 \leq x \leq 1/2 + \alpha/2 \,. \end{cases}$$

This is a continuous locally monotone circle map of degree 1, so it has an associated rotation number $\rho(\alpha) \in \mathbb{R}/\mathbb{Z}$. The rotation number $\rho(\alpha)$ grows continuously and monotonically with α, but not strictly monotonically. We have a fixed point and thus $\rho(\alpha) = 0$ if and only if $\alpha = 0$ in $\mathbb{R}/\mathbb{Z}$; as α moves from 0 to 1, $\rho(\alpha)$ also moves continuously from 0 to 1 and is thus a monotone surjective map from $\mathbb{R}/\mathbb{Z}$ to itself.

Consider the set

$$S_\alpha := \{x \in \mathbb{R}/\mathbb{Z} \colon f_\alpha^{\circ n}(x) \in [\alpha/2, 1/2 + \alpha/2] \text{ for all } n \geq 0\} \,.$$

If $\rho(\alpha)$ is irrational, then S_α is a non-empty compact subset of $\mathbb{R}/\mathbb{Z}$, and it contains the points α, $\alpha/2$ and $(\alpha+1)/2$ (otherwise, α would be periodic), and S_α is forward invariant. Let G be the gap in $\overline{D^2}$ that intersects the boundary exactly in the points $e^{2\pi i x}$ for $x \in S_\alpha$. Consider the invariant lamination $I(e^{2\pi i \alpha})$ with degenerate minor leaf at angle α. Then G is a gap in $I(e^{2\pi i\alpha})$; in fact, G must be an invariant gap of Siegel type with rotation number $\rho(\alpha)$.

If $\rho(\alpha)$ is rational, say $\rho(\alpha) = p/q$ in lowest terms, then there is at least one periodic orbit with combinatorial rotation number p/q, and every point on $\mathbb{R}/\mathbb{Z}$ must converge to such an orbit. Since f_α has derivative 2 everywhere except at the interval where it is constant, it follows that every periodic cycle that does not contain α is repelling. This implies that α must be periodic, and its cycle absorbs the interval where f_α is constant. It follows that there is a non-trivial closed interval of values α with the same rational rotation number $\rho(\alpha) = p/q$. The endpoints of this interval, say $\alpha_{p/q}^-$ and $\alpha_{p/q}^+$, have the property that α is periodic with period q and combinatorial rotation number p/q for the original map $x \mapsto 2x \pmod 1$ on $\mathbb{R}/\mathbb{Z}$. These endpoints form the periodic minor leaf we are looking for.

$$\boxed{\text{II.6.10c}}$$

Now we can classify the gaps of QML.

Theorem II.6.11. *The gaps of QML are of two types:*

P-gap: *a polygon P that occurs as the image of the central gap in some quadratic invariant lamination. If m is the side of P that is least in the partial order, then the associated lamination $I(m)$ is clean, and it has a central polygon of 6 or more sides which maps with degree two to P.*

Conversely, for any lamination having a central gap of 6 or more sides, the image of the central gap is a gap of QML.

H-gap: *a gap with countably many boundary leaves without any common endpoints and so that the gap intersects the boundary of the disk in a Cantor set. The least side m of this gap (in the partial order) is a minor leaf with periodic endpoints. The associated lamination $I(m)$ is clean, and has a central gap which is periodic with degree two (it has Cantor double cover type). The other sides of the gap are the immediate successors of m, and they also have periodic endpoints; each is the smallest leaf of a similar gap. There is one special case, in which m is the degenerate minor 1.*

Every periodic minor leaf is the least side of an H-gap of QML whose other sides are the immediate successors of m.

Proof of II.6.11. If G is any gap of QML, then there is some least side m of G, and all the other sides are immediate successors of m. We shall call m the leading edge of G. Classifying the gaps of QML is equivalent to classifying their leading edges, which are minor leaves having immediate successors in the partial order.

First we eliminate the need to discuss laminations with collapsing quadrilaterals. If an invariant lamination L has a collapsing quadrilateral in which the shorter pair of sides is isolated, then there is a clean lamination with the same minor leaf (see the discussion after Proposition II.6.4). On the other hand, if the shorter pair of sides is non-isolated, then the minor m is a limit of leaves m_i in L that are shorter than m (and thus contained in $D(m)$). By Lemma II.6.10a part a), all m_i sufficiently close to m are leaves in QML, so that the leaf m does not have an immediate successor in QML.

The remaining possibilities for a leading edge m are that it is the minor leaf for a lamination L which has a central gap G, where either

(a) G is a polygon of $2n$ sides, where $n \geq 3$, or

(b) G is a gap of Cantor double cover type: it has infinitely many sides, and eventually returns to itself with degree two.

To complete the proof, we will show that in each of these cases, m indeed has immediate successors, and that m together with its successors forms a gap as described in the theorem.

The polygonal case (P-gaps) is easiest. In this case, G can be subdivided by adding a collapsing quadrilateral in n ways. One of these subdivisions has the same minor leaf as the original. The other subdivisions give new minor leaves. These n minor leaves form a polygon G' in QML which is exactly the same as the image of the gap G under $z \mapsto z^2$.

It logically follows that G' is a union of finitely many gaps of QML. However, under the forward iterates of $z \mapsto z^2$, G' eventually maps by a homeomorphism to a periodic gap, and the return map of the periodic gap acts transitively on its vertices. Every diagonal of G' has interior intersections among its forward images, so G' is the actual gap.

The case of H-gaps is trickier to understand, partly because in this case the leading edge of the gap has infinitely many successors.

The special case of the degenerate minor 1 has been treated in Proposition II.6.10b; it is the prototype for all the others. All successors of the degenerate leaf 1 in QML are the chord joining $1/3$ to $2/3$, as well as the (non-degenerate) shortest leaves of the fixed polygons and the (degenerate) image of the diagonal in the fixed irrational rotation gaps. The non-degenerate immediate successors in QML of the minor 1 are thus leaves with periodic endpoints. Each such periodic minor m has, within the lamination $I(m)$, a periodic gap of equal period and of Cantor double cover type so that m is the longest boundary leaf of one of the gaps.

Now we consider the more general case, that m is the minor for a lamination L having a gap G of Cantor double cover type. The return map is a blown up version of the map $z \mapsto z^2$ (these maps are semi-conjugate on the boundary; a boundary point z on the circle has an arc as preimage in the boundary of G if and only if z eventually maps to the fixed point 1).

It follows that any rotating polygon, irrationally-rotating gap, or flipped leaf for $z \mapsto z^2$ (as in Proposition II.6.10b) can be transplanted into G via the blowing up correspondence, refining the lamination $I(m)$ (this is a combinatorial version of the tuning process of Douady and Hubbard). Since the blown-up points are all eventually fixed in $z \mapsto z^2$, they are disjoint from the rotating gap.

This transplantation always yields a forward-invariant lamination; it can be backward-saturated to form a fully invariant lamination L'. We will prove that the minors for the lamination L' are the immediate successors of m.

The proof is much the same as in the case of period 1. Let P and Q be the two major leaves in L and let m_1 be any minor which follows m in the partial ordering. There is some lamination L_1 with minor m_1 which also contains m (by Lemma II.6.10a part b)) and hence the major leaves P and Q of L, since one of those (say P) is in the forward image of m.

Let E' be the open region between P and Q, and let q be the period of m. Let E be the subset of E' consisting of all points which return to E' after all multiples of q iterations. Let g be the first return map from E to itself. The domain E is bounded by P and Q and countably many preimages of these leaves, and it maps P and Q to the leaf P which is fixed by g (this is the Cantor double cover dynamics).

Reasoning as in Proposition II.6.10b, we conclude that g must have a fixed point which is not on P. This implies the existence (in L') of either a polygonal gap which is rotated by g, a gap touching the circle in a Cantor set which is mapped as an irrational rotation (the Siegel case), or a leaf whose ends are flipped by g. This gives us the description of all of the immediate successors of m under $<$ as above. $\boxed{\text{II.6.11}}$

Remark II.6.11a. Every H-gap G has leading edge m_0 with periodic endpoints, say of period n. There is a map $\rho(m, m_0) \to \mathbb{R}/\mathbb{Z}$ that assigns to each (possibly degenerate) leaf m of QML a rotation number as follows. If $m \leq m_0$ or m and m_0 are not comparable, then $\rho(m, m_0) = 0$. If $m > m_0$, then there is a unique trailing edge m'' of G with $m_0 < m'' \leq m$, and m'' has periodic endpoints. If m'' is non-degenerate, then its endpoints are periodic of some period kn for $k \geq 2$, and we have the following:

- if $k = 2$, then the leaf m'' maps onto itself in an orientation reversing way, and we set $\rho(m, m_0) = \rho(m'', m_0) = 1/2$;

- if $k \geq 3$, then the leaf m'' is the shortest side of a periodic polygonal gap in $I(m'')$, and we define $\rho(m, m_0) = \rho(m'', m_0)$ as the combinatorial rotation number of this polygonal gap under its first return map;

- finally, if m'' is degenerate, then $m = m''$, and $I(m)$ has a periodic gap of Siegel type; we define $\rho(m, m_0)$ as the combinatorial rotation number of this Siegel gap.

The map $\rho(\,\cdot\,, m_0)\colon \partial G \to \mathbb{R}/\mathbb{Z}$ is an orientation preserving continuous circle map of degree 1, sending boundary leaves of G to the corresponding rotation numbers (so that the preimage of a rational rotation number is a boundary side of G, while the preimage of an irrational rotation number is a degenerate leaf of QML). For the case of period 1 (i.e., $m_0 = 1$), this follows from Proposition II.6.10b and Lemma II.6.10c, and in the general case this follows by the tuning argument as in the proof of Theorem II.6.11.

Each H-gap (whose leading edge is periodic) gives birth to an infinite collection of other H-gaps, since it has a countable number of immediate successors, corresponding to rational rotation numbers, which are periodic. All immediate successors of any periodic leaf in QML are thus other periodic leaves. Successive application of this argument shows that each H-gap generates an entire tree of further H-gaps.

If m is any leaf of QML, then there are likely to be a countable number of different leading edges of H-gaps less than m. The next proposition shows that this does not give a countable number of possible combinatorial rotation numbers, however.

Proposition II.6.12.[7] *For any non-degenerate leaf m of QML, there are only a finite number of leading edges m' of H-gaps such that $m' < m$ and $\rho(m, m') \neq \frac{1}{2}$.*

Proof of II.6.12. Each leading edge of an H-gap which precedes m has a unique successor which also precedes m. Let M consist of all such successors for leading edges m' with $\rho(m, m') \neq 1/2$. Note that each minor n in M is periodic, and in fact the forward saturation $F(n)$ has periodic polygonal gaps (because $\rho(m, m') \neq 1/2$) so that n is the shortest boundary side of all polygons in the cycle; this implies that all these polygons are disjoint for different n. The length of n is at least the length of m, so all polygons occupy a Euclidean area in the disk that satisfies a positive lower bound in terms of the length of m. This yields an upper bound on the number of polygons and hence on the size of M.

$$\boxed{\text{II.6.12}}$$

The following fact is not hard to check: if for any leading edge of an H-gap, an infinite sequence of successors is constructed, always with combinatorial rotation number $1/2$, then the limiting minor is not degenerate. For

---

[7]The original manuscript also had a claim on P-gaps. In fact, leaves in QML may well be preceded by infinitely many P-gaps.

example, beginning with the leading edge 1, the leaf corresponding to rotation number $1/2$ joins the points $1/3$ and $2/3$, or in binary notation, it joins $.(01)$ to $.(10)$, where the parentheses denote infinite repetition. Its successor with rotation number $1/2$ is the line joining $2/5 = 6/15$ to $3/5 = 9/15$, or binary $.(0110)$ to $.(1001)$. Next comes the leaf joining $7/17 = 105/255$ to $10/17 = 150/255$, or binary $.(01101001)$ to $.(10010110)$. The pattern continues, with the repeating parts of the binary expansion for an endpoint of the next leaf obtained by concatenating the repeating parts of the two endpoints of the preceding leaf. The limit case is the one which has been analyzed by Feigenbaum and others, with leaf joining the irrational points .41245403... and .58754596..., or binary .0110100110010110... and .1001011001101001....

Let m and m' be the leading edge and some trailing edge of a P-gap. Then $I(m')$ is a lamination with a central collapsing quadrilateral having an isolated side. In other words, it is not clean (Proposition II.6.4), and the equivalence relation on the circle determined by identifying the endpoints of all leaves of the lamination $I(m')$ is the same as the equivalence relation for $I(m)$.

For all non-degenerate minor leaves m which are not leading edges of gaps in QML, the central gap of $I(m)$ is a quadrilateral. There are (unclean) laminations whose minor is one of the two endpoints of m obtained by subdividing the eventually collapsing quadrilaterals into two triangles. If m is also not a trailing edge of a gap in QML, so that it is not isolated on either side in QML, then none of the sides of the quadrilateral are isolated. The converse also follows: if none of the sides of the collapsing quadrilateral are isolated, then m is not isolated in QML on both sides.

The limitation on choices for gaps may make it seem that there are not many gaps. However, we have the following.

Theorem II.6.13. *Gaps are dense in QML.*

Proof of II.6.13. If gaps are not dense in QML, then there is a whole non-degenerate interval of angles that are endpoints of non-degenerate leaves of QML. But any such interval contains a dyadic rational angle, and this is a contradiction to Proposition II.6.7 (d). $\boxed{\text{II.6.13}}$

Corollary II.6.13a. *The lamination QML is the closure of its periodic and preperiodic leaves.*

Proof of II.6.13a. Any point on any non-degenerate leaf in QML is a limit point of gaps by Theorem II.6.13, and by Theorem II.6.11, the boundary leaves of these gaps have both endpoints periodic or both endpoints preperiodic. $\boxed{\text{II.6.13a}}$

In fact, the lamination QML is the closure of its periodic leaves alone: all preperiodic leaves are limits of periodic leaves.

Corollary II.6.13b (The Branch Theorem). *For any two non-degenerate leaves m and m' of QML, one of the following holds:*

(a) $m < m'$ or $m' < m$;

(b) *there is a P-gap or an H-gap with two trailing edges $n \neq n'$ so that $n \leq m$ and $n' \leq m'$.*

Proof of II.6.13a. Consider the set of all leaves l that satisfy $l \leq m$ and $l \leq m'$. The set of these leaves is non-empty (there is at least the degenerate leaf 1), and it is totally ordered. Therefore, this set has a maximum, say n_0.

If n_0 is not the leading edge of a gap, then there is a sequence of leaves $n_i > n_0$ that converge to n_0. If $m > n_0$, then $m > n_i$ for sufficiently large i, and the same holds for m'. By choice of n_0, it follows that $m = n_0$ or $m' = n_0$ and we are done.

We are thus left with the case that n_0 is the leading edge of a gap, say G, and we may assume that $m \neq n_0$, $m' \neq n_0$. There is a unique trailing edge n of this gap with $n \leq m$, and a unique trailing edge n' with $n' \leq m'$. By choice of n_0, we have $n \neq n'$ and we are done. $\boxed{\text{II.6.13a}}$

It is good to think of QML itself as defining an equivalence relation on the circle. Points on the circle which are joined by leaves of QML define laminations which can be derived from each other by a process of cleaning and subdividing. It is worth stating the following.

Theorem II.6.14. *The quadratic minor lamination QML is clean, and the quotient space of the circle by identifying endpoints of leaves of QML is a Hausdorff space which can be embedded as a closed subset of the plane.*

Proof of II.6.14. If QML is not clean, then there are three leaves with common endpoint p. Since gaps are dense by Theorem II.6.13, there are at least two gaps for which p is an isolated boundary point. By Theorem II.6.11, these two gaps must be P-gaps, so all their boundary leaves are strictly preperiodic. Let G_1 and G_2 be these gaps and let m_1 and m_2 be their leading edges. If m_1 and m_2 are not comparable with respect to the partial order, then there must be another P-gap with isolated vertex p, so possibly by changing notation we may assume that $m_2 > m_1$. By Lemma II.6.7 part d), there is a degenerate dyadic minor leaf $n > m_2 > m_1$, and $I(n)$ is clean. Therefore, by Lemma II.6.10a part b), both leaves m_1 and m_2 occur in the lamination $I(n)$. We claim that all boundary leaves of G_1 and G_2 occur in $I(n)$; but since $I(n)$ is clean, we obtain a contradiction that shows that QML is clean.

To show the claim, note that by Theorem II.6.11, each gap G_i has an invariant quadratic lamination L_i with a preperiodic central gap that maps to G_i with degree 2. Each periodic gap on the forward orbit of G_i has the property that its first return map permutes all sides transitively. This implies that any invariant quadratic lamination that contains any side of G_i must contain all sides of G_i. Since $I(n)$ contains m_1 and m_2, the claim is proved.

The quotient space is the quotient of a compact space by a closed equivalence relation, so it is Hausdorff. The equivalence classes do not disconnect the plane, so it follows from a classical theorem of Moore that the quotient of the plane is homeomorphic to the plane. $\boxed{\text{II.6.14}}$

Appendix: Laminations, Julia Sets, and the Mandelbrot Set

Dierk Schleicher

Invariant laminations were developed in order to describe the dynamics of polynomials and the structure of the parameter space of quadratic polynomials, i.e., the Mandelbrot set. In the introduction to Section II.4, Thurston writes: "we define the notion of an invariant lamination (which we will secretly think of as determining a potential set of identifications for a polynomial map)". He had originally planned, but never written, a Section II.7: "we will show how the theory of laminations ties in with the Julia sets and Mandelbrot sets, and we will illustrate this with computer pictures".

In this appendix, we will try to make explicit the relation between invariant quadratic laminations and filled-in Julia sets of quadratic polynomials, and between the quadratic minor lamination QML and the Mandelbrot set, and we give illustrating computer pictures as planned originally. This appendix is closely related to Douady's classical text [D93].

Invariant Laminations and Quadratic Polynomials. Invariant laminations model the dynamics of polynomials on their filled-in Julia sets, provided the latter are connected. The relation is as follows. Consider a polynomial p, say of degree $d \geq 2$ and monic. Let K be its filled-in Julia set. By Proposition II.3.1, the set K is connected if and only if the orbits of all critical points in $\mathbb{C}$ are bounded. In this case, there is a unique Riemann map $\phi \colon (\mathbb{C} - K) \to (\mathbb{C} - \overline{D^2})$ that is tangent to the identity near ∞. The map ϕ conjugates the dynamics of p on $\mathbb{C} - K$ to $z \mapsto z^d$ on $\mathbb{C} - \overline{D^2}$. The *dynamic ray at angle* $\alpha \in \mathbb{R}/\mathbb{Z}$ is defined as the preimage (under ϕ) of $(1, \infty) \cdot e^{2\pi i \alpha}$. This ray *lands* at a point $z \in K$ if the limit $\lim_{r \searrow 1} \phi^{-1}(r e^{2\pi i \alpha})$ exists and equals z (for background and more details, see [M06, Section 18]).

First suppose that K is locally connected, so all dynamic rays land and their landing points define a continuous surjection from the circle to the Julia set (compare Carathéodory's Theorem II.2.3): the boundary map $\psi = \phi^{-1} \colon \partial D^2 \to \partial K$ is called the *Carathéodory loop*. We define a lamination L as follows: for any pair of dynamic rays at angles α and α' that land together at a common point in K, we stipulate that the lamination L has a leaf connecting $e^{2\pi i \alpha}$ and $e^{2\pi i \alpha'}$. If three or more rays land together (always a finite number), then we give L only a leaf for any pair of adjacent angles (in the circle), so that all angles of rays landing at a common point form the vertices of a single polygonal gap in L. This defines a lamination L. It is

111

easy to check that this lamination is invariant in the sense of Section II.4; it models the dynamics of p. This lamination is always clean: if we had three leaves with a common endpoint, we would have at least four dynamic rays that land together, and by construction the corresponding angles would just form a single polygonal gap without diagonals.

From this lamination, we obtain a topological model of the Julia set as follows: form the quotient space $\widetilde{L}$ defined by the smallest equivalence relation on $\overline{D^2}$ in which every leaf, as well as each polygonal gap, collapses to a point; since every lamination is closed by definition, it follows that this equivalence relation is closed. The natural projection $\pi_L \colon \overline{D^2} \to \widetilde{L}$ is compatible with the dynamics, so it induces a continuous d-to-one self-map on $\widetilde{L}$ which models the dynamics of p on its filled-in Julia set K: the dynamics of p on K is topologically conjugate to the induced map on $\widetilde{L}$. The homeomorphism $\pi_K \colon K \to \widetilde{L}$ is defined using the Carathéodory loop $\psi \colon \partial D^2 \to \partial K$ as follows: by construction, $\psi(e^{2\pi i \alpha}) = \psi(e^{2\pi i \beta})$ if and only if the dynamic rays at angles α and β land at a common point, which means $\pi_L(e^{2\pi i \alpha}) = \pi_L(e^{2\pi i \beta})$. We thus obtain a homeomorphism $\pi_K \colon \partial K \to \pi_L(\partial D^2)$ via $\pi_K(z) := \pi_L(\psi^{-1}(z))$. This homeomorphism extends to a homeomorphism $\pi_K \colon K \to \widetilde{L}$. The space $\widetilde{L}$ is called the *pinched disk model of K*. More details on pinched disk models can be found in [D93].

Lemma A.1. *The space $\widetilde{L}$ is locally connected. More generally, let L' be any clean geodesic lamination of D^2, and let $\widetilde{L}'$ be the quotient of $\overline{D^2}$ with respect to an equivalence relation on $\overline{D^2}$ that identifies each leaf and each polygonal gap to a point. Then $\widetilde{L}'$ is locally connected.*

Proof of A.1. The quotient topology in $\widetilde{L}'$ is defined so that the projection map is continuous. The continuous image of a compact locally path connected space is locally path connected; this is Lemma II.2.2. $\boxed{\text{A.1}}$

If K is not locally connected, then the construction of the lamination L needs to be modified. We do not specify the details (compare Kiwi [Ki04]), but specifically for a quadratic polynomial p with connected Julia set we note the following: if p has an attracting or parabolic periodic orbit, then K is known to be locally connected [DH84, M06], so we only need to consider the case that p has a periodic Siegel disk or Cremer point, or all periodic orbits are repelling.

In the case when all periodic orbits are repelling, then the construction of L using pairs of dynamic rays landing at a common point still works, except that the set of leaves thus obtained need not be closed. Taking the closure of the set of leaves yields an invariant lamination L and a locally connected quotient space $\widetilde{L}$ with continuous two-to-one dynamics, together with a continuous surjection $\pi_K \colon K \to \widetilde{L}$. The map π_K cannot be injective (or it would be a homeomorphism, but $\widetilde{L}$ is locally connected while

K is not); it induces a topological semiconjugation between (p, K) and the dynamics on $\widetilde{L}$.

If p has an invariant Siegel disk, the critical value v is on the boundary of K, but it may or may not be the landing point of a dynamic ray; accordingly, the critical point may or may not be the landing point of a pair of opposite dynamic rays: these two rays landing at the critical point would correspond to the diameter in L, which in turn would form the collapsing boundary leaf of the periodic gap of Siegel type. If no dynamic rays land at the critical value and hence at the critical point, the diameter would be missing; since this would be an isolated leaf in L, it could not be recovered by taking a limit of other leaves. In particular, if the Siegel disk has period 1, then two dynamic rays can land together only if they eventually map to the ray landing at the critical value (Proposition II.6.10b); if there is no such ray, the lamination constructed from ray pairs would be trivial. In this case, there is a unique angle α such that $I(e^{\sigma\pi i\alpha})$ has a Siegel gap of period 1 with the same combinatorial rotation as the Siegel disk of p. The lamination $L = I(e^{e\pi i\alpha})$ yields a locally connected model $\widetilde{L}$ for the filled-in Julia set K of p with the Siegel disk. Of course, the model $\widetilde{L}$ is not homeomorphic to K. Similar remarks apply to Siegel disks of higher period (here pairs of rays landing together define a nontrivial lamination L_0; now there is a unique α so that $I(e^{e\pi i\alpha})$ contains L_0 and has a Siegel disk with the right period and rotation number).

Finally, if p has a Cremer point, then the multiplier of the Cremer point is $e^{2\pi i\alpha}$ for some $\alpha \in \mathbb{R}/\mathbb{Z}$, and we obtain a lamination in the same way as in the Siegel case; of course, this lamination has a gap of Siegel type, so it models a locally connected Julia set with a Siegel disk: invariant quadratic laminations cannot distinguish between Siegel disks and Cremer points (recall the classification of periodic gaps in Theorem II.5.3: a Cremer point must be modeled by a periodic gap without periodic boundary leaf, and the only such gaps are those of Siegel type). In all these cases, the pinched disk models are locally connected by Lemma A.1 and model Julia sets with Siegel disk and "nice" topology, even if the actual Julia sets are quite different. More results on pinched disk models of Julia sets in all these cases can be found in [D93].

Gaps of Laminations and Polynomial Dynamics. Theorem II.5.3 describes all possible gaps of invariant quadratic laminations. Proposition II.6.4 rules out the collapsing chain case for clean laminations (and collapsing triangles also cannot happen for clean laminations). The remaining cases actually occur for Julia sets of quadratic polynomials; these are the following:

(a) A periodic polygonal gap arises when three or more periodic dynamic rays land at a common periodic point (all rays have necessarily the same period, so their number is finite); see Figure A.1.

(b) A periodic gap of Siegel type intersects the circle in a Cantor set and the first return map acts on the boundary as an irrational rotation blown up along the backward orbit of some point. Such a gap models a Siegel disk; see Figure A.3. Note that for any quadratic polynomial with a Siegel disk or a Cremer point, the corresponding invariant quadratic lamination has a periodic gap of Siegel type.

(c) A periodic gap of Cantor double cover type intersects the circle in a Cantor set and the first return map acts on the boundary as a shift map on two symbols. Such a gap models the immediate basin of an attracting or parabolic periodic point; see Figure A.4.

(d) A preperiodic gap eventually maps to a periodic gap: it may do so as a homeomorphism to any of the three cases above, or as a degree two cover of a finite-side polygon. The latter case happens if $n \geq 3$ dynamic rays land at the critical value, so $2n$ rays land at the critical point and form a gap with $2n$ sides; see Figure A.2.

(e) An eventually collapsing quadrilateral corresponds to the case when exactly four dynamic rays land at the critical point, and the two image rays land at the critical value; see Figure A.5.

The Mandelbrot Set. In order to describe the relation between the Quadratic Minor Lamination QML and the Mandelbrot set $\mathbf{M}$, we review some background on $\mathbf{M}$ that will be used later on.

Similarly as for filled-in Julia sets with non-escaping critical points, Douady and Hubbard proved that the Mandelbrot set $\mathbf{M}$ is connected, and there is a unique Riemann map $\Phi \colon (\mathbb{C} - \mathbf{M}) \to (\mathbb{C} - \overline{D^2})$ that is tangent to the identity near ∞. The *parameter ray at angle* $\alpha \in \mathbb{R}/\mathbb{Z}$ is defined as the preimage of $(1, \infty) \cdot e^{2\pi i \alpha}$. (Dynamic rays and parameter rays are often jointly called *external rays*.) The following fundamental result of Douady and Hubbard will be needed for all further arguments; for a proof, we refer the reader to [DH84, M00, Sch00].

Theorem A.2 (Parameter Ray Pairs at Periodic Angles). *For every periodic angle $\alpha \in \mathbb{R}/\mathbb{Z}$, there is a unique periodic angle $\beta \neq \alpha$ of equal period so that the parameter rays at angles α and β land at a common boundary point $c \in \mathbf{M}$. For the polynomial $p_c(z) = z^2 + c$, the dynamic rays at angles α and β land at a common periodic point z with the following property: the angles α and β minimize the distance on $\mathbb{R}/\mathbb{Z}$ of all pairs of angles of periodic rays landing at a common point on the orbit of z.*

Periodic angles α and β as in Theorem A.2 for which the corresponding parameter rays land at a common point are called *conjugate angles*.

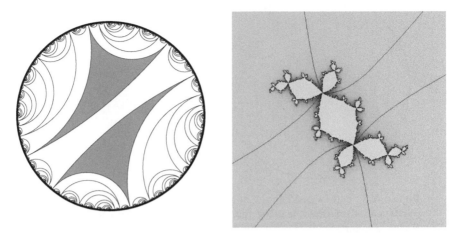

Figure A.1. Left: periodic (and preperiodic) gaps of polygonal type model several periodic (or preperiodic) dynamic rays that land at a common point. Right: a quadratic Julia set modeled by this lamination has the corresponding periodic dynamic rays landing at the same point.

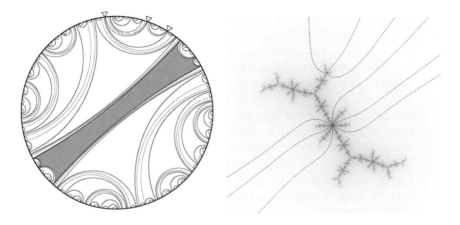

Figure A.2. Left: an invariant lamination with a central gap that is a preperiodic polygon. The central gap is shaded, the vertices of its image gap are marked. Right: a quadratic Julia set modeled by this lamination: the critical value is the landing point of $n \geq 3$ dynamic rays, and the critical point is the landing point of $2n$ dynamic rays.

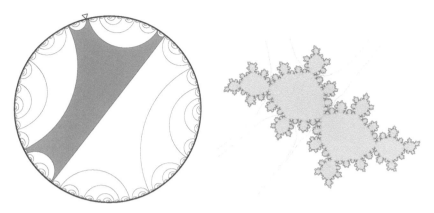

Figure A.3. Left: a periodic gap of Siegel type models a Siegel disk; the degenerate minor leaf is indicated near the top of the circle, and the invariant gap is shaded. Right: a quadratic Julia set modeled by this lamination; the large grey component on the upper left is an invariant Siegel disk. The leaves bounding the invariant gap in the lamination correspond to pairs of dynamic rays landing on the boundary of the Siegel disk (courtesy of Arnaud Chéritat).

A *Misiurewicz-Thurston parameter* is a point $c \in \mathbf{M}$ with the property that for $z \mapsto z^2 + c$, the critical point is strictly preperiodic. Such parameters are the landing point of a finite number $k \geq 1$ of parameter rays, all at strictly preperiodic angles $\alpha_1, \ldots, \alpha_k$. Moreover, in the dynamic plane of $z \mapsto z^2 + c$, the dynamic rays at the same angles $\alpha_1, \ldots, \alpha_k$ all land at the critical value, and these are the only such rays (it follows that all α_j have the same period and preperiod; this implies that there are only finitely many such angles). Conversely, if a Misiurewicz-Thurston parameter c has the property that the dynamic ray at some angle α lands at the critical value of p_c, then α is preperiodic and the parameter c is the landing point of the parameter ray at angle α. (For references, see [DH84, Sch00].)

A *hyperbolic component* of $\mathbf{M}$ is a connected component W of the interior of $\mathbf{M}$ so that for every $c \in \mathbf{M}$, the map $z \mapsto z^2 + c$ has an attracting periodic orbit of some period n; this period is necessarily constant throughout W. This n is called the *period of W*. Every hyperbolic component has a unique *root point*: this is a parameter $c \in \partial W$ so that $z \mapsto z^2 + c$ has an indifferent periodic point with period n and multiplier 1. The root is the landing point of two parameter rays at period n (if $n \geq 2$); the domain separated from the origin by these two parameter rays and their common landing point is called the *wake* of W, and it contains W. A *subwake of* W is a subset of the wake of W that is separated from W by a pair of parameter rays at periodic angles landing at a common point on ∂W. (For references, see [DH84, M00, Sch00].)

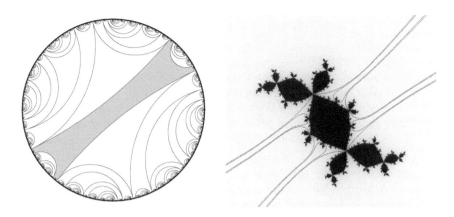

Figure A.4. Left: a periodic gap of Cantor double cover type models the immediate basin of an attracting or parabolic periodic point (in this case, of period 3). Right: the corresponding quadratic Julia set (the "Douady rabbit") with the some of the ray pairs corresponding to boundary leaves of the central gap drawn in.

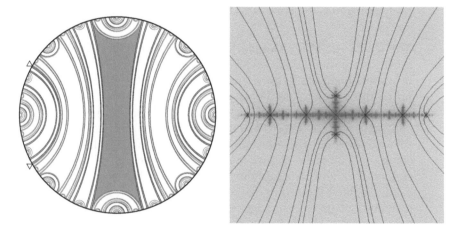

Figure A.5. Left: An invariant lamination with collapsing quadrilateral (the minor leaf connects the angles $5/12$ and $7/12$ and is marked). It models a quadratic Julia set in which two dynamic rays land at the critical value, so four dynamic rays land at the critical point. Right: the corresponding quadratic Julia set with dynamic ray pairs drawn in that land at the critical value, at its forward orbit, at the critical point, and at a number of preimages of the critical point.

Every parameter $c \in \mathbf{M}$ that is contained in the wake of W is contained either in $\overline{W}$ or in a subwake of W ("there are no ghost wakes"); similarly, every parameter ray in the wake of W either lands on ∂W (and no other parameter ray lands at the same point), or it is contained in a subwake of W [M00, Sch00, H93].

Parameter rays that land at common points give a lot of structure to $\mathbf{M}$, and such parameter rays are described by QML. The following discussion is taken from [Sch04]. We say that two parameter rays form a *ray pair* if they land at a common point. Such a ray pair, together with their common landing point, disconnects $\mathbb{C}$ into two complementary open components. If two points $z, z' \in \mathbb{C}$ are in different components, then we say that z and z' are *separated* by this ray pair.

A *fiber* of $\mathbf{M}$ is either a single point in the closure of a hyperbolic component, or an equivalence class of points in $\mathbf{M}$ so that none of these points is on the closure of any hyperbolic component, and no two of them can be separated by a ray pair of parameter rays at periodic or preperiodic angles. We say that a fiber is *trivial* if it consists of a single point. All parameter rays of $\mathbf{M}$ at rational rays land at parameters with trivial fibers (this fact is required for the proof that fibers indeed form an equivalence relation). Every fiber is compact and connected: it is the nested intersection of compact and connected subsets of $\mathbf{M}$ bounded by parameter rays at rational angles.

The Quadratic Minor Lamination and the Mandelbrot Set. The quadratic minor lamination QML models the Mandelbrot set $\mathbf{M}$ in a similar way as invariant quadratic laminations model quadratic Julia sets. We will describe this relation in more detail.

Recall from Theorem II.6.11 that QML has two kinds of gaps: P-gaps and H-gaps: P-gaps are polygons with finitely many vertices, all of them preperiodic with equal preperiod and equal period, while H-gaps have countably many non-degenerate boundary leaves, all of them with periodic endpoints, and uncountably many degenerate leaves at not eventually periodic angles. We define the *period* of an H-gap to be the period of the endpoints of its leading edge. Now we will describe the relation of these gaps with the Mandelbrot set: **P**-gaps correspond to single Misiurewicz-Thurston **P**oints of $\mathbf{M}$, while **H**-gaps correspond to **H**yperbolic components of $\mathbf{M}$.

The first relation that we describe is between landing properties of parameter rays and QML; it goes as follows.

Theorem A.3 (Landing of Parameter Rays and QML). *Parameter rays of the Mandelbrot set have the following properties:*

(a) *Two parameter rays at rational angles α and β land at a common point in $\partial \mathbf{M}$ if and only if QML has a leaf connecting the angles*

α and β, or QML has a P-gap the boundary of which contains the points α and β.

(b) *If two parameter rays at irrational angles α and β land at a common point, then QML has a leaf connecting the angles α and β; the converse holds provided* **M** *is locally connected.*

Proof of A.3. Theorem A.2 translates to quadratic laminations as follows: given two conjugate periodic angles $\alpha, \beta \in \mathbb{R}/\mathbb{Z}$, the leaf m connecting $e^{2\pi i \alpha}$ to $e^{2\pi i \beta}$ is shorter than all image leaves, it is disjoint from the interior of all its image leaves, and there is a preimage leaf of m of length at least $1/3$ that is disjoint from the interiors of all leaves on the forward orbit of m. Using Proposition II.6.5, it follows that there is a unique invariant quadratic lamination with minor leaf m, so m is a leaf in QML. All pairs of parameter rays at periodic angles landing at a common point in ∂**M** are thus represented by a leaf in QML with periodic angles. Since parameter rays of **M** at periodic angles land in pairs by Theorem A.2, it follows that QML also connects all periodic angles in pairs of equal period (see also Remark II.6.6a). If there were extra leaves in QML connecting periodic angles that do not correspond to pairs of parameter rays of **M** that land together, then QML would have gaps with boundary leaves that have common periodic endpoints, and this is not the case by the classification of gaps in Theorem II.6.11. This proves Theorem A.3 in the periodic case.

The proof in the preperiodic case is similar: if two parameter rays at preperiodic angles α and β land at a parameter $c \in \partial$**M**, then in the dynamic plane of $p(z) = z^2 + c$, the dynamic rays at angles α and β land together at the critical value, and conversely. The number of dynamic rays landing at the critical value is finite. The corresponding angles form a preperiodic polygonal gap in $\overline{D^2}$ (if there are just two angles, we obtain a preperiodic leaf). It is not hard to see that each of the boundary leaves of this gap satisfies the conditions of Proposition II.6.5, so it forms the minor of an invariant quadratic lamination. Therefore, α and β are connected by a leaf in QML or are boundary points of a common P-gap in QML. For the converse, it suffices to show that if QML has a leaf connecting two preperiodic angles α and β, then the corresponding parameter rays land at the same point in ∂**M**. Let c be the landing point of the parameter ray at angle α; for the polynomial $p_c(z) = z^2 + c$, the critical value is preperiodic and is the landing point of the dynamic ray at angle α; moreover, the Julia set is locally connected; in particular, all dynamic rays land. The two dynamic rays at angle $\alpha/2$ and $(1 + \alpha)/2$ land at the critical point and partition $\mathbb{C}$ into two open domains. It is not hard to show, using a hyperbolic contraction argument, that two dynamic rays at angles β and β' have a common landing point if and only if the following combinatorial condition is satisfied: for every $k \geq 0$, the two dynamic rays at angles $2^k \beta$ and $2^k \beta'$ are in the closure of the same component of the partition. If there is an invariant quadratic lamination with minor leaf connecting α and β, then this implies that this combinatorial condition is satisfied, and the dynamic rays at angles α and β land together at the critical value. It follows that the parameter rays at angles α and β both land at c.

Now we prove part b) of the theorem. Consider two irrational angles α and β so that the parameter rays at angles α and β land at a common point in $\partial \mathbf{M}$. Then these two parameter rays cannot be separated by a pair of parameter rays at rational angles: in other words, using the first part of the theorem, every leaf m of QML with periodic or preperiodic endpoints has the property that one component of $\overline{D^2} - m$ contains both $e^{2\pi i \alpha}$ and $e^{2\pi i \beta}$. Let m' be the supremum of all leaves in QML that are smaller than $e^{2\pi i \alpha}$ and $e^{2\pi i \beta}$ in the partial order of leaves in QML. We distinguish three cases, depending on whether m' is the leading edge of a P-gap, of an H-gap, or of no gap at all.

If m' is the leading edge of a P-gap, then both $e^{2\pi i \alpha}$ and $e^{2\pi i \beta}$ cannot be boundary points of this gap (because α and β) are irrational), so they are both greater than some trailing edge of this P-gap. In fact, since these angles are not separated by any preperiodic leaf of QML, they must be greater than the same trailing edge of the P-gap, and this contradicts maximality of m'.

If m' is the leading edge of an H-gap, then for similar reasons, neither $e^{2\pi i \alpha}$ and $e^{2\pi i \beta}$ can be greater than any trailing edge of the gap (which would have periodic endpoints), so by the "no ghost wake" theorem, the two parameter rays at angles α and β land at different boundary points of the same hyperbolic component; this is again a contradiction.

It follows that m' is not the leading edge of any gap, so it is approximated by greater leaves (in the partial order) that have endpoints disjoint from m' (because QML is clean). By maximality of m', at least one of $e^{2\pi i \alpha}$ and $e^{2\pi i \beta}$ is an endpoints of m'. If only one of them is, then the two points $e^{2\pi i \alpha}$ and $e^{2\pi i \beta}$ are separated from each other by some of the leaves approximating m', and this is a contradiction. Therefore, the endpoints of m' are $e^{2\pi i \alpha}$ and $e^{2\pi i \beta}$, and we are done.

For the converse, suppose there is a leaf l in QML connecting two angles $\alpha < \beta$ that are both neither periodic nor preperiodic. By Corollary II.6.13a, there is a sequence of leaves m_k with endpoints at rational angles α_k and β_k so that $\alpha_k \to \alpha$ and $\beta_k \to \beta$. We have already shown that the parameter rays at angles α_k and β_k land together at a point in $\partial \mathbf{M}$. If $\mathbf{M}$ is locally connected, then all parameter rays land and their landing points depend continuously on the angles (Carathéodory's theorem), so the claim follows.

A.3

Now we discuss the properties of the Mandelbrot set that correspond to the two types of gaps of QML.

Theorem A.4 (P-Gaps in QML). *Given a Misiurewicz-Thurston parameter in* **M** *that is the landing point of k parameter rays at angles $\alpha_1, \ldots, \alpha_k$, then in QML the angles $\alpha_1, \ldots, \alpha_k$ form a P-gap (if $k \geq 3$), a leaf (if $k = 2$) or a degenerate leaf (if $k = 1$).*

Conversely, for every P-gap in QML there is a Misiurewicz-Thurston parameter c of **M** *with the following property: if $\alpha_1, \ldots, \alpha_k$ are the vertices of the P-gap, then all parameter rays of* **M** *at angles $\alpha_1, \ldots, \alpha_k$ (but no others) land at c.*

Proof of A.4. The boundary of any P-gap consists of finitely many leaves at strictly preperiodic angles (Theorem II.6.11) so that any two vertices of the gap are connected by a finite chain of such leaves. If α and α' are the two angles forming the endpoints of such a leaf, then by Theorem A.3 the two parameter rays at angles α and α' land at a common point, and the landing point is a Misiurewicz-Thurston parameter.

The converse follows directly from Theorem A.3. $\boxed{\text{A.4}}$

Theorem A.5 (H-Gaps in QML). *There is a bijection between H-gaps of QML and hyperbolic components of* **M** *that preserves periods and that has the following property: if m is a non-degenerate boundary leaf of an H-gap G connecting the angles α and β, then the parameter rays of* **M** *at angles α and β land together on the boundary of the associated hyperbolic component W of* **M**, *and conversely; and if m is a degenerate boundary leaf of G at angle α, then the parameter ray of* **M** *at angle α lands on the boundary of W, and conversely.*

Proof of A.5. We only discuss the case of gaps and hyperbolic components of period $n \geq 2$; the case $n = 1$ is somewhat special and can be treated similarly. Every H-gap of any period $n \geq 2$ has a leading edge, also of period n. Denote this leading edge by m and let α and α' be its endpoints. Then by Theorem A.3, the parameter rays at angles α and α' land together at a common point in **M**; this landing point is the root point of a hyperbolic component W of **M**, and the two parameter rays at angles α and α', together with their landing point, separate W from the origin [DH84, M00, Sch00]. Conversely, every hyperbolic component of any period n has a root, and this is the landing point of two parameter rays at periodic angles. This establishes a bijection between H-gaps and hyperbolic components of equal period.

If G is an H-gap with leading edge m, then all non-degenerate immediate successors of m in the partial order of QML are exactly the trailing edges on the boundary of G: if m' is such a trailing edge with boundary angles

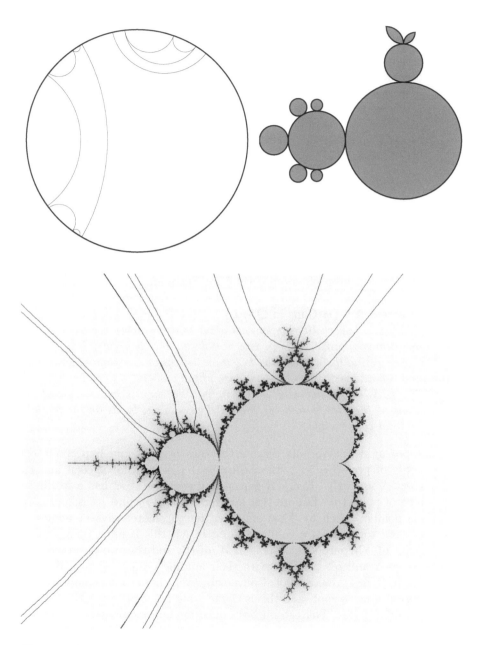

Figure A.6. Left top: several leaves from QML. Right top: the quotient of this partial lamination: each gap corresponds to a topological disk (except that polygonal gaps correspond to points). Bottom: the Mandelbrot set with corresponding parameter rays drawn in.

β and β', then the parameter rays at angles β and β' land together by Theorem A.3, and they bound the wake of some hyperbolic component W'. If the landing point of these parameter rays is not on ∂W, then it must be in a subwake of W, but then — again by Theorem A.3 — the edge m' could not be an immediate successor of m, a contradiction.

The converse follows directly from Theorem A.3. The argument about degenerate leaves is similar. $\boxed{\text{A.5}}$

The Pinched Disk Model of the Mandelbrot Set. Theorem II.6.14 says that the quotient space of the circle by identifying endpoints of leaves is a Hausdorff space. Equivalently, the quotient of the closed disk by identifying leaves or P-gaps yields another Hausdorff space, say $\widetilde{\mathbf{M}}$ (more precisely, $\widetilde{\mathbf{M}}$ is the quotient space of $\overline{D^2}$ with respect to the equivalence relation in which two points are equivalent if they are equal or if they are on the same leaf or P-gap in QML; recall that leaves and gaps are closed sets by definition). Denote by π_Q the natural projection map from $\overline{D^2}$ to the quotient $\widetilde{\mathbf{M}}$. The space $\widetilde{\mathbf{M}}$ is what Douady calls the *pinched disk model* of $\mathbf{M}$; see Douady [D93] for more details. The pinched disk model $\widetilde{\mathbf{M}}$ is locally connected (see Lemma A.1).

Note that any H-gap G in QML projects (under π_Q) to a topological disk in the quotient $\widetilde{\mathbf{M}}$ because G is bounded by countably many non-degenerate minors without common endpoints, and by uncountably many degenerate minors; the restriction of π_Q to G (the closed gap) is a homeomorphism onto its image in $\widetilde{\mathbf{M}}$. Conversely, any interior component of $\widetilde{\mathbf{M}}$ has the property that its closure is the homeomorphic image of an H-gap.

Theorem A.6 (The Projection Map $\pi_{\mathbf{M}}$). *There is a continuous surjection $\pi_{\mathbf{M}} \colon \mathbf{M} \to \widetilde{\mathbf{M}}$ with the following properties.*

(PM1) $\pi_{\mathbf{M}}$ *sends the closure of every hyperbolic component homeomorphically and in an orientation preserving way to its image, which is the π_Q-image of an H-gap of equal period.*

(PM2) $\pi_{\mathbf{M}}$ *sends the landing point of any pair of parameter rays at conjugate angles α and β to the quotient in $\widetilde{\mathbf{M}}$ of the leaf connecting the angles α and β.*

(PM3) *For every $x \in \widetilde{\mathbf{M}}$, the preimage $\pi_{\mathbf{M}}^{-1}(x)$ equals a single fiber of $\mathbf{M}$. In particular, $\pi_{\mathbf{M}}$ is constant on any non-hyperbolic component of $\mathbf{M}$.*

(PM4) $\pi_{\mathbf{M}}$ *is monotone in the sense that for every $x \in \widetilde{\mathbf{M}}$, the preimage $\pi_{\mathbf{M}}^{-1}(x)$ is connected.*

Such a map π_M is uniquely determined by (PM2) *on* $\partial \mathbf{M}$, *and it is unique on* $\mathbf{M}$ *up to postcomposition with homotopies on the union of hyperbolic components.*

Proof of A.6. Condition (PM2) defines $\pi_\mathbf{M}$ uniquely on landing points of parameter rays at periodic angles. Their landing points are roots of hyperbolic components, and these roots are dense in $\partial \mathbf{M}$ (see Theorem I.7.4a, as well as [DH84, Sch04]). Therefore, if $\pi_\mathbf{M}$ exists at all, then it is uniquely defined on $\partial \mathbf{M}$.

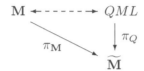

Figure A.7. The projection maps $\pi_\mathbf{M}$ and π_Q. There is a correspondence between $\mathbf{M}$ and QML as described in Theorems A.3, A.4, and A.4 (for instance, hyperbolic components of $\mathbf{M}$ correspond homeomorphically to H-gaps in QML), but there is no mapping between $\mathbf{M}$ and QML: P-gaps of QML correspond to Misiurewicz-Thurston points of $\mathbf{M}$, so a single point in $\mathbf{M}$ corresponds to a whole gap in QML; conversely, any non-hyperbolic components of $\mathbf{M}$ would correspond to a single (possibly degenerate) leaf of QML. Of course, if $\mathbf{M}$ is locally connected, then $\pi_\mathbf{M}$ is a homeomorphism, and $\pi_\mathbf{M}^{-1} \circ \pi_Q$ maps QML to $\mathbf{M}$.

Let W be any hyperbolic component of $\mathbf{M}$, let G be the corresponding H-gap in QML, and let $G' := \pi_Q(G)$. Then $\pi_\mathbf{M}$ extends as an orientation preserving homeomorphism from ∂W to $\partial G'$; and since both are topological circles, $\pi_\mathbf{M}$ extends as a homeomorphism from $\overline{W}$ to G'. We thus know that $\pi_\mathbf{M}$ can be defined on the union of the closures of all hyperbolic components and so that the restriction to each $\overline{W}$ separately is continuous. This proves property (PM1); the fact that the periods of H-gaps and corresponding hyperbolic components are equal follows from the construction, using property (PM2).

The next step is to define $\pi_\mathbf{M}$ for points $z \in \mathbf{M}$ that are not on the closure of any hyperbolic component. Let Y be the fiber of z. We will map all of Y to a single point in $\widetilde{\mathbf{M}}$. In fact, consider the set of all parameter ray pairs at rational angles that separate Y from the origin; these are totally ordered (the smaller ray pair separates the greater one from the origin). Every such parameter ray pair at angles α and β corresponds to a leaf of QML with periodic or preperiodic endpoints α and β (Theorem A.3): the order that these leaves have in QML is compatible with this correspondence. We thus obtain a totally ordered collection of leaves in QML. Since QML is closed, these leaves converge either to a non-degenerate leaf m in QML connecting two angles, say $\tilde{\alpha}$ and $\tilde{\beta}$, or they converge to a degenerate leaf at some angle $\tilde{\alpha}$. In both cases, the limit leaf projects to a single point $x \in \widetilde{\mathbf{M}}$. We set $\pi(z') := x$ for all $z' \in Y$. It follows from the construction that different points in Y define the same image x, so we obtain a well-defined map $\pi_{\mathbf{M}} : \mathbf{M} \to \widetilde{\mathbf{M}}$ that is constant on each fiber. It follows from Theorem A.3 that $\pi_{\mathbf{M}}$ is surjective.

Now we prove continuity of $\pi_{\mathbf{M}}$ at a point z not on the closure of a hyperbolic component. Set $x := \pi_M(z)$ and $q := \pi_Q^{-1}(x)$; then q is a non-degenerate or degenerate leaf of QML or a P-gap. A basis of open neighborhoods of x in $\widetilde{\mathbf{M}}$ can be constructed using π_Q-images of rational leaves in QML converging to q: since leaves with rational endpoints are dense in QML (Corollary II.6.13a), it follows that rational leaves bound a family of neighborhoods of q whose intersection is q alone. The π_Q-images of these boundary leaves bound arbitrarily small neighborhoods of x. The $\pi_{\mathbf{M}}$-preimage of such a neighborhood is a neighborhood of z bounded by finitely many parameter ray pairs at rational angles, and these neighborhoods (excluding the endpoints of the bounding ray pairs) are open in $\mathbf{M}$. Continuity of $\pi_{\mathbf{M}}$ at boundary points of hyperbolic components is shown similarly.

This shows that $\pi_{\mathbf{M}} : \mathbf{M} \to \widetilde{\mathbf{M}}$ is a continuous surjection; and it satisfies (PM1), as well as (PM2) that was part of the definition of $\pi_{\mathbf{M}}$.

Now we prove that for every $x \in \widetilde{\mathbf{M}}$, the preimage $\pi_M^{-1}(x)$ equals a single fiber. We already know that $\pi_{\mathbf{M}}$ is constant on each fiber. Conversely, consider two points $z, z' \in \mathbf{M}$ in different fibers. If z and z' are on the closure of the same hyperbolic component, then their images are different because the restriction of $\pi_{\mathbf{M}}$ to the closure of any hyperbolic component is injective. Otherwise, there is a parameter ray pair at rational angles that separates z from z', and there is a non-degenerate leaf m, say, in QML that corresponds to this ray pair. The construction of $\pi_{\mathbf{M}}$ assures that $\pi_Q(m)$ separates $\pi_{\mathbf{M}}(z)$ and $\pi_{\mathbf{M}}(z')$, so each $\pi_M^{-1}(x)$ indeed equals a single fiber. Since fibers of $\mathbf{M}$ are connected, this shows condition (PM4).

Every non-hyperbolic component Q of $\mathbf{M}$ (if any) cannot be disconnected by a parameter ray pair at rational angles, so it must be contained in a single fiber, and $\pi_{\mathbf{M}}$ must be constant on Q. This shows (PM3). $\boxed{\text{A.6}}$

One of the principal conjectures in complex dynamics says the following: *hyperbolic dynamics is dense in the space of quadratic polynomials* (compare the end of Section I.8). Equivalently, every interior component of $\mathbf{M}$ is a hyperbolic component. Any non-hyperbolic component of $\mathbf{M}$ would correspond to a (possibly degenerate) leaf of QML; see below. A stronger conjecture, called MLC, claims that *the Mandelbrot set $\mathbf{M}$ is locally connected*. It is well known that this conjecture implies the first one on density of hyperbolicity [DH84, D93, Sch04]. One way to see this is to consider the relation between the model space $\widetilde{\mathbf{M}}$ and the Mandelbrot set, as follows (compare [D93]).

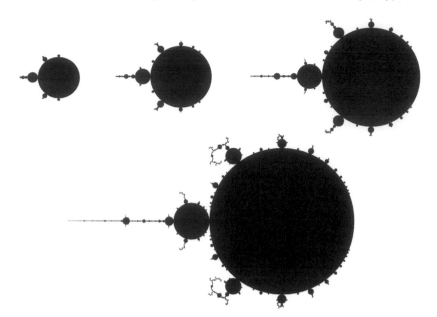

Figure A.8. The pinched disk model of the Mandelbrot set: the set $\widetilde{\mathbf{M}}$ is drawn so that every H-gap in QML becomes a round disk (every P-gap in QML must collapse to a point). Shown are approximations using hyperbolic components up to period 4, 6, 8, and 12. (Pictures courtesy of Michael Stoll.)

Theorem A.7 (The Projection Map $\pi_{\mathbf{M}}$ and the Two Conjectures).

The projection map $\pi_{\mathbf{M}} : \mathbf{M} \to \widetilde{\mathbf{M}}$ has the following properties:

(a) *Density of hyperbolicity in the space of quadratic polynomials is equivalent to the statement that for each $x \in \widetilde{\mathbf{M}}$, the preimage $\pi_{\mathbf{M}}^{-1}(x)$ has no interior.*

(b) *Local connectivity of $\mathbf{M}$ is equivalent to the statement that for each $x \in \widetilde{\mathbf{M}}$, the preimage $\pi_{\mathbf{M}}^{-1}(x)$ consists of a single point, or equivalently that $\pi_M : \mathbf{M} \to \widetilde{\mathbf{M}}$ is a homeomorphism.*

It follows that local connectivity of $\mathbf{M}$ implies density of hyperbolicity.

Proof of A.7. Every hyperbolic component of $\mathbf{M}$ maps homeomorphically to a single interior component of $\widetilde{\mathbf{M}}$, so if $\pi_{\mathbf{M}}^{-1}(x)$ has interior for an $x \in \widetilde{\mathbf{M}}$, then $\mathbf{M}$ must have a non-hyperbolic interior component. Conversely, $\pi_{\mathbf{M}}$ is constant on any non-hyperbolic component. Therefore, there exists some $x \in \widetilde{\mathbf{M}}$ for which $\pi_{\mathbf{M}}^{-1}(x)$ has interior if and only if $\mathbf{M}$ has a non-hyperbolic component, which is equivalent to saying that hyperbolic dynamics is *not* dense in the space of quadratic polynomials.

If each fiber of $\mathbf{M}$ is a single point, then $\pi_{\mathbf{M}} \colon \mathbf{M} \to \widetilde{\mathbf{M}}$ is a continuous bijection between compact Hausdorff spaces and thus a homeomorphism; this implies that $\mathbf{M}$ is locally connected because $\widetilde{\mathbf{M}}$ is (Lemma A.1). Conversely, if $\mathbf{M}$ is locally connected and $Y = \pi_{\mathbf{M}}^{-1}(x)$ is a fiber of $\mathbf{M}$ consisting of more than a single point, then Y is uncountable. By Carathéodory's Theorem II.2.3, at least three parameter rays at irrational angles $\alpha_1, \alpha_2, \alpha_3$ must land on ∂Y. No two of these parameter rays can be separated by ray pairs at rational angles, so the angle α_i cannot be separated by any rational leaf in QML and hence, by Corollary II.6.13a, by any leaf at all. Therefore, the three angles α_i are on the boundary of the same gap of QML. Since these angles are irrational, this must be an H-gap. But π_Q is injective on the set of irrational boundary angles of any H-gap. Therefore local connectivity of $\mathbf{M}$ is equivalent to the condition that all fibers of $\mathbf{M}$ consist of single points, i.e., that $\pi^{-1}(x)$ is a point for every $x \in \widetilde{\mathbf{M}}$. $\boxed{\text{A.7}}$

Remark. The original proof that local connectivity of $\mathbf{M}$ implies density of hyperbolicity is due to Douady and Hubbard [DH84]; it is based on their Branch Theorem (see [DH84, Exposé XXII] or [Sch04, Theorem 3.1]) that states, roughly speaking, that branch points in the Mandelbrot set are hyperbolic components or Misiurewicz-Thurston parameters. A combinatorial version of this branch theorem is Corollary II.6.13b, and it allows to describe the proof of Douady and Hubbard as follows: suppose $\mathbf{M}$ is locally connected and its interior has a non-hyperbolic component Q. Similarly as in the proof of Theorem A.7, we use the fact that $\partial Q \subset \partial \mathbf{M}$ is uncountable, so it contains three different points q_1, q_2, q_3. By local connectivity, there are three parameter rays at angles $\alpha_1, \alpha_2, \alpha_3$ that land at q_1, q_2, q_3, respectively, and we may suppose that $0 < \alpha_1 < \alpha_2 < \alpha_3 < 1$. Consider two periodic angles $\beta_1 \in (\alpha_1, \alpha_2)$ and $\beta_2 \in (\alpha_2, \alpha_3)$. Recall that parameter rays at periodic angles land in pairs; let β_1', β_2' the conjugate angles of β_1 and β_2, and relabel if necessary so that $\beta_1 < \beta_1'$ and $\beta_2 < \beta_2'$. We then have

$$0 < \alpha_1 < \beta_1 < \beta_1' < \alpha_2 < \beta_2 < \beta_2' < \alpha_3 < 1 .$$

By Theorem A.3, QML has two leaves m and m' with endpoints (β_1, β_1') and (β_2, β_2'), respectively. These clearly do not satisfy $m < m'$ or $m' < m$. By the Branch Theorem (Corollary II.6.13b), there is a gap G with two different trailing edges n, n' so that $n \leq m$ and $n' \leq m'$. But if all parameter rays at angles α_i land at the same non-hyperbolic component, then there must be a single trailing edge of G that precedes all α_i, and this is a contradiction.

The Lavaurs Algorithm. The quadratic minor lamination QML can be constructed by a simple algorithm, called the *Lavaurs algorithm* [La86]. It is based on the following three facts:

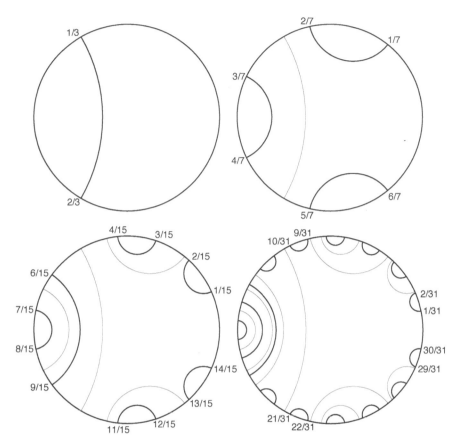

Figure A.9. The first steps in Lavaurs' algorithm, for periods up to 2, 3, 4, 5. After period 8, the resulting lamination is visually almost indistinguishable from the complete QML as shown in Figure II.9.

(a) all periodic angles are endpoints of non-degenerate leaves in QML with periodic endpoints of equal period (every periodic angle has a unique conjugate angle);

(b) every leaf of QML is the limit of leaves with periodic endpoints;

(c) if two leaves in QML have equal period and are comparable with respect to the partial order in QML, then these two leaves are separated (in $\overline{D^2}$) by a leaf of lower period.

In view of the first two facts, it suffices to construct all leaves of QML that have periodic endpoints, and this is what Lavaurs' algorithm does. Before describing it, we briefly justify these facts.

Fact (a) was mentioned in Remark II.6.6a; it also follows from Theorem A.2 (which is the analogous result for $\mathbf{M}$) and Theorem A.3 (which establishes the transfer to QML).

For Fact (b), compare Corollary II.6.13a: every leaf in QML is a limit of leaves with periodic or preperiodic endpoints. In order to show the second fact, all we need to do is to prove that every leaf with preperiodic endpoints is the limit of leaves with periodic endpoints. One proof of this fact uses symbolic dynamics in the following form: every angle $\alpha \in \mathbb{R}/\mathbb{Z}$ determines kneading sequence (in general, this is a sequence of symbols 0 and 1), and every kneading sequence has an associated internal address. If two angles α and α' have different kneading sequences, say with the first difference occurring at the nth position, then the parameter rays at angles α and α' are separated by a parameter ray pair at periodic angles of period n. If a leaf in QML with preperiodic endpoints is not approximated by a sequence of periodic leaves, then it must be approximated by a sequence of leaves with preperiodic endpoints so that all their angles have identical kneading sequences. But only finitely many preperiodic angles have the same kneading sequence. For details, see [Sch2].

Fact (c) is known as Lavaurs' lemma; for its proof, we refer to Lavaurs [La86].

Now we describe Lavaurs' algorithm: it constructs conjugate periodic angles in the order of increasing periods. Consider the boundary points of $\overline{D^2}$ that have period 2 under $z \mapsto z^2$ (these are the points at angles $\frac{1}{3}$ and $\frac{2}{3}$). Connect them by a leaf. Next, consider the boundary points of period 3 (at angles $k/7$ for $k = 1, 2, 3, 4, 5, 6$), and connect them pairwise by leaves. Do this in the order of increasing angles, always taking the lowest unused angle and connect it to the lowest angle that can be reached without crossing leaves that have already been constructed. Continue to do this for all further periods in increasing order. The first few steps in Lavaurs' algorithm are illustrated in Figure A.9.

Specifically for period 3, we obtain leaves connecting the pairs of angles

$$\left(\frac{1}{7}, \frac{2}{7}\right), \quad \left(\frac{3}{7}, \frac{4}{7}\right), \quad \left(\frac{5}{7}, \frac{6}{7}\right).$$

For period 4, we get the pairs of angles

$$\left(\frac{1}{15}, \frac{2}{15}\right), \quad \left(\frac{3}{15}, \frac{4}{15}\right), \quad \left(\frac{6}{15}, \frac{9}{15}\right), \quad \left(\frac{7}{15}, \frac{8}{15}\right), \quad \left(\frac{11}{15}, \frac{12}{15}\right), \quad \left(\frac{13}{15}, \frac{14}{15}\right);$$

the angles $\frac{5}{15} = \frac{1}{3}$ and $\frac{10}{15} = \frac{2}{3}$ are omitted because they have been used already for period 2. Note that the angle $\frac{6}{15}$ cannot be connected with $\frac{7}{15}$ or $\frac{8}{15}$ because the leaf $(\frac{3}{7}, \frac{4}{7})$ must not be crossed.

So far, every periodic angle had only a single angle of equal period available that could be reached without crossing leaves of lower periods.

This is different for period 5: the angles $\frac{1}{31}$, $\frac{2}{31}$, $\frac{9}{31}$, $\frac{10}{31}$, $\frac{21}{31}$, $\frac{22}{31}$, $\frac{29}{31}$ and $\frac{30}{31}$ are all not separated by leaves of lower period. This is where the rule is important that we connect the angles starting in increasing order and connect each one to the least angles that is still available. In this case, we connect $\frac{1}{31}$ to $\frac{2}{31}$, then (after taking care of six more angles for which there is no ambiguity) we connect $\frac{9}{31}$ to $\frac{10}{31}$, and so on. This is the only way that does not contradict Fact (c) mentioned above.

Having constructed leaves of all periods, taking the closure of leaves yields the lamination QML.

Index

Bibliography

[A06]   L. Ahlfors, *Lectures on Quasiconformal Mappings*, second edition. Univ. Lecture Ser. **38**. Amer. Math. Soc. 2006.

[A73]   L. Ahlfors, *Conformal Invariants: Topics in Geometric Function Theory*, McGraw-Hill, 1973.

[BFH92] B. Bielefeld, Y. Fisher, and J. Hubbard, *The classification of critically preperiodic polynomials as dynamical systems*, Jour. Amer. Math. Soc. **5** 4 (1992), 721–762.

[BL02]   A. Blokh, G. Levin, *An inequality for laminations, Julia sets and "growing trees"*, Ergod. Th. Dynam. Syst. 22 **1** (2002), 63–97.

[BO04]   A. Blokh, L. Oversteegen, *Wandering triangles exist*, C. R. Acad. Sci. Paris **339** 5 (2004), 365–370.

[BO08]   A. Blokh, L. Oversteegen, "Wandering gaps for weakly hyperbolic polynomials", in: *Complex Dynamics: Families and Friends* (this volume), Chapter 2. A K Peters, Wellesley, MA, 2009, 139–168.

[BKS]   H. Bruin, A. Kaffl, D. Schleicher, *Symbolic dynamics of quadratic polynomials*, monograph in preparation; see: Report **7**, Institute Mittag-Leffler, Djursholm, 2001/02.

[BS94]   S. Bullett, P. Sentenac, *Ordered orbits of the shift, square roots, and the devil's staircase*, Math. Proc. Cambridge Philos. Soc. **115** 3 (1994), 451–481.

[De94]   R. Devaney, *The complex dynamics of quadratic polynomials*, in: *Complex dynamical systems*, Proc. Symp. Appl. Math. **49**, Amer. Math. Soc., 1994, 1–27.

[D93]   A. Douady, *Descriptions of compact sets in* $\mathbb{C}$, in: Topological Methods in Modern Mathematics, L. Goldberg, A. Phillips Eds., Publish or Perish, Houston (1993), 429–465.

[DH84]   A. Douady and J. Hubbard, *Etude dynamique des polyômes complexes*. Prépublications mathématiques d'Orsay 1984/85.

[DH93]   A. Douady and J. Hubbard, *A proof of Thurston's topological characterization of rational functions*, Acta Math. **171** (1993), 263–297.

[Ep]   A. Epstein, *Infinitesimal Thurston rigidity and the Fatou-Shishikura inequality*, Stony Brook IMS Preprint **1**, 1999.

[FLP79]   A. Fathi, F. Laudenbach, and V. Poénaru, *Travaux de Thurston sur les surfaces*, Astérisque **66/67**, 1979.

[GM93]   L. Goldberg and J. Milnor, *Fixed points of polynomial maps. Part II*, Ann Sci. Ec. Norm. Sup., 4e série, **26** (1993), 51–98.

[HY88]   J. G. Hocking, G. S. Young, *Topology*, Dover Publications, New York, 1988.

[H93]   J. Hubbard, *Local connectivity of Julia sets and bifurcation loci: three theorems of J.-C. Yoccoz*, in: Topological Methods in Modern Mathematics, A. Philipps and L. Goldberg Eds., Publish or Perish, Houston, TX, 1993, 467–511.

[HS94]   J. Hubbard and D. Schleicher, *The spider algorithm*, in: Proc. Sympos. Appl. Math. **49**, Amer. Math. Soc. 1994, 155–180.

[HSS01]   J. Hubbard, D. Schleicher, and S. Sutherland, *How to find all roots of complex polynomials with Newton's method*, Invent. Math. **146** (2001), 1–33.

[Ke00]   K. Keller, *Invariant factors, Julia equivalences and the (abstract) Mandelbrot set*, Lect. Notes in Math. **1732**, Springer, New York, 2000.

[Ki02]   J. Kiwi, *Wandering orbit portraits*. Trans. Amer. Math. Soc. **354** (2002), 1473–1485.

[Ki04]   J. Kiwi, *Real laminations and the topological dynamics of complex polynomials*. Adv. Math. **184** (2002), 207–267.

[La86]   P. Lavaurs, *Une description combinatoire de l'involution définie par M sur les rationnels à dénominateur impair*, C. R. Acad. Sci. Paris Sér. I Math. **303** (1986) (4), 143–146.

[Le98]   G. Levin, *On backward stability of holomorphic dynamical systems*, Fund. Math. **158** (1998), 97–107.

[L83]   M. Lyubich, *Some typical properties of the dynamics of rational mappings* (Russian), Uspekhi Mat. Nauk **38** 5 (1983), 197–198. English translation: Russian Math. Surveys **38** 5 (1983), 154–155. *See also:* [L84]

[L84]   M. Lyubich, *Investigation of the stability of the dynamics of rational functions* (Russian), Teor. Funktsii Funktsional. Anal. i Prilozhen. **42** (1984), 72–91; English translation: Selecta Math. Soviet. **9** 1 (1990), 69–90.

[MSS83] R. Mañé, P. Sad, D. Sullivan, *On the dynamics of rational maps*, Ann. Sci. École Norm. Sup. (4) **16** 2 (1983), 193–217.

[MS98]   C. T. McMullen, D. Sullivan, *Quasiconformal homeomorphisms and dynamics III. The Teichmüller space of a holomorphic dynamical system*. Adv. Math. **135** 2 (1998), 351–395.

[Mc94]   C. McMullen, *Complex Dynamics and Renormalization*, Ann. Math. Studies **135**, Princeton University Press, 1994.

[M93]   J. Milnor, *Geometry and dynamics of quadratic rational maps*, with an appendix by the author and Tan Lei. Experiment. Math. **2** 1 (1993), 37–83.

[M00]   J. Milnor, *Periodic orbits, external rays and the Mandelbrot set: an expository account*, in: Géométrie Complexe et Systèmes Dynamiques, Astérisque **261** (2000), 277–333.

[M06]   J. Milnor, *Dynamics in one complex variable*, 3rd ed., Ann. Math, Study, Princeton University Press, 2006.

[MT88]   J. Milnor and W. P. Thurston, *On iterated maps of the interval*, in: Dynamical systems, Lecture Notes in Math. **1342**, Springer, Berlin, 1988, 465–563.

[Mo38]   C. Morrey, *On the solutions of quasi-linear elliptic partial differential equations*, Trans. Amer. Math. Soc. **43** (1938), 126–166.

[PM97]   R. Pérez-Marco, *Fixed points and circle maps*, Acta Math. **179** 2 (1997), 243–294.

[Po]   A. Poirier, *The classification of postcritically finite polynomials I/II*, Stony Brook IMS Preprints **5/7** (1993).

[Rü08]   J. Rückert, *Rational and transcendental Newton maps*, In: Holomorphic Dynamics and Renormalization, in honour of John Milnor's 75th birthday, M. Lyubich and M. Yampolsky Eds., Fields Inst. Commun. **53**, 2008.

[Sch00] D. Schleicher, *External rays of the Mandelbrot set*. In: Géométrie complexe et systèmes dynamiques. Astérisque **261** (2000), 409–447.

[Sch04] D. Schleicher, *On fibers and local connectivity of Mandelbrot and Multibrot sets*, in: Fractal Geometry and Applications, M. Lapidus and M. van Frankenhuysen Eds., Proc. Symp. Pure Math. **72**, Amer. Math. Soc., 2004, 477–507.

[Sch1]   D. Schleicher, *On fibers and local connectivity of compact sets in* $\mathbb{C}$, Stony Brook IMS Preprint **12**, 1998.

[Sch2]   D. Schleicher, *Internal addresses in the Mandelbrot set and Galois groups of polynomials.* Manuscript, ArXiv math.DS/9411238.

[Sh87]   M. Shishikura, *On the quasiconformal surgery of rational functions*, Ann. Sci. École Norm. Sup. (4) **20** 1 (1987), 1–29.

[Si42]   C. L. Siegel, *Iteration of analytic functions.* Ann. of Math. (2) **43** (1942), 607–612.

[S85a]   D. Sullivan, *Quasiconformal homeomorphisms and dynamics I. Solution of the Fatou-Julia problem on wandering domains.* Ann. of Math. (2) **122** 3 (1985), 401–418.

[S85b]   D. Sullivan, *Quasiconformal homeomorphisms and dynamics II. Structural stability implies hyperbolicity for Kleinian groups.* Acta Math. **155** 3–4 (1985), 243–260.

[S86]   D. Sullivan, *Quasiconformal homeomorphisms in dynamics, topology, and geometry*, in: Proc. Intern. Congr. Math. 1986, Vol. 1, 2, Amer. Math. Soc., Providence, RI, 1987, 1216–1228.

[T88]   W. Thurston, *On the geometry and dynamics of diffeomorphisms of surfaces*, Bull. Amer. Math. Soc. **19** (1988), 417–431.

[T97]   W. Thurston, *Three-dimensional geometry and topology*, Vol. 1. Edited by Silvio Levy. Princeton Mathematical Series **35**. Princeton University Press, Princeton, NJ, 1997.

2 Wandering Gaps for Weakly Hyperbolic Polynomials

Alexander Blokh and Lex Oversteegen

1 Introduction

The topological properties of Julia sets play an important role in the study of the dynamics of complex polynomials. For example, if the Julia set J is locally connected, then it can be given a nice combinatorial interpretation via relating points of J and angles at infinity [DH84]. Moreover, even in the case when J is connected but not locally connected, it often admits a nice locally connected model — the so-called *topological Julia set* with an induced map on it — which is always locally connected, similar to polynomial locally connected Julia sets, and has the same combinatorial interpretation as they do (Kiwi [Ki04] proved this for all polynomials with connected Julia sets but without Cremer or Siegel periodic points). This connection makes the study of both locally connected polynomial Julia sets and topological Julia sets important.

A striking result in this direction is the No Wandering Triangle Theorem due to Thurston [Th]. To state it, let us call a point with infinite forward orbit *wandering*. The theorem claims the non-existence of wandering non-precritical branch points of induced maps on quadratic topological Julia sets and extends a simple property of continuous maps of finite graphs, according to which all branch points of graphs are either preperiodic or precritical (which is quite surprising, since most quadratic topological Julia sets are complicated topologically and in general have infinitely many branch points).

The No Wandering Triangle Theorem is a beautiful result by itself. In addition, it is a central ingredient in the proofs of the main results of [Th]: namely, a locally connected model $\mathcal{M}_c$ of the Mandelbrot set $\mathcal{M}$ was suggested. It turns out that all branch points of $\mathcal{M}_c$ correspond to topological Julia sets whose critical points are (pre)periodic, and an important ingredient in proving this is the No Wandering Triangle Theorem. This motivates the study of the dynamics of branch points in topological Julia sets.

Thurston [Th] already posed the problem of extending the No Wandering Triangle Theorem to the higher-degree case and emphasized its importance.

The aim of this paper is to show that the No Wandering Triangle Theorem does *not* extend onto higher degrees. To state our main result, we recall that polynomials with the *Topological Collet-Eckmann property* (*TCE-polynomials*) are usually considered as having weak hyperbolicity.

Theorem 1.1. *There is an uncountable family* $\{P_\alpha\}_{\alpha \in \mathcal{A}}$ *of cubic TCE-polynomials* P_α *such that for every* α *the Julia set* J_{P_α} *is a dendrite containing a wandering branch point* x *of* J_{P_α} *of order 3, and the maps* $P_\alpha|_{J_{P_\alpha}}$ *are pairwise non-conjugate.*

Thus, the weak hyperbolicity of cubic polynomials does not prevent their Julia sets from exhibiting the "pathology" of having wandering branch points. Theorem 1.1 may be considered as a step towards the completion of the description of the combinatorial portrait of topological Julia sets. The main tool we use are *laminations*, introduced in [Th].

Let us describe how we organize the paper. In Section 2, we introduce laminations and discuss some known facts about them. In Section 3, we study (discontinuous) self-mappings of sets $A \subset S^1$ and give sufficient conditions under which such sets can be seen as invariant subsets of the circle under the map $\sigma_3 : S^1 \to S^1$. In Section 4, a preliminary version of the main theorem is proven; in this version, we establish the existence of an uncountable family of cubic laminations containing a wandering triangle. The proof was inspired by [BL02] and [OR80]; the result was announced in [BO04b]. The main idea is as follows: we construct a countable set $A \subset S^1$ and a function $g : A \to A$ so that (1) A is the g-orbit of a triple T such that all g-images of T have disjoint convex hulls in the unit disk, and (2) the set A can be embedded into S^1 by means of a one-to-one and order preserving function $\varphi : A \to S^1$ so that the map induced by g on $\varphi(A)$ is σ_3. Flexibility in our construction allows us to fine tune it in Section 5 to prove our main theorem (see [BO04b] for a detailed sketch of the proof).

Acknowledgments.   The results of this paper were discussed with participants of the Lamination Seminar at UAB, whom we thank for their interest in our work. The first author was partially supported by NSF grant DMS-0456748; the second author was partially supported by NSF grant DMS-0405774.

2   Laminations

As explained above, we use laminations for our construction. In this section, we give an overview of related results on laminations and describe properties of wandering gaps.

Laminations were introduced by Thurston [Th] as a tool for studying both individual complex polynomials and the space of all of them, especially in the degree 2 case. The former is achieved as follows. Let $P : \mathbb{C}^* \to \mathbb{C}^*$ be a degree d polynomial with a connected Julia set J_P acting on the complex sphere $\mathbb{C}^*$. Denote by K_P the corresponding filled-in Julia set. Let $\theta = z^d : \overline{\mathbb{D}} \to \overline{\mathbb{D}}$ ($\mathbb{D} \subset \mathbb{C}$ is the open unit disk). There exists a conformal isomorphism $\Psi : \mathbb{D} \to \mathbb{C}^* \setminus K_P$ such that $\Psi \circ \theta = P \circ \Psi$ [DH84]. If J_P is locally connected, then Ψ extends to a continuous function $\overline{\Psi} : \overline{\mathbb{D}} \to \overline{\mathbb{C}^* \setminus K_P}$, and $\overline{\Psi} \circ \theta = P \circ \overline{\Psi}$. Let $S^1 = \partial \mathbb{D}$, $\sigma_d = \theta|_{S^1}$, and $\psi = \overline{\Psi}|_{S^1}$. Define an equivalence relation $\sim_P$ on S^1 by $x \sim_P y$ if and only if $\psi(x) = \psi(y)$. The equivalence $\sim_P$ is called the *(d-invariant) lamination (generated by P)*. The quotient space $S^1/\sim_P = J_{\sim_P}$ is homeomorphic to J_P, and the map $f_{\sim_P} : J_{\sim_P} \to J_{\sim_P}$ induced by σ_d is topologically conjugate to P.

Kiwi [Ki04] extended this construction to *all* polynomials P with connected Julia sets and no irrational neutral cycles. For such polynomials, he obtained a d-invariant lamination $\sim_P$ on S^1. Then $J_{\sim_P} = S^1/\sim_P$ is a locally connected continuum and the induced map $f_{\sim_P} : J_{\sim_P} \to J_{\sim_P}$ is semi-conjugate to $P|_{J_P}$ under a monotone map $m : J_P \to J_{\sim_P}$ (by *monotone* we mean a map whose point preimages are connected). The lamination $\sim_P$ generated by P provides a combinatorial description of the dynamics of $P|_{J_P}$.

One can introduce laminations abstractly as equivalence relations on S^1 having certain properties similar to those of laminations generated by polynomials as above (we give detailed definitions below); in the case of such an abstract lamination $\sim$, we call $S^1/\sim = J_\sim$ a *topological Julia set* and denote the map *induced* by σ_d on $J_\sim$ by $f_\sim$. By the *positive* direction on S^1, we mean the *counterclockwise* direction and by the arc $(p, q) \subset S^1$ we mean the positively oriented arc from p to q. Consider an equivalence relation $\sim$ on the unit circle S^1 such that:

(E1) $\sim$ is *closed*: the graph of $\sim$ is a closed set in $S^1 \times S^1$,

(E2) $\sim$ defines a *lamination*, i.e., it is *unlinked*: if g_1 and g_2 are distinct equivalence classes, then the convex hulls of these equivalence classes in the unit disk $\mathbb{D}$ are disjoint,

(E3) each equivalence class of $\sim$ is *totally disconnected*.

We always assume that $\sim$ has a class of at least two points. Equivalence classes of $\sim$ are called *($\sim$-)classes*. A class consisting of two points is called a *leaf*; a class consisting of at least three points is called a *gap* (this is more restrictive than Thurston's definition in [Th]). Fix an integer $d > 1$. The equivalence relation $\sim$ is called *(d-)invariant* if:

(D1) $\sim$ is *forward invariant*: for a class g, the set $\sigma_d(g)$ is a class too,

(D2) $\sim$ is *backward invariant*: for a class g, its preimage $\sigma_d^{-1}(g) = \{x \in S^1 : \sigma_d(x) \in g\}$ is a union of classes,

(D3) for any gap g, the map $\sigma_d|_g : g \to \sigma_d(g)$ is a *covering map with positive orientation*, i.e., for every connected component (s,t) of $S^1 \setminus g$, the arc $(\sigma_d(s), \sigma_d(t))$ is a connected component of $S^1 \setminus \sigma_d(g)$.

Call a class g *critical* if $\sigma_d|_g : g \to \sigma(g)$ is not one-to-one and *precritical* if $\sigma_d^j(g)$ is critical for some $j \geq 0$. Call g *preperiodic* if $\sigma_d^i(g) = \sigma_d^j(g)$ for some $0 \leq i < j$. A gap g is *wandering* if g is neither preperiodic nor precritical. Let $p : S^1 \to J_\sim = S^1/\sim$ be the standard projection of S^1 onto its quotient space $J_\sim$, and let $f_\sim : J_\sim \to J_\sim$ be the map induced by σ_d.

Let us describe some properties of wandering gaps. J. Kiwi [Ki02] extended Thurston's theorem by showing that a wandering gap in a d-invariant lamination is at most a d-gon. In [Le98], G. Levin showed that laminations with one critical class have no wandering gaps. Let $k_\sim$ be the maximal number of critical $\sim$-classes g with pairwise disjoint infinite σ_d-orbits and $|\sigma_d(g)| = 1$. In [BL02], Theorem B, it was shown that if $\sim$ is a d-invariant lamination and Γ is a non-empty collection of wandering d_j-gons $(j = 1, 2, \dots)$ with distinct grand orbits, then $\sum_j (d_j - 2) \leq k_\sim - 1 \leq d - 2$.

Call laminations with wandering k-gons WT-*laminations*. The above and results of [BL02, Bl05, Ch07] show that cubic WT-laminations must satisfy a few necessary conditions. First, by [BL02], Theorem B, if $\sim$ is a cubic WT-lamination, then $k_\sim = 2$. The two critical classes of $\sim$ are leaves, and $J_\sim$ is a *dendrite*, i.e., a locally connected continuum without subsets homeomorphic to the circle. The two critical leaves in $\sim$ correspond to two critical points in $J_\sim$; by [Bl05] (see also [Ch07] for laminations of any degree), these critical points must be recurrent with the same limit set under the induced map $f_\sim$.

3 Circular Maps That Are σ-Extendable

In this section, we introduce the notion of a *topologically exact dynamical system* $f : A \to A, A \subset S^1$ of *degree n*. A dynamical system that can be embedded into $\sigma_n : S^1 \to S^1$ is said to be σ-*extendable (of degree n)*. We show that a topologically exact countable dynamical system of degree 3 without fixed points is σ-extendable of degree 3.

A subset of S^1 is said to be a *circular* set. An ordered circular triple $\{x, y, z\}$ is *positive* if $y \in (x, z)$. Given $X \subset S^1$, a function $f : X \to S^1$ is *order preserving* if for any positive triple $\{x, y, z\} \subset X$ the triple $\{f(x), f(y), f(z)\}$ is positive too. Given sets $A, B \subset S^1$, a (possibly discontinuous) function $f : A \to B$ is *of degree d* if d is the minimal number for which there exists a partition $x_0 < x_1 < \cdots < x_d = x_0$ of S^1 such that for each i, $f|_{[x_i,x_{i+1})\cap A}$ is order preserving. If A is finite, one can extend f to a map on S^1 that maps each arc complementary to A forward as an increasing map and is one-to-one inside the arc — the degree of the extension is equal to that of f. Thus, if A is finite, then $d < \infty$, but d may be finite even if A is infinite. If $d < \infty$, we denote it by $\deg(f)$.

If $A = B$ and $\deg(f) < \infty$, we call f a *circular map*. An order-preserving bijection $h : X \to Y$ (with $X, Y \subset S^1$) is called an *isomorphism*. Two circular maps $f : A \to A$ and $g : A' \to A'$ are *conjugate* if they are conjugate in the set-theoretic sense by an isomorphism $h : A \to A'$. The degree of a circular map is invariant under conjugacy. A circular map $f : A \to A$ is σ-*extendable* if for some $\sigma_{\deg(f)}$-invariant set $A' \subset S^1$, the map $f|_A$ is conjugate to the function $\sigma_{\deg(f)}|_{A'} : A' \to A'$. We prove that a version of *topological exactness* (i.e., the property that all arcs eventually expand and "cover" the entire circle) implies that a circular map is σ-extendable.

We need a few other definitions. An *arc in a circular set X (or X-arc)* is the intersection of an arc in S^1 and X. Every arc I in X (or in S^1) has the *positive* order $<_I$ determined by the positive orientation on S^1 (if it is clear from the context what I is, we omit the subscript I). Given sets A and B contained in an arc $J \subset S^1$, we write $A < B$ if $a < b$ for each $a \in A$ and each $b \in B$. Arcs in the circle may be open, closed, or include only one of the two endpoints and will be denoted (a, b), $[a, b]$, etc. Corresponding arcs in a circular set X will be denoted by $(a, b)_X$, $[a, b]_X$, etc. If $X, Y \subset S^1$ are two disjoint closed arcs, then by (X, Y) we mean the open arc enclosed between X and Y so that the movement from X to Y within this arc is in the positive direction. We always assume that a circular set A contains at least two points.

Definition 3.1. Let $f : A \to A$ be a circular map. Then f is said to be *topologically exact* if for each $x \neq y$ in A there exists an $n \geq 1$ such that either $f^n(x) = f^n(y)$ or $f([f^n(x), f^n(y)]_A) \not\subset [f^{n+1}(x), f^{n+1}(y)]_A$.

A circular map $f : A \to A$ may not admit a continuous extension over $\overline{A}$. However we define a class of set-valued functions which help in dealing with f anyway. Namely, a set-valued function $F : S^1 \to S^1$ is called an *arc-valued map* if for each $x \in S^1$, $F(x) = [a_x, b_x]$ (with $a_x \leq b_x \in S^1$ in the positive order) and for each sequence $z_i \to z$ in S^1, $\limsup F(z_i) \subset F(z)$; clearly, this is equivalent to the fact that the graph of F is closed as a subset of the 2-torus $\mathbb{T}^2 = S^1 \times S^1$.

Definition 3.2. We say that an arc-valued map $F : S^1 \to S^1$ is *locally increasing* if for *any* $z \in S^1$ there exists an arc $I = [x_z, y_z]$, $x_z <_I z <_I y_z$ with (1) $F(x_z) \cap F(y_z) = \emptyset$, and (2) for each $u <_I w \in (x_z, y_z)$, the arcs

$F(u), F(w)$ are contained in the open arc $(F(x_z), F(y_z))$ and $F(u) < F(w)$. The *degree* of a locally increasing arc-valued map F, denoted by $\deg(F)$, is the number of components of $F^{-1}(z)$ (by $F^{-1}(z)$ we mean the set of all $y \in S^1$ such that $z \in F(y)$). It is easy to see (by choosing a finite cover of S^1 by intervals (x_z, y_z)) that $\deg(F)$ is well defined and finite.

Let F be a locally increasing arc-valued map and $f : A \to A$, $A \subset S^1$, be a circular map; we say that F is an *arc-valued extension* of f if $f(a) \in F(a)$ for each $a \in A$. Now we prove the main result of this subsection; the statement is far from the most general one, but sufficient for our purpose.

Theorem 3.3. *Suppose that $f : A \to A$ is a topologically exact circular map of degree 3 such that A is countable and does not contain a fixed point. Then f is σ-extendable.*

Proof: We may assume that points of A are isolated (otherwise, replace each point of A with a small enough interval and put the point of A in the middle of it) and, hence, that limit points of A do not belong to A. Define an arc-valued extension F of f as follows. First, for each $z \in \overline{A}$ define

$$
L(z) = \begin{cases} f(z), & \text{if } z \in A \\ \lim f(a_i), & \text{if there exists } a_i \in A \text{ such that } a_i \nearrow z \\ \lim f(b_i), & \text{for a sequence } b_i \in A \text{ such that } b_i \searrow z, \text{ otherwise.} \end{cases}
$$

The map $L(z)$ is well defined, and it is easy to see that $L(z)$ maps $\overline{A}$ into $\overline{A}$ and that $L(z)$ is still of degree 3. Given a map $g : S^1 \to S^1$ defined at points a, b, let the *linear extension* of g on (a, b) be the map that maps the interval (a, b) linearly onto the interval $(g(a), g(b))$. Extend $L(z)$ on each component of $S^1 \setminus \overline{A}$ linearly. For each point $x \in S^1$, define $F(x)$ to be the interval $[\lim_{t \nearrow x} L(t), \lim_{t \searrow x} L(t)]$. Then F is a locally increasing arc-valued map with $f(z) \in F(z)$ for each $z \in A$ and $\deg(f) = \deg(F) = 3$. Note that for each $a \in A$, $F(a) = \{f(a)\}$, and the set of points with non-degenerate image is countable.

Let $p : \mathbb{R} \to S^1 = \mathbb{R}/\mathbb{Z}$ be a standard projection of the real line onto the circle. We may assume that $F(0)$ is a point. Choose a lifting G of F such that $G(0)$ is a point between 0 and 1. Then the graph of $G|_{[0,1]}$ stretches from the point $(0, G(0))$ to $(1, G(1))$ and $G(1) = G(0) + 3$. Hence the graph of $G|_{[0,1]}$ intersects the graph of $y = x + 1$, and we can change the projection p so that $0 \in G(0)$. Then $0 \notin A'$ (A' is the lifting of A) because otherwise $a = p(0)$ would be a fixed point in A.

Since the graph of G intersects each horizontal line at exactly one point, there are two points $0 < b' < c' < 1$ with $1 \in G(b'), 2 \in G(c')$. Let $b = p(b') \in A, c = p(c') \in A$. Note that $b', c' \notin A'$, since otherwise $b \in A$ or $c \in A$ and so $a \in A$, a contradiction. Hence $a \in F(a) \cap F(b) \cap F(c)$.

Consider the arcs $[a, b] = I_0, [b, c] = I_1$ and $[c, a] = I_2$ and associate to every point $x \in A$ its itinerary $\mathrm{itin}(x)$ in the sense of this partition. In fact, to each point $x \in A$ we may associate a well-defined itinerary, since then $F^k(x), k \geq 0$, is a point, and $F^k(x) \neq a, b, c, k \geq 0$, because $f|_A$ has no fixed points.

Let us show that any two points of A have distinct itineraries. Define pullbacks of the arcs I_0, I_1, I_2 by taking preimages of points a, b, c inside I_0, I_1, and I_2 appropriately and considering arcs between these preimages. This can be done arbitrarily many times, and hence every point $x \in A$ belongs to the intersection $I(\mathrm{itin}(x))$ of the appropriate pullbacks of I_0, I_1, and I_2. If points $x, y \in A$ had the same itinerary $\bar{r}$, then they would both belong to the same interval $I(\bar{r})$. Let J be the arc between x and y contained in $I(\bar{r})$. Then (1) J and all its F-images have well-defined endpoints (i.e., the endpoints of every F-image of J are such that their F-images are points, not intervals), and (2) every F-image of J is contained in I_0, or I_1, or I_2. This contradicts the topological exactness of $f|_A$ and shows that $\mathrm{itin}(x) \neq \mathrm{itin}(y)$. Hence no point $z \in A$ can have itinerary $\mathrm{itin}(z) = (iii \dots)$ for some $i = 0, 1, 2$ (otherwise, z and $f(z) \neq z$ would have the same itinerary).

The same construction applies to σ_3. Set $K_0 = [0, 1/3], K_1 = [1/3, 2/3]$, and $K_2 = [2/3, 1]$ (here 0 and 1 are identified and the full angle is assumed to be 1) and use the notation $K(\bar{r})$ for the point x with σ_3-itinerary $\bar{r}$. Given $x \in A$, define $h(x)$ as $K(\mathrm{itin}(x))$. Then h is a one-to-one map from A onto a σ-invariant set $B \subset S^1$. Since on each finite step the circular order among the F-pullbacks of I_0, I_1, and I_2 is the same as the circular order among the σ-pullbacks of K_0, K_1, and K_2, the map h is an isomorphism between the circular sets A and B, and hence h conjugates $f|_A$ and $\sigma|_B$. $\square$

4   Cubic Laminations with Wandering Triangles

In this section, we prove a preliminary version of Theorem 1.1. Set $\sigma_3 = \sigma$. The circle S^1 is identified with the quotient space $\mathbb{R}/\mathbb{Z}$; points of S^1 are denoted by real numbers $x \in [0, 1)$. Let $B = \{0 < c' < s_0 < u_0 < \frac{1}{2} < d' < v_0 < t_0 < 1\}$ with $v_0 - u_0 = 1/3$ and $t_0 - s_0 = 2/3$, $\bar{c}_0$ be the chord with endpoints u_0, v_0, and $\bar{d}_0$ be the chord with endpoints s_0, t_0. The intuition here is that the chords $\bar{c}_0$ and $\bar{d}_0$ will correspond to the critical points. Define the function g first on the endpoints of $\bar{c}_0, \bar{d}_0$ as $g(u_0) = g(v_0) = c'$, $g(s_0) = g(t_0) = d'$. Thus the points c' and d' will correspond to critical values. Also, set $g(0) = 0, g(\frac{1}{2}) = \frac{1}{2}$.

Our approach is explained in Section 1 and is based upon results of Section 3. The idea of the construction is as follows. First, we choose the location of the first seven triples $T_1, \dots, T_7$ on the circle. Naturally, we make the choice so that the corresponding triangles are pairwise disjoint. We also define the map g on these triples so that $g(T_1) = T_2, \ldots, g(T_6) = T_7$. According to our approach, we should only be interested in the relative location of the triples and the chords $\bar{c}_0, \bar{d}_0$. We then add more triples and define the map g on them in a step-by-step fashion. In so doing, we postulate from the very beginning of the construction that g restricted on each of the arcs complementary to B is monotonically increasing in the sense of the positive order on the circle; e.g., consider the arc $[0, s_0]$ in Figure 1. Then g restricted onto the set "under construction" should be monotonically increasing in the sense of the positive order on $[0, s_0]$ and $[0, d']$ (recall that g is already defined at 0 and s_0).

In some cases, the above assumptions force the location of images of certain points, and hence the location of the "next" triples. Indeed, suppose that $T_1 = \{x_1, y_1, z_1\}$ is a triple such that $t_0 < z_1 < 0 < x_1 < y_1 < s_0$ and that $g(T_1) = T_2$ is already defined and is such that $u_0 < g(x_1) = x_2 < \frac{1}{2} < g(y_1) = y_2 < d' < g(z_1) = z_2 < v_0$ (see Figure 1). Suppose that there is also a triple $T_7 = \{x_7, y_7, z_7\}$ such that $t_0 < z_7 < z_1$ and $y_1 < x_7 < y_7 < s_0$ (again, see Figure 1). Then it follows from the monotonicity of g on the arcs complementary to B that $g(T_7) = T_8$ must be located so that $y_2 < g(x_7) = x_8 < g(y_7) = y_8 < d' < g(z_7) = z_8 < z_2$.

This example shows that in the process of constructing the g-orbit of a triple some steps (in fact a lot of them) are forced and the relative location of the next triple is well defined. However it also shows the notational challenge which one faces in trying to describe the location of all of the triples on the circle. Below, we develop a specific "language" for the purpose of such a description.

First, we extend our definition of the function g onto a countable subset of the circle, which we construct. To do so, let u_{-k} be the point such that $u_{-k} \in (u_0, v_0)$, $\sigma(u_{-k}) \in (u_0, v_0), \ldots, \sigma^k(u_{-k}) = u_0$ and set $g(u_{-k}) = \sigma(u_{-k})$. Similarly, we define points v_{-k}, s_{-k}, t_{-k} and the map g on them. Then $\lim u_{-n} = \frac{1}{2}$ and $\sigma(u_{-i}) = u_{-i+1}$; analogous facts hold for v_{-n}, s_{-n}, and t_{-n}. All these points together with the set B form the set B'. This is obvious if $(a, b) \supset [s_0, u_0]$, or $(a, b) \supset [u_0, v_0]$, or $(a, b) \supset [v_0, t_0]$, or $(a, b) \supset [t_0, s_0]$.

The chord connecting u_{-k}, v_{-k} is denoted by $\bar{c}_{-k}$, and the chord connecting s_{-k}, t_{-k} is denoted by $\bar{d}_{-k}$. These chords will correspond to the appropriate preimages of critical points. Also, let $d' \in (v_{-1}, t_{-1})$. This gives a function $g : B' \setminus \{c', d'\} \to B'$. It acts on this set just like σ, mapping chords to the right except that at the endpoints of $\bar{c}_0$ and $\bar{d}_0$ the map g differs from σ. This, together with the above-postulated properties of g, introduces certain restrictions on the behavior of the triple whose orbit we want to construct.

Below, we define a triple $T_1 = \{\mathbf{x}_1, \mathbf{y}_1, \mathbf{z}_1\}$ and the set $X_1 = B' \cup T_1$. At each step a new triple $T_n = \{\mathbf{x}_n, \mathbf{y}_n, \mathbf{z}_n\}$ is added and the set $X_n = X_{n-1} \cup \{\mathbf{x}_n, \mathbf{y}_n, \mathbf{z}_n\}$ is defined. In describing the next step of the construction, denote *only* new points by boldface letters, while using standard font for the already defined points. This explains the following notation: the function g on points x_{n-1}, y_{n-1}, z_{n-1} is defined as $g(x_{n-1}) = \mathbf{x}_n$, $g(y_{n-1}) = \mathbf{y}_n$, $g(z_{n-1}) = \mathbf{z}_n$. Below, a "triple" means one of the sets T_i, and a "triangle" means the convex hull of a triple. By "the triangle (of the triple) T_i" we mean "the convex hull of the triple T_i". Define A as $\bigcup_i T_i$ and A' as $A \cup B' \setminus \{c', d'\}$.

In making the next step of our construction, we need to describe the location of the new triple T_i. We do this by describing the location of its points relative to the points of X_{i-1} (essentially, X_{i-1} is the set which has been constructed so far). The location of the ith triple T_i is determined by points $p, q, r \subset X_{i-1}$ with $p < \mathbf{x}_i < q < \mathbf{y}_i < r < \mathbf{z}_i$ and $[(p, \mathbf{x}_i) \cup (q, \mathbf{y}_i) \cup (r, \mathbf{z}_i)] \cap X_{i-1} = \emptyset$; then we write $T_i = T(p, \mathbf{x}_i, q, \mathbf{y}_i, r, \mathbf{z}_i)$. If 2 or 3 points of a triple lie between two adjacent points of X_{i-1}, we need fewer than 6 points to denote T_i — e.g., $T(p, \mathbf{x}_i, \mathbf{y}_i, q, \mathbf{z}_i)(p, q \in X_{i-1})$ means that $p < \mathbf{x}_i < \mathbf{y}_i < q < \mathbf{z}_i$, and $[(p, \mathbf{y}_i) \cup (q, \mathbf{z}_i)] \cap X_{i-1} = \emptyset$. The function g is constructed step by step to satisfy Rule A below.

Rule A. *All triples T_i are pairwise unlinked and disjoint from the set B'. The map g is order preserving on $[s_0, u_0]_{A'}, [u_0, v_0]_{A'}, [v_0, t_0]_{A'}, [t_0, s_0)_{A'}$ (which implies that the degree of $g|_{A'}$ is 3).*

Now we introduce locations of the initial triples:

$$T_1 = T(0, \mathbf{x}_1, c', \mathbf{y}_1, t_0, \mathbf{z}_1), \qquad T_2 = T(s_{-1}, \mathbf{x}_2, v_{-1}, \mathbf{y}_2, d', \mathbf{z}_2),$$
$$T_3 = T(s_0, \mathbf{x}_3, v_0, \mathbf{y}_3, \mathbf{z}_3), \qquad T_4 = T(x_1, \mathbf{x}_4, c', \mathbf{y}_4, \mathbf{z}_4),$$
$$T_5 = T(u_{-1}, \mathbf{x}_5, \mathbf{y}_5, t_{-2}, \mathbf{z}_5), \qquad T_6 = T(u_0, \mathbf{x}_6, \mathbf{y}_6, t_{-1}, \mathbf{z}_6),$$
$$T_7 = T(y_1, \mathbf{x}_7, \mathbf{y}_7, t_0, \mathbf{z}_7).$$

Rule A forces the location of some triples. For two disjoint chords $\bar{p}, \bar{q}$, denote by $S(\bar{p}, \bar{q})$ the strip enclosed by $\bar{p}, \bar{q}$ and arcs of the circle. Then the boundary A'-arcs of the strip $S(\bar{d}_{-1}, \bar{c}_0)$ must map one-to-one into the arcs $(t_0, c')_{A'}$ and $(c', s_0)_{A'}$. Also, the boundary A'-arcs of the strip $S(\bar{c}_{-i}, \bar{d}_{-i})$ map into the boundary arcs of the strip $S(\bar{c}_{-i+1}, \bar{d}_{-i+1})$ one-to-one, and the boundary A'-arcs of the strip $S(\bar{d}_{-i-1}, \bar{c}_{-i})$ map into the boundary arcs of the strip $S(\bar{d}_{-i}, \bar{c}_{-i+1})$ one-to-one. Observe that $T_2 \subset S(\bar{c}_{-1}, \bar{d}_{-1})$, and so by Rule A, $T_3 \subset S(\bar{c}_0, \bar{d}_0)$ (the point x_3 must belong to (s_0, u_0), whereas the points y_3, z_3 must belong to (v_0, t_0)). The segment of triples $T_1, \ldots, T_7$ is the basis of induction (see Figure 1).

Clearly, T_7 separates the chord $\bar{d}_0$ from T_1. Our rules then force the location of forthcoming triples $T_8, T_9, \ldots$ with respect to $X_7, X_8, \ldots$ for some time. More precisely, $T_8 = T(y_2, \mathbf{x}_8, \mathbf{y}_8, d', \mathbf{z}_8)$, $T_9 = T(y_3, \mathbf{x}_9, \mathbf{y}_9, \mathbf{z}_9)$,

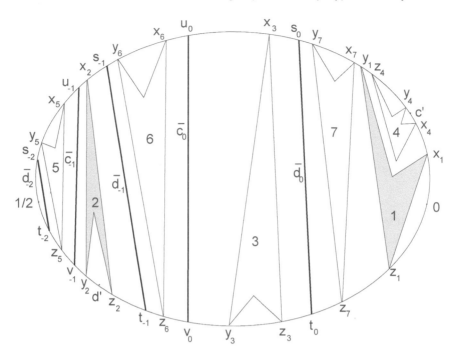

Figure 1. The first seven triangles.

and $T_{10} = T(y_4, \mathbf{x}_{10}, \mathbf{y}_{10}, \mathbf{z}_{10})$. The first time the location of a triple with respect to previously constructed triples and points of $B' \setminus \{c', d'\}$ is not forced is when T_{10} is mapped onto T_{11}. At this moment, Rule A guarantees that T_{11} is located in the arc (y_5, z_5), but otherwise the location of T_{11} is not forced. In particular, the location of the triangle T_{11} with respect to $\frac{1}{2}$ is not forced. The freedom of choice of the location of T_{11} at this moment, and the similar variety of options available later on at similar moments, is the reason why the construction yields not just one, but uncountably many types of behavior of a wandering triangle.

Next, we introduce another general rule which will be enforced throughout the construction and will help us determine the location of the triples.

Rule B. *Points of any triple T_i are ordered in the arc $(0,0)$ as follows:* $\mathbf{x}_i < \mathbf{y}_i < \mathbf{z}_i$. *All triangles are disjoint from the chords $\bar{c}_0, \bar{d}_0$.*

Since Rule B deals with the order of points on the arc $(0,0)$, it establishes more than the mere fact that the cyclic order among points $\mathbf{x}_i, \mathbf{y}_i, \mathbf{z}_i$ is kept. Denote by UP the upper semicircle $(0, \frac{1}{2})$ and by LO the lower semicircle $(\frac{1}{2}, 0)$. By Rule B, there are three types of triples:

1. *up triples*, or *triples of $\triangle$-type*: triples with $\mathbf{x}_i \in UP, \mathbf{y}_i < \mathbf{z}_i \in LO$, denoted by $\triangle(\cdot)$ (the standard notation is T with a subscript);

2. *down triples*, or *triples of $\triangledown$-type*: triples with $\mathbf{x}_i < \mathbf{y}_i \in UP, \mathbf{z}_i \in LO$, denoted by $\triangledown(\cdot)$;

3. *horizontal triples*, or *triples of $\triangleleft$-type*: triples with $\mathbf{x}_i < \mathbf{y}_i < \mathbf{z}_i$ contained entirely either in UP or in LO, denoted by $\triangleleft(\cdot)$.

Up triples and down triples are called *vertical triples*. Convex hulls of up, down, vertical, and horizontal triples are said to be *up, down, vertical,* and *horizontal* triangles. Chords with endpoints in UP and LO are *vertical* (e.g., $\bar{c}_0$ and $\bar{d}_0$ are vertical); otherwise, they are *horizontal* (all sides of a horizontal triangle are horizontal). Let us discuss properties of vertical triples. A *proper* arc is an arc that contains none of the points $0, \frac{1}{2}, s_0, t_0, u_0, v_0$. Given a triple $T_i = \{x_i, y_i, z_i\}$, call the arcs $(x_i, y_i), (y_i, z_i)$, and (z_i, x_i) *xy-arc*, *yz-arc*, and *zx-arc*; all such arcs are said to be *generated* by the corresponding triples (or simply arcs *of* that triple). An up triple generates only one proper arc contained in LO, and a down triple generates only one proper arc contained in UP. Also, if vertical triples T', T'' are unlinked, then none of them contains the other in its proper arc (this is not true for horizontal triples).

In our construction, there will be *crucial moments* at which the rules leave open the choice for the location of a new triple T_{n+1} with respect to X_n; the dynamics of a triangle at a crucial moment is called a *crucial event*. Crucial events are of 4 types: an $h_{\triangledown}$-event (the next closest approach of the triple to $\frac{1}{2}$ while the triple is of $\triangledown$-type), a d-event (the next closest approach to the *entire* chord $\bar{d}_0$ from the right), an $h_{\triangle}$-event (the next closest approach to $\frac{1}{2}$ while the triple is of $\triangle$-type), and a c-event (the next closest approach to the $\bar{c}_0$-chord from the right). The crucial moments of these types are denoted $h_{\triangledown}(i), d(i), h_{\triangle}(i)$, and $c(i)$; the number i indicates that the crucial event takes place at the corresponding crucial moment during the ith inductive step of the construction. We are now ready to state Rule C.

Rule C. *Vertical triples have the following properties:*

1. *up triples can only be contained in the strips* $S(\bar{c}_0, \bar{d}_0), S(\bar{c}_{-1}, \bar{d}_{-1}),$ $\ldots, S(\bar{c}_{-i}, \bar{d}_{-i}), \ldots$.

2. *down triples can only be located to the right of the chord* $\bar{d}_0$ *as well as in the strips* $S(\bar{d}_{-1}, \bar{c}_0), S(\bar{d}_{-2}, \bar{c}_{-1}), \ldots, S(\bar{d}_{-i-1}, \bar{c}_{-i}), \ldots$.

The rules allow us to explain how we choose the location of a triple; giving the order of the points without mentioning the rules would significantly lengthen the verification. Crucial moments always happen in the order $d(i) < h_\triangle(i) < c(i) < h_\triangledown(i) < d(i+1) < \ldots$. A triple T_k is *minimal* if it contains no triples $T_i, i < k$, in its proper arcs.

Let us pass on to the induction. It depends on a sequence of natural numbers $n_1 < m_1 < n_2 < m_2 < \ldots$ (each pair of numbers n_i, m_i corresponds to the ith step of induction), which can be chosen *arbitrarily*. The above-defined collection of triples $T_i, i = 1, \ldots, 7$, with the choice of crucial moments $d(0) = 1, h_\triangle(0) = 2, c(0) = 3, h_\triangledown(0) = 5$, and $d(1) = 7$, serves as the basis of induction. The inductive assumptions are of a dynamical nature and deal with the locations of triples on the circle. They are listed below as properties (a) through (h), where we describe the ith segment of the triples in the set A from the moment $d(i)$ through the moment $d(i+1) - 1$. It is easy to see that the basis of induction has properties (a)–(h). Now we can make our main claim concerning the construction.

Main Claim. *Suppose that $T_1, \ldots, T_7$ are the above-given triples and $d(0) = 1, h_\triangle(0) = 2, c(0) = 3, h_\triangledown(0) = 5, d(1) = 7$ are the above-given crucial moments. Suppose also that an arbitrary sequence of positive integers $n_1 < m_1 < n_2 < m_2 < \ldots$ is given. Then these finite sequences of triples and crucial moments can be extended to infinity so that the infinite sequences of triples $\{T_i\}, i = 1, 2, \ldots$ and crucial moments $d(i) < h_\triangle(i) < c(i) < h_\triangledown(i) < d(i+1) < \ldots$ satisfy the conditions listed below in* (a)–(h).

Inductive Assumptions for Step i.
(a) The i-th segment begins at the crucial moment $d(i)$, when the triple $T_{d(i)}$ is a down triple closest from the right to the chord $\bar{d}_0$:

$$T_{d(i)} = \triangledown(y_{d(i-1)}, \mathbf{x}_{d(i)}, \mathbf{y}_{d(i)}, t_0, \mathbf{z}_{d(i)}).$$

(b) Between the moments $d(i) + 1$ and $h_\triangle(i) - 1$, all triples are horizontal and minimal. Their location is determined by our rules and existing triples.

(c) At the crucial moment $h_\triangle(i)$, the triple $T_{h_\triangle(i)}$ is an up triple closest to $\frac{1}{2}$ and contained in the strip $S(\bar{c}_{-n_i}, \bar{d}_{-n_i})$:

$$T_{h_\triangle(i)} = \triangle(s_{-n_i}, \mathbf{x}_{h_\triangle(i)}, v_{-n_i}, \mathbf{y}_{h_\triangle(i)}, \mathbf{z}_{h_\triangle(i)}).$$

(d) We set $c(i) = h_\triangle(i) + n_i.$ (1)
For each $1 \leq j \leq n_i - 1$, we have

$$T_{h_\triangle(i)+j} = \triangle(s_{-n_i+j}, \mathbf{x}_{h_\triangle(i)+j}, v_{-n_i+j}, \mathbf{y}_{h_\triangle(i)+j}, \mathbf{z}_{h_\triangle(i)+j})$$

if the triple $T_{h_\triangle(i)+j}$ is the first triple entering the strip $S(\bar{c}_{-n_i+j}, \bar{d}_{-n_i+j})$. If this triple enters a strip of type $S(\bar{c}_{-r}, \bar{d}_{-r})$ already containing other triples, then we place it so that it becomes an up triple closest to $\bar{c}_{-r}$.

(e) At the crucial moment $c(i)$, the triple $T_{c(i)}$ is an up triple closest from the right to the chord $\bar{c}_0$:

$$T_{c(i)} = \triangle(x_{c(i-1)}, \mathbf{x}_{c(i)}, v_0, \mathbf{y}_{c(i)}, \mathbf{z}_{c(i)}).$$

(f) Between the moments $c(i) + 1$ and $h_\triangledown(i) - 1$, all triples are horizontal and minimal. Their location is determined by our rules and existing triples.

(g) At the crucial moment $h_\triangledown(i)$, the triple $T_{h_\triangledown(i)}$ is a down triple closest to $\frac{1}{2}$ and contained in the strip $S(\bar{d}_{-m_i}, \bar{c}_{-m_i+1})$:

$$T_{h_\triangledown(i)} = \triangledown(u_{-m_i+1}, \mathbf{x}_{h_\triangledown(i)}, \mathbf{y}_{h_\triangledown(i)}, t_{-m_i}, \mathbf{z}_{h_\triangledown(i)}).$$

(h) We set
$$d(i+1) = h_\triangledown(i) + m_i. \tag{2}$$

For each $1 \le j \le m_i - 1$, we have

$$T_{h_\triangledown(i)+j} = \triangledown(u_{-m_i+j+1}, \mathbf{x}_{h_\triangledown(i)+j}, \mathbf{y}_{h_\triangledown(i)+j}, t_{-m_i+j}, \mathbf{z}_{h_\triangledown(i)+j})$$

if $T_{h_\triangledown(i)+j}$ is the first triple entering the strip $S(\bar{d}_{-m_i+j}, \bar{c}_{-m_i+j+1})$. If this triple enters a strip of type $S(\bar{d}_{-r}, \bar{c}_{-r+1})$ already containing other triples, then we place it so that it becomes a down triple closest to $\bar{d}_{-r}$.

The properties (a)–(h) are exhibited at the basic step from $d(0)$ to $d(1)$. The inductive step can be made to satisfy the same properties. This can be easily verified, so except for part (a) of the inductive step, we leave the verification to the reader.

The Inductive Step.

(a) The $(i+1)$st segment begins at the crucial moment $d(i+1)$ when the triple $T_{d(i+1)}$ is a down triple closest from the right to the chord $\bar{d}_0$:

$$T_{d(i+1)} = \triangledown(y_{d(i)}, \mathbf{x}_{d(i+1)}, \mathbf{y}_{d(i+1)}, t_0, \mathbf{z}_{d(i+1)}).$$

It is easy to verify that this choice of $T_{d(i+1)}$ satisfies our rules. Indeed, by the inductive assumption (h), the triple $T_{h_\triangledown(i)+m_i-1} = T_{d(i+1)-1}$ precedes $T_{d(i+1)}$ and is a down triple located between $\bar{d}_{-1}$ and a down triple $T_{d(i)-1}$. Hence placing its image as described above, we satisfy all of our rules. Thus, $T_{d(i+1)}$ lies between $T_{d(i)}$ and $\bar{d}_0$. The rules and inductive assumptions determine the next few locations of the triple. We call $T_{d(i)}$ the *forcing* triple and $T_{d(i+1)}$ the *current* triple (this terminology applies to their images too).

(b) By our rules, on the next step the current triple $T_{d(i+1)+1}$ is contained in the arc $(y_{d(i)+1}, z_{d(i)+1})$, and for some time the triples $T_{d(i+1)+j}$ are contained inside yz-arcs of the images of the forcing triple. The containment holds at least until, at the crucial moment $h_\triangle(i)$, the crucial event of $h_\triangle$-type takes place for the forcing triple. However since the yz-arc of the forcing triple is then not exposed to $\frac{1}{2}$, we see that, for a while yet, the current triple stays inside the yz-arcs of the forcing triple and remains minimal. In fact, it remains minimal until, at the crucial moment $h_\bigtriangledown(i)$, the ith crucial event of type $h_\bigtriangledown$ takes place for the forcing triple. Then the location of the current triple with respect to B' and existing triples is not fully determined because the yz-arc of the forcing triple is "exposed" to $\frac{1}{2}$ for the first time. Choose this to be the crucial moment $h_\bigtriangleup(i+1)$ for our current triple. Then

$$h_\bigtriangleup(i+1) = d(i+1) + h_\bigtriangledown(i) - d(i). \tag{3}$$

(c) We explained in (a) that the choice which we make there can be made to comply with our rules. Similarly, one can easily check that the choice which we make below, in (c), can also be made to comply with our rules.

At the crucial moment $h_\bigtriangleup(i+1)$, the triple $T_{h_\bigtriangleup(i+1)}$ is an up triple closest to $\frac{1}{2}$ and contained in the strip $S(\bar{c}_{-n_{i+1}}, \bar{d}_{-n_{i+1}})$:

$$T_{h_\bigtriangleup(i+1)} = \bigtriangleup\left(s_{-n_{i+1}}, \mathbf{x}_{h_\bigtriangleup(i+1)}, v_{-n_{i+1}}, \mathbf{y}_{h_\bigtriangleup(i+1)}, \mathbf{z}_{h_\bigtriangleup(i+1)}\right).$$

(d) We set $c(i+1) = h_\bigtriangleup(i+1) + n_{i+1}$ (see (1)). Between the crucial moments $h_\bigtriangleup(i+1)$ and $c(i+1)$, the locations of the triples are almost completely determined by the rules. For each $1 \leq j \leq n_{i+1} - 1$, we have

$$T_{h_\bigtriangleup(i+1)+j} = \bigtriangleup\left(s_{-n_{i+1}+j}, \mathbf{x}_{h_\bigtriangleup(i+1)+j}, v_{-n_{i+1}+j}, \mathbf{y}_{h_\bigtriangleup(i+1)+j}, \mathbf{z}_{h_\bigtriangleup(i+1)+j}\right)$$

if the triple $T_{h_\bigtriangleup(i+1)+j}$ is the first triple entering the corresponding strip. If this triple enters a strip of type $S(\bar{c}_{-r}, \bar{d}_{-r})$ already containing other triples, then we locate it to be an up triple closest to $\bar{c}_{-r}$.

(e) At the crucial moment $c(i+1)$, the triple $T_{c(i+1)}$ is an up triple closest from the right to the chord $\bar{c}_0$:

$$T_{c(i+1)} = \bigtriangleup\left(x_{c(i)}, \mathbf{x}_{c(i+1)}, v_0, \mathbf{y}_{c(i+1)}, \mathbf{z}_{c(i+1)}\right).$$

Then $T_{c(i+1)}$ lies between $T_{c(i)}$ and $\bar{c}_0$. The rules and inductive assumptions determine the next few locations of the triple. We call $T_{c(i)}$ the *forcing* triple and $T_{c(i+1)}$ the *current* triple (this applies to their images too).

(f) By our rules, on the next step the current triple $T_{c(i+1)+1}$ is contained in the arc $(x_{c(i)+1}, y_{c(i)+1})$, and for some time the triples $T_{c(i+1)+j}$ are contained inside the xy-arcs of the images of the forcing triple. The containment holds at least until, at the crucial moment $h_\bigtriangledown(i)$, the crucial event of $h_\bigtriangledown$-type takes place for the forcing triple. However since the xy-arc of the forcing triple is then not exposed to $\frac{1}{2}$, we see that yet for a while the current triple stays inside the the xy-arcs of the forcing triple and remains minimal. In fact, it remains minimal until, at the crucial moment $h_\bigtriangleup(i)$, the ith crucial event of type $h_\bigtriangleup$ takes place for the forcing triple. Then the location of the current triple with respect to B' and existing triples is not fully determined because the xy-arc of the forcing triple is "exposed" to $\frac{1}{2}$ for the first time. Choose this to be the crucial moment $h_\nabla(i+1)$ for our current triple. Then

$$h_\nabla(i+1) = c(i+1) + h_\triangle(i+1) - c(i). \tag{4}$$

(g) At the crucial moment $h_\nabla(i+1)$, the triple $T_{h_\nabla(i+1)}$ is a down triple closest to $\frac{1}{2}$ and contained in the strip $S(\bar{d}_{-m_{i+1}}, \bar{c}_{-m_{i+1}+1})$:

$$T_{h_\nabla(i+1)} = \nabla(u_{-m_{i+1}+1}, \mathbf{x}_{h_\nabla(i+1)}, \mathbf{y}_{h_\nabla(i+1)}, t_{-m_{i+1}}, \mathbf{z}_{h_\nabla(i+1)}).$$

(h) We set $d(i+2) = h_\nabla(i+1) + m_{i+1}$ (see (2)). Between the crucial moments $h_\nabla(i+1)$ and $d(i+2)$, the locations of the triples are almost completely determined by the rules. For each $1 \le j \le m_{i+1} - 1$, we have

$$T_{h_\nabla(i+1)+j} - \nabla(u_{-m_{i+1}+j+1}, \mathbf{x}_{h_\nabla(i+1)+j}, \mathbf{y}_{h_\nabla(i+1)+j}, t_{-m_{i+1}+j}, \mathbf{z}_{h_\nabla(i+1)+j})$$

if the triple $T_{h_\nabla(i+1)+j}$ is the first triple entering the corresponding strip. If this triple enters a strip of type $S(\bar{d}_{-r}, \bar{c}_{-r+1})$ already containing other triples, then we locate it to be a down triple closest to $\bar{d}_{-r}$.

This concludes the induction. It is easy to check that the time between two consecutive crucial events grows to infinity. Let us check whether these examples generate an uncountable family of cubic WT-laminations with pairwise non-conjugate induced maps.

Lemma 4.1. *The map $g|_A$ is σ-extendable of degree 3 (here $A = \bigcup_{i=1}^\infty T_i$).*

Proof: It is easy to see that the degree of g is 3. By Theorem 3.3, we need to check that for $a \ne b \in A$ there exists an $n \ge 0$ such that $g([g^n(a), g^n(b)]_A) \not\subset [g^{n+1}(a), g^{n+1}(b)]$. This is obvious if $(a, b) \supset [s_0, u_0]$, or $(a, b) \supset [u_0, v_0]$, or $(a, b) \supset [v_0, t_0]$, or $(a, b) \supset [t_0, s_0]$. Suppose first that a and b are in the same triangle T_i. If $a = x_i$, $b = y_i$, and the next crucial moment of c-type is $c(j)$, then the arc $(f^{c(j)-i}(x_i), f^{c(j)-i}(y_i))$ contains (u_0, v_0) as desired. If $a = y_i$, $b = z_i$, and the next crucial moment of d-type is $d(l)$, then the arc $(g^{d(l)-i}(y_i), g^{d(l)-i}(z_i))$ contains (u_0, v_0) as desired. Now, let $a = z_i$ and $b = x_i$. Then it is enough to choose $T_j, j \ge i$, located to the left of $\bar{d}_0$ and observe that then $[z_j, x_j] \supset [t_0, s_0]$. Now assume that $a \in T_p$ and $b \in T_q$ with $p < q$. Since $q - p$ is finite and $m_i \to \infty$, we may assume that there exist k and i with $h_\nabla(i) \le p + k < q + k < d(i+1)$ and both T_{p+k} and T_{q+k} are down triples located in the arc (u_0, v_0). Then the set $[s_0, u_0] \cup [v_0, t_0]$ separates the points $g^{d(i+1)-q}(b)$ and $g^{d(i+1)-q}(a)$ in S^1, and the result follows. $\qquad\square$

By Lemma 4.1, from now on we assume that $T_1, T_2, \dots$ is the σ-orbit of a triple T_1 with the order among points of $A = \bigcup_{i=1}^\infty T_i$, exactly as before.

Lemma 4.2. *Let* $\hat{s}_0 = \lim_{i\to\infty} y_{d(i)}, \hat{t}_0 = \lim_{i\to\infty} z_{d(i)}, \hat{u}_0 = \lim_{i\to\infty} x_{c(i)}$, *and* $\hat{t}_0 = \lim_{i\to\infty} y_{c(i)}$. *Then the points* $\hat{s}_0, \hat{u}_0, \hat{v}_0, \hat{t}_0$ *are all distinct,* $\sigma(\hat{s}_0) = \sigma(\hat{t}_0)$, *and* $\sigma(\hat{u}_0) = \sigma(\hat{v}_0)$.

Proof: The limits in the lemma are well defined, and for every i there are points of A in the arcs $(y_{d(i)}, x_{c(i)})$, $(x_{c(i)}, y_{c(i)})$, $(y_{c(i)}, z_{d(i)})$, $(z_{d(i)}, y_{d(i)})$. Hence the points $\hat{s}_0, \hat{u}_0, \hat{v}_0, \hat{t}_0$ are all distinct. To see that $\sigma(\hat{s}_0) = \sigma(\hat{t}_0)$, we show that $\alpha = \lim_{i\to\infty} y_{d(i)+1}$ and $\beta = \lim_{i\to\infty} z_{d(i)+1}$ are the same. Indeed, if not, the arc $[\alpha, \beta]$ is non-degenerate, and there exists a least $l \geq 0$ such that $\sigma^l[\alpha, \beta] = [\sigma^l(\alpha), \sigma^l(\beta)] = I$ is an arc of length at least $1/3$. The chord connecting the endpoints of I is the limit of chords connecting $y_{d(i)+1+l}, z_{d(i)+1+l} \in T_{d(i)+1+l}$, and the endpoints of $T_{d(i)+1+l}$ are outside I. Let us show that $A \cap I = \emptyset$. Suppose otherwise. Then there is a triple $T_k \subset I$, because if a point of T_k is in I, then $T_k \subset I$ (if $T_k \not\subset I$, then a chord connecting points of T_k intersects chords connecting $y_{d(i)+1+l}$ and $z_{d(i)+1+l}$ with large i, a contradiction). Now we choose a big i so that $d(i) + l + 1 > k$ is between the crucial moments $d(i)$ and $h_\triangle(i)$. Then the triple $T_{d(i)+l+1}$ must be minimal among the already existing triples, contradicting the fact that $T_k \subset I$. Thus, I contains no points of A, which contradicts the fact that $g|_A$ is of degree 3 and implies that $\sigma(\hat{s}_0) = \sigma(\hat{t}_0)$. Similarly, $\sigma(\hat{u}_0) = \sigma(\hat{v}_0)$. $\square$

Let $\bar{c}_0$ be the chord connecting $\hat{s}_0$ with $\hat{t}_0$ and $\bar{d}_0$ be the chord connecting $\hat{v}_0$ with $\hat{u}_0$. To associate a lamination with $\Xi = \{\bar{c}_0, \bar{d}_0\}$ we rely on Kiwi [Ki05]. A collection $\Theta = \{X_1, \ldots, X_{d-1}\}$ of pairwise disjoint σ_d-critical chords (whose endpoints form a set $R = R_\Theta$) is called a *critical portrait* (e.g., Ξ is a critical portrait). The chords X_1, ..., X_{d-1} divide $\overline{\mathbb{D}}$ into components $B_1, \ldots, B_d$ whose intersections with S^1 are finite unions of open arcs with endpoints in R. Given $t \in S^1$, its *itinerary* $i(t)$ is the sequence $I_0, I_1, \ldots$ of sets B_1, ..., B_d, R with $\sigma_d^n(t) \in I_n (n \geq 0)$. A critical portrait Θ such that $i(t), t \in R_\Theta$, is not preperiodic is said to have a *non-periodic kneading*. Denote the family of all critical portraits with non-periodic kneadings by $\mathcal{Y}_d$. A lamination $\sim$ is Θ-*compatible* if the endpoints of every chord from Θ are $\sim$-equivalent. Theorem 4.3 is a particular case of Proposition 4.7 [Ki04].

Theorem 4.3. *To each* $\Theta \in \mathcal{Y}_d$ *one can associate a* Θ-*compatible lamination* $\sim$ *such that all* $\sim$-*classes are finite,* $J_\sim$ *is a dendrite, and the following holds:* (1) *any two points with the same itinerary, which does not contain* R, *are* $\sim$-*equivalent;* (2) *any two points whose itineraries are different at infinitely many places are not* $\sim$-*equivalent.*

Denote the family of laminations from Theorem 4.3 by $\mathcal{K}_d$.

Lemma 4.4. *We have $\Xi \in \mathcal{Y}_3$. There is a lamination $\sim$ from $\mathcal{K}_3$ compatible with Ξ such that T_1 forms a $\sim$-class, and $\bar{c}_0, \bar{d}_0$ are the $\sim$-critical leaves.*

Proof: We prove that $\hat{s}_0, \hat{u}_0, \hat{v}_0, \hat{t}_0$ have non-preperiodic itineraries and never map into one another. We have $\sigma^l(\hat{s}_0) \in (\hat{u}_0, \hat{v}_0), h_{\triangle(i)} \le l \le c_i - 1$. Since $c_i - h_{\triangle(i)} \to \infty$, the only way $i(\hat{s}_0)$ can be preperiodic is if $\hat{s}_0$ eventually stays in $(\hat{u}_0, \hat{v}_0)$ forever, a contradiction to the construction. Assume that $\hat{s}_0$ maps into $\hat{v}_0$ by σ^r. Choose j with $h_{\triangle}(j) - d(j) > r$. Then the triangle $T_{d(j)+r}$ intersects $\bar{c}_0$, contradicting the construction. The claim for $\hat{s}_0$ is proven; the claims for other points of R_Ξ can be proven similarly.

By Theorem 4.3, there exists a lamination $\sim$ in $\mathcal{K}_3$ compatible with Ξ; since points in any T_i have the same itinerary, which avoids R_Ξ, they are $\sim$-equivalent. Let us show that $\{\hat{u}_0, \hat{v}_0\}$ is a $\sim$-class. Were it not, the $\sim$-class of $\sigma(\hat{u}_0)$ would be non-degenerate. Since, by construction, the point $\sigma(\hat{u}_0)$ belongs to all arcs $(x_{c(j)+1}, y_{c(j)+1})$ that converge to it, the unlinked property (E2) of laminations implies that the $\sim$-class of $\sigma(\hat{u}_0)$ includes all triples $T_{c(j)+1}$ with big enough j and is infinite, contradicting Theorem 4.3. Similarly, $\{\hat{s}_0, \hat{t}_0\}$ is a $\sim$-class, and these are the only two critical $\sim$-classes. It follows that T_1 is a $\sim$-class. If not, T_1 is a proper subset of a $\sim$-class Q. Then Q contains more than 3 points, and by [Ki02], Q is preperiodic or precritical. If for some $i \ge 0$ the class $f^i(Q)$ is periodic, then, since the triple T_1 is wandering, $f^i(Q)$ must be infinite, contradicting Theorem 4.3. If for some minimal $i \ge 0$ the class $f^i(Q)$ is critical, then it has to consist of $|Q| > 3$ elements, contradicting the above.   □

By Lemma 4.4, for a sequence $\mathcal{T} = n_1 < m_1 < \ldots$, we construct a WT-lamination $\sim$ in $\mathcal{K}_3$; the family $\mathcal{W}$ of all such laminations is uncountable.

Theorem 4.5. *Laminations $\sim$ in $\mathcal{W}$ have pairwise non-conjugate induced maps $f_\sim|_{J_\sim}$.*

Proof: Let the sequences $\mathcal{T} = n_1 < m_1 < \ldots, \mathcal{T}' = n_1' < m_1' < \ldots$ be distinct, $\sim$ and $\sim'$ be the corresponding laminations from Lemma 4.4, p and p' be the corresponding quotient maps, and let the topological Julia sets with induced maps be $f : J \to J$ and $f' : J' \to J'$, respectively. All the points and leaves from our construction are denoted as before for f (e.g., $\hat{u}_0, \hat{v}_0, \ldots, \bar{c}_0, \bar{d}_0, \ldots$), whereas in the case of f' we add an apostrophe to the notation (e.g., $\hat{u}_0', \hat{v}_0', \ldots$).

Assume that the homeomorphism $\varphi : J \to J'$ conjugates f and f'. The critical points $p(\hat{u}_0) = C, p(\hat{s}_0) = D \in J$ of f are cut points, each of which cuts J into 2 pieces. Moreover the set $J \setminus (C \cup D)$ consists of 3 components: $L = p((\hat{u}_0, \hat{v}_0)), M = p((\hat{s}_0, \hat{u}_0) \cup (\hat{v}_0, \hat{t}_0))$, and $R = p((\hat{t}_0, \hat{s}_0))$. Similarly, the critical points $p'(\hat{u}_0') = C', p'(\hat{s}_0') = D' \in J'$ of f' are cut points each of which cuts J' into 2 pieces. Moreover, the set $J' \setminus (C' \cup D')$

consists of 3 components: $L' = p'((\hat{u}'_0, \hat{v}'_0))$, $M' = p'((\hat{s}'_0, \hat{u}'_0) \cup (\hat{v}'_0, \hat{t}'_0))$ and $R' = p'((\hat{t}'_0, \hat{s}'_0))$. Clearly, φ maps points C, D onto points C', D'.

For a $\sim$-class g, the point $p(g) \in J$ divides J into $|g|$ components (the same holds for $\sim'$). Let us show that the $\sim$-class of 0 is $\{0\}$. If not, by properties (D1)–(D3) of laminations, the $\sim$-class of 0 is $\{0, \frac{1}{2}\}$ contradicting the fact that the leaf $\bar{d}_0$ separates the points 0 and $\frac{1}{2}$. Similarly, $\{0\}$ is a $\sim'$-class, and $\{\frac{1}{2}\}$ is a $\sim$-class and a $\sim'$-class. Hence $a = p(\frac{1}{2}), b = p(0)$ are non-dividing f-fixed points, and $a' = p'(\frac{1}{2}), b' = p'(0)$ are non-dividing f'-fixed points. These are all the non-dividing fixed points, so φ maps the points a, b onto the points a', b'. By construction, a is the only non-dividing f-fixed point belonging to the limit sets of f-critical points. To show this, let us show that all triples from the original construction are contained in $[x_1, z_1]$. Indeed, the first seven triples $T_1, \ldots, T_7$ are located inside the arc $[x_1, z_1]$. After that, all the triples are either vertical (and then contained in $[x_1, z_1]$ by Rule C), or horizontal (and then contained in the appropriate arcs of vertical triples and again in $[x_1, z_1]$). This proves the claim and shows that a is indeed the only non-dividing f-fixed point belonging to the limit sets of f-critical points. Similarly, a' is the unique non-dividing f'-fixed point belonging to the limit sets of the f'-critical points. Hence $\varphi(a) = a'$, which implies that $\varphi(b) = b'$, and therefore $\varphi(C) = C', \varphi(D) = D'$. Thus, $\varphi(L) = L, \varphi(M) = M', \varphi(R) = R'$.

Assume that the first time the sequences $\mathcal{T}, \mathcal{T}'$ are different is at $n_i > n'_i$. Then $h_\triangle(i) = h'_\triangle(i) = h$, and up until that moment all corresponding crucial moments for the two laminations are equal: $d(r) = d'(r), h_\triangle(r) = h'_\triangle(r), c(r) = c'(r), h_\triangledown(r) = h'_\triangledown(r)(0 \le r \le i - 1)$, and $d(i) = d'(i)$. Before the crucial moment h, the behavior of the triples relative to the chords $\bar{c}_0, \bar{d}_0$ (respectively $\bar{c}'_0, \bar{d}'_0$) is the same. Consider the triple $T_{d(i)}$ (the closest approach to $\bar{d}_0$ preceding h), and the corresponding triple $T'_{d'(i)}$. The dynamics of $T_{d(i)}$ $(T'_{d'(i)})$ forces the same dynamics on $\bar{d}_0$ $(\bar{d}'_0)$ until $T_{d(i)}$ $(T'_{d'(i)})$ maps onto $T_{c(i)}$ $(T'_{c'(i)})$. Hence $\sigma^{h-d(i)+n'_i}(\hat{s}'_0)$ already belongs to the arc $(\hat{v}'_0, \hat{t}'_0)$, while $\sigma^{h-d(i)+n_i}(\hat{s}_0)$ still belongs to the arc $(\frac{1}{2}, \hat{v}_0)$. Therefore $f^{h-d(i)+n'_i}(D) \in L$, whereas $(f)'^{h-d(i)+n'_i}(D') \in M$. Since $\varphi(D) = D'$ and $\varphi(M) = M'$, we get a contradiction which shows that φ does not exist and the maps $f|_J$ and $f'|_{J'}$ are not conjugate.

5　TCE-Polynomials with Wandering Branch Points

In this section, we show that there exists an uncountable family of TCE-polynomials P whose induced laminations $\sim_P$ are WT-laminations (since, by [Pr00], if the Julia set of a TCE-polynomial is locally connected, then the polynomial on its Julia set and the induced map on the corresponding topological Julia set are conjugate). The *Topological Collet-Eckmann (TCE)* condition is studied in a number of papers (e.g., [GS98,Pr00,PRLS03,PR98, Sm00]; a list of references can be found in the nice recent paper [PRLS03]). It is considered a form of non-uniform (weak) hyperbolicity. By [PRLS03], several standard conditions of non-uniform hyperbolicity of rational maps, including the TCE condition, are equivalent. By Proposition 5.2 [Pr00] (see also [GS98, PR98]), the Julia set of a TCE-polynomial is Hölder (i.e., the Riemann map extends over the boundary as Hölder), and hence locally connected.

The plan is to construct WT-laminations $\sim$ from $\mathcal{W}$ corresponding to specific sequences $\mathcal{T}$ whose induced maps $f_\sim|_{J_\sim}$ satisfy the TCE condition (the definitions are below). Since $\mathcal{W} \subset \mathcal{K}_3$, by results of Kiwi [Ki04, Ki05], to each such lamination $\sim$ a polynomial $P_\sim$ is associated, and $P_\sim|_{J_{P_\sim}}$ is monotonically semiconjugate to the induced map $f_\sim|_{J_\sim}$. This implies that $P_\sim$ satisfies the TCE condition; by [Pr00] its Julia set is locally connected (actually Hölder), and $P_\sim|_{J_{P_\sim}}$ is in fact conjugate to $f_\sim : J_\sim \to J_\sim$.

A continuum $K \subset S^2$ is *unshielded* if it is the boundary of one of its complementary domains (see, e.g., [BO04a]). Below K is either S^2 or a locally connected unshielded continuum in S^2 (we then choose a metric in K such that all balls are connected; the existence of such a metric is proven in [Bi49], see also [MMOT92]). Given a set $A \subset K$ and a point $z \in A$, we denote by $\text{Comp}_z A$ the component of A containing z. Consider a branched covering map $f : K \to K$. Then the set of critical points Cr_f is finite.

Take a point $x \in K$ and the ball $B(f^n(x), r)$. For each $i, 0 \leq i \leq n$, consider $\text{Comp}_{f^i(x)} f^{-(n-i)}(B(f^n(x), r))$ and call it a *pull-back of $B(f^n(x), r)$ (along the orbit of x)*. Denote by $\Delta_f(x, r, n)$ the total number all moments i such that $\text{Comp}_{f^i(x)} f^{-(n-i)}(B(f^n(x), r)) \cap \text{Cr}_f \neq \emptyset$. A map $f : K \to K$ is said to satisfy the *TCE condition* (or to be a *TCE-map*, or just *TCE*) if and only if there are $M > 0, P > 1$, and $r > 0$ such that for every $x \in K$, there is an increasing sequence $n_j \leq Pj$ of numbers with $\Delta_f(x, r, n_j) \leq M$. Therefore, if a map is not TCE, then for *any* $M > 0, P > 1$, and $r > 0$, there exist $x \in K$ and $N > 0$ with

$$\frac{|\{n \in [0, N] \mid \Delta_f(x, r, n) > M\}|}{N + 1} > 1 - \frac{1}{P}.$$

Observe that in the case when f is a rational function and $K = S^2$, it is sufficient to consider only points $x \in J_f$ in the definition of TCE-maps.

A continuous map $f : X \to X$ of a metric space is *backward stable* at $x \in X$ if for any δ there is an ε such that for any connected set $K \subset B(x, \varepsilon)$, any $n \geq 0$, and any component M of $f^{-n}(K)$, $\text{diam}(M) \leq \delta$; f is *backward stable* if it is backward stable at all points. If X is compact, then f is *backward stable* if and only if, for any δ, there is an ε such that for any continuum T with $\text{diam}(T) \leq \varepsilon$, any $n \geq 0$, and any component M of $f^{-n}(T)$, $\text{diam}(M) \leq \delta$. The notion is essentially due to Fatou. Classic results (see, e.g., Fatou, [CG93]) imply that R is backward stable outside the critical limit sets and is not backward stable at parabolic or attracting periodic points. In an important paper [Le98], Levin showed that polynomials with one critical point and locally connected Julia sets are backward stable on their Julia sets. Later [BO04a], this result was extended to all induced maps on their topological Julia sets.

Orbit segments $\{z, f(z), \ldots, f^n(z)\}$ and $\{y, f(y), \ldots, f^n(y)\}$ δ-shadow (each other) if $d(f^i(z), f^i(y)) \leq \delta$ for $0 \leq i \leq n$. Denote the orbit of z by $\text{orb}(z)$; an $([i, j]$-$)$segment of $\text{orb}(z)$ is the set $\{f^i(z), \ldots, f^j(z)\}$. Given a point z, an integer n, and an $\varepsilon > 0$, we say that $f^n(z)$ is critically ε-shadowed of order k if there are precisely k distinct pairs (each pair consists of a critical point u and an iteration s) such that $f^s(z), \ldots, f^n(z)$ is ε-shadowed by $u, \ldots, f^{n-s}(u)$. If this is the case, we call n a critical ε-shadowing time of order k (for z). Lemma 5.1 is inspired by Lemma 2.2 of the paper [Sm00] by Smirnov.

Lemma 5.1. *Suppose that $f : K \to K$ is a branched covering, backward stable map, and there exist $\varepsilon' > 0$, s', and $\tau' > 0$ such that for any critical point u and any integer $N > 0$, there are more than $\tau'(N + 1)$ critically ε'-shadowed times of order less than s' in $[0, N]$ for u. Then f satisfies the TCE condition.*

Proof: We prove that if f does not satisfy the TCE condition, then for any given $\varepsilon > 0$, s, and $\tau > 0$, there is an $N > 0$ and a critical point u such that there are less than $\tau(N+1)$ critically ε-shadowed times of order less than s in $[0, N]$ for u. Since f is not TCE, for any $P > 1, r > 0$, and $M > 0$, there exist $x \in K$ and $N > 0$ such that for a set H of more than $\frac{(P-1)(N+1)}{P}$ integers $l \in [0, N]$, we have $\Delta_f(x, r, l) > M$. Let the distance between any two critical points be more than $R > 0$, and choose $M > \frac{sP}{(P-1)\tau}$. Since f is backward stable, we can find a $\delta < \min\{\varepsilon/2, R/2\}$ and $r > 0$ so that any pull-back of an r-ball is of diameter less than δ. For $x \in K$, let $c(x)$ be a closest to x critical point.

We define a collection $\mathcal{I}$ of *intervals of integers*. For an integer $j, 0 \leq j \leq N$, define (if possible) the largest number $k = k_j, j \leq k \leq N$, such that

$$\text{Comp}_{f^j(x)} f^{-(k-j)}(B(f^k(x), r)) \cap \text{Cr}_f \neq \emptyset.$$

Let A be the set of all j for which k_j exists, and let $\mathcal{I}$ be the family of all intervals of integers $\{[j, k_j] : j \in A\}$. The $[j, k_j]$-segment of $\text{orb}(x)$ is δ-shadowed by the $[0, k_j - j]$-segment of the critical point $c(f^j(x))$. If a critical point belongs to the pullback $U = \text{Comp}_{f^i(x)} f^{-(l-i)}(B(f^l(x), r))$ of $B(f^l(x), r)$ along the orbit of x, then $i \in A$ and $l \in [i, k_i]$. Hence, if $l \in H$, then more than M intervals from $\mathcal{I}$ contain l. Since $|H| \geq \frac{(P-1)(N+1)}{P}$, we have

$$\sum_{I \in \mathcal{I}} |I| \geq \frac{(P-1)(N+1)M}{P} > \frac{s(N+1)}{\tau}.$$

Let $i, j \in A$, $u = c(f^i(x))$, $v = c(f^j(x))$. If $j \geq i$ and $[i, k_i] \cap [j, k_j] = [j, l]$ ($l = k_i$ or $l = k_j$), then the $[j - i, l - i]$-segment of $\mathrm{orb}(u)$ and the $[0, l - j]$-segment of $\mathrm{orb}(v)$ 2δ-shadow each other. Since $2\delta < \varepsilon$, if $t \in [i, k_i]$ is covered by at least s intervals of the form $[j, k_j] \in \mathcal{I}$ with $i \leq j$, then $f^{t-i}(u)$ is critically ε-shadowed of order at least s. Let us show that in some interval $I = [i, k_i] \in \mathcal{I}$, there are $h > (1 - \tau)|I|$ integers $t_1, \ldots, t_h$ covered by at least s intervals of the form $[j, k_j] \in \mathcal{I}$, with $i \leq j$.

Let us show that such an interval $[i, k_i] \in \mathcal{I}$ exists. If not, then in each interval $I = [i, k_i] \in \mathcal{I}$, at most $(1 - \tau)|I|$ points are covered by s intervals of the form $[j, k_j] \in \mathcal{I}$, with $i \leq j$. Let us call a pair (I, l) admissible if $I \in \mathcal{I}, l \in I$, and there are at least s intervals $[j, k_j] \in \mathcal{I}$ with $i \leq j \leq l \leq k_j$. Denote the number of all admissible pairs by L and count it in two ways: over intervals I from $\mathcal{I}$, and over points l. If we count L over intervals from $\mathcal{I}$, then, since by assumption each interval $I \in \mathcal{I}$ contains at most $(1-\tau)|I|$ numbers l such that (I, l) is admissible, we see that $L \leq (1 - \tau) \sum_{I \in \mathcal{I}} |I|$. For each $l \in [0, N]$, let $m(l)$ be the number of intervals from $\mathcal{I}$ containing l. Then $\sum_{I \in \mathcal{I}} |I| = \sum m(l)$. Define two sets $A \subset [0, N], B \subset [0, N]$ as follows: A is the set of all integers $l \in [0, N]$ with $m(l) \leq s - 1$, and B is the set of all integers $l \in [0, N]$ with $m(l) \geq s$. Then it is easy to see that $L = \sum_{l \in B}(m(l) - s + 1)$. Hence

$$\sum_{I \in \mathcal{I}} |I| = \sum_{l=0}^{N} m(l) = (s-1)|B| + \sum_{l \in B}[m(l) - s + 1] + \sum_{l \in A} m(l) =$$

$$= (s-1)|B| + L + \sum_{l \in A} m(l).$$

Since $L \leq (1 - \tau) \sum_{I \in \mathcal{I}} |I|$ and $m(l) \leq s - 1$ for $l \in A$,

$$\sum_{I \in \mathcal{I}} |I| \leq (s-1)(|B|+|A|) + (1-\tau) \sum_{I \in \mathcal{I}} |I| = (s-1)(N+1) + (1-\tau) \sum_{I \in \mathcal{I}} |I|,$$

which implies that

$$\sum_{I \in \mathcal{I}} |I| \leq \frac{(s-1)(N+1)}{\tau},$$

a contradiction. Hence there exists an interval $I = [i, k_i] \in \mathcal{I}$ with $h > (1 - \tau)|I|$ integers $t_1, \ldots, t_h$ covered by at least s intervals of the form $[j, k_j] \in \mathcal{I}$

with $i \leq j$. Set $N = k_i - i$; then h integers $t_1 - i \in [0, N], \ldots, t_h - i \in [0, N]$ are critically ε-shadowing times of order at least s for u. Hence there are less than $N + 1 - h < \tau(N+1)$ integers in $[0, N]$ which are critically ε-shadowing times of order less than s for u. Doing this for $\varepsilon = \varepsilon', s = s', \tau = \tau'$ from the lemma, we get a contradiction to the assumptions of the lemma and complete its proof.   □

Let $\sim$ be a lamination, constructed as in Section 4, for a sequence $\mathcal{T} = n_1 < m_1 < \ldots$, and let $f|_J$ be its induced map. Let us state some facts about the construction in terms of the map f. Let $p : S^1 \to J$ be the corresponding quotient map, and let $I \subset J$ be the arc connecting $p(1/2) = b$ and $p(0) = a$. A $\sim$-class g contains points of the upper semicircle UP and the lower one LO if and only if $p(g) \in I$. Put $p(T_i) = t_i, p(\hat{u}_0) = C, p(\hat{d}_0) = D, f^i(C) = C_i$, and $f^i(D) = D_i$.

We assume that $J \subset \mathbb{C}$ and that the orientation of J agrees with that of the unit circle. Moreover, we visualize I as a subsegment of the x-axis such that b is the "leftmost" point of J (its x-coordinate is the smallest), a is the "rightmost" point of J (its x-coordinate is the greatest), the points of J corresponding to angles from UP belong to the upper half-plane, and the points of J corresponding to angles from LO belong to the lower half-plane.

By construction, $d(0) = 1, h_\triangle(0) = 2, c(0) = 3, h_\triangledown(0) = 5, d(1) = 7$. The crucial moments $d(i), h_\triangle(i), c(i), h_\triangledown(i)$ are the *moments of closest approach of images of* t_1 (or just the *closest approaches of* t_1) to $D, b, C, b, \ldots$, in this order. To explain the term "closer", we need the following notation: if $m, n \in J$, then $S(m, n)$ is the component of $J \setminus \{m, n\}$ which contains the unique arc in J connecting m and n.

Definition 5.2. A point $x \in J$ is *closer* to a point $w \in J$ than a point $y \in J$ if $y \notin S(x, w)$.

This notion is specific to the closest approaches of t_1 to C, D, b that take place on I. We distinguish between two types of closest approach to b depending on which critical point is approached next (equivalently, depending on the type of the triangle which approaches $1/2$). Thus, $h_\triangle(i)$ is a closest approach to b, after which t_1 will have the next closest approach to C ($h_\triangle(i)$ is the ith such closest approach to b). Similarly, $h_\triangledown(i)$ is a closest approach to b, after which t_1 will have the next closest approach to D ($h_\triangledown(i)$ is the ith such closest approach to b).

We apply Lemma 5.1 to f, choosing a collection of integers $\mathcal{T}$ appropriately. The behavior of C, D is forced by that of t_1. The three germs of J at t_1 corresponding to the arcs $(x_1, y_1), (y_1, z_1)$, and (z_1, x_1) in S^1 are denoted X, Y, Z; call their images X-germs, Y-germs, or Z-germs, respectively (at t_k). Thus, X points up, Y points to the left, and Z points to the right. Also, set $\sigma^k(x_1) = x_k, \sigma^k(y_1) = y_k, \sigma^k(z_1) = z_k, k \geq 1$. Because of the connection between the map $f|_J$ and the map σ at the circle at infinity, the dynamics of the arcs is reflected by the behavior of the germs. This helps one see where in J images of C, D are located.

We use the expressions "the X-germ (at t_k) points up", "the Y-germ (at t_k) points to the left", etc., which are self-explanatory if $t_k \in I$. The components $C_X(t_k), C_Y(t_k), C_Z(t_k)$ of $J \setminus t_k$ containing the corresponding germs at t_k correspond to the X-, Y-, and Z-germs at t_k; the components are called the X-, Y-, Z-components (of J at t_k), respectively.

For $t_k \in I$, the Z-germ at t_k always points to the right, so we only talk about X- and Y-germs at points $t_k \in I$. At the moment $d(i)$, the point $t_{d(i)} \in I$ is to the right of D in $S(D, t_{d(i-1)})$, its X-germ points up, and its Y-germ points to the left. Then the point $t_{d(i)}$ leaves I, and between the moments $d(i) + 1$ and $h_\triangle(i) - 1$, all its images avoid $I \cup S(D, t_{d(i)}) \cup S(C, t_{c(i-1)}) \cup S(b, t_{h_\triangledown(i-1)})$ (its images are farther away from D, C, b than the three previous closest approaches to these points).

The next crucial moment is $h_\triangle(i)$, when t_1 maps into $I \cap S(b, t_{h_\triangledown(i-1)})$ (so it is the next closest approach to b), its X-germ points to the left, and its Y-germ points down. The map locally "rotates" J: the X-germ, which was pointing up, now points to the left, and the Y-germ, which was pointing to the left, now points down. Moreover, it follows that, along the way, the Y-components of images of $T_{d(i)}$ never contain a critical point, and hence points that used to belong to the Y-component at $t_{d(i)}$ still belong to the Y-component at $t_{h_\triangle(i)}$. Observe that D belongs to the Y-component at $t_{d(i)}$. Hence the following holds.

Claim $\triangle$. *D maps by* $f^{h_\triangle(i)-d(i)}$ *to the point* $D_{h_\triangle(i)-d(i)}$ *inside the Y-component at* $t_{h_\triangle(i)}$, *which points down.*

For the next n_i steps, t_1 stays in I while being repelled from b to the right with no "rotation" (the X-germ points to the left, the Y-germ points down). For these n_i steps, the images of t_1 and D stay close while being repelled "together" from b. At the next crucial moment $c(i) = h_\triangle(i) + n_i$, the point $t_{c(i)} \in I$ is to the right of C in $S(c, t_{c(i-1)})$, its X-germ points to the left, and its Y-germ points down. As with the above, we conclude that the following claim holds.

Claim Θ. *D maps by* $f^{c(i)-d(i)}$ *to the point* $D_{c(i)-d(i)}$ *inside the Y-component at* $t_{c(i)}$, *which points down.*

Now $t_{c(i)}$ leaves I, and between the moments $c(i) + 1$ and $h_\triangledown(i) - 1$ all its images avoid $I \cup S(D, t_{d(i)}) \cup S(C, t_{c(i)}) \cup S(b, t_{h_\triangle(i)})$ (its images are farther away from D, C, b than the three previous closest approaches to these points). For a while after this moment, the behavior of t_1 is mimicked by both images of $D_{c(i)-d(i)}$ *and* images of C. The next crucial moment is

$h_\nabla(i)$, when t_1 maps into $I \cap S(b, t_{h_\triangle(i)})$ (so it is the next closest approach to b), its X-germ points up, and its Y-germ points to the left. The map locally "rotates" J: the X-germ, which pointed to the left, now points up, and the Y-germ, which pointed down, now points to the left. Observe that $D_{c(i)-d(i)}$ belongs to the Y-component at $t_{c(i)}$ and on the next step maps by $f^{h_\nabla(i)-c(i)}$ to the point $D_{h_\nabla(i)-d(i)}$. Hence, as with the explanation prior to Claim $\triangle$, we conclude that the following holds.

Claim Γ. *D maps by $f^{h_\nabla(i)-d(i)}$ to the point $D_{h_\nabla(i)-d(i)}$ inside the Y-component at $t_{h_\nabla(i)}$.*

By construction, so far all the points from the appropriate segments of the orbits of $t_{d(i)}$ and D are very close, because the appropriate images of the arc $[y_{d(i)}, z_{d(i)}]$, which correspond to the images of the Y-component at $t_{d(i)}$, are very small. However now the behaviors of t_1 and D differ. In terms of t_1, for the next m_i steps $t_{h_\nabla(i)}$, stays in I while being repelled from b to the right with no rotation (the X-germ points up, and the Y-germ points to the left). At the next crucial moment $d(i+1) = h_\nabla(i) + m_i$, the point t_1 maps inside $S(D, t_{d(i)})$ (this is the next closest approach to D), and the process for t_1 is repeated inductively (the segments of the constructed orbit repeat the same structure as the one described above). However the dynamics of D is more important.

At the moment when D maps by $f^{h_\nabla(i)-d(i)}$ to the point $D_{h_\nabla(i)-d(i)}$, the point $D_{h_\nabla(i)-d(i)}$ is still associated with the $f^{h_\nabla(i)-d(i)}$-image of $t_{d(i)}$, i.e., with the point $t_{h_\nabla(i)}$. Since by formula (3) (see "The Inductive Step", Part (b), Page 152) $h_\nabla(i) - d(i) = h_\triangle(i+1) - d(i+1) = q$, by Claim $\triangle$ applied to $i+1$ rather than to i, the point D_q belongs to the Y-component at $t_{h_\triangle(i+1)}$.

Now the next segment of the orbit of D begins, which, according to Claim Θ and Claim Γ applied to $i+1$, includes n_{i+1} steps when D is repelled away from b (while the appropriate images of t_1 are also repelled from b on I), and then $h_\nabla(i+1) - c(i+1)$ steps when D is shadowed by the orbit of C. Thus, the orbit of D can be divided into countably many pairs of segments, described below.

(d1) Segment D_i', from the $h_\triangle(i) - d(i) = h_\nabla(i-1) - d(i-1)$th to the $c(i) - d(i) - 1$th iteration of D of length n_i when D is repelled from b with the images $t_{h_\triangle(i)}, \ldots, t_{c(i)-1}$ of t_1, so that the images of D belong to the Y-components of the appropriate images of t_1, which belong to I and stay to the left of C while the images of D are below the images of t_1.

(d2) Segment D_i'', from the $c(i)-d(i)$th to the $h_\nabla(i)-d(i)-1 = h_\triangle(i+1) - d(i+1) - 1$th iteration of D of length $h_\nabla(i) - c(i) = h_\triangle(i) - c(i-1)$, when D is closely shadowed by the orbit of C and has no closest approaches to b, C, D; $h_\nabla(i) - c(i) = h_\triangle(i) - c(i-1)$ by (4).

Since the construction is symmetric with respect to D and C, the orbit of C can be divided into countably many pairs of segments, described below.

(c1) Segment C_i', from the $h_\triangledown(i) - c(i) = h_\triangle(i) - c(i-1)$th to the $d(i+1) - c(i) - 1$th iteration of C of length m_i, when C is repelled from b with the images $t_{h_\triangledown(i)}, \ldots, t_{d(i+1)-1}$ of t_1 so that the images of C belong to the X-components of the appropriate images of t_1, which belong to I and stay to the left of C while the images of C are above the images of t_1.

(c2) Segment C_i'', from the $d(i+1) - c(i)$th to the $h_\triangledown(i+1) - c(i+1) - 1 = h_\triangle(i+1) - c(i) - 1$th iteration of C of length $h_\triangle(i+1) - d(i+1) = h_\triangledown(i) - d(i)$, when C is closely shadowed by the orbit of D and has no closest approaches to b, C, D.

By (c1), the segment C_i' begins at $h_\triangledown(i) - c(i) = h_\triangle(i) - c(i-1)$; since $c(i-1) < d(i)$, $h_\triangle(i) - d(i) < h_\triangle(i) - c(i-1)$, and the segment C_i' begins after the segment D_i'. By (d1), the segment D_{i+1}' begins at $h_\triangle(i+1) - d(i+1) = h_\triangledown(i) - d(i)$; since $d(i) < c(i)$, $h_\triangledown(i) - c(i) < h_\triangledown(i) - d(i)$, and the segment D_{i+1}' begins after the segment C_i'.

The length of the segment D_i'' does not depend on n_i, m_i. Indeed, the length of D_i'' is $h_\triangledown(i) - c(i) = h_\triangle(i) - c(i-1)$ by (4). However both $h_\triangle(i)$ and $c(i-1)$ are defined before n_i, m_i need to be defined. Likewise, the length of C_i'' equals $h_\triangle(i+1) - d(i+1) = h_\triangledown(i) - d(i)$ (see (3)). Since both $h_\triangledown(i), d(i)$ are defined before m_i, n_{i+1} need to be defined, the length of the segment C_i'' does not depend on m_i and n_{i+1}.

Lemma 5.3. *Suppose that $\mathcal{T} = n_1 < m_1 < \ldots$ is such that $n_i > 9h_\triangle(i)$ and $m_i > 9h_\triangledown(i)$. Then the corresponding map f is TCE.*

Proof: By Lemma 5.1, we need to show that there exist $\varepsilon > 0, s$, and $\tau < 1$ such that for any N and any critical point u there are more than $\tau(N+1)$ critically ε-shadowed times of order less than s in $[0, N]$ for u. Set $\tau = .4$ and $s = 2$; ε will be chosen later.

The segment D_{i+1}' begins at $h_\triangledown(i) - d(i)$, whereas the segment C_i' ends at $m_i + (h_\triangledown(i) - c(i)) - 1$; since $m_i > 9h_\triangledown(i)$, C_i' ends after D_{i+1}' begins. The segment C_{i+1}' begins at $h_\triangle(i+1) - c(i)$, whereas the segment D_{i+1}' ends at $n_{i+1} + (h_\triangledown(i) - d(i)) - 1$; since $n_{i+1} > 9h_\triangle(i+1)$, D_{i+1}' ends after C_{i+1}' begins. Thus, C_i' ends inside D_{i+1}'. Likewise, D_i' ends inside C_i'. All these segments form a "linked" sequence in which (1) each D'-segment begins and ends inside the appropriate consecutive C' segments, (2) each C'-segment begins and ends inside the appropriate consecutive D'-segments, (3) $D_i'' \subset C_i'$, and (4) $C_i'' \subset D_{i+1}'$

The segment D_i' is at least 9 times longer than any segment $D_q'', q \le i$, (the length of D_i'' is $h_\triangle(i) - c(i-1)$, and the length of D_i' is n_i); D_i' is also at least 9 times longer than any segment $C_q'', q < i$, since all these segments are shorter than $h_\triangle(i)$ by construction. Similarly, the segment C_i' is at least 9 times longer than any C''-segment before it *and* the segment C_i'' (the length of C_i'' is $h_\triangledown(i) - d(i)$ and the length of C_i' is m_i); C_i' is also at least 9 times longer than any segment $D_q'', q \leq i$, since all these segments are shorter than $h_\triangledown(i)$ by construction.

It is easy to check that the construction and the choice of the constants imply the following. Let $u = C$ or $u = D$. Each D''-segment begins when the image of D is to the right of C, close to C, and ends when the image of D is to the right of C, close to a preimage of b not equal to b. Each C''-segment begins when the image of C is to the right of C, close to D, and ends when the image of C is to the right of C, close to a preimage of b not equal to b.

Within segments D_i' and C_i', critical points are repelled from b while staying to the left of C. In the beginning of a segment, the appropriate image of a critical point is close to b, whereas at the first step after the end of a segment, it maps very close to either C or D. Hence there exists an $\varepsilon > 0$ such that within any segment D_i', C_i' the images of critical points are more than 3ε-distant from the closure of the component of $J \setminus \{C\}$ located to the right of C, and in particular from both critical points. Assume also that 3ε is less than the distance between any two points from the set $\{C, D, f(C), f(D)\}$. This completes the choice of constants.

We consider the critical point D and show that all times in the subsegment $E_i = [h_\triangle(i) - d(i) + n_{i-1}, c(i) - d(i) - 1]$ of $D_i' = [h_\triangle(i) - d(i), c(i) - d(i) - 1]$ are critically ε-shadowed of order at most 2. One such shadowing is trivial — the point D shadows itself. Let us show that there is no more than 1 *non-trivial* shadowing for the times described above. Choose a $t \in E_i$. Suppose that for some q and a critical point u the $[q, t]$-segment of orb(D) is shadowed by the $[0, t - q]$-segment of u. Then $f^q(D)$ is ε-close to u. Hence $1 \leq q < h_\triangle(i) - d(i)$ by our choice of ε.

Thus, u stays to the left of C for $t - [h_\triangle(i) - d(i)] + 1 > n_{i-1}$ consecutive iterations of f as it shadows $f^{h_\triangle(i) - d(i)}(D), \ldots, f^t(D)$ within the $[h_\triangle(i) - d(i) - q, t - q]$-segment Q of its orbit. The segment Q begins before the segment D_i', consists of images of u located to the left of C, and is at least $n_{i-1} + 1$ long. Hence it must be contained in a segment of one of the four types of length at least $n_{i-1} + 1$ listed above. There is only one such segment, namely the C_{i-1}'-segment of the orbit of C, and so $u = C$ and $Q \subset C_{i-1}'$.

Let us show that $q = c(i-1) - d(i-1)$ coincides with the beginning of D_{i-1}''. If $q < c(i-1) - d(i-1)$, then as the orbit of C ε-shadows the orbit of $f^q(D)$, an iteration of C from the C_{i-1}'-segment of the orbit of C will correspond to the last iteration of D in the segment D_{i-1}'', which is impossible, since this image of D is to the right of C and is therefore more than ε-distant from any image of C from C_{i-1}'. On the other hand, if

$q > c(i-1) - d(i-1)$, then as the orbit of C ε-shadows the orbit of $f^q(D)$, the last iteration of C in the segment C''_{i-2} of the orbit of C will correspond to an iteration of D from D'_i, which is a contradiction, since this iteration of C is to the right of C and is therefore more than ε-distant from any image of D from D'_i. Thus, the only non-trivial critical ε-shadowing which may take place for a time $t \in E_i$ is by the orbit of C, which ε-shadows the $[f^{c(i-1)-d(i-1)}, t]$-segment of the orbit of D, and so any $t \in E_i$ is critically ε-shadowed of order at most 2.

Let us estimate which part of any segment $[0, N]$ is occupied by the times that are critically ε-shadowed of order at most 2 for D. Assume that N belongs to $F_i = [h_\triangle(i) - d(i) + n_{i-1}, h_\triangle(i+1) - d(i+1) + n_i - 1]$ for some i. The segment E_i lies in the beginning of F_i and forms a significant portion of F_i. Indeed, $n_{i-1} < h_\triangle(i) < 9h_\triangle(i) < n_i$. Hence $|E_i| > \frac{8}{9}n_i$. After E_i, the segment $D''_i \subset F_i$ follows, and by (d2), we have $|D''_i| < h_\triangle(i) < \frac{n_i}{9}$. Finally, the last part of F_i is occupied by $n_i - 1$ initial times from D'_{i+1}. Hence $\frac{|E_i|}{|F_i|} > \frac{4}{9}$, which implies that the times that are critically ε-shadowed of order at most 2 for D form at least $\frac{4}{9}$ of the entire number of times in $[0, N]$. Similar arguments show that the times that are critically ε-shadowed of order at most 2 for C form at least $\frac{4}{9}$ of the entire number of times in $[0, N]$. By Lemma 5.1, this implies that f is TCE, as desired. $\square$

So far, we have dealt with the dynamics of induced maps $f = f_\sim$ of laminations $\sim$. However our goal is to establish corresponding facts concerning polynomials. To "translate" our results from the language of induced maps of laminations into that of polynomials, we need an important result of Kiwi [Ki04, Ki05]. In Section 3, we defined the family $\mathcal{Y}_d$ of collections of σ_d-critical chords whose endpoints have non-preperiodic itineraries and the corresponding family $\mathcal{K}_d$ of laminations whose properties are described in [Ki04, Ki05] (see Theorem 4.3 in Section 3). The following theorem is a version of results of Kiwi [Ki04, Ki05] which is sufficient for our purpose.

Theorem 5.4. *Let $\sim$ be a lamination from $\mathcal{K}_d$. Then there exists a polynomial P of degree d such that its Julia set J_P is a non-separating continuum on the plane and $P|_{J_P}$ is monotonically semiconjugate to $f_\sim|_{J_\sim}$ by a map ψ_P. Moreover, $J_\sim$ is a dendrite, ψ_P-images of critical points of P are critical points of $f_\sim$, ψ_P-preimages of preperiodic points of $f_\sim$ are preperiodic points, and J_P is locally connected at all its preperiodic points.*

We combine Lemma 5.3 and Theorem 5.4 to prove Theorem 1.1.

Proof of Theorem 1.1: Let a sequence $\mathcal{T}$ satisfy the conditions of Lemma 5.3. By Lemma 5.3, the induced map $f_\sim = f$ of the corresponding lamination $\sim$ is TCE. The lamination $\sim$ belongs to $\mathcal{W} \subset \mathcal{K}_3$; hence by Theorem 5.4

there is a polynomial P such that the Julia set J_P is a non-separating continuum on the plane and $P|_{J_P}$ is monotonically semiconjugate to $f|_{J_\sim}$ by a map ψ_P. Let $M \geq 0, L \geq 1, r' > 0$ be constants for which f exhibits the TCE property, i.e., such that for every $x \in J_\sim$ and every positive integer N we have

$$\frac{|\{n \in [0, N] \mid \Delta_f(x, r', n) \leq M\}|}{N + 1} \geq \frac{1}{L}.$$

Clearly, for some $r > 0$ and any point $z \in J_P$, we have $\psi_P(B(z, r)) \subset B(\psi_P(z), r')$. Let $z \in J_P$. To estimate the number of integers $n \in [0, N]$ with $\Delta_P(z, r, n) \leq M$, take $x = \psi_P(z)$. The number of integers $n \in [0, N]$ with $\Delta_f(x, r', n) \leq M$ is at least $(N + 1)/L$. Let n be one such number, and estimate $\Delta_P(z, r, n)$. Observe that if $\mathrm{Comp}_{f^i(x)} f^{-(n-i)}(B(f^n(x), r')) \cap \mathrm{Cr}_f = \emptyset$, then $\mathrm{Comp}_{P^i(z)} P^{-(n-i)}(B(f^n(z), r)) \cap \mathrm{Cr}_P = \emptyset$ because ψ_P maps critical points of P to critical points of f. Hence $\Delta_P(z, r, n) \leq \Delta_f(x, r', n) \leq M$, and there are at least $(N + 1)/L$ numbers $n \in [0, N]$ with $\Delta_P(z, r, n) \leq M$. Thus, P is TCE, and by Proposition 5.2 [Pr00] (cf. [GS98, PR98]), it follows that the Julia set of P is Hölder and hence locally connected.

By Carathéodory theory, this means that for any sequence $\mathcal{T}$ satisfying the conditions of Lemma 5.3 and the corresponding lamination $\sim$, there exists a TCE-polynomial P such that J_P and $J_\sim$ are homeomorphic and $P|_{J_P}$ and $f_\sim|_{J_\sim}$ are topologically conjugate. It is easy to see that there are uncountably many sequences $\mathcal{T}$ inductively constructed so that $n_i > 9h_\triangle(i), m_i > 9h_\triangledown(i)$, i.e., so that they satisfy the conditions of Lemma 5.3. This completes the proof of Theorem 1.1.   □

Bibliography

[Bi49]   R. H. Bing, *A convex metric for a locally connected continuum*, Bull. AMS **55** (1949), 812–819.

[Bl05]   A. Blokh, *Necessary conditions for the existence of wandering triangles for cubic laminations*, Disc. Cont. Dyn. Syst. **13** (2005), 13–34.

[BL02]   A. Blokh and G. Levin, *An inequality for laminations, Julia sets and 'growing trees'*, Ergod. Th. Dynam. Syst. **22** (2002), 63–97.

[BO04a]   A. Blokh and L. Oversteegen, *Backward stability for polynomial maps*, Trans. Amer. Math. Soc. **356** (2004), 119–133.

[BO04b] A. Blokh and L. Oversteegen, *Wandering triangles exist*, C. R. Acad. Sci. Paris, Ser. I **339** (2004), 365–370.

[CG93] L. Carleson and T. W. Gamelin, *Complex dynamics*, Universitext: Tracts in Mathematics, Springer-Verlag, New York, 1993.

[Ch07] D. K. Childers, *Wandering polygons and recurrent critical leaves*, Ergod. Th. Dynam. Syst. **27** (2007), 87–107.

[DH84] A. Douady and J. H. Hubbard, *Étude dynamique des polynomes complexes*, Public. Math. d'Orsay (1984), 1–75.

[GS98] J. Graczyk and S. Smirnov, *Collet, Eckmann and Hölder*, Invent. Math. **131** (1998), 69–96.

[Ki02] J. Kiwi, *Wandering orbit portraits*, Trans. Amer. Math. Soc. **354** (2002), 1473–1485.

[Ki04] J. Kiwi, *Real laminations and the topological dynamics of complex polynomials*, Adv. Math. **184** (2004), 207–267.

[Ki05] J. Kiwi, *Combinatorial continuity in complex polynomial dynamics*, Proc. Lond. Math. Soc. **91** (2005), 215–248.

[Le98] G. Levin, *On backward stability of holomorphic dynamical systems*, Fund. Math. **158** (1998), 97–107.

[MMOT92] J. C. Mayer, L. Mosher, L. G. Oversteegen, and E. D. Tymchatyn, *Characterization of separable metric $\mathbb{R}$-trees*, Proc. Amer. Math. Soc. **115** (1992), 257–264.

[OR80] L. G. Oversteegen and J. T. Rogers, Jr., *An inverse limit description of an atriodic tree-like continuum and an induced map without the fixed point property*, Houston J. Math. **6** (1980), 549–564.

[Pr00] F. Przytycki, *Hölder implies Collet-Eckmann*, Asterisque **261** (2000), 385–403.

[PRLS03] F. Przytycki, J. Rivera-Letelier, and S. Smirnov, *Equivalence and topological invariance of conditions for non-uniform hyperbolicity in the iteration of rational maps*, Invent. Math. **151** (2003), 29–63.

[PR98] F. Przytycki and S. Rohde, *Porosity of Collet–Eckmann Julia sets*, Fund. Math. **155** (1998), 189–199.

[Sm00]   S. Smirnov, *Symbolic dynamics and Collet-Eckmann conditions*, Internat. Math. Res. Notices **7** (2000), 333–351.

[Th]   W. P. Thurston, "On the geometry and dynamics of iterated rational maps", in: *Complex Dynamics: Families and Friends* (this volume), Chapter 1. A K Peters, Wellesley, MA, 2009, 3–137.

3 Combinatorics of Polynomial Iterations

Volodymyr Nekrashevych

1 Introduction

The topics of these notes are group-theoretical and combinatorial aspects of iteration of post-critically finite polynomials, including topological polynomials and post-critically finite non-autonomous backward iterations.

As the main objects encoding the combinatorics of iterations we use the associated iterated monodromy groups and permutational bimodules. These algebraic structures encode in a condensed form all topological information about the corresponding dynamical systems. For more on iterated monodromy groups and their applications in symbolic dynamics see [Ne05, Ne08, BGN03].

In the context of polynomial iterations the iterated monodromy groups are analogs or generalizations of the classical tools of symbolic dynamics of quadratic and higher-degree polynomials: kneading sequences, internal addresses, Hubbard trees, critical portraits etc. In some cases the transition from the classical objects to iterated monodromy groups and permutational bimodules is very straightforward, but in some cases it is more involved. For more on symbolic dynamics of post-critical polynomials, see the works [BFH92, BS02, Ke00, Po93a, Po93b]. For relations of kneading sequences and iterated monodromy groups of quadratic polynomials, see [BN08].

We answer some basic questions about iterated monodromy groups of polynomials and formulate some problems for further investigations. This area is fresh and many questions are open, even though they might be not so hard to answer.

We give in our paper a complete description of the iterated monodromy groups of *post-critically finite backward iterations* of topological polynomials. Here a post-critically finite backward iteration is a sequence $f_1, f_2, \ldots$ of complex polynomials (or orientation-preserving branched coverings of planes) such that there exists a finite set P such that all critical values of $f_1 \circ f_2 \circ \cdots \circ f_n$ belong to P for every n.

The *iterated monodromy group* of such a sequence is the automorphism group of the *tree of preimages*

$$T_t = \bigsqcup_{n \geq 0} (f_1 \circ f_2 \circ \cdots \circ f_n)^{-1}(t),$$

induced by the monodromy actions of the fundamental group $\pi_1(\mathbb{C} \setminus P, t)$. Here t is an arbitrary basepoint.

We prove that a group of automorphisms of a rooted tree is the iterated monodromy group of a backward polynomial iteration if and only if it is generated by a set of automorphisms satisfying a simple planarity condition.

Namely, if $A \subset \mathfrak{S}(\mathsf{X})$ is a set of permutations, then its *cycle diagram* $D(A)$ is the oriented 2-dimensional CW-complex in which for every cycle of each permutation we have a 2-cell such that the corresponding cycle is read on the boundary of the cell along the orientation. Two cycles of different permutations are not allowed to have common edges in $D(A)$.

We say that a set $A \subset \mathfrak{S}(\mathsf{X})$ is *dendroid* (*tree-like* in [Ne05]) if the diagram $D(A)$ is contractible. A set of automorphisms $A = \{a_1, a_2, \ldots, a_n\}$ of a rooted tree T is said to be *dendroid* if A acts on every level of the tree T as a dendroid set of permutations.

The main result of Section 5 is the following complete description of the iterated monodromy groups of polynomial iterations (given in Propositions 5.2 and 5.3, which contain more details).

Theorem 1.1. *An automorphism group of a rooted tree T is an iterated monodromy group of a post-critically finite backward iteration of polynomials if and only if it is generated by a dendroid set of automorphisms of T.*

Very little is known about the class of groups generated by dendroid sets of automorphisms. This class contains many interesting examples of groups (especially in relation with questions of amenability and growth of groups see [GZ02, BV05, BP06, Ne07, BKN08]), and further study of such groups is of great interest for group theory and dynamics.

We also give in our paper a general recipe for construction of dendroid sets of automorphisms of rooted trees using automata. We define a class of *dendroid automata* and prove in Theorem 4.8 that dendroid sets of automorphisms of a rooted tree are defined using sequences of dendroid automata. General definitions of automata are given in Subsection 3.5; dendroid automata are described in Definition 4.6. See the paper [Ne07], where three dendroid automata were used to construct an uncountable set of groups with unusual properties.

We will see that combinatorics of post-critically finite backward iterations is described by the corresponding sequences of dendroid automata and that composition of polynomials corresponds to a certain composition of automata.

It is natural in the case of iterations of a single post-critically finite polynomial f to ask if it is possible to describe the iterated monodromy group by one automaton of some special kind (i.e., by a special constant sequence of automata). It was shown in [Ne05] that for any post-critically finite polynomial f there exists n such that the iterated monodromy group *of the nth iteration* of f is described by a particularly simple *kneading automaton* (one can take $n = 1$ if f is hyperbolic). The kneading automaton is found using the technique of *external angles* to the Julia set, i.e., using analytic techniques.

The last section of the paper deals with the problem of finding the kneading automaton associated with a post-critically finite polynomial using purely topological information and group-theoretical techniques. We propose an iterative algorithm, which seems to work in many cases, which we call a *combinatorial spider algorithm*. It takes as input the dendroid automaton associated with the polynomial (called *twisted kneading automaton*), which can be easily found from the action of the topological polynomial on the fundamental group of the punctured plane, and simplifies it until we get the associated kneading automaton, or a twisted kneading automaton close to it. The associated automaton together with a cyclic ordering of its states uniquely determines the Thurston combinatorial class of the polynomial (hence determines the complex polynomial uniquely, up to an affine conjugation), see Proposition 6.2.

The algorithm seems to work in most examples, but the general question of convergence of the combinatorial spider algorithm is still not very clear. We show that this algorithm is equivalent to some simple computations in a permutational bimodule over a subgroup of the outer automorphisms of the free group. Convergence problem of the combinatorial spider algorithm is closely related to the (sub)-hyperbolicity of this bimodule, which is also open.

The structure of the paper is as follows. Section 2 studies elementary properties of dendroid subsets of the symmetric group. Section 3 is an overview of notions and basic results of the theory of groups acting on rooted trees, including permutational bimodules, wreath recursions and automata. The techniques and language of this section are used throughout the paper.

Section 4 gives a complete description of dendroid sets of automorphisms of a rooted tree. We introduce the notion of a dendroid automaton (Definition 4.6) and prove that dendroid sets of automorphisms of a rooted tree are exactly automorphisms defined by sequences of dendroid automata (Theorem 4.8).

In Section 5 we apply the developed techniques and give a complete description of iterated monodromy groups of post-critically finite backward polynomial iterations, as described in Theorem 1.1 above. In particular, we give an explicit description of the associated sequence of dendroid automata (Proposition 5.7), and discuss cyclic ordering of states of a dendroid automaton and its relation with topology (Section 5.3).

The last section, "Iterations of a Single Polynomial", deals with iterations of post-critically finite polynomials. We describe the *twisted kneading automata*, which are the automata obtained when applying the general techniques of dendroid automata to the case of iterations of a single polynomial. This automaton, in principle, already describes the iterated monodromy group of the polynomial, but it may be too complicated. So, a natural goal is to transform it to the simplest possible form. This can be done analytically (using external angles and invariant spiders), as in Chapter 6 of [Ne05], but this approach may be not available, if the polynomial is given by purely topological information.

The suggested algorithm for simplification of the twisted kneading automata is described in Section 6.3. The underlying algebraic structure of a permutational bimodule over a subgroup of the outer automorphism group of the free group is described in Section 6.4. In the last subsection we give a criterion of absence of obstructions (i.e., realizability as complex polynomials) for topological polynomials given by kneading automata.

Acknowledgments. The author wishes to thank Dierk Schleicher for encouraging him to write this paper and Laurent Bartholdi and Kevin Pilgrim for useful discussions. This paper is based upon work supported by the National Science Foundation under Grant DMS-0605019.

2   Dendroid Sets of Permutations

Let X be a finite set and denote by $\mathfrak{S}(X)$ the symmetric group of all permutations of X. Let $A = (a_i)_{i \in I}$ be a sequence of elements of $\mathfrak{S}(X)$. Draw an oriented 2-dimensional CW-complex with the set of 0-cells X in which for every cycle $(x_1, x_2, \ldots, x_n)$ of every permutation a_i we have a 2-cell with the vertices $x_1, x_2, \ldots, x_n$ so that their order on the boundary of the cell and in the cycle of the permutation coincide. We label each cycle by the corresponding permutation. Two different 2-cells are not allowed to have common 1-cells.

The constructed CW-complex is called the *cycle diagram* of the sequence A and is denoted $D(A)$. For example, Figure 1 shows the cycle diagram for $X = \{1, 2, 3, 4\}$ and $A = \{(12)(34), (1234), (123)\}$.

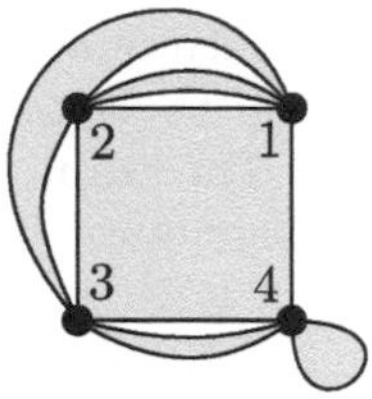

Figure 1. A cycle diagram.

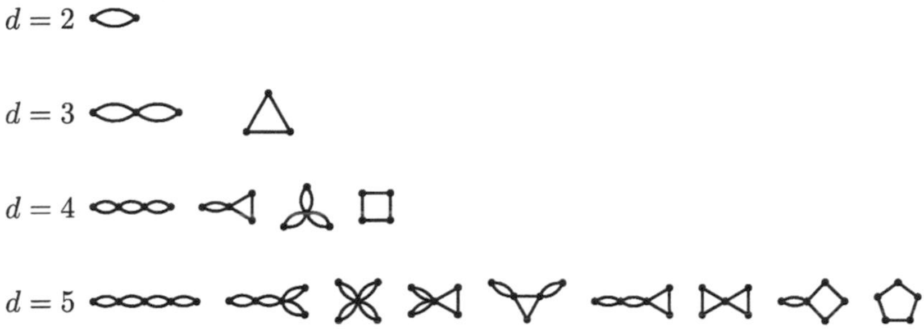

$d = 2$

$d = 3$

$d = 4$

$d = 5$

Figure 2. Dendroid sets of permutations.

Definition 2.1. A sequence A of elements of $\mathfrak{S}(\mathsf{X})$ is said to be *dendroid* if its cycle diagram $D(A)$ is contractible.

Note that if A is dendroid, then only trivial cycles can appear twice as cycles of elements of A. In particular, only the trivial permutation can appear more than once in the sequence A. Moreover, any two cycles of A are either disjoint (hence commute) or have only one common element.

See Figure 2 for all possible types of cycle diagrams of dendroid sets of permutations of X for $2 \leq d = |\mathsf{X}| \leq 5$. We do not show the trivial cycles there.

Proposition 2.2. *Let A be a sequence of elements of $\mathfrak{S}(\mathsf{X})$ generating a transitive subgroup. Denote by N the total number of cycles of the elements of A (including the trivial ones).*

Then the sequence A is dendroid if and only if

$$N - 1 = |\mathsf{X}| \cdot (|A| - 1).$$

Here and later $|A|$ denotes the length of the sequence A (the size of the index set).

Proof: Choose a point in each face of the cycle diagram $D(A)$ and replace each face by a star, i.e., by the graph connecting the chosen point with the vertices of the face. The obtained one-dimensional complex Γ is homotopically equivalent to the diagram $D(A)$, hence the sequence is dendroid if and only if the graph Γ is a tree. The graph Γ has $|\mathsf{X}| + N$ vertices and $|\mathsf{X}| \cdot |A|$ edges (every permutation contributes $|\mathsf{X}|$ edges). It is well known that a graph is a tree if and only if it is connected and the number of vertices minus one is equal to the number of edges. Hence, A is dendroid if and only if $D(A)$ is connected and $|\mathsf{X}| + N - 1 = |\mathsf{X}| \cdot |A|$. $\square$

Corollary 2.3. *Let* $A = (a_i \in \mathfrak{S}(\mathsf{X}))_{i \in I}$ *be a dendroid sequence. Suppose that the sequence* $B = (b_i \in \mathfrak{S}(\mathsf{X}))_{i \in I}$ *is such that* b_i *is conjugate to* a_i *or to* a_i^{-1} *in* $\mathfrak{S}(\mathsf{X})$ *for every* $i \in I$ *and that the permutations* b_i *generate a transitive subgroup of* $\mathfrak{S}(\mathsf{X})$. *Then* B *is also dendroid.*

Proof: The numbers of cycles in A and B are the same and $|A| = |B|$. $\square$

Corollary 2.4. *Let* $A = (a_1, a_2, \ldots, a_m)$ *be a dendroid sequence. Then the sequences* $(a_1^{-1}, a_2, \ldots, a_m)$ *and* $(a_1^g, a_2, \ldots, a_m)$ *for* $g \in \langle a_2, \ldots, a_m \rangle$ *are dendroid.*

Proposition 2.5. *Suppose that* $a_1, a_2, \ldots, a_m$ *is a dendroid sequence of elements of* $\mathfrak{S}(\mathsf{X})$. *Then the product* $a_1 a_2 \cdots a_m$ *is a transitive cycle.*

Proof: It is sufficient to prove the proposition for the case when each a_i is a cycle, since we can replace a_i by the sequence of its cycles without changing the cycle diagram.

Suppose that two cycles $(x_1, x_2, \ldots, x_n) \cdot (y_1, y_2, \ldots, y_m)$ have only one common point, i.e., $|\{x_1, x_2, \ldots, x_m\} \cap \{y_1, y_2, \ldots, y_m\}| = 1$. Without loss of generality we may assume that $y_m = x_1$. Then

$$(y_m, x_2, \ldots, x_n) \cdot (y_1, y_2, \ldots, y_m) = (y_1, y_2, \ldots y_{m-1}, x_2, \ldots, x_n, y_m).$$

Thus, the product of two cycles having one common point is a cycle involving the union of the points moved by the cycle. This finishes the proof, since any two cycles in a dendroid sequence of permutations have at most one common point. If two cycles are disjoint, then they commute; if we replace two cycles having a common point by their product, which is a cycle on their union, then we will not change the homotopy type of the cycle diagram. $\square$

Corollary 2.6. *Suppose that* $A = (a_1, a_2, \ldots, a_m)$ *is a dendroid sequence and let*

$$(a_{i_{1,1}}, \ldots a_{i_{1,k_1}}), \quad (a_{i_{2,1}}, \ldots, a_{i_{2,k_2}}), \quad \ldots, \quad (a_{i_{l,1}}, \ldots, a_{i_{l,k_l}})$$

be any partition of the sequence A into disjoint sub-sequences. Then the sequence of products

$$(a_{i_{1,1}} \cdots a_{i_{1,k_1}}, \quad a_{i_{2,1}} \cdots a_{i_{2,k_2}}, \quad \ldots, \quad a_{i_{l,1}} \cdots a_{i_{l,k_l}})$$

is dendroid.

Proof: By Proposition 2.5 the sets of vertices of the connected components of the cycle diagram of $(a_{i_{j,1}}, \ldots, a_{i_{j,k_j}})$ are exactly the sets of vertices of the cycles of the product $a_{i_{j,1}} \cdots a_{i_{j,k_j}}$, hence the cycle diagrams of A and of the sequence of products are homotopically equivalent. $\qquad\square$

3 Automorphisms of Rooted Trees and Bimodules

3.1 Rooted Trees

A *rooted tree* is a tree T with a fixed vertex, called the *root* of the tree. We consider only locally finite trees.

The *level number* n (or the *nth level*) of a rooted tree T is the set of vertices on distance n from the root. The set of vertices of T is then a disjoint union of its levels $L_0, L_1, \ldots$, where the 0th level contains only the root. Two vertices may be connected by an edge only if they belong to consecutive levels.

We say that a vertex u is *below* a vertex v if the path from the root to u passes through v. The set of vertices below v together with v as a root form a *rooted subtree* T_v.

An *automorphism* of a rooted tree T is an automorphism of the tree T fixing the root. Every automorphism of a rooted tree preserves the levels.

A group acting on a rooted tree is said to be *level transitive* if it is transitive on the levels of the tree. A *level-homogeneous rooted tree* is a rooted tree admitting a level transitive automorphism group.

A level-homogeneous rooted tree is uniquely determined, up to an isomorphism of rooted trees, by its *spherical index* $(d_1, d_2, \ldots)$, where d_1 is degree of the root and $d_k + 1$ is the degree of a vertex of the kth level for $k \geq 2$. In other words, d_k is the number of vertices of the kth level adjacent to a common vertex of the $(k-1)$st level.

For a given sequence $\kappa = (d_1, d_2, \ldots)$ a rooted tree with spherical index κ can be constructed as a *tree of words* in the following way. Choose a sequence of finite sets (*alphabets*) $\mathsf{X} = (\mathsf{X}_1, \mathsf{X}_2, \ldots)$ such that $|\mathsf{X}_k| = d_k$. The nth level of the tree of words X^* is equal to

$$\mathsf{X}^n = \mathsf{X}_1 \times \mathsf{X}_2 \times \cdots \times \mathsf{X}_n = \{x_1 x_2 \ldots x_n \ : \ x_k \in \mathsf{X}_k\},$$

so that $X^* = \bigsqcup_{k \geq 0} X^k$. Here X^0 consists of a single *empty word* $\varnothing$, which will be the root of the tree X^*.

We connect two words $v \in X^n$ and $u \in X^{n+1}$ if and only if $u = vx$ for some $x \in X_{n+1}$.

If X is a finite alphabet, then the *regular tree* X^* is defined as above for the constant sequence $(X, X, \ldots)$, i.e., X^* is the set of all finite words over the alphabet X (in other terms, it is the free monoid generated by X).

For an arbitrary rooted tree T the *boundary* ∂T is the set of simple infinite paths in T starting in the root. We denote by ∂T_v the subset of paths passing through a given vertex v. The set of subsets ∂T_v for all vertices v of T is a basis of a natural topology on ∂T.

If T is a tree of words X^* over a sequence of alphabets $(X_1, X_2, \ldots)$, then the boundary ∂X^* is naturally identified with the direct product

$$X^\omega = X_1 \times X_2 \times \cdots .$$

For more on groups acting on rooted trees and related notions, see the papers [GNS00, Si98, Ne05].

3.2   Permutational Bimodules

Let G be a level-transitive automorphism group of a rooted tree T. Let v be a vertex of the nth level L_n of T and let G_v be the stabilizer of v in G. The subtree T_v is invariant under G_v. Denote by $G|_v$ the automorphism group of T_v equal to the restriction of the action of G_v onto T_v.

Since we assume that G is level transitive, the conjugacy class of G_v in G, and hence the isomorphism class of the action of $G|_v$ on T_v depend only on the level of the vertex v and do not depend on the choice of v.

Denote by $\mathfrak{M}_{\varnothing, v}$ the set of isomorphisms $\phi : T_v \longrightarrow T_u$ for $u \in L_n$ induced by the action of an element $g \in G$ (different u give different elements of $\mathfrak{M}_{\varnothing, v}$).

Note that for every $\phi \in \mathfrak{M}_{\varnothing, v}$ and $g \in G|_v$ the composition $\phi \cdot g$ belongs to $\mathfrak{M}_v$, since g is an automorphism of T_v induced by the action of an element of G.

Similarly, for every $g \in G$ the composition $g \cdot \phi$ (restricted to T_v) is an element of $\mathfrak{M}_{\varnothing, v}$. It is easy to see that in this way we get a right action of $G|_v$ and a left action of G on $\mathfrak{M}_{\varnothing, v}$ and that these actions commute.

Note that the right action of $G|_v$ on $\mathfrak{M}_{\varnothing, v}$ is *free*, i.e., that $\phi \cdot g = \phi$ implies that g is trivial, and that two elements $\phi_1 : T_v \longrightarrow T_{u_1}$ and $\phi_2 : T_v \longrightarrow T_{u_2}$ belong to one $G|_v$-orbit if and only if $u_1 = u_2$.

Let us formalize the obtained structure in the following definition.

Definition 3.1. Let G and H be groups. A *permutational $(G-H)$-bimodule* is a set $\mathfrak{M}$ with a left action of G and a right action of H, which commute.

More explicitly, we have maps $G \times \mathfrak{M} \longrightarrow \mathfrak{M} : (g, x) \mapsto g \cdot x$ and $\mathfrak{M} \times H \longrightarrow \mathfrak{M} : (x, h) \mapsto x \cdot h$ satisfying the following conditions:

(1) $1 \cdot x = x \cdot 1 = x$ for all $x \in \mathfrak{M}$;

(2) $g_1 \cdot (g_2 \cdot x) = (g_1 g_2) \cdot x$ and $(x \cdot h_1) \cdot h_2 = x \cdot (h_1 h_2)$ for all $x \in \mathfrak{M}$, $g_1, g_2 \in G$ and $h_1, h_2 \in H$;

(3) $(g \cdot x) \cdot h = g \cdot (x \cdot h)$ for all $x \in \mathfrak{M}$, $g \in G$ and $h \in H$.

A *(finite) covering* bimodule is a bimodule in which the right group action is free and has a finite number of orbits.

We have seen that $\mathfrak{M}_{\varnothing, v}$ is a *covering permutational* $(G - G|_v)$-*bimodule*. In general, if v_1, v_2 are vertices of T such that v_2 is below v_1, then $\mathfrak{M}_{v_1, v_2}$ is the $(G|_{v_1} - G|_{v_2})$-bimodule consisting of the isomorphisms $T_{v_2} \longrightarrow T_u$ induced by elements of $G|_{v_1}$ (or, equivalently, of G_{v_1}).

We will say that $\mathfrak{M}$ is a *G-bimodule* if it is a $(G - G)$-bimodule.

Definition 3.2. Two $(G - H)$-bimodules $\mathfrak{M}_1$ and $\mathfrak{M}_2$ are *isomorphic* if there exists a bijection $F : \mathfrak{M}_1 \longrightarrow \mathfrak{M}_2$ which agrees with the actions, i.e., such that

$$F(g \cdot x \cdot h) = g \cdot F(x) \cdot h$$

for all $g \in G$, $x \in \mathfrak{M}$ and $h \in H$.

3.3   Tensor Products and Bases

Definition 3.3. Let $\mathfrak{M}_1$ and $\mathfrak{M}_2$ be a $(G_1 - G_2)$-bimodule and a $(G_2 - G_3)$-bimodule, respectively. Then the *tensor product* $\mathfrak{M}_1 \otimes \mathfrak{M}_2 = \mathfrak{M}_1 \otimes_{G_2} \mathfrak{M}_2$ is the $(G_1 - G_3)$-bimodule equal as a set to the quotient of $\mathfrak{M}_1 \times \mathfrak{M}_2$ by the identification

$$x_1 \otimes g \cdot x_2 = x_1 \cdot g \otimes x_2$$

for $x_1 \in \mathfrak{M}_1$, $x_2 \in \mathfrak{M}_2$, $g \in G_2$. The actions are defined by the rule

$$g_1 \cdot (x_1 \otimes x_2) \cdot g_3 = (g_1 \cdot x_1) \otimes (x_2 \cdot g_3)$$

for $g_i \in G_i$ and $x_i \in \mathfrak{M}_i$.

It is an easy exercise to prove that $\mathfrak{M}_1 \otimes \mathfrak{M}_2$ is a well-defined $(G_1 - G_3)$-bimodule. Moreover, if $\mathfrak{M}_1$ and $\mathfrak{M}_2$ are covering bimodules, then $\mathfrak{M}_1 \otimes \mathfrak{M}_2$ is also a covering bimodule. It is also not hard to prove that tensor product of bimodules is an associative operation, i.e., that $(\mathfrak{M}_1 \otimes \mathfrak{M}_2) \otimes \mathfrak{M}_3$ is isomorphic to $\mathfrak{M}_1 \otimes (\mathfrak{M}_2 \otimes \mathfrak{M}_3)$, where the isomorphism is induced by the natural identification of the direct products of sets.

Proposition 3.4. *Let v_1 and v_2 be vertices of T such that v_2 is below v_1. Then the bimodule $\mathfrak{M}_{\varnothing,v_2}$ is isomorphic to $\mathfrak{M}_{\varnothing,v_1} \otimes \mathfrak{M}_{v_1,v_2}$.*

Proof: Let $\phi_1 : T_{v_1} \longrightarrow T_{u_1}$ and $\phi_2 : T_{v_2} \longrightarrow T_{u_2}$ be elements of $\mathfrak{M}_{\varnothing,v_1}$ and $\mathfrak{M}_{v_1,v_2}$. Then u_2 is below v_1, since ϕ_2 is restriction of an element of G_{v_1} onto $T_{v_2} \subset T_{v_1}$. Define

$$F(\phi_1 \otimes \phi_2) = \phi_1 \circ \phi_2 : T_{v_2} \longrightarrow T_{\phi_1(u_2)}.$$

Since both ϕ_1 and ϕ_2 are restrictions of elements of G, the isomorphism $F(\phi_1 \otimes \phi_2)$ belongs to $\mathfrak{M}_{\varnothing,v_2}$. We leave to the reader to prove that F is a well-defined bijection preserving the actions. $\square$

Corollary 3.5. *Let $\varnothing, v_1, v_2, \ldots$ be a path in the tree T starting at the root, such that v_n belongs to the nth level of T. Denote $G_n = G|_{v_n}$ (called the nth upper companion group in [Gr00]). Then the $(G - G_n)$-bimodule $\mathfrak{M}_{\varnothing,v_n}$ is isomorphic to the tensor product*

$$\mathfrak{M}_{\varnothing,v_1} \otimes \mathfrak{M}_{v_1,v_2} \otimes \cdots \otimes \mathfrak{M}_{v_{n-1},v_n}.$$

In particular, the action of G on the nth level of the tree T is conjugate with the action of G on the right G_n-orbits of this tensor product.

Covering bimodules can be encoded symbolically using the notion of a basis of the bimodule.

Definition 3.6. Let $\mathfrak{M}$ be a covering $(G - H)$-bimodule. A *basis* of $\mathfrak{M}$ is an orbit transversal for the right H-action, i.e., a set $\mathsf{X} \subset \mathfrak{M}$ such that every H-orbit of $\mathfrak{M}$ contains exactly one element of X.

If X is a basis of $\mathfrak{M}$, then every element of $\mathfrak{M}$ can be written uniquely as $x \cdot h$ for some $x \in \mathsf{X}$ and $h \in H$.

Then for every $g \in G$ and $x \in \mathsf{X}$ there exists a unique pair $y \in \mathsf{X}$, $h \in H$ such that

$$g \cdot x = y \cdot h.$$

If X_1 and X_2 are bases of the bimodules $\mathfrak{M}_1$ and $\mathfrak{M}_2$, respectively, then the set $\mathsf{X}_1 \otimes \mathsf{X}_2 = \{x_1 \otimes x_2 : x_1 \in \mathsf{X}_1, x_2 \in \mathsf{X}_2\}$ is a basis of the bimodule $\mathfrak{M}_1 \otimes \mathfrak{M}_2$ (see Proposition 2.3.2 of [Ne05]).

By induction, if $\mathfrak{M}_i$ for $i = 1, 2, \ldots$ is a $(G_{i-1} - G_i)$-bimodule, and X_i is its basis, then

$$\mathsf{X}_1 \otimes \mathsf{X}_2 \otimes \cdots \otimes \mathsf{X}_n \subset \mathfrak{M}_1 \otimes \mathfrak{M}_2 \otimes \cdots \otimes \mathfrak{M}_n$$

is also a basis of the $(G_0 - G_n)$-bimodule $\mathfrak{M}_1 \otimes \cdots \otimes \mathfrak{M}_n$.

If the sequence $X = (X_1, X_2, \ldots)$ of bases is fixed, then we denote by X^n the basis $X_1 \otimes \cdots \otimes X_n$ and by X^* the disjoint union of X^n for $n \geq 0$. Here X^0 contains only one element $\varnothing$ with the property $\varnothing \otimes v = v$ for all $v \in X^*$. We will often write elements $x_1 \otimes x_2 \otimes \cdots \otimes x_n \in X^n$ just as words $x_1 x_2 \ldots x_n$.

Then every element of the bimodule $\mathfrak{M}_1 \otimes \cdots \otimes \mathfrak{M}_n$ can be uniquely written in the form $v \cdot h$ for $v \in X^n$ and $h \in G_n$. In particular, for any $v \in X^n$ and every $g \in G_0$ there exist a unique $u \in X^n$ and $h \in G_n$ such that

$$g \cdot v = u \cdot h.$$

We denote $u = g(v)$ and $h = g|_v$. We have then the following properties, which are easy corollaries of the definitions:

$$g_1(g_2(u)) = (g_1 g_2)(u), \quad 1(u) = u,$$
$$g(u \otimes v) = g(u) \otimes g|_u(v).$$

Consequently, we get in this way a natural action of G_0 on the rooted tree X^*. The natural action of G_0 on X^* does not depend, up to a conjugacy of the actions, on the choice of the bases X_i. The most direct way to see this is to note that this action coincides with the natural action of G_0 on the set $\bigsqcup_{n \geq 0} \mathfrak{M}_1 \otimes \cdots \otimes \mathfrak{M}_n / G_n$ of the right orbits of the tensor product bimodules. An orbit is mapped by the conjugacy to the unique element of X^* contained in it. For more on the tree of right orbits see [Ne08].

3.4 Wreath Recursions

Let $\mathfrak{M}$ be a covering $(G - H)$-bimodule and let X be its basis. Then for every $g \in G$ and $x \in X$ there exist unique elements $h \in H$ and $y \in X$ such that

$$g \cdot x = y \cdot h.$$

Recall that we denote $y = g(x)$ and $h = g|_x$.

For a fixed $g \in G$ we get then a permutation $\sigma_g : x \mapsto g(x)$ of X induced by g and a sequence of *sections* $(g|_x)_{x \in X}$. For $g_1, g_2 \in G$ we have $\sigma_{g_1 g_2} = \sigma_{g_1} \sigma_{g_2}$ and

$$(g_1 g_2)|_x = g_1|_{\sigma_{g_2}(x)} g_2|_x.$$

Hence we get a homomorphism from G to the wreath product $\mathfrak{S}(X) \wr H = \mathfrak{S}(X) \ltimes H^X$:

$$\psi : g \mapsto \sigma_g (g|_x)_{x \in X},$$

since the elements of the wreath product are multiplied by the rule

$$\sigma_1 (g_x)_{x \in X} \cdot \sigma_2 (h_x)_{x \in X} = \sigma_1 \sigma_2 (g_{\sigma_2(x)} h_x)_{x \in X}.$$

We call the homomorphism ψ the *wreath recursion* associated with the bimodule $\mathfrak{M}$ and the basis X.

We will usually order the basis $\mathsf{X} = \{x_1, x_2, \ldots, x_d\}$ and write the elements of the wreath product as sequences

$$\sigma(g_x)_{x \in \mathsf{X}} = \sigma(g_{x_1}, g_{x_2}, \ldots, g_{x_d}),$$

thus implicitly identifying the wreath product $\mathfrak{S}(\mathsf{X}) \wr H$ with $\mathfrak{S}(d) \wr H$ (where $\mathfrak{S}(d) = \mathfrak{S}(\{1, 2, \ldots, d\})$).

If we change the basis X, then we compose the wreath recursion with an inner automorphism of the group $\mathfrak{S}(d) \wr H$. More explicitly, if $\mathsf{Y} = \{y_1, \ldots, y_d\}$ is another basis of $\mathfrak{M}$, then $y_i = x_{\pi(i)} \cdot h_i$ for some $\pi \in \mathfrak{S}(d)$ and $h_i \in H$. Then the wreath recursions ψ_X and ψ_Y are related by

$$\psi_\mathsf{Y}(g) = \alpha^{-1} \psi_\mathsf{X}(g) \alpha,$$

where $\alpha = \pi(h_1, h_2, \ldots, h_d)$, see [Ne05, Proposition 2.3.4] and [Ne08, Proposition 2.12].

In particular, conjugation of all coordinates of H^X in the wreath product by an element of the group H is equivalent to a change of the basis of the bimodule.

If G is a finitely generated group, then the wreath recursion, and hence the bimodule, are determined by a finite number of equations of the form

$$g_i = \sigma_i(h_{1,i}, h_{2,i}, \ldots, h_{d,i}),$$

where $\{g_i\}$ is a finite generating set of G and $h_{j,i}$ are elements of H. Equations of this form is the main computational tool in the study of coverings using iterated monodromy groups.

3.5   Automata

Permutation bimodules and wreath recursions can be encoded by *automata*. We interpret equalities

$$g \cdot x = y \cdot h$$

in a bimodule as a work of an automaton, which being in a state g and reading an input letter x, gives on the output the letter y and goes to the state h, ready to process further letters.

More formally, we adopt the following definition.

Definition 3.7. An *automaton* over the alphabet X is a set of *internal states* A together with a map

$$\tau : A \times \mathsf{X} \longrightarrow \mathsf{X} \times A.$$

For $a \in A$ and $x \in X$, the first and second coordinates of $\tau(a, x)$ as functions from $A \times X$ to X and A are called the *output* and the *transition* functions, respectively.

If $\mathfrak{M}$ is a covering G-bimodule and X is a basis of $\mathfrak{M}$, then the *associated automaton* $\mathcal{A}(G, X, \mathfrak{M})$ is the automaton with the set of internal states G, defined by

$$\tau(g, x) = (y, h), \text{ iff } g \cdot x = y \cdot h,$$

i.e., by

$$\tau(g, x) = (g(x), g|_x).$$

We will use similar notation for all automata, so that for a state q and a letter x we have

$$\tau(q, x) = (q(x), q|_x),$$

i.e., $q(x)$ denotes the value of the output function and $q|_x$ denotes the value of the transition function.

We will also usually write

$$g \cdot x = y \cdot h$$

instead of

$$\tau(g, x) = (y, h).$$

An automaton is conveniently described by its *Moore diagram*. It is an oriented graph with the set of vertices equal to the set A of internal states of the automaton. For every pair $q \in A$ and $x \in X$ there is an arrow going from q to $q|_x$ labeled by the pair $(x, q(x))$.

We will need to deal also with the $(G-H)$-bimodules for different groups G and H. Therefore, we will also use the following generalized notion.

Definition 3.8. An *automaton* over alphabet X is given by its *input set* A_1, *output set* A_2 and a map

$$\tau : A_1 \times X \longrightarrow X \times A_2.$$

We also use the notation $q(x)$ for the first coordinate of $\tau(q, x)$ and $q|_x$ for the second coordinate of $\tau(q, x)$.

Definition 3.9. An automaton is called a *group automaton* if for every element a of the input set the transformation $x \mapsto a(x)$ of the alphabet is a permutation.

We assume implicitly that the input and the output sets of a group automaton contain special *trivial states* denoted 1 with the property that

$$1 \cdot x = x \cdot 1$$

for all letters x of the alphabet.

Perhaps it would be less confusing to dualize Definition 3.8 and say that X, A_1 and A_2 are the set of internal states, input and output alphabets, respectively. But since we think of the groups acting on words, and not words acting on groups, we stick to the terminology of Definition 3.8.

Moore diagrams is not an appropriate way of describing such generalized automata. Therefore, we will usually describe them using *dual Moore diagrams*. It is a directed graph with the set of vertices X in which for every $x \in X$ and q in the input set there is an arrow from x to $q(x)$ labeled by $(q, q|_x)$.

Products of automata correspond to tensor products of bimodules and are described in the following way.

Definition 3.10. Let $\mathcal{A}_1$ and $\mathcal{A}_2$ be automata over the alphabets X_1 and X_2, input sets A_1 and A_2, output sets A_2 and A_3, respectively. Their *product* $\mathcal{A}_1 \otimes \mathcal{A}_2$ is the automaton over the alphabet $X_1 \times X_2$ with the input set A_1 and the output set A_3, with the output and transition functions given by the rules

$$q_1(x_1, x_2) = (q_1(x_1), q_1|_{x_1}(x_2))$$

and

$$q_1|_{(x_1, x_2)} = (q_1|_{x_1})|_{x_2}.$$

The product of automata is different from the dual notion of *composition*, which is defined only for automata with coinciding input and output sets. Namely, if $\mathcal{A}_1$ and $\mathcal{A}_2$ are automata over the alphabet X with the sets of internal states A_1 and A_2, respectively, then their *composition* is the automaton with the set of internal states $A_1 \times A_2$ over the alphabet X in which the output and transition functions are defined by the rules

$$(q_1, q_2)(x) = q_1(q_2(x))$$

and

$$(q_1, q_2)|_x = (q_1|_{q_2(x)}, q_2|_x).$$

Definition 3.11. Let $\mathcal{A}_1, \mathcal{A}_2, \ldots$ be a sequence of group automata over alphabets $X_1, X_2, \ldots$, respectively. Suppose that $\mathcal{A}_i$ has input set A_{i-1} and output set A_i. Then the *action of A_0 on X^**, for $X = (X_1, X_2, \ldots)$, is the action defined on the nth level $X^n = \prod_{i=1}^{n} X_i$ of X^* as the action of the input set A_0 of the product automaton $\mathcal{A}_1 \otimes \mathcal{A}_2 \otimes \cdots \otimes \mathcal{A}_n$ on its alphabet $X^n = \prod_{i=1}^{n} X_i$.

It is not hard to see that the action defined by a sequence of group automata is an action by automorphisms of the rooted tree X^*.

3.6   Hyperbolic and Sub-Hyperbolic Bimodules

Definition 3.12. Let $\mathfrak{M}$ be a covering G-bimodule and let X be its basis. We say that the bimodule $\mathfrak{M}$ is *hyperbolic* if there exists a finite set $\mathcal{N} \subset G$ such that for every $g \in G$ there exists $n_0 \in \mathbb{N}$ such that $g|_v \in \mathcal{N}$ whenever $v \in X^n$ and $n \geq n_0$.

It is proved in [Ne05, Corollary 2.11.7] that the property of being hyperbolic does not depend on the choice of the basis X (though the set $\mathcal{N}$ does).

If the group G is finitely generated, then hyperbolicity can be expressed in terms of a uniform contraction of the length of the group elements under restriction.

Definition 3.13. Let G be a finitely generated group and let $l(g)$ denote the word length of $g \in G$ with respect to some fixed finite generating set of G (i.e., the minimal length of a representation of g as a product of the generators and their inverses).

Then the number

$$\rho = \limsup_{n \to \infty} \sqrt[n]{\limsup_{l(g) \to \infty} \max_{v \in X^n} \frac{l(g|_v)}{l(g)}}$$

is called the *contraction coefficient* of the bimodule $\mathfrak{M}$ (with respect to the basis X).

It is not hard to prove that the contraction coefficient does not depend on the choice of the generating set. It is also proved in [Ne05, Proposition 2.11.11] that it does not depend on the basis X, if $\rho < 1$. Moreover, the following holds.

Proposition 3.14. *The bimodule $\mathfrak{M}$ is hyperbolic if and only if its contraction coefficient is less than 1.*

It is possible that the action of G on X^* associated with $\mathfrak{M}$ is not faithful. Then the kernel K of the action is uniquely determined as the maximal subgroup with the property that it is normal and if $g \cdot x = y \cdot h$ for $x, y \in \mathfrak{M}$ and $g \in K$, then $h \in K$. Then the set $\mathfrak{M}/K$ of the right K-orbits of $\mathfrak{M}$ is naturally a G/K-bimodule such that the action of G/K on the tree X^* is faithful (and coincides with the action of G/K on X^* induced by G). The G/K-bimodule $\mathfrak{M}/K$ is called the *faithful quotient* of the bimodule $\mathfrak{M}$.

Definition 3.15. A bimodule is said to be *sub-hyperbolic* if its faithful quotient is hyperbolic.

In general, we say that a normal subgroup $N \lhd G$ is $\mathfrak{M}$-invariant, if $g|_x \in N$ for all $g \in N$ and $x \in \mathsf{X}$. This property does not depend on the choice of X. The kernel of the induced action on X^* is the maximal $\mathfrak{M}$-invariant normal subgroup. If N is a normal $\mathfrak{M}$-invariant subgroup, then the set $\mathfrak{M}/N$ of the right N-orbits is naturally a G/N-bimodule. It is easy to see that if $\mathfrak{M}$ is hyperbolic, then $\mathfrak{M}/N$ is also hyperbolic and that the faithful quotient of $\mathfrak{M}/N$ coincides with the faithful quotient of $\mathfrak{M}$. Consequently, we have the following version of the definition of a sub-hyperbolic bimodule.

Proposition 3.16. *A G-bimodule $\mathfrak{M}$ is sub-hyperbolic if and only if there exists a $\mathfrak{M}$-invariant normal subgroup $N \lhd G$ such that the G/N-bimodule $\mathfrak{M}/N$ is hyperbolic.*

Examples. If f is a hyperbolic post-critically finite rational function, then the bimodule $\mathfrak{M}_f$ over the fundamental group of the sphere minus the post-critical set is hyperbolic. For the definition of $\mathfrak{M}_f$ see Section 5.2. It follows from uniform expansion of f on a neighborhood of the Julia set of f, which does not contain the post-critical points.

On the other hand, if f is sub-hyperbolic, then $\mathfrak{M}_f$ is hyperbolic only as a bimodule over the fundamental group of the associated *orbifold*. The bimodule over the fundamental group of the punctured sphere is *sub-hyperbolic*. For more on the bimodules associated with post-critically finite rational functions, see [Ne05, Section 6.4].

4   Dendroid Automata

4.1   Dendroid Sets of Automorphisms of a Rooted Tree

Definition 4.1. A sequence $(a_1, a_2, \ldots, a_m)$ of automorphisms of a rooted tree T is said to be *dendroid* if for every n the sequence of permutations defined by $(a_1, a_2, \ldots, a_m)$ on the nth level of T is dendroid.

If $A = (a_1, a_2, \ldots, a_m)$ is a dendroid sequence of automorphisms of a rooted tree T, then it generates a level-transitive subgroup of $\mathrm{Aut}\,(T)$. Consequently, the tree T is spherically homogeneous.

We will assume that A does not contain trivial automorphisms of T. Then all elements of A are different, and we may consider it as a set.

Denote by D_n the cycle diagram of the action of the sequence A on the nth level L_n of the tree T. We say that an oriented edge e of D_n *corresponds to* $a_i \in A$ if it is an edge of a 2-cell corresponding to a cycle of a_i.

If $\gamma = (e_1, e_2, \ldots, e_k)$ is a path in the 1-skeleton of D_n (we are allowed to go against the orientation), then we denote by $\pi(\gamma)$ the element $g_k g_{k-1} \cdots g_1$, where g_i is the element of A corresponding to the edge e_i, if we pass it along the orientation; and its inverse, if we go against the orientation. In particular, if e is an oriented edge of D_n, then $\pi(e)$ is the corresponding element of A.

Proposition 4.2. *Let $A = \{a_1, a_2, \ldots, a_m\}$ be a dendroid set of automorphisms of a rooted tree T generating a group $G < \mathrm{Aut}\,(T)$. Then for every vertex v of T the group $G|_v$ is also generated by a dendroid set.*

Recall, that $G|_v$ is restriction of the action of the stabilizer G_v on the sub-tree T_v.

Proof: The stabilizer G_v of v in G is generated by $\pi(\gamma)$, where γ runs through a generating set of the fundamental group of the 1-skeleton of the cycle diagram D_n of the action of A on the nth level $L_n \ni v$ of T.

Choose one edge in each 2-cell of D_n. Let E be the set of chosen arrows (called *marked edges*). Let D'_n be the union of non-marked edges. It is a spanning tree of the 1-skeleton of D_n.

For every edge $e \in E$ let γ_e be the path going from v to the beginning of e through edges belonging to D'_n, then along e and then back to v using only the edges from D'_n. Since D'_n is a tree, this description gives a unique automorphism $\pi(\gamma_e)$ of T.

It is well known that the set of the loops $\{\gamma_e\}_{e \in E}$ is a free generating set of the fundamental group of the 1-skeleton of D_n. Hence the elements of the form $\pi(\gamma_e)$ give a generating set of G_v.

Let us denote by b_e the restriction of $\pi(\gamma_e)$ onto the sub-tree T_v. Then $\{b_e\}_{e \in E}$ is a generating set of $G|_v$. Let us prove that it is dendroid.

Let L_m be a level of T below the level L_n of v. Denote by D_m the cycle diagram of the action of A on L_m.

The 1-skeleton of D_m covers the 1-skeleton of D_n by the natural map $L_m \longrightarrow L_n$ sending a vertex u to its ancestor (i.e., to the vertex of L_n below which u is). Let D''_m be the inverse image of D'_n under this covering map. Since D'_n is a tree, the graph D''_m is a disjoint union of $|L_m|/|L_n|$ trees, which are mapped isomorphically onto D'_n by the covering map, see Figure 3. In particular, every connected component of D''_m contains exactly one vertex, which is below v. We get hence a natural bijection between the set of connected components of D''_m and $T_v \cap L_m$.

For every marked $e \in E$ and every connected component C of D''_m there exists exactly one preimage of e in D_m starting in a vertex v_1 of C and exactly one preimage ending in a vertex v_2 of C, since the projection $D_m \longrightarrow D_n$ is a covering of the 1-skeletons. The vertex v_2 is connected to

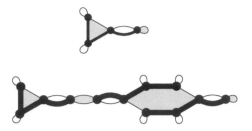

Figure 3. The sets D'_n and D''_m in the proof of Proposition 4.2.

the vertex v_1 by a directed chain of edges belonging to C and corresponding to the same element of A as e does, since e must belong to a cycle.

It follows from the definition of the generators γ_e that the inverse images of the edges $e \in E$ connect the components of D''_m exactly in the same way that the generators b_e act on $T_v \cap L_m$ (if we identify the components with the vertices of $T_v \cap L_m$ that they contain).

Contract the components of D''_m in D_m to points. We will obtain a CW-complex homotopically equivalent to D_m, i.e., contractible. Every preimage of $e \in E$ in D_m belongs to boundary of a disc, which after contraction becomes a disc corresponding to a cycle of b_e. Hence the obtained CW-complex is isomorphic to the cycle diagram of $\{b_e\}_{e \in E}$ and is contractible. $\qquad \square$

Note that some of the generators b_e from the proof may be trivial. In this case we can remove them from the generating set.

Let us call the generating set $\{b_e\}_{e \in E}$ (with trivial elements removed) the *induced generating set* of $G|_v$.

The induced generating set of $G|_v$ depends on the choice of marked edges, however it is not hard to see that the conjugacy classes in $\mathrm{Aut}\,(T_v)$ of its elements depend only on A.

In particular, the cardinality of the induced generated set depends only on A and the level number of v. It can be also found in the following way.

Definition 4.3. A *support* of a group $H < \mathrm{Aut}\,(T)$ is the set of vertices $v \in T$ such that the stabilizer H_v acts non-trivially on the sub-tree T_v.

It is easy to see that the support of a non-trivial group H is a sub-tree containing the root of the tree.

Definition 4.4. We call a cycle of the action of an automorphism $a \in \mathrm{Aut}\,(T)$ on vertices *active* if it is contained in the support of $\langle a \rangle$.

Proposition 4.5. *Let $A \subset \mathrm{Aut}\,(T)$ be a dendroid set and let $G = \langle A \rangle$ be the group it generates. The cardinality of the induced generating set of $G|_v$ is equal to the sum over $a \in A$ of the number of active cycles of a on the level of v.*

Proof: Let $x_1, x_2, \ldots, x_k$ be a cycle of the action of $a \in A$ on the level of v. If a^k acts trivially on the subtrees T_{x_i}, then the vertices x_i do not belong to the support of $\langle a \rangle$ and also the corresponding generator b_e of $G|_v$ will be trivial, since it is conjugate to the restriction of a^k onto T_v. Otherwise, if b_e is not trivial, then the action of a^k on the subtrees T_{x_i} will be conjugate to b_e, and thus non-trivial. In this case the points x_i will belong to the support of $\langle a \rangle$.   □

4.2 Dendroid Automata

Let $A = \{a_1, a_2, \ldots, a_n\}$ be a dendroid set of automorphisms of a rooted tree T. Denote by D_n the cycle diagram of the action of A on the nth level L_n of T. Recall, that a *marking* of D_n is a set E of (*marked*) edges containing exactly one edge from the boundary of every 2-cell.

Choose a vertex $v \in L_n$ of D_n. Our aim is to describe the bimodule $\mathfrak{M}_{\varnothing,v}$.

Denote, for $u \in L_n$, by $\gamma_{v,u}$ a path in D_n from v to u not containing marked edges. Let $h_u = \pi(\gamma_{v,u})$. The automorphisms h_u does not depend on the choice of $\gamma_{v,u}$, since the complement D_n' of the set of marked edges is a spanning tree of the 1-skeleton of D_n.

Denote by $x_u : T_v \longrightarrow T_u$ the element of $\mathfrak{M}_{\varnothing,v}$ equal to the restriction of h_u onto the sub-tree T_v. The set $\{x_u\}_{u \in L_n}$ is a basis of the bimodule $\mathfrak{M}_{\varnothing,v}$.

Let us describe now the bimodule $\mathfrak{M}_{\varnothing,v}$ with respect to the basis $\mathsf{X} = \{x_u\}_{u \in L_n}$, the generating set $A = \{a_1, \ldots, a_m\}$ of G and the induced generating set $\{b_e\}_{e \in E}$ of $G|_v$, constructed in the proof of Proposition 4.2.

Recall that b_e is the restriction of $h_{r(e)}^{-1} \pi(e) h_{s(e)}$ onto T_v, where $s(e)$ is the beginning and $r(e)$ is the end of the edge e.

Let $a_i \in A$ and $x_u \in \mathsf{X}$ be arbitrary. Denote by e_{a_i,x_u} the edge of D_n starting in u and corresponding to a_i. If e_{a_i,x_u} is not marked, then $a_i \cdot h_u = h_{a_i(u)}$, hence

$$a_i \cdot x_u = x_{a_i(u)}. \tag{4.1}$$

If e_{a_i,x_u} is marked, then $b_{e_{a_i,x_u}} = h_{a_i(u)}^{-1} a_i h_u$, hence

$$a_i \cdot x_u = x_{a_i(u)} \cdot b_{e_{a_i,x_u}}. \tag{4.2}$$

Let us generalize the properties of the obtained automaton in the following definition.

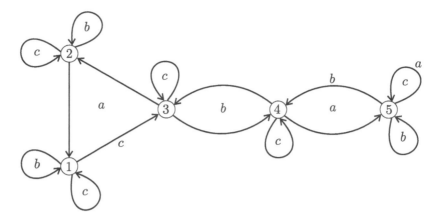

Figure 4. A dendroid automaton.

Definition 4.6. A group automaton with the set of states X, input set A and output set B is said to be a *dendroid* automaton if the following conditions are satisfied:

(1) The set of permutations defined by A on X is dendroid.

(2) For every $b \in B \setminus \{1\}$ there exists a unique pair $a \in A, x \in X$ such that $a \cdot x = y \cdot b$ for some $y \in X$.

(3) For any cycle $(x_1, x_2, \ldots, x_k)$ of the action of $a \in A$ on X we have $a \cdot x_i = x_{i+1} \cdot 1$ for all but possible one index i. (Here indices are taken modulo k.)

We will draw the dual Moore diagram of dendroid automata as the cycle diagram of A in which a cell is labeled by the corresponding element of A. If $a \cdot x = y \cdot b$ for $a \in A$ and $b \in B$, then we label the edge from x to y corresponding to a by $b \in B$. If $a \cdot x = y \cdot 1$, then we do not label the corresponding edge (since we typically ignore trivial states). This labeled diagram completely describes the automaton. It coincides (up to a different labeling convention) with its dual Moore diagram.

Condition (1) of Definition 4.6 is equivalent to contractibility of the diagram; condition (2) means that every element $b \in B$ appear exactly once as a label of an edge; and condition (3) means that every 2-cell of the diagram has at most one label of its boundary edge. See an example of a diagram of a dendroid automaton on Figure 4.

By Equations (4.1) and (4.2), the automaton describing the bimodule $\mathfrak{M}_{\varnothing,v}$ with respect to the basis $\{x_u\}_{u \in L_n}$, the (input) generating set A and the (output) induced generating set $\{b_e\}$ is a dendroid automaton. The third condition follows from the fact that every cycle of a has exactly one marked edge. (The element b_e might be trivial, though).

Our aim now is to show that dendroid sets of automorphisms of a rooted tree are exactly the sets defined by sequences of dendroid automata.

Proposition 4.7. *The product of two dendroid automata is a dendroid automaton.*

The proof of this proposition also gives a description of a procedure of constructing the diagram of the product of dendroid automata.

Proof: The proof essentially coincides with the proof of [Ne05, Proposition 6.7.5]. We rewrite it here, making the necessary changes.

Let $\mathcal{A}_1$ be a dendroid automaton over the alphabet X_1, input set A_1 and output set A_2. Let $\mathcal{A}_2$ be a dendroid automaton over X_2 with input and output sets A_2 and A_3, respectively.

If we have $a_1 \cdot x_1 = y_1 \cdot 1$ in $\mathcal{A}_1$, then $a_1 \cdot x_1 x_2 = y_1 x_2 \cdot 1$ in $\mathcal{A}_1 \otimes \mathcal{A}_2$. If $a_1 \cdot x_1 = y_1 \cdot a_2$ in $\mathcal{A}_2$, then $a_1 \cdot x_1 x_2 = y_1 \cdot a_2 \cdot x_2$.

Consequently, the diagram of the automaton $\mathcal{A}_1 \otimes \mathcal{A}_2$ can be described in the following way.

Take $|X_2|$ copies of the diagram D_1 of $\mathcal{A}_1$. Each copy will correspond to a letter $x_2 \in X_2$ and the respective copy will be denoted $D_1 x_2$. If $x_1 \in X_1$ is a vertex of D_1, then the corresponding vertex of the copy $D_1 x_2$ will become the vertex $x_1 x_2$ of the diagram of $\mathcal{A}_1 \otimes \mathcal{A}_2$.

If we have an arrow labeled by a_2 in the copy $D_1 x_2$, (i.e., if we have $a_1 \cdot x_1 = y_1 \cdot a_2$) then we detach it from its end $y_1 x_2 \in D_1 x_2$ and attach it to the vertex $y_1 a_2(x_2) \in D_1 a_2(x_2)$. If $a_2|_{x_2} \neq 1$, then we label the obtained arrow by $a_2|_{x_2}$. The rest of the arrows of $\bigsqcup_{x_2 \in X_2} D_1 x_2$ are not changed.

It is easy to see that in this way we get the diagram of $\mathcal{A}_1 \otimes \mathcal{A}_2$. We see that the copies of D_1 are connected in the same way as the vertices of D_2 are. See, for example, in Figure 5 the diagram of $\mathcal{A} \otimes \mathcal{A}$, where $\mathcal{A}$ is the automaton from Figure 4.

It follows immediately that every element $a_3 \in X_3$ is a label of exactly one arrow of the diagram of $\mathcal{A}_1 \otimes \mathcal{A}_2$.

Let us reformulate the procedure of construction of the diagram of $\mathcal{A}_1 \otimes \mathcal{A}_2$ in a more geometric way. The diagram is obtained by gluing discs, corresponding to the cells of D_2, to the copies of D_1 along their labeled edges. Namely, if the edge (a_1, x_1) is labeled in D_1 by $a_2 = a_1|_{x_1}$ and $x_2 \in X_2$ belongs to a cycle $\left(x_2, a_2(x_2), \ldots, a_2^{k-1}(x_2)\right)$ of length k under the action of a_2, then we have to take a $2k$-sided polygon and glue its every other side to the copies of the edge (a_1, x_1) in the diagrams $D_1 x_2$, $D_1 a_2(x_2)$, $\ldots$, $D_1 a_2^{k-1}(x_2)$ in the given cyclic order. We will glue in this way the k copies of a cell of D_1 together and get a cell of the diagram of

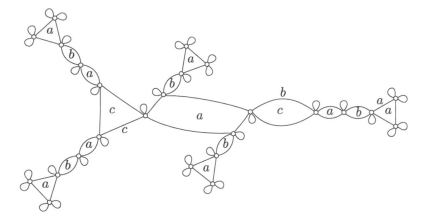

Figure 5. Diagram of $\mathcal{A} \otimes \mathcal{A}$.

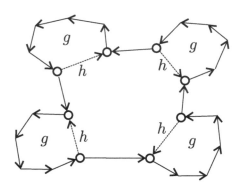

Figure 6. A cell of the diagram of $\mathcal{A}_1 \otimes \mathcal{A}_2$.

$\mathcal{A}_1 \otimes \mathcal{A}_2$. See, for example, Figure 6, where the case $k = 4$ is shown. It is easy to see that in this procedure is equivalent to the one described before.

It follows that we can contract the copies of D_1 in the diagram of $\mathcal{A}_1 \otimes \mathcal{A}_2$ to single points, and get a cellular complex homeomorphic to D_2, which is contractible. This proves that the diagram of $\mathcal{A}_1 \otimes \mathcal{A}_2$ is contractible.

We also see that every cell of this diagram has at most 1 labeled side, since the labels come only from the attached $2k$-sided polygons, whose sides are labeled in the same way as the corresponding cell of D_2. $\square$

Theorem 4.8. *A set $A = \{a_1, \ldots, a_m\}$ is a dendroid set of automorphisms of T if and only if there exists a sequence of automata $\mathcal{A}_n$, $n \geq 1$, and sequences of finite sets B_n and X_n such that $B_0 = A$ and*

- *$\mathcal{A}_n$ is a dendroid automaton over alphabet X_n, input set B_{n-1} and output set B_n;*

- *the action of the elements a_i on the rooted tree $\mathsf{X}^* = \bigcup_{n \geq 0} \mathsf{X}_1 \times \cdots \times \mathsf{X}_n$ defined by the sequence of automata $\mathcal{A}_n$ is conjugate to the action of a_i on T.*

In other words, all dendroid sets of automorphisms of a rooted tree are defined taking products of kneading automata.

The actions defined by sequences of automata is described in Definition 3.11.

Proof: Sequence of dendroid automata define dendroid sets of automorphisms of a rooted tree by Proposition 4.7.

On the other hand, we have seen in Proposition 4.5 that the automaton describing the bimodules $\mathfrak{M}_{\varnothing,v}$ with respect to the basis x_v and the generating set A is a dendroid automaton with a dendroid output set B, which is the generating set of $\langle A \rangle|_v$. It follows now from Corollary 3.5 that the action of A on the tree T is conjugate to the action defined by a sequence of dendroid automata. $\square$

Figure 7 shows an example of products of a sequence of dendroid automata. It shows two dendroid automata $\mathcal{A}_1$ and $\mathcal{A}_2$ over the binary alphabet and their products $\mathcal{A}_1 \otimes \mathcal{A}_1$, $\mathcal{A}_1 \otimes \mathcal{A}_1 \otimes \mathcal{A}_2$ and $\mathcal{A}_1 \otimes \mathcal{A}_1 \otimes \mathcal{A}_2 \otimes \mathcal{A}_1$. We use colors "grey" and "white" instead of letters to mark the cells and arrows of the dual Moore diagrams (arrows are marked by dots near them). The diagrams of the products are shown without marking of the edges and orientation.

5 Backward Polynomial Iteration

5.1 Iterated Monodromy Groups

Let f_n be a sequence of complex polynomials and consider them as an inverse sequence of maps between complex planes:

$$\mathbb{C} \xleftarrow{f_1} \mathbb{C} \xleftarrow{f_2} \mathbb{C} \xleftarrow{f_3} \cdots .$$

We call such sequences *backward iterations*. More generally, we may consider backward iterations of *topological polynomials*. Here a topological

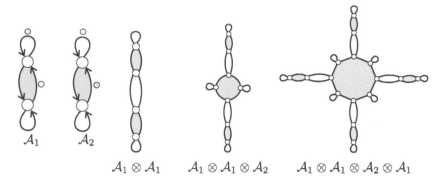

$$\mathcal{A}_1 \qquad \mathcal{A}_2 \qquad\qquad \mathcal{A}_1 \otimes \mathcal{A}_1 \qquad \mathcal{A}_1 \otimes \mathcal{A}_1 \otimes \mathcal{A}_2 \qquad \mathcal{A}_1 \otimes \mathcal{A}_1 \otimes \mathcal{A}_2 \otimes \mathcal{A}_1$$

Figure 7. Products of dendroid automata.

polynomial is a continuous map $f : \mathbb{R}^2 \longrightarrow \mathbb{R}^2$ which is an orientation preserving local homeomorphism at all but a finite number of (*critical*) points of $\mathbb{R}^2$. For more on topological polynomials, see [BFH92] and [BN06].

A backward iteration $(f_n)_{n \geq 1}$ is said to be *post-critically finite* if there exists a finite set $P \subset \mathbb{C}$ such that the set of critical values of the composition $f_1 \circ f_2 \circ \cdots \circ f_n$ is contained in P for all n. The smallest such set P is called the *post-critical set* of the iteration.

Example. Every sequence $(f_1, f_2, \dots)$ such that f_i is either z^2 or $1 - z^2$ is post-critically finite with post-critical set a subset of $\{0, 1\}$.

If the backward iteration $f_1, f_2, \dots$ is post-critically finite, then the shifted iteration $f_2, f_3, \dots$ is also post-critically finite. Let us denote by P_n the post-critical set of the iteration $f_n, f_{n+1}, \dots$.

Choose a basepoint $t \in \mathbb{C} \setminus P_1$ and consider the rooted *tree of preimages*

$$T_t = \{t\} \sqcup \bigsqcup_{n \geq 1} (f_1 \circ f_2 \circ \cdots \circ f_n)^{-1}(t),$$

where a vertex $z \in (f_1 \circ f_2 \circ \cdots \circ f_n)^{-1}(t)$ is connected by an edge with the vertex $f_n(z) \in (f_1 \circ f_2 \circ \cdots \circ f_{n-1})^{-1}(t)$ and the vertex t is considered to be the root.

The fundamental group $\pi_1(\mathbb{C} \setminus P_1, t)$ of the punctured plane acts on the tree T_t by the monodromy action. The image of a point $z \in (f_1 \circ f_2 \circ \cdots \circ f_n)^{-1}(t)$ under the action of a loop $\gamma \in \pi_1(\mathbb{C} \setminus P_1, t)$ is the endpoint of the unique lift of γ by $f_1 \circ f_2 \circ \cdots \circ f_n$ starting at z. This is clearly an action by automorphisms of the tree T_t. This action is called the *iterated monodromy action*.

The *iterated monodromy group* of the iteration $f_1, f_2, \dots$ is the quotient of the fundamental group by the kernel of the iterated monodromy action, i.e., the group of automorphisms of T_t defined by the loops $\gamma \in \pi_1(\mathbb{C} \setminus P_1, t)$.

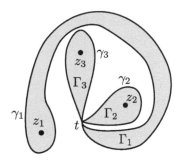

Figure 8. A planar generating set.

Example. This is a slight generalization of a construction due to Richard Pink (see also [AHM05] and [BJ07]). Consider the field of rational functions $\mathbb{C}(t)$ and the polynomial over $\mathbb{C}(t)$

$$F_n(z) = f_1 \circ f_2 \circ \cdots \circ f_n(z) - t.$$

Let Ω_n be the splitting field of F_n in an algebraic closure of $\mathbb{C}(t)$. It is not hard to see that $\Omega_{n+1} \supset \Omega_n$. Denote by Ω the field $\bigcup_{n \geq 1} \Omega_n$. Then the Galois group of the extension $\Omega/\mathbb{C}(t)$ is isomorphic to the closure of the iterated monodromy group of the iteration $f_1, f_2, \ldots$ in the automorphism group of the tree of preimages T_t (see [Ne05, Proposition 6.4.2]).

Definition 5.1. Let $P = \{z_1, z_2, \ldots, z_n\}$ be a finite subset of the plane $\mathbb{R}^2$. A *planar* generating set of the fundamental group $\pi_1(\mathbb{R}^2 \setminus P, t)$ is a collection of simple (i.e., without self-intersections) loops γ_i such that the region Γ_i bounded by γ_i contains only one point z_i of P, γ_i goes around the region in the positive direction and Γ_i are disjoint for different i.

See Figure 8 for an example of a planar generating set of the fundamental group of a punctured plane.

There exists a unique cyclic order $\gamma_1, \gamma_2, \ldots, \gamma_n$ on a planar generating set such that the loop $\gamma_1 \gamma_2 \cdots \gamma_n$ has no transversal self-intersections. We call this order the *natural cyclic order* of the planar generating set.

Proposition 5.2. *Let $f_1, f_2, \ldots$ be a post-critically finite backward iteration of topological polynomials. Let P be its post-critical set and let $t \in \mathbb{R}^2 \setminus P$ be a basepoint. Let $\{\gamma_z\}_{z \in P}$ be a planar generating set of $\pi_1(\mathbb{R}^2 \setminus P, t)$, where γ_z goes around z. Let a_z be the corresponding elements of the iterated monodromy group of the backward iteration. Then the set $(a_z)_{z \in P}$ of automorphisms of the tree of preimages T_t is dendroid and generates the iterated monodromy group of the backward iteration.*

Proof: The loops γ_z generate the fundamental group, hence the elements a_z generate the iterated monodromy group. Denote $F = f_1 \circ f_2 \circ \cdots \circ f_n$. Let us prove that the generators a_z define a dendroid set of permutations of the nth level $F^{-1}(t)$ of the tree T_t.

Let $\Delta = \overline{\bigcup_{z \in P} \Gamma_z}$ be the closed part of the plane bounded by the curves γ_z. Then the set $\mathbb{C} \setminus \Delta$ contains no critical values of F and is a homeomorphic to an annulus. Consequently, the map $F : F^{-1}(\mathbb{C} \setminus \Delta) \longrightarrow \mathbb{C} \setminus \Delta$ is a $\deg F$-covering. In particular, $F^{-1}(\mathbb{C} \setminus \Delta)$ is also an annulus, hence the set $F^{-1}(\Delta)$ is contractible. But it is easy to see that $F^{-1}(\Delta)$ is homeomorphic to the cycle diagram of the action of the set $(a_z)_{z \in P}$ on the nth level $F^{-1}(t)$ of the tree T_t. Hence the set of permutations of $F^{-1}(t)$ defined by $(a_z)_{z \in P}$ is dendroid. □

Consequently, the iterated monodromy group of any post-critically finite backward iteration is defined by a sequence of dendroid automata, due to Theorem 4.8.

Example. Consider, as in the first example in Section 5.1, a backward iteration (f_n), where $f_i(z) = z^2$ or $f_i(z) = 1 - z^2$ for every i. Then the iterated monodromy group is defined by the sequence of automata $(\mathcal{A}_{i_1}, \mathcal{A}_{i_2}, \ldots)$, where $\mathcal{A}_{i_j}$ is the automaton $\mathcal{A}_1$ from Figure 7, if $f_j(z) = 1 - z^2$ and $\mathcal{A}_2$ if $f_j(z) = z^2$.

Let us show that the converse to Proposition 5.2 is also true.

Proposition 5.3. *Let $(a_1, \ldots, a_m)$ be a cyclically ordered dendroid set of automorphisms of a rooted tree T. Then there exists a post-critically finite backward iteration of complex polynomials $f_1, f_2, \ldots$ with post-critical set $P = \{z_1, z_2, \ldots, z_m\}$, a planar generating set $\{\gamma_i\}_{i=1,\ldots,m}$ of $\pi_1(\mathbb{C} \setminus P, t)$ with the natural order $(\gamma_1, \ldots, \gamma_m)$ (where γ_i goes around the point z_i), and an isomorphism of T with the preimage tree T_t conjugating the iterated monodromy action of the loops γ_i with the automorphisms a_i.*

Proof: Denote by D_n the cycle diagram of the sequence of permutations induced by $(a_1, \ldots, a_m)$ on the nth level L_n of the tree T. The product $a_m \cdots a_1$ acts on L_n as a transitive cycle by Proposition 2.5. Choose a vertex $x \in L_n$ of D_n and consider the path γ_n starting at x and passing along the arrows

$$e_1, e_2, \ldots, e_m, e_{m+1}, e_{m+2}, \ldots, e_{m \cdot |L_n|},$$

where e_i is the edge corresponding to the action of a_i, where the indices of a_i are taken modulo m. The path γ_n is closed and contains every arrow of D_n exactly once.

Embed D_n into the plane so that γ_n is embedded as the path going around the image of D_n in the positive direction (without transverse self-intersections). Choose a point in each of the (images of the) 2-cells of D_n, which we will call *centers*.

The 1-skeleton of D_{n+1} covers of the 1-skeleton of D_n by the map carrying a vertex $v \in L_{n+1}$ to the adjacent vertex $u \in L_n$. This covering induces an $|L_{n+1}|/|L_n|$-fold covering of the curve γ_n by the curve γ_{n+1}. Let us extend this covering to a covering of the complement of the image of D_n in the plane by the complement of the image of D_{n+1}, so that we get a covering of the union of the complements with the images of the 1-skeletons of diagrams. We can then extend this covering inside the image of the diagrams so that we get an orientation-preserving branched covering of the planes, mapping D_{n+1} onto D_n so that the 1-skeletons are mapped as before, while the branching points are the centers of the 2-cells of D_{n+1} (not all centers must be branching points). We also require that images of centers are centers. See a more explicit construction in [Ne05, Theorem 6.10.4], which can be easily adapted to our more general situation.

We get in this way a sequence of branched coverings $\mathbb{R}^2 \xleftarrow{f_1} \mathbb{R}^2 \xleftarrow{f_2} \cdots$ with embedded diagrams $D_0, D_1, \ldots$. Note that D_0 is a bouquet of m circles going around m centers. The backward iteration $f_1, f_2, \ldots$ is post-critically finite with the post-critical set equal to the set of centers of D_0. The set of preimages of the basepoint of D_0 under $f_1 \circ \cdots \circ f_n$ is equal to the image of $L_n \subset D_n$. It follows from the construction of the maps f_i that the action of the loops of D_n on the tree of preimages of the basepoint of D_0 coincides with the action of the corresponding elements a_i.

It remains now to introduce a complex structure on the first plane and pull it back by the branched coverings $f_1 \circ \cdots \circ f_n$ so that the maps f_i become complex polynomials. $\qquad\qquad\square$

5.2 Bimodules Associated with Coverings

We multiply paths (in particular in the fundamental groups) as functions: in a product $\gamma_1\gamma_2$ the path γ_2 is passed first.

Definition 5.4. Let $P_1, P_2 \subset \mathbb{R}^2$ be finite sets of points and let $f : \mathbb{R}^2 \longrightarrow \mathbb{R}^2$ be a continuous map such that $f : \mathbb{R}^2 \setminus f^{-1}(P_2) \longrightarrow \mathbb{R}^2 \setminus P_2$ is a d-fold covering and $f^{-1}(P_2) \supseteq P_1$. Denote $\mathcal{M}_i = \mathbb{R}^2 \setminus P_i$. We say that f is a *partial covering* of $\mathcal{M}_2$ by (a subset of) $\mathcal{M}_1$. Choose a basepoint $t_2 \in \mathcal{M}_2$ and let $t_1 \in f^{-1}(t_2)$ be some of its preimages.

The *bimodule* $\mathfrak{M}_f$ *associated with the covering* f is the $(\pi_1(\mathcal{M}_2, t_2) - \pi_1(\mathcal{M}_1, t_1))$-bimodule consisting of homotopy classes of paths in $\mathcal{M}_2$ starting in t_1 and ending in any point of $f^{-1}(t_2)$. The right action $\pi_1(\mathcal{M}_1, t_1)$ is appending the loop $\gamma \in \pi_1(\mathcal{M}_1, t_1)$ at the beginning of the path $\ell \in \mathfrak{M}_f$.

The left action of $\pi_1(\mathcal{M}_2, t_2)$ is appending the f-lift of the loop $\gamma \in \pi_1(\mathcal{M}_2, t_2)$ to the endpoint of the path $\ell \in \mathfrak{M}_f$.

It is not hard to prove that the bimodule $\mathfrak{M}_f$ does not depend (up to an isomorphism) on the choice of the basepoints t_1 and t_2, if we identify the respective fundamental groups in the usual way using paths.

Moreover, if $P_1 = P_2$, then we may identify the fundamental groups $\pi_1(\mathcal{M}_1, t_1)$ and $\pi_1(\mathcal{M}_2, t_2)$ by a path from t_1 to t_2, and assume that $\mathfrak{M}_f$ is a $\pi_1(\mathcal{M})$-bimodule, where $\mathcal{M} = \mathcal{M}_1 = \mathcal{M}_2$. The isomorphism class of the $\pi_1(\mathcal{M})$-bimodule $\mathfrak{M}_f$ will not depend on the choice of the connecting path.

Note also that the bimodule $\mathfrak{M}_f$ has free right action and that two elements belong to the same orbit of the right action if and only if the ends of the corresponding paths coincide. Hence the number of the right orbits is equal to the degree of the covering f. The induced left action of the group $\pi_1(\mathcal{M}_2, t_2)$ on the right orbits coincides with the monodromy action on the fiber $f^{-1}(t_2)$, if we identify orbits with the corresponding endpoints.

Proposition 5.5. *Let $P_1, P_2, P_3 \subset \mathbb{R}^2$ be finite sets and let $f_i : \mathbb{R}^2 \longrightarrow \mathbb{R}^2$ for $i = 1, 2$ be branched coverings such that the set of critical values of f_i is contained in P_i and $f_i(P_{i+1}) \subseteq P_i$ for $i = 1, 2$. Let $\mathfrak{M}_{f_i}$ be the bimodules associated with the respective partial coverings f_i of $\mathbb{R}^2 \setminus P_i$ by a subset of $\mathbb{R}^2 \setminus P_{i+1}$.*

Then the bimodule $\mathfrak{M}_{f_1} \otimes \mathfrak{M}_{f_2}$ is isomorphic to the bimodule $\mathfrak{M}_{f_1 \circ f_2}$ associated with the partial covering of $\mathbb{R}^2 \setminus P_1$ by a subset of $\mathbb{R}^2 \setminus P_3$.

Proof: Denote $\mathcal{M}_i = \mathbb{R}^2 \setminus P_i$ and let the basepoints $t_i \in \mathcal{M}_i$ be chosen so that $f_i(t_{i+1}) = t_i$. Then $\mathfrak{M}_{f_i}$ consists of the homotopy classes of paths in $\mathcal{M}_{i+1}$ starting in t_{i+1} and ending in a point from the set $f_i^{-1}(t_i)$.

For $\ell_i \in \mathfrak{M}_{f_i}$ for $i = 1, 2$ be arbitrary elements. Then it is an easy exercise to prove that the map

$$\ell_1 \otimes \ell_2 \mapsto f_2^{-1}(\ell_1)_{\ell_2} \ell_2$$

is an isomorphism of bimodules, where $f_2^{-1}(\ell_1)_{\ell_2}$ is the lift of ℓ_1 by f_2 to a path starting at the endpoint of ℓ_2.   □

Proposition 5.6. *Let f_n be a post-critically finite backward iteration of topological polynomials. Let P_n be the post-critical set of the iteration $f_n, f_{n+1}, \ldots$. Then the iterated monodromy action of $\pi_1(\mathbb{R}^2 \setminus P_1)$ is conjugate to the action associated with the infinite tensor product $\mathfrak{M}_{f_1} \otimes \mathfrak{M}_{f_2} \otimes \cdots$.*

Proof: The bimodule $\mathfrak{M}_{f_1} \otimes \cdots \otimes \mathfrak{M}_{f_n}$ is isomorphic to the bimodule $\mathfrak{M}_{f_1 \circ \cdots \circ f_n}$, hence the corresponding left action on the right orbits coincides with the monodromy action associated with the covering $f_1 \circ \cdots \circ f_n$.   □

We have seen that the bimodules associated with dendroid sets of automorphisms of a rooted tree can be put in a simple form of dendroid automata. Let us describe how it is done for the bimodule $\mathfrak{M}_f$ associated with a partial covering directly and interpret the dendroid automata geometrically. The construction we describe is nothing more than a translation of the proof of Proposition 4.2 in terms of paths and monodromy actions.

As above, let P_1, P_2 be finite subsets of the plane and let $f : \mathbb{R}^2 \longrightarrow \mathbb{R}^2$ be an orientation-preserving branched covering such that P_2 contains all critical values of f and $f(P_1) \subseteq P_2$. Denote $\mathcal{M}_i = \mathbb{R}^2 \setminus P_i$ and choose a basepoint $t_2 \in \mathcal{M}_2$ and $f_1 \in f^{-1}(t_2)$. Let $\mathfrak{M}_f$ be the associated $(\pi_1(\mathcal{M}_2, t_2) - \pi_1(\mathcal{M}_1, t_1))$-bimodule. Let $\gamma_1, \ldots, \gamma_n$ be a planar generating set of $\pi_1(\mathcal{M}_2, t_2)$, where γ_i goes around a point $z_i \in P_2$.

The union of the preimages of γ_i under f is a oriented graph isomorphic to the 1-skeleton of the cycle diagram of the monodromy action of the sequence $\gamma_1, \gamma_2, \ldots, \gamma_n$ on the set $f^{-1}(t_2)$. Every connected component of the total inverse image $f^{-1}(\gamma_i)$ corresponds to a cycle of the monodromy action of γ_i and is a closed simple path containing a unique point of $f^{-1}(P_2)$. If this point belongs to P_1, then we call the component *active*. The topological disc bounded by a component of $f^{-1}(\gamma_i)$ is called a *cell*. Every cell corresponds to a 2-cell of the cycle diagram of the monodromy action.

Choose one lift e_i of γ_i in each active component of $f^{-1}(\gamma_i)$ and call the chosen arcs *marked*. The choice of the marked paths (together with the choice of the generators γ_i) determine a basis of $\mathfrak{M}_f$ and a generating set of $\pi_1(\mathcal{M}_1, t_1)$ with respect to which it is given by a dendroid automaton. Namely, for $z \in P_1$ define γ_z' as the loop going along unmarked paths of $f^{-1}(\gamma_1 \gamma_2 \cdots \gamma_n)$ to the beginning of the marked path e of the component containing z, then going back to t_1 along unmarked paths. Since the union of unmarked paths is contractible in $\mathcal{M}_1$, this description defines $\gamma_z' \in \pi_1(\mathcal{M}_1, t_1)$ uniquely. It is also easy to see that the loops γ_z' generate freely the fundamental group $\pi_1(\mathcal{M}_1, t_1)$.

For every $t \in f^{-1}(t_2)$ the corresponding element x_t of the basis of $\mathfrak{M}_f$ is the path connecting t_1 to t along the unmarked paths of $f^{-1}(\gamma_1 \gamma_2 \cdots \gamma_n)$.

The following proposition is checked directly.

Proposition 5.7. *The recursion defined by the generating sets γ_i, γ_z' and the basis x_t is given by a dendroid automaton. The generating set $\{\gamma_z'\}$ of $\pi_1(\mathcal{M}_1, t_1)$ is planar.*

5.3 Cyclically Ordered Dendroid Automata

Let $\mathcal{A}$ be a dendroid automaton over the alphabet X with the input set A and output set B. Let D be the dual Moore diagram of $\mathcal{A}$. Let $(a_1, a_2, \ldots, a_m)$ be a cyclic order on the input set A. Dendroid automata with a cyclic order of the input set are called *cyclically ordered*.

It follows from Proposition 2.5 that the product $a = a_m \cdots a_2 a_1$ of the permutations of X defined by the states a_i acts transitively on the alphabet X.

Let us denote by (a_i, y) for $y \in X$ the edge of D corresponding to a_i and starting in y. Pick a letter $x \in X$. Since the product $a = a_m \cdots a_2 a_1$ is transitive on X, the oriented edge path

$$(a_1, x), (a_2, a_1(x)), \ldots, (a_m, a_{m-1} \cdots a_2 a_1(x))$$
$$(a_1, a(x)), (a_2, a_1 a(x)), \ldots, (a_m, a_{m-1} \cdots a_1 a(x))$$
$$\vdots$$
$$(a_1, a^{d-1}(x)), (a_2, a_1 a^{d-1}(x)), \ldots, (a_m, a_{m-1} \cdots a_1 a^{d-1}(x)),$$

where $d = |X|$, is a closed cycle containing every edge of D exactly once. Consequently, the sequence of their labels

$$a_1|_x, a_2|_{a_1(x)}, \ldots, a_m|_{a_{m-1} \cdots a_2 a_1(x)}$$
$$a_1|_{a(x)}, a_2|_{a_1 a(x)}, \ldots, a_m|_{a_{m-1} \cdots a_1 a(x)}$$
$$\vdots$$
$$a_1|_{a^{d-1}(x)}, a_2|_{a_1 a^{d-1}(x)}, \ldots, a_m|_{a_{m-1} \cdots a_1 a^{d-1}(x)},$$

contains every element of the output set B of the dendroid automaton $\mathcal{A}$ once (while the rest of its elements are trivial). Let $(b_1, b_2, \ldots, b_k)$ be the cyclic order on B of appearance in the above sequence. We call it the *induced cyclic ordering of the output set*. The cyclic ordering of the output set depends only on the cyclic ordering of the original set $(a_1, a_2, \ldots, a_m)$. Note that different choice of the initial letter x changes only the initial point of the sequence $(b_1, b_2, \ldots, b_k)$, but does not change the cyclic ordering.

Suppose that we have a sequence $\mathcal{A}_1, \mathcal{A}_2, \ldots$ of dendroid automata over a sequence of alphabets $(X_1, X_2, \ldots)$, such that $\mathcal{A}_i$ has an input set A_{i-1} and output set A_i. Then every cyclic order on A_0 induced a cyclic order on the output set A_1 of $\mathcal{A}_1$, which in turn induces a cyclic order on the output set A_2 of $\mathcal{A}_2$ (since A_1 is the input set of $\mathcal{A}_1$), and so on. Thus, every cyclic order on A_0 induces cyclic orders on each of the sets A_i.

We leave the proof of the following proposition to the reader.

Proposition 5.8. *Let $\mathfrak{M}_f$ be the $(\pi_1(\mathcal{M}_2), \pi_1(\mathcal{M}_1))$-bimodule of a partial covering f of punctured planes. Let $A = (\gamma_1, \gamma_2, \ldots, \gamma_n)$ be a planar generating set of $\pi_1(\mathcal{M}_2)$ in its natural order. Choose a marking of the inverse image of this generating set and let $\mathcal{A}$ be the corresponding dendroid automaton. Then the cyclic ordering on the output set of $\mathcal{A}$ induced by the natural ordering of A coincides with its natural cyclic ordering as a planar generating set of $\pi_1(\mathcal{M}_1)$.*

5.4   Action of the Braid Groups

Let us denote the set of all dendroid sequences of length m of elements of Aut (T) by $\Delta_m(T)$. If $(a_1, a_2, \ldots, a_m) \in \Delta_m(T)$ and g belongs to the group generated by $\{a_1, \ldots, a_m\}$, then the sequence $(a_1^g, a_2^g, \ldots, a_m^g)$ belongs to $\Delta_m(T)$ and generates the same group.

We also know that the map

$$\beta_i : (a_1, a_2, \ldots, a_i, a_{i+1}, \ldots, a_m) \mapsto (a_1, a_2, \ldots, a_{i+1}, a_i^{a_{i+1}}, \ldots, a_m)$$

is an invertible transformation of $\Delta_m(T)$ (by Corollary 2.4), which does not change the group generated by the sequence.

The transformations β_i satisfy the usual defining relations for the generators of the braid group B_m on m strands, hence we get an action of B_m on $\Delta_m(T)$.

We will need the action of the braid group on dendroid sequence due to the following fact.

Proposition 5.9. *If* $(\gamma_1, \gamma_2, \ldots, \gamma_n)$ *and* $(\delta_1, \delta_2, \ldots, \delta_n)$ *are planar generating sets of the fundamental group of a punctured plane in their natural orders, then there exists an element of the braid group* $\alpha \in B_n$ *and an element* γ *of the fundamental group such that*

$$(\gamma_1^\gamma, \gamma_2^\gamma, \ldots, \gamma_n^\gamma)^\alpha = (\delta_1, \delta_2, \ldots, \delta_n).$$

Proof: By a classical result, the braid group is the mapping class group of the n-punctured disc. Conjugating by γ, we may achieve that

$$(\gamma_1 \gamma_2 \cdots \gamma_n)^\gamma = \delta_1 \delta_2 \cdots \delta_n.$$

After that α is the braid representing the isotopy class of the homeomorphisms mapping γ_i^γ to δ_i. $\qquad\square$

6   Iteration of a Single Polynomial

6.1   Planar Generating Sets

Let $f : \mathbb{R}^2 \longrightarrow \mathbb{R}^2$ be a post-critically finite topological, so that the backward iteration $(f, f, \ldots)$ is post-critically finite. Let P be the post-critical set of this iteration. Denote $\mathcal{M} = \mathbb{R}^2 \setminus P$ and let $t \in \mathcal{M}$ be an arbitrary basepoint.

Consider a planar generating set $\{\gamma_z\}_{z \in P}$ of $\pi_1(\mathcal{M}, t)$. Then construction of Proposition 5.7 gives a basis $\{x_p\}_{p \in f^{-1}(t)}$ of $\mathfrak{M}_f$ and the induced planar generating set $\{\gamma_z'\}_{z \in P}$ of the fundamental group $\pi_1(\mathcal{M}, t_1)$, where $t_1 \in f^{-1}(t)$.

Let us identify the fundamental groups $\pi_1(\mathcal{M}, t)$ and $\pi_1(\mathcal{M}, t_1)$ by a path, i.e., by the isomorphism

$$L : \pi_1(\mathcal{M}, t) \longrightarrow \pi_1(\mathcal{M}, t_1) : \gamma \mapsto \ell^{-1}\gamma\ell,$$

where ℓ is a path starting at t_1 and ending in t. This identification is well defined, up to inner automorphisms of the fundamental groups.

After the identification by the isomorphism L the bimodule $\mathfrak{M}_f$ becomes a bimodule over one group. More formally, denote by F_n the free group with the free generating set $g_1, g_2, \ldots, g_n$. Let us identify F_n with the fundamental group $\pi_1(\mathcal{M}, t)$ by the isomorphism $\phi_0 : g_i \mapsto \gamma_{z_i}$ and with $\pi_1(\mathcal{M}, t_1)$ by the isomorphism $L \circ \phi_0$. Then, by Proposition 5.9,

$$(L \circ \phi_0)^{-1}(\gamma'_{z_{i_k}}) = g_k^{g\alpha}$$

for some $g \in F_n$ and braid $\alpha \in B_n < \mathrm{Aut}\,(F_n)$. Composing the wreath recursion with an inner automorphism of the wreath product $\mathfrak{S}\,(d) \wr F_n$, we may assume that $g = 1$.

Then the bimodule $\mathfrak{M}_f$ is described by an ordered automaton with the input set $(g_1, g_2, \ldots, g_n)$ and the output set $(g_1, g_2, \ldots, g_n)^\alpha$. Here the ordering of the output set is induced by the ordering of the input set and corresponds, by Proposition 5.8, to the natural ordering of the respective planar generating set.

Definition 6.1. A *twisted kneading automaton* is a dendroid automaton with a cyclically ordered input set $(g_1, \ldots, g_n)$ and output set $(g_1, \ldots, g_n)^\alpha$ for some $\alpha \in B_n$.

Question 1. Is every bimodule given by a twisted kneading automaton sub-hyperbolic?

The twisted kneading automaton comes with a cyclic orderings of the input and output sets agreeing with the natural order on the corresponding planar generating sets of the fundamental groups. The cyclic order on the output set coincides with the order induced by the cyclic order on the input set (see Proposition 5.8).

For the definition of Thurston's combinatorial equivalence of post-critically finite branched coverings of the sphere, see [DH93].

Proposition 6.2. *The twisted kneading automaton associated to the topological polynomial f together with the cyclic order $g_1, g_2, \ldots, g_n$ of the generators of F_n uniquely determines the Thurston combinatorial class of the polynomial f.*

Proof: The Proposition is a direct corollary of the following fact, which is [Ne05, Theorem 6.5.2].

Theorem 6.3. *Let f_1, f_2 be post-critically finite orientation preserving self-coverings of the sphere S^2 with post-critical sets P_{f_1}, P_{f_2} and let $\mathfrak{M}_{f_i}$, $i = 1, 2$, be the respective $\pi_1\left(S^2 \setminus P_{f_i}\right)$-bimodules.*

Then the maps f_1 and f_2 are combinatorially equivalent if and only if there exists an isomorphism $h_* : \pi_1\left(S^2 \setminus P_{f_1}\right) \longrightarrow \pi_1\left(S^2 \setminus P_{f_2}\right)$ conjugating the bimodules $\mathfrak{M}(f_1)$ and $\mathfrak{M}(f_2)$ and induced by an orientation preserving homeomorphism $h : S^2 \longrightarrow S^2$ such that $h\left(P_{f_1}\right) = P_{f_2}$.

Here we say that an isomorphism conjugates the bimodules if the bimodules become isomorphic if we identify the groups by the isomorphism.$\square$

Of course, it would be nice to be able to find the simplest possible twisted kneading automaton describing $\mathfrak{M}_f$. In particular, we would like to know if α can be made trivial. We will call the generating set $\{\gamma_{z_i}\}$ *invariant* (for a given marking), if $\alpha = 1$, i.e., if $L(\gamma_{z_i}) = \gamma'_{z_i}$. In this case the twisted kneading automaton describing the bimodule $\mathfrak{M}_f$ is called a *kneading automaton* (see [Ne05, Section 6.7]).

The following theorem is proved in [Ne05, Theorems 6.8.3 and 6.9.1], where more details can be found.

Theorem 6.4. *If a post-critically finite polynomial f is hyperbolic (i.e., if every post-critical cycle contains a critical point), then $\pi_1(\mathcal{M}, t)$ has an invariant generating set for some choice of marking.*

In general, if f is post-critically finite, then there exists n such that there exists an invariant generating set of $\pi_1(\mathcal{M}, t)$ for the nth iteration of f.

Sketch of the proof. The idea of the proof is to find an *invariant spider* of the polynomial using external angles, i.e., a collection of disjoint curves p_{z_i} connecting the post-critical points to infinity such that $\bigcup_{z_i \in P} f^{-1}(p_{z_i})$ contains $\bigcup_{z_i \in P} p_{z_i}$ (up to homotopies). If we can find such a spider, then the generators γ_{z_i} are uniquely defined by the condition that γ_{z_i} is a simple loop going around z_i in the positive direction and intersects only p_{z_i} and only once.

Every active component of $f^{-1}(\gamma_{z_i})$ will go around one post-critical point, and hence will intersect only one path p_{z_j}. The arc intersecting p_{z_j} will be marked. It is easy to show that the chosen generating set will be invariant with respect to the given marking, if we identify $\pi_1(\mathcal{M}, t)$ with $\pi_1(\mathcal{M}, t_1)$ by a path disjoint with the legs (i.e., paths) of the spider.

If f is hyperbolic, then there exists an invariant spider constructed using "external-internal" rays. In the general case there will be no way to choose an invariant collection of external rays, but any such collection will be periodic, hence after passing to some iteration of f, we may find an invariant spider.

6.2 Example: Quadratic Polynomials

If f is a post-critically finite hyperbolic quadratic polynomial, then $\mathfrak{M}_f$ can be described by a kneading automaton (see Theorem 6.4). The following description of such kneading automata is proved in [BN08] (see also [Ne05, Section 6.11]).

If $(x_1 x_2 \ldots x_n *)^\omega$ is the kneading sequence of the polynomial f (for a definition of the kneading sequence see the above references and [BS02]), then the bimodule $\mathfrak{M}_f$ is described by the following wreath recursion:

$$
\begin{aligned}
a_1 &= \sigma(1, a_n) \\
a_{i+1} &= \begin{cases} (a_i, 1) & \text{if } x_i = 0 \\ (1, a_i) & \text{if } x_i = 1 \end{cases} , \quad i = 1, \ldots, n-1,
\end{aligned}
$$

where $\sigma \in \mathfrak{S}(2)$ is the transposition.

In general, if one deletes the trivial state and all arrows coming into it in the Moore diagram of a kneading automaton and inverts the direction of all arrows, then one will get a graph isomorphic to the graph of the action of f on its post-critical set.

Not for every sub-hyperbolic polynomial f the bimodule $\mathfrak{M}_f$ can be represented by a kneading automaton. However, it follows from the results of [BN08] that for every sub-hyperbolic iteration $(f, f, \ldots)$ of the same post-critically finite quadratic polynomial there exists a constant sequence of kneading automata $(\mathcal{A}, \mathcal{A}, \ldots)$, such that the iterated monodromy action is conjugate to the action defined by this sequence of automata. However, the input-output set of the automaton $\mathcal{A}$ will correspond to different generating sets of the fundamental group for different instances of the bimodule in the sequence (the sequence of generating sets is periodic though).

6.3 Combinatorial Spider Algorithm

The proof of Theorem 6.4 is analytic, which seems to be not satisfactory. At least it would be interesting to understand invariance of generating sets (or, which is equivalent, of spiders) in purely algebraic terms. For instance, it would be nice to be able to find algorithmically the simplest automaton describing the bimodule $\mathfrak{M}_f$ starting from a given twisted kneading automaton (which can be easily found for a given topological polynomial). This will give a way to decide when two given topological polynomials are combinatorially equivalent (due to Proposition 6.2 below) and possibly will give a better understanding of Thurston obstructions of topological polynomials.

We propose here an algorithm, inspired by the "spider algorithm" by J. Hubbard and D. Schleicher from [HS94] and by the solution of the "Hubbard's twisted rabbit problem" in [BN06]. Note that this algorithm provides only combinatorial information about the polynomial and it lacks an important part of the original spider algorithm of J. Hubbard and D. Schleicher: numerical values of coefficients of the polynomial.

Suppose that we have a post-critically finite topological polynomial f and suppose that we have chosen an arbitrary planar generating set $\{\gamma_{z_i}\}_{z_i \in P}$ of $\pi_1(\mathcal{M}, t)$ and a marking, so that we have a bimodule over the free group F_n given by a twisted kneading automaton with the input generating set $(g_1, \ldots, g_n)$ and the output generating set $(g_1, \ldots, g_n)^\alpha$ for some braid $\alpha \in B_n$. Let $\psi : F_n \longrightarrow \mathfrak{S}(\mathsf{X}) \wr F_n$ be the associated wreath recursion. We write it as a list:

$$\psi(g_i) = \sigma_i(h_{1,i}, h_{2,i}, \ldots, h_{d,i}).$$

The elements $h_{k,i} \in F_n$ are either trivial or of the form g_j^α.

Our aim is to simplify α by changing the generating set of F_n. Namely, we replace the ordered generating set (g_i) of F_n by the generating set $(g_{i,1}) = (g_i)^\alpha$, computing the images of g_i^α under ψ and rewriting the coordinates of the wreath product as words in the generating set $\{g_{i,1}\}$.

The new wreath recursion on the ordered generating set $(g_{i,1})$ will not correspond to a dendroid automaton, but since the corresponding generating set of the fundamental group is planar we can post-conjugate the recursion (i.e., change the basis of the bimodule), using Proposition 5.7, so that the new recursion will correspond to a twisted kneading automaton, i.e., to a dendroid automaton with the output set (i.e., the induced generating set) of the form $(g_{i,1})^{\alpha_1}$ for some new element $\alpha_1 \in B_n$. Note that the element α_1 is defined only up to an inner automorphism of F_n. This means that we actually work with the quotient $\overline{B}_n$ of the braid group B_n by its center.

The idea is that α_1 will be shorter than α and therefore iterations of this procedure will give us simple wreath recursions. In many cases an invariant generating set can be found in this way. We will formalize the algorithm and the question of its convergence in more algebraic terms in the next subsection. Here we present examples of work of this algorithm.

Example. A detailed analysis of this example (in a more general setting) is described in [BN06]. Consider the "rabbit polynomial", which is a quadratic polynomial f for which the bimodule $\mathfrak{M}_f$ is described by the wreath recursion

$$a = \sigma(1, c),$$
$$b = (1, a),$$
$$c = (1, b),$$

with the cyclic order (a, b, c), where $\sigma \in \mathfrak{S}(2)$ is the transposition. Note that $abc = \sigma(1, cab)$, hence this cyclic order agrees with the structure of the kneading automaton.

Let us pre-compose now this polynomial with the Dehn twist around the curve bc. The obtained post-critically finite topological polynomial will correspond to the following wreath recursion:

$$a = \sigma(1, c^{bc}),$$
$$b = (1, a),$$
$$c = (1, b^{bc}),$$

which is now a twisted kneading automaton with the same cyclic order of the generators. Let us run the combinatorial spider algorithm on this example. With respect to the new generating set $a_1 = a, b_1 = b^{bc} = b^c, c_1 = c^{bc}$ the wreath recursion is

$$a_1 = \sigma(1, c_1),$$
$$b_1 = (1, a)^{(1, b^c)} = (1, a_1^{b_1}),$$
$$c_1 = (1, b^c)^{(1,a)(1,b^c)} = (1, b_1^{a_1 b_1}).$$

This is still not a kneading automaton, but a dendroid automaton with the output set $a_2 = a_1^{b_1}, b_2 = b_1^{a_1 b_1}, c_2 = c_1$. We have to rewrite now the last recursion in terms of the new generating set a_2, b_2, c_2 (with the cyclic order (a_2, b_2, c_2)):

$$a_2 = (1, a_1^{b_1})^{-1}\sigma(1, c_1)(1, a_1^{b_1}) = \sigma((a_1^{b_1})^{-1}, c_1 a_1^{b_1}) = \sigma(a_2^{-1}, c_2 a_2),$$
$$b_2 = (1, a_1^{b_1})^{\sigma(1,c_1)(1,a_1^{b_1})} = (a_1^{b_1}, 1) = (a_2, 1),$$
$$c_2 = (1, b_2).$$

This is not a dendroid automaton, but composing the wreath recursion with conjugation by $(a_2, 1)$, we get

$$a_2 = \sigma(1, c_2^{a_2}),$$
$$b_2 = (a_2, 1),$$
$$c_2 = (1, b_2).$$

Again, we have to change the generating set to $a_3 = a_2, b_3 = b_2, c_3 = c_2^{a_2}$. Note that this generating set is ordered (a_3, c_3, b_3), since we have applied one generator of the braid group and $c_2 a_2 b_2 = a_2 c_2^{a_2} b_2 = a_3 c_3 b_3$. Then

$$a_3 = \sigma(1, c_3),$$
$$b_3 = (a_3, 1),$$
$$c_3 = (1, b_2)^{\sigma(1,c_3)} = (b_2, 1) = (b_3, 1).$$

Hence this topological polynomial is described by the kneading automaton

$$a_3 = \sigma(1, c_3), \quad b_3 = (a_3, 1), \quad c_3 = (b_3, 1),$$

with the cyclic order (a_3, c_3, b_3) of the generators. Conjugating the wreath recursion by σ, we get

$$a_3 = \sigma(c_3, 1), \quad b_3 = (1, a_3), \quad c_3 = (1, b_3),$$

which is exactly the recursion for a^{-1}, b^{-1}, c^{-1}. This implies that this recursion corresponds to the polynomial, which is complex conjugate to the original "rabbit polynomial".

6.4 The Bimodule of Twisted Kneading Automata

We will need some more technical notions related to permutational bimodules.

Definition 6.5. If α is an automorphism of a group G, then the associated bimodule $[\alpha]$ is the set of expressions of the form $\alpha \cdot g$ for $g \subset G$, where the actions are given by

$$(\alpha \cdot g) \cdot h = \alpha \cdot gh$$

and

$$h \cdot (\alpha \cdot g) = \alpha \cdot h^\alpha g.$$

We denote the element $(\alpha \cdot 1)$ just by α.

Proposition 6.6. *If α is an inner automorphism, then the bimodule $[\alpha]$ is isomorphic to the trivial bimodule G with the natural left and right actions of G on itself. In particular, if $\mathfrak{M}$ is a G-bimodule, then $\mathfrak{M} \otimes [\alpha]$ and $[\alpha] \otimes \mathfrak{M}$ are isomorphic to $\mathfrak{M}$.*

Proof: Suppose that α is conjugation by g, i.e., that $h^\alpha = h^g$ for all $h \in G$. Then the map $x \mapsto \alpha \cdot g^{-1}x$ is an isomorphism of the bimodules G and $[\alpha]$, since

$$h \cdot (\alpha \cdot g^{-1}x) = \alpha \cdot g^{-1}hg \cdot g^{-1}x = \alpha \cdot g^{-1} \cdot hx$$

and

$$(\alpha \cdot g^{-1}x) \cdot h = \alpha \cdot g^{-1} \cdot (xh)$$

for all $h, x \in G$.

The rest of the proposition follows from the fact that $\mathfrak{M} \otimes G$ and $G \otimes \mathfrak{M}$ are isomorphic to $\mathfrak{M}$ by the isomorphisms $x \otimes g \mapsto x \cdot g$ and $g \otimes x \mapsto g \cdot x$. $\qquad \square$

As above, let f be a post-critically finite topological polynomial with the post-critical set P. Fix a basepoint $t \in \mathcal{M} = \mathbb{R}^2 \setminus P$ and $t_1 \in f^{-1}(t)$. Let $\mathfrak{M}_f$ be the associated $(\pi_1(\mathcal{M}, t) - \pi_1(\mathcal{M}, t_1))$-bimodule.

Denote by F_n the free group of rank $n = |P|$ with the basis $g_1, g_2, \ldots, g_n$. Let us choose, as in the previous subsections, a planar generating set $\{\gamma_i\}$

of $\pi_1(\mathcal{M}, t)$. Choose marked lifts of γ_i and let $\gamma'_1, \gamma'_2, \ldots, \gamma'_m$ be the in-duced generating set of $\pi_1(\mathcal{M}, t_1)$. Connect t and t_1 by a path and let $L : \pi_1(\mathcal{M}, t) \longrightarrow \pi_1(\mathcal{M}, t_1)$ be the corresponding isomorphism of funda-mental groups.

We get then a pair of isomorphisms $\phi : F_n \longrightarrow \pi_1(\mathcal{M}, t)$ and $\phi_1 : F_n \longrightarrow \pi_1(\mathcal{M}, t_1)$ given by

$$\phi(g_i) = \gamma_i, \qquad \phi_1(g_i) = L(\gamma_i).$$

Let $\mathfrak{M}_0$ be the F_n-bimodule obtained from $\mathfrak{M}_f$ by identification of the group F_n with $\pi_1(\mathcal{M}, t)$ and $\pi_1(\mathcal{M}, t_1)$ by the isomorphisms ϕ and ϕ_1. More formally, it is the bimodule equal to $\mathfrak{M}_f$ as a set, with the actions given by

$$g \cdot x \cdot h = \phi(g) \cdot x \cdot \phi_1(h),$$

where on the right-hand side the original action of $\pi_1(\mathcal{M}, t)$ and $\pi_1(\mathcal{M}, t_1)$ are used.

Note that it follows from Proposition 6.6 that the isomorphism class of $\mathfrak{M}_0$ does not depend on a particular choice of the connecting path defining the isomorphism L.

Recall that the *outer automorphism group* of the free group F_n is

$$\mathrm{Out}(F_n) = \mathrm{Aut}\,(F_n)\,/\,\mathrm{Inn}(F_n),$$

where $\mathrm{Inn}(F_n)$ is the group of inner automorphisms of F_n. Let $\mathcal{G}_n$ be the image in $\mathrm{Out}(F_n)$ of the group generated by the automorphisms of F_n of the form

$$g_k^{a_{ij}} = \begin{cases} g_k & \text{if } k \neq i, \\ g_i^{g_j} & \text{if } k = i. \end{cases}$$

Note that it follows from Corollary 2.4 that image of a dendroid sequence of automorphisms of a tree under the action of the elements of $\mathcal{G}_n$ is a dendroid sequence.

Denote by $\mathfrak{G}_f$ the set of isomorphism classes of F_n-bimodules $[\alpha] \otimes \mathfrak{M}_0 \otimes [\beta]$ for all pairs $\alpha, \beta \in \mathcal{G}_n$.

Recall that for every $\alpha \in \mathrm{Out}(F_n)$ and every F_n-bimodule $\mathfrak{M}$ the bi-modules $[\alpha] \otimes \mathfrak{M}$ and $\mathfrak{M} \otimes [\alpha]$ are well defined, up to an isomorphism of bimodules, by Proposition 6.6.

Note also that the automorphisms of F_n coming from the braid group B_n become also elements of $\mathcal{G}_n$, if we permute the images of the generators, so that the image of every generators g_i is conjugate to g_i.

Proposition 6.7. *Every element of $\mathfrak{G}_f$ is equal to $\mathfrak{M} \otimes [\alpha]$ for some $\alpha \in \mathcal{G}_n$ and an F_n-bimodule $\mathfrak{M}$ given with respect to the input set $\{g_i\}$ by a kneading automaton (in some basis of $\mathfrak{M}$).*

Recall that a kneading automaton is a dendroid automaton with the same input and output sets.

Proof: It is sufficient to show that for any kneading automaton $\mathcal{A}$ defining a bimodule $\mathfrak{M}$ and every generator a_{ij} of $\mathcal{G}_n$ the bimodule $[a_{ij}] \otimes \mathfrak{M}$ is of the form $\mathfrak{M}' \otimes [\alpha]$ for some $\alpha \in \mathcal{G}_n$ and a bimodule $\mathfrak{M}'$ given by a kneading automaton.

We can find an ordering of the generating set of F_n such that a_{ij} is a generator of the braid group (one just has to put the generators g_i and g_j next to each other). We can assume then that $[a_{ij}] \otimes \mathfrak{M}$ is a bimodule associated with a post-critically finite topological polynomial. But any such a bimodule can be represented as a twisted kneading automaton, i.e., is of the form $\mathfrak{M}' \otimes [\alpha]$ for some bimodule $\mathfrak{M}'$ given by a kneading automaton and an element $\alpha \in \mathcal{G}_n$. □

The combinatorial spider algorithm can be formalized now in the following way. The cyclically ordered generating set $(\phi_1^{-1}(\gamma_1'), \dots, \phi_1^{-1}(\gamma_n'))$ of F_n corresponding to the induced generating set of $\pi_1(\mathcal{M}, t_1)$ is the image of the standard generating set $(g_i = \phi^{-1}(\gamma_i))$ of F_n under an element of the braid group, which modulo permutation of the images of the generators is equal to some $\alpha \in \mathcal{G}_n$. Consequently, $\mathfrak{M}_0 = \mathfrak{M}_0' \otimes [\alpha]$, where $\mathfrak{M}_0'$ is an F_n-bimodule given by a kneading automaton. Our aim is to find a planar generating set of $\pi_1(\mathcal{M}, t)$ for which α is trivial, or as short as possible. Since all planar generating sets are obtained from any planar generating set by application of elements of the braid group, we will change the generating set of F_n applying elements of $\mathcal{G}_n$.

Changing the generating set of F_n from (g_i) to $(g_i)^\alpha$ corresponds to conjugation of the bimodule $\mathfrak{M}_0$ by α^{-1}, i.e., to passing from $\mathfrak{M}_0' \otimes [\alpha]$ to $\mathfrak{M}_1 = [\alpha] \otimes \mathfrak{M}_0'$. The bimodule $[\alpha] \otimes \mathfrak{M}_0'$ can be also written in the form $\mathfrak{M}_1' \otimes [\alpha_1]$, where $\mathfrak{M}_1'$ is a kneading bimodule and $\alpha_1 \in \mathcal{G}_n$. Our hope is that α_1 will be shorter than α and iterating this procedure we will find a simple representation of the bimodule $\mathfrak{M}_f$. Keeping track of the cyclic order of the generators of F_n (and passing each time to the induced order) will hopefully provide an ordered kneading automaton.

Question 2. Is the bimodule $\mathfrak{G}_f$ always sub-hyperbolic?

If the bimodule $\mathfrak{G}_f$ is sub-hyperbolic, then the combinatorial spider algorithm will always converge to a finite cycle of kneading automata.

If the bimodule $\mathfrak{G}_f$ is always sub-hyperbolic, then the answer on Question 1 is positive, since then for every bimodule $\mathfrak{M}$ given by a twisted kneading automaton there will exist, by [Ne05, Proposition 2.11.5], a finite subset $N \subset \mathcal{G}_n$ and a number $m \in \mathbb{N}$, such that $\mathfrak{M}^{\otimes km}$ is isomorphic to $\mathfrak{M}_1^k \otimes \alpha_k$ for some $\alpha_k \in N$ and a bimodule $\mathfrak{M}_1$ not depending on k and given by a kneading automaton. Since bimodules given by kneading automata are sub-hyperbolic (as kneading automata are *bounded*, see [BoN03] and [Ne05, Section 3.9]), this implies sub-hyperbolicity of $\mathfrak{M}$.

6.5   An Example of the Bimodule $\mathfrak{G}_f$

The following example is considered in [Ne07]. Let F_3 be the free group on three free generators a, b, c. Denote by $\mathfrak{M}_0$ and $\mathfrak{M}_1$ the bimodules given by the recursions

$$a = \sigma(1, c), \quad b = (1, a), \quad c = (1, b)$$

and

$$a = \sigma(1, c), \quad b = (a, 1), \quad c = (1, b),$$

respectively.

Consider the following elements of the group $\mathcal{G}_3$:

$$a^\alpha = a, \quad b^\alpha = b^a, \quad c^\alpha = c,$$
$$a^\beta = a, \quad b^\beta = b, \quad c^\beta = c^b,$$
$$a^\gamma = a^c, \quad b^\gamma = b, \quad c^\gamma = c.$$

Note that these three automorphisms of F_3 generate $\mathcal{G}_3$ (for instance, the automorphism $a \mapsto a, b \mapsto b, c \mapsto c^a$ is equal to α, modulo conjugation by a, i.e., modulo an inner automorphism).

The bimodule $[\alpha] \otimes \mathfrak{M}_0$ is given by the recursion

$$a = \sigma(1, c),$$
$$b = (1, a)^{\sigma(1,c)} = (a, 1),$$
$$c = (1, b),$$

hence $[\alpha] \otimes \mathfrak{M}_0 = \mathfrak{M}_1$.

The bimodule $[\alpha] \otimes \mathfrak{M}_1$ is given by

$$a = \sigma(1, c),$$
$$b = (a, 1)^{\sigma(1,c)} = (1, a^c),$$
$$c = (1, b),$$

hence $[\alpha] \otimes \mathfrak{M}_1 = \mathfrak{M}_0 \otimes [\gamma]$. Similarly, $[\beta] \otimes \mathfrak{M}_0 = \mathfrak{M}_0 \otimes [\alpha]$ and $[\beta] \otimes \mathfrak{M}_1 = \mathfrak{M}_1$.

The bimodule $[\gamma] \otimes \mathfrak{M}_0$ is given by

$$a = (1, b^{-1})\sigma(1, c)(1, b) = \sigma(b^{-1}, cb),$$
$$b = (1, a),$$
$$c = (1, b),$$

composing with conjugation by $(b,1)$, we get

$$a = \sigma(1, c^b),$$
$$b = (1, a),$$
$$c = (1, b),$$

hence $[\gamma] \otimes \mathfrak{M}_0$ is isomorphic $\mathfrak{M}_0 \otimes [\beta]$. Similarly, $[\gamma] \otimes \mathfrak{M}_1$ is isomorphic to $\mathfrak{M}_1$.

We see that the bimodule $\mathfrak{G}_f$ is given by the recursion

$$\alpha = \sigma(1, \gamma),$$
$$\beta = (\alpha, 1),$$
$$\gamma = (\beta, 1).$$

Note that in this case $\mathcal{G}_3$ is isomorphic to the free group on 3 generators, and the bimodule $\mathfrak{G}_f$ is conjugate with the bimodule $\mathfrak{M}_0$, i.e., with the bimodule associated with the "rabbit polynomial".

The computations in the example of Section 6.3 can be written now as a the following sequence of equalities in $\mathfrak{G}_f$. We have started with the twisted kneading automaton $\mathfrak{M}_0 \otimes [\beta\gamma^{-1}]$, since

$$a^{\beta\gamma^{-1}c} = a, \quad b^{\beta\gamma^{-1}c} = b^c, \quad c^{\beta\gamma^{-1}c} = c^{bc}.$$

Then we have run through the following sequence

$$[\beta\gamma^{-1}] \otimes \mathfrak{M}_0 = \mathfrak{M}_0 \otimes [\alpha\beta^{-1}],$$
$$[\alpha\beta^{-1}] \otimes \mathfrak{M}_0 = \mathfrak{M}_1 \otimes [\alpha^{-1}],$$
$$[\alpha^{-1}] \otimes \mathfrak{M}_1 = \mathfrak{M}_0.$$

Looking at the parity of the corresponding braids, we see that the cyclic order of the input set has been changed.

Now it is easy to run the combinatorial spider algorithm for any composition of the rabbit polynomial with a homeomorphism of the plane fixing the post-critical set pointwise. The corresponding computations is the essence of a solution of J. Hubbard's "twisted rabbit problem", given in [BN06].

6.6 The Bimodule over the Pure Braid Group

The quotient $\overline{\mathcal{P}}_n$ of the pure braid group P_n by the center (i.e., its image in $\mathrm{Out}(F_n)$) is a subgroup of $\mathcal{G}_n$. Consequently, instead of looking at the bimodule $\mathfrak{G}_f$ of the isomorphism classes of the bimodules $[\alpha] \otimes \mathfrak{M}_0 \otimes [\beta]$ for $\alpha, \beta \in \mathcal{G}_n$, one can consider the $\overline{\mathcal{P}}_n$-bimodule of the isomorphism classes of the bimodules $[\alpha] \otimes \mathfrak{M}_0 \otimes [\beta]$ for $\alpha, \beta \in \overline{\mathcal{P}}_n$. This bimodule was considered in [Ne05] (see Proposition 6.6.1 about the bimodule $\mathfrak{F}$) and in [BN06]. It is isomorphic to the bimodule associated with a correspondence on the moduli space of the puncture sphere coming from the pull-back map of complex structures by the topological polynomial f.

In particular, in the above example of the rabbit polynomial, the pure braid group $\overline{\mathcal{P}}_3 < \mathcal{G}_3$ is the sub-group generated by the automorphisms $T = \beta^{-1}\alpha$ and $S = \gamma^{-1}\beta$. It follows from the recursion defining α, β, γ, that

$$T = (\alpha^{-1}, 1)\sigma(1, \gamma) = \sigma(1, \alpha^{-1}\gamma) = \sigma(1, T^{-1}S^{-1})$$

and

$$S = (\beta^{-1}, 1)(\alpha, 1) = (\beta^{-1}\alpha, 1) = (T, 1).$$

This recursion was used in [BN06] and it is the recursion associated with the post-critical rational function $1 - 1/z^2$, which is the map on the moduli space induced by the rabbit polynomial. For more details see [BN06, Ne07] and [Ne05, Section 6.6].

Question 3. Does there exist an analytic interpretation of the bimodule $\mathfrak{G}_f$ similar to the description of the bimodule over the pure braid group? In particular, are they always associated with some post-critically finite multidimensional rational maps (correspondences)?

6.7   Limit Space and Symbolic Presentation of the Julia Set

Let $\mathfrak{M}$ be a hyperbolic G-bimodule. Fix a basis X of $\mathfrak{M}$. By $X^{-\omega} = \{\ldots x_2 x_1 : x_i \in X\}$ we denote the space of the left-infinite sequences with the direct product topology.

Definition 6.8. We say that two sequences $\ldots x_2 x_1$ and $\ldots y_2 y_1$ are *equivalent* if there exists a finite set $N \subset G$ and a sequence $g_k \in N$, $k = 1, 2, \ldots$, such that

$$g_k(x_k \ldots x_1) = y_k \ldots y_1$$

for all k.

The quotient of $X^{-\omega}$ by the equivalence relation is called the *limit space* of the bimodule $\mathfrak{M}$ and is denoted $\mathcal{J}_{\mathfrak{M}}$.

Note that the equivalence relation on $X^{-\omega}$ is invariant under the shift $\ldots x_2 x_1 \mapsto \ldots x_3 x_2$, hence the shift induces a continuous self-map $\mathsf{s} : \mathcal{J}_{\mathfrak{M}} \longrightarrow \mathcal{J}_{\mathfrak{M}}$ of the limit space. The obtained dynamical system $(\mathcal{J}_{\mathfrak{M}}, \mathsf{s})$ is called the *limit dynamical system* of the bimodule.

In some cases the following description of the equivalence relation on $X^{-\omega}$ may be more convenient. For its proof see [Ne05, Proposition 3.2.7].

Proposition 6.9. *Let S be a* state-closed generating set *of G, i.e., such a generating set that for every $g \in S$ and $x \in \mathsf{X}$ we have $g|_x \in S$.*

Let $\mathcal{S} \subset \mathsf{X}^{-\omega} \times \mathsf{X}^{-\omega}$ be the set of pairs of sequences read on the labels of the left-infinite paths in the Moore diagram of S, i.e.,

$$\mathcal{S} = \{(\ldots x_2 x_1, \ldots y_2 y_1) \ : \ there\ exist\ g_k \in S\ such\ that\ g_k \cdot x_k = y_k \cdot g_{k-1}\}.$$

Then the asymptotic equivalence relation on $\mathsf{X}^{-\omega}$ is the equivalence relation generated by $\mathcal{S}$.

If $\mathfrak{M}$ is a sub-hyperbolic bimodule, then its limit dynamical system $(\mathcal{J}_{\mathfrak{M}}, \mathsf{s})$ is defined to be the limit dynamical system of its faithful quotient.

More about the limit spaces of hyperbolic bimodules, see [Ne05, Chapter 3].

A corollary of [Ne05, Theorem 5.5.3] is the following description of a topological model of the Julia set of a post-critically finite rational map.

Theorem 6.10. *Let f be a post-critically finite rational function with postcritical set P. Then the $\pi_1(\mathbb{PC}^2 \setminus P)$-bimodule $\mathfrak{M}_f$ is sub-hyperbolic and the limit dynamical system $(\mathcal{J}_{\mathfrak{M}}, \mathsf{s})$ is topologically conjugate with the restriction of f on its Julia set.*

Question 4. Find an interpretation of the limit space of the bimodule $\mathfrak{G}_f$, defined in Section 6.4. An interpretation of the limit space of the bimodule over the pure braid group can be found in [BN06, Ne07].

6.8 Realizability and Obstructions

Proposition 6.11. *Every twisted kneading automaton over F_n is associated with some post-critically finite topological polynomial.*

Proof: It is proved in [Ne05, Theorem 6.10.4] that every cyclically ordered kneading automaton can be realized by a topological polynomial (see also the proof of Proposition 5.3 in this paper). It remains to compose it with the homeomorphism realizing the respective element of the braid group.□

A natural question now is which twisted kneading automata can be realized by *complex* polynomials. This question for *kneading automata* was answered in [Ne05] in Theorem 6.10.7. We reformulate this theorem here perhaps in slightly more accessible terms.

Let $\mathcal{A}$ be an ordered kneading automaton over the alphabet X and input-output set A. Denote by D the set of sequences $x_1 x_2 \ldots \in \mathsf{X}^{\omega}$ such that there exists a non-trivial element $g \in A \cup A^{-1}$ such that $g|_{x_1 \ldots x_m}$ is non-trivial for every $k \geq 1$.

Note that then $g(x_1 x_2 \ldots)$ also belongs to D. Connect $x_1 x_2 \ldots$ to $g(x_1 x_2 \ldots)$, if $g \in A$, by an arrow, thus transforming D into a graph (multiple edges and loops are allowed, since elements of A may have fixed points or act in the same way on some sequences). We call D the *boundary graph* of the kneading automaton $\mathcal{A}$.

It also follows directly from the definition of a kneading automaton that every element of D is periodic and there is no more than one sequence $x_1 x_2 \ldots \in D$ for every $g \in A \cup A^{-1} \setminus \{1\}$. More explicitly, if m is sufficiently big (e.g., bigger than the lengths of cycles in the Moore diagram of $\mathcal{A}$), then D is isomorphic to the subgraph of the *dual* Moore diagram of $\mathcal{A}^{\otimes m}$ consisting of the edges marked by non-finitary elements of $\mathcal{A}$. (Here a state g of $\mathcal{A}$ is called finitary if there exists k such that $g|_v = 1$ in $\mathcal{A}^{\otimes k}$ for all $v \in \mathsf{X}^k$.) In particular, every component of the boundary graph is a tree with loops.

Theorem 6.12. *The kneading automaton $\mathcal{A}$ can be realized by a complex polynomial if and only if every connected component of its boundary graph has at most one loop.*

We leave to the reader to check that the condition of this theorem is equivalent to the condition of Theorem 6.10.7 of [Ne05].

Question 5. Find a simple criterion of absence of obstruction for a topological polynomial given by an arbitrary twisted kneading automaton.

Bibliography

[AHM05] W. Aitken, F. Hajir, and C. Maire, *Finitely ramified iterated extensions,* Int. Math. Res. Not. **14** (2005), 855–880.

[BGN03] L. Bartholdi, R. Grigorchuk, and V. Nekrashevych, *From fractal groups to fractal sets,* in: Fractals in Graz 2001. Analysis–Dynamics–Geometry–Stochastics, P. Grabner and W. Woess Eds., Birkhäuser Verlag, Basel, Boston, Berlin, 2003, 25–118.

[BKN08] L. Bartholdi, V. Kaimanovich, and V. Nekrashevych, *On amenability of automata groups* (2008), Preprint, arXiv:0802.2837.

[BN06] L. Bartholdi and V. Nekrashevych, *Thurston equivalence of topological polynomials,* Acta Math. **197**:1 (2006), 1–51.

[BN08] L. Bartholdi and V. Nekrashevych, *Iterated monodromy groups of quadratic polynomials I,* Groups, Geometry, and Dynamics **2** 3 (2008), 309–336.

[BV05]   L. Bartholdi and B. Virág, *Amenability via random walks*, Duke Math. J. **130**:1 (2005), 39–56.

[BFH92] B. Bielefeld, Y. Fisher, and J. H. Hubbard, *The classification of critically preperiodic polynomials as dynamical systems*, Jour. Amer. Math. Soc. 5:4 (1992), 721–762.

[BoN03] E. Bondarenko and V. Nekrashevych, *Post-critically finite self-similar groups*, Algebra and Discrete Mathematics 2:4 (2003), 21–32.

[BJ07]   N. Boston and R. Jones, *Arboreal Galois representations*, Geom Dedic. **124** (2007), 27–35.

[BKS]   H. Bruin, A. Kaffl, and D. Schleicher, *Symbolic dynamics of quadratic polynomials*, Manuscript, in preparation.

[BS02]   H. Bruin and D. Schleicher, *Symbolic dynamics of quadratic polynomials*, Institut Mittag-Leffler, Report No. 7, (2001/2002). To be published as [BKS].

[BP06]   K. U. Bux and R. Pérez. *On the growth of iterated monodromy groups*, in: Topological and Asymptotic Aspects of Group Theory, Contemp. Math., **394**, Providence, RI: Amer. Math. Soc., 2006, 61–76.

[DH93]   A. Douady and J. H. Hubbard, *A proof of Thurston's topological characterization of rational functions*, Acta Math. **171**:2 (1993), 263–297.

[Gr00]   R. I. Grigorchuk, *Just infinite branch groups*, in: New Horizons in Pro-p Groups, A. Shalev, M. P. F. du Sautoy, and D. Segal Eds. Progress in Mathematics **184**. Birkhäuser Verlag, Basel, 2000, 121–179.

[GNS00] R. I. Grigorchuk, V. Nekrashevich, and V. I. Sushchanskii, *Automata, dynamical systems and groups*, Proceedings of the Steklov Institute of Mathematics **231** (2000), 128–203.

[GZ02]   R. I. Grigorchuk and A. Żuk, *On a torsion-free weakly branch group defined by a three state automaton*, Internat. J. Algebra Comput., 12:1 (2002), 223–246.

[HS94]   J. H. Hubbard and D. Schleicher, *The spider algorithm*, in: Complex Dynamical Systems: The Mathematics Behind the Mandelbrot and Julia Sets, R. L. Devaney Ed., Proc. Symp. Appl. Math **49** (1994), 155–180.

[Ke00]   K. Keller, *Invariant factors, Julia equivalences and the (abstract) Mandelbrot set*, Lecture Notes in Mathematics **1732**, Springer, New York, 2000.

[Ne05]   V. Nekrashevych, *Self-similar groups*, Mathematical Surveys and Monographs, **117**, Amer. Math. Soc., Providence, RI, 2005.

[Ne07]   V. Nekrashevych, *A minimal Cantor set in the space of 3-generated groups*, Geom. Dedic. 124:2 (2007), 153–190.

[Ne08]   V. Nekrashevych, *Symbolic dynamics and self-similar groups*, in: Holomorphic Dynamics and Renormalization, in honour of John Milnor's 75th birthday, M. Lyubich and M. Yampolsky Eds., Fields Institute Communications Series **53**, 2008, 25–73.

[Po93a]   A. Poirier, *The classification of postcritically finite polynomials I: Critical portraits*, Stony Brook IMS Preprint 5, 1993.

[Po93b]   A. Poirier, *The classification of postcritically finite polynomials II: Hubbard trees*, Stony Brook IMS Preprint 7, 1993.

[Si98]   S. N. Sidki, *Regular trees and their automorphisms*, Monografias de Matematica, **56**, IMPA, Rio de Janeiro, 1998.

4   The Unicritical Branner-Hubbard Conjecture

Tan Lei and Yin Yongcheng

1   Introduction

The Branner-Hubbard conjecture has recently been proved in its full generality (see [KS, QY]):

Theorem (Qiu-Yin, Kozlovski-van Strien). *Let $f : \mathbb{C} \to \mathbb{C}$ be a complex polynomial. Denote by K_f its filled Julia set. Assume that every component of K_f containing a critical point is aperiodic. Then K_f is a Cantor set.*

A particular case of this result is the unicritical one:

Theorem 1.1. *Assume that $f : \mathbb{C} \to \mathbb{C}$ is a polynomial with all but one critical point c escaping to ∞. The non-escaping critical point c might have multiplicity. Assume that the K_f-component of c is not periodic. Then K_f is a Cantor set.*

The case when c is a simple critical point was originally proved by Branner-Hubbard [BH92], which led them to the conjecture for a general polynomial.

In this short note we will illustrate the new techniques involved in the proof of this conjecture by proving a key statement (Theorem 2.1 below) that leads easily to the unicritical Theorem 1.1. The proof we present here follows essentially the same line as [QY].

These new techniques, namely the Kozlovski-Shen-Strien nest and the Kahn-Lyubich covering lemma, when combined with various other techniques, have led to several interesting results in the field. For a detailed account, see for example [KS, QY].

Acknowledgments. The authors would like to thank J. Cannizzo, L. De Marco, P. Roesch and D. Schleicher for their helpful comments.

2   Results

We set up a puzzle by taking an equipotential of f so that the bounded pieces contain no other critical points than c. These pieces consist of our puzzle of depth 0. A connected component under f^{-n} is a puzzle piece of depth n.

The tableau $T(c)$ is by definition an array indexed by $-\mathbb{N} \times \mathbb{N}$, where $\mathbb{N} = \{0, 1, \cdots\}$, with the 0th column representing the nest of the consecutive puzzle pieces containing the critical point c. Now let I be a critical puzzle piece of depth, say, m. It therefore occupies the $(-m, 0)$th entry of $T(c)$. We use $f^j(I)$, $j = 1, \cdots, m$ to form the north-east diagonal segment starting from I, so that $f^j(I)$ occupies the $(-m + j, j)$th entry of $T(c)$.

A position in $T(c)$ is marked by $\circ$ if it is non-critical, by $\bullet$ if it is a critical puzzle piece, and by $\times$ if unknown.

A *child* of a critical puzzle piece I is defined to be a critical puzzle piece J with deeper depth so that, setting $k = |I - J|$, $f^j(J)$, $j = 1, \cdots, k-1$, is not a critical piece, and $f^k(J) = I$. We will use the symbol $J \xrightarrow{f^k} I$. In other words, $f : J \to f(J)$ is non-univalent but $f^{k-1} : f(J) \to I$ is univalent. We will use δ to denote the multiplicity of c. Then $\deg(J \xrightarrow{f^k} I) = \delta$.

Viewed from $T(c)$, a child J of a critical piece I (say at depth m) is a position on the column 0, say of depth m', so that, marching from J diagonally up and to the right until reaching the depth m, one meets only non-critical positions, except the last one, which is critical.

The critical tableau $T(c)$ is said to be *persistently recurrent* if

- every horizontal line contains infinitely many critical positions,

- every vertical line contains at most finitely many critical positions,

- every critical piece has at least one and at most finitely many children.

The key in proving Theorem 1.1 is the following:

Theorem 2.1. *Assume that the critical tableau $T(c)$ is persistently recurrent. Then in the critical nest there are disjoint annuli with moduli μ_n such that $\mu_n \geq C > 0$ for some C independent of n.*

The implication Theorem 2.1 $\Longrightarrow$ Theorem 1.1 is quite classical. For details, see [QY].

To start with, we recall the three tableau rules:

Rule 1 (Vertical Segment). In each column there is either no critical position or the critical positions form a single vertical segment on the top of the column.

An *upper triangle* in a tableau is by definition a triangle with a vertical side on the left, a horizontal side on top and a diagonal segment as the third side. Its *size* is simply the length of any of its sides, and its *depth* is the depth of its lowest vertex.

Rule 2 (Double Triangle Rule). Two upper triangles in $T(c)$ of the same size and depth such that both vertical sides are critical are identical.

A *puzzle parallelogram* is a parallelogram with two vertical sides and with the top and lower sides made of diagonal segments.

Rule 3 (Double Parallelogram Rule). Given a puzzle parallelogram D in $T(c)$ whose two vertical sides are critical and whose other pieces are non-critical, then for any other puzzle parallelogram D' of the same size and depth, with the two top vertices being critical, either the lower side of D' is completely non critical, or D' is identical to D.

Lemma 2.2. *Every critical puzzle piece I has at least two children.*

Proof: This is similar to Lemma 1.3 (b) in [M00]. Since the convention there is slightly different from ours (we look at the pieces rather than the annuli), we include a proof here (see Figure 1 for the construction).

Start from a piece I on the column 0 of $T(c)$, say of depth d. March to the right until first meeting a critical position, say at column $k \geq 1$. Denote this position by I' (we know that I and I' represent the same critical piece). Now march diagonally south-west (down and to the left) to reach a position S on the column 0. By Rule 1, every position on the open diagonal $diag]S, I'[$ is non-critical. Therefore S is a child of I, and is in fact the first child.

By the hypothesis of the persistent recurrence of $T(c)$ and by Rule 1, the top of the kth-column consists of a segment of finitely critical positions (containing I') and then becomes non-critical from some depth, say d' (with $d' > d$). Denote by P this first non-critical position.

Start now from P. Follow first the south-west diagonal to reach a piece L on the column 0. By Rule 1, every position on $diag]L, P[$ is non-critical. Now start from P again and proceed up and to the right along the diagonal until reaching the depth of I. There one hits a position, denoted by W. Consider the closed diagonal segment $diag[P, W]$. Repeatedly apply Rule 2 and Rule 3 to each block of k steps starting from P (by comparing with a left-most triangle/parallelogram), and one concludes that the entire diagonal from P to W, except possibly the last block (of length $< k$), is non-critical. Then, use Rule 2 (by comparing with a left-most triangle) and then Rule 1 to conclude that every position on the last block, except possibly W, is non-critical. Therefore $diag]L, W[$ is entirely non-critical.

If W is critical then it represents the critical piece I, and L is a second child of I. If not, march right from W until the first hit of a critical position I'' (representing again the critical piece I). This position exists by assumption. Then by Rule 1, I'' leads to a second child Z of I by following its south-west diagonal until the 0th column. $\qquad \square$

Let $P_f := \bigcup_{n \geq 1} f^n(c)$ be the postcritical set of f.

Lemma 2.3. *For any two nested pieces $L \subset\subset K$ in the 0-column of $T(c)$, $(K \smallsetminus L) \cap P_f = \emptyset$ iff the horizontal strip in $T(c)$ from K to L does not contain semi-critical positions, i.e., no critical vertical segment ends at a depth m with $\mathrm{depth}(L) > m \geq \mathrm{depth}(K)$.*

3   The KSS Nest

Definition 3.1. Let I_0 be a critical puzzle piece. We will inductively define a nested sequence of critical puzzle pieces

$$I_0 \supsetneq \cdots \supsetneq I_n \supsetneq L_n \supsetneq K'_n \supsetneq K_n \supsetneq I_{n+1} \supsetneq \cdots ,$$

all descendents of I_0, as follows: Assume I_n is already defined. Then its first and last child are named L_n and K'_n, respectively. The piece K_n is the critical puzzle piece contained in K'_n such that the depth difference $|K_n - K'_n|$ is equal to $|L_n - I_n|$. Finally, I_{n+1} is the last child of K_n. By Lemma 2.2, all these pieces exist and are mutually distinct.

$$
\begin{array}{ccc}
 & & K_{n-1} \\
 & & {}^{\nearrow}{}_{f^{r_{n-1}}} \\
 & I_n & & & I_n \\
 & {}^{\nearrow}{}_{f^{s_n}} & & & {}^{\nearrow}{}_{f^{t_n}} \\
L_n & & & K'_n \\
{}^{\nearrow}{}_{f^{t_n}} \\
K_n
\end{array}
$$

This nested sequence of puzzle pieces is called the *KSS nest* (Kozlovski-Shen-Strien nest). See [KSS07] for its original construction.

Here are some basic combinatorial properties of the KSS nest. Set $s_n := |L_n - I_n| = |K_n - K'_n|$, $t_n := |K'_n - I_n|$, $r_n := |I_{n+1} - K_n|$, and $p_n := |K_n - K_{n-1}|$.

Lemma 3.2. *We have $f^{r_{n-1}}(I_n) = K_{n-1}$, $f^{s_n}(L_n) = I_n$, $f^{t_n}(K'_n) = I_n$, $f^{t_n}(K_n) = L_n$; each is of degree δ, where δ is the local degree of f at c, and $f^{p_n}(K_n) = K_{n-1}$, $\deg(f^{p_n} : K_n \to K_{n-1}) = \delta^3$.*

Proof: The first equalities $f^{r_{n-1}}(I_n) = K_{n-1}$, $f^{s_n}(L_n) = I_n$, $f^{t_n}(K'_n) = I_n$ are almost trivial from the construction: a child S of I is, in particular, a pullback of I, and the number of pullback iterates is equal to the depth difference.

However the equality $f^{t_n}(K_n) = L_n$ is a lot less trivial. We know that $f^{t_n}(K_n)$ is a piece at the same depth as L_n.

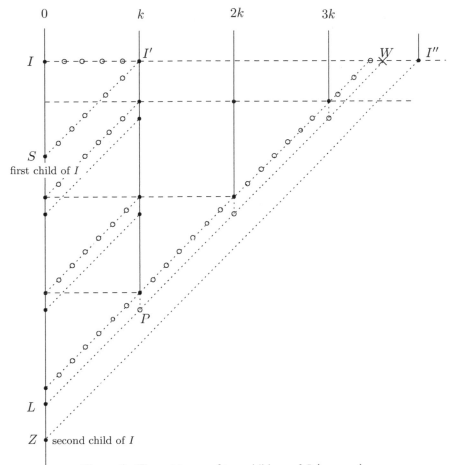

Figure 1. The existence of two children of I (or more).

Looking in the tableau $T(c)$, by the definition of a child, there are no critical positions on the diagonal between K'_n and $f^{t_n}(K'_n)$, except the ends. By Rule 1, there is no critical position on the diagonal between K_n and $f^{t_n}(K_n)$, except probably the north-east end.

Assume by contradiction that $f^{t_n}(K_n)$ is not critical.

Start from $f^{t_n}(K_n)$ and go to the west (left) until reaching the diagonal between K'_n and $f^{t_n}(K'_n)$; this horizontal segment has no critical position by Rule 1 and the fact that K'_n is a child.

One goes now from $f^{t_n}(K_n)$ to the east (right) until one hits a critical position for the first time, and then turns north and continues until reaching the depth of I_n, to reach a piece, which we name I'. Now I' must be critical by Rule 1, i.e., it is equal to I_n. Follow now the south-west diagonal from

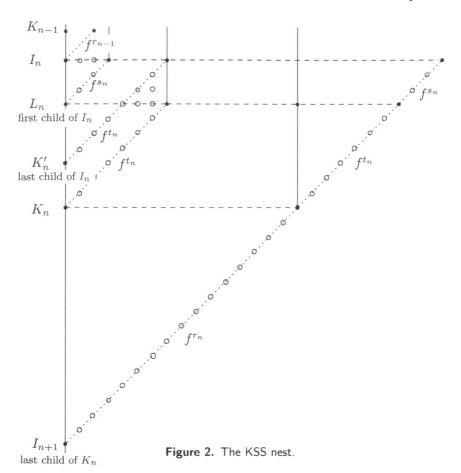

Figure 2. The KSS nest.

I' until reaching the 0th column; we will find another child of I_n by Rule 2 and the fact that L_n is the first child and the pieces of the diagonal from I' up to the depth above that of L_n have no critical positions. When it reaches the depth of L_n, it is again non-critical, and then it continues to avoid critical positions all the way to the 0th column due to Rule 1. This contradicts the choice of K'_n as the last child. Consequently, $f^{t_n}(K_n) = L_n$.

Finally, the equality $f^{p_n}(K_n) = K_{n-1}$, which says that K_n is a pullback of K_{n-1}, follows: as a child, I_n is a pullback of K_{n-1}, and L_n is a pullback of I_n, and finally K_n is a pullback of L_n by the above.

The degree estimates are simple consequences. $\square$

The following properties are essential:

Lemma 3.3. *For any $n \geq 1$,*

(a) *we have $t_n > s_n \geq r_{n-1}$, $p_n = t_n + s_n + r_{n-1}$, $(K_n' \setminus K_n) \cap P_f = \emptyset$;*

(b) *any two consecutive critical positions on the horizontal line of K_n have width difference in $\{t_n, t_n + 1, \ldots, r_n\}$;*

(c) *we have $r_n \geq 2t_n$, $3t_n \geq p_n \geq 2p_{n-1}$, and $p_n \geq t_n$;*

(d) *for any $m > m^-$, and for $M := |K_m - K_{m^-}|$, we have $f^M(K_m) = K_{m^-}$ of degree $\delta^{3(m-m^-)}$, and $M < 2p_m$.*

Proof: (a) The relations $t_n > s_n \geq r_{n-1}$, $p_n = t_n + s_n + r_{n-1}$ are trivial. To prove $(K_n' \setminus K_n) \cap P_f = \emptyset$, we apply Lemma 2.3. Assume that there is a semi-critical position $Q \supset E$ on the horizontal strip of K_n' and K_n. As Q is critical, we have $Q = K_n'$. As E is not critical, we have $E \neq K_n$. By Rule 2, $f^{t_n}(Q) =: I'$ is critical and therefore equal to I_n, and there are no other critical positions on the diagonal. By Rule 3, $f^{t_n}(E)$ is not critical. Now follow the argument as in the proof of $f^{t_n}(K_n) = L_n$, by going at first to the right of $f^{t_n}(E)$ until reaching the first critical position, then upwards up to the depth of I_n, and then follow the south-west diagonal until reaching the first critical position. We then find a child of I_n with depth deeper than $\text{depth}(Q) = \text{depth}(K_n')$, contradicting the fact that K_n' is the last child.

(b) At first we look at the distance between K_n and the first critical position E' to the east (right) of K_n. As the diagonal between K_n and $f^{t_n}(K_n)$ has no critical positions (except the ends), by Rule 1 the width difference $|E' - K_n|$ is at least t_n.

Now let E'', E''' be two consecutive critical positions at the depth of K_n. As E'' is critical, it is equal to K_n. By Rule 2, the diagonal between E'' and $f^{t_n}(E'')$ has no critical positions (except the ends); therefore the width difference from E'' to E''' is at least t_n.

Assume that the width difference between E'' and E''' exceeds r_n, and that E''' is on the right of E''. Then start from E''' and follow the south-west diagonal until reaching a critical position for the first time; we find a child of K_n that is deeper than I_{n+1}, contradicting the fact that I_{n+1} is the last child of K_n.

(c) Proof of $r_n \geq 2t_n$: As I_{n+1} is a child of K_n, but is not the first child, we have that $f^{r_n}(I_{n+1})$ is some kth critical position to the east (right) of K_n with $k \geq 2$. So $r_n \geq 2t_n$. The fact that $p_n = t_n + s_n + r_{n-1} \leq 3t_n$ is due to (a). Now $t_n \geq s_n \geq r_{n-1} \geq 2t_{n-1} \geq 2s_{n-1} \geq 2r_{n-2}$, so

$$p_n = t_n + s_n + r_{n-1} \geq 2t_{n-1} + 2s_{n-1} + 2r_{n-2} = 2p_{n-1}.$$

(d) $M = p_m + p_{m-1} + \cdots + p_{m^- + 1}$. So $M + p_{m^- + 1} \leq 2p_m$. The rest follows from Lemma 3.2. $\hfill\square$

4 From the KSS Nest to the Conditions of the KL Lemma

Lemma 4.1 (Kahn-Lyubich Covering Lemma [KL]). *Fix $D, d \in \mathbb{N}$ and $\eta > 0$. There is an $\varepsilon(\eta, D) > 0$ such that the following holds: given any two nests of three hyperbolic discs $A \subset\subset A' \subset\subset U$ and $B \subset\subset B' \subset\subset V$, and any proper holomorphic map $g : U \to V$ of degree at most D such that $g|_{A'} : A' \to B'$ and $g|_A : A \to B$ are both proper, $\deg(g|_{A'}) \leq d$, and $\mathrm{mod}(B' \setminus B) \geq \eta \cdot \mathrm{mod}(U \setminus A)$, then one of the following holds:*

$$\mathrm{mod}(U \setminus A) > \varepsilon(\eta, D) \quad or \quad \mathrm{mod}\,(U \setminus A) > \frac{\eta}{2d^2}\,\mathrm{mod}\,(V \setminus B)\,.$$

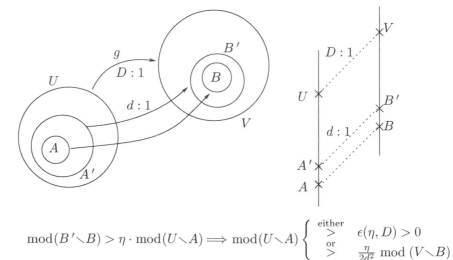

$$\mathrm{mod}(B' \setminus B) > \eta \cdot \mathrm{mod}(U \setminus A) \Longrightarrow \mathrm{mod}(U \setminus A) \begin{cases} \text{either} \\ > \\ \text{or} \\ > \end{cases} \begin{array}{l} \epsilon(\eta, D) > 0 \\ \\ \frac{\eta}{2d^2}\,\mathrm{mod}\,(V \setminus B) \end{array}$$

Figure 3. The Kahn-Lyubich Covering Lemma.

This lemma will be our main analytic tool (along with the Grötzsch inequality) to estimate the moduli of annuli.

Fix $N > 0$ (which will be $2\delta^{34} + 1$ later on). Fix $m > N$. We will construct a KL-map $g : (U, A', A) \to (V, B', B)$ depending on the pair (N, m). Set $U = K_m$ and $V = K_{m-N}$.

Let σ, M be the integers such that $f^\sigma(K'_{m+4}) = K_m = U$, $f^M(K_m) = K_{m-N} = V$. View U, V as sitting on the σth and the $(\sigma + M)$th column of $T(c)$. Set $g = f^M|_U : U \to V$.

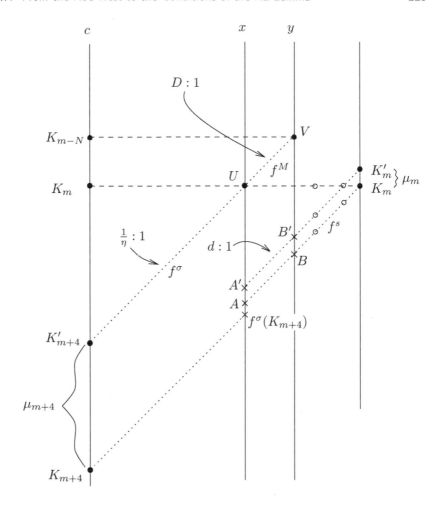

Figure 4. Applying the Kahn-Lyubich Covering Lemma.

Note that $c \in K'_{m+4}$. Set $x = f^{\sigma}(c)$ and $y = f^M(x)$. As c is persistently recurrent, there is a minimal $s \geq 0$ such that $f^s(y) \in K_m$. This corresponds in $T(c)$ to starting from the column of y at depth(K_m) and going to the right $s \geq 0$ steps until first hitting a critical position. If $s = 0$, i.e., if $y \in K_m$, set $B = K_m$ and $B' = K'_m$. If $s > 0$, then $f^{-s}(K_m)$ has a component containing y, denoted B. As B never visits a critical piece before reaching K_m, the map $f^s : B \to K_m$ is actually conformal. Now $f^{-s}(K'_m)$ has a component containing y, denoted B'. As $(K'_m \setminus K_m) \cap P_f = \emptyset$ by Lemma 3.3, we know that $f^s : B' \to K'_m$ is also conformal. As B, B' are puzzle pieces of depth deeper than that of $K_{m-N} = V$, we have $B \subset\subset B' \subset\subset V$. Pull back B, B' by $g = f^M$ to get components A, A' containing x.

In order to apply the KL-lemma, we need to control $D := \deg(g|_U)$ and $d := \deg(g|_{A'})$.

Lemma 4.2. *Set* $\mu_j := \mathrm{mod}(K'_j \setminus K_j)$.

(a) $\sigma = t_{m+4} + r_{m+3} + p_{m+3} + p_{m+2} + p_{m+1}$ *and* $\deg(K'_{m+4} \xrightarrow{f^\sigma} K_m) = \delta^{11}$;

(b) $D := \deg(g : U \to V) = \delta^{3N}$ *(independent of m) and* $M < 2p_m$;

(c) $\#\{0 \leq j < M + s, f^j(A) \subset K_m\} \leq 6$;

(d) $d := \deg(g : A \to B) = \deg(g : A' \to B')$ *is at most* δ^6;

(e) $f^\sigma(K_{m+4}) \subset A$ *and* $\mathrm{mod}(U \setminus A) \leq \delta^{11} \cdot \mu_{m+4}$;

(f) $\mathrm{mod}(V \setminus B) \geq \sum_{j=m-N+1}^{m} \mu_j$;

(g) $\mathrm{mod}(B' \setminus B) = \mu_m$.

Proof: (a) Note that the number of iterates from K_{m+4} to K_m equals $p_{m+4} + p_{m+3} + p_{m+2} + p_{m+1}$. But $|K_{m+4} - K'_{m+4}| = s_{m+4}$. So

$$\begin{aligned} \sigma &= -s_{m+4} + p_{m+4} + p_{m+3} + p_{m+2} + p_{m+1} \\ &= t_{m+4} + r_{m+3} + p_{m+3} + p_{m+2} + p_{m+1} . \end{aligned}$$

Now the total degree is $\delta^{1+1+3+3+3} = \delta^{11}$, by Lemma 3.2.

(b) This is due to Lemma 3.3.

(c) and (d) Note that A is on the σth column of $T(c)$, and $f^M(A) = B$. As $f^s(B) = K_m$, we have $\mathrm{depth}(B) - \mathrm{depth}(K_m) = s$ and $\mathrm{depth}(A) - \mathrm{depth}(B) = M$. As $A \subset K_m$, the critical segment on the column of A reaches at least the depth of K_m.

By Lemma 3.3, two adjacent critical positions on the depth K_m are at least t_m apart. Each visit of $A, f(A), \cdots f^{M-1}(A)$ to K_m corresponds to a critical position. Hence

$$\#\{0 \leq j < M, f^j(A) \subset K_m\} \leq \frac{M}{t_m} + 1 \overset{b)}{<} \frac{2p_m}{t_m} + 1 \overset{Le.3.3}{\leq} \frac{2 \cdot 3t_m}{t_m} + 1 = 7 .$$

Note that $f^M(A)$ (which is B) first visits K_m after s-iterates. So

$$\#\{0 \leq j < M + s, f^j(A) \subset K_m\} = \#\{0 \leq j < M, f^j(A) \subset K_m\} \leq 6.$$

Now for $j \in \{0, 1, \cdots, M + s - 1\}$, if $f^j(A)$ is a critical piece, then $f^j(A) \subset K_m$. It follows that there are at most six such j and $\deg(f^{M+s} : A \to K_m) \leq \delta^6$. But $\deg(f^{M+s} : A \to K_m) = \deg(f^M : A \to B) = d$. We have $d \leq \delta^6$.

Now B' is a pullback by f^s of K'_m and B is a pullback by f^s of K_m. As $(K'_m \smallsetminus K_m) \cap P_f = \emptyset$, we have that $B' \smallsetminus B$ is a pullback by f^s of $K'_m \smallsetminus K_m$ and is also disjoint from P_f. This implies that $A' \smallsetminus A$ is a pullback by f^M of $B' \smallsetminus B$ and $\deg(f^M|_{A'}) = \deg(f^M|_A)$.

(e) Note that $f^\sigma(K_{m+4})$ and A are both puzzle pieces containing x. Thus one just needs to estimate the depths to see which is contained in which. As they are both pullbacks of K_m, we will estimate the number of visits to K_m through iteration:

$$K_{m+4} \xrightarrow{\ f^\sigma\ } f^\sigma(K_{m+4}) \xrightarrow[a)]{\ f^{s_{m+4}}\ } K_m.$$

By Lemma 3.3, two adjacent critical positions at the depth of K_m are at most r_m apart. Thus, denoting by $[x]$ the integer part of x,

$$\#\{0 \le j < s_{m+4}, f^{j+\sigma}(K_{m+4}) \subset K_m\} \ge \left[\frac{s_{m+4}}{r_m}\right] \overset{Le.3.3}{\ge} \left[\frac{r_{m+3}}{r_m}\right] \overset{Le.3.3}{\ge} 8 \ .$$

Comparing with (c), we get $f^\sigma(K_{m+4}) \subset A$. Now

$$
\begin{aligned}
\mathrm{mod}(U \smallsetminus A) \quad &\overset{\text{Grötzsch}}{\le} \quad \mathrm{mod}(U \smallsetminus f^\sigma(K_{m+4})) \\
&= \quad \mathrm{mod}(f^\sigma(K'_{m+4}) \smallsetminus f^\sigma(K_{m+4})) \\
&\overset{*}{\le} \quad \deg(f^\sigma|_{K'_{m+4}}) \cdot \mathrm{mod}(K'_{m+4} \smallsetminus K_{m+4}) \\
&= \quad \deg(f^\sigma|_{K'_{m+4}}) \cdot \mu_{m+4} \overset{a)}{=} \delta^{11} \mu_{m+4},
\end{aligned}
$$

where the inequality marked by $*$ is an equality if f^σ has no critical point on $K'_{m+4} \smallsetminus K_{m+4}$, and a strict inequality otherwise (this can be proved by the Grötzsch inequality: see [M00], Problem 1-b).

(f) Note that B is a piece on the $(\sigma + M)$th column of $T(c)$. Its critical vertical segment goes down at least to the depth of K_{m-N}. For each $j = m - N + 1, \cdots, m$, denote by $l_j \ge 0$ the integer so that $f^{l_j}(y)$ visits K_j for the first time. This corresponds in $T(c)$ to starting from the column of y at $\mathrm{depth}(K_j)$ and going to the right $l_j \ge 0$ steps until the first hit of a critical position. We have $l_j \le s$, by Rule 1. Use f^{l_j} to pull back the annulus $K'_j \smallsetminus K_j$ to get an annulus C_j surrounding B. It is quite easy, using Rule 1 and the fact that $(K'_j \smallsetminus K_j) \cap P_f = \emptyset$, to see that the C_j's are mutually disjoint, and $\mathrm{mod}(C_j) = \mu_j$. Therefore

$$\mathrm{mod}(V \smallsetminus B) \ge \sum_{j=m-N+1}^{m} \mu_j \ .$$

(g) $\mathrm{mod}(B' \smallsetminus B) = \mathrm{mod}(f^s(B') \smallsetminus f^s(B)) = \mathrm{mod}(K'_m \smallsetminus K_m) = \mu_m$. $\qquad \square$

5   Final Arguments

Proof of Theorem 2.1: Fix a critical puzzle piece I_0. Construct the corresponding KSS nest. In particular, the annuli $K'_n \smallsetminus K_n$ are mutually disjoint and are nested. Set $\mu_n = \mathrm{mod}(K'_n \smallsetminus K_n)$. We want to show that there is a $C > 0$ so that $\mu_n \geq C$ for all n.

Set $N = 2\delta^{34} + 1$ and construct $(U, V, \sigma, M, x, y, g, B', B, A', A)$ as above, for all $m > N$. Then by Lemma 4.2,

$$
(*) \begin{cases}
\eta := \dfrac{1}{\deg(f^\sigma|_{K'_{m+4}})} &= \dfrac{1}{\delta^{11}}, \\[2mm]
D := \deg(f^M|_U) &= \delta^{3N}, \\[2mm]
d := \deg(f^M|_{A'}) &\leq \delta^6 \quad \text{and} \quad f^\sigma(K_{m+4}) \subset A .
\end{cases}
$$

Assume there is a sequence $k_n \to \infty$ such that $\mu_{m'} \geq \mu_{k_n}$ for $m' < k_n$ (otherwise, there is an n_0 with $\inf \mu_n = \mu_{n_0} > 0$). Fix n so that $k_n - 4 > N$. Set $m = k_n - 4$. Then $\mu_{m'} \geq \mu_{m+4}$ for any $m' \leq m + 4$. Hence,

$$
\mathrm{mod}(B' \smallsetminus B) \overset{g)}{=} \mu_m \overset{\text{choice of } m}{\geq} \mu_{m+4} \geq \frac{\mathrm{mod}(U \smallsetminus f^\sigma(K_{m+4}))}{\deg(f^\sigma)}
$$

$$
\overset{e)}{\geq} \eta \cdot \mathrm{mod}(U \smallsetminus A) .
$$

We can then apply the Kahn-Lyubich Covering Lemma to the mapping $g = f^M \colon (U, A', A) \to (V, B', B)$ and obtain:

$$
\frac{1}{\eta} \mu_{m+4} \geq \mathrm{mod}\,(U \smallsetminus A) \begin{cases} \overset{\text{either}}{>} \varepsilon(\eta, D) > 0 ; \\[2mm] \overset{\text{or}}{>} \dfrac{\eta}{2d^2} \ \mathrm{mod}\,(V \smallsetminus B) \overset{f)}{\geq} \dfrac{\eta}{2d^2}(\mu_m + \cdots + \mu_{m-N+1}) . \end{cases}
$$

But

$$
\frac{\eta}{2d^2}(\mu_m + \cdots + \mu_{m-N+1}) \overset{\text{choice of } m}{\geq} \frac{\eta N}{2d^2}\mu_{m+4} .
$$

As $N = 2\delta^{34} + 1 > \frac{2d^2}{\eta^2}$, we get $\mu_{m+4} > \eta \cdot \varepsilon(\eta, D) > 0$.

Therefore $\inf \mu_n = \lim \mu_{k_n} \geq \eta \cdot \varepsilon(\eta, D) > 0$. (As one can see from the proof, the actual constants in $(*)$ are not important. One needs the following: D is independent of m, the constant η is independent of m, N, and the degree d is bounded by a number independent of m, N). $\square$

Bibliography

[BH92]   B. Branner and J. H. Hubbard, *The iteration of cubic polynomials, Part II: patterns and parapatterns*, Acta Math. **169** (1992), 229–325.

[KL]   J. Kahn and M. Lyubich, *The quasi-additivity law in conformal geometry*, ArXiv: math.DS/0505194, Annals of Math., to appear.

[KSS07] O. Kozlovski, W. Shen, and S. van Strien, *Rigidity for real polynomials*, Ann. of Math. **165** (2007), 749–841.

[KS]   O. Kozlovski and S. van Strien, *Local connectivity and quasiconformal rigidity of non-renormalizable polynomials*, ArXiv: math.DS/0609710.

[M00]   J. Milnor, *Local connectivity of Julia sets: expository lectures*, in: The Mandelbrot set, Theme and Variations, Tan Lei Ed., London Math. Soc. Lecture Note Ser. **274**, Cambridge Univ. Press, 2000, 67–116.

[QY]   Qiu Weiyuan and Yin Yongcheng, *Proof of the Branner-Hubbard conjecture on Cantor Julia sets*, ArXiv: math.DS/0608045.

5 A Priori Bounds for Some Infinitely Renormalizable Quadratics, III. Molecules

Jeremy Kahn and Mikhail Lyubich

1 Introduction

The most prominent component of the interior of the Mandelbrot set M is the component bounded by the main cardioid. There are infinitely many secondary hyperbolic components of int M attached to it. In turn, infinitely many hyperbolic components are attached to each of the secondary components, etc. Let us take the union of all hyperbolic components of int M obtained this way, close it up and fill it in (i.e., add all bounded components of its complement[1]). We obtain the set called the *molecule* $\mathcal{M}$ of M; see Figure 1.[2] In this paper, we consider infinitely primitively renormalizable quadratic polynomials satisfying a *molecule condition*, which means that the combinatorics of the primitive renormalization operators involved stays away from the molecule (see Section 2.2 for the precise definition in purely combinatorial terms).

An infinitely renormalizable quadratic map f is said to have *a priori bounds* if its renormalizations can be represented by quadratic-like maps $R^n f : U'_n \to U_n$ with $\mathrm{mod}(U_n \smallsetminus U'_n) \geq \mu > 0$, $n = 1, 2 \ldots$.

Our goal is to prove the following result:

Main Theorem. *Infinitely renormalizable quadratic maps satisfying the molecule condition have a priori bounds.*

By [Ly97], this implies:

Corollary 1.1. *Let $P_c : z \mapsto z^2 + c$ be an infinitely renormalizable quadratic map satisfying the molecule condition. Then the Julia set $J(P_c)$ is locally connected, and the Mandelbrot set M is locally connected at c.*

Given an $\eta > 0$, let us say that a renormalizable quadratic map satisfies the η-*molecule* condition if the combinatorics of the renormalization operators involved stays η-away from the molecule of $\mathcal{M}$.

---

[1] These bounded components could be only queer components of int M.

[2] It is also called the *cactus*.

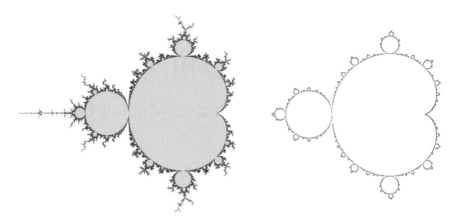

Figure 1. The Mandelbrot set (left) and its central molecule (right; courtesy of Nils Becker).

In this paper, we will deal with the case of renormalizations with sufficiently high periods. Roughly speaking, we show that if a quadratic-like map is nearly degenerate, then its geometry improves under such a renormalization. The precise statement requires the notion of a "pseudo-quadratic-like map" f, defined in Section 3, and its modulus, $\mathbf{mod}\,(f)$.

Theorem 1.2. *Given $\eta > 0$ and $\rho \in (0,1)$, there exist $\bar{\mu} > 0$ and $\underline{p} \in \mathbb{N}$ with the following property. Let f be a renormalizable quadratic-like map with period p that satisfies the η-molecule condition. If $\mathbf{mod}\,(Rf) < \bar{\mu}$ and $p \geq \underline{p}$, then $\mathbf{mod}\,(f) < \rho\,\mathbf{mod}\,(Rf)$.*

The complementary case of "bounded periods" was dealt with in [Ka06].

Remark 1. Theorem 1.2 is proved in a similar way as a more special result of [KL06]. The main difference occurs on the top level of the Yoccoz puzzle, which is modified here so that it is associated with an appropriate periodic point rather than with the fixed point of f.[3] We will focus on explaining these new elements, while only outlining the parts that are similar to [KL06].

Let us now outline the structure of the paper.

---

[3] It is similar to the difference between "non-renormalizable" and "not infinitely renormalizable" cases in the Yoccoz Theorem.

In the next section, Section 2, we lay down the combinatorial framework for our result, the Yoccoz puzzle associated to dividing cycles, and formulate the Molecule Condition precisely.

In Section 3, we summarize necessary background about *pseudo-quadratic-like maps* introduced in [Ka06], and the *pseudo-puzzle* introduced in [KL06]. From now on, the usual puzzle will serve only as a combinatorial frame, while all the geometric estimates will be made for the pseudo-puzzle. Only at the end (Section 5.6) will we return to the standard quadratic-like context.

In Section 4, we formulate a Transfer Principle that will allow us to show that if $\mathbf{mod}\,(Rf)$ is small, then the pseudo-modulus in-between appropriate puzzle pieces is even smaller.

In Section 5, we apply the Transfer Principle to the dynamical context. It implies that the extremal pseudo-distance between two specific parts of the Julia set (obtained by removing from the Julia set the central puzzle piece Y^1) is much smaller than $\mathbf{mod}\,(Rf)$ (provided the renormalization period is big). On the other hand, we show that under the Molecule Condition, this pseudo-distance is comparable with $\mathbf{mod}\,(f)$. This yields Theorem 1.2.

1.1   Terminology and Notation

Let $\mathbb{N} = \{1, 2, \dots\}$ denote the set of natural numbers, and set $\mathbb{Z}_{\geq 0} = \mathbb{N} \cup \{0\}$. Moreover, let $\mathbb{D} = \{z : |z| < 1\}$ be the unit disk and $\mathbb{T}$ the unit circle.

A *topological disk* means a simply connected domain in $\mathbb{C}$. A *continuum* K is a connected closed subset in $\mathbb{C}$. It is called *full* if all components of $\mathbb{C} \setminus K$ are unbounded.

For subsets K, Y of a topological space X, the notation $K \Subset Y$ will mean (in a slightly non-standard way) that the closure of K is compact and contained in $\operatorname{int} Y$.

We let $\operatorname{orb}(z) \equiv \operatorname{orb}_g(z) = (g^n z)_{n=0}^{\infty}$ be the *orbit* of z under a map g.

Given a map $g : U \to V$ and an open topological disk $D \subset V$, components of $g^{-1}(D)$ are called *pullbacks* of D under g. If the disk D is closed, we define pullbacks of D as the closures of the pullbacks of $\operatorname{int} D$.[4] In either case, given a connected set $X \subset g^{-1}(\operatorname{int} D)$, we let $g^{-1}(D)|X$ be the pullback of D containing X.

Acknowledgments. This work has been partially supported by the NSF and NSERC.

---

[4]Note that the pullbacks of a closed disk D can touch one another, so they are not necessarily connected components of $g^{-1}(D)$.

2 Dividing Cycles, Yoccoz Puzzles, and Renormalization

Let $f : U' \to U$ be a quadratic-like map. We assume that the domains U' and U are smooth disks and that f is even, and we normalize f so that 0 is its critical point. We let $U^m = f^{-m}(U)$. The boundary of U^m is called the *equipotential of depth* m.

By means of straightening, we can define external rays for f. They form a foliation of $U \smallsetminus K(f)$ transversal to the equipotential ∂U. Each ray is labeled by its *external angle*. These rays will play a purely combinatorial role, so a particular choice of the straightening is not important.

2.1 Dividing Cycles and Associated Yoccoz Puzzles

Let us consider a repelling periodic point γ of period $\mathbf{t}$ and the corresponding cycle $\boldsymbol{\gamma} = \{\gamma_k \equiv f^k\gamma\}_{k=0}^{\mathbf{t}-1}$. This point (and the cycle) is called *dividing* if there exist at least two rays landing at it. For instance, the landing point of the zero ray is a non-dividing fixed point, whereas the other fixed point is dividing (if repelling).

In what follows, we assume that γ *is dividing.* Let $\mathcal{R}(\gamma)$ (respectively $\mathcal{R}(\boldsymbol{\gamma})$) stand for the family of rays landing at γ (respectively $\boldsymbol{\gamma}$). Let $\mathbf{s} = \#\mathcal{R}(\gamma)$ and let $\mathbf{r} = \mathbf{ts} = \#\mathcal{R}(\boldsymbol{\gamma})$. These rays divide U into $\mathbf{t}(\mathbf{s}-1)+1$ closed topological disks $Y^0(j) \equiv Y_{\boldsymbol{\gamma}}^0(j)$ called *Yoccoz puzzle pieces of depth* 0.

Yoccoz puzzle pieces $Y^m(j) \equiv Y_{\boldsymbol{\gamma}}^m(j)$ *of depth* m are defined as the pullbacks of $Y^0(i)$ under f^m. They tile the neighborhood of $K(f)$ bounded by the equipotential ∂U^m. Each of them is bounded by finitely many arcs of this equipotential and finitely many external rays of $f^{-m}(\mathcal{R}(\boldsymbol{\gamma}))$. We will also use notation $Y^m(z)$ for the puzzle piece $Y^m(j)$ containing z in its interior. If $f^m(0) \notin \boldsymbol{\gamma}$, then there is a well-defined *critical* puzzle piece $Y^m \equiv Y^m(0)$. The critical puzzle pieces are nested around the origin:

$$Y^0 \supset Y^1 \supset Y^2 \cdots \ni 0.$$

Notice that all Y^m, $m \geq 1$, are symmetric with respect to the origin.

Let us take a closer look at some puzzle piece $Y = Y^m(i)$. Different arcs of ∂Y meet at the *corners* of Y. The corners where two external rays meet will be called *vertices* of Y; they are f^m-preimages of $\boldsymbol{\gamma}$. Let $K_Y = K(f) \cap Y$. This is a closed connected set that meets the boundary ∂Y at its vertices. Moreover, the external rays meeting at a vertex $v \in \partial Y$ chop off from $K(f)$ a continuum S_Y^v, the component of $K(f) \smallsetminus \mathrm{int}\, Y$ containing v. Let $\mathcal{Y}_{\boldsymbol{\gamma}}$ stand for the family of all puzzle pieces $Y_{\boldsymbol{\gamma}}^m(j)$. Let us finish with an obvious observation that will be constantly exploited:

Lemma 2.1. *If a puzzle piece* $Y^m(z)$ *of* $\mathcal{Y}_{\boldsymbol{\gamma}}$ *does not touch the cycle* $\boldsymbol{\gamma}$*, then* $Y^m(z) \Subset Y^0(z)$.

2.2 Renormalization Associated with a Dividing Cycle

Lemma 2.2 (see [Th, M00a]). *The puzzle piece $X^0 \equiv Y^0(f(0))$ of $\mathcal{Y}_\gamma$ containing the critical value has only one vertex, and thus is bounded by only two external rays (and one equipotential).*

In what follows, γ will denote the point of the cycle γ such that $f(\gamma)$ is the vertex of X^0. Notice that $f(0) \in \text{int } X^0$, for otherwise $f(0) = f(\gamma)$, which is impossible since 0 is the only preimage of $f(0)$.

Since the critical puzzle piece Y^1 is the pullback of X^0 under f, it has two vertices, γ and $\gamma' = -\gamma$, and is bounded by four rays, two of them landing at γ and two landing at γ'.[5] See Figure 2.

Lemma 2.3 (see [Do86]). *Let $X^r \equiv Y^r(j) \subset X^0$ be the puzzle piece attached to the boundary of X^0. Then $f^r : X^r \to X^0$ is a double branched covering.*

Proof: Let $\mathcal{C}^0$ be the union of the two rays that bound X^0, and let $\mathcal{C}^r$ be $\mathcal{C}^0$ cut by the equipotential ∂U^r. Let us orient $\mathcal{C}^0$ and then induce an orientation in $\mathcal{C}^r$. Since $f^r : \mathcal{C}^r \to \mathcal{C}^0$ is an orientation-preserving homeomorphism, it maps X^r onto X^0.

Since for $m = 1, \ldots, r - 1$, the arcs $f^m(\mathcal{C}^r) \subset \partial(f^m X^r)$ are disjoint from int X^0, the puzzle pieces $f^m X^r$ are not contained in X^0. Since they have greater depth than X^0, they are disjoint from int X^0. It follows that all the puzzle pieces $f^m(X^r)$, $m = 0, 1, \ldots, r - 1$, have pairwise disjoint interiors. (Otherwise $f^m(X^r) \supset f^n(X^r)$ for some $r > m > n \geq 0$, and applying f^{r-m}, we would conclude that $X^0 \supset f^{r-m+n}(X^r)$.)

Moreover, $f^{r-1}X^r = Y^1 \ni 0$, since Y^1 is the only pullback of X^0 under f. Hence the puzzle pieces $f^m(X^r)$, $m = 0, 1, \ldots, r - 2$, do not contain 0. It follows that $\deg(f^r : X^r \to X^0) = 2$. $\qquad\square$

Corollary 2.4. *If $f(0) \in \text{int } X^r$, then the puzzle piece Y^{r+1} has four vertices, and the map $f^r : Y^{r+1} \to Y^1$ is a double branched covering.*

Let $\Theta(\gamma) \subset \mathbb{T}$ be the set of external angles of the rays of $\mathcal{R}(\gamma)$. There is a natural equivalence relation on $\Theta(\gamma) \subset \mathbb{T}$: two angles are equivalent if the corresponding rays land at the same periodic point. Let us consider the hyperbolic convex hulls of these equivalence classes (in the disk $\mathbb{D}$ viewed as the hyperbolic plane). The union of the boundaries of these convex hulls is a finite lamination $\mathcal{P} = \mathcal{P}(\gamma)$ in $\mathbb{D}$ which is also called the *periodic ray portrait*. One can characterize all possible ray portraits that appear in this way (see [M00a]).

---

[5] We will usually say that "a puzzle piece is bounded by several external rays" without mentioning equipotentials that also form part of its boundary.

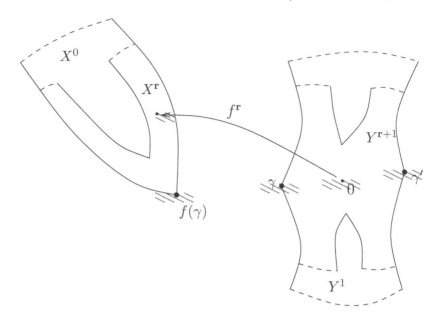

Figure 2. The critical puzzle pieces Y^1 and $Y^{\mathbf{r}+1}$.

Definition 2.5. A map f is called $\mathcal{P}$-*renormalizable* (or "renormalizable with combinatorics $\mathcal{P}$") if $f(0) \in \text{int}\, X^{\mathbf{r}}$ and $f^{\mathbf{r}m}(0) \in Y^{\mathbf{r}+1}$ for all $m = 0, 1, 2, \ldots$. In this case, the double covering $f^{\mathbf{r}} : Y^{\mathbf{r}+1} \to Y^1$ is called the *renormalization* $R_{\mathcal{P}} f = R_\gamma f$ of f (associated with the cycle γ). The corresponding *little (filled) Julia set* $\mathcal{K} = K(Rf)$ is defined as

$$\{z : f^{\mathbf{r}m} z \in Y^{\mathbf{r}+1}, \quad m = 0, 1, 2 \ldots \}.$$

If the little Julia sets $f^m \mathcal{K}$, $k = 0, 1 \ldots, \mathbf{r} - 1$, are pairwise disjoint, then the renormalization is called *primitive*; otherwise it is called *satellite*.

In case γ is the dividing fixed point of f, the map is also called *immediately renormalizable*. (This is a particular case of satellite renormalization.)

Remark 2. The above definition of renormalization is not quite standard, since the map $R_\gamma f$ is not quadratic-like. To obtain the usual notion of renormalization, one should thicken the domain of $R_\gamma f$ a bit to make it quadratic-like (see [Do86, M00b]). This thickening does not change the Julia set, so $K(Rf)$ possesses all the properties of quadratic-like Julia sets. In particular, it has two fixed points, one of which (different from γ) is either non-repelling or dividing.

Note also that in the case when $f^r(0) = \gamma'$ the puzzle piece $Y^{\mathbf{r}+1}$ degenerates (is pinched at 0), but this does not affect any of our further considerations.

Given a periodic ray portrait $\mathcal{P}$, the set of parameters $c \in \mathbb{C}$ for which the quadratic polynomial P_c is $\mathcal{P}$-renormalizable form a *little copy* $M_{\mathcal{P}}$ of the Mandelbrot set ("M-copy"). Thus, there is a one-to-one correspondence between admissible ray portraits and little M-copies. Consequently, one can encode the combinatorics of the renormalization by the little M-copies themselves.

2.3 The Molecule Condition

The molecule $\mathcal{M}$ defined in the introduction consists of the quadratic maps which are:

- either finitely many times renormalizable; all these renormalizations are satellite, and the last renormalization has a non-repelling cycle;

- or infinitely many times renormalizable, with all the renormalizations satellite.

The molecule condition that we are about to introduce will ensure that our map f has frequent "qualified" primitive renormalizations. Though f is allowed to be satellite renormalizable once in a while, *we will record only the primitive renormalizations*. They are naturally ordered according to their periods, $1 = p_0 < p_1 < p_2 < \ldots$, where p_i is a multiple of p_{i-1}.

Along with these "absolute" periods of the primitive renormalizations, we will consider *relative periods* $\tilde{p}_i = p_i/p_{i-1}$ and the corresponding M-copies $\tilde{M}_i$ that encode the combinatorics of $R^i f$ as the renormalization of $R^{i-1} f$.

Given an $\eta > 0$, we say that a sequence of primitively renormalizable quadratic-like maps f_i satisfies the *η-molecule condition* if the corresponding M-copies M_i stay η-away from the molecule $\mathcal{M}$ (the latter is defined in the introduction). We say that $\{f_i\}$ satisfies the *molecule condition* if it satisfies it for some $\eta > 0$.

An infinitely primitively renormalizable map f satisfies the η-molecule condition if the sequence of its primitive renormalizations $R^i f$ does (i.e., if the corresponding relative copies $\tilde{M}_i$ stay η-away from the molecule $\mathcal{M}$). We say the map f satisfies the non-quantified molecule condition under analogous circumstances.

There is, however, a more specific combinatorial way to describe the molecule condition.

Let us consider a quadratic-like map f with straightening P_c, $c \in M$. We are going to associate to f (in some combinatorial region, and with some choice involved) three combinatorial parameters, $(\mathbf{r}, \mathbf{q}, \mathbf{n})$ ("period, valence, and escaping time") the boundedness of which will be equivalent to the molecule condition.

Assume first that f admits a dividing cycle γ with the ray portrait $\mathcal{P}$ with $\mathbf{r}$ rays. This happens if and only if c belongs to the parabolic limb of M cut off by two external rays landing at an appropriate parabolic point.

On the central domain of $\mathbb{C} \smallsetminus (\mathcal{R}(\gamma) \cup \mathcal{R}(\gamma'))$, $f^{\mathbf{r}}$ has a unique fixed point α. *Next, we assume that α is repelling and there are $\mathbf{q}$ rays landing at it.*

Assume next that the finite orbit $f^{\mathbf{r}j}(0)$, $j = 1, \ldots, \mathbf{q}\mathbf{n} - 1$, does not escape the central domain of $\mathbb{C} \smallsetminus (\mathcal{R}(\gamma) \cup \mathcal{R}(\gamma'))$. This happens if and only if c lies outside certain *decorations* of small Mandelbrot sets (see [KL06]). In particular, this happens if f is $\mathcal{P}(\gamma)$-renormalizable.

Finally, assume that $\mathbf{n}$ is the first moment n such that $f^{\mathbf{r}\mathbf{q}n}(0)$ escapes the central domain of $\mathbb{C} \smallsetminus (\mathcal{R}(\alpha) \cup \mathcal{R}(\alpha'))$. This happens if and only if c belongs to the union of $2^{\mathbf{n}}$ *decorations* inside the above parabolic limb. In particular, the map f is not $\mathcal{P}(\alpha)$-renormalizable.

Under the above assumptions, we say that f satisfies the $(\bar{\mathbf{r}}, \bar{\mathbf{q}}, \bar{\mathbf{n}})$-*molecule condition* if there is a choice of $(\mathbf{r}, \mathbf{q}, \mathbf{n})$ with $\mathbf{r} \leq \bar{\mathbf{r}}$, $\mathbf{q} \leq \bar{\mathbf{q}}$ and $\mathbf{n} \leq \bar{\mathbf{n}}$.

Lemma 2.6. *The $(\mathbf{r}, \mathbf{q}, \mathbf{n})$-molecule condition is equivalent to the η-molecule condition.*[6]

Proof: If f satisfies the $(\mathbf{r}, \mathbf{q}, \mathbf{n})$-molecule condition, then c belongs to the finite union of decorations. Each of them does not intersect the molecule $\mathcal{M}$, so c stays some distance η away from $\mathcal{M}$.

Conversely, assume there is a sequence of maps f_i that satisfies the η-molecule condition, but with $(\mathbf{r}, \mathbf{q}, \mathbf{n}) \to \infty$ for any choice of $(\mathbf{r}, \mathbf{q}, \mathbf{n})$. Let us select a convergent subsequence $c_i \to c$. Since $c \notin \mathcal{M}$, there can be only finitely many hyperbolic components $H_0, H_1, \ldots, H_m$ of $\mathcal{M}$ such that H_0 is bounded by the main cardioid, H_{k+1} bifurcates from H_k, and f is m times immediately renormalizable with the corresponding combinatorics. Let us consider the two rays landing at the last bifurcation point (where H_m is attached to H_{m-1}), and the corresponding parabolic limb of the Mandelbrot set.[7] This parabolic point has a certain period $\mathbf{r}$.

Since the quadratic polynomial P_c is not immediately renormalizable any more, the corresponding cycle α is repelling with $\mathbf{q}$ rays landing at each of its periodic points, and there is some escaping time $\mathbf{n}$. Thus, P_c satisfies the $(\mathbf{r}, \mathbf{q}, \mathbf{n})$-molecule condition. Since this condition is stable under perturbations, the P_{c_i} satisfy it as well — a contradiction. $\qquad\square$

---

[6]In the sense that if f satisfies the $(\mathbf{r}, \mathbf{q}, \mathbf{n})$-molecule condition, then it satisfies the η-molecule condition with some $\eta = \eta(\mathbf{r}, \mathbf{q}, \mathbf{n})$, and vice versa.

[7]If $m = 0$, then we consider the whole Mandelbrot set.

In what follows, we assume that parameters $\mathbf{r}, \mathbf{q}, \mathbf{n}$ *are well defined for a map f under consideration; in particular, we have two dividing cycles,* γ *and* α. We let $\mathbf{k} = \mathbf{rqn}$. Let us state for the record the following well-known combinatorial property:

Lemma 2.7. *The point* $\zeta = f^k(0)$ *is separated from* α *and* 0 *by the rays landing at* α'.[8]

Remark 3. The molecule condition can also be formulated in terms of the existence of "combinatorial horseshoes of bounded type". To define the latter, let us say that a sequence $a_1, \ldots, a_n$ of pre-periodic points in $K(f)$ is "linearly ordered" if the rays landing at a_j separate a_i from a_k whenever $i < j < k$. A *combinatorial horseshoe* in $K(f)$ is a linearly ordered sequence v, v_1, w_1, w_2, v_2, w such that $f^k(v_i) = v$ and $f^k(w_i) = w$ for some k. One can show that *the* $(\mathbf{r}, \mathbf{q}, \mathbf{n})$*-molecule condition is equivalent to the existence of a combinatorial horseshoe with bounded k* (meaning that k is bounded in terms of $\mathbf{r}, \mathbf{q}, \mathbf{n}$ and vice versa).

2.4   Combinatorial Separation Between γ and α

Along with the puzzle $\mathcal{Y}_\gamma$ associated with γ, let us consider the puzzle $\mathcal{Y}_\alpha$ associated with α. The critical puzzle piece Y_α^1 has two vertices, α and α', and is bounded by four external rays landing at these vertices. Let $\mathcal{C}$ be the union of the two external rays of ∂Y_α^1 landing at α, and let $\mathcal{C}'$ be the symmetric pair of external rays landing at α'.

Recall that $\mathbf{t}$ stands for the period of γ and $\mathbf{s}$ stands for $\#\mathcal{R}(\gamma)$.

Lemma 2.8. *There exist inverse branches* $f^{-tm} | \mathcal{C}'$, $m = 0, 1, \ldots, \mathbf{s} - 1$, *such that the union of the arcs* $f^{-tm}(\mathcal{C}')$ *separates* γ *from the cycle* α *and the co-cycle* α' *(except when* $\alpha' \in \mathcal{C}'$*).*

Proof: Let us pull the puzzle piece Y_γ^1 along the orbit γ (or equivalently, along the orbit α). By Corollary 2.4, the corresponding inverse branches

$$f^{-m} : Y_\gamma^1 \to f^{-m}(Y_\gamma^1) = f^{\mathbf{r}-m} Y_\gamma^{\mathbf{r}+1}, \quad m = 0, 1, \ldots, \mathbf{r} - 1,$$

have disjoint interiors. Hence each of these puzzle pieces contains exactly one point of α. Moreover, none of these puzzle pieces except Y_γ^1 may intersect α' (for otherwise, its image would contain two points of α).

It follows from standard properties of quadratic maps that the arc $\mathcal{C}'$ *separates* γ (which is the non-dividing fixed point of $R_\gamma f$) from γ' and α (which is the dividing fixed point of $R_\gamma f$). Hence the arcs $f^{-tm}(\mathcal{C}')$ separate γ from $f^{-tm}(\gamma')$ and $f^{-tm}(\alpha)$.

---

[8]While the latter two points are not separated.

Since each of the puzzle pieces $f^{-tm}(Y_\gamma^1)$, $m = 0, 1, \ldots, s - 1$, has two vertices (γ and $f^{-tm}(\gamma')$) and their union forms a neighborhood of γ, the rest of the Julia set is separated from γ by the union of arcs $f^{-tm}(\mathcal{C}')$, $m = 0, 1, \ldots, s - 1$. It follows that this union separates γ from the whole cycle α, and from the co-cycle α'. $\qquad\qquad\square$

2.5 A Non-degenerate Annulus

In what follows, we will be dealing only with the puzzle $\mathcal{Y}_\alpha$, so we will skip the label α in our notation.

Since by Lemma 2.7 the point $\zeta = f^k(0)$ is separated from α by $\mathcal{C}'$, the union of arcs $f^{-tm}(\mathcal{C}')$, $m = 0, 1, \ldots, s - 1$, from Lemma 2.8 separates ζ from the whole cycle α. By Lemma 2.1, $Y^r(\zeta) \Subset Y^0(\zeta)$. Pulling this back by f^k, we conclude that:

Lemma 2.9. *We have $Y^{k+r} \Subset Y^k$.*

Let $E^0 = Y^{k+r}$.

2.6 Buffers Attached to the Vertices of $P = Y^{rq(n-1)+1}$

Let us consider a nest of critical puzzle pieces

$$Y^1 \supset Y^{rq+1} \supset Y^{2rq+1} \supset \cdots \supset Y^{rq(n-1)+1} = P.$$

Since $f^{rq} : Y^{rq+1} \to Y^1$ is a double branched covering such that

$$f^{rqm}(0) \in Y^1, \quad m = 0, 1, \ldots, n - 1,$$

the puzzle piece Y^{rqk+1} is mapped by f^{rq} onto $Y^{rq(k-1)+1}$ as a double branched covering, $k = 1 \ldots, n - 1$. However, since $f^k(0) = f^{rqn}(0) \notin Y^1$, there are two non-critical puzzle pieces of depth $k + 1$ mapped univalently onto P under f^{rq}. One of these puzzle pieces, called Q_L, is attached to the point α; another one, called Q_R, is attached to α'. The following lemma is similar to [KL06, Lemma 2.1]:

Lemma 2.10. *For any vertex of P, there exists a puzzle piece $Q^v \subset P$ of depth $rq(2n-1)+1$ attached to the boundary rays of P landing at v, which is a univalent f^k-pullback of P. Moreover, these puzzle pieces are pairwise disjoint.*

2.7 The Modified Principal Nest

Until now, the combinatorics of the puzzle depended only on the parameters (r, q, n). Now we will dive into deeper waters.

Let l be the first return time of 0 to int E^0 and let $E^1 = Y^{k+r+l}$ be the pullback of E^0 along the orbit $\{f^m(0)\}_{m=0}^l$. Then $f^l : E^1 \to E^0$ is a double branched covering.

Corollary 2.11. *We have $E^1 \Subset E^0$.*

Proof: Since $\{f^m(0)\}_{m=1}^{\mathbf{r}-1}$ is disjoint from $Y_\gamma^1 \supset Y_\alpha^1 \supset E^0$, we have $l \geq \mathbf{r}$. Hence

$$f^l(E^0) \supset Y^{\mathbf{k}} \ni E^0 = f^l(E^1),$$

and the conclusion follows. $\qquad\qquad\qquad\qquad\qquad\qquad\qquad\qquad\qquad\qquad\square$

Given two critical puzzle pieces $E^1 \subset \operatorname{int} E^0$, we can construct the *(Modified) Principal Nest* of critical puzzle pieces

$$E^0 \Supset E^1 \Supset \ldots \Supset E^{\chi-1} \Supset E^\chi$$

as described in [KL05b]. It comes together with quadratic-like maps $g_n : E^n \to E^{n-1}$.

If the map f is renormalizable, then the Principal Nest terminates at some level χ. In this case, the last quadratic-like map $g_\chi : E^\chi \to E^{\chi-1}$ has a connected Julia set that coincides with the Julia set of the renormalization $R_\beta f$, where β is the f-orbit of the non-dividing fixed point β of g_χ. The renormalization level χ is also called the *height* of the nest.

Lemma 2.12. *Let f be renormalizable with period p, and let χ be the height of the Principal Nest. Then* $\operatorname{depth} E^{\chi-1} \leq 4p + \mathbf{k} + \mathbf{r}$.

Proof: Following the notation of [KL05b], we have $V^n = E^{2n}$, $W^n = E^{2n+1}$.

The level $\chi - 1$ is even, so let $\chi - 1 = 2n$. Then $E^{\chi-1} = V^n$, and $E^\chi = W^n$.

Given two puzzle pieces Y and Z such that $f^t Y = Z$, we let $t = \operatorname{Time}(Y, Z)$. Then

$$\operatorname{depth} V^n = \operatorname{Time}(V^n, V^0) + \operatorname{depth} V^0 \leq \operatorname{Time}(W^n, W^{n-2}) + \mathbf{k} + \mathbf{r},$$

where the last estimate follows from Lemma 2.9 of [KL05b]. But

$$\operatorname{Time}(W^n, W^{n-2}) \leq 4 \operatorname{Time}(W^n, V^n) = 4p.$$

2.8 Stars

Given a vertex v of some puzzle piece of depth n, let $S^n(v)$ stand for the union of the puzzle pieces of depth n attached to v (the "star" of v). Given a finite set $\mathbf{v} = \{v_j\}$ of vertices v_j, we let

$$S^n(\mathbf{v}) = \bigcup_j S^n(v_j).$$

Let us begin with an obvious observation that follows from Lemma 2.1:

Lemma 2.13. *If a puzzle piece $Y^n(z)$ is not contained in $S^n(\alpha)$, then we have $Y^n(z) \Subset Y^0(z)$.*

Lemma 2.14. *For $\lambda = \mathbf{k} + 1 + 2\mathbf{r}$, the stars $S^\lambda(\alpha_j)$ and $S^\lambda(\alpha'_j)$ do not overlap and do not contain the critical point.*

Proof: Let us consider the curves $\mathcal{C}$ and $f^{-\mathbf{t}m}\mathcal{C}'$, $m = 1, \ldots, \mathbf{s} - 1$, from Lemma 2.8. They separate α' from all other points of $\boldsymbol{\alpha} \cup \boldsymbol{\alpha'} \smallsetminus \{\alpha\}$. Furthermore, since $f^{\mathbf{k}}(0)$ is separated from 0 by $\mathcal{C}'$, there is a lift Γ of $\mathcal{C}'$ under $f^{\mathbf{k}}$ that separates α' from 0 and hence from α. It follows that the curves Γ and $f^{-\mathbf{t}m}\mathcal{C}'$, $m = 1, \ldots, \mathbf{s} - 1$, separate α' from all other points of $\boldsymbol{\alpha} \cup \boldsymbol{\alpha'}$. Since the maximal depth of these curves is $\mathbf{k} + 1$ (which is the depth of Γ), the star $S^{\mathbf{k}+1}(\alpha')$ does not overlap with the interior of the stars $S^{\mathbf{k}+1}(a)$ for all other $a \in \boldsymbol{\alpha} \cup \boldsymbol{\alpha'}$.

By symmetry, the same is true for the star $S^{\mathbf{k}+1}(\alpha)$. Since these stars do not contain 0, the pullback of $S^{\mathbf{k}+1+\mathbf{r}}(\alpha)$ under $f^{\mathbf{r}}$ (along α) is compactly contained in its interior, int $S^{\mathbf{k}+1+\mathbf{r}}(\alpha)$. It follows that $S^{\mathbf{k}+1+\mathbf{r}}(\alpha)$ does not overlap with the stars $S^{\mathbf{k}+1}(a)$ for all other $a \in \boldsymbol{\alpha} \cup \boldsymbol{\alpha'}$.

Pulling this star once more around α, we obtain a disjoint family of stars. Hence all the stars $S^{\mathbf{k}+1+2\mathbf{r}}(a)$, $a \in \boldsymbol{\alpha} \cup \boldsymbol{\alpha'}$, are pairwise disjoint. $\square$

2.9   Geometric Puzzle Pieces

In what follows we will deal with more general puzzle pieces.

Given a puzzle piece Y, of depth m, let $Y[l]$ stand for a Jordan disk bounded by the same external rays as Y and arcs of equipotentials of level l (so $Y[m] = Y$). Such a disk will be called a puzzle piece of *bidepth* (m, l).

A *geometric puzzle piece* of bidepth (m, l) is a closed Jordan domain that is the union of several puzzle pieces of the same bidepth. As for ordinary pieces, a pullback of a geometric puzzle piece of bidepth (m, l) under some iterate f^k is a geometric puzzle piece of bidepth $(m + k, l + k)$. Note also that if P and P' are geometric puzzle pieces with[9] bidepth $P \geq$ bidepth P' and $K_P \subset K_{P'}$ then $P \subset P'$.

The family of geometric puzzle pieces of bidepth (m, l) will be called $\mathcal{Y}^m[l]$.

Stars give examples of geometric puzzle pieces. Note that the pullback of a star is a geometric puzzle piece as well, but it is a star only if the pullback is univalent.

Lemma 2.15. *Given a point $z \in$ int $S^1(\alpha) \cup$ int $S^1(\alpha')$ such that $f^{\mathbf{k}}z \in$ int $S^1(\alpha)$, let $P = f^{-\mathbf{k}}(S^1(\alpha))|z$. Then $P \subset S^1(\alpha)$ or $P \subset S^1(\alpha')$.*

---

[9] The inequality between bidepths is understood componentwise.

Proof: Notice that α and α' are the only points of $\boldsymbol{\alpha} \cup \boldsymbol{\alpha'}$ contained in int $S^1(\alpha) \cup$ int $S^1(\alpha')$. Since $f^{\mathbf{k}}$ maps $\tilde{\boldsymbol{\alpha}} = \boldsymbol{\alpha} \cup \boldsymbol{\alpha'} \smallsetminus \{\alpha, \alpha'\}$ to $\boldsymbol{\alpha} \smallsetminus \{\alpha\}$, no point of $\tilde{\boldsymbol{\alpha}}$ is contained in int P. Hence P is contained in $S^1(\alpha) \cup S^1(\alpha')$.

But by our construction of $Y^{\mathbf{k}+1}$, the interior of $f^{\mathbf{k}}(Y^{\mathbf{k}+1})$ does not overlap with $S^1(\alpha)$. Hence int $Y^{\mathbf{k}+1}$ does not overlap with P. But $S^1(\alpha) \cup S^1(\alpha') \smallsetminus$ int $Y^{\mathbf{k}+1}[1]$ consists of two components, one inside $S^1(\alpha)$ and the other inside $S^1(\alpha')$. Since P is connected, it is contained in one of them. $\qquad\square$

3 Pseudo-Quadratic-Like Maps and the Pseudo-Puzzle

In this section, we will summarize the needed background on pseudo-quadratic-like maps and pseudo-puzzles. The details can be found in [Ka06, KL06].

3.1 Pseudo-Quadratic-Like Maps

Suppose that $\mathbf{U'}$, $\mathbf{U}$ are disks, $i : \mathbf{U'} \to \mathbf{U}$ is a holomorphic immersion, and $f : \mathbf{U'} \to \mathbf{U}$ is a degree d holomorphic branched cover. Suppose further that there exist full continua $K \Subset \mathbf{U}$ and $K' \Subset \mathbf{U'}$ such that $K' = i^{-1}(K) = f^{-1}(K)$. Then we say that $F = (i, f) : \mathbf{U'} \to \mathbf{U}$ is a ψ-*quadratic-like* (ψ-*ql*) map with filled Julia set K. We let

$$\mathbf{mod}\,(F) = \mathbf{mod}\,(f) = \mathrm{mod}(\mathbf{U} \smallsetminus K).$$

Lemma 3.1. *Let $F = (i, f): \mathbf{U'} \to \mathbf{U}$ be a ψ-ql map of degree d with filled Julia set K. Then i is an embedding in a neighborhood of $K' \equiv f^{-1}(K)$, and the map $g \equiv f \circ i^{-1}: \mathbf{U'} \to \mathbf{U}$ near K is quadratic-like.*

Moreover, the domains U and U' can be selected in such a way that

$$\mathrm{mod}(U \smallsetminus i(U')) \geq \mu(\mathbf{mod}\,(F)) > 0.$$

There is a natural ψ-ql map $\mathbf{U}^n \to \mathbf{U}^{n-1}$, the "restriction" of (i, f) to $\mathbf{U}^n$. Somewhat loosely, we will use the same notation $F = (i, f)$ for this restriction.

Let us normalize the ψ-quadratic-like maps under consideration so that diam $K' = $ diam $K = 1$, both K and K' contain 0 and 1, 0 is the critical point of f, and $i(0) = 0$. Let us endow the space of ψ-quadratic-like maps (considered up to independent rescalings in the domain and the range) with the Carathéodory topology. In this topology, a sequence of normalized maps $(i_n, f_n) : \mathbf{U'}_n \to \mathbf{U}_n$ converges to $(i, f) : \mathbf{U'} \to \mathbf{U}$ if the pointed domains $(\mathbf{U'}_n, 0)$ and $(\mathbf{U}_n, 0)$ converge to $\mathbf{U'}$ and $\mathbf{U}$, respectively, and the maps i_n, f_n converge uniformly to i, f, respectively, on compact subsets of $\mathbf{U'}$.

Lemma 3.2. *Let $\mu > 0$. Then the space of ψ-ql maps F with connected Julia set and $\mathrm{mod}\,(F) \geq \mu$ is compact.*

To simplify notation, we will often refer to f as a "ψ-ql map", keeping i in mind implicitly.

3.2   The Pseudo-Puzzle

Definitions.   Let $(i, f) : \mathbf{U}' \to \mathbf{U}$ be a ψ-ql map. By Lemma 3.1, it admits a quadratic-like restriction $U' \to U$ to a neighborhood of its (filled) Julia set $K = K_\mathbf{U}$. Here U' is embedded into U, so we can identify U' with $i(U')$ and $f : U' \to U$ with $f \circ i^{-1}$.

Assume that both fixed points of f are repelling. Then we can cut U by external rays landing at the α-fixed point and consider the corresponding Yoccoz puzzle.

Given a (geometric) puzzle piece Y of bidepth (m, l), recall that K_Y stands for $Y \cap K(f)$. Let us consider the topological annulus $A = \mathbf{U}^l \setminus K(f)$ and its universal covering $\hat{A}$. Let Y_i be the components of $Y \setminus K_Y$. There are finitely many of them, and each Y_i is simply connected. Hence they can be embedded into $\hat{A}$. Select such an embedding $e_i : Y_i \to \hat{A}_i$, where $\hat{A}_i$ stands for a copy of $\hat{A}$. Then glue the $\hat{A}_i$ to Y by means of e_i, i.e., let $\mathbf{Y} = Y \sqcup_{e_i} \hat{A}_i$. This is the *pseudo-piece* ("ψ-piece") associated with Y. Note that the Julia piece K_Y naturally embeds into $\mathbf{Y}$.

Lemma 3.3.

(i) *Consider two puzzle pieces Y and Z such that the map $f : Y \to Z$ is a branched covering of degree k (where $k = 1$ or $k = 2$ depending on whether Y is off-critical or not). Then there exists an induced map $\mathbf{f} : \mathbf{Y} \to \mathbf{Z}$ that is a branched covering of the same degree k.*

(ii) *Given two puzzle pieces $Y \subset Z$, the inclusion $i : Y \to Z$ extends to an immersion $\mathbf{i} : \mathbf{Y} \to \mathbf{Z}$.*

Boundary of Puzzle Pieces.   The ideal boundary of a ψ-puzzle piece $\mathbf{Y}$ is tiled by (finitely many) arcs $\lambda_i \subset \partial \hat{A}_i$ that cover the ideal boundary of $\mathbf{U}^l$ (where $(m, l) = \mathrm{bidepth}\, Y$) and arcs $\xi_i, \eta_i \subset \partial \hat{A}_i$ mapped onto the Julia set $J(f)$. The arc λ_i meets each ξ_i, η_i at a single boundary point corresponding to a path $\delta : [0, 1) \mapsto A$ that wraps around $K(f)$ infinitely many times, while η_i meets ξ_{i+1} at a vertex $v_i \in K_Y$. We say that the arcs λ_i form the *outer boundary* (or "O-boundary") $\partial_O \mathbf{Y}$ of the puzzle piece $\mathbf{Y}$, while the arcs ξ_i and η_i form its *J-boundary* $\partial_J \mathbf{Y}$. Given a vertex $v = v_i$ of a puzzle piece Y, let $\partial^v \mathbf{Y} = \eta_i \cup \xi_{i+1}$ stand for the part of the J-boundary of $\mathbf{Y}$ attached to v.

4 Transfer Principle

Let us now formulate two analytic results which will play a crucial role in what follows. The first one appears in [KL05a, Section 2.10.3]:

Quasi-Additivity Law. *Fix some $\eta \in (0,1)$. Let $\mathbf{V}$ be a topological disk, let $K_i \Subset \mathbf{V}$, $i = 1, \ldots, m$, be pairwise disjoint full compact continua, and let $\phi_i : \mathbb{A}(1, r_i) \to \mathbf{V} \setminus \cup K_j$ be holomorphic maps such that each ϕ_i is an embedding of some proper collar of $\mathbb{T}$ to a proper collar of ∂K_i. Then there exists a $\delta_0 > 0$ (depending on η and m) such that:*

If for some $\delta \in (0, \delta_0)$, $\mathrm{mod}(\mathbf{V}, K_i) < \delta$ while $\log r_i > 2\pi\eta\delta$ for all i, then

$$\mathrm{mod}(\mathbf{V}, \cup K_i) < \frac{2\eta^{-1}\delta}{m}.$$

The next result appears in [KL05a, Section 3.1.5]:

Covering Lemma. *Fix some $\eta \in (0,1)$. Let us consider two topological disks $\mathbf{U}$ and $\mathbf{V}$, two full continua $A' \subset \mathbf{U}$ and $B' \subset \mathbf{V}$, and two full compact continua $A \Subset A'$ and $B \Subset B'$.*

Let $f : \mathbf{U} \to \mathbf{V}$ be a branched covering of degree D such that A' is a component of $f^{-1}(B')$, and A is a component of $f^{-1}(B)$. Let $d = \deg(f : A' \to B')$.

Let B' also be embedded into another topological disk $\mathbf{B}'$. Assume $\mathbf{B}'$ is immersed into $\mathbf{V}$ by a map i in such a way that $i| B' = \mathrm{id}$, $i^{-1}(B') = B'$, and $i(\mathbf{B}') \setminus B'$ does not contain the critical values of f. Suppose the following "Collar Assumption" holds:

$$\mathrm{mod}(\mathbf{B}', B) > \eta\,\mathrm{mod}(\mathbf{U}, A) .$$

Then

$$\mathrm{mod}(\mathbf{U}, A) < \varepsilon(\eta, D)$$

implies

$$\mathrm{mod}(\mathbf{V}, B) < 2\eta^{-1}d^2 \,\mathrm{mod}(\mathbf{U}, A).$$

We will now apply these two geometric results to a dynamical situation. Recall from Section 2.7 that χ stands for the height of the Principal Nest, so that the quadratic-like map $g_\chi : E^\chi \to E^{\chi-1}$ represents the renormalization of f with the filled Julia set $\mathcal{K}$.

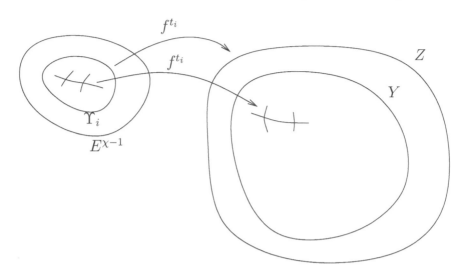

Figure 3. The Transfer Principle.

Transfer Principle. *Suppose there are two geometric puzzle pieces $Y \Subset Z$ with* $\operatorname{depth} Z < \operatorname{depth} E^{\chi-1}$, *and a sequence of moments of time* $0 < t_1 < t_2 < \cdots < t_m$ *such that:*

- $t_m - t_1 < p$ *and* $t_m < (l+1)p$ *where* $l \in \mathbb{N}$;
- $f^{t_i}(\mathcal{K}) \subset Y$;
- $\Upsilon_i = f^{-t_i}(Z)| \mathcal{K} \subset E^{\chi-1}$;
- $\deg(f^{t_i} : \Upsilon_i \to Z) \leq D$.

Then there exist a constant $C = C(l)$ and $\varepsilon = \varepsilon(D)$ such that

$$\operatorname{\mathbf{mod}}(Z, Y) < \frac{C}{m} \operatorname{\mathbf{mod}}(E^\chi, \mathcal{K}),$$

provided $\operatorname{\mathbf{mod}}(E^\chi, \mathcal{K}) < \varepsilon$.

Proof: Let $\mathcal{K}_j = f^{t_j}(\mathcal{K})$. We want to apply the Covering Lemma to the maps $f^{t_j} : (\Upsilon_j, \mathcal{K}) \to (Z, \mathcal{K}_j)$. We take $E^{\chi+l}$ as the buffer around $\mathcal{K}$. Since

$$\operatorname{depth} E^{\chi+l} = \operatorname{depth} E^{\chi-1} + (l+1)p > \operatorname{depth} Z + t_j = \operatorname{depth} \Upsilon_j,$$

we have $E^{\chi+l} \subset \Upsilon_j$ for any $j = 1, \ldots, m$.

We let $\Omega_j = f^{t_j}(E^{\chi+l})$ be the corresponding buffer around $\mathcal{K}_j$. Then $\deg(f^{t_j} : E^{\chi+l} \to \Omega_j) \leq 2^{l+1}$ since $t_j < (l+1)p$. Moreover,

$$\mathbf{mod}\,(\Omega_j, \mathcal{K}_j) \geq 2\,\mathbf{mod}\,(E^{\chi+l}, \mathcal{K}) = \frac{1}{2^l}\mathbf{mod}\,(E^{\chi-1}, \mathcal{K}) \geq \frac{1}{2^l}\mathbf{mod}\,(\Upsilon_j, \mathcal{K}),$$

(4.1)

which puts us in a position to apply the Covering Lemma with $\eta = 1/2^l$ and $d = 2^{l+1}$. It yields:

$$\mathbf{mod}\,(Z, \mathcal{K}_j) \leq 2^{3(l+1)}\,\mathbf{mod}\,(E^{\chi-1}, \mathcal{K}) = 2^{3l+4}\,\mathbf{mod}\,(E^\chi, \mathcal{K}),$$

(4.2)

provided $\mathbf{mod}\,(E^\chi, \mathcal{K}) < \varepsilon(D)$.

Let k_j be the residue of t_j mod p. Then the k_j's are pairwise different numbers in-between 0 and $p-1$, and hence the sets $f^{k_j}(E^\chi)$ are pairwise disjoint. Since $\Omega_j \subset f^{k_j}(E^\chi)$, the buffers Ω_j are pairwise disjoint as well. Moreover, by (4.1),

$$\mathbf{mod}\,(\Omega_j, \mathcal{K}_j) \geq \frac{1}{2^{l-1}}\mathbf{mod}\,(E^\chi, \mathcal{K}),$$

which, together with (4.2), puts us in a position to apply the Quasi-Additivity Law with $\eta = 1/2^{4l+3}$. It yields

$$\mathbf{mod}\,(Z, Y) \leq \frac{2^{7l+8}}{m}\mathbf{mod}\,(E^\chi, \mathcal{K}),$$

provided $\mathbf{mod}\,(E^\chi, \mathcal{K}) < \varepsilon(D)$. $\qquad\qquad\qquad\qquad\qquad\square$

5 Improving the Moduli

In this section, we will prove Theorem 1.2 for ψ-ql maps.

Let $f_i : \mathbf{U}'_i \to \mathbf{U}_i$ be a sequence of renormalizable ψ-ql maps satisfying the η-molecule condition. Let $p_i \to \infty$ stand for the renormalization periods of the f_i, and let $\mathbf{mod}\,(Rf_i) \to 0$. We need to show that

$$\mathbf{mod}\,(Rf_i)/\mathbf{mod}\,(f_i) \to \infty.$$

Let $P_{c_i} : z \mapsto z^2 + c_i$ be the straightenings of the f_i. Without loss of generality, we can assume that $c_i \to c$. Then the η-molecule condition implies that the quadratic polynomial P_c satisfies the $(\bar{\mathbf{r}}, \bar{\mathbf{q}}, \bar{\mathbf{n}})$-condition, with $\bar{\mathbf{r}}$, $\bar{\mathbf{q}}$ and $\bar{\mathbf{n}}$ depending only on η. Hence all nearby maps satisfy the $(\bar{\mathbf{r}}, \bar{\mathbf{q}}, \bar{\mathbf{n}})$-condition as well. In what follows, we will fix one of these maps, $f = f_i$, with parameters $(\mathbf{r}, \mathbf{q}, \mathbf{n}) \leq (\bar{\mathbf{r}}, \bar{\mathbf{q}}, \bar{\mathbf{n}})$, and consider its puzzle as described in Section 2. All of the objects under consideration (e.g., the principal nest $E^0 \supset E^1 \supset \dots$) will be associated with f without our making it notationally explicit.

5.1   From the Bottom to the Top of the Principal Nest

The following result, proved in [KL06, Lemma 5.3], shows that if χ is big while the modulus $\mathbf{mod}\,(E^{\chi-1}, E^\chi)$ is small, then $\mathbf{mod}\,(E^0, E^1)$ is even smaller:

Lemma 5.1. *For any* $\mathbf{k} \in \mathbb{N}$ *and* $\rho \in (0,1)$, *there exist* $\varepsilon > 0$ *and* $\underline{\chi} \in \mathbb{N}$ *such that if* $\chi \geq \underline{\chi}$ *and* $\mathbf{mod}\,(E^{\chi-1}, E^\chi) < \varepsilon$, *then*

$$\mathbf{mod}\,(E^0, E^1) < \rho\,\mathbf{mod}\,(E^{\chi-1}, E^\chi) \leq \rho\,\mathrm{mod}(\mathbf{E}^\chi, \mathcal{K}).$$

Corollary 5.2. *For any* $\mathbf{k} \in \mathbb{N}$ *and* $\rho \in (0,1)$, *there exist* $\varepsilon > 0$ *and* $\underline{\chi} \in \mathbb{N}$ *such that if* $\chi \geq \underline{\chi}$, *and* $\mathrm{mod}(\mathbf{E}^{\chi-1}, \mathcal{K}) < \varepsilon$, *then for some puzzle piece* $Y^m(z)$ *we have:*

$$0 < \mathbf{mod}\,(Y^0(z), Y^m(z)) < \rho\,\mathrm{mod}(\mathbf{E}^\chi, \mathcal{K}).$$

Proof: We apply $f^{\mathbf{r}(\mathbf{qn}+1)}$ to the pair (E^0, E^1). It maps E^1 onto some puzzle piece $Y^m(z)$, and maps E^0 onto $Y^0(z)$ with degree at most $2^{\mathbf{r}(\mathbf{qn}+1)}$. By Lemma 3.3,

$$\mathbf{mod}\,(Y^0(z), Y^m(z)) \leq 2^{\mathbf{r}(\mathbf{qn}+1)}\mathbf{mod}\,(E^0, E^1).$$

Together with Lemma 5.1, this yields the assertion.   □

5.2   Around the Stars

Recall that p stands for its renormalization period, $\mathbf{k} = \mathbf{rqn}$, and $\boldsymbol{\lambda}$ is introduced in Lemma 2.14.

Lemma 5.3. *For any* $\mathbf{k} \in \mathbb{N}$ *and* $\rho \in (0,1)$, *there exist* $\varepsilon > 0$ *and* $\underline{p} \in \mathbb{N}$ *such that if* $p \geq \underline{p}$, *and* $\mathrm{mod}(\mathbf{E}^{\chi-1}, \mathcal{K}) < \varepsilon$, *then either for some puzzle piece* $Y^\lambda(z)$,

$$0 < \mathbf{mod}\,(Y^0(z), Y^\lambda(z)) < \rho\,\mathrm{mod}(\mathbf{E}^\chi, \mathcal{K}),$$

or for some periodic point $\alpha_\mu \in \boldsymbol{\alpha}$,

$$0 < \mathbf{mod}\,(S^1(\alpha_\mu), S^\lambda(\alpha_\mu)) < \rho\,\mathrm{mod}(\mathbf{E}^\chi, \mathcal{K}).$$

Proof: By Corollary 5.2, the lemma is true when $\chi \geq \underline{\chi}$, so assume $\chi \leq \underline{\chi}$. We will derive the desired result from the Transfer Principle of Section 4. Let $E^{\chi-1} = Y^{\tau_0}$. Note that

$$\deg(f^{\tau_0} \,|\, E^{\chi-1}) \leq 2^{\underline{\chi}+\mathbf{r}(\mathbf{qn}+1)},$$

so it is bounded in terms of $\mathbf{k}$ and ρ.

Let us then select the first moment $\tau \geq \tau_0$ such that $f^\tau(\mathcal{K}) \not\subset S^\lambda(\alpha)$. Note that $\tau \leq 5p + \mathbf{k}$. Indeed, $f^{5p+\mathbf{k}}(\mathcal{K}) = f^{\mathbf{k}}(\mathcal{K}) \not\subset S^\lambda(\alpha)$ and $5p + \mathbf{k} > 4p + \mathbf{k} + \mathbf{r} \geq \tau_0$ by Lemma 2.12.

Let $m = [\frac{C}{\rho}] + 1$, where C is the constant from the Transfer Principle. Let $s = s(\mathbf{r}, \mathbf{q}, \mathbf{n})$ be the number of puzzle pieces $Y^\lambda(z)$ in the complement of the star $S^\lambda(\alpha)$ (see Section 2.8), and let $N = \mathbf{k}sm^2$. Let us consider the piece of orb $\mathcal{K}$ of length N,

$$f^t(\mathcal{K}), \quad t = \tau + 1, \ldots, \tau + N. \tag{5.3}$$

Then one of the following takes place:

(i) m sets $\mathcal{K}_j = f^{t_j}\mathcal{K}$, $j = 0, \ldots, m-1$, in the orbit (5.3) belong to some puzzle piece $Y^\lambda(z)$ in the complement of the star $S^\lambda(\alpha)$;

(ii) $\mathbf{k}m$ consecutive sets $f^t(\mathcal{K})$, $t = i + 1, \ldots i + \mathbf{k}m$, in the orbit (5.3) belong to the star $S^\lambda(\alpha)$. Here i is selected so that that $f^i(\mathcal{K})$ does not belong to the star $S^\lambda(\alpha)$.

Assume the first possibility occurs. We consider the puzzle piece $Y^0(z)$ of depth 0 containing $Y^\lambda(z)$. By Lemma 2.13, $Y^\lambda(z) \Subset Y^0(z)$. Let us consider the pullback $\Upsilon_j = f^{-t_j}(Y^0(z))|\mathcal{K}$ containing $\mathcal{K}$. Then $\Upsilon_j \subset E^{\chi-1}$ (since $t_j \geq \tau_0$).

Moreover, the map $f^{t_j} : \Upsilon_j \to Y^0(z)$ has bounded degree in terms of $\mathbf{k}$ and ρ. Indeed, $\deg(f^{\tau_0}| E^{\chi-1})$ and N, λ, $\mathbf{k}$ are bounded in these terms. Hence it is enough to show that the trajectory

$$f^m(\Upsilon_j), \quad \tau_0 < m < \tau - \lambda - \mathbf{k},$$

does not hit the critical point. But if $f^m(\Upsilon_j) \ni 0$, then $f^{m+\mathbf{k}}(\Upsilon_j)$ would land outside $S^\lambda(\alpha)$ (since $f^{\mathbf{k}}(0) \not\subset S^\lambda(\alpha)$ and depth $f^{m+\mathbf{k}}(\Upsilon_j) \geq \lambda$). Then $f^{m+\mathbf{k}}(\mathcal{K})$ would land outside $S^\lambda(\alpha)$ as well, contradicting the definition of τ as the first landing moment of orb $\mathcal{K}$ in $S^\lambda(\alpha)$ after τ_0.

Now, selecting p bigger than $\mathbf{k} + N$, we bring ourselves in the position to apply the Transfer Principle with $Y = Y^\lambda(z)$, $Z = Y^0(z)$. It yields

$$\mathbf{mod}\,(Y^0(z), Y^\lambda(z)) \leq \frac{C}{m}\mathbf{mod}\,(E^\chi, \mathcal{K}) \leq \rho\,\mathbf{mod}\,(E^\chi, \mathcal{K}).$$

Assume now that the second possibility occurs. Then there is a point $\alpha_\mu \in \alpha$ such that $f^{i+\mathbf{k}j}(\mathcal{K}) \subset S^\lambda(\alpha_\mu)$ for $j = 1, \ldots, m$, while $f^i(\mathcal{K}) \subset S^\lambda(\alpha'_\mu)$. Let us pull the star $S^1(\alpha_\mu)$ back by $f^{i+\mathbf{k}j}$:

$$\Upsilon_j = f^{-(i+\mathbf{k}j)}(S^1(\alpha_\mu))|\,\mathcal{K}, \quad j = 0, \ldots, m - 1.$$

We show that

$$\Upsilon_j \subset E^{\chi-1}. \tag{5.4}$$

We fix some j and let $\Upsilon = \Upsilon_j$. We claim that int $f^i(\Upsilon)$ does not contain any point α_ν of $\boldsymbol{\alpha}$. Otherwise, $\alpha_\nu \in f^{kj}(\text{int } f^i(\Upsilon)) = \text{int } S^1(\alpha_\mu)$. Since α_μ is the only point of $\boldsymbol{\alpha}$ inside int $S^1(\alpha_\mu)$, we conclude that $\nu = \mu$. If $\mu = 0$, then $f^i(\mathcal{K}) \subset S^\lambda(\alpha')$, so by Lemma 2.15, int $f^i(\Upsilon) \subset$ int $S^1(\alpha')$, which does not contain α. If $\mu \neq 0$, then the points α_μ and α'_μ are separated by α in the filled Julia set. Since $f^i(\Upsilon)$ is a geometric puzzle piece intersecting both $S^\lambda(\alpha_\mu)$ and $S^\lambda(\alpha'_\mu)$, it must contain α as well — a contradiction. The claim follows.

Thus, $f^i(\Upsilon)$ is a geometric puzzle piece whose interior does not contain any points of $\boldsymbol{\alpha}$. Hence it is contained in some puzzle piece $Y^0(z)$ of zero depth. Then $\Upsilon \subset f^{-i}(Y^0(z)) | 0 = Y^i$. But since $i \geq \tau_0$, $Y^i \subset Y^{\tau_0} = E^{\chi-1}$, and (5.4) follows.

Other assumptions of the Transfer Principle (with $Y = S^\lambda(\alpha_\mu)$ and $Z = S^1(\alpha_\mu)$) are valid for the same reason as in the first case. The lemma follows. $\qquad \square$

5.3 Bigons

A geometric puzzle piece with two vertices is called a *bigon*, and the corresponding pseudo-puzzle piece is called a ψ-*bigon*.

Recall from Section 3.2 that the ideal boundary of the corresponding ψ-puzzle piece $\mathbf{Y}$ comprises the outer boundary $\partial_O \mathbf{Y}$ (in the case of a bigon consisting of two arcs) and the J-boundary $\partial_J \mathbf{Y} = \partial^v \mathbf{Y} \cup \partial^w \mathbf{Y}$ attached to the vertices. Let $\mathcal{G}_{\mathbf{Y}} = \mathcal{G}_{\mathbf{Y}}(v, w)$ stand (in the case of a bigon) for the family of horizontal curves in $\mathbf{Y}$ connecting $\partial^v \mathbf{Y}$ to $\partial^w \mathbf{Y}$, and let $\mathbf{d}_{\mathbf{Y}}(v, w)$ stand for its extremal length.

More generally, let us consider a puzzle piece Y whose vertices are bicolored, i.e., they are partitioned into two non-empty subsets, B and W. This induces a natural bicoloring of $K(f) \smallsetminus \text{int } Y$ and of $\partial_J \mathbf{Y}$: namely, a component of these sets attached to a black/white vertex inherits the corresponding color. Let $\mathcal{G}_{\mathbf{Y}}$ stand for the family of horizontal curves in $\mathbf{Y}$ connecting boundary components with different colors.

For a geometric puzzle piece Y, let $v(Y) \subset K$ denote the set of vertices of Y. Suppose that $Y \subset Z$ are (geometric) puzzle pieces with the same equipotential depth; we say that Y is *cut out* of Z if $v(Z) \subset v(Y)$, so that we have produced Y by cutting out pieces of Z. If the vertices of Y are bicolored, then the vertices of Z are as well.

Lemma 5.4. *If Y is cut out of Z, and $v(Y)$ is bicolored, then the family of curves $\mathcal{G}_Z$ overflows $\mathcal{G}_Y$.*

Proof: We prove the lemma by induction on the cardinality of $v(Y) \setminus v(Z)$. First suppose that $v(Y) = v(Z) \cup \{w\}$, where $w \notin v(Z)$. Let $\gamma \in \mathcal{G}_Z$; the two endpoints of γ lie in differently colored components $\partial^x \mathbf{Z}$, $\partial^y \mathbf{Z}$ of $\partial_J \mathbf{Z}$. If γ lifts to $\mathcal{G}_Y$, then we are finished. Otherwise, we can start lifting γ from the endpoint (of γ) that lies in the component (say $\partial^x Z$) of $\partial_J Z$ whose color is different from that of w. Then that partial lift of γ will connect $\partial^x Y$ and $\partial^w Y$ and hence will belong to $\mathcal{G}_Y$.

In general, if $|v(Y)| > |v(Z)| + 1$, we can let Y' be such that $v(Y) = v(Y') \cup \{w\}$, and $v(Y') \supset v(Z)$. Then given $\gamma \in \mathcal{G}_Z$, we can lift part of γ to $\mathcal{G}_{Y'}$ by induction, and then to $\mathcal{G}_Y$ as before. $\qquad\square$

Given a puzzle piece Y with $v(Y)$ bicolored and a bigon B, we set $B \succ Y$ if the vertices of B belong to components of $K(f) \setminus \mathrm{int}\, Y$ with different colors, while the equipotential depths of B and Y are the same.

Lemma 5.5. *If $B \succ Y$, then the family $\mathcal{G}_B$ overflows $\mathcal{G}_Y$.*

Proof: Let B' be the bigon whose vertices are the vertices of Y that separate $\mathrm{int}\, Y$ from $v(B)$. Then $\mathcal{G}_B$ overflows $\mathcal{G}_{B'}$, and $\mathcal{G}_{B'}$ overflows $\mathcal{G}_Y$ by Lemma 5.5. $\qquad\square$

Let $\mathcal{W}_Y$ stand for the width of $\mathcal{G}_\mathbf{Y}$.

Lemma 5.6. *Let Y and B be two bigons such that the vertices of $f^n()$ are separated by Y for some n, and the equipotential depth of Y is 2^n times bigger than the equipotential depth of B. Then $\mathcal{W}_Y \geq 2^{-n}\mathcal{W}_B$.*

Proof: Since the vertices of $f^n(B)$ are separated by Y, there is a component Z of $f^{-n}(Y)$ such that $B \succ Z$. By Lemma 5.5, $\mathcal{W}_B \leq \mathcal{W}_Z$. On the other hand, the map $f^n : Z \to Y$ has degree at most 2^n and maps horizontal curves in Z to horizontal curves in Y. Hence $\mathcal{W}_Z \leq 2^n \mathcal{W}_Y$. $\qquad\square$

Lemma 5.7. *Let Y be a bigon with vertices u and v of depths l and m satisfying the following property: If $l = m$ then $f^l u \neq f^l v$. Then there exists an $n \leq \max(l, m) + \mathbf{r}$ such that $f^n u$ and $f^n v$ are separated by the puzzle piece $\mathrm{int}\, Y^1$.*

Proof: By symmetry, we can assume that $l \leq m$. Suppose that $f^m u \neq f^m v$; then we can find $0 \leq t < \mathbf{r}$ such that $f^{m+t} u$ and $f^{m+t} v$ are on opposite sides of the critical point, so they are separated by $\mathrm{int}\, Y^1$. Otherwise, we must have $l < m$, and then $f^{m-1} u = -f^{m-1} v$; hence they are separated by $\mathrm{int}\, Y^1$. $\qquad\square$

5.4   Amplification

We can now put together all the above results of this section as follows:

Lemma 5.8. *For any* $\mathbf{k} \in \mathbb{N}$ *and* $\rho \in (0,1)$*, there exist* $\varepsilon > 0$ *and* $\underline{p} \in \mathbb{N}$ *such that if* $p \geq \underline{p}$*, and* $\mathrm{mod}(\mathbf{E}^{\chi-1}, \mathcal{K}) < \varepsilon$*, then*

$$\mathbf{d}_{Y^1}(\alpha, \alpha') \leq \rho\,\mathrm{mod}(\mathbf{E}^{\chi}, \mathcal{K}).$$

Proof: Under our circumstances, Lemma 5.3 implies that there exist geometric puzzle pieces $Y \Subset Z$ with the bidepth of Z bounded by $(1,1)$ while the bidepth of Y is bounded in terms of $\mathbf{k}$, such that

$$\mathbf{mod}\,(Z, Y) \leq \rho\,\mathrm{mod}(\mathbf{E}^{\chi}, \mathcal{K}).$$

For any vertex v of Z, there exists a vertex v' of Y such that the rays of ∂Y landing at v' separate $\mathrm{int}\,Y$ from v. These two rays together with the two rays landing at v (truncated by the equipotential of Z) form a bigon B^v. By the Parallel Law, there exists a vertex v of Z such that

$$\mathbf{d}_{B^v}(v, v') \leq N\mathbf{mod}\,(Z, Y),$$

where $N \leq N(\mathbf{r})$ is the number of vertices of Z. By Lemma 5.7, there is an iterate $f^n(B^v)$ such that the vertices $f^n(v)$ and $f^n(v')$ are separated by $\mathrm{int}\,Y^1$. By Lemma 5.6,

$$\mathbf{d}_{Y^1}(\alpha, \alpha') \leq 2^n \mathbf{d}_{B^v}(v, v').$$

Putting the above three estimates together, we obtain the assertion.   □

Remark 4. The name "amplification" alludes to the extremal *width* that is amplified under the push-forward procedure described above.

5.5   Separation

The final step of our argument is to show that the vertices α and α' are well separated in the bigon Y^1.

Lemma 5.9. *There exists* $\kappa = \kappa(\mathbf{r}, \mathbf{q}, \mathbf{n}) > 0$ *such that*

$$\mathbf{d}_{Y^1}(\alpha, \alpha') \geq \kappa\,\mathrm{mod}(\mathbf{U}, K).$$

Idea of the proof: The proof is the same as the one of [KL06, Proposition 5.12], so we will only give an idea here.

Let $\mathbf{Y}$ be a ψ-puzzle piece, and let v and w be two vertices of it. A *multicurve in* $\mathbf{Y}$ *connecting* $\partial^v \mathbf{Y}$ *to* $\partial^w \mathbf{Y}$ is a sequence of proper paths γ_i, $i = 1, \ldots, n$, in $\mathbf{Y}$ connecting $\partial^{v_{i-1}} \mathbf{Y}$ to $\partial^{v_i} \mathbf{Y}$, where $v = v_0, v_1, \ldots, v_n = w$

is a sequence of vertices in $\mathbf{Y}$. Let $\mathbf{W_Y}(v, w)$ stand for the extremal width of the family of multicurves in $\mathbf{Y}$ connecting $\partial^v \mathbf{Y}$ to $\partial^w \mathbf{Y}$. Let

$$\mathbf{W}_Y = \sup_{v,w} \mathbf{W}_Y(v, w).$$

Let us estimate this width for the puzzle piece P introduced in Section 2.6. To this end, let us consider puzzle pieces Q^v from Lemma 2.10. Let r be the depth of these puzzle pieces, $T^{vw} = \text{cl}(K_P \smallsetminus (Q^v \cup Q^w))$, and let $v' = Q^v \cap T^{vw}$, $w' = Q^w \cap T^{vw}$. For any multicurve γ in $\mathbf{P}$ connecting $\partial^v \mathbf{P}$ to $\partial^w \mathbf{P}$, one of the following things can happen:

(i) γ skips over T^{vw};

(ii) γ contains an arc γ' connecting an equipotential of depth r to T^{vw};

(iii) γ contains two disjoint multicurves, δ^v and δ^w, that do not cross this equipotential and such that δ^v connects $\partial^v \mathbf{Q}^v$ to $\partial^{v'} \mathbf{Q}^v$, while δ^w connects $\partial^{w'} \mathbf{Q}^w$ to $\partial^w \mathbf{Q}^w$.

It is not hard to show that the width of the first two families of multicurves is $O(\text{mod}(\mathbf{U}, K)^{-1})$ (see Section 5.5-5.6 of [KL06]). Concerning each family of multicurves δ^v or δ^w that appear in (iii), it is conformally equivalent to a family of multicurves connecting two appropriate vertices of P (since Q^v and Q^w are conformal copies of P). By the Series and Parallel Laws,

$$\mathbf{W}_P \leq \frac{1}{2} \mathbf{W}_P + O(\text{mod}(\mathbf{U}, K)^{-1}),$$

which implies the desired estimate.

5.6 Conclusion

Everything is now prepared for the main results. Lemmas 5.8 and 5.9 imply:

Theorem 5.10 (Improving of the Moduli). *For any parameters $\bar{\mathbf{r}}, \bar{\mathbf{q}}, \bar{\mathbf{n}}$ and any $\rho > 0$, there exist $\underline{p} \in \mathbb{N}$ and $\varepsilon > 0$ with the following property. Let f be a renormalizable ψ-quadratic-like map with renormalization period p satisfying the $(\bar{\mathbf{r}}, \bar{\mathbf{q}}, \bar{\mathbf{n}})$-molecule condition, and let g be its first renormalization. Then*

$$\{p \geq \underline{p} \ \text{ and } \ \textbf{mod}\,(g) < \varepsilon\} \Rightarrow \textbf{mod}\,(f) < \rho\,\text{mod}(g).$$

Theorem 5.10, together with Lemma 3.1, implies Theorem 1.2 from the Introduction. The Main Theorem follows from Theorem 5.10 combined with the following result [Ka06, Theorem 9.1]:

Theorem 5.11 (Improving of the Moduli: Bounded Period).   *For any $\rho \in (0,1)$, there exists $\underline{p} = \underline{p}(\rho)$ such that for any $\bar{p} \geq \underline{p}$, there exists $\varepsilon = \varepsilon(\bar{p}) > 0$ with the following property. Let f be a primitively renormalizable ψ-quadratic-like map, and let g be the corresponding renormalization. Then*

$$\{\underline{p} \leq p \leq \bar{p} \text{ and } \mathbf{mod}\,(g) < \varepsilon\} \Rightarrow \mathbf{mod}\,(f) < \rho\,\mathbf{mod}\,(g).$$

Putting the above two theorems together, we obtain:

Corollary 5.12. *For any $(\bar{\mathbf{r}}, \bar{\mathbf{q}}, \bar{\mathbf{n}})$, there exist an $\varepsilon > 0$ and $l \in \mathbb{N}$ with the following property. For any infinitely renormalizable ψ-ql map f satisfying the $(\bar{\mathbf{r}}, \bar{\mathbf{q}}, \bar{\mathbf{n}})$-molecule condition with renormalizations $g_n = R^n f$, if $\mathbf{mod}\,(g_n) < \varepsilon$, $n \geq l$, then $\mathbf{mod}\,(g_{n-l}) < \frac{1}{2}\mathbf{mod}\,(g_n)$.*

This implies the Main Theorem, in an important refined version. We say that a family $\mathcal{M}$ of little Mandelbrot copies (and the corresponding renormalization combinatorics) has *beau*[10] *a priori* bounds if there exists an $\varepsilon = \varepsilon(\mathcal{M}) > 0$ and a function $N : \mathbb{R}_+ \to \mathbb{N}$ with the following property. Let $f : U \to V$ be a quadratic-like map with $\mathrm{mod}(V \smallsetminus U) \geq \delta > 0$ that is at least $N = N(\delta)$ times renormalizable. Then for any $n \geq N$, the n-fold renormalization of f can be represented by a quadratic-like map $R^n f : U_n \to V_n$ with $\mathrm{mod}(V_n \smallsetminus U_n) \geq \varepsilon$.

Beau Bounds. *For any parameters $(\bar{\mathbf{r}}, \bar{\mathbf{q}}, \bar{\mathbf{n}})$, the family of renormalization combinatorics satisfying the $(\bar{\mathbf{r}}, \bar{\mathbf{q}}, \bar{\mathbf{n}})$-molecule condition has beau a priori bounds.*

5.7   Table of Notations

p is the renormalization period of f;
γ is a dividing periodic point of period $\mathbf{t}$,
$\boldsymbol{\gamma}$ is its cycle;
$\mathbf{s}$ is the number of rays landing at γ;
α is a dividing periodic point of periods $\mathbf{r} = \mathbf{ts}$;
$\alpha' = -\alpha$, $\alpha_j = f^j \alpha$;
$\boldsymbol{\alpha} = \{\alpha_j\}_{j=0}^{\mathbf{r}-1}$ is the cycle of α;
$\mathbf{q}$ is the number of rays landing at α;
$\mathbf{n}$ is the first moment such that $f^{\mathbf{rqn}}(0)$ is separated from 0 by the rays
    landing at α',
$\mathbf{k} = \mathbf{rqn}$;
$\boldsymbol{\lambda} = \mathbf{k} + 1 + 2\mathbf{r}$ is a depth such that the stars $S^\lambda(\alpha_j)$, $j = 0, 1, \ldots, \mathbf{r} - 1$,
    are all disjoint.

---

[10] According to Dennis Sullivan, "beau" stands for "bounded and eventually universal".

Bibliography

[Ah73] L. Ahlfors, *Conformal invariants: Topics in geometric function theory*, McGraw Hill, New York, 1973.

[AKLS05] A. Avila, J. Kahn, M. Lyubich, and W. Shen, *Combinatorial rigidity for unicritical polynomials*, Stony Brook IMS Preprint 5, 2005. Annals of Math., to appear.

[Ch06] D. Cheraghi, *Combinatorial rigidity for some infinitely renormalizable unicritical polynomials*, Stony Brook IMS Preprint 6, 2007.

[Do86] A. Douady, *Chirurgie sur les applications holomorphes*, in: Proceedings of the International Congress of Mathematicians, Berkeley (1986), 724–738.

[DH85] A. Douady and J. H. Hubbard, *On the dynamics of polynomial-like maps*, Ann. Sc. Éc. Norm. Sup., **18** (1985), 287–343.

[Ka06] J. Kahn, *A priori bounds for some infinitely renormalizable quadratics, I: Bounded primitive combinatorics*, Stony Brook IMS Preprint 5, 2006.

[KL05a] J. Kahn and M. Lyubich, *Quasi-additivity law in conformal geometry*, Stony Brook IMS Preprint 2, 2005. Annals of Math., to appear. arXiv.math.DS/0505191v2.

[KL05b] J. Kahn and M. Lyubich, *Local connectivity of Julia sets for unicritical polynomials*, Stony Brook IMS Preprint 3, 2005. Annals of Math., to appear.

[KL06] J. Kahn and M. Lyubich, *A priori bounds for some infinitely renormalizable quadratics, II: Decorations*, Ann. Scient. Éc. Norm. Sup., **41** (2008), 57–84.

[Ly97] M. Lyubich, *Dynamics of quadratic polynomials, I-II*, Acta Math., **178** (1997), 185–297.

[Mc94] C. McMullen, *Complex dynamics and renormalization*, Princeton University Press, 1994.

[M00a] J. Milnor, *Periodic orbits, external rays, and the Mandelbrot set: an expository account*, in: "Géométrie complexe et systémes dynamiques", volume in honor of Douady's 60th birthday, Astérisque **261** (2000), 277–333.

[M00b]   J. Milnor, *Local connectivity of Julia sets: expository lectures*, in: The Mandelbrot Set, Themes and Variations, Tan Lei Ed., London Math. Soc. Lecture Note Ser. **274**, Cambridge University Press, 2000, 67–116.

[Th]   W. P. Thurston, "On the geometry and dynamics of iterated rational maps", in: *Complex Dynamics: Families and Friends* (this volume), Chapter 1. A K Peters, Wellesley, MA, 2009, 3–137.

II. Beyond Polynomials: Rational and Transcendental Dynamics

6 The Connectivity of the Julia Set and Fixed Points

Mitsuhiro Shishikura

We study the relationship between the connectivity of the Julia sets of rational maps and their fixed points that are repelling or parabolic with multiplier 1. It is proved that if a rational map has only one such fixed point, then its Julia set is connected. This is the case for Newton's method applied to polynomials. The proof is given in terms of the method of quasiconformal surgery.

Introduction

For a non-constant complex polynomial $P(z)$, the rational map $N_P : \widehat{\mathbb{C}} \to \widehat{\mathbb{C}}$ defined as

$$N_P : z \mapsto z - \frac{P(z)}{P'(z)}$$

is called *Newton's method* for P. The iteration of N_P gives an algorithm for finding the roots of P. All fixed points except one (namely ∞) are roots of P and are attracting for N_P. One of the purposes of this paper is to prove that the Julia set of N_P is connected.

More generally, we will prove:

Theorem I. *If the Julia set of a rational map f is disconnected, there exist two fixed points of f such that each of them is either repelling or parabolic with multiplier 1, and they belong to different components of the Julia set.*

Corollary II. *If a rational map has only one fixed point which is repelling or parabolic with multiplier 1, then its Julia set is connected. In other words, every component of the complement of the Julia set is simply connected. In particular, the Julia set of Newton's method for a non-constant polynomial is connected.*

Remark. There are several works concerning the connectivity of the Julia set of Newton's method. Przytycki [Pr89] proved that every root of P has a simply connected immediate basin. The author then proved connectivity using the theory of trees associated with configurations of the complement of the Julia set [Sh89] and quasiconformal surgery. The proof has been simplified several times, and we present here the latest version. While writing this paper, the author learned that Meier [Me89] proved connectivity in the case of degree 3 under an additional assumption, independently and by a completely different method. Tan Lei [Ta97] has a generalization of Meier's result to higher degrees. According to Hubbard (with whom the author communicated personally), he also knows a proof of connectivity in the cubic case using J. Head's result [He87]. The author has another proof in the cubic case which is much simpler than the proof of Theorem I (see the end of this section).

For simplicity, we call a *weakly repelling fixed point* a fixed point which is either repelling or parabolic with multiplier 1. We say a set $X \subset \widehat{\mathbb{C}}$ *separates* two sets $A, B, \subset \widehat{\mathbb{C}}$ if A and B are contained in different connected components of $\widehat{\mathbb{C}} - X$.

Theorem 1 follows from the following theorem, together with Sullivan's Theorem 1.2 (see Claim 1.3).

Theorem III. *Let f be a rational map of degree greater than one. Then:*

(i) *If f has an attractive basin or a parabolic basin which is not simply connected, then there exist two weakly repelling fixed points which are separated by a forward orbit of the basin.*

(ii) *If f has a Herman ring, then there exist two weakly repelling fixed points which are separated by a forward orbit of the ring.*

(iii) *If D is a connected component of the complement of the Julia set such that D is not simply connected and f(D) is simply connected, then every component of $\widehat{\mathbb{C}} - D$ contains a weakly repelling fixed point.*

Remark. Przytycki [Pr89] proved that if an attractive basin of *period one* is not simply connected, then its boundary contains two fixed points in different components. However, his proof does not seem to generalize to higher periods.

Theorem III and its proof not only give connectivity results (Theorem I) but suggest a connection between the structure (or the *configuration*) of the Fatou set $F_f = \widehat{\mathbb{C}} - J_f$ and the weakly repelling fixed points. If there are not too many such fixed points, then the possible configuration of Fatou components (connected components of F_f) will be restricted. See, for example, Proposition 4.1, Proposition 7.1 and Remark 7.5.

The outline of the proof of Theorem III is as follows. If the rational map f restricted to a region of $\widehat{\mathbb{C}}$ satisfies certain conditions, it can be conjugated to another rational map, which is simpler than f (Theorem 2.1). This will be done by the method of quasiconformal surgery (Section 3); gluing a different map (non-analytic) to the outside of the region, one obtains a new map of $\widehat{\mathbb{C}}$, which is proved to be conjugate to a rational map via the measurable Riemann mapping theorem. Using Fatou's Theorem 1.1, one can deduce the existence of a weakly repelling fixed point in this region (Corollary 2.2). The problem is then to find appropriate regions, as above, for a given rational map, according to the cases stated in Theorem III. The cases of a strictly preperiodic component, attractive basin, parabolic basin and Herman ring are treated separately in Section 4, Section 5, Section 6 and Section 7, respectively.

Acknowledgments. The author would like to thank many people who have discussed this subject with him, especially S. Ushiki, M. Yamaguti, F. Przytycki, J.H. Hubbard, A. Douady, H.-G. Meier and Tan Lei. He also thanks the Max-Planck-Institut für Mathematik (1988), the Institute for Advanced Study (1989-90) and the Institut des Hautes Etudes Scientifiques (1989-90) and Akizuki project (sponsored by T. Taniguchi, 1989-90) for having supported him. Finally, the author would like to thank D. Schleicher, for pushing him to publish this paper, and Y. Mikulich, for valuable comments about the paper.

Historical Remark. This paper has been circulating as IHES Preprint M-90-37 since 1990. Since then, there have been several works related to this paper. Bergweiler and Terglane [BT96] applied our method to study the relation between weakly repelling fixed points and multiply connected wandering domains of entire functions. Mayer and Schleicher [MS06] proved that Newton's methods for entire functions, the immediate basins of roots and also the virtual immediate basin of ∞ are simply connected. It is still unknown whether the Julia set of Newton's methods for entire functions are connected.

1 Preliminaries

Let f be a rational map of degree greater than one.

Periodic Points. The *multiplier* of a periodic point z of period p is $\lambda = (f^p)'(z)$ if $z \neq \infty$, and $\lambda = (A \circ f^p \circ A^{-1})'(0)$ if $z = \infty$, where $A(z) = 1/z$. The periodic point is called *attracting, repelling, indifferent* or *parabolic*, if $|\lambda| < 1$, $|\lambda| > 1$, $|\lambda| = 1$ or $\lambda = e^{2\pi i \theta}$ ($\theta \in \mathbb{Q}$), respectively. It is called *weakly repelling*, if $|\lambda| > 1$ or $\lambda = 1$, as defined in the introduction.

The *Julia set* J_f of f is the closure of the set of repelling periodic points. Its complement is the Fatou set $F_f = \widehat{\mathbb{C}} - J_f$. An equivalent definition is that $z \in F_f$ if and only if z has a neighborhood on which $\{f^n | n \geq 0\}$ is a normal family. See, for example, [Mi06, Be91].

Theorem 1.1 (Fatou [Fa19] vol. 47 (1919), Section 52, p. 167–168).
A rational map of degree greater than one has at least one weakly repelling fixed point.

Proof (Outline): Suppose f does not have a parabolic fixed point with multiplier 1. By residue calculus on $1/(f(z) - z)$, one obtains *Fatou's formula*:

$$\sum_z \frac{1}{\lambda_z - 1} = -1,$$

where the sum is taken over all fixed points z and λ_z is the multiplier. Note the fact that $|\lambda| \leq 1$ ($\lambda \neq 1$) if and only if $Re\left(\frac{1}{\lambda-1}\right) \leq -\frac{1}{2}$. Since there are $d + 1$ fixed points, the theorem follows immediately. See [Mi06] for details.   □

   Because of this theorem, weakly repelling fixed points play a special role in this paper.

Theorem 1.2 (Sullivan [Su85, MS83]). *Every connected component of the Fatou set F_f is preperiodic under f as a set. A periodic component is of one of the following types: attractive basin, parabolic basin (the orbits are attracted to an attracting or parabolic periodic cycle), Siegel disk or Herman ring (f is conformally conjugate to an irrational rotation on the disk or on an annulus).*

Claim 1.3. *Theorem III implies Theorem I.*

Proof: Suppose J_f is not connected. Then there exists a non-simply connected component of $F_f = \widehat{\mathbb{C}} - J_f$. If a periodic component is not simply connected, then either (i) or (ii) of Theorem III applies. If all the periodic components are simply connected, then there is a strictly preperiodic component U such that U is not simply connected and $f(U)$ is simply connected. Hence (iii) of Theorem III applies. In any case, there exist two weakly repelling fixed points of f which are separated by F_f.   □

Newton's Method.   Let $P(z)$ be a non-constant polynomial of degree d with n distinct roots. Then the Newton's method $N_P(z)$ is of degree n. The fixed points of N_P are the roots of P and ∞. A root of P with multiplicity m is an attracting fixed point of N_P with multiplier $(m-1)/m$, and ∞ is a repelling fixed point with multiplier $d/(d-1)$. Hence N_P has only one weakly repelling fixed point, and this justifies the second half of Corollary II. See [Ma92] for more details. If $\deg N_P = 1$, then $J_{N_P} = \{\infty\}$ for any definition of the Julia set.

A Short Proof of the Connectivity of J_{N_P} When $\deg P = 3$.

Proof: By an inequality in [Sh87] (Corollary 2), for a rational map f,

$$\text{(number of attracting periodic cycles)}$$
$$+ 2 \times \text{(number of cycles of Herman rings)} \ \leq \ 2 \deg f - 2 \,.$$

For $f = N_P$, there are at least 3 attracting fixed points and $\deg f \leq 3$; hence there is no Herman ring. By Przytycki [Pr89], the immediate basins of the roots are simply connected. Other components of F_{N_P} can contain at most one critical point. Note that if U is simply connected and $f^{-1}(U)$ contains at most one critical point, then each component of $f^{-1}(U)$ is simply connected. Hence all the components of the pre-images of the immediate basins of the roots are simply connected. Let B be an attractive basin which is not a parabolic basin or an attracting basin of the roots. Then B can be written as a component of $\bigcup_{n \geq 0} f^{-n}(D)$, where D is a simply connected set satisfying $f^p(D) \subset D$. By the same reasoning as above, $f^{-n}(D)$ consists of simply connected sets, so B and their pre-images are simply connected. Thus all the components of F_{N_P} are simply connected. □

This proof does not generalize to higher degrees because it relies on the number of critical points. Instead of critical points, we shall look at the weakly repelling fixed points — this is the key to our proof of the connectivity of J_{N_P} for higher degrees.

2   The Surgery Theorem and Its Corollary

Theorem 2.1. *Let V_0, V_1 be connected and simply connected open sets in $\widehat{\mathbb{C}}$ such that $V_0 \neq \widehat{\mathbb{C}}$ and f is an analytic mapping from a neighborhood $\mathcal{N}$ of $\widehat{\mathbb{C}} - V_0$ to $\widehat{\mathbb{C}}$ such that $f(\partial V_0) = \partial V_1$ and $f(V_0 \cap \mathcal{N}) \subset V_1$. Suppose that up to some $k \geq 1$, the iterates $f^j(V_1)$ are defined and satisfy*

$$f^j(V_1) \cap V_0 \neq \emptyset \quad (0 \leq j < k - 1),$$

and assume that one of the following holds ($k < \infty$ for (b) (c) (d)):

(a) *$k = \infty$, i.e., for any $j \geq 0$, $f^j(V_1) \cap V_0 = \emptyset$.*

(b) *$f^{k-1}(\overline{V}_1) \subset V_0$.*

(c) *$f^{k-1}(\overline{V}_1) \subset V_0 \cup \{\alpha\}$ for some $\alpha \in \partial V_0$ satisfying $f^k(\alpha) = \alpha$, and for any neighborhood $\mathcal{N}_0 \ (\subset \mathcal{N})$ of α, there exists a connected open set $W \subset V_0 \cap \mathcal{N}_0$ such that $f^k(\overline{W}) \subset W \cup \{\alpha\}$, $\partial W \cap V_0$ is an arc and $\alpha \notin \overline{V_0 - W}$.*

(d) $f^{k-1}(V_1) = V_0$, ∂V_0 is a real analytic Jordan curve, $f^k(\partial V_0) = \partial V_0$ and $f^k|_{\partial V_0}$ is real analytically conjugate to an irrational rotation $z \to e^{2\pi i\theta} z$ on $S^1 = \{|z| = 1\}$ $(\theta \in \mathbb{R} - \mathbb{Q})$.

Then there exists a quasiconformal mapping $\varphi : \widehat{\mathbb{C}} \to \widehat{\mathbb{C}}$ and a rational map g with $\deg g \geq 1$ such that

- $\varphi \circ f \circ \varphi^{-1} = g$ on $\widehat{\mathbb{C}} - V_0^*$, where $V_0^* = \varphi(V_0)$;

- $g(V_0^*) = V_1^*$, where $V_1^* = \varphi(V_1)$;

- V_0^* contains at most one critical point of g (counted without multiplicity);

- $\frac{\partial \varphi}{\partial \bar{z}} = 0$ a.e. on $\{z \in \widehat{\mathbb{C}} - V_0|\ f^n(z)$ is defined and in $\widehat{\mathbb{C}} - V_0$ for $n \geq 0\}$;

- $V_0^* \subset F_g$.

Moreover, in Case (b), g has an attracting periodic point α^* of period k in V_0^*, and its immediate basin contains V_0^*. In Case (c), $\alpha^* = \varphi(\alpha)$ is an attracting or parabolic periodic point of g with multiplier 1 and its basin contains V_0^*. In Case (d), g has a Siegel disk containing V_0^*.

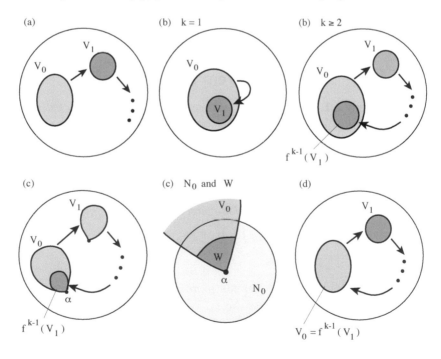

Corollary 2.2. *Assuming the hypotheses of Theorem 2.1, f has a weakly repelling fixed point in $\widehat{\mathbb{C}} - V_0$, except if $k = 1$ and $f(\widehat{\mathbb{C}} - V_0) = \widehat{\mathbb{C}} - V_0$ in Case* (d).

Remark 2.3. In Case (b), if $k = 1$, and if $f(\widehat{\mathbb{C}} - V_0) = \widehat{\mathbb{C}} - V_1$, then f is a polynomial-like mapping [DH85] (after a slight modification of the region). More generally, the surgery for Cases (b) and (d) can be considered as a very special case of the surgery in [Sh87] (Proposition 2) if ∂V_0 and ∂V_1 are real analytic Jordan curves.

Theorem 2.1 means that, under its hypotheses, one can replace the dynamics of f in V_0 by something simpler, or, in other words, one can ignore what happens in V_0.

Proof of Corollary 2.2: Let φ, g be as in Theorem 2.1. The rational map g has a weakly repelling fixed point z_0, except if $\deg g = 1$ and g is an elliptic Möbius transformation (See Fatou's Theorem 1.1). However, g cannot be an elliptic transformation for the following reasons: In Case (a), V_0^* is non-recurrent ($g^n(V_0^*) \cap V_0^* = \emptyset$ for $n \geq 1$). In Cases (b) and (c), g has an attracting or parabolic point α^* (in the parabolic case, α^* has a basin). In Case (d), with $k > 1$, g has a periodic point of period k in the Siegel disk and $g^k \neq id$. Hence $\deg g > 1$. In Case (d) with $k = 1$ and $f(\widehat{\mathbb{C}} - V_0) \neq \widehat{\mathbb{C}} - V_0$, since $\partial f(\widehat{\mathbb{C}} - V_0) \subset f(\partial(\widehat{\mathbb{C}} - V_0)) = f(\partial V_0)$, we have $f(\widehat{\mathbb{C}} - V_0) \supset V_0$ by the open mapping property; hence g cannot be injective and $\deg g > 1$.

We have $z_0 \notin V_0^*$, since z_0 is in the Julia set of g and $V_0^* \subset F_g$. Finally, the following lemma implies that $\varphi^{-1}(z_0)$ is a weakly repelling fixed point of f in $\widehat{\mathbb{C}} - V_0$. □

Lemma 2.4. *Let f, g be germs of analytic functions with fixed points x_0, y_0, respectively. Suppose f and g are conjugated by a local homeomorphism that sends x_0 to y_0. Then*
$|f'(x_0)| < 1$ (*respectively* > 1, $= 1$) *if and only if* $|g'(y_0)| < 1$ (*respectively* > 1, $= 1$); $f'(x_0) = 1$ *if and only if* $g'(y_0) = 1$.

Proof: For the first statement, note that $|f'(x_0)| < 1$ (respectively > 1) if and only if there exists an arbitrarily small open neighborhood W of x_0 such that $\overline{f(W)} \subset W$ (respectively $\overline{W} \subset f(W)$).

For the second statement, if $f'(x_0) = 1$, then there exists an arbitrarily small connected open set W such that $x_0 \in \partial W$, $f(W) \subset W$ and $f^n(z) \to x_0$ in W as $n \to \infty$ (See Lemma 6.2). On the other hand, we have:

Theorem 2.5 (Fatou [Fa19] vol. 48, Sullivan [MS83], Proposition 3.5).
Let f be an analytic function defined in a connected open set $W \subset \mathbb{C}$ and in a neighborhood of $x_0 \in \partial W \cap \mathbb{C}$. Suppose $f(W) \subset W$ and $f^n \to x_0$ in W ($n \to \infty$). Then either $|f'(x_0)| < 1$ or $f'(x_0) = 1$.

Therefore if $f'(x_0) = 1$, then $|g'(y_0)| < 1$ or $g'(y_0) = 1$. By the first statement, we have $g'(y_0) = 1$. □

3 Surgery-Proof of Theorem 2.1

We prove Theorem 2.1 by the method of quasiconformal surgery (see [Sh87], Section 3). First, we will extend $f : \widehat{\mathbb{C}} - V_0 \to \widehat{\mathbb{C}}$ to a quasiregular mapping $f_1 : \widehat{\mathbb{C}} \to \widehat{\mathbb{C}}$. A *quasiregular mapping* is a mapping which is a composition of an analytic mapping and a quasiconformal mapping with uniformly bounded dilatation at any point. We need the following:

Lemma 3.1. *Let V, V' be simply connected open sets in $\widehat{\mathbb{C}}$ such that $V \neq \widehat{\mathbb{C}}$, f is an analytic mapping from a neighborhood $\mathcal{N}$ of ∂V to $\widehat{\mathbb{C}}$ such that $f(\partial V) = \partial V'$ and $f(V \cap \mathcal{N}) \subset V'$. Then there exists a quasiregular mapping $f_1 : V \to V'$ such that*

- $f_1 = f$ *in* $V \cap \mathcal{N}'$, *where* $\mathcal{N}'$ *is a neighborhood of* ∂V *with* $\mathcal{N}' \subset \mathcal{N}$,

and

- f_1 *has at most one critical point in* V.

Moreover, if a compact set K in V, $a \in V$ and $b \in V'$ are given, then f_1 can be chosen so that

- f_1 *is analytic in a neighborhood of K, $f_1(a) = b$ and there is no critical point in V except at a.*

The proof will be given at the end of this section.

Now suppose f satisfies the hypotheses of Theorem 2.1. In Cases (a), (b) and (c), by applying Lemma 3.1 to f with $V = V_0$ and $V' = V_1$, we can extend f to V_0 as a quasiregular mapping $f_1 : \widehat{\mathbb{C}} \to \widehat{\mathbb{C}}$ such that $f_1 = f$ in a neighborhood of $\widehat{\mathbb{C}} - V_0$, $f_1(V_0) \subset V_1$ and f_1 has at most one critical point in V_0. (We will require more properties of f_1 in (b).)

Case (a). Define $E = \bigcup_{n \geq 0} f^n V_1$ and $X = V_0$. We have

(∗) f_1 is analytic in $\widehat{\mathbb{C}} - X$; $f_1(E) \subset E$, $X \cap E = \emptyset$ and $f_1(X) \subset E$.

Let σ_0 be the standard conformal structure on $\widehat{\mathbb{C}}$ (i.e., the conformal class of the standard metric $\frac{|dz|}{1+|z|^2}$), and let $f_1^* \sigma_0$ denote the pull-back of σ_0 by f_1 (defined almost everywhere). (See Appendix of [Ah06] for a discussion on conformal structures.) Define a new measurable conformal structure σ on $\widehat{\mathbb{C}}$ by

$$\sigma = \begin{cases} (f_1^n)^* \sigma_0 & \text{on } f_1^{-n}(E) \ (n \geq 0), \\ \sigma_0 & \text{on } \widehat{\mathbb{C}} - \bigcup_{n \geq 0} f_1^{-n}(E). \end{cases}$$

Then σ is measurable and f_1-invariant, i.e., $f_1^*\sigma = \sigma$ a.e. In fact, the invariance is obvious on $f_1^{-n}(E)$ $(n \geq 0)$ and follows from the analyticity of f_1 in $\widehat{\mathbb{C}} - \bigcup_{n \geq 0} f_1^{-n}(E) \subset \widehat{\mathbb{C}} - X$. Moreover σ has a bounded dilatation with respect to σ_0 (in the sense that it has a representative $|dz + \mu(z)d\bar{z}|$ with $\|\mu\|_\infty < 1$), since the analytic map $f_1|_{\widehat{\mathbb{C}} - X}$ does not change the dilatation.

Thus we can apply the measurable Riemann mapping theorem (see [Ah06] Chapter V) to σ to obtain a quasiconformal mapping $\varphi : \widehat{\mathbb{C}} \to \widehat{\mathbb{C}}$ such that $\varphi^*\sigma_0 = \sigma$ a.e. Let $g = \varphi \circ f_1 \circ \varphi^{-1}$. Then we have $g^*\sigma_0 = \sigma_0$ a.e. This implies that $g : \widehat{\mathbb{C}} \to \widehat{\mathbb{C}}$ is analytic and hence a rational map.

Note that $\sigma = \sigma_0$ a.e. on $\{z \in \widehat{\mathbb{C}} - V_0 \,|\, f^n(z)$ is defined and in $\widehat{\mathbb{C}} - V_0$ for any $n \geq 0\}$ by definition. Hence $\varphi^*\sigma_0 = \sigma_0$, and then $\frac{\partial \varphi}{\partial \bar{z}} = 0$ a.e. on this set.

The iterates $g^n|_{V_0^*}$ $(n \geq 1)$ omit V_0^*, so they form a normal family and $V_0^* \subset F_g$.

It is easy to check the rest of the conclusions of the theorem for Case (a). (See [Sh87] Section 3, Lemma 1 and its proof for details of the surgery procedure).

Case (b). Construct $f_1 : \widehat{\mathbb{C}} \to \widehat{\mathbb{C}}$ as above, with the additional properties (for this purpose, choose $b \in V_1$, $a = f^{k-1}(b) \in V_0$ and $K = f^{k-1}(\overline{V}_1)$ in Lemma 3.1:

- f_1 is analytic in a neighborhood of $f^{k-1}(\overline{V}_1)$;

- there is a point $a \in V_0$ such that $f_1^k(a) = a$.

Define $E = \bigcup_{j=0}^{k-1} f^j(V_1)$ and $X = V_0 - f^{k-1}(V_1)$. Then f_1 satisfies $(*)$ from Case (a), so by the same argument, we obtain a quasiconformal mapping φ and a rational map $g = \varphi \circ f_1 \circ \varphi^{-1}$.

Consider $g^k : V_0^* \to V_0^*$, where $V_0^* = \varphi(V_0)$. Since $g^k(V_0^*) = \varphi(f_1^{k-1}(f_1(V_0))) \subset \varphi(f_1^{k-1}(V_1)) \subsetneqq V_0^*$, g^k is strictly contracting on any compact set in V_0^*, with respect to the Poincaré metric of V_0^*. The point $a^* = \varphi(a)$ must be an attracting periodic point of g of period k, and $g^{nk} \to a^*$ on V_0^*, uniformly on compact sets. So V_0^* is contained in the immediate basin of a^*.

Case (c). Let f_1 be as above. There is a neighborhood $\mathcal{N}_0$ of α on which $f_1 = f$. Let W be as in the statement of (c). Note that $V_0 - \overline{W}$ is simply connected, $\partial W \cap V_0$ is a part of its boundary and $f^{k-1}(\overline{V}_1) - W$ is compact and contained in V_0. There exists a quasiconformal mapping $h : \widehat{\mathbb{C}} \to \widehat{\mathbb{C}}$ such that $h = id$ in a neighborhood of $(\widehat{\mathbb{C}} - V_0) \cup f^k(\overline{W})$, and $h(f^{k-1}(\overline{V}_1)) \subset W$.

In order to construct the map h, first send the set $f^{k-1}(\overline{V}_1) - W$ to a neighborhood of $\partial W \cap V_0$ by a quasiconformal mapping of $V_0 - \overline{W}$, then push it into W across $\partial W \cap V_0$ by another quasiconformal mapping.

Now define $f_2 = f_1 \circ h$, $E = \bigcup_{j=1}^{k} f^j(W)$ and $X = V_0 - f^k(W)$. Then $f_2 = f$ is analytic in $\widehat{\mathbb{C}} - X$, $E \cap X = \emptyset$, $h(E) = E$ and $f_2(E) = f(E) \subset E$. Since $h(V_0) = V_0$, we have $f_2(X) \subset f_2(V_0) = f_1(V_0) \subset V_1$ and $h \circ f_2^k(X) = h \circ f_2^{k-1}(f_2(X)) \subset h \circ f_2^{k-1}(V_1) \subset W$. Finally, $f_2^{k+1}(X) \subset E$. Hence each orbit of f_2 passes X a bounded number of times (in fact at most twice). Thus the argument in Case (a) applies, and there exist φ and $g = \varphi \circ f_2 \circ \varphi^{-1}$.

Since $g^k(\overline{V_0^*}) \subset V_0^* \cup \{\alpha^*\}$, by Sullivan's Theorem 1.2, V_0^* must be contained in a basin of an attracting or parabolic periodic point β, and $f^{nk} \to \beta$ on V_0 ($n \to \infty$). By the assumption of the existence of arbitrarily small W, we conclude that $\alpha^* = \beta$.

Case (d). Note that the analytic map $f^{k-1} : \overline{V_1} \to \overline{V_0} = f^{k-1}(\overline{V_1})$ must be a homeomorphism, since V_0, V_1 are simply connected, $f^k|_{\partial V_0} = (f^{k-1}|_{\partial V_1}) \circ (f|_{\partial V_0})$ is a homeomorphism and so is $f^{k-1}|_{\partial V_1}$.

By our assumption, there exists a real analytic diffeomorphism $h : \partial V_0 \to S^1 = \{|z| = 1\}$ such that $h \circ f^k(z) = e^{2\pi i\theta} h(z)$ for $z \in \partial V_0$. We can extend h to $h : \overline{V_0} \to \overline{D} = \{|z| \leq 1\}$, replacing h by $\overline{h}$ and θ by $-\theta$ if necessary. Define $f_1 : \widehat{\mathbb{C}} \to \widehat{\mathbb{C}}$ as

$$f_1(z) = (f^{k-1}|_{\overline{V_1}})^{-1} \circ h^{-1}(e^{2\pi i\theta} h(z)) \text{ for } z \in \overline{V_0},$$
$$f_1 = f \text{ on } \widehat{\mathbb{C}} - V_0.$$

Then f_1 is well defined, continuous at ∂V_0 and quasiregular on $\widehat{\mathbb{C}}$. Let $E = \bigcup_{j=0}^{k-1} f^j(V_1)$ and $X = f^{-1}(E) - E$.

Define a measurable conformal structure σ_1 on $\widehat{\mathbb{C}}$ by

$$\sigma_1 = \begin{cases} h^* \sigma_0 & \text{on } V_0 = f^{k-1}(V_1), \\ (f^{k-1-j})^* (h^* \sigma_0) & \text{on } f^j(V_1) \quad (0 \leq j < k-1), \\ \sigma_0 & \text{on } \widehat{\mathbb{C}} - E. \end{cases}$$

Note that f_1 is analytic a.e. (i.e., $f_1^* \sigma_0 = \sigma_0$) on $\widehat{\mathbb{C}} - (E \cup X)$. On the other hand, we have $f_1^* \sigma_1 = \sigma_1$ on E. This is trivial on $f^j(V_1)$ ($0 \leq j < k-1$); that it holds on V_0 follows from the fact that $h \circ f_1^k = e^{2\pi i\theta} h$ and $(f_1^k)^* h^* \sigma_0 = f_1^* (f^{k-1})^* h^* \sigma_0 = h^* \sigma_0$ on $V_0 = f^{k-1}(V_1)$.

Now f_1 satisfies (∗) in Case (a), with the first condition replaced by

$$f_1 \text{ is analytic in } \widehat{\mathbb{C}} - (X \cup E) \text{ and } f_1^* \sigma_1 = \sigma_1 \text{ on } E.$$

Since σ_1 is bounded, we can apply a similar argument with σ_0 replaced by σ_1 in the definition of σ and obtain a quasiconformal mapping φ and a rational map $g = \varphi \circ f_1 \circ \varphi^{-1}$.

Since $\left(h \circ \varphi^{-1}\right)^* \sigma_0 = \left(\varphi^{-1}\right)^* h^* \sigma_0 = \left(\varphi^{-1}\right)^* \sigma = \sigma_0$, the mapping $h \circ \varphi^{-1} : V_0^* = \varphi(V_0) \to \Delta$ is conformal and conjugates $g^k : V_0^* \to V_0^*$ to $z \to e^{2\pi i \theta} z$ on Δ. Hence V_0^* is contained in a Siegel disk of g. Thus Theorem 2.1 is proved. $\qquad\square$

Now let us prove Lemma 3.1 to complete the proof of Theorem 2.1.

Proof of Lemma 3.1: Let γ be a Jordan curve in V. Then there exists a component A of $V - \gamma$ which is an annulus such that $\partial A = \gamma \cup \partial V$. Choose γ so that $\overline{A} \subset \mathcal{N}$ and f has no critical point in A. The image $f(\gamma)$ is compact, connected and contained in V'. Thus there exists a component A' of $V' - f(\gamma)$ which is an annulus such that $\partial V' \subset \partial A'$. It is easy to see that $\#f^{-1}(z) \cap A$ is finite and constant (> 0) for $z \in A'$. Hence $f : A'' = f^{-1}(A') \cap A \to A'$ is a covering of finite degree, and A'' is a disjoint union of annuli in A. However $f^{-1}(\partial V') \cap \overline{A} = \partial V$; it follows that A'' consists of a single annulus.

Let γ' be a real analytic Jordan curve in A' which is homotopically non-trivial in A'. Then $\gamma'' = f^{-1}(\gamma') \cap A''$ is also a real analytic Jordan curve, and there exist annuli A_0, A_1 such that $A_0 \subset A''$, $\partial A_0 = \gamma'' \cup \partial V$, $A_1 \subset A'$, $\partial A_1 = \gamma' \cup \partial V'$ and $f : A_0 \to A_1$ is a covering. Let $\Psi_0 : V - \overline{A}_0 \to \Delta = \{|z| < 1\}$ and $\Psi_1 : V' - \overline{A}_1 \to \Delta$ be conformal mappings. Define f_1 on $V - \overline{A}_0$ by $f_1(z) = \Psi_1^{-1}\left([\Psi_0(z)]^m\right)$ $(z \in V - \overline{A}_0)$, where $m = \deg(f : A'' \to A')$. The map f_1 extends analytically to a neighborhood of γ'' without a critical point on γ''. Both f and f_1 are real analytic coverings from γ'' to γ' of the same degree without a critical point, and hence they are homotopic.

We can extend f_1 to $\overline{A}_0$, modifying f only in a neighborhood of γ'', so that f_1 is a C^∞-local diffeomorphism, f_1 on γ'' is defined as above and $f_1 = f$ in a neighborhood of ∂V.

When K and a, b are given, γ, Ψ_0 and Ψ_1 can be chosen so that $K \cap \overline{A}_0 = \emptyset$, $a \notin \overline{A}_0$, $b \notin \overline{A}_1$ and $\Psi_0(a) = \Psi_1(b) = 0$.

It is clear that f_1 has the required properties. $\qquad\square$

4 Strictly Preperiodic Components

In this section, we prove Theorem III (iii), which is the easiest case. Proposition 4.1 is a convenient reformulation of Theorem 2.1, Case (a), Case (b) and Corollary 2.2. Case (β) of the proposition will be used in the next section for attractive basins.

Note that if U is a component of F_f, then $f : U \to f(U)$ is proper, or, equivalently, U is a component of $f^{-1}(f(U))$. Hence the proof of Theorem III (iii) reduces to the following proposition, Case (α) :

Proposition 4.1. *Let f a rational map of degree ≥ 1 and U a connected open set in $\widehat{\mathbb{C}}$ such that $U \neq \widehat{\mathbb{C}}$, $f(U)$ is simply connected and U is a connected component of $f^{-1}(f(U))$. Suppose either*

(α) for any $j \geq 1$, $f^j(U) \cap U = \emptyset$, or

(β) there exists a $p \geq 1$ such that $f^j(U) \cap U = \emptyset$ $(1 \leq j < p)$ and $f^p(\overline{U}) \subset U$.

Then each connected component of $\widehat{\mathbb{C}} - U$ contains a weakly repelling fixed point of f.

Proof: Let E be a component of $\widehat{\mathbb{C}} - U$. Define $V_0 = \widehat{\mathbb{C}} - E$, $V_1 = f(U)$ and $\mathcal{N} = U \cup E$. Then, obviously, V_0, V_1 are simply connected open sets, $\mathcal{N}$ is a neighborhood of $E = \widehat{\mathbb{C}} - V_0$ and $f(V_0 \cap \mathcal{N}) \subset V_1$. Since $f : U \to f(U)$ is proper, we have $f(\partial V_0) = \partial V_1$. For each $j \geq 0$, $f^j(V_1) = f^{j+1}(U)$ must be contained in E or another component of $\widehat{\mathbb{C}} - U$, which is also closed. There are two possibilities:

(a) for any $j \geq 0$, $f^j(V_1) \subset E$, or

(b) there exists a $k \geq 1$ such that $f^j(V_1) \subset E$ $(0 \leq j < k - 1)$ and $f^{k-1}(V_1) \subset \widehat{\mathbb{C}} - E = V_0$.

In Case (β), only (b) can occur and $k \leq p$.

Now we can apply Theorem 2.1, Cases (a), (b) and Corollary 2.2, and conclude that f has a weakly repelling fixed point in $E = \widehat{\mathbb{C}} - V_0$. □

Remark 4.2. If the condition $f^p(\overline{U}) \subset U$ in (β) is replaced by $f^p(U) \subset U$ (without closure), then the following is still true:

If E is a component of $\widehat{\mathbb{C}} - U$ such that $f^j(\overline{U}) \cap E = \emptyset$ for some j with $0 \leq j \leq p$, then E contains a weakly repelling fixed point of f.

The proof is the same as above.

5   Attractive Basins

Proposition 5.1. *Suppose that a rational map f $(\deg f > 1)$ has a non-simply connected immediate attractive basin. Then there exists a connected open set U such that*

- *U is contained in the forward orbit of the immediate attractive basin,*

- *U is not simply connected and $f(U)$ is simply connected,*

- *U is a connected component of $f^{-1}(f(U))$,*

- *$f^p(\overline{U}) \subset U$, where p is the period of the basin.*

Proof of Theorem III (i) for attractive basins: Let U be as in Proposition 5.1. Note that $f^j(U) \cap U = \emptyset$ $(0 < j < p)$. It follows from Case (β) of Proposition 4.1 that each component of $\widehat{\mathbb{C}} - U$ contains a weakly repelling fixed point of f. Therefore there are at least two weakly repelling fixed points which are separated by U, and they are separated by the basin containing U, since weakly repelling fixed points belong to the Julia set. $\square$

Proof of Proposition 5.1: Let α be an attracting periodic point of period p of f such that its immediate attractive basin $A^*(\alpha)$ is not simply connected. Take a small disk neighborhood U_0 of α such that $f^p(\overline{U}_0) \subset U_0$.

For $n \geq 0$, let U_n be the connected component of $f^{-n}(U_0)$ containing $f^j(\alpha)$, where $0 \leq j < p$, $n \equiv -j \pmod{p}$.

Lemma 5.2. *We have*

$$A^*(\alpha) = \bigcup_{n \geq 0} U_{np}.$$

Proof: The inclusion $\bigcup_{n \geq 0} U_{np} \subset A^*(\alpha)$ is obvious. Let z be any point in $A^*(\alpha)$. Take an arc δ connecting z to α. Then there exists an $n \geq 0$ such that $f^{np}(\delta) \subset U_0$. Hence $\delta \subset f^{-np}(U_0)$, and z is in U_{np} by definition. $\square$

Since $f(U_{n+1}) \subset f^{-n}(U_0)$, $f(U_{n+1})$ is connected and contains $f^j(\alpha)$ (for some $j \geq 0$ with $n \equiv -j \pmod{p}$), and it follows that $f(U_{n+1}) \subset U_n$. Hence we have

$$U_{n+1} \subset f^{-1}(U_n) \subset f^{-1}(f^{-n}(U_0)) = f^{-(n+1)}(U_0).$$

It follows that U_{n+1} is a component of $f^{-1}(U_n)$, and $U_n = f(U_{n+1})$, since f is proper on $\widehat{\mathbb{C}}$.

Since $f^n(f^p(\overline{U}_n)) = f^p(f^n(\overline{U}_n)) \subset f^p(\overline{U}_0) \subset U_0$ and $f^p(\overline{U}_n)$ is connected, we have $f^p(\overline{U}_n) \subset U_n$. Hence $U_{np} = f^p(U_{(n+1)p}) \subset U_{(n+1)p}$.

There exists an $n > 0$ such that U_{np} is not simply connected; if not, the union of the increasing simply connected open sets U_{np} would also be simply connected. Therefore there exists an $n_0 > 0$ such that U_{n_0} is not simply connected and U_{n_0-1} is simply connected. Then $U = U_{n_0}$ satisfies the conditions of the proposition. $\square$

6 Parabolic Basins

The proof of Theorem III (i) in the case of parabolic basins is very similar to the case of an attractive basin. However we need to be careful with the details, since we touch parabolic periodic points, which are more delicate.

Proposition 6.1. *Suppose that a rational map f (deg $f > 1$) has a non-simply connected parabolic basin B. Let p be its period and α the corresponding parabolic periodic point. Then there exists a connected open set U and an i $(0 \leq i < p)$ such that, putting $\alpha' = f^i(\alpha)$,*

- *$U \subset f^i(B)$ and $\alpha' \in \partial U$,*

- *U is not simply connected and $f(U)$ is simply connected,*

- *U is a connected component of $f^{-1}(f(U))$,*

- *$f^p(\overline{U}) \subset U \cup \{\alpha'\}$,*

- *for any neighborhood $\mathcal{N}_0$ of α', there exists a simply connected open set $W \subset U \cap \mathcal{N}_0$ such that $f^p(\overline{W}) \subset W \cup \{\alpha'\}$, $\partial W \cap U$ is an arc and $\alpha' \notin \overline{U - W}$.*

Proof of Theorem III (i) for parabolic basins: It suffices, as in Section 5, to show that each connected component of $\widehat{\mathbb{C}} - U$ contains a weakly repelling fixed point, where U is as in Proposition 6.1. For simplicity, we may assume that $i = 0$ and $\alpha = \alpha'$. Let $\widehat{\mathbb{C}} - U = \bigcup_{\ell=0}^{L} E_\ell$, where the E_ℓ are the connected components and $\alpha \in E_0$. Since U is contained in a parabolic basin of period p, each $f^j(\overline{U})$ $(1 \leq j \leq p-1)$ must be contained in one of the E_ℓ's. Fix E_ℓ, and set $V_0 = \widehat{\mathbb{C}} - E_\ell$, $V_1 = f(U)$ and $\mathcal{N} = U \cup E_\ell$. Then V_0, V_1 are simply connected and $\mathcal{N}$ is a neighborhood of $\widehat{\mathbb{C}} - V_0$.

If $\ell \neq 0$, then $f^p(\overline{U}) \cap E_\ell = \emptyset$. Hence for some k with $1 \leq k \leq p$,

$$f^j(V_1) \cap V_0 = \emptyset \quad (0 \leq j \leq k-1) \text{ and } f^{k-1}(\overline{V}_1) \subset V_0.$$

By Theorem 2.1 Case (b) and Corollary 2.2, f has a weakly repelling fixed point in E_ℓ. (See Remark 4.2.)

Similarly, if $\ell = 0$ and if $f^k(U) \subset E_m$ for some k, m with $1 \leq k < p$, $m \neq 0$, then for the minimal such k, V_0 and V_1 satisfy the above condition, and E_0 contains a weakly repelling fixed point. (See also Remark 4.2.)

The remaining case is E_0 with $f^k(U) \subset F_0$ $(1 \leq k < p)$. Then $f^j(V_1) \cap V_0 = \emptyset$ $(0 \leq j < p-1)$ and $f^{p-1}(\overline{V}_1) \subset V_0 \cup \{\alpha\}$, and by the last assertion of Proposition 6.1, all the conditions in Case (c) of Theorem 2.1 are satisfied, with $k = p$. Hence E_0 contains a weakly repelling fixed point by Corollary 2.2. □

Proof of Proposition 6.1: Assume that f, B and α are as in the hypotheses of Proposition 6.1.

Lemma 6.2. (i) *There exists a simply connected open set $U_0 \subset B$ such that $f^p(\overline{U}_0 - \{\alpha\}) \subset U_0$, ∂U_0 is a Jordan curve containing α, $f^p|_{\overline{U}_0}$ is injective and $B \subset \bigcup_{n \geq 0} f^{-np}(U_0)$.*
(ii) *For any neighborhood $\mathcal{N}_0$ of α, there exists a simply connected open set $W \subset U_0 \cap \mathcal{N}_0$ such that $f^p(\overline{W}) \subset W \cup \{\alpha\}$, $\partial W \cap U_0$ is an arc and $\alpha \notin \overline{U}_0 - W$.*

Proof: (See, for example [Mi06, Be91] Flower theorem.) By a coordinate change, we may assume that $\alpha = 0$ and

$$f^p(z) = z(1 - z^\nu + O(z^{\nu+1})) \quad (z \to 0),$$

where ν is the number of parabolic basins attached to α. Let $H(z) = 1/z^\nu$, $S_j = \{z \in \mathbb{C} - \{0\} | \,|\arg z - 2\pi j/\nu| \leq \pi/\nu\}$ and $H_j = H|_{S_j} : S_j \to \mathbb{C} - \mathbb{R}_-$ ($j = 0, 1, \ldots, \nu - 1$), where $\mathbb{R}_-$ is the negative real axis. Then $H_j \circ f^p \circ H_j^{-1}(w) = w + 1 + O\left(\left(\frac{1}{|w|}\right)^{\frac{1}{\nu}}\right)$, $w \in \mathbb{C} - \mathbb{R}_-$, as $w \to \infty$. For large $L > 0$, f^p is injective on $U_0^{(j)} = H_j^{-1}\{w | Re\, w > L\}$ and $f^p(\overline{U}_0^{(j)} - \{\alpha\}) \subset U_0^{(j)}$. The parabolic basin B is the component of $\bigcup_{n \geq 0} f^{-np}(U_0^{(j)})$ containing $U_0^{(j)}$ for some j. Let $U_0 = U_0^{(j)}$ and $W_R = H_j^{-1}\{w | Re\, w > L, Re\, w + |Im\, w| > R\}$. Note that for any neighborhood $\mathcal{N}_0$ of α, $W_R \subset \mathcal{N}_0$ for large R. It is easy to check the conditions on U_0 and to see that $W = W_R$. □

Fix U_0 as in Lemma 6.2. For $n \geq 0$, let U_n be the connected component of $f^{-n}(U_0)$ containing $f^j(U_0)$, where $0 \leq j < p$, $n \equiv -j \pmod{p}$. For $n < 0$, let $U_n = f^{-n}(U_0)$.

Lemma 6.3. *We have*

$$B = \bigcup_{n \geq 0} U_{np}.$$

The proof is the same as that of Lemma 5.2, using the arc δ connecting z to U_0.

Now we continue the proof of Proposition 6.1. It can be shown, as in Section 5, that U_{n+1} is a component of $f^{-1}(U_n)$, $U_n = f(U_{n+1})$ and $U_{np} \subset U_{(n+1)p}$. Thus there exists an $n_0 > 0$ such that U_{n_0} is not simply connected, whereas U_{n_0-1} is simply connected. Let us write $n_0 = mp - i$ ($0 \leq i < p$) and let $\alpha' = f^i(\alpha)$. If we set $U = U_{n_0}$, U satisfies all conclusions of the proposition except $f^p(\overline{U}) \subset U \cup \{\alpha'\}$. As for the last point, we use Lemma 6.2 and the fact that f is locally injective along parabolic periodic orbits. It is clear that U_{n_0} satisfies $f^p(U_{n_0}) \subset U_{n_0}$, but the problem is that $f^p(\overline{U}_{n_0})$ may intersect ∂U_{n_0} at points other than α'. In order to achieve $f^p(\overline{U}) \subset U \cup \{\alpha'\}$, we need to modify U_{n_0-1} near the boundary.

By replacing B and U_0 by their forward iterates, we may assume that $n_0 = mp + 1$ and $U_{n_0-1} = U_{mp}$. Note that if $z \in \partial U_{mp}$ and $f^{mp}(z) \neq \alpha$, then $f^{mp}(z) \in \partial U_0$ and $f^{mp}(f^p(z)) \in f^p(\overline{U}_0 - \{\alpha\}) \subset U_0$. This implies that $f^p(z) \in U_{mp}$ and not in its boundary. Therefore the intersection of $f^p(\overline{U}_{mp})$ and ∂U_{mp} can only occur at $f^p(f^{-mp}(\alpha)) = f^{-(m-1)p}(\alpha)$.

Now we are going to modify $U' = U_{mp}$ in a small neighborhood of $f^{-mp}(\alpha) - \{\alpha\}$ so that $f^p(\overline{U}' - \{\alpha\}) \subset U'$. The boundary U' locally looks like an injective arc, except possibly at $f^{-mp}(\alpha)$, where the domain looks like several sectors touching at a vertex (when a critical orbit lands on α). Near each point z of $f^{-mp}(\alpha) - \{\alpha\}$, we "push" the boundary curve into the domain U', so that the domain becomes slightly smaller near z. It then has the effect of making its image $f^p(U')$ slightly smaller near $f^p(z)$. (Moreover possible self-intersections of $\partial U'$ will also be removed, and the boundary becomes a Jordan curve.)

In order to remove the intersection of $f^p(\overline{U}')$ and $\partial U'$, we proceed as follows. First do the modification in a small neighborhood of

$$\left(f^{-mp}(\alpha) - f^{-(m-1)p}(\alpha) \right) \cap \partial U' ,$$

where there is no intersection of $f^p(\overline{U}')$ and $\partial U'$ ($f^p(\overline{U}')$ does not reach here). This will shrink $f^p(\overline{U}')$ near $\left(f^{-(m-1)p}(\alpha) - f^{-(m-2)p}(\alpha) \right) \cap \partial U'$ and U' itself is not touched here, so the intersection is removed. Next, we modify $\partial U'$ near

$$\left(f^{-(m-1)p}(\alpha) - f^{-(m-2)p}(\alpha) \right) \cap \partial U' ;$$

the modification should be so small that the previously removed intersection does not regenerate. On the other hand, $f^p(\overline{U}')$ will be shrunk near $\left(f^{-(m-2)p}(\alpha) - f^{-(m-3)p}(\alpha) \right) \cap \partial U'$ and the intersection will be removed there. Continuing this process, we can remove the intersection up to $(f^{-p}(\alpha) - \{\alpha\}) \cap \partial U'$. The final modification near $(f^{-p}(\alpha) - \{\alpha\}) \cap \partial U'$ will give us $f^p(\overline{U}' - \{\alpha\}) \subset U'$.

Let U' denote the domain left after all modifications have been made, and define U to be the connected component of $f^{-1}(U')$ containing a forward orbit of U_0. Let us show that $f^p(\overline{U}) \subset U \cup \{\alpha'\}$ (under the above assumption, $\alpha' = f^{p-1}(\alpha)$). Suppose there exists a $z \in \partial U$ such that $f^p(z) \in \partial U$ (hence $f^p(z) \in f^p(\overline{U}) \cap \partial U$). Then $f(z) \in \partial U'$, and $f^p(f(z)) \in \partial U'$. By the above property of U', we have $f(z) = \alpha$. Hence $f^p(z) = f^{p-1}(\alpha) = \alpha'$. This shows that $f^p(\overline{U}) \cap \partial U = \{\alpha'\}$ and $f^p(\overline{U}) \subset U \cup \{\alpha'\}$. The modification does not change the topological types of U' and U. Hence $U' = f(U)$ is simply connected and U is multiply connected. The last conclusion regarding W follows from Lemma 6.2 (ii).

Thus Proposition 6.1 is proved. $\square$

7   Herman Rings

For Theorem III (ii), we need a study of the configuration of Herman rings. Suppose a rational map f has a Herman ring A of period p. Choose a periodic curve γ_0 in A which does not intersect the orbit of the critical points. Hence γ_0 is a real analytic Jordan curve, $f^p(\gamma_0) = \gamma_0$ and $f^p|_{\gamma_0}$ is real analytically conjugate to an irrational rotation. Let $\Gamma = \{f^j(\gamma_0)|\, 0 \leq j < p\}$.

By Fatou's Theorem 1.1, there exists at least one weakly repelling fixed point z_0. Denote by E_γ the connected component of $\widehat{\mathbb{C}} - \gamma$ not containing z_0 ($\gamma \in \Gamma$). For Theorem III (ii), we need to show that there is a $\gamma \in \Gamma$ such that E_γ contains a weakly repelling fixed point. In fact, we prove the following:

Proposition 7.1.

(i) If $f(\gamma) \not\subset E_\gamma$ and $f(E_\gamma) \cap E_\gamma \neq \emptyset$, then E_γ contains a weakly repelling fixed point.

(ii) If $f(\gamma) \subset E_\gamma$, then E_γ contains a weakly repelling fixed point.

(iii) Either (i) or (ii) occurs for some $\gamma \in \Gamma$.

Remark 7.2. (0) If $\gamma_1, \gamma_2 \in \Gamma$ and $\gamma_1 \neq \gamma_2$, then exactly one of the following holds: $\gamma_1 \subset E_{\gamma_2}$, $\gamma_2 \subset E_{\gamma_1}$ or $\overline{E}_{\gamma_1} \cap \overline{E}_{\gamma_2} = \emptyset$.

(1) By the open mapping property, $f(E_\gamma)$ is either $E_{f(\gamma)}$, $\widehat{\mathbb{C}} - \overline{E}_{f(\gamma)}$ or $\widehat{\mathbb{C}}$.

(2) Let D_1, D_1' (respectively D_2, D_2') be the connected components of $\widehat{\mathbb{C}} - \gamma$ (respectively $\widehat{\mathbb{C}} - f(\gamma)$). Then there exists a neighborhood $\mathcal{N}$ of γ such that either $f(D_1 \cap \mathcal{N}) \subset D_2$ and $f(D_1' \cap \mathcal{N}) \subset D_2'$, or $f(D_1 \cap \mathcal{N}) \subset D_2'$ and $f(D_1' \cap \mathcal{N}) \subset D_2$.

Lemma 7.3. Let γ_0, $\gamma_1 = f(\gamma_0) \in \Gamma$, let D_i be a component of $\widehat{\mathbb{C}} - \gamma_i$ $(i = 0, 1)$ and let $z^* \in D_0 \cap f^{-1}(D_1)$. Then there exists a component (a real analytic Jordan curve) γ_0^* of $f^{-1}(\gamma_1)$ (which may be γ_0), a component D_0^* of $\widehat{\mathbb{C}} - \gamma_0^*$ and a neighborhood $\mathcal{N}$ of γ_0^* such that $\gamma_0^* \subset \overline{D}_0$, $z^* \in D_0^* \subset D_0$ and $f(D_0^* \cap \mathcal{N}) \subset D_1$, $f((\widehat{\mathbb{C}} - \overline{D}_0^*) \cap \mathcal{N}) \subset \widehat{\mathbb{C}} - \overline{D}_1$.

Proof: If the conclusion does not hold for $\gamma_0^* = \gamma_0$ and $D_0^* = D_0$ (this implies $f(D_0 \cap \mathcal{N}) \subset \widehat{\mathbb{C}} - \overline{D}_1$, etc. for a small neighborhood $\mathcal{N}$), then there is a component U of $f^{-1}(\widehat{\mathbb{C}} - \overline{D}_1)$ such that $U \subset D_0$ and $\gamma_0 \subset \partial U$. By $z^* \in D_0 \cap f^{-1}(D_1)$, we have $z^* \notin U$. Take a component D_0^* of $\widehat{\mathbb{C}} - U$ containing z^* and let $\gamma_0^* = \partial D_0^*$. All properties can be easily checked.   □

Proof of Proposition 7.1: Our goal is to find an appropriate subset $\widehat{\mathbb{C}} - V_0$ within E_γ so that we can apply Theorem 2.1 (b) or (d). To do this, we have to divide the proof into several cases, according to the configuration of Γ.

(i) If $f(\gamma) = \gamma$, i.e., $p = 1$, then $V_0 = V_1 = \widehat{\mathbb{C}} - E_\gamma$ satisfy condition (d) of Theorem 2.1 with $k = 1$. If $f(E_\gamma) \subset E_\gamma$, then $f^p|_{E_\gamma}$ is normal, and hence $E_\gamma \subset F_f$, and this contradicts the fact that A is a Herman ring. Therefore $f(E_\gamma) \neq E_\gamma$, and by Corollary 2.2, E_γ contains a weakly repelling fixed point.

From now on, suppose $p > 1$, so that $f(\gamma') \neq \gamma'$ ($\gamma' \in \Gamma$). Let $D_0 = E_\gamma$, and let D_1 be the component of $\widehat{\mathbb{C}} - f(\gamma)$ containing $\overline{E}_\gamma$. Pick $z^* \in E_\gamma \cap f^{-1}(E_\gamma)$, which is non-empty by assumption. Applying Lemma 7.3, we obtain γ_0^*, D_0^* and $\mathcal{N}$. Then $V_0 = \widehat{\mathbb{C}} - \overline{D}_0^*$ and $V_1 = \widehat{\mathbb{C}} - \overline{D}_1$ satisfy $\widehat{\mathbb{C}} - V_0 = \overline{D}_0^* \subset \overline{D}_0 = \overline{E}_\gamma$, $\overline{V}_1 = \widehat{\mathbb{C}} - D_1 \subset \widehat{\mathbb{C}} - \overline{E}_\gamma = V_0$ and $f(V_0 \cap \mathcal{N}) \subset V_1$. Hence condition (b) of Theorem 2.1 is satisfied with $k = 1$. By Corollary 2.2, $\widehat{\mathbb{C}} - V_0$ and then E_γ contain a weakly repelling fixed point.

(ii) It is enough to show the assertion for an innermost curve among $\gamma \in \Gamma$ satisfying $f(\gamma) \subset E_\gamma$, where "innermost" is defined by the inclusion between the E_γ's. Let us call this curve γ_0, renumbering γ's, and write $E_0 = E_{\gamma_0}$. Hence we assume that $f(\gamma_0) \subset E_0$, and if $\gamma \subset E_0$ ($\gamma \in \Gamma$), then $f(\gamma) \not\subset E_\gamma$.

If there is a $\gamma \subset E_0$ such that $f(E_\gamma) \cap E_\gamma \neq \emptyset$, then by (i) we have a weakly repelling fixed point in $E_\gamma \subset E_0$. So we now assume that if $\gamma \subset E_0$, then $f(E_\gamma) \cap E_\gamma = \emptyset$. In fact, in this case, we have $f(\overline{E}_\gamma) \cap \overline{E}_\gamma = \emptyset$, because $f(\gamma) \neq \gamma$.

Lemma 7.4. *If $\gamma \subset E_0$, then either $f(\gamma) \subset E_0$ and $f(E_\gamma) = E_{f(\gamma)} \subset \overline{E}_{f(\gamma)} \subset E_0$, or $f(\gamma) \not\subset E_0$ and $f(E_\gamma) \subset \widehat{\mathbb{C}} - E_0$.*

Proof: Let $\gamma \in \Gamma$ be such that $\gamma \subset E_0$. First assume that $\gamma \subset E_{f(\gamma)}$. Then $f(\gamma) \subset f(E_{f(\gamma)})$, and $E_{f(\gamma)}$ accumulates to $f(\gamma)$, and therefore we have $f(E_{f(\gamma)}) \cap E_{f(\gamma)} \neq \emptyset$. This implies that $f(\gamma) \not\subset E_0$ by the above assumption. In other words, if $f(\gamma) \subset E_0$, then $\gamma \not\subset E_{f(\gamma)}$, and hence $\overline{E}_{f(\gamma)} \cap \overline{E}_\gamma = \emptyset$, by Remark 7.2 (0), and $f(\gamma) \not\subset E_\gamma$. In this case, it follows from Remark 7.2 (1) and the fact that $f(E_\gamma) \cap E_\gamma = \emptyset$ that $f(E_\gamma) = E_{f(\gamma)}$. It is obvious that $\overline{E}_{f(\gamma)} \subset E_0$.

On the other hand, if $f(\gamma) \not\subset E_0$, $f(E_\gamma)$ cannot intersect $E_\gamma (\subset E_0)$. Then $f(E_\gamma)$ is the component of $\widehat{\mathbb{C}} - f(\gamma)$ not containing E_0, and therefore $f(E_\gamma) \subset \widehat{\mathbb{C}} - E_0$. $\qquad \square$

Let us write $\gamma_j = f^j(\gamma_0)$ and $E_j = E_{\gamma_j}$ ($j = 1, 2, \dots$). The assumptions $\gamma_1 \subset E_0$ and $f(\overline{E}_{\gamma_1}) \cap \overline{E}_{\gamma_1} = \emptyset$ imply that $E_1 \subset E_0 \cap f^{-1}(\widehat{\mathbb{C}} - \overline{E}_1)$. Apply Lemma 7.3 to $D_0 = E_0$, $D_1 = \widehat{\mathbb{C}} - E_1$ and $z^* \in E_1$ to obtain γ_0^*, D_0^* and $\mathcal{N}$. We have $z^* \in D_0^* \subset E_0$ and $f((\widehat{\mathbb{C}} - \overline{D}_0^*) \cap \mathcal{N}) \subset E_1$. Since

$\partial E_1 = \gamma_1$ and $\partial D_0^* = \gamma_0^* \subset f^{-1}(\gamma_1)$, which is disjoint from $\gamma_1 (\neq \gamma_0)$, we have $z^* \in \overline{E}_1 \subset D_0^*$.

The periodicity of γ_0 implies that there exists a k with $2 \leq k \leq p$ such that $\gamma_k \not\subset D_0^*$. Take the minimal such k. Then we have $\gamma_j \subset D_0^*$ $(1 \leq j < k)$ and $\gamma_k \cap D_0^* = \emptyset$. By Lemma 7.4, we have

- $E_j \subset D_0^*$ $(1 \leq j < k)$, $f(E_j) = E_{j+1}$ $(1 \leq j < k - 1)$ and

- either (α) $f(E_{k-1}) = E_k \subset \overline{E}_k \subset E_0 - \overline{D}_0^*$, or (β) $f(E_{k-1}) \subset \widehat{\mathbb{C}} - E_0 \subset \widehat{\mathbb{C}} - \overline{D}_0^*$.

- If $f(E_{k-1}) = \widehat{\mathbb{C}} - \overline{D}_0^*$ in Case (β), then $D_0^* = E_0$ and $k = p$. Otherwise $f(\overline{E}_{k-1}) \subset \widehat{\mathbb{C}} - \overline{D}_0^*$.

Setting $V_0 = \widehat{\mathbb{C}} - \overline{D}_0^*$ and $V_1 = E_1$, we can apply Theorem 2.1 (b) when $f^{k-1}(\overline{V}_1) = f(\overline{E}_{k-1}) \subset \widehat{\mathbb{C}} - \overline{D}_0^* = V_0$ and Theorem 2.1 (d) when $f^{k-1}(V_1) = f(E_{k-1}) = \widehat{\mathbb{C}} - \overline{D}_0^* = \widehat{\mathbb{C}} - E_0 = V_0$ and $k = p$. Hence by Corollary 2.2, in either case, there is a weakly repelling fixed point in $\widehat{\mathbb{C}} - V_0 = \overline{D}_0^* \subset \overline{E}_0$. This completes the proof of (ii).

(iii) Suppose neither (i) nor (ii) occur for any $\gamma \in \Gamma$. This means that for every $\gamma \in \Gamma$, $f(\gamma) \not\subset E_\gamma$ and $f(E_\gamma) \cap E_\gamma = \emptyset$. If moreover $f(E_\gamma) = E_\gamma$ for all $\gamma \in \Gamma$, then $\{f^n\}$ is normal in E_γ, and this contradicts the assumption that γ is a periodic curve within a Herman ring. Therefore there exists a $\gamma \in \Gamma$ such that $f(E_\gamma) = \widehat{\mathbb{C}} - \overline{E}_{f(\gamma)}$, and the right-hand side should not intersect E_γ. Then the period p cannot be 1, and $\overline{E}_\gamma \subset E_{f(\gamma)}$. On the other hand, $E_{f(\gamma)}$ should also be mapped into a proper subset of $\widehat{\mathbb{C}} - \overline{E}_{f(\gamma)}$ due to the fact that $f(E_{f(\gamma)}) \cap E_{f(\gamma)} = \emptyset$. Hence E_γ is also mapped into a proper subset of $\widehat{\mathbb{C}} - \overline{E}_{f(\gamma)}$. This is a contradiction. $\square$

Remark 7.5. One can ask what happens if there are more than one cycle of Herman rings. Should there be $n + 1$ weakly repelling fixed points in different components of the Julia set if there are n cycles of Herman rings?

Bibliography

[Ah06]   L. V. Ahlfors, *Lectures on quasiconformal mappings, Second edition, with supplemental chapters by C. J. Earle, I. Kra, M. Shishikura, and J. H. Hubbard*, University Lecture Series **38**, Amer. Math. Soc., Providence, RI, 2006.

[Be91]   A. F. Beardon, *Iteration of rational functions*, Graduate Texts in Mathematics **132**, Springer-Verlag, New York, 1991.

[BT96]   W. Bergweiler and N. Terglane, *Weakly repelling fixpoints and the connectivity of wandering domains*, Trans. Amer. Math. Soc. **348** 1 (1996), 1–12.

[DH85]   A. Douady and J. H. Hubbard, *On the dynamics of polynomial-like mappings*, Ann. Scient. Ec. Norm. Sup., 4e série, t. 18 (1985), 287–343.

[Fa19]   P. Fatou, *Sur les équations fonctionnelles*, Bull. Soc. Math. France **47** (1919), 161–271; **48** (1920), 33–94, 208–304.

[He87]   J. Head, *The combinatorics of Newton's method for cubic polynomials*, Ph. D. Thesis, Cornell University, 1987.

[Ma92]   A. Manning, *How to be sure of finding a root of a complex polynomial using Newton's method*, Bol. Soc. Brasil. Mat. (N.S.) **22** 2 (1992), 157–177.

[MS06]   S. Mayer and D. Schleicher, *Immediate and virtual basins of Newton's method for entire functions*, Ann. Inst. Fourier **56** 2 (2006), 325–336.

[MS83]   C. T. McMullen and D. Sullivan, *Quasiconformal homeomorphisms and dynamics III*, Adv. Math. **135** (1998), 351–395, Original version by D. Sullivan, Preprint IHES 1983.

[Me89]   H. G. Meier, *On the connectedness of the Julia-set for rational functions*, Preprint RWTH Aachen, 1989.

[Mi06]   J. Milnor, *Dynamics in one complex variable, Third edition*, Ann. Math. Studies **160**, Princeton University Press, 2006.

[Pr89]   F. Przytycki, *Remarks on the simple connectedness of basins of sinks for iterations of rational maps*, in: Dynamical systems and ergodic theory, K. Krzyzewski Ed., PWN-Polish Scientific Publishers, 1989, 229–235.

[Sh87]   M. Shishikura, *On the quasiconformal surgery of rational functions*, Ann. Scient. Ec. Norm. Sup, 4e série, t. 20 (1987), 1–29.

[Sh89]   M. Shishikura, *Trees associated with the configuration of Herman rings*, Ergod. Th. Dynam. Syst. **9** (1989), 543–560.

[Su85]   D. Sullivan, *Quasiconformal homeomorphisms and dynamics I*, Ann. Math. **122** (1985), 401–418.

[Ta97]   L. Tan, *Branched coverings and cubic Newton maps*, Fund. Math. **154** (1997), 207–260.

7 The Rabbit and Other Julia Sets Wrapped in Sierpiński Carpets

Paul Blanchard, Robert L. Devaney,
Antonio Garijo, Sebastian M. Marotta,
and Elizabeth D. Russell

1 Introduction

In this paper we consider complex analytic rational maps of the form

$$F_\lambda(z) = z^2 + c + \frac{\lambda}{z^2}$$

where $\lambda, c \in \mathbb{C}$ are parameters. For this family of maps, we fix c to be a parameter that lies at the center of a hyperbolic component of the Mandelbrot set, i.e., a parameter such that, for the map

$$F_0(z) = z^2 + c,$$

0 lies on a periodic orbit. We then perturb F_0 by adding a pole at the origin. Our goal is to investigate the structure of the Julia set of F_λ, which we denote by $J(F_\lambda)$, when λ is non-zero.

For these maps, the point at ∞ is always a superattracting fixed point, so we have an immediate basin of attraction of ∞ that we denote by B_λ. As a consequence, we may also define the filled Julia set for these maps to be the set of points whose orbits remain bounded. We denote this set by $K(F_\lambda)$.

In the case where c is chosen so that the map has a superattracting cycle of period 1, the structure of $J(F_\lambda)$ has been well studied [BDLSS05], [DG], [DL06], [DLU05]. In this case, $c = 0$ and the map is $z^2 + \lambda/z^2$. This map has four "free" critical points at the points $c_\lambda = \lambda^{1/4}$, but there is essentially only one critical orbit, since one checks easily that $F_\lambda^2(c_\lambda) = 4\lambda + 1/4$. Hence all four of the critical orbits land on the same point after two iterations. Then the following result is proved in [BDLSS05].

Theorem. *Suppose the free critical orbit of $z^2 + \lambda/z^2$ tends to ∞ but the critical points themselves do not lie in B_λ. Then the Julia set of this map is a Sierpiński curve. In particular, there are infinitely many disjoint open sets in any neighborhood of 0 in the parameter plane (the λ-plane) for which this occurs. Furthermore, two maps drawn from different open sets in this collection are not topologically conjugate on their Julia sets.*

A *Sierpiński curve* is any planar set that is homeomorphic to the well-known Sierpiński carpet fractal. By a result of Whyburn [Wh58], a Sierpiński curve may also be characterized as any planar set that is compact, connected, locally connected, nowhere dense, and has the property that any pair of complementary domains are bounded by simple closed curves that are pairwise disjoint. A Sierpiński curve is an important object from the topological point of view because it is a universal plane continuum among all topologically one-dimensional sets, i.e., it contains a homeomorphic copy of any planar one-dimensional set.

When $|\lambda|$ is small, we consider the map $z^2 + \lambda/z^2$ to be a singular perturbation of the simple map $F_0(z) = z^2$. As is well known, the Julia set of F_0 is the unit circle. When $|\lambda|$ is small, it is known that the Julia set of F_λ is also bounded by a simple closed curve that moves continuously as λ varies. Note that, when the hypothesis of the above theorem is met, the Julia set suddenly changes from a simple closed curve to a much more complicated Sierpiński curve.

We remark that the situation for families of the form

$$z^n + \frac{\lambda}{z^d}$$

where $n, d \geq 2$ but not both are equal to 2 is quite different. For these families, it is known [Mc88] that the Julia sets for $|\lambda|$ small consist of a Cantor set of simple closed curves, each of which surrounds the origin. Sierpiński curves do occur as Julia sets for these maps, but only for larger parameter values.

Our goal in this paper is to describe a related but somewhat different phenomenon that occurs in the family

$$F_\lambda(z) = z^2 + c + \frac{\lambda}{z^2}$$

when the singular perturbation occurs at c-values that are the centers of other hyperbolic components of the Mandelbrot set. A similar explosion in the Julia set takes place when $\lambda \neq 0$. Unlike the case $c = 0$, the boundaries of the components of the basin of ∞ are no longer simple closed curves. Rather, these domains are usually bounded by "doubly" inverted copies of the Julia set of $z^2 + c$. By removing the attachments on these components, we then find infinitely many disjoint Sierpiński curves that now lie in the Julia set whenever the critical orbits eventually escape. In addition, there is a collection of other points in the perturbed Julia set when $c \neq 0$.

To be more precise, suppose now that $n = d = 2$ and c lies at the center of some hyperbolic component with period $k > 1$ in the Mandelbrot set. When $\lambda = 0$, the Julia set and filled Julia set of F_0 are connected sets whose structure is also well understood: the interior of $K(F_0)$ consists of countably many simply connected open sets, each of which is bounded by a simple closed curve that lies in the Julia set. Let C_0 denote the closure of the component of this set that contains 0. Let C_j be the closure of the component that contains $F_0^j(0)$ for $1 \leq j \leq k-1$. Then F_0^k maps each C_j to itself. Moreover, F_0^k on C_j is conjugate to the map $z \mapsto z^2$ on the closed unit disk. For example, the center of the hyperbolic component of period 2 in the Mandelbrot set occurs when $c = -1$; this Julia set is known as the basilica and is displayed in Figure 1. Similarly, when $c \approx -0.12256+0.74486i$, c lies at the center of a period 3 hyperbolic component and the corresponding Julia set is the Douady rabbit. See Figure 1.

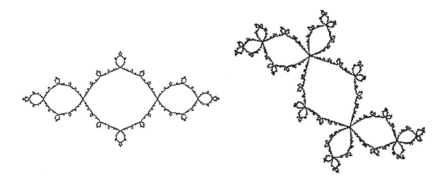

Figure 1. The basilica ($c = -1$) and the Douady rabbit ($c = -0.12256 + 0.74486i$) Julia sets.

Our first goal in this paper is to prove the following result.

Theorem 1.1. *There exists $\delta > 0$ such that, if $|\lambda| < \delta$, the boundary of B_λ is homeomorphic to $\partial B_0 = J(F_0)$ and F_λ restricted to ∂B_λ is conjugate to F_0 on $J(F_0)$.*

By this result the structure of the Julia set of $z^2 + c$ persists as ∂B_λ when $|\lambda|$ is small. However, the structure of $J(F_\lambda)$ inside ∂B_λ is quite a bit more complex. In Figure 2, we display perturbations of the basilica and the Douady rabbit. Note that the boundary of B_λ in these cases is a

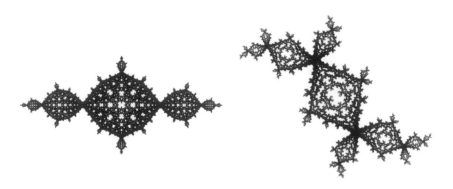

Figure 2. Perturbations of the basilica ($\lambda = -0.001$) and the Douady rabbit ($\lambda = 0.0013 - 0.002i$) Julia sets.

copy of the original basilica or rabbit, but that there are infinitely many "doubly inverted" basilicas or rabbits inside this set. See Figures 3 and 4. By a double inversion of the rabbit, for example, we mean the following. Choose one of the C_j and translate the rabbit linearly so that the periodic point inside C_j moves to the origin. Then invert the set via the two-to-one map $z \mapsto 1/z^2$. This map moves all the components of the filled Julia set that lie in the exterior of C_j so that they now lie inside the image of the boundary of C_j, and the external boundary of this set is now a simple closed curve that is the image of the boundary of C_j. A homeomorphic copy of this set is what we called a doubly inverted rabbit. See Figure 4.

As a consequence of Theorem 1.1, there is a region $C_j(\lambda)$ that corresponds to the region C_j for F_0. Consider the set of points in $C_j(\lambda)$ whose orbits travel through the $\cup C_i(\lambda)$ in the exact order that the point c_j travels through the $\cup C_i$ under F_0. Call this set Λ_λ^j. Note that Λ_λ^j is contained in the disk $C_j(\lambda)$ and is invariant under F_λ^k. We shall prove:

Theorem 1.2. *Suppose that $|\lambda|$ is sufficiently small and that all of the free critical orbits of F_λ escape to ∞ but the critical points themselves do not lie in B_λ. Then, for $j = 0, \ldots, k-1$, the set Λ_λ^j is a Sierpiński curve.*

Corollary 1.3. *Inside every component of the interior of $\mathbb{C} - \partial B_\lambda$ that corresponds to an eventually periodic component of the interior of $K(F_0)$ there is a similar copy of a Sierpiński curve that eventually maps to the Sierpiński curves inside the $C_j(\lambda)$. Furthermore, each interior complementary domain of all of these Sierpiński curves contains an inverted copy of the Julia set of F_0, and then each interior component of this set also contains a Sierpiński curve, and so forth.*

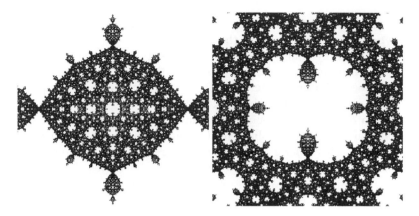

Figure 3. Several magnifications of the perturbed basilica Julia set.

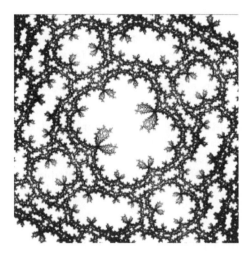

Figure 4. A magnification of the doubly inverted rabbit. Note that there are some "quadruply" inverted copies of the rabbit surrounding this set. These bound regions that contain critical points of F_λ or their preimages.

Thus, when λ becomes non-zero and the critical orbits eventually escape to ∞, we see a similar phenomenon as in the case of z^2: suddenly each component of filled Julia set inherits the structure of a Sierpiński curve while ∂B_λ remains homeomorphic to ∂B_0. However, there is actually much more to the structure of the full Julia set of F_λ than that described in the above theorems.

By Theorems 1.1 and 1.2, when $|\lambda|$ is sufficiently small, we have two types of invariant subsets of the Julia set of F_λ: ∂B_λ and the Λ_λ^j. Moreover, we completely understand both the topology of and the dynamics on these sets. However, each of these sets has infinitely many preimages and each of these preimages lies in the Julia set but contains no periodic points. So there are many other points in $J(F_\lambda)$.

To describe these points, we shall assign in Section 5 an itinerary to each such point in the Julia set (excluding those in the various preimages of ∂B_λ). This itinerary will be an infinite sequence of non-negative integers that specifies how the orbit of the given point moves through the various preimages of the $C_j(\lambda)$. For example, the itinerary of any point in Λ_λ^0 will be $\overline{012\ldots n-1}$ and the itinerary of any point in Λ_λ^j will be the j-fold shift of this sequence. Similarly, any point in a preimage of any of the sets Λ_λ^j will be a sequence that terminates in such a sequence. The itinerary of any other point in the Julia set will not have this property. Then we shall prove:

Theorem 1.4. *Let $\Gamma_s(\lambda)$ denote the set of points whose itinerary is the sequence of non-negative integers $s = (s_0 s_1 s_2 \ldots)$. Then, if s ends in a repeating sequence of the form $\overline{012\ldots n-1}$, $\Gamma_s(\lambda)$ is a Sierpiński curve. Otherwise, $\Gamma_s(\lambda)$ is a Cantor set.*

Acknowledgments. The third author would like to thank to Department of Mathematics at Boston University for their hospitality while this work was in progress. The third author was supported by MTM2005-02139/Consolider (including a FEDER contribution) and CIRIT 2005 SGR 01028.

2 Preliminaries

Let c be a center of a hyperbolic component of the Mandelbrot set with period greater than 1. Let

$$F_\lambda(z) = z^2 + c + \frac{\lambda}{z^2}$$

where $\lambda \in \mathbb{C}$. When $\lambda \neq 0$, these maps have critical points at 0, ∞, and the four points $\lambda^{1/4}$. Since 0 maps to ∞, which is a superattracting fixed point, we call the remaining four critical points the *free critical points*. There are really only two free critical orbits for this family, since $F_\lambda(-z) = F_\lambda(z)$, so $\pm\lambda^{1/4}$ both map onto the same orbit after one iteration. Thus we have only two critical values for F_λ, namely $c \pm 2\lambda^{1/2}$.

One checks easily that the circle of radius $|\lambda|^{1/4}$ centered at the origin is mapped four-to-one onto the straight line segment connecting the two critical values $c \pm 2\lambda^{1/2}$. We call this circle the *critical circle* and its image the *critical segment*. The points $(-\lambda)^{1/4}$ on the critical circle are all mapped to c. Also, the straight lines from the origin to ∞ passing through each of the four critical points are mapped two-to-one onto straight line segments extending from one of the two the critical values to ∞ and extending the critical segment so that these lines together with the critical segment form a single straight line in the plane. One also checks easily that any circle centered at the origin (except the critical circle) is mapped by F_λ two-to-one onto an ellipse whose foci are the two critical values. As these circles tend to the critical circle, the image ellipses tend to the critical segment. Thus the exterior (resp., interior) of the critical circle in $\mathbb{C}$ is mapped as a two-to-one covering of the complement of the critical segment.

For these maps, recall that we always have an immediate basin of attraction of ∞ denoted by B_λ. For each $\lambda \neq 0$, there is a neighborhood of 0 that is mapped into B_λ. If this neighborhood is disjoint from B_λ, we call the component of the full basin of ∞ that contains the origin the *trap door* and denote it by T_λ. We will be primarily concerned with the case where B_λ and T_λ are disjoint in this paper.

The *Julia set* of F_λ, denoted by $J(F_\lambda)$, is the set of points at which the family of iterates of F_λ fails to be a normal family in the sense of Montel. The complement of the Julia set is the *Fatou set*. It is known that $J(F_\lambda)$ is the closure of the set of repelling periodic points of F_λ. The Julia set is also the boundary of the full basin of attraction of ∞. When T_λ and B_λ are disjoint, there are infinitely many distinct components of the entire basin of ∞, so the Julia set surrounds infinitely many disjoint open sets in which orbits eventually escape into B_λ. These holes all lie in the Fatou set.

Since $F_\lambda(-z) = F_\lambda(z)$, it follows that $J(F_\lambda)$ is symmetric under $z \mapsto -z$. There is a second symmetry for these maps: let $H_\lambda(z) = \sqrt{\lambda}/z$. Then we have $F_\lambda(H_\lambda(z)) = F_\lambda(z)$, so $J(F_\lambda)$ is also symmetric under each of the involutions H_λ.

Recall that we have assumed that 0 lies on a cycle of period $k > 1$ for F_0. Let $c_j = F_0^j(0)$ for $j = 1, \ldots, k-1$. The set C_j is the closure of the component of the interior of $\mathbb{C} - J(F_0)$ that contains c_j. As is well known, the interior of C_j is the immediate basin of attraction of F_0^k surrounding c_j. Also, F_0^k maps C_j to itself as a two-to-one branched covering with c_j acting as the only branch point. On the boundary of C_j, F_0^k is conjugate to the map $z \mapsto z^2$ on the unit circle. All other components of $\mathbb{C} - J(F_0)$ eventually map to the C_j (with the exception of the basin of attraction of ∞, which is mapped to itself).

3   The Boundary of the Basin of ∞

Our goal in this section is to prove Theorem 1.1.

Theorem 1.1. *There exists $\delta > 0$ such that, if $|\lambda| < \delta$, the boundary of B_λ is homeomorphic to $\partial B_0 = J(F_0)$ and F_λ restricted to ∂B_λ is conjugate to F_0 on $J(F_0)$.*

Proof: We shall use quasiconformal surgery to modify each of the maps F_λ so that the resulting maps are all conjugate to F_0 via a conjugacy h_λ, at least for $|\lambda|$ small enough. Then h_λ will be shown to be a homeomorphism taking ∂B_λ to $\partial B_0 = J(F_0)$.

Let $\mathcal{O}_0$ be the closed disk of radius r about the origin. We choose r small enough so that $\mathcal{O}_0$ lies in the interior of the Fatou component of $K(F_0)$ that contains the origin. For $i = 1, \ldots, k$, let $\mathcal{O}_i = F_0^i(\mathcal{O}_0)$. Note that $\mathcal{O}_k$ is strictly contained in the interior of $\mathcal{O}_0$. Let β_i denote the boundary of $\mathcal{O}_i$. There is a simple closed curve γ_0 that lies outside of β_0 in the component of $K(F_0)$ containing the origin and that is mapped two-to-one onto β_0 by F_0^k. We may then choose $\delta > 0$ small enough so that, if $|\lambda| < \delta$, there is a similar curve $\gamma_0(\lambda)$ lying outside β_0 that is mapped two-to-one onto β_0 by F_λ^k. This follows since, for $|\lambda|$ small enough, $F_\lambda \approx F_0$ outside of $\mathcal{O}_0$. Let $\gamma_i(\lambda) = F_\lambda^i(\gamma_0(\lambda))$ for $i = 1, \ldots, k$ so that $\gamma_k(\lambda) = \beta_0$. Let $A_i(\lambda)$ denote the closed annulus bounded by β_i and $\gamma_i(\lambda)$ for each $i \le k$.

For $|\lambda| < \delta$, we define a new map G_λ on $\overline{\mathbb{C}}$ as follows. We first set $G_\lambda = F_0$ on each of the $\mathcal{O}_i$. Then we set $G_\lambda = F_\lambda$ on the region outside the union of all the $\mathcal{O}_i$ and $A_i(\lambda)$. We now only need to define G_λ on the $A_i(\lambda)$ for $i = 0, \ldots, k - 1$. To do this, recall that F_0 maps β_i to β_{i+1} while F_λ maps $\gamma_i(\lambda)$ to $\gamma_{i+1}(\lambda)$. For $i = 0, \ldots, k - 1$, we then define $G_\lambda : A_i(\lambda) \to A_{i+1}(\lambda)$ to be a smooth map that:

1. G_λ agrees with F_0 on β_i and with F_λ on $\gamma_i(\lambda)$;

2. $G_0 = F_0$ on each $A_i(\lambda)$;

3. G_λ is a two-to-one covering map on $A_0(\lambda)$ and one-to-one on $A_i(\lambda)$ for $1 \le \lambda \le k - 1$;

4. G_λ varies continuously with λ.

According to this definition, we have that $G_0 = F_0$ everywhere on $\overline{\mathbb{C}}$. Furthermore, G_λ is holomorphic at all points outside of the $A_i(\lambda)$ and G_λ has a superattracting cycle of period k at 0. Finally, $G_\lambda = F_\lambda$ on B_λ and its boundary, so the immediate basin of ∞ for G_λ is just B_λ.

We now construct a measurable ellipse field ξ_λ that is invariant under G_λ. Define ξ_λ to be the standard complex structure on the union of the $\mathcal{O}_i$, i.e., the circular ellipse field. Now we begin pulling back this structure by successive preimages of G_λ. The first k preimages defines ξ_λ on the union of the $A_i(\lambda)$ (and elsewhere). Each of these pullbacks yields an ellipse field on the $A_i(\lambda)$ since we are pulling back by a map that is not necessarily holomorphic on these annuli. However, since G_λ is a smooth map on these annuli, this new portion of the ellipse field has bounded dilatation. Then all subsequent pullbacks of the ellipse field are done by holomorphic maps since $G_\lambda = F_\lambda$ outside of the $A_i(\lambda)$. This defines ξ_λ on the union of all of forward and backward images of the $\mathcal{O}_i$. As defined so far, ξ_0 is just the standard complex structure on the union of all the bounded components of $K(F_0)$. Furthermore, each G_λ preserves ξ_λ. To complete the definition of ξ_λ, we set ξ_λ to be the standard complex structure on all remaining points in $\overline{\mathbb{C}}$. Since $G_\lambda = F_\lambda$ on this set of points, it follows that G_λ preserves ξ_λ everywhere and, in particular, ξ_λ has bounded dilatation on the entire Riemann sphere.

By the Measurable Riemann Mapping Theorem, there is then a quasiconformal homeomorphism h_λ that converts ξ_λ to the standard complex structure on $\overline{\mathbb{C}}$. We may normalize h_λ so that $h_\lambda(\infty) = \infty$, $h_\lambda(c) = c$, and $h_\lambda(0) = 0$. Since ξ_λ depends continuously on λ, so too does h_λ. Moreover, h_0 is the identity map. Thus h_λ conjugates each G_λ to a holomorphic map that is a polynomial of degree two with a superattracting cycle of period k. This polynomial must therefore be F_0 for each λ. We therefore have that h_λ is a homeomorphism that takes ∂B_λ to $J(F_0)$. This completes the proof.   □

4   Sierpiński Carpets

Our goal in this section is to prove Theorem 1.2.

Proof: For the rest of this section, we fix a λ value with $|\lambda| \leq \delta$ so that, by Theorem 1.1, ∂B_λ is homeomorphic to $J(F_0)$. Hence we have the k regions $C_j(\lambda)$ for F_λ that correspond to the periodic regions C_j for F_0. By Theorem 1.1, each of the $C_j(\lambda)$ is a closed disk that is bounded by a simple closed curve. We shall prove that there is an F_λ^k-invariant set $\Lambda_\lambda \subset J(F_\lambda)$ that is contained in $C_0(\lambda)$, is homeomorphic to the Sierpiński carpet, and has the property that all points in this set have orbits that remain for all iterations in $\cup C_i(\lambda)$ and travel through these sets in the same order as the orbit of 0 does under F_0. The other parts of Theorem 1.2 and its Corollary then follow immediately from this result by taking appropriate preimages of Λ_λ.

So consider the region $C_0(\lambda)$ and its boundary curve $\nu_0(\lambda)$. Similarly, let $\nu_j(\lambda)$ denote the boundary of $C_j(\lambda)$. Since $|\lambda| \leq \delta$, the critical segment lies inside $C_1(\lambda)$, so the critical circle lies in the interior of $C_0(\lambda)$. Now recall that F_λ maps the interior of the critical circle as a two-to-one covering onto the exterior of the critical segment in $\overline{\mathbb{C}}$. It follows that there is another simple closed curve in $C_0(\lambda)$ that lies inside the critical circle (and hence inside $\nu_0(\lambda)$), and, like $\nu_0(\lambda)$, this curve is mapped two-to-one onto $\nu_1(\lambda)$. Call this curve $\xi_0(\lambda)$. The region between $\xi_0(\lambda)$ and $\nu_0(\lambda)$ is therefore an annulus that is mapped by F_λ as a four-to-one branched covering onto the interior of the disk $C_1(\lambda)$. Call this annulus $\mathcal{A}_\lambda$. Note that all four of the free critical points of F_λ lie in $\mathcal{A}_\lambda$ since the critical values reside in the interior of $C_1(\lambda)$.

The complement of $\mathcal{A}_\lambda$ in $C_0(\lambda)$ is therefore a closed disk that is mapped by F_λ two-to-one to the complement of the interior of $C_1(\lambda)$ in $\overline{\mathbb{C}}$. Hence there is a subset of this disk that is mapped two-to-one onto ∂B_λ. This subset includes the boundary curve $\xi_0(\lambda)$ that is mapped two-to-one to $\nu_1(\lambda)$ in ∂B_λ and the preimages of all of the other points in ∂B_λ lie strictly inside the curve $\xi_0(\lambda)$. Since F_λ is two-to-one inside $\xi_0(\lambda)$, the preimage of ∂B_λ is thus a doubly inverted copy of ∂B_λ. Note that there is a component of the complement of this inverted copy of ∂B_λ that is an open set containing the origin that is mapped two-to-one onto B_λ. This set is the trap door, T_λ.

To prove that the set Λ_λ in $C_0(\lambda)$ is homeomorphic to the Sierpiński carpet, we use quasiconformal surgery. We shall construct a quasiconformal map $L_\lambda : \overline{\mathbb{C}} \to \overline{\mathbb{C}}$ that agrees with F_λ^k in $\mathcal{A}_\lambda$. The set of points whose orbits under L_λ are bounded will be exactly the set Λ_λ. So we then show that L_λ is conjugate to a rational map of the form

$$Q_{\lambda,\alpha}(z) = z^2 + \frac{\lambda}{z^2} + \alpha$$

where α is a complex parameter and that, with the given assumptions on the critical orbits of F_λ, the Julia set of $Q_{\lambda,\alpha}$ is a Sierpiński curve. This will show that Λ_λ is a Sierpiński curve.

To construct L_λ, first recall that F_λ^k maps $\nu_0(\lambda)$ two-to-one onto itself and is hyperbolic in a neighborhood of this set. Also, $\nu_0(\lambda)$ is symmetric under $z \mapsto -z$. Hence we may choose a simple closed curve $\zeta_1(\lambda)$ having the following properties:

1. $\zeta_1(\lambda)$ lies close to but strictly outside $\nu_0(\lambda)$ and surrounds $\nu_0(\lambda)$;

2. $\zeta_1(\lambda)$ is symmetric under $z \mapsto -z$;

3. there is a preimage of $\zeta_1(\lambda)$ under F_λ^k, namely $\zeta_0(\lambda)$, that lies between $\nu_0(\lambda)$ and $\zeta_1(\lambda)$ and F_λ^k maps $\zeta_0(\lambda)$ to $\zeta_1(\lambda)$ as a two-to-one covering, so $\zeta_0(\lambda)$ is also symmetric under $z \mapsto -z$;

4. all points in the open annulus between $\nu_0(\lambda)$ and $\zeta_1(\lambda)$ eventually leave this annulus under iteration of F_λ^k.

We remark that the curve $\zeta_1(\lambda)$ does not lie in B_λ; indeed, $\zeta_1(\lambda)$ passes through portions of $J(F_\lambda)$ close to but outside the curve $\nu_0(\lambda)$.

We now define L_λ in stages. We first define $L_\lambda(z) = F_\lambda^k(z)$ if z is in the closed annulus bounded on the outside by $\zeta_0(\lambda)$ and on the inside by $\xi_0(\lambda)$. To define L_λ outside $\zeta_0(\lambda)$, we proceed in two stages. First consider the region V_λ outside $\zeta_1(\lambda)$ in the Riemann sphere. In this region we "glue in" the map $z \mapsto z^2$. More precisely, since $\zeta_1(\lambda)$ is invariant under $z \mapsto -z$, it follows that V_λ also has this property. Let ϕ_λ be the exterior Riemann map taking V_λ onto the disk

$$D_2 = \{z \in \overline{\mathbb{C}} \,|\, |z| \geq 2\}$$

in $\overline{\mathbb{C}}$ and fixing ∞ with $\phi_\lambda'(\infty) > 0$. Because of the $z \mapsto -z$ symmetry in V_λ, we have that $\phi_\lambda(-z) = -\phi_\lambda(z)$. Let $f(z) = z^2$, so f takes D_2 to

$$D_4 = \{z \in \overline{\mathbb{C}} \,|\, |z| \geq 4\}.$$

We then define L_λ on V_λ by

$$L_\lambda(z) = \phi_\lambda^{-1}(\phi_\lambda(z))^2.$$

Note that $L_\lambda(z) = L_\lambda(-z)$ since $\phi_\lambda(-z) = -\phi_\lambda(z)$.

So we now have L_λ defined on the region inside $\zeta_0(\lambda)$ (but outside $\xi_0(\lambda)$) and also outside $\zeta_1(\lambda)$. We next need to define L_λ on the open annulus U_λ between $\zeta_0(\lambda)$ and $\zeta_1(\lambda)$. On the boundary curve $\zeta_0(\lambda)$ of U_λ, we have that L_λ is the two-to-one covering map F_λ^k and $L_\lambda(\zeta_0(\lambda)) = \zeta_1(\lambda)$; on the other boundary curve $\zeta_1(\lambda)$, L_λ is the map above that is conjugate to $z \mapsto z^2$. So we define a smooth map q_λ on U_λ such that:

1. q_λ takes U_λ to the annulus bounded by $\zeta_1(\lambda)$ and $L_\lambda(\zeta_1(\lambda))$ as a two-to-one covering;

2. q_λ agrees with L_λ on both boundaries of U_λ;

3. $q_\lambda(-z) = q_\lambda(z)$;

4. q_λ varies smoothly with λ.

We now have L_λ defined everywhere outside $\xi_0(\lambda)$. Inside $\xi_0(\lambda)$, we then set $L_\lambda(z) = L_\lambda(H_\lambda(z))$. Since H_λ maps the disk bounded on the outside by $\xi_0(\lambda)$ to the exterior of $\nu_0(\lambda)$ and then L_λ maps this region to itself, it follows that L_λ takes the disk bounded by $\xi_0(\lambda)$ onto the exterior of $\nu_0(\lambda)$ in two-to-one fashion. Note that L_λ is continuous along $\xi_0(\lambda)$, since, on this curve, the "exterior" definition of L_λ, namely F_λ^k, and the interior definition, $L_\lambda \circ H_\lambda = F_\lambda^k \circ H_\lambda$ agree. Also, we have that $L_\lambda(-z) = L_\lambda(z)$. See Figure 5.

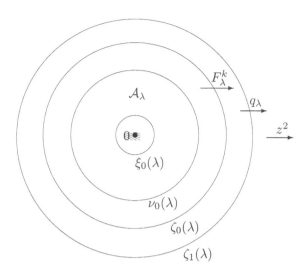

Figure 5. Construction of L_λ.

Proposition 4.1. *The set of points whose entire orbits under L_λ lie in $\mathcal{A}_\lambda$ is precisely the set Λ_λ.*

Proof: Suppose $z \in \mathcal{A}_\lambda$. Then we have that $F_\lambda^j(z) \in C_j(\lambda)$ for $j = 0, \ldots, k-1$ since $F_\lambda^j(\mathcal{A}_\lambda) = C_j(\lambda)$ for each such j. If $L_\lambda(z)$ also lies in $\mathcal{A}_\lambda$, then we have that $F_\lambda^k(z)$ also lies in $C_0(\lambda)$ and the first k points on the orbit of z under F_λ travel around the $C_j(\lambda)$ in the correct fashion. Similarly, if $L_\lambda^j(z)$ lies in $\mathcal{A}_\lambda$ for all j, then the entire orbit of z visits the $C_j(\lambda)$ in the correct order and we have that $z \in \Lambda_\lambda$.

Conversely, if $L_\lambda^i(z) \in \mathcal{A}_\lambda$ for $0 \le i < j$ but $L_\lambda^j(z) \notin \mathcal{A}_\lambda$, then $L_\lambda^j(z)$ must lie inside the disk bounded by $\xi_0(\lambda)$. But F_λ takes this disk to the exterior of $C_1(\lambda)$, so the orbit of z does not follow the orbit of 0 and so $z \notin \Lambda_\lambda$. $\square$

Proposition 4.2. *The map L_λ is quasiconformally conjugate to a rational map of the form*

$$Q_{\lambda,\alpha}(z) = z^2 + \alpha + \frac{\lambda}{z^2}.$$

Proof: We first construct an L_λ-invariant ellipse field in $\overline{\mathbb{C}}$. First define this field to be the standard complex structure in the region outside $\zeta_1(\lambda)$. Then pull this structure back by q_λ to define the ellipse field in the annulus between $\zeta_0(\lambda)$ and $\zeta_1(\lambda)$. Since q_λ is a smooth map, the ellipse field in this region has bounded dilatation. To define the ellipse field in the annulus between $\nu_0(\lambda)$ and $\zeta_0(\lambda)$, we keep pulling the already defined ellipses back by the appropriate branch of L_λ, which, in this region, equals F_λ^k. Since F_λ^k is holomorphic, the ellipse field continues to have bounded dilatation under these pull-backs.

We next define the ellipse field inside $\xi_0(\lambda)$ by pulling the given field back by L_λ. This is possible since L_λ maps the region inside $\xi_0(\lambda)$ as a two-to-one covering of the exterior of $\nu_0(\lambda)$.

Finally, we extend the ellipse field to the annulus between $\xi_0(\lambda)$ and $\nu_0(\lambda)$ as follows. We first use the iterates of the map F_λ^k to define the ellipses at any point whose orbit eventually enters the disk bounded by $\xi_0(\lambda)$. If a point never enters this region, then we put the standard structure at this point. This defines the ellipse field almost everywhere. Note that this field is preserved by L_λ, has bounded dilatation, is symmetric under $z \mapsto -z$, and is also preserved by H_λ.

By the Measurable Riemann Mapping Theorem, there exists a quasiconformal homeomorphism ψ_λ that straightens this ellipse field. We may normalize ψ_λ so that $\psi_\lambda(0) = 0$ and $\psi_\lambda(\infty) = \infty$. Because of the symmetries in the ellipse field, we have that $\psi_\lambda(-z) = -\psi_\lambda(z)$ and $\psi_\lambda(H_\lambda(z)) = H_\lambda(\psi_\lambda(z))$. Therefore the map $\psi_\lambda \circ L_\lambda \circ \psi_\lambda^{-1}$ is a rational map of degree 4 that fixes ∞ and has a pole of order 2 at the origin. So we have

$$\psi_\lambda \circ L_\lambda \circ \psi_\lambda^{-1}(z) = \frac{a_4 z^4 + \ldots + a_1 z + a_0}{z^2}.$$

Since $\psi_\lambda(-z) = -\psi_\lambda(z)$ and $L_\lambda(-z) = L_\lambda(z)$, we must have $a_3 = a_1 = 0$ and this map simplifies to

$$\psi_\lambda \circ L_\lambda \circ \psi_\lambda^{-1}(z) = \frac{a_4 z^4 + a_2 z^2 + a_0}{z^2} = a_4 z^2 + a_2 + \frac{a_0}{z^2}.$$

We may scale this map so that $a_4 = 1$ and then the H_λ-symmetry shows that $a_0 = \lambda$. Therefore, L_λ is conjugate to the rational map

$$Q_{\lambda,\alpha}(z) = z^2 + \frac{\lambda}{z^2} + \alpha,$$

and this is what we claimed.   □

We now complete the proof that Λ_λ is a Sierpiński curve. It suffices to show that Λ_λ is compact, connected, locally connected, nowhere dense, and has the property that any two complementary domains are bounded by simple closed curves that are pairwise disjoint.

By the previous propositions, we know Λ_λ is the set of points whose orbits are bounded under iteration of L_λ. Also, the set Λ_λ contains no critical points by assumption. Moreover, this set is homeomorphic to the filled Julia set of $Q_{\lambda,\alpha}$. Thus all of the critical points of $Q_{\lambda,\alpha}$ must escape to ∞ as well and so the Fatou set of $Q_{\lambda,\alpha}$ is the union of all of the preimages of the basin of ∞. Standard facts about the Julia sets of rational maps then yields the fact that $J(Q_{\lambda,\alpha})$ is compact and nowhere dense. In particular, the filled Julia set of this map is equal to $J(Q_{\lambda,\alpha})$. By a result of Yin [Yi99], since all the critical orbits of $Q_{\lambda,\alpha}$ tend to ∞, the Julia set is locally connected.

Now we know that the set of L_λ-bounded orbits is bounded on the outside by the simple closed curve $\nu_0(\lambda)$ and on the inside by $\xi_0(\lambda)$, and all the other complementary domains are preimages of these sets. Therefore this set is connected and all of the complementary domains are bounded by simple closed curves. These curves must be pairwise disjoint for, otherwise, a point of intersection would necessarily be a critical point whose orbit would then be bounded. But this cannot happen since all of the critical orbits escape to ∞. This completes the proof.   □

5   Dynamics on the Rest of the Julia Set

In this section we turn our attention to the dynamical behavior of all other points in $J(F_\lambda)$. As in the previous sections, we continue to assume that $|\lambda|$ is sufficiently small and that all of the critical orbits of F_λ tend to ∞ (but the critical points themselves do not lie in B_λ).

By Theorem 1.1 there exists the invariant set ∂B_λ on which F_λ is conjugate to F_0 on $J(F_0)$. Let $\mathcal{B}_\lambda$ denote the set consisting of ∂B_λ together with all of its preimages under F_λ^j for each $j \geq 0$. Similarly, by Theorem 1.2, there exist Sierpiński curves Λ_λ^j for $j = 0, \ldots, k-1$ on each of which F_λ^k is conjugate to a map of the form $z^2 + \alpha(\lambda) + \lambda/z^2$. Let Ω_λ denote the union of the Λ_λ^j together with all of the preimages of these sets under all iterates of F_λ. By our earlier results, we understand the topology of and dynamics on each of these sets.

We therefore consider points in the set

$$\mathcal{O}_\lambda = J(F_\lambda) - (\mathcal{B}_\lambda \cup \Omega_\lambda).$$

Note that there must be infinitely many points in this set since none of the preimages of the Λ_λ^j or ∂B_λ contain periodic points and, as is well known, repelling periodic points must be dense in $J(F_\lambda)$.

To describe the structure of the set $\mathcal{O}_\lambda$, we first assign an itinerary to each point in this set. For $0 \leq j \leq k-1$, let I_j denote the interior of the disk C_j in $K(F_0)$ that contains $F_0^j(0)$. The interior of $K(F_0)$ consists of infinitely many other such open disks. So for each $j > k-1$, we let

I_j denote a unique such disk. How these I_j are indexed is not important. Then let $I_j(\lambda)$ denote the corresponding open disk for F_λ.

Given $z \in \mathcal{O}_\lambda$, we define the *itinerary* of z to be the sequence of non-negative integers $S(z) = (s_0 s_1 s_2 \ldots)$ where, as usual, $s_j = \ell$ if and only if $F_\lambda^j(z) \in I_\ell(\lambda)$. The itinerary is said to be *allowable* if it actually corresponds to a point in $\mathcal{O}_\lambda$. Note the following:

1. We do not assign an itinerary to any point in $\mathcal{B}_\lambda$ since the $I_j(\lambda)$ are disjoint from this set (and we already understand the dynamics on this set anyway).

2. The itinerary of any point in $\mathcal{O}_\lambda$ necessarily contains infinitely many zeroes. This follows immediately from the fact that each I_j with $j > 0$ must eventually be mapped to I_0 by some iterate of F_0 and so the same must be true for $I_j(\lambda)$ and F_λ.

3. The itinerary of $z \in \mathcal{O}_\lambda$ cannot end in an infinite string of the form $(\overline{0\, 1\, \ldots\, k-1})$ since we have assumed that $z \notin \Omega_\lambda$.

Now suppose $z \in \mathcal{O}_\lambda$. If some entry of $S(z) = 0$, say $s_j = 0$, then either $s_{j+1} = 1$, in which case we have that

$$s_{j+2} = 2, s_{j+3} = 3, \ldots, s_{j+k} = 0,$$

or else $s_{j+1} \neq 1$. In the latter case, we call the index j a *departure index*, since these are the points on the orbit of z where this orbit "deviates" from a similar orbit for F_0. As above, there must be infinitely many departure indices for any orbit in $\mathcal{O}_\lambda$, since this itinerary cannot end in the repeating sequence $(\overline{0\, 1\, \ldots\, k-1})$.

Before turning to the proof of Theorem 1.4, we give several illustrative examples of why the set of points in $\mathcal{O}_\lambda$ with a given itinerary is a Cantor set. For clarity, we restrict to the case where $F_\lambda(z) = z^2 - 1 + \lambda/z^2$, i.e., the case where c is drawn from the center of the period two bulb in the Mandelbrot set. We let $I_2(\lambda) = -I_1(\lambda)$, so $I_2(\lambda)$ is the other preimage of $I_0(\lambda)$.

Example 1 (The itinerary $(\overline{0})$). In this case, each index j is a departure index since $s_{j+1} \neq 1$. Let $V_n(\lambda)$ be the set of points in $I_0(\lambda)$ whose itinerary begins with $n+1$ consecutive zeroes. Then $V_0(\lambda) = I_0(\lambda)$ and $V_1(\lambda)$ is a pair of disjoint open disks in $I_0(\lambda)$, each of which lies inside the curve $\xi_0(\lambda)$ that is mapped two-to-one to the boundary of $I_1(\lambda)$. Each of these disks is mapped univalently over $I_0(\lambda)$ since the critical points of F_λ lie outside the curve $\xi_0(\lambda)$ and are mapped to $I_1(\lambda)$. But then $V_2(\lambda)$ consists of 4 disjoint open disks, two in each component of $V_1(\lambda)$ that are mapped onto the two components of $V_1(\lambda)$. Continuing in this fashion, we see that $V_n(\lambda)$ consists of 2^n disjoint open disks, and $V_n(\lambda) \subset V_{n-1}(\lambda)$ for each n. Since F_λ maps $V_n(\lambda)$ to $V_{n-1}(\lambda)$ as above, standard arguments from complex dynamics then show that the set of points in $\mathcal{O}_\lambda$ whose itinerary is $(\overline{0})$ is a Cantor set.

From now on, we let $W_{s_0 s_1 \ldots s_n}(\lambda)$ denote the set of points in $\mathcal{O}_\lambda$ whose itinerary begins with $s_0 s_1 \ldots s_n$.

Example 2 (The itinerary $(\overline{02})$). In this case we again have that the critical points are mapped into $I_1(\lambda)$, so, as above, $W_{02}(\lambda)$ is a pair of open disks in $I_0(\lambda)$, each of which is mapped univalently onto $I_2(\lambda)$. Thus $W_{202}(\lambda)$ is also pair of disks lying in $I_2(\lambda)$. But then, since F_λ maps each disk in $W_{02}(\lambda)$ univalently onto $I_2(\lambda)$, we have that $W_{0202}(\lambda)$ consists of four disks, two in each of the disks comprising $W_{02}(\lambda)$. Continuing, we see that every second iterate produces double the number of disks contained in the previous $W(\lambda)$, so again we see that the set of points on $\mathcal{O}_\lambda$ with itinerary $(\overline{02})$ is a Cantor set.

Example 3 (The itinerary $(\overline{0102})$). As in the previous example, $W_{02}(\lambda)$ is a pair of disks, since each of the critical points in $I_0(\lambda)$ is mapped into $I_1(\lambda)$ for small $|\lambda|$. Then $W_{102}(\lambda)$ is a pair of disks in $I_1(\lambda)$ since $I_1(\lambda)$ is mapped univalently onto $I_0(\lambda)$. But now $W_{0102}(\lambda)$ consists of at least 4 and at most 8 disjoint disks in $I_0(\lambda)$. To see this, note that the preimage of $I_1(\lambda)$ in $I_0(\lambda)$ is the annulus bounded by the curves $\xi_0(\lambda)$ and $\nu_0(\lambda)$ defined earlier, and F_λ takes this annulus four-to-one onto $I_1(\lambda)$. There are four critical points in this annulus, and it could be the case that one of the critical points map into one of the two disks in $W_{102}(\lambda)$. If that happens, then the negative of this critical point (also a critical point) maps to the same disk. So the preimage of this particular disk has either one or two components since the map is four-to-one. But, if this preimage has only one component, by the $z \mapsto -z$ symmetry, this component would necessarily surround the origin. Now this preimage must be disjoint from the Sierpiński curve invariant set in $\overline{I}_0(\lambda)$ and also separate $\xi_0(\lambda)$ from $\nu_0(\lambda)$. This then gives a contradiction to the connectedness of the Sierpiński curve. Hence each of these disks would have at least two preimages for a total of at least four and at most eight preimages of $W_{102}(\lambda)$ in $I_0(\lambda)$. But then $W_{20102}(\lambda)$ also consists of at least 4 and at most 8 disks, while $W_{020102}(\lambda)$ now consists of double this number of disks, since F_λ maps $I_0(\lambda)$ two-to-one onto $I_2(\lambda)$, but the critical points map into $I_1(\lambda)$. That is, each of the two original disks in $W_{02}(\lambda)$ acquires from 4 to 8 preimages when we pull them back by the four appropriate inverses of F_λ. Then, continuing in this fashion, each time we pull back each of these disks, again by the four appropriate preimages, we find at least 4 preimages for each one. Again, the set of points with this itinerary is a Cantor set.

We now complete the proof of Theorem 1.4. Consider the allowable itinerary $(s_0 s_1 s_2 \ldots)$. We may assume at the outset that $s_0 = 0$ and that 0 is a departure index. So say that the itinerary is given by $(0 s_1 \ldots s_i 0 \ldots)$ where $s_j \neq 0$ for $1 \leq j \leq i$. Then the set of points whose itinerary begins this way is a pair of disks in $I_0(\lambda)$, and each of these disks is mapped univalently onto $I_0(\lambda)$ by F_λ^{i+1}. Then there are two cases: either $i + 1$ is a departure index or it is not. In the former case, the itinerary may be continued $(0 s_1 \ldots s_i 0 s_{i+2} \ldots s_{i+\ell} 0 \ldots)$ where again $s_j \neq 0$ for $i + 2 \leq j \leq i + \ell$. Just as in Examples 1 and 2, the set of points in $I_0(\lambda)$ whose itinerary begins in this fashion now consists of 4 disks. In the other case, we have that $s_{i+1} = 1, \ldots, s_{i+k-1} = k - 1, s_{i+k} = 0$. Arguing as in Example 3, we have that the set of points whose itinerary now begins in this fashion consists of between four and eight disjoint open disks, all contained in the original pair of disks, and each mapped onto $I_0(\lambda)$ (at most two-to-one) by $F_\lambda^{i+\ell+1}$. In any event, the number of disks that correspond to this initial itinerary has at least doubled. Continuing in this fashion, we see that at each index for which $s_j = 0$, we find at least double the number of disjoint open disks in $I_0(\lambda)$ that begin with this itinerary. These disks are nested and converge to points. Hence the set of points with the given itinerary is a Cantor set.

6   Several Examples

For completeness, we give several examples of parameters in the family

$$F_\lambda(z) = z^2 - 1 + \frac{\lambda}{z^2}$$

for which both critical orbits eventually escape to ∞ and the above theorems hold. We shall choose λ real and negative. Note that, in this case, we have $F_\lambda(\bar{z}) = \overline{F_\lambda(z)}$, so the Julia sets are symmetric under $z \mapsto \bar{z}$. More importantly, the two critical values are symmetric under this map, so if one critical value eventually escapes, then so does the other one.

The first example occurs when $\lambda = -.0025$ so that one of the critical values is $v_\lambda = -1 + 0.1i$. We then compute

$$
\begin{aligned}
F_\lambda(v_\lambda) &= -.0124262 - .20049i, \\
F_\lambda^2(v_\lambda) &= -.978559 + .0126334i, \\
F_\lambda^3(v_\lambda) &= -.0451908 - .0247924i, \\
F_\lambda^4(v_\lambda) &= -1.50415 + .795835i, \\
F_\lambda^5(v_\lambda) &= .628637 - 2.39483i.
\end{aligned}
$$

We have that v_λ and $F_\lambda^2(v_\lambda)$ belong to $I_1(\lambda)$ whereas $F_\lambda(v_\lambda)$ and $F_\lambda^3(v_\lambda)$ belong to $I_0(\lambda)$. One checks easily that $F_\lambda^4(v_\lambda)$ then lies outside ∂B_λ. Hence both critical orbits escape at this iteration. See Figure 6 for a picture of this Julia set.

Figure 6. The Julia sets for $z^2 - 1 - \lambda/z^2$ when $\lambda = -.0025$ and also $\lambda = -.0001$.

The second example occurs when $\lambda = -.0001$ so that one of the critical values is $-1 + .02i$. We then compute

$$
\begin{aligned}
F_\lambda(v_\lambda) &= .00049988 - .040004i, \\
F_\lambda^2(v_\lambda) &= -.939142 + .00160116i, \\
F_\lambda^3(v_\lambda) &= -.118129 - .00300783i, \\
F_\lambda^4(v_\lambda) &= -.993207 + .00107508i, \\
F_\lambda^5(v_\lambda) &= -.0136424 - .00213578i, \\
F_\lambda^6(v_\lambda) &= -1.449917 + .160339i, \\
F_\lambda^7(v_\lambda) &= 1.22177 - .48076i, \\
F_\lambda^8(v_\lambda) &= .261538 - 1.17479i, \\
F_\lambda^9(v_\lambda) &= -2.31167 - .614536i.
\end{aligned}
$$

so that $F_\lambda^6(v_\lambda)$ now is the first point on the critical orbit to lie outside ∂B_λ. See Figure 6.

Bibliography

[BDLSS05] P. Blanchard, R. L. Devaney, D. M. Look, P. Seal, and Y. Shapiro, *Sierpiński curve Julia sets and singular perturbations of complex polynomials*, Ergod. Th. Dynam. Syst. **25** (2005), 1047–1055.

[De05] R. L. Devaney, *Structure of the McMullen domain in the parameter planes for rational maps*, Fund. Math. **185** (2005), 267–285.

[DG]   R. L. Devaney and A. Garijo Real, *Julia sets converging to the unit disk*, Proc. Amer. Math. Soc., to appear.

[DL06]   R. L. Devaney and D. M. Look, *A criterion for Sierpiński curve Julia sets*, Topology Proceedings **30** (2006), 163–179.

[DLU05]   R. L. Devaney, D. M. Look, and D. Uminsky, *The escape trichotomy for singularly perturbed rational maps*, Indiana Univ. Math. J. **54** (2005), 1621–1634.

[DM06]   R. L. Devaney and S. M. Marotta, *Evolution of the McMullen domain for singularly perturbed rational maps*, Topology Proc., to appear.

[DH84]   A. Douady and J. H. Hubbard, *Ètude dynamique des polynômes complexes*, Publ. Math. D'Orsay (1984).

[Mc88]   C. McMullen, *Automorphisms of rational maps*, in: Holomorphic Functions and Moduli I, Math. Sci. Res. Inst. Publ. **10**, Springer, New York, 1988.

[Mc95]   C. McMullen, *The classification of conformal dynamical systems*, Current Developments in Mathematics, Internat. Press, Cambridge, MA, 1995, 323–360.

[Mi06]   J. Milnor, *Dynamics in one complex variable*, Third Edition, Princeton University Press, 2006.

[MT93]   J. Milnor and L. Tan, *A "Sierpiński carpet" as Julia set*, Appendix F in: Geometry and Dynamics of Quadratic Rational Maps, Experiment. Math., **2** (1993), 37–83.

[PR00]   C. Petersen and G. Ryd, *Convergence of rational rays in parameter spaces*, in: The Mandelbrot Set: Theme and Variations, London Math. Soc. Lecture Note Series **274**, Cambridge University Press, 2000, 161–172.

[Ro06]   P. Roesch, *On capture zones for the family $f_\lambda(z) = z^2 + \lambda/z^2$*, in: Dynamics on the Riemann Sphere, European Mathematical Society, 2006, 121–130.

[Wh58]   G. T. Whyburn, *Topological characterization of the Sierpiński curve*, Fundamenta Mathematicae **45** (1958), 320–324.

[Yi99]   Y. Yin, *Julia sets of semi-hyperbolic rational maps*, Chinese Ann. Math. Ser. A **20** (1999), 559–566. English translation in: Chinese J. Contemp. Math. **20** (2000), 469–476.

Color Plates

Plate I. Top: the mathematical *family tree* of John Hubbard (see Page xiv). Bottom: *Family and friends* of complex dynamics (see Page xv).

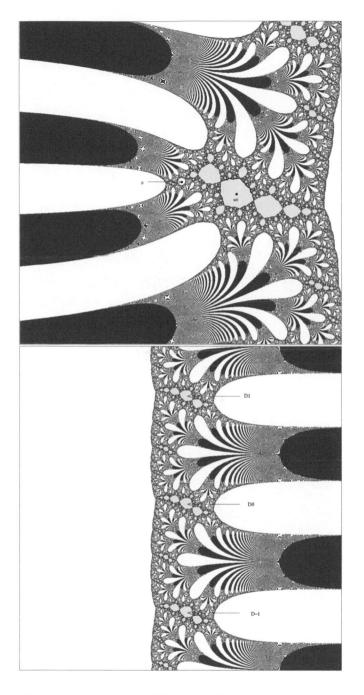

Plate II. Figures 2 and 3 from Chapter 8 (Fagella and Henriksen).

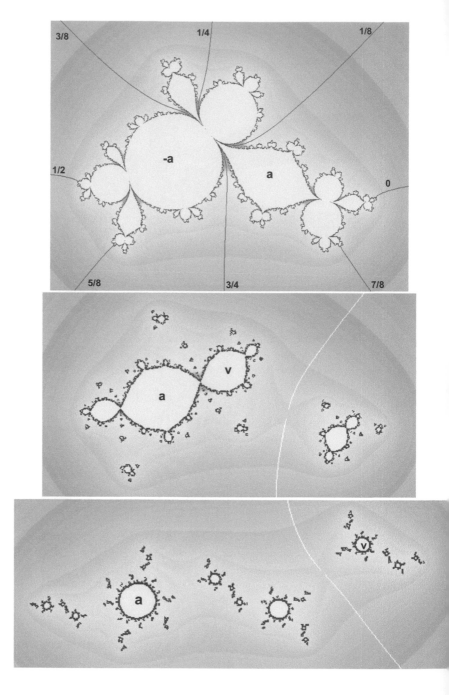

Plate III. Figures 17, 18, and 19 from Chapter 9 (Milnor).

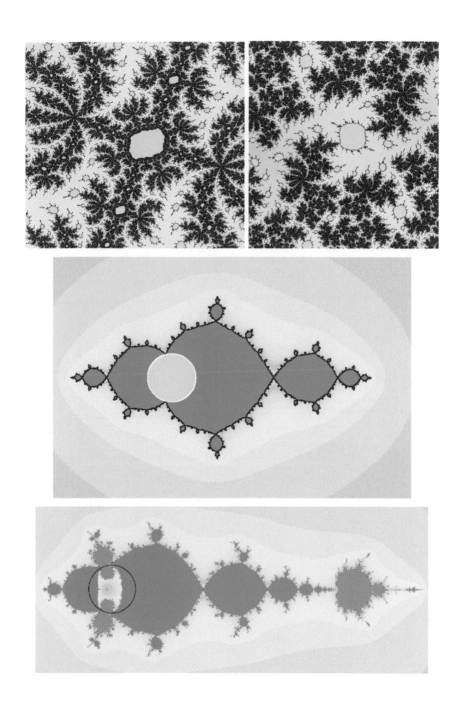

Plate IV. Figures 20, 21, 24, and 25 from Chapter 9 (Milnor).

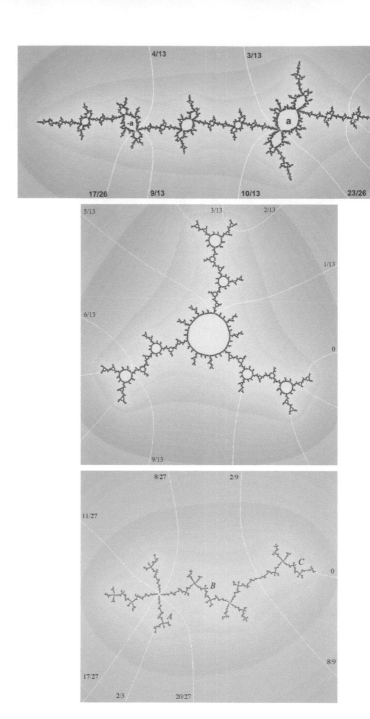

Plate V. Figures 31, 32, and 33 from Chapter 9 (Milnor).

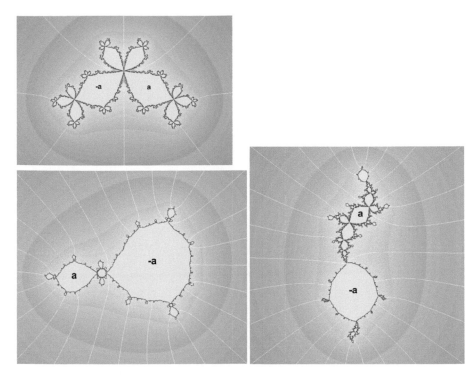

Plate VI. Top three pictures: Figure 36 from Chapter 9 (Milnor).
Bottom two pictures: Figure 1 from Chapter 13 (Bedford and Diller).

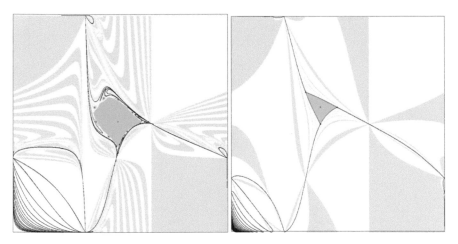

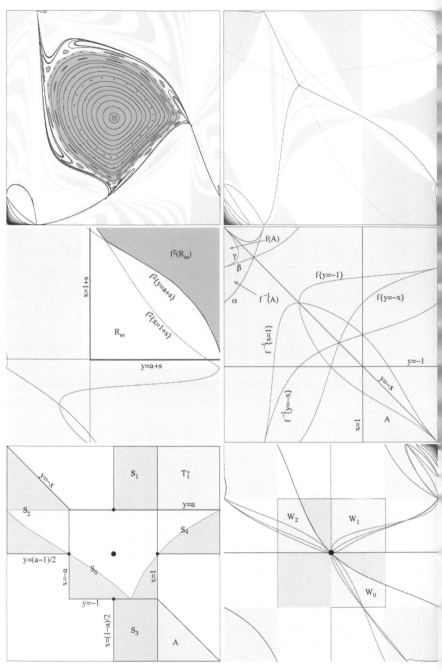

Plate VII. Figures 2, 3, 5, 6, 7, 8 from Chapter 13 (Bedford and Diller).

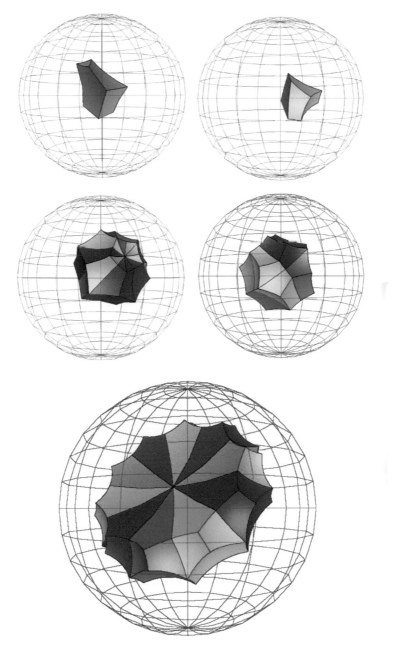

Plate VIII. Figure 1 (top four pictures) and Figure 7 (bottom picture) from Chapter 17 (Antolín-Camarena, Maloney, and Roeder).

8   The Teichmüller Space of an Entire Function

Núria Fagella and Christian Henriksen

We consider the Teichmüller space of a general entire transcendental function $f : \mathbb{C} \to \mathbb{C}$ regardless of the nature of the set
of singular values of f, i.e., critical and asymptotic values. We
prove that, as in the known case of periodic points and critical
values, asymptotic values are also fixed points of any quasiconformal automorphism that commutes with f and which is
homotopic to the identity relative to the ideal boundary of the
domain. As a consequence, the general framework of McMullen
and Sullivan [MS98] for rational functions applies also to entire
functions and we can apply it to study the Teichmüller space of
f, analyzing each type of Fatou component separately. Baker
domains were already considered in [FH06], but the considerations of wandering domains are new. We provide different
examples of wandering domains; each of them adding a different quantity to the dimension of the Teichmüller space. In
particular, we give examples of rigid wandering domains.

1   Introduction

Given a holomorphic endomorphism $f : S \to S$ on a Riemann surface S,
we consider the dynamical system generated by the iterates of f, denoted
$f^n = f \circ \overset{n}{\ldots} \circ f$. There is a dynamically natural partition of the phase
space S into the *Fatou set* $\mathcal{F}(f)$, restricted to which the iterates of f form a
normal family, and the *Julia set* $\mathcal{J}(f) = S - \mathcal{F}(f)$, which is the complement
of the Fatou set.

If $S = \widehat{\mathbb{C}} = \mathbb{C} \cup \{\infty\}$, then f is a rational map. If $S = \mathbb{C}$ and f does
not extend to $\widehat{\mathbb{C}}$, then f is an entire transcendental mapping, i.e., infinity is an essential singularity. There are fundamental differences between
transcendental functions and rational maps.

One of these differences concerns the set of singularities of the inverse
function of the respective map. For a rational map f, all branches of
the inverse function are locally well defined except on the set of *critical*

values, i.e., points $v = f(c)$, where $f'(c) = 0$. If f is transcendental, inverse branches may not be well defined for another reason: Some inverse branches may not be well defined in any neighborhood of an *asymptotic value.* A point $a \in \mathbb{C}$ is called an asymptotic value if there exists a path $\gamma(t) \to \infty$ as $t \to \infty$ such that $f(\gamma(t)) \to a$ as $t \to \infty$. We call a point $z \in \mathbb{C}$ *regular* if there exists a neighborhood U of z such that $f : V \to U$ is a conformal isomorphism for every component V of $f^{-1}(U)$. The complement of the set of regular values is called the set of *singular values,* and it can be shown to be equal to the closure of the union of the set of critical values and the set of asymptotic values. The set of critical values, even when infinite, always forms a discrete set, unlike the set of asymptotic values. In fact there exists an entire map whose set of asymptotic values coincide with $\mathbb{C}$ [Gr18]. However, it follows from a theorem of Denjoy, Carleman and Ahlfors that entire functions of finite order may have only a finite number of asymptotic values (see, e.g., [Ne70] or Theorem 4.11 in [HY98]).

This fact motivated the definition and study of special classes of entire transcendental maps, such as the class $\mathcal{S}$ of functions of *finite type,* or function with a finite number of singular values (see, e.g., [EL92] and [DT86]). The space of entire functions with a given finite number of singular values is finite dimensional and therefore share many properties with the set of rational maps, such as the fact that every component of the Fatou set of a function from either space is eventually periodic [EL92, GK86]. There is a classification of the periodic components of the Fatou set of a rational map or a map in $\mathcal{S}$: A periodic component can either be a cycle of rotation domains or the basin of attraction of an attracting, superattracting or indifferent periodic point.

If we allow f to have infinitely many singular values, then there are more possibilities. For example, a component of $\mathcal{F}(f)$ may be *wandering,* that is, such that it will never be iterated to a periodic component. The first example of a wandering domain was given by Baker [Ba76], and elementary examples were given by Herman in [He81]. It is easy to check, for instance, that the function $f(z) = z + 2\pi + \sin(z)$ has a sequence of wandering domains, each of which contains a critical point (see example 2). In general, however, wandering domains need not contain any singular value. It was shown in [Ba75] that wandering domains are the only Fatou components of an entire transcendental function that may be multiply connected.

There is a classification of the eventually periodic components of the Fatou set of a map of finite type according to the dynamical behavior of the map (see [Be93]). An entire map f with infinitely many singular values allow for one more type of dynamical behavior: An invariant connected component U of the Fatou set is called a Baker domain if for all $z \in U$ we have $f^n(z) \to \infty$ as $n \to \infty$. We call U a period p Baker domain if $f^k(U)$ is an invariant Baker domain under f^p, for some $k \leq p$. The first example of an entire function with a Baker domain was given by Fatou in [Fa19], where he considered the function $f(z) = z + 1 + e^{-z}$ and showed that the right half-plane is contained in an invariant Baker domain. Since then, many other examples have been considered, which exhibit various properties that are possible for this type of Fatou components (see for example [EL87], [Be95], [BD99], [RS99a], [RS99b], [Kö99] and [BF00]). It follows from [Ba75] that a Baker domain of an entire function is simply connected.

As is the case with basins of attraction and rotation domains, there is a relation between Baker domains and the singularities of the inverse map. In particular, it is shown in [EL92] that Baker domains do not exist for a map f for which the set $\mathrm{Sing}(f^{-1})$ is bounded. Here $\mathrm{Sing}(f^{-1})$ denotes the closure in $\mathbb{C}$ of the set of singular values of f. The actual relationship between this set and a Baker domain U has to do with the distance of singular orbits to the boundary of U (see [Be01] for a precise statement). We remark that it is not necessary, however, that any of the singular values of f lie inside the Baker domain. Indeed, there are examples of Baker domains which contain an arbitrary finite number of singular values (including none).

We consider two holomorphic endomorphisms $f, g : S \to S$ to be equivalent if they are conjugate via a conformal isomorphism of S, i.e., if there exists a holomorphic bijection $c : S \to S$ such that $g = c^{-1} \circ f \circ c$. In the case of entire maps, the conjugacy needs to be affine, so our maps live in the space of entire functions modulo affine conjugations. From now on we will mostly consider the case $S = \mathbb{C}$.

Our goal in this paper is to study the possible quasiconformal deformations of an entire transcendental map f, i.e., those entire maps $g : \mathbb{C} \to \mathbb{C}$ such that $g = h^{-1} \circ f \circ h$ for some quasiconformal homeomorphism h. This analysis was started in [HT97] for entire maps whose set of singular values is a countable set. Later, in [FH06], the case of entire maps possessing an invariant Baker domain was studied, without imposing restrictions on the set of singular values. In this paper, we extend the study to all entire maps also without imposing restrictions on the set of singular values.

More precisely, we will consider the Teichmüller space $\mathcal{T}(\mathbb{C}, f)$ of an entire function f using the very general framework given by [MS98] (see Section 2.2 for an introduction). Another approach, applied to entire functions of finite type, can be found in [Ep95]. Morally, the Teichmüller space of f is precisely the space of quasiconformal deformations of f provided with a nice manifold structure. It is shown in [MS98] that under some conditions the Teichmüller space can be split into several smaller parts, which correspond to completely invariant subsets of the dynamical plane. In Section 3 we will see how these splitting theorems can be applied to our setting. To do this, we show the following technical result, which is the main difficulty we need to overcome in order to prove our results for *all* entire functions. We first define some necessary concepts. By the *grand orbit* of a point $z \in \mathbb{C}$, we mean the set of all points $w \in \mathbb{C}$ such that $f^n(z) = f^m(w)$ for some $n, m \in \mathbb{N}$. Alternatively, if we define the grand orbit relation as the finest equivalence relation on $\mathbb{C}$ such that $z \simeq f(z)$, then the grand orbit of z is the equivalence class of z. Let $\mathrm{QC}(U, f)$ be the group of quasiconformal automorphisms of U that commute with f, and let $\mathrm{QC}_0(U, f) \subset \mathrm{QC}(U, f)$ be the subgroup of automorphisms which are homotopic to the identity rel. the ideal boundary of U, through a uniformly K-quasiconformal subset of $\mathrm{QC}(U, f)$ (see Section 2.2). We now can give a precise statement (see also Theorem 3.2).

Theorem. *Let f be an entire transcendental function and $\widehat{\mathcal{J}}$ the closure of the grand orbits of all marked points of f (periodic points and singular values), and let ω be an element of $\mathrm{QC}_0(\mathbb{C}, f)$. Then ω restricts to the identity on $\widehat{\mathcal{J}}$.*

This theorem is well known in the case of rational maps (critical values, like periodic points of a given period, always form a discrete set) and entire maps of finite type (for the same reason), and it extends to the case of functions with a countable set of singularities [HT97]. In our case, however, it is possible that the set of singular values forms a continuum, and even a set of positive measure.

The main consequence of the previous theorem is that marked points play a special role when studying the Teichmüller space of a given function (see Section 3). In other words, we can split the dynamical plane into two sets, namely $\widehat{\mathcal{J}}$ and its complement, and study the Teichmüller space separately in each of them. Likewise, we can then split $\mathbb{C} - \widehat{\mathcal{J}}$ into several completely invariant subsets, namely the different grand orbits of Fatou components (or subsets of them).

For basins of attracting, superattracting or parabolic periodic points, and for rotation domains, the results are practically the same as those for rational maps: if U is a Fatou component of f and $\mathcal{U}$ is its grand orbit, then $\mathcal{U}$ contributes a certain finite quantity to the dimension of the Teichmüller space, plus the number of distinct singular grand orbits that belong to it, which may be infinite. In the case of rotation domains, this is not a precise statement, although the moral is the same.

New situations arise with the presence of Baker and wandering domains, since these are not present in the dynamical planes of rational maps. For Baker domains, we recall the results obtained in [FH06] (see Proposition 4.2), where we classify different types of Baker domains and show that the corresponding Teichmüller spaces can have 0, finite non-zero or infinite dimension depending on the class they fall into and the number of singular grand orbits they contain.

We perform a similar study for wandering domains. By giving explicit examples, we show that there exist maps with a certain type of dynamics on a wandering domain. To our knowledge, this was not known.

Finally, in Section 5, we address the problem of studying Teichmüller space supported on the Julia set, which of course can give rise to several different possibilities.

Acknowledgments. The first author was partially supported by grants MTM2005-02139, MTM2006-05849/Consolider (including a FEDER contribution) and CIRIT 2005/SGR01028.

2   Preliminaries

In this section, we briefly recall the relevant definitions and results related to quasiconformal mappings and to Teichmüller spaces. For quasiconformal mappings, the standard references are [Ah66] and [LV73]. For Teichmüller spaces, we will use the general framework of [MS98]. We also refer to [H06] as a reference for Riemann surfaces and Teichmüller theory.

2.1   Quasiconformal Mappings

Let $V, V' \subset \mathbb{C}$ be open subsets of the complex plane or, more generally, one dimensional complex manifolds.

Definition 2.1. Given a measurable function $\mu : V \to \mathbb{C}$, we say that μ is a k-*Beltrami coefficient* of V if $\|\mu(z)\|_\infty \leq k < 1$ almost everywhere in V. Two Beltrami coefficients of V are equivalent if they coincide almost everywhere in V. We say that μ is a Beltrami coefficient if μ is a k-Beltrami coefficient for some $k < 1$.

Definition 2.2. A homeomorphism $\phi : V \to V'$ is said to be *quasiconformal* if it has locally square integrable generalized derivatives and

$$\mu_\phi(z) = \frac{\frac{\partial \phi}{\partial \bar{z}}(z)}{\frac{\partial \phi}{\partial z}(z)} = \frac{\bar{\partial}\phi(z)}{\partial\phi(z)}$$

is a k-Beltrami coefficient for some $k < 1$. In this case, we say that ϕ is K-quasiconformal, where $K = \frac{1+k}{1-k} < \infty$, and that μ_ϕ is the *complex dilatation* or the *Beltrami coefficient* of ϕ.

If we drop the condition that ϕ be a homeomorphism, we call ϕ a K-*quasiregular map*.

Definition 2.3. Given a Beltrami coefficient μ of V and a quasiregular map $f : V \to V'$, we define the *pull-back* of μ by f as the Beltrami coefficient of V defined by

$$f^*\mu = \frac{\frac{\partial f}{\partial \bar{z}} + (\mu \circ f)\overline{\frac{\partial f}{\partial z}}}{\frac{\partial f}{\partial z} + (\mu \circ f)\overline{\frac{\partial f}{\partial \bar{z}}}}.$$

We say that μ is f-invariant if $f^*\mu = \mu$. If $\mu = \mu_g$ for some quasiregular map g, then $f^*\mu = \mu_{g \circ f}$.

It follows from Weyl's Lemma that a quasiregular map f is holomorphic if and only if $f^*\mu_0 = \mu_0$, where $\mu_0 \equiv 0$.

Definition 2.4. Given a Beltrami coefficient μ, the partial differential equation

$$\frac{\partial \phi}{\partial \bar{z}} = \mu(z)\frac{\partial \phi}{\partial z} \tag{2.1}$$

is called the *Beltrami equation*. By *integration* of μ we mean the construction of a quasiconformal map ϕ, which solves 2.1 almost everywhere, or equivalently, the construction of ϕ such that $\mu_\phi = \mu$ almost everywhere.

The famous *Measurable Riemann Mapping Theorem* by Morrey, Bojarski, Ahlfors and Bers states that every k-Beltrami coefficient is integrable.

Theorem 2.5 (Measurable Riemann Mapping Theorem). *Let μ be a k-Beltrami coefficient of V, for some $k < 1$, where $V = \mathbb{C}$ or is isomorphic to $\mathbb{D}$. Then there exists a quasiconformal map $\phi : V \to \mathbb{C}$ such that $\mu_\phi = \mu$ almost everywhere. Moreover, ϕ is unique up to post-composition with conformal isomorphisms of V.*

2.2   Teichmüller Space

In this section, we set up the machinery necessary for defining the Teichmüller space of a dynamical system, using the general framework of McMullen and Sullivan. The classical concept of the Teichmüller space of a Riemann surface is a special case of the general definition.

Let V be an open subset of the complex plane or, more generally, a one dimensional complex manifold, and let f be a holomorphic endomorphism of V. Define an equivalence relation $\sim$ on the set of quasiconformal homeomorphisms on V by identifying $\phi : V \to V'$ with $\psi : V \to V''$ if there exists a conformal isomorphism $c : V' \to V''$ such that $c \circ \phi = \psi$, i.e., the following diagram commutes.

It then follows that $\phi \circ f \circ \phi^{-1}$ and $\psi \circ f \circ \psi^{-1}$ are conformally conjugate (although the converse is not true in general). The deformation space of f on V is

$$\mathrm{Def}(V, f) = \{\phi : V \to V' \text{ quasiconformal} \mid \mu_\phi \text{ is } f\text{-invariant }\}/ \sim .$$

From the Measurable Riemann Mapping Theorem (see [Ah66] or Theorem 2.5), one obtains a bijection between $\mathrm{Def}(V, f)$ and

$$\mathcal{B}_1(V, f) = \{f\text{-invariant Beltrami forms } \mu \in L^\infty \text{ with } \|\mu\|_\infty < 1\},$$

and this is used to endow $\mathrm{Def}(V, f)$ with the structure of a complex manifold. Indeed, $\mathcal{B}_1(V, f)$ is the unit ball in the Banach space of f-invariant Beltrami forms equipped with the infinity norm.

We denote by $\mathrm{QC}(V, f)$ the group of quasiconformal automorphisms of V that commute with f. Given $K < \infty$, a family of quasiconformal mappings is called *uniformly K-quasiconformal* if each element of the family is K-quasiconformal.

A hyperbolic Riemann surface V is covered by the unit disk; in fact V is isomorphic to $\mathbb{D}/\Gamma$, where Γ is a Fuchsian group. Let $\Omega \subseteq \mathbb{S}^1$ denote the complement of the limit set of Γ. Then $(\mathbb{D} \cup \Omega)/\Gamma$ is a bordered surface and Ω/Γ is called the ideal boundary of V. A homotopy $\omega_t : V \to V, 0 \leq t \leq 1$, is called *rel. ideal boundary* if there exists a lift $\hat\omega_t : \mathbb{D} \to \mathbb{D}$ that extends continuously to Ω as the identity. If V is not hyperbolic, then the ideal boundary is defined to be the empty set.

We denote by $\mathrm{QC}_0(V, f) \subseteq \mathrm{QC}(V, f)$ the subgroup of automorphisms which are homotopic to the identity rel. the ideal boundary of V through a uniformly K-q.c. subset of $\mathrm{QC}(V, f)$, for some $K < \infty$.

Earle and McMullen [EM88] proved the following result for hyperbolic subdomains of the Riemann sphere.

Theorem 2.6. *Suppose $V \subseteq \widehat{\mathbb{C}}$ is a hyperbolic subdomain of the Riemann sphere. Then a uniformly quasiconformal homotopy $\omega_t : V \to V, 0 \leq t \leq 1$, rel. ideal boundary can be extended to a uniformly quasiconformal homotopy of $\widehat{\mathbb{C}}$ by letting $\omega_t = \mathrm{Id}$ on the complement of V. Conversely, a uniformly quasiconformal homotopy $\omega_t : V \to V$ such that each ω_t extends continuously as the identity to the topological boundary $\partial V \subseteq \widehat{\mathbb{C}}$ is a homotopy rel the ideal boundary.*

Proof: The proof can be found in [EM88]: Proposition 2.3 and the proof of Corollary 2.4 imply the first statement. Theorem 2.2 implies the second.□

The group $\mathrm{QC}(V, f)$ acts on $\mathrm{Def}(V, f)$ by $\omega_* \phi = \phi \circ \omega^{-1}$. Indeed if ϕ and ψ represent the same element in $\mathrm{Def}(V, f)$, then $\omega_* \phi = \omega_* \psi$ as elements of $\mathrm{Def}(V, f)$.

Definition 2.7. The Teichmüller space $\mathcal{T}(V, f)$ is the deformation space $\mathrm{Def}(V, f)$ modulo the action of $\mathrm{QC}_0(V, f)$, i.e.,

$$\mathcal{T}(V, f) = \mathrm{Def}(V, f)/\mathrm{QC}_0(V, f).$$

If V is a one dimensional complex manifold, we denote by $\mathcal{T}(V)$ the Teichmüller space $\mathcal{T}(V, \mathrm{Id})$.

Teichmüller space can be equipped with the structure of a complex manifold and a (pre)metric (we refer to [MS98]).

Let us give a rough idea of Teichmüller space and the motivation for studying it. In holomorphic dynamics one is often interested in studying the set $\mathbf{F}$ of holomorphic mappings that are quasiconformally conjugate to a given holomorphic map $f : V \to V$ modulo conjugacy by conformal isomorphisms. Such a mapping can be written as $\phi \circ f \circ \phi^{-1}$ for some $\phi \in \mathrm{Def}(V, f)$. Now $\phi \circ f \circ \phi^{-1}$ and $\psi \circ f \circ \psi^{-1}$ are conformally conjugate exactly when they represent the same element in $\mathrm{Def}(V, f)/\mathrm{QC}(V, f)$. So we can study $\mathbf{F}$ by looking at $\mathrm{Def}(V, f)/\mathrm{QC}(V, f)$. Clearly the Teichmüller space is related to this space, and it can be shown to be, at least morally, a covering of it. Because of the nice properties it has, it is often more convenient to study Teichmüller space than to study $\mathbf{F}$.

As we will see in the next section, it is often useful to split Teichmüller space into several smaller parts that are easier to study. In this direction Sullivan and McMullen proved stronger versions of the following two theorems. To state them, we need the definition of a restricted product of Teichmüller spaces.

Definition 2.8. Let f be an entire function, and suppose that $\{U_\alpha\}$ is a family of pairwise disjoint, completely invariant open subsets of $\mathbb{C}$. For an element $u \in \mathcal{T}(U_\alpha, f) = \mathcal{B}_1(U_\alpha, f)/\mathrm{QC}_0(U_\alpha, f)$, we set

$$|u| = \inf\{\|\mu\|_\infty \mid \mu \in \mathcal{B}_1(U_\alpha, f) \text{ is a representative of } u\}.$$

Then $0 \le |u| < 1$, and we define the *restricted product of Teichmüller spaces*

$$\prod_\alpha^* \mathcal{T}(U_\alpha, f)$$

as the set of sequences $\{u_\alpha \mid u_\alpha \in \mathcal{T}(U_\alpha, f)\}$ that satisfy $\sup_\alpha |u_\alpha| < 1$. In particular, if the family $\{U_\alpha\}$ is finite, the restricted product coincides with the usual product of spaces.

The motivation for this definition is that we could have a collection of invariant Beltrami forms μ_α, each with complex dilatation $K_\alpha < 1$, defined on each U_α. However, if we tried to put them all together to form a global invariant Beltrami form μ on $\bigcup_\alpha U_\alpha$, with $\mu \mid_{U_\alpha} = \mu_\alpha$, we would obtain a complex dilatation of μ whose infinity norm would be bounded by $k = \sup_\alpha k_\alpha$ which, given infinitely many α's, could be equal to 1. In such a case, μ would not belong to $\mathcal{B}_1(\bigcup_\alpha U_\alpha)$. By considering the restricted product, we allow precisely the sequences for which this problem does not arise.

With this concept, we are now ready to state the following splitting theorem.

Theorem 2.9. *Let f be an entire function, and suppose that U_α is a family of pairwise disjoint, completely invariant open subsets of $\mathbb{C}$. Then*

$$\mathcal{T}\left(\bigcup_\alpha U_\alpha, f\right) \simeq \prod_\alpha^* \mathcal{T}(U_\alpha, f).$$

Proof: This follows from Theorem 5.5 in [MS98].   □

The next theorem characterizes completely the Teichmüller space of an open, invariant, hyperbolic subset of the complex plane, as long as it has no singular values and it is in some sense minimal. To properly state the theorem, we need the notions of a grand orbit equivalence relation and of discreteness.

Definition 2.10. Let U be a subset of $\mathbb{C}$ and $f : U \to U$ a holomorphic endomorphism. The *grand orbit* of a point $z \in U$ is the set of $w \in U$ such that $f^n(z) = f^m(w)$ for some $n, m \in \mathbb{N}$. We denote the grand orbit equivalence relation by $z \sim w$, and we denote by U/f the quotient of U by the grand orbit equivalence relation. We call this relation *discrete* if all grand orbits are discrete; otherwise we call it *indiscrete*.

As an example, if U is a basin of attraction, the grand orbit relation is discrete. On the other hand, Siegel disks or basins of superattraction give rise to indiscrete grand orbit relations. See also Section 4.

Theorem 2.11 (McMullen and Sullivan, [MS98] Thm. 6.1). *Let U be an open subset of $\mathbb{C}$ and suppose every connected component of U is hyperbolic. Assume $f : U \to U$ is a holomorphic covering map and U/f is connected.*

(a) *If the grand orbit equivalence relation is discrete, then U/f is a Riemann surface or orbifold and*

$$\mathcal{T}(U, f) \simeq \mathcal{T}(U/f).$$

(b) *If the grand orbit equivalence relation is indiscrete, and some component A of U is an annulus of finite modulus, then*

$$\mathcal{T}(U, f) \simeq \mathcal{T}(A, \mathrm{Aut}_0(A)) \simeq \mathbb{H}.$$

(c) *Otherwise, $\mathcal{T}(U, f)$ is a point.*

3 The Structure of Teichmüller Space

Let $f : \mathbb{C} \to \mathbb{C}$ be an entire function. By using the decomposition results at the end of the previous section, we partition the complex plane into dynamically meaningful subsets in such a way that the Teichmüller space $\mathcal{T}(\mathbb{C}, f)$ can be studied separately in each of them. First we note the following.

Proposition 3.1. *Suppose $U \subset \mathbb{C}$ is a completely invariant open set and $K \subset U$ is a closed, completely invariant subset with the property that each element of $\mathrm{QC}_0(U, f)$ fixes K pointwise. Then*

$$\mathcal{T}(U, f) \simeq \mathcal{B}_1(K, f) \times \mathcal{T}(U - K, f).$$

Proof: Since K is measurable, we have $\mathcal{B}_1(U, f) \simeq \mathcal{B}_1(K, f) \times \mathcal{B}_1(U - K, f)$. If we write $\mathrm{QC}_0(U, f)|_{U-K}$ for the set comprised by the elements of $\mathrm{QC}_0(U, f)$ restricted to $U - K$, it follows that $\mathrm{QC}_0(U, f)|_{U-K} = \mathrm{QC}_0(U - K, f)$ by Theorem 2.6. Hence, we get that

$$
\begin{aligned}
\mathcal{T}(U, f) &\simeq (\mathcal{B}_1(K, f) \times \mathcal{B}_1(U - K, f))/\mathrm{QC}_0(U, f) \\
&\simeq \mathcal{B}_1(K, f) \times (\mathcal{B}_1(U - K, f)/\mathrm{QC}_0(U - K, f)) \\
&\simeq \mathcal{B}_1(K, f) \times \mathcal{T}(U - K, f). \qquad \Box
\end{aligned}
$$

The completely invariant set we take as K is the set of *marked points*, which consists of all periodic points of f together with the singular values of the map (i.e., critical and asymptotic values). More precisely, we define the set $\widehat{\mathcal{J}}$ as the closure of the grand orbit of the set of marked points of f, that is,

$$\widehat{\mathcal{J}} = \mathrm{Cl}\,(\mathrm{GO}\{\text{ marked points of } f\})\,.$$

Observe that $\widehat{\mathcal{J}}$ always contains the Julia set of f. We will see that points in the grand orbit of the set of marked points are dynamically distinguished, in the sense that every automorphism of $\mathbb{C}$ isotopic to the identity that commutes with f must have them as fixed points.

Theorem 3.2. *Let f be an entire transcendental function, $\widehat{\mathcal{J}}$ the closure of the grand orbits of all marked points of f and ω an element of $\mathrm{QC}_0(\mathbb{C}, f)$. Then ω restricts to the identity on $\widehat{\mathcal{J}}$.*

The proof of this theorem is involved, mainly because we do not know what the set of asymptotic values looks like. It may, for instance, have non-empty interior; recall that every point in the plane may be an asymptotic value. For the sake of exposition we leave the proof for the end of the section.

Combining this theorem with Proposition 3.1, we obtain

$$T(\mathbb{C}, f) = T(\hat{\mathcal{J}}, f) \times T(\mathbb{C} - \hat{\mathcal{J}}, f) = \mathcal{B}_1(\hat{\mathcal{J}}, f) \times T(\mathbb{C} - \hat{\mathcal{J}}, f).$$

We now would like to further decompose the space $T(\mathbb{C} - \hat{\mathcal{J}}, f)$, in order to apply Theorem 2.11 to each of its parts. Let S denote the set of singular values and periodic points of f that lie in $\mathcal{F}(f)$, and let $\hat{S}$ denote the grand orbit of the elements of S. Clearly $\hat{S}$ is a subset of $\mathcal{F}$ (in fact $\hat{\mathcal{J}}$ equals the disjoint union $\mathcal{J} \coprod \hat{S}$) and therefore $\mathbb{C} - \hat{\mathcal{J}} = \mathcal{F} - \hat{S} := \hat{\mathcal{F}}$, which is an open hyperbolic set. This set splits naturally into completely invariant subsets. Indeed let us define an equivalence relation in the set of connected components of $\hat{\mathcal{F}}$ by identifying all components that have the same grand orbit. Since they are all open, we may choose one representative from each class and thereby obtain a countable collection of connected open sets U_i. We put $V_i = \mathrm{GO}(U_i)$ and $\hat{\mathcal{F}} = \bigcup_i V_i$, where the V_i are pairwise disjoint, completely invariant, hyperbolic and open. It then follows from Theorem 2.9 that

$$T(\mathbb{C} - \hat{\mathcal{J}}, f) = \prod_i^* T(V_i, f).$$

In summary, we have proved the following theorem.

Theorem 3.3 (First Structure Theorem). *Let f be an entire transcendental function. Let $\hat{S}$ denote the closure of the grand orbits of the marked points in $\mathcal{F}(f)$ and $\hat{\mathcal{J}} = \mathcal{J} \cup \hat{S}$. Let V_i denote the collection of pairwise disjoint grand orbits of the connected components of $\mathbb{C} - \hat{\mathcal{J}}$. Then*

$$T(\mathbb{C}, f) = \mathcal{B}_1(\hat{\mathcal{J}}, f) \times \prod_i^* T(V_i, f).$$

We would now like to determine the Teichmüller space of each of the completely invariant sets V_i using Theorem 2.11. Recall that if X is an open subset of the plane which is completely invariant under f, we defined X/f to be the quotient of X by the grand orbit equivalence relation. We will split up the collection of $V_i's$ into two classes. Given a fixed i, we will say that V_i belongs to $\hat{\mathcal{F}}^{dis}$ if all grand orbits in V_i are discrete. Otherwise, the grand orbit relation in V_i is indiscrete, and we say that V_i belongs to $\hat{\mathcal{F}}^{ind}$. In the latter case, we rename it $V_i' = V_i$ and modify the indices as necessary, so that

$$\hat{\mathcal{F}}^{dis} = \bigcup_i V_i, \qquad \hat{\mathcal{F}}^{ind} = \bigcup_j V_j', \qquad \hat{\mathcal{F}} = \hat{\mathcal{F}}^{dis} \cup \hat{\mathcal{F}}^{ind}. \qquad (3.2)$$

In the same fashion, we distinguish between sets U_i and U_j', so that $V_i = \mathrm{GO}(U_i)$ and $V_j' = \mathrm{GO}(U_j')$.

Theorem 3.4 (Second Structure Theorem). *Let $f : \mathbb{C} \to \mathbb{C}$ be an entire transcendental function, and define $\widehat{\mathcal{J}}, \widehat{\mathcal{F}}^{dis}, \widehat{\mathcal{F}}^{ind}, V_i, U_i, V'_j$ and U'_j as previously. Then for every i, the quotient $U_i/f \simeq V_i/f$ is a connected Riemann surface or orbifold and*

$$\mathcal{T}(V_i, f) \simeq \mathcal{T}(U_i/f).$$

Consequently, we can write

$$\mathcal{T}(\mathbb{C}, f) \simeq \mathcal{B}_1(\widehat{\mathcal{J}}, f) \times \prod_i^* \mathcal{T}(U_i/f) \times \prod_j^* \mathcal{T}(V'_j, f).$$

Moreover, for every j such that a component of V'_j is an annulus of finite modulus, we have

$$\mathcal{T}(V'_j, f) \simeq \mathbb{H};$$

and otherwise, $\mathcal{T}(V'_j, f)$ is trivial.

Proof: This theorem is in fact Theorem 2.11 adapted to our setting. One thing we must check, though, is that the quotients V_i/f and V'_j/f are indeed connected sets, so we can apply the aforementioned theorem. We first check that they coincide with U_i/f and U'_j/f, respectively, and then see that the latter are connected. Let U be U_i or U'_j and V be V_i or V'_j, for some i or j.

By definition, V is the grand orbit of U, so any point in V has an image or a preimage in U. In other words, all grand orbits must have at least one representative in U. This implies that $V/f \subset U/f$; the other inclusion follows trivially from the fact that $U \subset V$. To see that U/f is connected, we first recall that U is connected (by definition). Let $\pi : U \to U/f$ denote the natural projection, and endow U/f with the quotient topology. Then if U/f could be split into two disjoint open sets, so could U.

We are now ready to apply Theorem 2.11 and obtain the first and last parts of the theorem. The middle decomposition comes from (3.2) and Theorem 2.9. □

The remainder of this section will be dedicated to the proof of Theorem 3.2. The proof is divided in two different parts. The first part concerns the set of periodic points and critical values, and the corresponding proof is relatively simple, given the fact that critical values and periodic points of a given period form a discrete set. The second part concerns asymptotic values, and the proof is much more delicate since, a priori, we do not know what this set looks like. For the proof of the second part, we need some preliminary results, which we include in the following subsection.

3.1 Preliminary Results

The first ingredient is due to Lindelöf.

Lindelöf's Theorem (e.g., Theorem 5.4 in [Co87]). *Let* $g : \mathbb{D} \to \mathbb{D}$ *be a holomorphic map and* $\Gamma : [0,1] \to \mathbb{D}$ *a curve such that* $\Gamma(1) = 1$. *If* $\lim_{t \to 1} g(\Gamma(t))$ *exists and equals* L, *then the map* g *has radial limit* L *at* 1.

What we will actually use is the following immediate corollary.

Corollary 3.5. *Let* $g : \mathbb{D} \to \mathbb{D}$ *be a holomorphic map, and let* $\Gamma_1, \Gamma_2 : [0,1] \to \mathbb{D}$ *be two curves such that* $\Gamma_1(1) = \Gamma_2(1) = 1$. *If the curves* $g(\Gamma_1(t))$ *and* $g(\Gamma_2(t))$ *has a limit as* $t \to 1$, *then they must land at the same point.*

Proof: If any image curve lands, then there exists a radial limit L, and the landing must occur at L. Hence any other landing image curve must do so at the same point L. $\square$

The second ingredient is the following. The statement can be found in Theorem 1.1 of [EM88], but we include a proof for completeness.

Proposition 3.6. *Let* $V \subset \mathbb{C}$ *be an open hyperbolic subset of the plane and* $\omega : V \to V$ *an element of* $\mathrm{QC}_0(V, \mathrm{Id})$. *Then there exists a lift* $\tilde{\omega} : \mathbb{D} \to \mathbb{D}$ *which extends as the identity to* $\partial\mathbb{D}$.

Proof: By assumption, we know that there exists a lift $\tilde{\omega}$ which extends as the identity to a subset of $\partial\mathbb{D}$, namely, the complement Ω of the limit set of the Fuchsian group Γ. We must check that the extension also exists at the remaining points and equals the identity.

Let μ be the Beltrami form on $\mathbb{D}$ induced by the quasiconformal map $\tilde{\omega}$. We can extend μ to $\mathbb{C}$ by setting $\mu = 0$ on the complement of the unit disk.

Let φ be the quasiconformal map that integrates μ, given by the Measurable Riemann Mapping Theorem. Then $\varphi(\partial\mathbb{D})$ is a quasicircle. We define the composition $\psi = \tilde{\omega} \circ \varphi^{-1}$ on $\varphi(\mathbb{D})$ as shown in the following diagram.

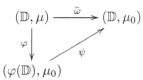

The map ψ is conformal, since it transports the standard complex structure $\mu_0 = 0$ to itself. Hence it extends continuously to the boundary of $\varphi(\mathbb{D})$. Let us denote this extension by $\widehat{\psi}$.

Now define

$$\widehat{\omega} = \begin{cases} \widetilde{\omega} & \text{on } \mathbb{D} \\ \widehat{\psi} \circ \varphi & \text{on } \partial\mathbb{D}. \end{cases}$$

This is a continuous extension of $\widetilde{\omega}$ to the boundary of the unit disk. By assumption, we know that $\widetilde{\omega}$ extends as the identity to $\Omega \subset \partial\mathbb{D}$, the complement of the limit set of γ. It follows that $\widehat{\omega}$ is the identity on Ω and, since Ω is dense in the circle, that $\widehat{\omega}$ is the identity on the whole unit circle. □

3.2   Proof of Theorem 3.2

We first show that ω restricts to the identity on the set of periodic points and on the Julia set of f, as well as on the set of critical points of f.

Let ω_t be a homotopy, i.e., a path in $\mathrm{QC}(\mathbb{C}, f)$ that connects $\omega_0 = \mathrm{Id}$ to $\omega_1 = \omega$. Since ω_t commutes with f, the set of periodic points of a given period is ω_t–invariant. Thus if $p \in \mathbb{C}$ is a periodic point of period, say, N, then the path $t \mapsto \omega_t(p)$ is a subset of the periodic points of period N. Since this set is discrete, $\omega_t(p) = p$ for all t. Since ω_t commutes with f we immediately see that ω_t fixes all periodic points for all t.

By continuity, every ω_t fixes any point in the closure of the set of periodic points, and, in particular, any point in the Julia set.

The same argument proves that ω_t fixes all critical points, since these also form a discrete set.

Since any automorphism ω in $\mathrm{QC}_0(\mathbb{C}, f)$ restricts to the identity on the Julia set, it follows that ω restricts to an automorphism of the Fatou set, which is a completely invariant open set whose connected components are hyperbolic. For this reason, the remaining cases (singular values and their grand orbits, and the closure of these sets) follow from the below proposition.

Proposition 3.7. *Let f be an entire function and $\mathcal{U}$ a totally invariant open set whose connected components are hyperbolic. Denote by S the set of singular values of f in $\mathcal{U}$. Then any ω in $\mathrm{QC}_0(U, f)$ restricts to the identity on the closure of the grand orbit of S in U.*

We remark that this result is very similar to Proposition 3 in [FH06], except that here we do not require $\mathcal{U}$ to be simply connected. The proof, however, is simplified.

Proof: The case of critical points was dealt with above. Now assume we know also that ω fixes every asymptotic value of f.

Since every singular value is in the closure of the set of asymptotic and critical values (see Section 1), we get by continuity that ω_t fixes the singular values of f in $\mathcal{U}$. Since ω_t commutes with f, we get that ω_t restricts to the identity on the forward orbit of this set. Now suppose $\omega_t(y) = y$ for all t

and that $f^n(x) = y$. Then $\omega_t(x)$ must map into $f^{-n}(\{y\})$. Since this set is discrete, we get that $\omega_t(x) = x$ for all x. It follows that ω_t restricts to the identity on the grand orbit of S for all t; by continuity, this is also true on the closure.

It remains to prove that ω fixes every asymptotic value of f. Let $z_0 \in \mathcal{U}$ be an asymptotic value of f, and let $\gamma = \gamma_0 \subset U$ be an associated asymptotic path, that is, let $\gamma : [0, \infty) \to \mathcal{U}$ be a curve such that $\gamma(s) \to \infty$ and $f(\gamma(s)) \to z_0$ as $s \to \infty$. Our goal is to show that $\omega_t(f(\gamma(s)))$ also tends to z_0 as $s \to \infty$, which would imply that z_0 is fixed by ω_t.

Let U and U' be the connected components of $\mathcal{U}$ containing z_0 and γ, respectively (we could have $U = U'$), and denote by $\pi' : \mathbb{D} \to U'$ the universal covering of U'. Let $\tilde{\omega}_t : \mathbb{D} \to \mathbb{D}$ be the lift of $\omega_t : U' \to U'$ that extends as the identity on the boundary of $\mathbb{D}$ (see Proposition 3.6) and depends continuously on t. Then, $\gamma_t = \omega_t(\gamma)$ is a family of curves in U', all of which tends to infinity, since ω_t extends as the identity on the boundary of U' (see Theorem 2.6).

We now lift the family of curves $\{\gamma_t\}$ to a family $\{\tilde{\gamma}_t\}$ in the unit disk in the following way: We choose any lift $\tilde{\gamma}_0$ of γ_0 and then define $\tilde{\gamma}_t = \tilde{\omega}_t(\tilde{\gamma}_0)$ (see Figure 1). Observe that these are lifts of $\tilde{\gamma}_t$, since

$$\pi'(\tilde{\gamma}_t) = \pi' \circ \tilde{\omega}_t(\tilde{\gamma}_0) = \omega_t \circ \pi'(\tilde{\gamma}_0) = \omega_t(\gamma_0) = \gamma_t.$$

Notice that all curves $\tilde{\gamma}_t$ must land at the same point, say, u, in the boundary of $\mathbb{D}$, given that $\tilde{\omega}_t$ extends as the identity to $\partial\mathbb{D}$, and $\tilde{\gamma}_t = \tilde{\omega}_t(\tilde{\gamma}_0)$.

In addition, observe that the image curves $f(\pi'(\tilde{\gamma}_t))$ also land at some point $z_t = \omega_t(z_0) \in U$, since the ω_t are continuous maps and

$$f(\pi'(\tilde{\gamma}_t)) = f(\omega_t(\gamma_0)) = \omega_t(f(\gamma_0)).$$

We would like to show that $z_t = z_0$ for all t.

To that end, we observe that we are working under the hypotheses of Corollary 3.5, except that U is not necessarily a disk. To fix this problem, let $\pi : \mathbb{D} \to U$ be the universal covering of U satisfying, for example, $\pi(0) = z_0$. Denote by $g : \mathbb{D} \to U$ the composition $g = f \circ \pi'$ and by $\tilde{g} : \mathbb{D} \to \mathbb{D}$ one of its lifts to the disk, as shown in the following diagram.

Then $\tilde{g}$ is holomorphic, and the curves $\tilde{\gamma}_t$ are mapped to curves in $\mathbb{D}$ that land at interior points. Indeed, if this were not the case, their images under π could not be the γ_t landing at interior points of U. Hence we assume the

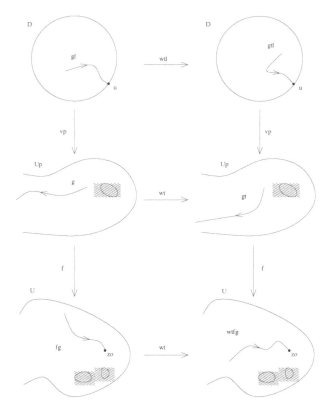

Figure 1. Commutative diagram of the construction in the proof of Proposition 3.7.

hypotheses of Lindelöf's Theorem and conclude that all of the curves $\tilde{g}(\tilde{\gamma}_t)$ land at the same point. It follows that their images under π must also land at the same point, and we conclude that $z_t = z_0$ for all t.   □

4   Teichmüller Space Supported on the Fatou Set

In this section, we treat each of the possible types of Fatou components separately in order to study how they might contribute to the Teichmüller space of f, in view of the structure theorems in the previous section. We will put more emphasis on Fatou components that cannot occur for rational maps (Baker and wandering domains) and which, consequently, were not treated in [MS98]. In general, we will use calligraphic letters to denote Fatou components and their grand orbits and regular letters to denote these objects with their marked points removed.

4.1 Fatou Components with Discrete Grand Orbit Relations

Attracting Basins. Suppose f has a periodic attracting basin of period p. Let $\mathcal{U}$ be a connected component of the immediate basin, and $\mathcal{V}$ the grand orbit of $\mathcal{U}$. In such a Fatou component, each grand orbit is a discrete set (points are isolated in the linearizing domain and then one can just pull back). It is well known that we can choose (using linearizing coordinates) a simple closed curve γ in the linearizable domain of the periodic orbit such that after p iterations γ is mapped to a curve γ' strictly inside γ. It follows that all orbits in $\mathcal{V}$ (except the periodic points and their preimages) pass exactly once through the annulus bounded by γ and γ' (including only one of the two curves). By identifying these two boundaries, we obtain a fundamental domain T isomorphic to a torus.

Now we must remove the points of $\widehat{S}$ from T (we did not consider the grand orbit of the periodic orbit which is part of $\widehat{S}$). Suppose that the periodic basin contains only one singular value v, the minimum number possible. Then the grand orbit of v has exactly one element in T. Thus if we set $V = \mathcal{V} - \widehat{S}$, we find that $V/f = (\mathcal{V}/f) - (\widehat{S}/f)$ is isomorphic to a torus with one marked point. It follows that the complex dimension of $\mathcal{T}(V/f)$ is 1.

In practice, the basin of attraction could contain many singular values (even a set with non-empty interior). The number to take into account is the number of *singular grand orbits*, i.e., the number of distinct grand orbits that contain a singular value. Indeed, if this number is finite, each of the singular grand orbits will add a marked point to the torus T and increase the dimension of $\mathcal{T}(V/f)$ by one.

If the number of distinct singular grand orbits is infinite, the dimension of $\mathcal{T}(\mathcal{V}, f)$ is infinite. If V is non-empty, then a component of V/f has a infinitely many punctures or non-empty ideal boundary (or both). If $V = \emptyset$, then

$$\mathcal{T}(\mathcal{V}, f) = B_1(\mathcal{V}/f, f) \, ,$$

which is again infinite dimensional.

In summary, we have the following statement.

Proposition 4.1. *Let $\mathcal{V}$ denote the grand orbit of a periodic attracting basin, and let $V = \mathcal{V} - \widehat{S}$, where $\widehat{S}$ is the closure of the grand orbits of all singular values. Let $1 \leq N \leq \infty$ be the number of distinct grand orbits in $\widehat{S} \cap \mathcal{V}$ (other than the attracting periodic point). Then if N is finite, V/f is a torus with N punctures and*

$$\dim \mathcal{T}(\mathcal{V}, f) = \dim_{\mathbb{C}} \left(\mathcal{T}(V/f) \right) = N.$$

If N is infinite, $\mathcal{T}(\mathcal{V}, f)$ has infinite dimension.

It is easy to produce examples of functions with periodic attracting basins that add any finite number or infinity to the dimension of Teichmüller space.

Parabolic Basins. Parabolic basins can be treated in the same way as attracting basins. Passing to Fatou coordinates, we find that V/f is an infinite cylinder with as many punctures as distinct singular grand orbits lying in V, a quantity which must be at least one. The conclusions are then the same as those in Proposition 4.1.

Baker Domains. An extensive treatment of Baker domains was given in [FH06]; we refer the reader there for details. Here we summarize the main ideas and results.

Suppose that f has an invariant Baker domain $\mathcal{U}$ (as before, the periodic case can be handled in the same way). Then $\mathcal{U}$ is simply connected [Ba75], and we can choose a Riemann map $\varphi : \mathcal{U} \to \mathbb{D}$ that maps $\mathcal{U}$ conformally onto the unit disk. Such a map conjugates f to a self-mapping of $\mathbb{D}$, which we denote by $B_{\mathcal{U}}$, and we call $B_{\mathcal{U}}$ the *inner function* associated to $\mathcal{U}$. It follows from the Denjoy-Wolf Theorem that there exists a point $z_0 \in \partial\mathbb{D}$ such that $B_{\mathcal{U}}^n$ converges to the constant mapping z_0 locally uniformly in $\mathbb{D}$ as $n \to \infty$.

It can be deduced from the work of Cowen in [Co81] that the grand orbit relation in $\mathcal{U}$ is always discrete. Hence, by Theorem 3.4, we have to analyze the set $(\mathcal{U} - \widehat{S})/f$ or, equivalently, $(\mathcal{U}/f) - (\widehat{S}/f)$. In [FH06], we have the following classification, following Cowen.

Proposition 4.2 (Fagella and Henriksen [FH06], Prop. 1). *Let f be entire and $\mathcal{U}$ an invariant Baker domain. Then $\mathcal{U}/f$ is a Riemann surface conformally isomorphic to one of the following cylinders:*

(a) $\{-s < \mathrm{Im}(z) < s\}/\mathbb{Z}$ *for some $s > 0$, and we call $\mathcal{U}$ hyperbolic;*

(b) $\mathbb{H}/\mathbb{Z}$, *and we call $\mathcal{U}$ simply parabolic;*

(c) $\mathbb{C}/\mathbb{Z}$, *and we call $\mathcal{U}$ doubly parabolic. In this case $f : \mathcal{U} \to \mathcal{U}$ is either not proper or has degree at least 2.*

Once we know what kind of Riemann surface $\mathcal{U}/f$ is, we need to remove the elements of $\widehat{S}$. Observe that in the hyperbolic and simply parabolic cases, the cylinders we obtain have a non-empty ideal boundary. This means that their Teichmüller spaces are infinite dimensional, even before adding any of the punctures given by $\widehat{S}$. In the doubly parabolic case, we have a doubly infinite cylinder, and hence the Teichmüller space will depend on the number of punctures created when removing $\widehat{S}$. In summary, we have the following theorem.

Theorem 4.3 (Fagella and Henriksen [FH06], Main Theorem). *Let* $\mathcal{U}$ *be a fixed Baker domain of the entire function* f, *and let* $V = \mathrm{GO}(\mathcal{U}) - \widehat{S}$. *Then* $\mathcal{T}(V, f) \simeq \mathcal{T}(V/f)$, *which is infinite dimensional except if* $\mathcal{U}$ *is doubly parabolic and* $\widehat{S}/f$ *is finite. In the latter case, the dimension of* $\mathcal{T}(V, f)$ *equals* $\#\widehat{S}/f - 1$.

Examples of all three types of Riemann surfaces can be found in [FH06], which includes an example of a Baker domain whose Teichmüller space is trivial, and an example where the dimension is one. By introducing more critical points in a doubly parabolic Baker domain, one can construct examples of Baker domains which contribute any finite number to the dimension of the global Teichmüller space.

4.2   Fatou Components with Indiscrete Grand Orbit Relations

In the case of indiscrete grand orbit relations, it is not important how many distinct singular grand orbits we have, but how many of them have distinct closures. It is in this spirit that we introduce the following concept (following [MS98]).

Definition 4.4. A singular value is *acyclic* if its forward orbit is infinite. Two points z and w in the Fatou set belong to the same *foliated equivalence class* if the closures of their grand orbits agree.

Two singular values whose grand orbits are not discrete may belong to the same foliated equivalence class although their grand orbits are disjoint. This is the case, for example, if their grand orbits intersect the same equipotential in a superattracting basin or the same invariant curve in a Siegel disk. If their orbits are discrete, then they belong to the same foliated equivalence class if and only if their grand orbits coincide.

Superattracting Basins. Suppose f has a superattracting periodic basin $\mathcal{V}'$. For such a Fatou component, the grand orbit equivalence relation is indiscrete (excepting the grand orbit of the periodic point). To see this, let us consider the map $z \mapsto z^2$ and choose $z_0 = re^{i\theta}$ for some fixed r and θ. It is easy to check that all points of the form

$$\{r^{2^s} \exp(i(2^s\theta + \frac{n}{2^k})) \mid s \in \mathbb{Z}, 1 \le k < \infty, 0 \le n < 2^k\}$$

belong to the same grand orbit. Clearly, the closure of these points is an infinite set of concentric circles of radii r^{2^s} for all $s \in \mathbb{Z}$. So the grand orbit relation is indiscrete. This is also true in the general case; it can be easily seen using Böttcher coordinates which locally conjugate the dynamics to $z \mapsto z^d$.

In order to apply Theorem 3.4, we need to remove the points of $\widehat{S}$ which lie in the superattracting basin $\mathcal{V}'$. Suppose first that there are no singular values in the basin. Then the set $V' = \mathcal{V}' - \widehat{S}$ consists of a collection of punctured disks (the punctures corresponding to the grand orbit of the periodic point), all belonging to the same grand orbit. No component of this set is an annulus of finite modulus (for it to be mapped to a disk, it would have to contain a critical point, and we are assuming there are none) and it therefore follows from Theorem 3.4 that $\mathcal{T}(V', f)$ is trivial.

Now suppose that $\mathcal{V}'$ contains singular values. We need to remove from $\mathcal{V}'$ the closure of their grand orbits. Assume that $\mathcal{V}'$ contains a finite number of singular grand orbits, each belonging to different foliated equivalence classes. Denote their closures by $\widehat{S}_1, \ldots, \widehat{S}_N$. Let $\mathcal{U}'$ be a connected component of the immediate basin, which we assume for simplicity to be fixed (if not, one only needs to consider f^p). We start by removing $\widehat{S}_1$. Close enough to the fixed (and critical) point, we can assume that there are no critical points, so $\widehat{S}_1$ consists of infinitely many concentric simple closed curves (equipotentials). After removing these, we are left (close enough to the fixed point) with infinitely many annuli, all belonging to the same grand orbit. Let U' be one of these annuli. Then U' is a fundamental domain for $V' = \mathcal{V}' - \widehat{S}_1$. Let us now consider the remaining $N - 1$ singular orbits. Since U' is a fundamental domain, each of them will intersect U' in a single equipotential. Therefore, removing $\widehat{S}_2, \ldots, \widehat{S}_N$ will partition U' into exactly N disjoint open annuli $U_1', \ldots, U_n'$, each of which gives rise to a distinct grand orbit. Denote these grand orbits by $V_1', \ldots, V_n'$; they form a partition of V' into completely invariant, open subsets.

We can apply Theorem 3.4 to each V_i'. Since each V_i' has a connected component which is an annulus of finite modulus, i.e., U_i', we conclude that

$$\mathcal{T}(V_i', f) = \mathbb{H} \, .$$

It follows that

$$\mathcal{T}(V', f) \simeq \prod_{i=1}^{N} \mathcal{T}(V_i', f) \simeq \mathbb{H}^N,$$

and hence the superattracting basin adds N dimensions to the global Teichmüller space.

Finally, we consider the case where there are infinitely many singular grand orbits in $\mathcal{V}'$, each of them in a distinct foliated equivalence class. In this case, either the set $\widehat{S} \cap \mathcal{V}'$ has nonempty interior or $\mathcal{V}' - \widehat{S}$ contains infinitely many distinct grand orbits of annuli. In both cases, $\mathcal{T}(\mathcal{V}', f)$ has infinite dimension.

We summarize the results as follows.

Proposition 4.5. *Let $\mathcal{V}'$ denote the grand orbit of a superattracting periodic attracting basin, and let $V' = \mathcal{V}' - \widehat{S}$, where $\widehat{S}$ is the closure of the grand orbits of all singular values and periodic points. Let $0 \leq N \leq \infty$ equal the number of grand orbits in $\widehat{S} \cap \mathcal{V}'$, excluding the superattracting periodic point, that belong to distinct foliated equivalence classes. Then the grand orbit relation is indiscrete, and*

(a) *If $N = 0$, the set V' is a collection of punctured disks, and therefore $\mathcal{T}(V', f)$ is trivial.*

(b) *If $N > 0$ is finite, V' splits into N completely invariant subsets $V'_1, \ldots, V'_N$, and each of them has a connected component which is an annulus of finite modulus. Then*

$$\dim_{\mathbb{C}} (\mathcal{T}(V', f)) = N.$$

(c) *Finally, if $N = \infty$, either V' is empty or $\mathcal{T}(V', f)$ has infinite dimension. In both cases, $\dim \mathcal{T}(\mathcal{V}', f) = \infty$.*

Siegel Disks. Suppose that f has a fixed Siegel point z_0 with an invariant Siegel disk Δ (the periodic case is analogous), and let $\mathcal{V}'$ denote the grand orbit of Δ. Observe that the grand orbit relation is indiscrete since, given any point in the Siegel disk (except for the fixed point), the accumulation of its grand orbit is exactly one invariant simple closed curve in Δ, and all its preimages in the other components of $\mathcal{V}'$.

As before, let us first assume that $\mathcal{V}'$ contains no singular grand orbits, so that $V' = \mathcal{V}' - \widehat{S} = \mathcal{V}' - \mathrm{GO}(z_0)$. Then V'/f is isomorphic to $\Delta - \{z_0\}$ and therefore connected (equivalently, V' has no completely invariant proper subsets). Hence we can apply Theorem 2.11 or 3.4 to V' directly. Since no component of V' is an annulus of finite modulus (for it to be mapped to Δ, it would have to contain a critical point, and we assume there are none, it follows that the Teichmüller space $\mathcal{T}(V', f)$ is trivial.

Let us now suppose we have a finite number, say N, of singular grand orbits (other than the Siegel point), each of which belongs to a distinct foliated equivalence class. Then the set $\widehat{S} \cap \Delta$ consists of the fixed Siegel point together with N disjoint invariant closed curves around z_0. Hence $\Delta - \widehat{S}$ consists of N annuli of finite modulus $U'_1, \ldots, U'_N$ together with a punctured disk U'_{N+1}. For $1 \leq i \leq N + 1$, define $V'_i = \mathrm{GO}(U'_i)$. Then the sets $V'_1, \ldots, V'_{N+1}$ are open and completely invariant and form a partition of V'. Applying Theorem 3.4 to each of them, we find that $\mathcal{T}(V'_i, f) = \mathbb{H}$ for $1 \leq i \leq N$ whereas $\mathcal{T}(V'_{N+1}, f)$ is trivial.

If $\mathcal{V}'$ contains infinitely many foliated equivalence classes, arguments identical to those used for attracting and superattracting basins apply.

We summarize the results as follows.

Proposition 4.6. *Let* $\mathcal{V}'$ *denote the grand orbit of a periodic cycle of Siegel disks, and let* $V' = \mathcal{V}' - \widehat{S}$, *where* $\widehat{S}$ *is the closure of the grand orbits of all singular values and periodic points. Let* $0 \leq N \leq \infty$ *equal the number of grand orbits in* $\widehat{S} \cap \mathcal{V}'$ *that belong to distinct foliated equivalence classes. Then the grand orbit relation is indiscrete, and*

(a) *If* $N = 0$, *the set* V' *is a collection of punctured disks, and therefore* $\mathcal{T}(V', f)$ *is trivial.*

(b) *If* $N > 0$ *is finite,* V' *splits into* $N + 1$ *completely invariant subsets* $V'_1, \ldots, V'_{N+1}$, *and each of them, except one, has a connected component which is an annulus of finite modulus. Then*

$$\dim_{\mathbb{C}} (\mathcal{T}(V', f)) = N.$$

(c) *Finally, if* $N = \infty$, *either* V' *is empty or* $\mathcal{T}(V', f)$ *has infinite dimension. In both cases,* $\dim_{\mathbb{C}} (\mathcal{T}(\mathcal{V}', f)) = \infty$.

4.3   Wandering Domains: Discrete and Indiscrete Grand Orbit Relations

We will treat the case of wandering domains by giving several examples that illustrate the different situations that can occur. In particular, some wandering domains have discrete grand orbit relations whereas others do not. They may also increase the dimension of global Teichmüller space by zero, a finite amount, or infinity. To our knowledge, no example of a wandering domain having a finite (non-zero) dimensional Teichmüller space was previously known. We construct such an example (see Example 3) by surgery, and obtain a wandering domain with an indiscrete grand orbit relation. We have not been able to construct a wandering domain with a discrete grand orbit relation, and a finite dimensional Teichmüller space. A summary of examples is given in Table 4.1.

Example 1 (Discrete Relation, $V/f \simeq \mathbb{D}$, $\dim(\mathcal{T}(\mathcal{V}, f)) = \infty$).
Let $f(w) = aw^2 e^{-w}$, and set $a = e^{2-\lambda}/(2 - \lambda)$, with

$$\lambda = \exp(2\pi i(1 - \sqrt{5})/2) = -0.737369\ldots + i0.675490\ldots.$$

It is easy to check that f has a superattracting fixed point at $w = 0$ (for all values of a) and that for this particular value of the parameter, f has an indifferent fixed point at $w_0 = 2 - \lambda$. The multiplier of w_0 is precisely

Example	GO rel.	V/f	$\dim(\mathcal{T}(V, f))$
1. Lift of a Siegel disk	disc.	$\mathbb{D}$	∞
2. Lift of a superattracting basin	indisc.	$-$	0
3. Surgery on 2	indisc.	$-$	$1 \leq N < \infty$
4. Lift of a basin of attraction	disc.	$\mathbb{C}^* - \{a_i\}_{i \in \mathbb{N}}$	∞
5. Lift of a superattracting basin with an additional critical point	indisc.	$-$	∞
Unknown	disc.	$-$	$0 \leq N < \infty$

Table 4.1. Examples of wandering domains.

λ and thus w_0 is linearizable and has a Siegel disk Δ around it. Given the presence of a superattracting basin for $w = 0$, we know that Δ does not intersect a neighborhood of this point. Just to make the picture complete, notice that f has two critical points: the origin and $c = 2$, whose orbit accumulates on the boundary of Δ. See Figure 2.

We are now going to take a logarithmic lift of f in such a way that the Siegel disk lifts to a wandering domain. More precisely, let

$$F(z) = A + 2z - e^z, \qquad A = \log(a) = 2 - \lambda - \log(2 - \lambda),$$

and observe that the function $w = e^z$ is a semiconjugacy between f and F. It follows that the superattracting basin of f lifts to a (hyperbolic) Baker domain of F, as studied originally in [Be95] and later in [BF00] and in [FH06] for some other parameter values. Let us see what happens to the Siegel disk after lifting. See Figure 3.

An easy computation shows that all points of the form

$$z_k = \log(2 - \lambda) + 2k\pi i$$

project down to the fixed Siegel point. However, these points are not all fixed by f, but instead satisfy $F(z_k) = z_{2k}$. Since the Siegel disk Δ omits a neighborhood of zero and is simply connected, it must lift to infinitely

Figure 2. Dynamical plane of $f(w) = aw^2e^{-w}$ for $a = e^{2-\lambda}/(2 - \lambda)$, with $\lambda = \exp(2\pi i(1 - \sqrt{5})/2) = -0.737369\ldots + i0.675490\ldots$. In white we see the basin of attraction of $w = 0$ (the superattracting fixed point is at the center of the image). In gray are the Siegel disk around w_0 and its preimages. (See Plate II.)

many disjoint domains Δ_k each of which surrounds the point z_k. Since the points z_k map to z_{2k}, it follows that the domains Δ_k map to Δ_{2k}, and hence we obtain a Siegel disk Δ_0 and infinitely many distinct grand orbits of wandering domains. Since the original Δ was a Siegel disk, it follows that F is one to one on each of the domains. In fact, F morally still acts like a rotation on the wandering domains, the only difference being that it moves to a different fiber of the Siegel disk every time it is applied.

Let us choose one of these sequences, say $\Delta_1, \Delta_2, \Delta_4, \ldots$, and denote by V its grand orbit. Notice that V does not contain any singular value nor any periodic point; in other words, V contains no element of $\widehat{S}$. Since F is one to one in each of the domains, the grand orbit relation is clearly discrete. In view of Theorem 3.4 we must then look at V/f, which is isomorphic to, say, Δ_1, since every grand orbit in V passes once and only once through Δ_1. Hence V/f is a disk, and its Teichmüller space is infinite dimensional because of the presence of a non-empty ideal boundary.

Figure 3. The dynamical plane of $F(z) = A + 2z - e^z$ as in Example 1. In white we see the Baker domain, which is the lift of the superattracting immediate basin for f and all the preimages of the Baker domain. In gray are the domains Δ_k and their preimages; they correspond to lifting the Siegel disk in Figure 2. (See Plate II.)

Example 2 (Indiscrete Relation, $\dim(\mathcal{T}(\mathcal{V}', f)) = 0$).

Consider the map $F(z) = z + 2\pi + \sin(z)$. Observe that F has infinitely many (double) critical points on the real line located at $c_k = (2k + 1)\pi$, which satisfy $F(c_k) = c_{k+1}$. Consider also the vertical lines whose real part is an even multiple of π, which we denote by $l_k = 2k\pi + it$, $t \in \mathbb{R}$, and observe that they satisfy $F(l_k) = l_{k+1}$. Moreover, these lines belong to the Julia set, since all their points (with non-zero imaginary part) tend to infinity exponentially fast under iteration.

It follows that there is a sequence of distinct Fatou domains W_k containing the critical points c_k in their interiors and satisfying $F(W_k) = W_{k+1}$. These are wandering domains which map to each other with degree 3 since each of them contains a double critical point. See Figure 4.

Observe that F projects under $w = e^{iz}$ to $f(w) = w \exp(\frac{1}{2}(w - \frac{1}{w}))$, which has a superattracting fixed point at $w = -1$ and a repelling fixed point at $w = 1$. The wandering domains W_k are in fact the logarithmic

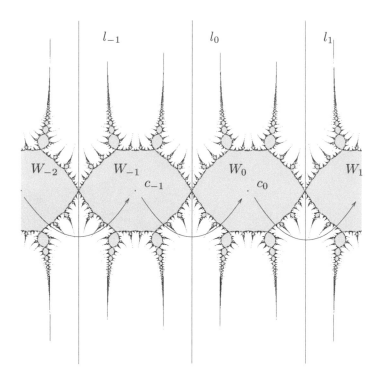

Figure 4. The dynamics of $F(z) = z + 2\pi + \sin z$ used in example 2. The vertical lines are part of the Julia set and cut the plane into vertical strips each of which contains a wandering domain W_k.

lifts of the superattracting basin around $w = -1$. Abusing language, we will call *equipotential curves* those simple closed curves in W_k that are lifts of the equipotentials in the superattracting basin.

Returning to F, observe that the grand orbit relation is indiscrete. Just as with the superattracting basins, the grand orbit of any point accumulates in one of the equipotential curves, a different one in each W_k. This is easy to check by picking any point, mapping it forward several times and then consider *all* its preimages in each of the preceding domains. Taking this procedure to its limit, we see that we fill up an equipotential curve in each of the W_k's.

Let $\mathcal{V}'$ denote the grand orbit of W_k (for any k), and let $V' = \mathcal{V}' - \widehat{S}$, which in this case is a collection of punctured disks (the only elements in $\widehat{S}$ are the critical points c_k and their preimages). Since the orbit relation is indiscrete and no component of V' is an annulus of finite modulus, we conclude from Theorem 3.4 that $T(V', f)$ is trivial. Now since $\hat{S}$ has measure zero, we have from Proposition 3.1 that $T(V', f) \simeq T(V', f)$ and thus that $T(V', f)$ has dimension 0.

Example 3 (Indiscrete Relation, $\dim(T(V, f)) = 1$).
We modify the previous example in such a way that, instead of one grand orbit of singular values, we get two. Let F and W_k, $k \in \mathbb{Z}$ be defined as before. Notice that each W_k is simply connected and that $F : W_k \to W_{k+1}$ is a degree 3 proper map. Let U be an open disk neighborhood of π that is compactly contained in W_0. Let U' be the component of $F^{-1}(U)$ that contains $-\pi$. Then $U' = F^{-1}(U) \cap W_{-1}$, and more importantly $F : U' \to U$ is a proper degree 3 mapping.

Construct $h : \overline{U} \to \overline{U}$ so that h satisfies $h|_{\partial U} = \text{Id}_{\partial U}$, $h(\pi) \neq \pi$ and $h|_U$ is quasiconformal. Set

$$\tilde{F} = \begin{cases} h \circ F & \text{on } \overline{U'} \\ F & \text{on } \mathbb{C} - U'. \end{cases}$$

Then $\tilde{F}$ is a quasiregular mapping. Since, as for F, $\tilde{F}$ maps W_k onto W_{k+1} for all k, we find that a grand orbit contains at most one point in U'.

We now define an almost complex structure with respect to which $\tilde{F}$ is holomorphic. Let μ_0 denote the standard complex structure. Define μ by letting it equal the pullback $(\tilde{F}^n)^*\mu_0$ on $\tilde{F}^{-n}(U)$, $n = 1, 2, \ldots$ and letting it equal μ_0 on the complement, i.e., the set $\{z \mid \tilde{F}^n(z) \notin U\}$.

Then $\|\mu\|_\infty = \|h^*(\mu_0)\|_\infty < 1$. By the Measurable Riemann Mapping Theorem, there exists a map $\phi : \mathbb{C} \to \mathbb{C}$ that satisfies $\phi^*\mu_0 = \mu$. By Weyl's Lemma, the composition $G = \phi \circ \tilde{F} \circ \phi^{-1}$ is a holomorphic mapping.

Notice that ϕ conjugates F to G everywhere except on U', and that $G(\phi(W_{-1})) = \phi(W_0)$. It follows that ϕ maps the Julia set of F to the Julia set of G, and the Fatou set of F to the Fatou set of G. By construction, $\phi(W_0)$ is a simply connected wandering domain. The critical points of G are $p_k = \phi((2k+1)\pi)$, $k \in \mathbb{Z}$. Again by the construction of G, we see that G maps p_k to p_{k+1} for all $k \in \mathbb{Z} - \{-1\}$, and that $F(p_{-1}) \neq p_0$.

We conclude that we have at most two distinct grand orbits of singular values, namely $\text{GO}(p_{-1})$ and $\text{GO}(p_0)$. That these two grand orbits are distinct follows from an easy induction using the fact that the only point in W_k that is mapped to p_{k+1} is p_k, $k = 0, 1, \ldots$.

Let $S_i = \overline{\text{GO}(p_i)}$, $i = -1, 0$. We can argue as in the previous example that $\phi(W_0) \cap S_{-1}$ is a simple closed analytic curve encircling $\{p_0\} = \phi(W_0) \cap S_0$. Thus $\phi(W_0) - \tilde{S}$ is composed of an annulus X_A and a pointed disk X_B. A connected component of the preimage of a pointed disk by an entire function cannot be an annulus of finite modulus. It follows that no component of $\text{GO}(X_B)$ is an annulus of finite modulus. We can therefore apply Theorem 2.11 and Proposition 3.1 and get $T(\text{GO}(X_A), G) \simeq \mathbb{H}$, whereas $T(\text{GO}(X_B), G)$ is trivial. Since $\text{GO}(\phi(W_0)) \cap \widehat{S}$ has measure zero, we conclude that $T(\text{GO}(\phi(W_0)), G)$ has dimension 1.

Remark. We could repeat the surgery and modify G so that $G(p_0) \neq p_1$. In this way, we would obtain a mapping with three distinct grand orbits of singular values and add one to the dimension of the Teichmüller space. In fact, given any n, we could do this a finite number of times and obtain a map with a wandering domain on which the grand orbit relation is indiscrete and where the corresponding Teichmüller space has dimension n.

Example 4 (Discr. Relation, $V/f \simeq \mathbb{C} - \{a_i\}_{i \in \mathbb{Z}}$, $\dim(T(V, f)) = \infty$).
Let $f(w) = \lambda w e^w$, where we set $\lambda = e^{\frac{1}{2}}$. Then f has a repelling fixed point at the asymptotic value $w = 0$ and an attracting fixed point at $w = -1/2$. By plotting the map on the real line, we can see that the semi-axis $[0, \infty)$ belongs to the Julia set (all its points except the origin escape to ∞), while $(-\infty, 0)$ is contained in the basin of attraction Ω of the attracting fixed point. The map has only one critical point at $w_0 = -1$, which is of course contained in Ω.

We now choose a lift of f by $w = e^z$ and obtain a new function

$$F(z) = z + \frac{1}{2} + 2\pi i + e^z.$$

The basin of attraction Ω lifts to infinitely many band-like Fatou domains $\{\Omega_k\}_{k \in \mathbb{Z}}$, each of which contains the horizontal line $\text{Im}(z) = (2k+1)\pi$ and a critical point at $c_k = (2k+1)\pi i$. They are separated by the lifts of the positive real line, i.e., by the horizontal lines $\text{Im}(z) = 2k\pi$, which belong to the Julia set. Because of the choice of the lift, the domains are mapped in such a way that $F(\Omega_k) = \Omega_{k+1}$ with degree 2. Hence they are all wandering domains and belong to the same grand orbit, which we denote by V. Observe that these sets are morally basins of attraction with one critical point inside, except that as the orbits move closer to a fiber of the fixed point, they also jump to the next domain.

Since the grand orbit relation of f in $\Omega - \widehat{S}$ is discrete, so is the grand orbit relation of F in V. We would now like to know what the set $V/f = (V - \widehat{S})/f$ looks like. Observe that the grand orbit of every point must have at least one element in each of the Ω_k. Hence $V/f = \Omega_k/f$ for any k. Let us take for example $k = 0$.

Proposition 4.7.

(a) *Let $z_1, z_2 \in \Omega_0$ not belong to any singular grand orbit. Then z_1 and z_2 belong to the same grand orbit if and only if $F^n(z_1) = F^n(z_2)$ for some $n \geq 0$.*

(b) *There exists a holomorphic bijection $\phi : \Omega_0/f \to \mathbb{C}$.*

(c) *There are infinitely many distinct singular grand orbits in $\mathcal{V}$, and*

$$V/f \simeq (\Omega_0 - \widehat{S})/f \simeq \mathbb{C}^* - \{a_i\}_{i \in \mathbb{Z}},$$

where $a_i \in \mathbb{C}$ for all $i \in \mathbb{Z}$. We conclude that $\dim(\mathcal{T}(\mathcal{V}, f)) = \dim(\mathcal{T}(\mathcal{V}/f)) = \infty$.

Proof: (a) The "if" part is clear by definition. To see the "only if" part, note that if $z_1, z_2 \in \Omega_0$ belong to the same grand orbit, there exist $n, m \geq 0$ such that $f^n(z_1) = f^m(z_2)$. Since $f^n(z_1) \in \Omega_n$ and $f^m(z_2) \in \Omega_m$, we must have $n = m$.

(b) Recall that F was obtained as the logarithmic lift of a function f having an attracting basin Ω. Let $\varphi : \Omega \to \mathbb{C}$ denote the extended linearizing coordinates conjugating f in Ω to $M_\rho(w) = \rho w$, where $\rho = 1/2$ is the multiplier of the fixed point. Define the maps

$$\phi_k : \quad \Omega_k \overset{\exp}{\longrightarrow} \Omega \overset{\varphi}{\longrightarrow} \mathbb{C}$$
$$z \longmapsto \exp(z) \longmapsto \varphi(\exp(z)),$$

which are holomorphic. For any $n \geq 0$, the following diagram commutes:

$$
\begin{array}{ccc}
\Omega_0 & \overset{F^n}{\longrightarrow} & \Omega_n \\
\exp \downarrow & & \downarrow \exp \\
\Omega & \overset{f^n}{\longrightarrow} & \Omega \\
\varphi \downarrow & & \downarrow \varphi \\
\mathbb{C} & \overset{M_\rho^n}{\longrightarrow} & \mathbb{C},
\end{array}
$$

and therefore $M_\rho^{-n} \circ \phi_n \circ F^n = \phi_0$ on Ω_0. It follows that if $F^n(z_1) = F^n(z_2)$, then $\phi_0(z_1) = \phi_0(z_2)$, so the map $\phi : \Omega_0/f \to \mathbb{C}$ defined by $\phi[z] = \phi_0(z)$ is well defined. To see that it is a bijection, observe that the linearizing coordinate satisfies $\varphi(w_1) = \varphi(w_2)$ if and only if $f^n(w_1) = f^n(w_2)$ for some $n \geq 0$. Now suppose that z_1 and z_2 are in different classes in Ω_0/f but $\phi(z_1) = \phi(z_2)$. Since the exponential is bijective on Ω_0, we have that $\varphi(\exp(z_1)) = \varphi(\exp(z_2))$, and therefore $f^n(\exp(z_1)) = f^n(\exp(z_2))$ for some $n \geq 0$. Lifting to Ω_n, this implies that $F^n(z_1) = F^n(z_2)$, and thus z_1 and z_1 belong to the same class. This proves the injectivity of ϕ. The surjectivity of ϕ follows immediately from the surjectivity of φ.

(c) There is one critical point c_k in each wandering domain Ω_k, and each of them belongs to a distinct grand orbit, since they all project to the same point $c \in \Omega$, which is not fixed nor periodic. Given the absence of asymptotic values, this implies that $\phi(\widehat{S}/f)$ consists of a countable sequence of points $\{a_i\} \cup \{0\}$ which are the points in the full orbit of $\phi(c)$

under M_ρ, together with the point 0, which is the image of the former fixed point in Ω_0 under ϕ. In other words, $\phi(\widehat{S}/f) = \{\rho^i \varphi(c)\}_{i \in \mathbb{Z}} \cup \{0\}$, and the statement follows. $\square$

Example 5 (Indiscrete Relation, $\dim(\mathcal{T}(\mathcal{V}, f)) = \infty$).
As in other examples, we get the mapping by lifting. First consider the entire map f given by $f_{a,b}(z) = az^2 \exp(bz^2 + z)$, $a \neq 0$. We claim there exist real values of a and b such that $f_{a,b}$ has three real critical points $\omega_2 < \omega_1 < \omega_0 = 0$, and such that ω_1 is a fixed point having ω_2 in its immediate basin of attraction. Indeed, $\omega_0 = 0$ is a critical point, and for $0 < b \leq 1/16$, the two other critical points are given by $\omega_1 = \frac{-1 + \sqrt{1 - 16b}}{4b}$ and $\omega_2 = \frac{-1 - \sqrt{1 - 16b}}{4b}$. For $0 < b \leq 1/16$, there exists a function $a(b)$ such that $f_{a(b),b}(\omega_1) = \omega_1$. Abusing notation, we write $f_b = f_{a(b),b}$. For $b = \frac{1}{16}$, ω_1 and ω_2 coincide, and for $b < \frac{1}{16}$ they are distinct. By the semicontinuity of the immediate basin of attraction of ω_1, it follows that for values of b slightly less than $1/16$, ω_2 stays in the immediate basin of ω_1. This proves the claim.

Fix a and b as in the claim, and write f for $f_{a,b}$. Let W be the immediate basin of ω_1. The equipotential going through ω_2 is a figure eight that cuts the sphere into three simply connected regions; let U denote the one that contains ω_1. Now lift f by the exponential map to obtain a mapping F given by $F(w) = \log a + 2w + b \exp(2w) + \exp(w)$. Let $\tilde{W}$ and $\tilde{U}$ denote the preimages of W and U, respectively, under the exponential map. Since the positive real axis does not meet W, the preimage $\tilde{W}$ contains countably many components W_k which we can number so that $F(\tilde{W}_k) = \tilde{W}_{2k}, k \in \mathbb{Z}$. Similarly, define $\tilde{\omega}_i(k) \in W_k$ as the fibers of $\omega_i, i = 1, 2$.

Let $\mathcal{V}$ denote the grand orbit of $\tilde{W}_1$. We leave it to the reader to verify that the grand orbit relation is indiscrete, and we sketch a proof that $\mathcal{T}(\mathcal{V}, f)$ is infinite dimensional. Number the components $\tilde{U}_k$ of $\log(U)$ so that $\tilde{U}_k \subset \tilde{W}_k$. Notice that W_1 is a wandering component. Now the closure of $\mathrm{GO}(\omega_1)$ cuts U into a countably infinite number of annuli, each delimited by equipotential curves of $f^k(\omega_2)$ and $f^{k+1}(\omega_2)$. For F, the grand orbit of $\widehat{S}$ is the grand orbit of the set $O = (\bigcup_k \omega_1(k)) \bigcup (\bigcup_k \omega_2(k))$. It follows that $\widehat{S}$ cuts $\tilde{U}_1$ into infinitely many nested annuli, and we conclude that on the wandering domain $\mathcal{V}$ the grand orbit relation is indiscrete and that the dimension of $\mathcal{T}(\mathcal{V}, f)$ is infinite.

5   Teichmüller Space Supported on the Julia Set

Whereas the dynamics of an entire function on the Fatou set is well understood, its dynamics on the Julia set is generally not. We survey here a few well known facts.

The Teichmüller space supported on the Julia set is isomorphic to $\mathcal{B}_1(J, f)$, because the Julia set is formed by dynamically distinguished points. In particular, the Julia set has to have positive Lebesgue measure in order to have non-trivial Teichmüller space. Thus in many examples $\mathcal{B}_1(J, f)$ can be understood if it can be shown that J has measure zero. A very general result used to show that J has measure zero is the following theorem by Eremenko and Lyubich.

Theorem 5.1 (Eremenko and Lyubich [EL92], Prop. 4 and Thm. 8).
Let f be an entire transcendental function of finite order and finite type such that f^{-1} has a logarithmic singularity $a \in \mathbb{C}$. Assume that the orbit of every singular point of f^{-1} is either absorbed by a cycle or converges to an attracting or parabolic cycle. Then either $J(f) = \mathbb{C}$ or $\mathrm{meas}(J(f)) = 0$.

Just because the Julia set has positive measure does not mean that $\mathcal{B}_1(J, f)$ is non-trivial. For instance, it is known that there exist quadratic polynomials whose Julia sets have positive measure (see [BC06]), but even if it is not known, it is generally believed that $\mathcal{B}_1(J, P)$ is trivial for every quadratic polynomial P. In particular, the Cremer polynomials constructed in [BC06] do not admit a one dimensional family of quasiconformal deformations since they belong to the boundary of the Mandelbrot set.

However, there are examples of entire transcendental mappings f for which $\mathcal{B}_1(J, f)$ is non-trivial. Such an example was given in [EL87]; it is stated as follows.

Theorem. *There is an entire transcendental map f whose Julia set $J(f)$ is nowhere dense, has positive measure and supports an invariant line field.*

By an invariant line field, one means a measurable function μ supported on the Julia set such that $\|\mu\|_\infty = 1$ and $f^*\mu = \mu$. Notice that the existence of an invariant line field implies that the measurable functions $\mu_t = t\mu$ are f-invariant, $|t|$-Beltrami forms on the Julia set for all $t \in \mathbb{D}$. It follows that $\mathcal{B}_1(J, f)$ has dimension at least 1.

In fact, in [EL87] Eremenko and Lyubich indicate how to modify this example so that it could support an infinite-dimensional family of measurable invariant line fields. There are several recent relevant results on this issue, such as [UZ07], [GKS04], [Re06] or [RvS07].

Bibliography

[Ah66]   L. Ahlfors, *Lectures on quasiconformal mappings*, Wadsworth & Brooks/Cole Mathematics Series, 1966.

[Ba75]   I. N. Baker, *The domains of normality of an entire function*, Ann. Acad. Sci. Fenn. Ser. A **1** (1975), 277–283.

[Ba76]   I. N. Baker, *An entire function which has wandering domains*, J. Austral. Math. Soc. Ser. A **22** (1976), 173–176.

[BD99]   I. N. Baker and P. Domínguez, *Boundaries of unbounded Fatou components of entire functions*, Ann. Acad. Sci. Fenn. Math. **24** (1999), 437–464.

[BF00]   K. Barański and N. Fagella, *Univalent Baker domains*, Nonlinearity **14** (2001), 411–429.

[Be93]   W. Bergweiler, *Iteration of meromorphic functions*, Bull. Amer. Math. Soc. (N.S.) **29** 2 (1993), 151–188.

[Be95]   W. Bergweiler, *Invariant domains and singularities*, Math. Proc. Camb. Phil. Soc. **117** (1995), 525–532.

[Be01]   W. Bergweiler, *Singularities in Baker domains*, Comput. Methods Funct. Theory **1** (2001), 41–49.

[BC06]   X. Buff and A. Chéritat, *Quadratic Julia sets with positive area*, arXiv math.DS/0605514.

[Co87]   J. B. Conway, *Functions of one complex variable II*, Springer, New York, 1995.

[Co81]   C. C. Cowen, *Iteration and the solution of functional equations for functions analytic in the unit disk.* Trans. Amer. Math. Soc. **265** 1 (1981), 69–95.

[DT86]   R. L. Devaney and F. Tangerman, *Dynamics of entire functions near an essential singularity*, Ergod. Th. Dynam. Syst. **6** (1986), 489–503.

[EM88]   C. J. Earle and C. T. McMullen, *Quasiconformal isotopies*, in: Holomorphic Functions and Moduli I, Math. Sci. Res. Inst. Publ. **10**, Springer, New York, 1988, 143–154.

[Ep95]   A. L. Epstein, *Towers of finite type complex analytic maps*, PhD Thesis, City University of New York, 1995.

[EL87]   A. Eremenko and M. Lyubich, *Examples of entire functions with pathological dynamics*, J. London Math. Soc. (2) **36** (1987), 458–468.

[EL92]   A. Eremenko and M. Lyubich, *Dynamical properties of some class of entire functions*, Ann. Inst. Fourier **42** (1992), 989–1020.

[FH06]   N. Fagella and C. Henriksen, *Deformation of entire functions with Baker domains*, Disc. Cont. Dynam. Syst. **15** (2006), 379–394.

[Fa19]   P. Fatou, *Sur les équations fonctionelles*, Bull. Soc. Math. France **47** (1919), 161–271, **48** (1920), 33–94, 208–314.

[GK86]   L. R. Goldberg and L. Keen, *A finiteness theorem for a dynamical class of entire functions*, Ergod. Th. Dynam. Syst. **6** (1986), 183–192.

[GKS04] J. Graczyk, J. Kotus, and G. Światek, *Non-recurrent meromorphic functions*, Fund. Math. **182** (2004), 269–281.

[Gr18]   W. Gross, *Eine ganze Funktion, für die jede komplexe Zahl Konvergenzwert ist*, Math. Ann. 79 1–2 (1918), 201–208.

[HT97]   T. Harada and M. Taniguchi, *On Teichmüller space of complex dynamics by entire functions*, Bull. Hong Kong Math. Soc. **1** (1997), 257–266.

[He81]   M. Herman, *Exemples des fractions rationelles ayant une orbite dense sur la sphère de Riemann*, Bull. Soc. Math. France **112** (1984), 93–142.

[HY98]   X. Hua and C. Yang, *Dynamics of transcendental functions*, Gordon and Breach Science Publishers, 1998.

[H06]   J. H. Hubbard, *Teichmüller theory and applications to geometry, topology, and dynamics, vol. I: Teichmüller theory*, Matrix Editions, Ithaca, 2006.

[Kö99]   H. König, *Conformal conjugacies in Baker domains*, J. London. Math. Soc. **59** (1999), 153–170.

[LV73]   O. Lehto and K. I. Virtanen, *Quasiconformal mappings in the plane*, Springer-Verlag, Berlin, Heidelberg, New York, 1973.

[MS98]   C. T. McMullen and D. P. Sullivan, *Quasiconformal homeomorphisms and dynamics III: The Teichmüller space of a holomorphic dynamical system*, Adv. Math. **135** (1998), 351–395.

[Ne70]   R. Nevanlinna, *Analytic functions*, Springer-Verlag, New York, 1970.

[Re06]   L. Rempe, *Rigidity of escaping dynamics for transcendental entire functions*, Acta Math., to appear.

[RvS07] L. Rempe and S. van Strien, *Absence of line fields and Mañés theorem for non-recurrent transcendental functions*, Manuscript (2007).

[RS99a] P. J. Rippon and G. M. Stallard, *Families of Baker domains I*, Nonlinearity **12** (1999), 1005–1012.

[RS99b] P. J. Rippon and G. M. Stallard, *Families of Baker domains II*, Conform. Geom. Dyn. **3** (1999), 67–78.

[UZ07] M. Urbański and A. Zdunik, *Instability of exponential Collet-Eckman maps*, Israel J. Math. **161** (2007), 347–371.

III. Two Complex Dimensions

9   Cubic Polynomial Maps with Periodic Critical Orbit, Part I

John Milnor

Dedicated to JHH:

> Julia sets looked peculiar—
> Unruly and often unrulier
>> Till young Hubbard with glee
>> Shrank each one to a tree
> And taught us to see them much trulier.

This will be a discussion of the dynamic plane and the parameter space for complex cubic maps which have a superattracting periodic orbit. It makes essential use of Hubbard trees to describe associated Julia sets.

1   Introduction

The parameter space for cubic polynomial maps has complex dimension 2. Its non-hyperbolic subset is a complicated fractal locus which is difficult to visualize or study. One helpful way of exploring this space is by means of complex 1-dimensional slices. This note will pursue such an exploration by studying maps belonging to the complex curve $\mathcal{S}_p$ consisting of all cubic maps with a superattracting orbit of period p. Here p can be any positive integer.

A preliminary draft of this paper, based on conversations with Branner, Douady and Hubbard, was circulated in 1991 but not published. The present version tries to stay close to the original; however, there has been a great deal of progress in the intervening years. (See especially Faught [Fa92], Branner and Hubbard [BH92], Branner [Br93], Hubbard [H93], Roesch [Ro99, 06], Kaffl [Ka06] and Kiwi [Ki06].) In particular, a number of conjectures in the original have since been proved; and new ideas have made sharper statements possible.

We begin with the period 1 case. Section 2 studies the dynamics of a cubic polynomial map F which has a superattracting fixed point, and whose Julia set $J(F)$ is connected. The filled Julia set of any such map consists of a central Fatou component bounded by a Jordan curve, together with various *limbs* sprouting off at internal angles which are explicitly described. (See Figures 2, 3. This statement was conjectured in the original manuscript and then proved by Faught.) Section 3 studies the parameter space $\mathcal{S}_1$ consisting of all monic, centered cubic maps with a specified superattractive fixed point, and provides an analogous description of the non-hyperbolic locus in $\mathcal{S}_1$. (See Figure 5.) Section 4 makes a more detailed study of hyperbolic components in $\mathcal{S}_1$. Section 5 begins the study of the period p case, describing the geometry of the complex affine curve $\mathcal{S}_p$ consisting of maps with a marked critical point of period p. This is a non-compact complex 1-manifold; but can be made into a compact complex 1-manifold $\overline{\mathcal{S}}_p$ by adjoining finitely many *ideal points*. There is a conjectured cell subdivision of $\overline{\mathcal{S}}_p$ with a 2-cell centered at each ideal point, and with the union of all *simple closed regulated curves* as 1-skeleton. To each quadratic map $Q(z) = z^2 + c$ with period p critical orbit, there is associated a 2-cell $\mathbf{e}_Q$. Section 6 describes a conjectural canonical embedding of the filled Julia set $K(Q)$, cut open along its minimal Hubbard tree, into this 2-cell. However, there are many other 2-cells which cannot be described in this way. This paper concludes with an appendix which discusses *Hubbard trees*, following Poirier [Po93], and also describes the slightly modified *puffed-out* Hubbard trees.

1A   Basic Concepts and Notations

Any polynomial map $F : \mathbb{C} \to \mathbb{C}$ of degree $d \geq 2$ is affinely conjugate to one which is *monic* and *centered*, that is, of the form

$$F(z) = z^d + c_{d-2}z^{d-2} + \cdots + c_0 .$$

This normal form is unique up to conjugation by a $(d-1)$-st root of unity, which replaces $F(z)$ by $G(z) = \omega F(z/\omega)$ where $\omega^{d-1} = 1$, and replaces the Julia set $J(F)$ by the rotated Julia set $J(G) = \omega J(F)$.

The set $\mathcal{P}(d)$ of all such monic, centered maps forms a complex $(d-1)$-dimensional affine space. A polynomial $F \in \mathcal{P}(d)$ belongs to the *connectedness locus* $\mathcal{C}(\mathcal{P}(d))$ if its Julia set $J(F)$ is connected, or equivalently if the orbit of every critical point is bounded. This connectedness locus is always a compact cellular subset of $\mathcal{P}(d)$. This was proved in [BH88] for the cubic case, and in [La89] for higher degrees. (See also [Br86]. By definition, following [Brn60], a subset of some Euclidean space $\mathbb{R}^n$ is *cellular* if its complement in the sphere $\mathbb{R}^n \cup \infty$ is an open topological cell.)

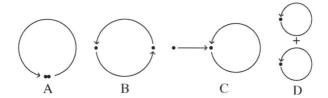

Figure 1. Schematic diagrams for the four classes of cubic hyperbolic components. Each dot represents a critical point (or the Fatou component containing it), and each arrow represents some iterate of F.

A polynomial map F is *hyperbolic* if the orbit of every critical point converges to an attracting cycle. (See for example [Mi06, Section 19].) The set $\mathcal{H}$ consisting of all hyperbolic maps in $\mathcal{C}(\mathcal{P}(d))$ is a disjoint union of open topological cells, each containing a unique post-critically finite map which will be called its *center*. (Compare [Mi92b].) Thus every critical orbit of such a center map is either periodic or eventually lands on a periodic critical orbit. One noteworthy special case is the *principal hyperbolic component* $\mathcal{H}_0 \subset \mathcal{C}(\mathcal{P}(d))$, centered at the map $z \mapsto z^d$, and consisting of all $F \in \mathcal{H} \subset \mathcal{P}(d)$ such that $J(F)$ is a Jordan curve. (For a study of $\mathcal{H}_0$ in the degree 3 case, see [PT04].)

Hyperbolic components in $\mathcal{C}(\mathcal{P}(3))$ fall into four distinct types as follows. (Compare [Mi92a].)

A. Adjacent Critical Points, with both critical points in the same periodic Fatou component.

B. Bitransitive, with the two critical points in different Fatou components belonging to the same periodic cycle.

C. "Capture", with just one critical point in the cycle of periodic Fatou components. The orbit of the other critical point must eventually land in (or be "captured by") this cycle.

D. Disjoint Attracting Orbits, with two distinct attracting periodic orbits, each of which necessarily attracts just one critical orbit.

For components of the first three types, the corresponding maps have just one attracting periodic orbit, and hence just one cycle of periodic Fatou components.

Remark 1.1 (Outside of the Connectedness Locus). There are many hyperbolic components in $\mathcal{P}(3)$ which belong to the complement of the connectedness locus. These will be called *escape components*, since they consist of hyperbolic maps for which at least one critical orbit "escapes to infinity", so that the Julia set is disconnected.

One such component, called the *shift locus*, has an extremely complicated topological structure. (Compare [BDK91].) It consists of maps for which the Julia set is isomorphic to a one-sided shift on three symbols. (More generally, a polynomial or rational map of degree d belongs to the *shift locus* if its Julia set is isomorphic to the one-sided shift on d symbols. A completely equivalent condition is that all of its critical points on the Riemann sphere belong to the immediate basin of a common attracting fixed point, which must be the point at infinity in the polynomial case.)

For maps in the remaining escape components in $\mathcal{P}(3) \setminus \mathcal{C}(\mathcal{P}(3))$, there is only one critical point in the basin of infinity, while the other critical point belongs to the immediate basin of a bounded attracting periodic orbit. We will give a rough classification of these components in Section 5.

Here is a rough picture of the complement $\mathcal{P}(3) \setminus \mathcal{C}(\mathcal{P}(3))$. (See [Br93].) Take a large sphere centered at the origin in the space $\mathcal{P}(3) \cong \mathbb{C}^2$. Then each escape hyperbolic component with an attracting orbit intersects this 3-sphere in an embedded solid torus, which forms one interior component of a "*Mandelbrot torus*", that is, a product of the form (Mandelbrot set)$\times$(circle). There are countably many such Mandelbrot-tori, and also many connected components without interior, for example solenoids or circles, corresponding to polynomials whose Julia set is a Cantor set which contains one critical point. If we remove all of these Mandelbrot-tori, solenoids, etc., from the 3-sphere, then what is left are points of the shift locus. (Compare Remark 4.1 and Figure 15.)

Remark 1.2 (Quadratic Rational Maps). (Compare [Re90, Re92, Re95] and [Mi93].) In a suitable parameter space for quadratic rational maps, there are again four different types of hyperbolic components. One of these is the shift locus as described above. (In the terminology of Rees, this is of Type I.) The remaining three are precise analogues of Types B, C, D (or in Rees's terminology, Types II, III, IV).

Caution. I have used the term "capture component" for components of Type C, even for quadratic rational maps. However, extreme care is needed, since the term "*capture*" is often used with a completely different meaning. See for example [Wi88], [Re92], and [Lu95], where this word refers instead to a procedure for modifying the dynamics of a quadratic polynomial to yield a quadratic rational map.

Definition 1.3 (The Moduli Space $\hat{\mathcal{P}}(3)/\mathcal{I}$). We are interested in cubic maps for which one of the two critical points has a periodic orbit. Hence it is convenient to work with the space $\hat{\mathcal{P}}(3)$ consisting of monic centered cubic maps together with a *marked critical point* a. Since there are two possible choices for the marked point, this space $\hat{\mathcal{P}}(3)$ is a 2-fold ramified covering of $\mathcal{P}(3)$. Each $F \in \hat{\mathcal{P}}(3)$ can be written in *Branner-Hubbard normal form* as

$$F(z) = z^3 - 3a^2 z + b, \qquad (1.1)$$

with critical points a and $-a$. Thus $\hat{\mathcal{P}}(3)$ could be identified with the complex coordinate space $\mathbb{C}^2$, using a, b as coordinates. However, for the purpose of this paper, it will be more convenient to use coordinates (a, v) where a is the marked critical point and

$$v = F(a) = b - 2a^3$$

is the corresponding critical value. We will write

$$F(z) = F_{a,v}(z) = z^3 - 3a^2 z + (2a^3 + v), \qquad (1.2)$$

and will use the notations $\hat{\mathcal{H}}_0 \subset \mathcal{C}(\hat{\mathcal{P}}(3)) \subset \hat{\mathcal{P}}(3)$ for the corresponding principal hyperbolic component and connectedness locus in the complex (a, v)-plane.

It is not hard to check that two distinct maps $F_{a,v}$ and $F_{a',v'}$ in $\hat{\mathcal{P}}(3)$ are affinely conjugate, in a conjugacy which carries the marked critical point a to the marked critical point a', if and only if $a' = -a$ and $v' = -v$, with conjugacy $z \mapsto -z$. Thus we define the *canonical involution* $\mathcal{I}$ of $\hat{\mathcal{P}}(3)$ to be the correspondence $F(z) \mapsto -F(-z)$, taking $F_{a,v}$ to $F_{-a,-v}$ and rotating the associated Julia set by $180°$ degrees. The quotient $\hat{\mathcal{P}}(3)/\mathcal{I}$ can be described as the *moduli space*, consisting of all affine conjugacy classes of cubic polynomials with marked critical point. (Thus I distinguish between a "parameter space", whose elements are actual maps, and a "moduli space" made up of conjugacy classes of maps.) A complete set of conjugacy class invariants for a polynomial F with marked critical point is provided by the numbers a^2 and v^2, together with a choice of square root $av = \pm\sqrt{a^2 v^2}$ which is needed to specify the choice of marking. Thus $\hat{\mathcal{P}}(3)/\mathcal{I}$ can be identified with an algebraic variety in $\mathbb{C}^3$, with coordinates a^2, v^2, av satisfying a homogeneous quadratic equation. This variety has a mild singularity at the origin.

One special feature of cubic maps is that for each critical point there is a uniquely defined *co-critical point* which has the same image under F. Using the normal form of Equations (1.1) or (1.2), the marked point a has co-critical point $-2a$, while $-a$ has co-critical point $+2a$. (Even if we don't use this normal form, the critical points, co-critical points, and their center of gravity will still lie in arithmetic progression along a straight line in the z-plane.)

2   Maps with Critical Fixed Point: The Julia Set

Before trying to understand configurations in parameter space, it is important to study the z-plane. We first consider the case of a superattracting orbit of period one. That is, using the normal form $F(z) = z^3 - 3a^2 z + 2a^3 + v$, we consider maps satisfying $F(a) = a$, or in other words[1]

$$v = a, \qquad F(z) = F_{a,a}(z) = z^3 - 3a^2 z + (2a^3 + a). \qquad (2.3)$$

The locus of all such maps is denoted by $\mathcal{S}_1$. Let U_a be the *immediate attracting basin* of the superattracting point a under this map F. Thus U_a is a simply connected bounded open neighborhood of a.

Let $\mathcal{C}(\mathcal{S}_1) = \mathcal{S}_1 \cap \mathcal{C}(\hat{\mathcal{P}}(3))$ be the connectedness locus within $\mathcal{S}_1$. If $F \in \mathcal{C}(\mathcal{S}_1)$, then the filled Julia set $K(F)$ (the complement of the attracting basin of infinity) is a compact connected subset of the z-plane. We divide the discussion into two cases, according as the *free critical point* $-a$ does or does not belong to the immediate basin $U_a \subset K(F)$.

Case 1 (Hyperbolic of Type A). Suppose that $F \in \mathcal{S}_1 \cap \hat{\mathcal{H}}_0$. In other words, suppose that F belongs to the unique hyperbolic component of Type A within $\mathcal{S}_1$, so that the other critical point $-a$ also belongs to the immediate basin U_a of the superattracting point a. In this case, the dynamics is quite well understood. The Julia set $J(F) = \partial K(F)$ is a Jordan curve. The bounded component of its complement is the attractive basin U_a, and the unbounded component is the attractive basin of infinity. Furthermore, the map F restricted to U_a is conformally conjugate to a Blaschke product of the form $\quad \Psi(w) = e^{2\pi i t} w^2 (r - w)/(1 - rw)$, with $t \in \mathbb{R}/\mathbb{Z}$ and $0 \leq r < 1$; and this conformal conjugacy extends homeomorphically over the closure $\overline{U}_a$. The map F is uniquely determined, up to affine conjugacy, by the parameter $e^{2\pi i t} r$, which varies over the open unit disk $\mathbb{D}$; however, F does not depend holomorphically on this parameter. (A holomorphic parametrization will be described in Lemma 3.6.)

Case 2 (Everything Else). For the rest of this section we will concentrate on the more difficult case where $\quad F \in \mathcal{C}(\mathcal{S}_1) \smallsetminus \hat{\mathcal{H}}_0$. *In other words, we will assume that both critical points have bounded orbits, and that the free critical point $-a$ lies outside the immediate basin U_a of the superattracting critical point a.* Then there is a unique *Böttcher isomorphism* from the basin U_a onto the open unit disk which conjugates F to the squaring map $w \mapsto w^2$, that is

$$\beta : U_a \xrightarrow{\ \cong\ } \mathbb{D} \qquad \text{with} \qquad \beta(F(z)) = \beta(z)^2. \qquad (2.4)$$

---

[1] An extra motive for studying this particular family of maps is the close relationship between this family of cubic maps and the family of rational maps which arise from cubic polynomial equations via Newton's method. See [Ta97].

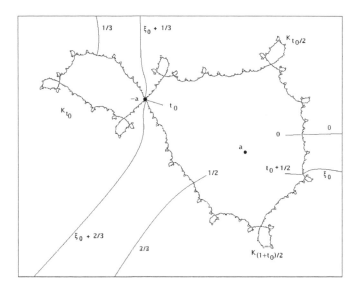

Figure 2. Julia set for an $F \in \mathcal{S}_1$ on the boundary of $\hat{\mathcal{H}}_0$ with non-periodic internal angle. Note that there is no limb at the critical value. The two rays which land at the critical point map to a single critical value ray. (In this particular example, the critical internal angle is $t_0 = .34326\cdots$, and the external angle at the co-critical point is $\xi_0 = .95884\cdots$.)

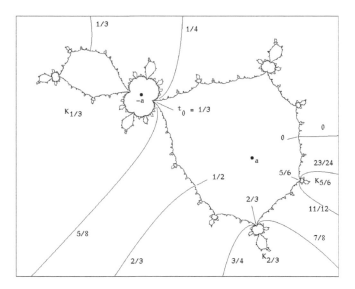

Figure 3. Julia set for a map $F \in \mathcal{S}_1$ on the boundary of $\hat{\mathcal{H}}_0$ with periodic internal angle $t_0 = 1/3$.

(See for example [Bl84] or [Mi06].) According to [Fa92] (see also [Ro99, 06], the boundary ∂U_a is locally connected, and in fact is a simple closed curve. By a well-known theorem of Carathéodory, this implies that the Böttcher map extends uniquely to a homeomorphism

$$\beta : \overline{U}_a \to \overline{\mathbb{D}}$$

from the closure of U_a to the closed unit disk. (See for example [Mi06, Section 17.16].) In particular, each point z of the boundary ∂U_a can be uniquely labeled by its *internal angle* $t \in \mathbb{R}/\mathbb{Z}$, where $z = \beta(e^{2\pi i t})$. (Angles are always measured in fractions of a full turn.)

Making use of Faught's result, we will prove the following.

Theorem 2.1. *If* $F \in \mathcal{C}(\mathcal{S}_1)$ *with* $-a \notin U_a$, *then the filled Julia set* $K(F)$ *is equal to the union of the topological disk* $\overline{U}_a$ *with a collection of compact connected sets* K_t, *where* t *ranges over a countable subset* Λ *of the circle* $\mathbb{R}/\mathbb{Z}$. *Furthermore*

(i) *The* K_t *are pairwise disjoint, and each* K_t *intersects* $\overline{U}_a$ *in the single boundary point* $\beta(e^{2\pi i t})$.

(ii) *There is a preferred element* $t_0 \in \Lambda$ *such that the free critical point* $-a$ *belongs to* K_{t_0}.

(iii) *The angle* t *belongs to this index set* $\Lambda \subset \mathbb{R}/\mathbb{Z}$ *if and only if* $2^n t \equiv t_0 \pmod{\mathbb{Z}}$ *for some integer* $n \geq 0$.

(iv) *For* $t \not\equiv t_0 \pmod{\mathbb{Z}}$, *the map* F *carries* K_t *homeomorphically onto* K_{2t}. *However,* F *carries* K_{t_0} *onto the entire filled Julia set* $K(F)$.

By definition, K_t is the *limb* which is attached to $\overline{U}_a$ at the point $\beta(e^{2\pi i t})$ with internal angle t, and K_{t_0} is the *critical limb*.

In fact, there are two rather different cases. In the simplest case (Figure 2), the critical angle t_0 is not periodic under angle doubling. In other words, t_0 is either irrational, or rational with even denominator. The critical limb K_{t_0} then maps homeomorphically onto the entire filled Julia set, and the free critical point $-a$ is precisely equal to the boundary point $\beta(e^{2\pi i t_0}) \in \partial U_a$. Furthermore, the map F itself, considered as a point in parameter space, belongs to the boundary $\partial \hat{\mathcal{H}}_0$ of the principal hyperbolic component.

On the other hand, if the critical angle t_0 is periodic under angle doubling, then $-a$ lies strictly outside of $\overline{U}_a$. In this case, the critical limb K_{t_0} is the union of an "inner" part which maps onto the critical value limb K_{2t_0} by a 2-fold branched covering, and an "outer" part which maps homeomorphically onto $K(F) \setminus K_{2t_0}$. The map F may belong to the boundary $\partial \hat{\mathcal{H}}_0$ (compare Figure 3), or it may lie completely outside the closure of $\hat{\mathcal{H}}_0$.

The proof of Theorem 2.1 will be based on a comparison between internal angles, measured at the critical fixed point a, and external angles, measured at infinity. Note that internal angles multiply by two under the map F, while external angles multiply by three. Equality between angles will be denoted by the symbol $\equiv$ with $(\bmod \, \mathbb{Z})$ understood.

Definition 2.2. Angles $t_1, \ldots, t_k$ are in *positive cyclic order* if it is possible to choose representatives $\hat{t}_j \in \mathbb{R}$ so that $\hat{t}_1 < \cdots < \hat{t}_k < \hat{t}_1 + 1$. For any $t_1 \not\equiv t_2$ in $\mathbb{R}/\mathbb{Z}$, the *open interval* (t_1, t_2) will mean the set of all angles $t \in \mathbb{R}/\mathbb{Z}$ for which t_1, t, t_2 are in positive cyclic order. The corresponding *closed interval* $[t_1, t_2]$ is defined to be the closure of (t_1, t_2). Note that these intervals have *length* equal to $\mathbf{frac}(t_2 - t_1)$, where $\mathbf{frac} : \mathbb{R}/\mathbb{Z} \to [0, 1)$ maps each point of the circle $\mathbb{R}/\mathbb{Z}$ to its unique representative in the half-open interval.

Basic Construction. For each rational angle $\tau \in \mathbb{Q}/\mathbb{Z}$, the internal ray of angle τ lands at a point

$$\beta(e^{2\pi i \tau}) \in \partial U_a \subset J(F),$$

which is periodic or preperiodic under F. It follows that $\beta(e^{2\pi i \tau})$ is also the landing point of at least one external ray $\mathcal{R}_\xi \subset \mathbb{C} \setminus K(F)$ which is periodic or preperiodic. (See for example [Mi06, Sections 18.11 and 18.12].) There can be at most finitely many such rays, so we can make an explicit choice $\xi = \xi(\tau)$ by choosing the largest one in cyclic order, measured from the internal ray which lands at this same point. The identity

$$\xi(2\tau) \equiv 3\xi(\tau) \tag{2.5}$$

then follows easily.

Definition 2.3. Let $G : \mathbb{C} \to [0, \infty)$ be the Green's function (= canonical potential function) which vanishes precisely on $K(F)$. Given two rational internal angles $\tau_0 \not\equiv \tau_1$ in $\mathbb{Q}/\mathbb{Z}$, and given some equipotential curve $G = G_0 > 0$, define the *quadrilateral* $\mathcal{Q} = \mathcal{Q}(\tau_0, \tau_1, G_0)$ to be the compact simply connected region in $\mathbb{C} \setminus U_a$ which is bounded by three edges in the Fatou set and one edge in the Julia set, as follows. (Figure 4.)

(a) The segments of the external rays $\mathcal{R}_{\xi(\tau_0)}$ and $\mathcal{R}_{\xi(\tau_1)}$ defined by the potential inequality $G \leq G_0$.

(b) The segment of the equipotential curve $G = G_0$ which lies between these two external rays so that the external angle lies in the closed interval $[\xi(\tau_0), \xi(\tau_1)]$.

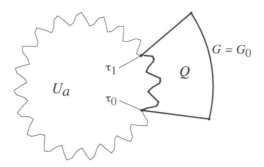

Figure 4.   Sketch of the quadrilateral $\mathcal{Q} = \mathcal{Q}(\tau_0, \tau_1, G_0)$.

(c) The segment of the boundary ∂U_a consisting of all $\beta(e^{2\pi i t})$ with $t \in [\tau_0, \tau_1]$.

Note that this quadrilateral $\mathcal{Q}$ contains all limbs which are attached to $\overline{U}_a$ at internal angles strictly between τ_0 and τ_1 in cyclic order.

Lemma 2.4. *Now suppose that the length* $\mathbf{frac}(\tau_1 - \tau_0)$ *of the interval* (τ_0, τ_1) *of internal angles is less than* $1/2$, *and also that the length of the corresponding interval* $(\xi(\tau_0), \xi(\tau_1))$ *of external angles satisfies*

$$\mathbf{frac}\Big(\xi(\tau_1) - \xi(\tau_0)\Big) \; < \; 1/3\,. \tag{2.6}$$

Then the quadrilateral $\mathcal{Q} = \mathcal{Q}(\tau_0, \tau_1, G_0)$ *contains no critical point, and maps biholomorphically onto the quadrilateral* $\mathcal{Q}' = \mathcal{Q}(2\,\tau_0, 2\,\tau_1, 3\,G_0)$. *But if*

$$\mathbf{frac}\Big(\xi(\tau_1) - \xi(\tau_0)\Big) \; > \; 1/3\,, \tag{2.7}$$

with $\mathbf{frac}(\tau_1 - \tau_0) < 1/2$ *as above, then* $\mathcal{Q}$ *contains the free critical point* $-a$, *and maps onto the entire region* $\{G \le 3\,G_0\}$. *In this case, points in* $\mathcal{Q}'$ *have two preimages in* $\mathcal{Q}$, *counted with multiplicity, while the remaining points in the region* $\{G \le 3\,G_0\}$ *have only one preimage.*

Proof: Let z_0 be any point of $\mathbb{C}$ which does not belong to the image $F(\partial \mathcal{Q})$ of the boundary of $\mathcal{Q}$. By the Argument Principle, the number of solutions to the equation $F(z) = z_0$ with $z \in \mathcal{Q}$, counted with multiplicity, is equal to the winding number of $F(\partial \mathcal{Q})$ around z_0. If we are in the case of Equation (2.6), then it is not hard to check that F maps this boundary homeomorphically onto the boundary $\partial \mathcal{Q}'$. Hence this winding number is $+1$ for z_0 in the interior of $\mathcal{Q}'$, and zero for z_0 outside. For the case of Equation (2.7), the argument is similar, but now the image $F(\partial \mathcal{Q})$

consists of a circuit around $\partial \mathcal{Q}'$, together with a circuit around the entire equipotential $G = 3\,G_0$.

To check for the presence of critical points, we use a form of the Riemann-Hurwitz formula. Choose a cell subdivision of $F(\mathcal{Q})$, with $\mathcal{Q}'$ as subcomplex (if it is not the entire image), and with $F(-a)$ as vertex if $-a \in \mathcal{Q}$. Then each cell in $F(\mathcal{Q})$ lifts up to either a single cell in $\mathcal{Q}$ if it lies outside of $\mathcal{Q}'$, or if it is equal to the vertex $F(-a)$, or to two cells in $\mathcal{Q}$. Computing the Euler characteristic,

$$\chi = \sum_{n=0}^{2} (-1)^n (\text{number of } n \text{ cells}) = +1,$$

for both $\mathcal{Q}$ and $F(\mathcal{Q})$, we find easily that $-a \in \mathcal{Q}$ if and only if we are in the second case (2.7). (Furthermore, the critical value $F(-a)$ then belongs to $\mathcal{Q}'$. A similar argument shows that the third case $\xi(\tau_1) \in (\xi(\tau_0) + \frac{2}{3}, \xi(\tau_0) + 1)$ cannot occur.)   □

Proof of Theorem 2.1: For each internal angle $t \in \mathbb{R}/\mathbb{Z}$, define K_t to be the intersection of all quadrilaterals $\mathcal{Q}(\tau_0, \tau_1, G_0)$ for which $t \in (\tau_0, \tau_1)$ and $G_0 > 0$. Then K_t can be described as the intersection of a nested sequence of compact, connected, non-vacuous sets, and hence is itself compact, connected, and non-vacuous.

For every t, it is easy to check that the intersection $K_t \cap \overline{U}_a$ consists of the single point $\beta(e^{2\pi i t})$. For countably many choices of t, we will see that K_t is much larger than this single intersection point. The following statement follows immediately from Lemma 2.4.

> *For each $t \not\equiv t_0$, the map F carries K_t homeomorphically onto K_{2t}. However, F carries K_{t_0} onto the entire filled Julia set $K(F)$.*

In particular, if $2^n t \equiv t_0$, then it follows that $F^{\circ(n+1)}$ maps K_t onto $K(F)$, so that K_t contains infinitely many points. To complete the proof of Theorem 2.1, we need only prove a converse statement:

> *If $2^n t \not\equiv t_0$ for all $n \neq 0$, then K_t consists of the single point $\beta(e^{2\pi i t})$.*

Define the *angular width* $\Delta\xi(\mathcal{Q})$ of a quadrilateral $\mathcal{Q} = \mathcal{Q}(\tau_0, \tau_1, G_0)$, to be the length of the interval $(\xi(\tau_0),\ \xi(\tau_1)) \subset \mathbb{R}/\mathbb{Z}$; and define the angular width $\Delta\xi(K_t)$ of the set K_t to be the infimum of $\Delta\xi(\mathcal{Q})$ over all quadrilaterals $\mathcal{Q}$ which contain K_t. In other words, this angular width

$$0 \leq \Delta\xi(K_t) < 1$$

is the infimum, over all open intervals (τ_0, τ_1) which contain t, of the length of the interval $(\xi(\tau_0), \xi(\tau_1))$. (Intuitively, this is just the infimum of $\xi(\tau_1) - \xi(\tau_0)$.)

Lemma 2.5. *This angular width satisfies*

$$\Delta\xi(K_{2t}) \;=\; 3\Delta\xi(K_t) \qquad for \qquad t \not\equiv t_0,$$

but

$$\Delta\xi(K_{2t_0}) \;=\; 3\,\Delta\xi(K_{t_0}) - 1.$$

Proof: The congruence $\Delta\xi(K_{2t_0}) \equiv 3\,\Delta\xi(K_{t_0}) \pmod{\mathbb{Z}}$ follows immediately from Equation (2.5), and the more precise statement then follows from Lemma 2.4. ☐

To complete the proof of Theorem 2.1, consider any angle t such that K_t contains more than one point. Since the boundary of the connected set K_t is contained in the Julia set, and since repelling periodic points are dense in the Julia set, it follows that K_t contains many repelling periodic points. Each of these must be the landing point of an external ray, and it follows easily that $\Delta\xi(K_t) > 0$.

But if this were true with $2^n t \not\equiv t_0$ for all $n \geq 0$, then it would follow inductively from Lemma 2.5 that

$$\Delta\xi(K_{2^n t}) \;=\; 3^n \Delta\xi(K_t) \;\to\; \infty \qquad as \qquad n \to \infty,$$

which is clearly impossible. This shows that K_t contains more than one point if and only if some forward image is equal to the critical limb K_{t_0}, which proves Theorem 2.1. ☐

Next we will show that each limb is separated from the rest of $K(F)$ by two external rays. Recall that $\Lambda \subset \mathbb{R}/\mathbb{Z}$ is the countable set consisting of all t such that $2^n t \equiv t_0$ for some $n \geq 0$. For each internal angle t, consider the closed intervals $[\xi(\tau_0), \xi(\tau_1)]$ of external angles which are associated with quadrilaterals $\mathcal{Q}(\tau_0, \tau_1, G_0)$ such that $t \in (\tau_0, \tau_1)$. If $t \in \Lambda$, then evidently these closed intervals intersect in an interval, to be called $[\xi^-(t), \xi^+(t)]$, with length equal to the angular width $\Delta\xi(K_t) > 0$. On the other hand, if $t \notin \Lambda$, so that $\Delta\xi(K_t) = 0$, then a similar argument show that the intervals $[\xi(\tau_0), \xi(\tau_1)]$ intersect in a single point $\xi(t)$.

Theorem 2.6. *For each $t \in \Lambda$, the two external rays $\mathcal{R}_{\xi^\pm(t)}$ both land at the point of attachment $\beta(e^{2\pi i t})$ for the limb K_t, and these rays together with their landing point, separate K_t from the rest of $K(F)$. For $t \notin \Lambda$, the external ray $\mathcal{R}_{\xi(t)}$ lands at $\beta(e^{2\pi i t})$, and no other ray accumulates at this point.*

Proof: First suppose that $t \notin \Lambda$. Then it follows from the proof of Theorem 2.1 that K_t consists of the single point $\beta(e^{2\pi it})$, and that $\Delta\xi(K_t) = 0$. This means that we can find quadrilaterals $\mathcal{Q}(\tau_0, \tau_1, G_0)$ such that the open interval (τ_0, τ_1) is arbitrarily small, and contains t, and such that the interval $[\xi(\tau_0), \xi(\tau_1)]$ of exterior angles is also arbitrarily small. Taking the intersection of $[\xi(\tau_0), \xi(\tau_1)]$ over all such quadrilaterals, we clearly obtain a single exterior angle $\xi(t)$. Let $X_t \subset J(F)$ be the set of all accumulation points for the corresponding external ray $\mathcal{R}_{\xi(t)}$. For every $t' \not\equiv t$, the set X_t is separated from $K_{t'}$ by some rational external ray. Hence X_t must consist of the singleton K_t, as required. Similarly, for any $\xi' \not\equiv \xi(t)$, the ray $\mathcal{R}_{\xi'}$ is separated from K_t by some rational external ray, and hence cannot accumulate on K_t.

Now suppose that $t \in \Lambda$. If t_0 is not periodic under angle doubling, then $2t_0 \notin \Lambda$, so that there is a single ray $\mathcal{R}_{\xi(2t_0)}$ landing at the critical value $F(-a)$. Since F carries a neighborhood of $-a$ to a neighborhood of $F(-a)$ by a 2-fold branched covering, it follows that exactly two rays land at $-a = \beta(e^{2\pi it_0})$.

On the other hand, if t_0 is periodic, then the point of attachment $\beta(e^{2\pi it_0})$ is a periodic point of rotation number zero in the Julia set, and there are exactly two ways of accessing $\beta(e^{2\pi it_0})$ from $\mathbb{C} \smallsetminus K(F)$, or in other words, exactly two prime ends of $\mathbb{C} \smallsetminus K(F)$ which map to $\beta(e^{2\pi it_0})$. Hence, again there must be exactly two external rays which land on $\beta(e^{2\pi it_0})$. (See for example [Mi06, Sections 17, 18].) In either case, these two rays, together with their common landing point, must separate at least one limb from U_a. Since no other limb can have this property, it follows that these two rays must separate K_{t_0} from the rest of $K(F)$. It is then easy to check that the two rays must be precisely $\mathcal{R}_{\xi^{\pm}(t_0)}$. The corresponding statement for an arbitrary limb K_t then follows, since $F^{\circ n} : K_t \xrightarrow{\cong} K_{t_0}$ for some $n \geq 0$. $\square$

Remark 2.7. It follows easily that there is a canonical retraction from $\mathbb{C} \smallsetminus \{a\}$ to the circle ∂U_a which carries each limb to its point of attachment, and which takes a constant value on each internal or external ray. In particular, there is a canonical map $T : \mathbb{R}/\mathbb{Z} \to \mathbb{R}/\mathbb{Z}$ from external angles to internal angles with the following two properties:

- For any limb K_t this map $\xi \mapsto T(\xi)$ takes the constant value $T(\xi) = t$ for ξ in the interval $[\xi^-(t), \xi^+(t)]$ of length $\Delta\xi(K_t)$.

- Furthermore, T is monotone of degree one, in the sense that it lifts to a monotone map $\hat{\xi} \mapsto \hat{T}(\hat{\xi})$ from $\mathbb{R}$ to $\mathbb{R}$, with $\hat{T}(\hat{\xi}+1) = \hat{T}(\hat{\xi})+1$.

It is not difficult to compute the lengths $\Delta\xi(K_t)$ of these intervals of constancy.

Lemma 2.8. *If $2^n t \equiv t_0$ with $n \geq 0$ minimal, then*

$$\Delta\xi(K_t) = \Delta\xi(K_{t_0})/3^n, \quad where \tag{2.8}$$

$$\Delta\xi(K_{t_0}) = \begin{cases} 1/3 & if \ t_0 \ is \ not \ periodic \ under \ angle \ doubling, \\ 3^{p-1}/(3^p - 1) & if \ 2^p t_0 \equiv t_0 \ (\mathrm{mod} \ \mathbb{Z}) \ with \ p \geq 1 \ minimal. \end{cases}$$

In both cases, the sum of $\Delta\xi(K_t)$ over all $t \in \mathbb{R}/\mathbb{Z}$ is precisely equal to $+1$. In other words, almost every external angle ξ belongs to such an interval of constancy.

Thus the set of ξ such that the external ray $\mathcal{R}_\xi$ lands on the boundary ∂U_a has measure zero.

Proof of Lemma 2.8: Equation (2.8) follows immediately from Lemma 2.5. Suppose first that t_0 is not periodic under angle doubling (Figure 2). Then for each $n \geq 0$ there are exactly 2^n distinct solutions t to the congruence $2^n t \equiv t_0$, and $\Delta\xi(K_t) = \Delta\xi(K_{t_0})/3^n$ for each one of these solutions. Summing over all n and all solutions, we get

$$\sum_t \Delta\xi(K_t) = \Delta\xi(K_{t_0}) \sum_{n \geq 0} 2^n/3^n = 3\,\Delta\xi(K_{t_0}). \tag{2.9}$$

We have $\Delta\xi(K_{t_0}) \geq 1/3$ by Lemma 2.5; but the sum (2.9) must be ≤ 1 since the sum of the lengths of subintervals of $\mathbb{R}/\mathbb{Z}$ cannot be greater than one. Thus $\Delta\xi(K_{t_0})$ is exactly $1/3$, and the sum is exactly one.

Now suppose that $2^p t_0 \equiv t_0$ with $p \geq 0$ minimal (Figure 3). Then by Lemma 2.5,

$$\Delta\xi(K_{2t_0}) = 3\Delta\xi(K_{t_0}) - 1,$$

hence

$$\Delta\xi(K_{2^n t_0}) = 3^{n-1}(3\Delta\xi(K_{t_0}) - 1) \ \text{for} \ 1 \leq n \leq p.$$

In particular,

$$\Delta\xi(K_{t_0}) = \Delta\xi(K_{2^p t_0}) = 3^{p-1}(3\Delta\xi(K_{t_0}) - 1),$$

hence we can solve for the required expression

$$\Delta\xi(K_{t_0}) - \frac{3^{p-1}}{3^p - 1}.$$

It then follows by Lemma 2.5 that

$$\Delta\xi(K_{2^n t_0}) = \frac{3^{n-1}}{3^p - 1} \quad \text{for} \quad 1 \leq n \leq p. \tag{2.10}$$

(Curiously enough, the sum of these angular widths (2.10) over all angles $2^n t_0$ in the periodic orbit is always precisely $1/2$.) For each t with

$\Delta\xi(K_t) > 0$, let $m \geq 0$ be the smallest integer such that $2^m t \equiv 2^n t_0$ for some angle $2^n t_0$ in the periodic orbit. Then $\Delta\xi(K_t) = 3^{n-m-1}/(3^p - 1)$. Summing over all such t, we see that

$$\sum_t \Delta\xi(K_t) = \sum_{n=1}^{p} \frac{3^{n-1}}{3^p - 1} \left(1 + 1/3 + 2/9 + 4/27 + \cdots\right) = 1,$$

as required. This proves Lemma 2.8.   □

The precise relationship between the internal argument t and the external argument or arguments ξ at a point of ∂U_a can be described more explicitly as follows. According to Remark 2.7, the correspondence $\xi \mapsto t = T(\xi)$ is a well-defined, continuous, and monotone map of degree one from the circle $\mathbb{R}/\mathbb{Z}$ to itself. However, it turns out to be easier to describe the inverse function $t \mapsto \xi = T^{-1}(t)$, which is monotone, but has a jump discontinuity at t for every limb K_t. Recall that the mapping F doubles internal arguments and triples external arguments. Hence it is often convenient to describe t by its base 2 expansion, but to describe ξ by its base 3 expansion, which we write as $\xi = .x_1 x_2 x_3 \cdots$ (base 3) $= \sum x_i/3^i$ with $x_i \in \{0, 1, 2\}$.

Suppose, to fix our ideas, that the internal argument t_0 of the principal limb satisfies $0 < t_0 < 1/2$. Let us start with the unique fixed point on the circle ∂U_a, with internal argument zero. Since U_a is mapped onto itself by F, the corresponding external argument must be either zero or $1/2$. Applying the involution $\mathcal{I} : (a, v) \mapsto (-a, -v)$ if necessary, we may assume that this point has external argument zero. (See Section 3.)

Lemma 2.9. *With these hypotheses, the correspondence*

$$t \mapsto \xi = \xi_{t_0}(t) = T^{-1}(t) = .x_1 x_2 x_3 \cdots$$

is obtained by setting x_m equal to either $0, 1$, or 2 according as $2^m t$ belongs to the interval $[0, t_0]$, $[t_0, 1/2]$, or $[1/2, 1]$ modulo one.

Thus there is a jump discontinuity whenever $2^m t$ lies exactly at the boundary between two of these intervals. When $1/2 < t_0 < 1$, the statement is similar, except that we use the intervals

$$[0, 1/2], \quad [1/2, t_0], \quad \text{and} \quad [t_0, 1].$$

In the case where the fixed point on ∂U_a has external argument $1/2$, we must add $1/2$ to the value of ξ described above.

Proof of Lemma 2.9: Consider the three pre-images of the fixed point which has internal and external arguments zero. One is the point itself, one must lie in the principal limb, by Theorem 2.1, and the third must be the unique point on ∂U_a which has internal argument $1/2$. The corresponding external arguments must be 0, $1/3$ and $2/3$, respectively. Given a completely arbitrary internal argument t, we can now compute the corresponding external argument $\xi = T^{-1}(t)$, simply by following its orbit under F. $\square$

Remark 2.10 (The Non-Periodic Case (Figure 2)). In the case of a critical angle t_0 which is not periodic under doubling, the map F is uniquely determined by t_0 (up to the involution $\mathcal{I}$), and we can give a much more precise description of $K(F)$. If $2^n t \equiv t_0$, then the map $F^{\circ(n+1)}$ carries the limb K_t homeomorphically onto $K(F)$. Let

$$ f_t \; : \; K(F) \; \xrightarrow{\;\cong\;} \; K_t $$

be the inverse homeomorphism. Then f_t carries each limb $K_{t'}$ onto a *secondary limb* $f_t(K_{t'}) \subset K_t$, to be denoted by $K_{tt'}$. More generally, for any finite sequence of limbs $K_{t_1}, K_{t_2}, \ldots, K_{t_m}$, we can form an *m-th order limb*

$$ f_{t_1} \circ f_{t_2} \circ \cdots \circ f_{t_m}(K(F)) \,, $$

which will be denoted briefly by

$$ K_{t_1 t_2 \cdots t_m} \; \subset \; K_{t_1 t_2 \cdots t_{m-1}} \; \subset \; \cdots \; \subset K_{t_1 t_2} \; \subset \; K_{t_1} \,. $$

Each of these higher order limbs contains an associated Fatou component

$$ f_{t_1} \circ f_{t_2} \circ \cdots \circ f_{t_m}(U_a) \,, $$

and every Fatou component within $K(F)$ is uniquely determined by such a list $t_1, t_2, \ldots t_m$ with $m \geq 0$. Note that

$$ F(K_{t_1 t_2 \cdots t_m}) \; = \; K_{2t_1 \, t_2 \cdots t_m} \quad \text{for} \quad t_1 \not\equiv t_0 $$

but

$$ F(K_{t_0 t_1 \cdots t_m}) \; = \; K_{t_1 \cdots t_m} \,, $$

with similar formulas for the associated Fatou components. (Here t_0 is the fixed critical angle, but $t_1, \ldots t_m$ can be the internal angles for arbitrary limbs.) It seems natural to conjecture that $K(F)$ is locally connected in this situation, and in particular that the diameter of the m-th order limb $K_{t_1 \cdots t_m}$ tends to zero as $m \to \infty$.

Remark 2.11 (The Periodic Case). For periodic t_0 the situation is much more complicated, since there may be Cremer points or other difficulties. However, if we consider only maps F which belong to the boundary $\partial \hat{\mathcal{H}}_0$

of the principal hyperbolic component, as in Figure 3, then the situation is well understood. In this case, the point of attachment $\beta(e^{2\pi i t_0})$ is parabolic, of period $p \geq 1$, with rotation number zero, and the Julia set is certainly locally connected. (Compare [TY96].) In fact, this parabolic F is the root point of a hyperbolic component which has Hubbard tree with an easily described topological model, consisting of the line segment between the two critical vertices 0 and $e^{2\pi i t_0}$, together with the images of this line segment under the map $z \mapsto z^2$. (See, for example, the top three examples in Figure 35, which represent "puffed-out" versions of three such trees.)

3   Parameter Space: The Curve $\mathcal{S}_1$

Consider the set of all cubics having a critical fixed point. Using the normal form (1.2), we define the *superattracting period one curve* $\mathcal{S}_1$ to be the one-parameter subspace of $\hat{\mathcal{P}}(3)$ consisting of all $F = F_{a,a} \in \hat{\mathcal{P}}(3)$ for which the critical value $v = F(a)$ is equal to a, so that the marked critical point a is a fixed point. (In Section 5 and Section 6, we will study the analogous curve $\mathcal{S}_p$, consisting of cubics with a marked critical point of period p.) Evidently the curve $\mathcal{S}_1 \subset \hat{\mathcal{P}}(3)$ is canonically biholomorphic to the complex a-plane. We will sometimes use the abbreviated notation F_a for a point in $\mathcal{S}_1$.

The boundary of the intersection $\mathcal{C}(\hat{\mathcal{P}}(3)) \cap \mathcal{S}_1$ (considered as a subset of the a-plane) is shown in Figure 5. Since there is only one free critical point in this family, much of the Douady-Hubbard theory concerning the parameter space for quadratic polynomials carries over with minor changes. However, there are new difficulties. [Fa92] proved locally connectivity, modulo local connectivity of the Mandelbrot set, and showed that all hyperbolic components in $\mathcal{S}_1$ are bounded by Jordan curves. See [Ro06] for a simplified proof, for a generalization of these results to higher degrees, and for a proof that the limbs which branch off from the principal hyperbolic component have diameters tending to zero.

Recall that the *canonical involution* $\mathcal{I}$ of $\hat{\mathcal{P}}(3)$ takes the pair (a, v) to $(-a, -v)$, preserving equation (2.3). It corresponds to the linear conjugation $F(z) \mapsto -F(-z)$; and clearly preserves the subsets $\hat{\mathcal{H}}_0 \subset \mathcal{C}(\hat{\mathcal{P}}(3))$ and $\mathcal{S}_1$. Geometrically, its effect is to rotate the Julia set of F by $180°$, and to add $1/2$ to all external angles. Note that the curve $\mathcal{S}_1$ has uniformizing parameter a, while the quotient curve $\mathcal{S}_1/\mathcal{I}$ has uniformizing parameter a^2. (Figures 5, 6.) We will often use the abbreviated notation $F = F_a$ to indicate the dependence of $F \in \mathcal{S}_1$ on the parameter a.

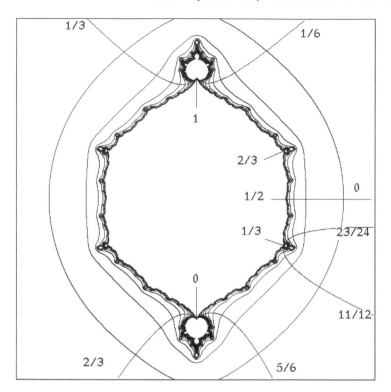

Figure 5. The non-hyperbolic locus in $\mathcal{S}_1$, projected into the a-plane. The connectedness locus $\mathcal{C}(\hat{\mathcal{P}}(3)) \cap \mathcal{S}_1$ consists of this non-hyperbolic locus together with the bounded components of its complement.

Remark 3.1. Alternatively, we could equally well work with the affinely conjugate normal form

$$z \mapsto F(z+a) - a = z^3 + 3az^2,$$

with superattracting fixed point at the origin. More generally, for any fixed constant μ, we can look at the complex curve $\mathrm{Per}(1; \mu)$ consisting of all cubic maps

$$z \mapsto z^3 + 3\alpha z^2 + \mu z,$$

having a fixed point of multiplier μ at the origin. The cases where $\mu \neq 1$ is a root of unity are of particular importance, since these curves contain regions which border on two different hyperbolic components within the ambient space $\hat{\mathcal{P}}(3)$. Note that the canonical involution, which maps the function $F(z)$ to $-F(-z)$, sends each $\mathrm{Per}(1; \mu)$ onto itself, changing the

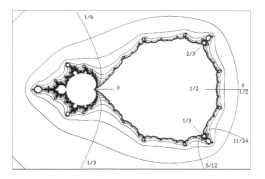

Figure 6. Non-hyperbolic locus in the quotient plane $\mathcal{S}_1/\mathcal{I}$, with parameter a^2.

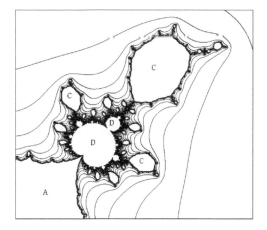

Figure 7. Detail of Figure 6 showing the $2/3$-limb. (For labels, see Section 1A.)

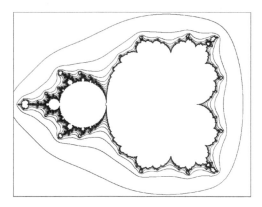

Figure 8. Configuration analogous to Figure 6 in the plane $\mathrm{Per}(1; 1)/\mathcal{I}$ of maps with a fixed point of multiplier $+1$.

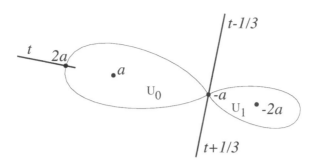

Figure 9. Sketch in the dynamic plane for a map with one escaping critical orbit. The equipotential and the external rays through the escaping critical point $-a$ and through its co-critical point $2a$ are shown, with the rays labeled by their angles.

sign of α. In general, the connectedness locus in $\mathrm{Per}(1; \mu)$ varies by an isotopy as μ varies within the open unit disk, but changes topology as μ tends to a limit on the unit circle. However, there is one noteworthy exception:

Conjecture 3.2. *The connectedness locus in the quotient* $\mathrm{Per}(1; \mu)/\mathcal{I}$ *tends to a limit without changing topology, as* $\mu \to 1$.

In fact the limiting configuration in $\mathrm{Per}(1; 1)/\mathcal{I}$, as shown in Figure 8, looks topologically very much like the corresponding configuration in Figure 6, although the geometrical shapes are different.

3A   Maps Outside of the Connectedness Locus

We begin the analysis of the curve $\mathcal{S}_1$ with the following analogue of a well-known result of [DH82].

Lemma 3.3. *The connectedness locus* $\mathcal{C}(\mathcal{S}_1) = \mathcal{S}_1 \cap \mathcal{C}(\hat{\mathcal{P}}(3))$ *in* $\mathcal{S}_1$ *is a cellular set. Furthermore, there is a canonical conformal diffeomorphism from the complement* $\mathcal{E} = \mathcal{S}_1 \setminus \mathcal{C}(\mathcal{S}_1)$ *onto* $\mathbb{C} \setminus \overline{\mathbb{D}}$.

By definition, $\mathcal{E}$ will be called the *escape region* in $\mathcal{S}_1$. (More generally, when discussing the curve $\mathcal{S}_p$ of maps with critical orbit of period $p > 1$, we will see that there are always two or more connected escape regions.)

Proof of Lemma 3.3: First consider some fixed polynomial $F = F_{a,a}$ in $\mathcal{S}_1$. Then $F : \mathbb{C} \to \mathbb{C}$ is conjugate, throughout some neighborhood of infinity, to the map $w \mapsto w^3$. In other words, there exists a *Böttcher diffeomorphism*

$z \mapsto \beta_F(z)$, defined and holomorphic throughout a neighborhood of infinity, satisfying the identity

$$\beta_F(F(z)) = \beta_F(z)^3 .$$

(See for example [Mi06].) In fact β_F is unique up to sign, and can be normalized by the requirement that $\beta_F(z) \sim z$ as $|z| \to \infty$. If we draw an equipotential through the free critical point $-a$, as illustrated in Figure 9, then β is well defined everywhere in the region outside this equipotential, and maps this exterior region diffeomorphically onto the complement of a suitable disk centered at the origin. It is not well defined at the critical point $-a$ itself, but does extend smoothly through a neighborhood of the co-critical point $2a$. Thus (following Branner and Hubbard) we can define the map $\widehat{\beta} : \mathcal{E} \to \mathbb{C} \backslash \overline{\mathbb{D}}$ by setting

$$\widehat{\beta}(F) = \beta_F(2a) \qquad \text{where} \qquad F = F_{a,a} .$$

(We may also use the alternate notation $\widehat{\beta}(a)$ for $\widehat{\beta}(F_{a,a})$.)

It is not hard to check that $\widehat{\beta}$ is holomorphic and locally bijective, and that $|\widehat{\beta}(F)|$ converges to $+1$ as F converges towards the connectedness locus. In order to show that it is a covering map, we must describe its behavior near infinity. Note that the orbit of $2a$ under F is given by

$$F : 2a \mapsto 4a^3 + a \mapsto \cdots \mapsto 4^{3^{k-1}} a^{3^k} + (\text{lower order terms}) \mapsto \cdots .$$

The asymptotic formula

$$\widehat{\beta}(F_{a,v}) \sim \sqrt[3]{4}\, a \qquad \text{as} \qquad |a| \to \infty \tag{3.11}$$

follows easily. Thus $\widehat{\beta} : \mathcal{E} \to \mathbb{C} \backslash \overline{\mathbb{D}}$ is proper and locally bijective. Since it has degree one near infinity, it follows that it is a conformal isomorphism, as required. $\qquad\square$

Using this description of the escape region $\mathcal{E} \subset \mathcal{S}_1$, we can talk about *external rays* $\mathcal{R}_\xi(\mathcal{E})$ within this escape region in parameter space. The reader should take care, since we will discuss external rays $\mathcal{R}'_\xi(F)$ for the Julia set $J(F)$ at the same time.

Definition 3.4. The map $F = F_{a,a}$ belongs to the ray $\mathcal{R}_t(\mathcal{S}_1)$ in parameter space if and only if the corresponding dynamic ray $\mathcal{R}'_t(F)$ passes through the co-critical point $2a$. It then follows that the ray $\mathcal{R}'_{3t}(F)$ passes through the critical value $F(2a) = F(-a)$, and hence that two rays $\mathcal{R}'_{t \pm 1/3}(F)$ must crash together at the free critical point $-a$ (Figure 9).

Note that the intersection $\mathcal{C}(\widehat{\mathcal{P}}(3)) \cap \mathcal{S}_1$ is not known to be locally connected,[2] so that we do not know that these external rays in parameter

---

[2] As noted at the beginning of this section, Faught showed that this set is locally connected if and only if the Mandelbrot set is locally connected.

space land at well-defined points of $\mathcal{C}(3)$. However, we can prove the following.

Lemma 3.5. *If the number $\xi \in \mathbb{Q}/\mathbb{Z}$ is periodic under tripling (or in other words if its denominator is relatively prime to 3), then the two external rays $\mathcal{R}_{\xi+1/3}(\mathcal{S}_1)$ and $\mathcal{R}_{\xi-1/3}(\mathcal{S}_1)$ both land at well-defined points of the connectedness locus. Furthermore, for either one of these two landing maps F, the Julia set $J(F)$ contains a parabolic periodic point, namely the landing point of the periodic external ray $\mathcal{R}'_\xi(F)$.*

Proof: (Compare [GM93, Appendices B, C].) Let $F \in \mathcal{C}(3)$ be any accumulation point of the ray $\mathcal{R}_{\xi\pm1/3}(\mathcal{S}_1)$. Then the ray $\mathcal{R}'_\xi(F)$ must land at a well-defined periodic point in $J(F)$, which a priori can be either repelling or parabolic. (See for example [Mi06].) If it were repelling, then for any nearby map $F_1 \in \mathcal{S}_1$ the corresponding ray $\mathcal{R}'_\xi(F_1)$ would land at a nearby periodic point. However, as noted above, for F_1 in the ray $\mathcal{R}_{\xi\pm1/3}(\mathcal{S}_1)$ this ray $\mathcal{R}'_\xi(F_1)$ must crash into the critical point $-a$, and hence cannot land. Thus F must have a parabolic cycle, with period dividing the period of ξ.

On the other hand, the set of all $F = F_a \in \mathcal{S}_1$ having a parabolic cycle of bounded period forms an algebraic variety. Since it is not the whole curve $\mathcal{S}_1$, it must be finite. But the collection of all accumulation points for the ray $\mathcal{R}_{\xi\pm1/3}(\mathcal{S}_1)$ must be connected, so this set of accumulation points can only be a single point.   □

3B   Maps in $\hat{\mathcal{H}}_0$

An argument quite similar to the proof of Lemma 3.3 applies to the principal hyperbolic component $\hat{\mathcal{H}}_0$, intersected with $\mathcal{S}_1$. In fact we will show that the quotient $(\hat{\mathcal{H}}_0 \cap \mathcal{S}_1)/\mathcal{I}$ is canonically biholomorphic to the unit disk.

We suppose that $F = F_a$ belongs to the principal hyperbolic component $\hat{\mathcal{H}}_0$, or in other words we suppose that the immediate basin U_a contains both critical points. If $a \neq 0$, then as in the discussion in Section 2 there is a unique Böttcher coordinate $w = \beta(z) = \beta_a(z)$ which maps some neighborhood of $z = a$ biholomorphically onto a neighborhood of $w = 0$, and which conjugates F to the squaring map $w \mapsto w^2$, so that, as in Equation (2.4)

$$\beta_a(F(z)) = \beta_a(z)^2 .$$

Since $F \in \hat{\mathcal{H}}_0$ we cannot extend this Böttcher coordinate throughout the basin U_a. For this basin will also contain the co-critical point $-2a$, which satisfies $F(-2a) = F(a) = a$. Evidently $\beta_a(z) = \pm\sqrt{\beta_a(F(z))}$ cannot be defined as a single valued function in a neighborhood of $-2a$. However an argument quite similar to the proof of Lemma 3.3 shows the following.

Lemma 3.6. *There is a canonical conformal isomorphism η from the quotient space $(\hat{\mathcal{H}}_0 \cap \mathcal{S}_1)/\mathcal{I}$ onto the open unit disk. More explicitly, if $F = F_a \in \hat{\mathcal{H}}_0 \cap \mathcal{S}_1$, then the Böttcher coordinate $z \mapsto w = \beta_a(z)$, which initially is defined only in a neighborhood of $z = a$, can be analytically continued to a neighborhood of the other critical point $z = -a$ in such a way that the resulting correspondence $a \mapsto \beta_a(-a) \in \mathbb{D}$ is well defined, holomorphic and even, as F_a varies through the region $\hat{\mathcal{H}}_0 \cap \mathcal{S}_1$. This correspondence induces the required conformal isomorphism $\eta : a^2 \mapsto \beta_a(-a)$.*

Thus the dynamical behavior of the critical point $-a$ under the map F_a is just like that of the point $w_a = \beta_a(-a)$ under the squaring map. Intuitively we can say that F_a is obtained from the squaring map by "enramifying" the point w_a.

Proof of Lemma 3.6: We continue to assume that $a \neq 0$. Note first that the absolute value $|\beta_a(z)|$ extends as a well-defined function of z throughout the basin U_{F_a}. This extended function will be smooth except at points which map precisely onto a under some iteration of F_a, and will have non-zero gradient except at points which map onto $-a$ under some iteration of F_a. For any $0 < r \leq 1$, let C_r be that component of the open set $\{z \in U_{F_a} : |\beta_a(z)| < r\}$ which contains the superattracting point a. Evidently there is a largest value of r so that β_a extends to a conformal diffeomorphism from C_r onto the open disk $\{w : |w| < r\}$. We claim that the boundary ∂C_r must contain the critical point $-a$. For if z is any non-critical boundary point of C_r, then using Equation (2.4) there exists a unique holomorphic extension of β_a to a neighborhood of z. Hence, if $|\beta_a(-a)| \neq r$ there would be no obstruction to a holomorphic extension to a larger neighborhood. In fact, we claim that β_a extends homeomorphically over the closure $\overline{C}_r$ (and holomorphically over a neighborhood of $\overline{C}_r$). Here we must rule out the possibility that ∂C_r consists of two loops, one inside the other, meeting at the point $-a$. But this configuration is easily excluded by the maximum modulus principle.

In this way, we see that the map $\beta = \beta_a$ takes a well-defined value at the critical point $-a$. Thus we obtain a well-defined point

$$a \mapsto w_a = \beta_a(-a) \in \mathbb{D}$$

whose dynamical properties under the squaring map are the same as those of $-a$ under F_a. Evidently this image point w_a will not be changed if we apply the involution $\mathcal{I}$. Hence it can be considered as a holomorphic function $w_a = \eta(a^2)$. Here a ranges over all non-zero parameters for which the associated map F_a belongs to $\mathcal{S}_1 \cap \hat{\mathcal{H}}_0$. As $a \to 0$, a brief computation shows that $\eta(a^2) \sim -\sqrt{12}\,a^2$. Hence the apparent singularity at $a = 0$ is removable. Since this correspondence $\eta : a^2 \mapsto \beta_a(-a)$ is well defined and holomorphic, it suffices to show that η is a proper map of degree one from a region in the a^2-plane onto the open unit disk. First consider a boundary point F_a of the region $\mathcal{S}_1 \cap \hat{\mathcal{H}}_0$. Then as noted earlier the Böttcher mapping from the immediate basin U_{F_a} onto the unit disk has no critical points, and in fact is a conformal diffeomorphism. In particular, β_a^{-1} can be defined as a single valued function on the disk of radius $1 - \epsilon$, for any $\epsilon > 0$. This last property must be preserved under any small perturbation of F_a, and it follows that $|\beta_b(-b)| > 1 - \epsilon$ for any $F_b \in \hat{\mathcal{H}}_0$ sufficiently close to F_a. Thus η is a proper map from $(\hat{\mathcal{H}}_0 \cap \mathcal{S}_1)/\mathcal{I}$ onto $\mathbb{D}$. Since $\eta^{-1}(0)$ is the single point 0, with $\eta'(0) = -\sqrt{12} \neq 0$, it follows that η is a conformal diffeomorphism. $\square$

3C   Maps Outside of $\hat{\mathcal{H}}_0$

In analogy with Lemma 2.4 in the dynamic plane, we have the following result in parameter space.

Lemma 3.7. *The conformal diffeomorphism* $\eta : (\mathcal{S}_1 \cap \hat{\mathcal{H}}_0)/\mathcal{I} \xrightarrow{\cong} \mathbb{D}$ *of Lemma 3.6 extends to a continuous map*

$$\overline{\eta} : \mathcal{S}_1/\mathcal{I} \to \overline{\mathbb{D}}$$

that maps each $F_{\pm a} \in \mathcal{S}_1/\mathcal{I}$ *outside of* $\hat{\mathcal{H}}_0/\mathcal{I}$ *to the point* $e^{2\pi i t_0}$, *where* t_0 *is the internal argument for the principal limb of* F_a *or of* F_{-a}.

Intuitively, each $F_{\pm a}$ outside of $\hat{\mathcal{H}}_0/\mathcal{I}$ should belong to a limb which is attached to the boundary of $\hat{\mathcal{H}}_0/\mathcal{I}$, and we want to map it to the corresponding point of the circle $\partial\mathbb{D}$.

Proof of Lemma 3.7: Fixing some $F \in \mathcal{S}_1 \cap (\mathcal{C}(3) \smallsetminus \hat{\mathcal{H}}_0)$, choose two rational angles $t_\ell < t_0 < t_r$ close to t_0. Then the critical point $-a$ is contained in the sector bounded by the two extended rays $\hat{\mathcal{R}}_{t_\ell}$ and $\hat{\mathcal{R}}_{t_r}$. Without loss of generality, we may assume that these extended rays meet ∂U_a at repelling periodic points, since there can be at most finitely many parabolic points. Evidently this situation will be preserved under a small perturbation of F. This proves that the correspondence $F \mapsto e^{2\pi i t_0}$ is continuous as F varies over $\mathcal{C}(3) \smallsetminus \hat{\mathcal{H}}_0$. (It is conjectured that this correspondence is not only continuous, but actually locally constant away from the boundary of $\hat{\mathcal{H}}_0$. Compare Lemmas 3.9 and 3.10.) If we perturb $F = F_a$ into $\hat{\mathcal{H}}_0$, then a similar argument, using the construction from Lemma 3.3, shows that $\overline{\eta}(F_{\pm a})$ depends continuously on a. $\square$

In analogy with the discussion above, let us define the *limb* $\mathcal{C}_t$, attached to $(\mathcal{S}_1 \cap \overline{\mathcal{H}}_0)/\mathcal{I}$ at internal angle t, to be the set $\overline{\eta}^{-1}(e^{2\pi i t})$. In other words, $F_{\pm a}$ belongs to $\mathcal{C}_t$ if and only if the principal limb of the filled Julia set $K(F_a)$ is attached at internal angle t, so that $-a \in K_t \subset K(F_a)$. *By abuse of language, we may say that the map F_a belongs to the limb $\mathcal{C}_t$, although properly speaking it is the unordered pair $\{F_a, F_{-a}\}$ which belongs to $\mathcal{C}_t$.*

According to Faught, the principal hyperbolic component $\mathcal{S}_1 \cap \hat{\mathcal{H}}_0$ in $\mathcal{S}_1$ is bounded by a Jordan curve, so that $\overline{\eta}$ maps $(\mathcal{S}_1 \cap \overline{\mathcal{H}}_0)/\mathcal{I}$ homeomorphically onto $\overline{\mathbb{D}}$. (We cannot be sure that the connectedness locus $\mathcal{C}(\mathcal{S}_1)$ in $\mathcal{S}_1$ is locally connected, since it contains many copies of the Mandelbrot set. However Faught showed that such Mandelbrot copies are the only possible source of non-local-connectivity.) It follows that the limb $\mathcal{C}_t \subset (\mathcal{S}_1 \cap \mathcal{C}(3))/\mathcal{I}$ has more than one point if and only if the angle t is periodic under doubling, or in other words if and only if t is rational with odd denominator. Compare Figure 6, in which the 0-limb to the left, the 1/3-limb to the lower right, and the 2/3-limb (Figure 7) to the upper right are clearly visible.

In analogy with Lemmas 2.8 and 2.9, let us describe the relationship between internal and external angles in parameter space. *It will be convenient to measure internal arguments t in the a^2-plane $\mathcal{S}_1/\mathcal{I}$, where we identify affinely conjugate polynomials, but to measure external arguments η in the a-plane $\mathcal{S}_1$ where we make no such identification.* (Compare Figures 2, 3.)

Lemma 3.8. *The correspondence $t \mapsto \eta(t)$ between internal and external angles in parameter space can be expressed in terms of the corresponding function $t \mapsto \xi_{t_0}(t)$ in the dynamic plane (Lemma 2.9), by the formula $\eta(t) = \xi_t(t + \frac{1}{2})$. This function $t \mapsto \eta(t)$ is strictly monotone, increasing by $1/2$ as t increases by 1, and has a jump discontinuity at t if and only if t is periodic under the doubling map mod 1. In fact if t has period p under doubling then the discontinuity at t is given by*

$$\Delta\eta(t) = \eta(t^+) - \eta(t^-) = \xi_t(\frac{1}{2} + t^+) - \xi_t(\frac{1}{2} + t^-) = \frac{1}{3(3^p - 1)}.$$

For example the jump from $\xi_{1/3}(\frac{5}{6}^-) = 11/12$ to $\xi_{1/3}(\frac{5}{6}^+) = 23/24$ in Figure 3 corresponds exactly to the jump from $\eta(\frac{1}{3}^-) = 5/12$ to $\eta(\frac{1}{3}^+) = 11/24$ in Figure 6. (In fact the corresponding shapes in the Julia set and in parameter space are very similar! It would be interesting to explore this phenomenon.)

Note that the sum of these discontinuities,

$$\sum \left\{ \frac{1}{3(3^p - 1)} \; : \; 2^p t \equiv t, \; 0 \le t < 1, \; p \; \text{minimal} \right\},$$

is equal to $1/2$. In fact, writing $(3^p - 1)^{-1}$ as $3^{-p} + 3^{-2p} + 3^{-3p} + \cdots$, we can express this sum as

$$\frac{1}{3} \sum \left\{ 3^{-p} \; : \; 2^p t \equiv t, \, 0 \le t < 1 \right\} = \frac{1}{3} \sum_{1}^{\infty} \frac{2^p - 1}{3^p} = \frac{1}{3} \left(2 - \frac{1}{2} \right) = \frac{1}{2}.$$

The proof of Lemma 3.8 is not difficult, and will be omitted.   □

Thus the correspondence $t \mapsto \xi_t(t + \frac{1}{2})$ is discontinuous precisely when $2^m(t + \frac{1}{2}) \equiv t \pmod 1$ for some $m \ge 0$, or in other words when $t = t_0$ is rational with odd denominator. *It is natural to conjecture that these are precisely the internal arguments at which some non-trivial limb $\mathcal{C}_t$ is attached to $\partial \hat{\mathcal{H}}_0 \cap \mathcal{S}_1$ within $\mathcal{C}(3) \cap \mathcal{S}_1$.* The points of attachment are of particular interest. These are the maps $F = F_a$ for which the periodic point $k(t) \in \partial U_a$ is parabolic, with multiplier equal to $+1$.

Caution. Although the boundary $\partial \hat{\mathcal{H}}_0 \cap \mathcal{S}_1$ is a topological circle, parametrized by the internal argument t_0, it definitely is not true that the corresponding Julia sets vary continuously with t_0. In fact the cases where $-a$ does or does not belong to $\overline{U}_a$ are presumably both everywhere dense along this circle.

Now suppose that we fix some angle t_0 which is periodic of order p under doubling.

Lemma 3.9. *The two angles $\eta(t_0^-) = \xi_{t_0}(\frac{1}{2} + t_0^-)$ and $\eta(t_0^+) = \xi_{t_0}(\frac{1}{2} + t_0^+)$ are consecutive angles of the form $\frac{i}{3(3^p-1)}$. The corresponding external rays in parameter space land at a common map F_0 which has the following property. In the dynamic plane $\mathbb{C} \setminus J(F_0)$, the external rays of argument $\eta(t_0^-)$ and $\eta(t_0^+)$ and the internal ray of argument $t_0 + \frac{1}{2}$ all land at a common pre-periodic point z_0 in the Julia set. Furthermore, the multiplier $F^{\circ p\prime}(F(z_0))$ is equal to $+1$.*

These two external rays $\mathcal{R}_{\eta(t_0^-)}(\mathcal{S}_1)$ and $\mathcal{R}_{\eta(t_0^+)}(\mathcal{S}_1)$ cut off an open region $W(t_0) \subset \mathcal{S}_1$ which (following [At92]) we may call the *wake* of the t_0-limb. It can be characterized as follows.

Lemma 3.10. *Every map $F \in W(t_0)$ has the property that the internal ray of argument $t_0 + \frac{1}{2}$ for F, as well as the external rays of argument $\eta(t_0^-)$ and $\eta(t_0^+)$, all land at a common pre-periodic point in the Julia set $J(F)$. However, for any map $F \notin \overline{W}(t_0)$, the two external rays of argument $\eta(t_0^-)$ and $\eta(t_0^+)$ for F land at distinct pre-periodic points.*

Proof of Lemmas 3.9 and 3.10 (Outline): To simplify the discussion and fix our ideas we will only describe the case $t_0 = 1/3$. The general case

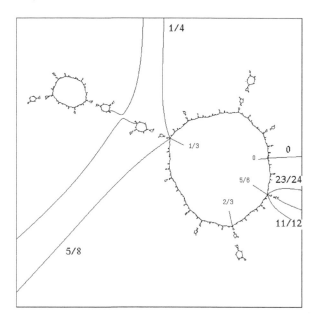

Figure 10. Julia set for a map which belongs to the wake $W(1/3)$, but not to the connectedness locus. (Compare Figure 5.) The unlabeled rays on the left pass close to the escaping critical point $-a$ and that on the right passes close to the co-critical point $2a$.

is not essentially different. As a first step, we must check that there exist maps $F \in \mathcal{S}_1$ which satisfy the condition that the internal ray $\mathcal{R}_{5/6}(F)$ and the two external rays $\mathcal{R}'_{11/12}(F)$ and $\mathcal{R}'_{23/24}(F)$ all land at a common point. For example any hyperbolic map in the $1/3$-rd limb will satisfy this condition. (Compare Figure 3.) Using the dynamics, it then follows that other triples such as $\mathcal{R}_{2/3}$, $\mathcal{R}'_{3/4}$, $\mathcal{R}'_{7/8}$ and $\mathcal{R}_{1/3}$, $\mathcal{R}'_{1/4}$, $\mathcal{R}'_{5/8}$ also have a common landing point. For a map satisfying this condition, since the two angles $1/4$ and $5/8$ differ by more than $1/3$, it follows that there must be a critical point, namely $-a$, lying in the region between the two rays $\mathcal{R}'_{1/4}(F)$ and $\mathcal{R}'_{5/8}(F)$.

On the other hand, there are also maps, such as $F(z) = z^3$, for which the two rays $\mathcal{R}'_{11/12}(F)$ and $\mathcal{R}'_{23/24}(F)$ land at distinct pre-periodic points. As we deform the map F along some path in $\mathcal{S}_1$, how can we pass from one type of behavior to the other? If F is a transition point which belongs to the connectedness locus, then at least one of these two rays must land at a pre-parabolic point (that is a pre-periodic point whose orbit falls onto a parabolic cycle). Compare [GM93, Appendix B]. Note that there are only finitely many maps in $\mathcal{S}_1$ which possess parabolic cycles of the appropriate period and multiplier. On the other hand, for a transition outside of the connectedness locus, the critical point $-a$ must pass out of the region bounded by $\mathcal{R}'_{1/4}(F)$ and $\mathcal{R}'_{5/8}(F)$. Hence, at the transition point, one of these two rays must crash into the critical point $-a$. But this is exactly the defining property of a map F that belongs to the external ray of angle $11/12$, respectively $23/24$, in parameter space. Thus the boundary between the two types of behavior is formed by these two external rays, each of which lands at a well-defined map by Lemma 3.5, together with a finite set. Hence these two rays must land at a common map, as asserted in Lemma 3.9. The rest of the proof is straightforward. $\square$

4   Hyperbolic Components in $\mathcal{S}_1$

This section will present a more detailed, but partially conjectural, picture of the connectedness locus intersected with $\mathcal{S}_1$. Recall that a map in $\mathcal{C}(3)$ is called *hyperbolic* if the orbits of both critical points converge to attracting periodic orbits. The set of hyperbolic points forms a union of components of the interior of $\mathcal{C}(3)$. Conjecturally it constitutes the entire interior. It is shown in [Mi92b] that each hyperbolic component is an open topological 4-cell, which is canonically biholomorphic to one of four standard models. Furthermore, each hyperbolic component contains one and only one post-critically finite map, called its *center*. (A map is *post-critically finite* if the forward orbit of every critical point is either periodic, or eventually falls onto a periodic cycle which may be either repelling or superattracting. However, in the hyperbolic case, such a post-critical cycle must necessarily be superattracting.)

In the case of a hyperbolic component which intersects $\mathcal{S}_1$, clearly Type B cannot occur, and Type A occurs only for the principal hyperbolic component $\hat{\mathcal{H}}_0$. However, we will see that Type C and D both occur infinitely often (and all four types are important in studying maps with a periodic critical orbit of higher period). It is not difficult to check that for *any* hyperbolic component in the connectedness locus which intersects $\mathcal{S}_1$, the intersection is an open topological 2-cell which contains the center point. (Compare Lemma 3.6.) All of these hyperbolic components in $\mathcal{S}_1$ are bounded by Jordan curves. (See [Fa92] or [Ro06].)

In the case of a *capture component*, we can be even more explicit. The closure $\overline{U}_a$ of the immediate basin of the fixed point $+a$ is homeomorphic to the disk $\overline{\mathbb{D}}$, using the Böttcher coordinate. There must be some first element in the orbit of the other critical point $-a$ which belongs to $\overline{U}_a$. *Using the Böttcher coordinate of this point, say $F^{\circ n}(-a)$, we obtain the* *required homeomorphism $a \mapsto \beta_a(F^{\circ n}(-a))$ from the closure of the capture component in $\mathcal{S}_1$ onto the closed unit disk.*

In the case of a component of type D (disjoint attracting orbits), we can make the much sharper statement. *If F_0 is the center map in the component, then by the Douady-Hubbard operation of "tuning", we obtain a copy $F_0 * M$ of the Mandelbrot set $M = \mathcal{C}(2)$ which is topologically embedded into $\mathcal{S}_1$.* (See [DH85] and compare the discussion in [Mi89].) In particular, there are infinitely many other hyperbolic components of type D which are canonically subordinated to the given one. When discussing such an embedded Mandelbrot set, we will always implicitly assume that it is maximal, i.e., that $F_0 * M$ is not a subset of some strictly larger embedded Mandelbrot set. In other words, we assume that F_0 cannot itself be obtained by tuning some other center point of lower period.

The "directions" in which we can proceed from one hyperbolic component or embedded Mandelbrot set to any other, measured around the boundary of the component or Mandelbrot set, can be described quite explicitly as follows. (Note that Case B is excluded, since it does not occur in $\mathcal{S}_1$.)

Case A. From the hyperbolic component $\hat{\mathcal{H}}_0$ in $\mathcal{S}_1$, as discussed in Section 1, we can proceed outward in any direction $t \in \mathbb{R}/\mathbb{Z}$ which is rational with odd denominator, or equivalently is periodic under doubling. Components which are attached in this direction are said to belong to the *limb $\mathcal{C}_t$*. In particular, there is one copy of the Mandelbrot set which is immediately attached to $\hat{\mathcal{H}}_0$ in each such direction. We will use the notation $F_t * M$ for this "satellite" of $\hat{\mathcal{H}}_0$ in the limb $\mathcal{C}_t \subset \mathcal{S}_1$.

Case C. If C is a capture component in the limb $\mathcal{C}_t$, then we can go out from C in any direction α which is a preimage of t under doubling. In other words, α must satisfy $2^k \alpha \equiv t \mod 1$ for some $k \geq 0$. (Compare Figure 11.) Here the "direction" from C is measured using the Böttcher parametrization of the boundary ∂C, as described above. One particular direction plays a special role: namely, the direction $\alpha = 2t$, which leads from the component C back towards the principal component $\hat{\mathcal{H}}_0$.

Case D. From each embedded Mandelbrot set $F * M$ we can go out in any dyadic direction $\delta = m/2^k$, measured around the Carathéodory loop $\delta \mapsto \gamma(\delta) \in \partial M$ which parametrizes the boundary of M. (The number $\delta \in \mathbb{R}/\mathbb{Z}$ can be described as an external argument with respect to M, but is certainly not an external argument with respect to the cubic connectedness locus.) Here the case $\delta = 0$ plays a special role, as the direction in which we must proceed from $F * M$ in order to get back to the principal hyperbolic component $\hat{\mathcal{H}}_0$.

In particular, if we start out on some immediate satellite $F_t * M$ of the principal hyperbolic component, then at each dyadic boundary point

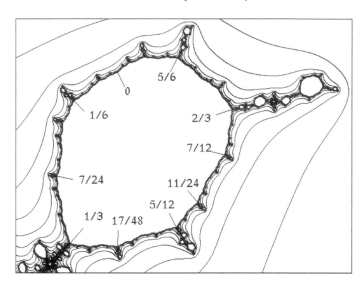

Figure 11. Detail of Figure 7, showing the capture component $C(2/3\,,1/2)$. (Here $2/3$ is the internal angle in $\hat{\mathcal{H}}_0$ at which a small Mandelbrot set is attached, and $1/2$ is the external angle with respect to this Mandelbrot set. The interior of this component $C(2/3\,,1/2)$ is parametrized by the Böttcher coordinate of $F^{\circ 3}(-a)$.

$F_t * \gamma(\delta)$, $\delta \neq 0$, there is a capture component, which we will denote by $C(t,\delta)$, immediately attached.

Thus the principal component $\hat{\mathcal{H}}_0$ has immediate satellites $F_t * M$, and these have immediate satellites $C(t,\delta)$. According to [Ro06]: *These are the only examples of hyperbolic components or Mandelbrot sets in $\mathcal{S}_1$ which are immediately contiguous to each other.* If we exclude these cases, and if we exclude contiguous components within an embedded Mandelbrot set, then it is conjectured that we can pass from one hyperbolic component to another only by passing through infinitely many components, both of Type C and of Type D.

4A   Hubbard Trees

(Compare Section 6A as well as the appendix.) In order to partially justify this picture, let us describe Hubbard trees for the various hyperbolic components. The Hubbard tree for the center point $z \mapsto z^3$ of $\hat{\mathcal{H}}_0$ is of course just a single doubly-critical vertex.

The Hubbard tree $T(t)$ for the center point F_t of the satellite $F_t * M$ can be described as follows. We assume that the argument $t \in \mathbb{Q}/\mathbb{Z}$ has

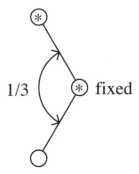

Figure 12. Tree for the center point $F_{1/3}$ of the satellite $F_{1/3} * M$ at internal angle $1/3$. Critical points are indicated by stars, and vertices in the Fatou set by small circles.

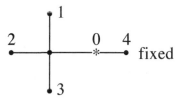

Figure 13. Tree for the quadratic map $\gamma(1/8)$ with external angle $1/8$ in the Mandelbrot set. The post-critical vertices are numbered so that $0 \mapsto 1 \mapsto 2 \mapsto \cdots$.

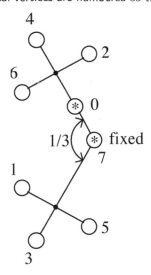

Figure 14. Tree for the center of the capture component $C(1/3, 1/8)$ in $\mathcal{S}_1$ which is attached to $F_{1/3} * M$ at the point $F_{1/3} * \gamma(1/8)$.

period $n \geq 1$ under doubling modulo 1. *Then $T(t)$ consists of n different edges radiating out from a central vertex at angles t, $2t$, $4t$, ... modulo 1, as measured from some fixed base direction.* Here the central vertex $\mathbf{v}_0$ and the other endpoint $\mathbf{w}_0$ of the edge at angle t are both critical, but all other vertices are non-critical. The canonical mapping τ from $T(t)$ to itself fixes the central vertex $\mathbf{v}_0$ and permutes the other vertices cyclically, carrying the vertex at angle α to the vertex at angle 2α.

Now choose some dyadic angle $\delta = m/2^k \not\equiv 0$ in $\mathbb{Q}/\mathbb{Z}$. Let $\gamma(\delta) \in M$ be the quadratic map at external argument δ in the Mandelbrot set, and let $T'(\delta)$ be its Hubbard tree. Thus the $(k+1)$-st forward image of the critical vertex in $T'(\delta)$ is a fixed vertex. If we tune F_t by $\gamma(\delta)$, or equivalently if we tune $T(t)$ by $T'(\delta)$, then we obtain a new tree $T(t) * T'(\delta)$ for which the $(nk + 1)$-st forward image $\mathbf{w}_{nk+1}$ of the "outer" critical point $\mathbf{w}_0$ is periodic of period n, lying at angle $2t$ from the central critical point $\mathbf{v}_0$. Thus for each edge of $T(t)$ it contains a complete copy of $T'(\delta)$, all of these copies being pasted together at the post-critical fixed point, which is now critical. (However, only the primary copy at angle t contains another critical point.)

In order to obtain the tree $T(t, \delta)$ for the center of the satellite $C(t, \delta)$, we modify this construction very slightly as follows. As an angled topological tree with two marked critical points, $T(t, \delta)$ is identical with $T(t) * T'(\delta)$. However, $T(t, \delta)$ has fewer post-critical points, hence fewer vertices, and the canonical mapping from the tree to itself is changed so that the nk-th forward image $\mathbf{w}_{nk}$ of the outer critical point $\mathbf{w}_0$ maps to the central critical point $\mathbf{w}_{nk+1} = \mathbf{v}_0$. In other words, the edge $\mathbf{e}$ in the t-limb which leads out to $\mathbf{w}_{nk}$ is now to be mapped to a path in the $2t$-limb which leads all the way in to $\mathbf{v}_0$. (Figure 12, 13, 14.)

More generally, consider the tree T for an arbitrary component of Type D in $\mathcal{S}_1$. Suppose that the outer critical point $\mathbf{w}_0$ has period n, and lies at angle t from the central vertex $\mathbf{v}_0$. For any dyadic angle δ as above, we can tune to obtain a tree $T * T'(\delta)$ for which the $(nk + 1)$-st image $\mathbf{w}_{nk+1}$ of $\mathbf{w}_0$ is periodic of period n, and lies in the $2t$-limb. Again we can stretch this $(nk + 1)$-st image in towards the central vertex, and thus construct other hyperbolic components. But in general, there does not seem to be a immediately contiguous component which can be constructed in this way.

Similarly, we can consider a completely arbitrary capture component in $\mathcal{S}_1$. The corresponding tree T has outer critical point $\mathbf{w}_0$ lying in a limb which has angle say t from the fixed central vertex $\mathbf{v}_0$. If the $(k + 1)$-st forward image $\mathbf{w}_{k+1}$ of $\mathbf{w}_0$ is equal to $\mathbf{v}_0$, then it is not difficult to see that the k-th forward image $\mathbf{w}_k$ must lie in the t-limb. (Every other limb, at angle say α, maps isomorphically into the limb at angle 2α.) Thus, the edge $\mathbf{e}$ in the t-limb which leads out to $\mathbf{w}_k$ must map to a path in the $2t$-

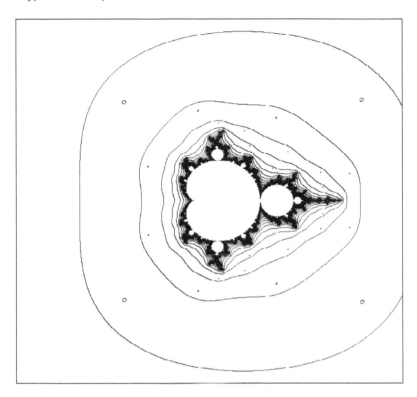

Figure 15.   A small slice of constant height $b \equiv 18$ through a Mandelbrot-torus in the plane $\hat{\mathcal{P}}(3)$, using coordinates (a, b). Here a varies over a box of width .04 centered at $a = 2$. This slice intersects the curve $\mathcal{S}_1$ transversally at the central point of the figure.

limb which leads in to v_0. Now suppose that we choose any angle $\alpha \not\equiv 2t$ which is a pre-image of t under doubling modulo 1. Then we can modify this tree, adding an α-limb if it does not already exist, so that the image of $\mathbf{e}$, leading from the $2t$-limb into the center, will extend on out into the α-limb. In this new tree, the $(k+1)$-st image of $\mathbf{w}_0$ will lie in the α-limb, hence some further iterated image will lie back in the t-limb. By such constructions, it is not difficult to obtain either a tree in which $\mathbf{w}_0$ is periodic, or a tree in which $\mathbf{w}_0$ eventually maps to the fixed point v_0. In either case, the associated hyperbolic component can be described as one which lies in the α-direction from our initial capture component with tree T.

Remark 4.1 (What Does a Neighborhood of $\mathcal{S}_1$ Look Like?). (Compare Remark 1.1.) Understanding the curve $\mathcal{S}_1$ should be a first step towards understanding the dynamics for maps in $\hat{\mathcal{P}}(3)$ which are close to $\mathcal{S}_1$. Perhaps the easiest points to understand are those in the escape locus. According to [BH92] or [Br93], each escape point in $\mathcal{S}_1$ is the center of a small Mandelbrot set in the transverse direction, with each period p center in this Mandelbrot set corresponding to an intersection with the period p curve $\mathcal{S}_p$. A transverse section, as shown in Figure 15, illustrates such a small Mandelbrot set. Note the small dots outside of the Mandelbrot copy. Each one seems to represent a small Cantor set of maps. The complementary region, outside of these Cantor sets and outside this Mandelbrot set, represents maps in the shift locus.

5   Topology and Geometry of $\mathcal{S}_p$

For any integer $p \geq 1$, let $\mathcal{S}_p \subset \hat{\mathcal{P}}(3)$ be the *period p superattracting curve* consisting of all $F \in \hat{\mathcal{P}}(3)$ for which the critical point $+a$ has period exactly p. In other words, $\mathcal{S}_p$ can be identified with the affine algebraic variety consisting of all pairs $(a, v) \in \mathbb{C}^2$ such that the critical point a has period exactly p under the map

$$F(z) \;=\; z^3 - 3a^2 z + (2a^3 + v). \tag{5.12}$$

Remark 5.1. It is important to work with this normal form, rather than with $F(z) = z^3 - 3a^2 z + b$, since it will allow a simpler description of the curve $\mathcal{S}_p$. As examples, the equations for $\mathcal{S}_1$ and $\mathcal{S}_2$ in the (a, v)-plane take the form

$$v - a \;=\; 0 \qquad \text{and} \qquad v^3 - 3a^2 v + 2a^3 + v - a \;=\; 0\,,$$

with degrees one and three, respectively. The corresponding equations in the (a, b)-plane, obtained by substituting $b - 2a^3$ in place of v, would have degrees three and nine.

Theorem 5.2. *Each $\mathcal{S}_p$ is a smooth affine algebraic curve.*

The proof will be given later in this section.

Question 5.3 (Is $\mathcal{S}_p$ Connected?). It seems quite possible that all of the curves $\mathcal{S}_p$ are irreducible, or equivalently that they are topologically connected. For example, we will see that $\mathcal{S}_1$ and $\mathcal{S}_2$ are connected curves of genus zero, while $\mathcal{S}_3$ is a connected curve of genus one. However, I don't know how to attack this question in general.

The degree of the affine curve $\mathcal{S}_p$ can be computed as follows. It will be convenient to first consider the disjoint union $\bigcup_{n|p} \mathcal{S}_n \subset \mathbb{C}^2$ of the curves $\mathcal{S}_n$ where n ranges over all divisors of p. This will be denoted by $\mathcal{S}_p^{\uplus}$.

Lemma 5.4. *The degree of this affine curve $\mathcal{S}_p^{\uplus}$ is*

$$\deg(\mathcal{S}_p^{\uplus}) \;=\; \sum_{n|p} \deg(\mathcal{S}_n) \;=\; 3^{p-1}.$$

Furthermore, the number of hyperbolic components of Type A in $\mathcal{S}_p^{\uplus}$ is also equal to 3^{p-1}.

Remark 5.5. Given this statement, it is easy to compute the degree of $\mathcal{S}_p$. Assuming inductively that we have computed the degree $\deg(\mathcal{S}_n)$ for all proper divisors of p, we simply need to subtract these numbers from 3^{p-1} to get the degree of $\mathcal{S}_p$. More generally, it will be convenient to define numbers $\nu_d(p)$ by the equation[3]

$$d^p \;=\; \sum_{n|p} \nu_d(n)\,.$$

For example, $\nu_d(p)$ can be interpreted as the number of period p points for a generic polynomial map of degree d, and $\nu_2(p)/2$ can be interpreted as the number of period p centers in the Mandelbrot set. With this notation, our conclusion is that

$$\deg(\mathcal{S}_p) \;=\; \nu_3(p)/3\,.$$

Here is a table listing $\nu_2(p)/2$ and $\nu_3(p)/3$ for small p.

p	1	2	3	4	5	6	7	8	9	10
$\nu_2(p)/2$	1	1	3	6	15	27	63	120	252	495
$\nu_3(p)/3$	1	2	8	24	80	232	728	2160	6552	19600

Proof of Lemma 5.4: Evidently $\mathcal{S}_p^{\uplus}$ can be defined by the polynomial equation $F^{\circ p}(a) - a = 0$. Since $F(z) = z^3 - 3a^2 z + 2a^3 + v$, we can write

$$\begin{aligned} F(a) &= v\,, \\ F^{\circ 2}(a) &= (v^3 - 3a^2 v + 2a^3) + v\,, \end{aligned}$$

---

[3]Equivalently, by the Möbius Inversion Formula, $\nu(p) = \sum_{n|p} \mu(n)\, d^{p/n}$, where the *Möbius function* $\mu(n)$ equals $(-1)^k$ if $n = p_1 p_2 \cdots p_k$ is a product of k distinct prime factors, with $\mu(1) = 1$, but with $\mu(n) = 0$ whenever n has a squared prime factor.

and in general

$$F^{\circ p}(a) \;=\; (v^3 - 3a^2 v + 2a^3)^{3^{p-2}} + \text{(lower order terms)} \tag{5.13}$$

for $p \geq 2$. Thus the equation $F^{\circ p}(a) = a$ has degree 3^{p-1} in the variables a, v, as asserted.

To count hyperbolic components of Type A, note that the center of each such component is a polynomial of the form $F(z) = z^3 + v$. More generally, for any degree d we can study the family of maps

$$g_v(z) \;=\; z^d + v, \tag{5.14}$$

counting the number of v such that the critical orbit $0 \mapsto v \mapsto v^d + v \mapsto \cdots$ has period p. The argument will be based on [Sch04] which studies the connectedness locus for for the family (5.14), known as the *Multibrot set*. In particular, Schleicher studies external rays in the v parameter plane. He shows that for each angle which has period p under multiplication by d, the corresponding parameter ray lands on the boundary of a hyperbolic component of period p. Furthermore, if $p \geq 2$, then exactly d such rays land on the boundary of any given period p component. In the period one case, the corresponding statement is that all $d - 1$ of the rays of period one land on the boundary of the unique period one component. Since there are exactly $d^p - 1$ rays which have period dividing p, a straightforward argument now shows that the number of period p components in the Multibrot set is $\nu_d(p)/d$, and the conclusion follows.   □

Remark 5.6. The centers of period dividing p in this Multibrot family are precisely the roots of the polynomial $g_c^{\circ p}(0)$, which has degree d^{p-1}. Thus an immediate corollary of Lemma 5.4 is the purely algebraic statement that this polynomial has d^{p-1} distinct roots.

These same numbers $\nu_3(p)/3$ can also be used to count hyperbolic components of Type B and D.

Definition 5.7. Let $\mathcal{S}_p' \subset \hat{\mathcal{P}}(3)$ be the *dual* superattractive period p curve, consisting of all maps $F(z) = z^3 - 3a^2 z + 2a^3 + v$ for which the critical point $-a$ has period exactly p.

Lemma 5.8. *For each $p, r \geq 1$, the curve $\mathcal{S}_p$ intersects $\mathcal{S}_r'$ transversally in $\nu_3(p)\nu_3(r)/3$ distinct points. These intersection points comprise precisely the center points of all hyperbolic components in $\mathcal{C}(3)$ which have Type A, B, or D.*

(On the other hand the center point of a component of Type C lies on only one of these curves $\mathcal{S}_p$ or $\mathcal{S}_r'$.) As examples, for $p = 1, 2, 3$, the intersection $\mathcal{S}_p \cap \mathcal{S}_1'$ consists of 3, 6, and 24 points, respectively, while $\mathcal{S}_2 \cap \mathcal{S}_2'$ has 12 points. Representative Hubbard trees are shown in Figures 34 and 35, while Julia sets illustrating three of these trees are shown in Figure 36.

Proof of Lemma 5.8: We will use Bezout's theorem, which states that if two curves in the complex projective plane intersect transversally, then the number of intersection points is equal to the product of the degrees of the two curves. As noted above, the curve $\mathcal{S}_p$ has degree $\nu_3(p)/3$. A similar computation shows that the curve $\mathcal{S}_r'$ has degree $\nu_3(r)$. (Note: The asymmetry between these two formulas arises from the fact that we are using coordinates (a, v) which are particularly adapted to studying the orbit of a rather than $-a$. The polynomial

$$F^{\circ r}(-a) - (-a) = (4a^3 + v)^{3^{r-1}} + (\text{lower order terms})$$

has degree 3^r rather than 3^{r-1}.) The curve $\mathcal{S}_p$ intersects the line at infinity in two (highly multiple) points where the ratios $(a : v : 1)$ take the values $(1 : 1 : 0)$ and $(1 : -2 : 0)$, respectively; while $\mathcal{S}_r'$ intersects the line at infinity in the single point $(0 : 1 : 0)$. Thus there are no intersections at infinity, and the conclusion follows. $\square$

5A   Escape Regions

The complement $\mathcal{S}_p \setminus \mathcal{C}(\mathcal{S}_p)$ of the connectedness locus will be called the *escape locus* in $\mathcal{S}_p$. Each connected component $\mathcal{E}$ of the escape locus will be called an *escape region*.

We will see that each escape region $\mathcal{E}$ is conformally isomorphic to a punctured disk (or equivalently to the region $\mathbb{C} \setminus \overline{\mathbb{D}}$). Thus $\mathcal{S}_p$ can be made into a smooth compact surface $\overline{\mathcal{S}}_p$ by adjoining finitely many ideal points, one for each escape region. We can then think of each connected component of $\mathcal{S}_p$ as a multiply punctured Riemann surface with its connectedness locus as a single connected "continent", and with the escape regions as the complementary "oceans", each centered at one of the puncture points. Assuming that this connected component of $\mathcal{S}_p$ is mapped to itself by the canonical involution $\mathcal{I}$, it is a 2-fold branched covering of the corresponding connected component of $\mathcal{S}_p/\mathcal{I}$.

Here is a precise statement.

Lemma 5.9. *Each escape region $\mathcal{E}$ is canonically isomorphic to the μ-fold cyclic covering of $\mathbb{C} \setminus \overline{\mathbb{D}}$ for some integer $\mu \geq 1$.*

By definition, this integer $\mu = \mu(\mathcal{E}) \geq 1$ will be called the *multiplicity* of the escape region $\mathcal{E}$.

Proof of Lemma 5.9: For any $F = F_{a,v} \in \hat{\mathcal{P}}(3)$, the associated Böttcher coordinate $\beta(z) = \beta_{a,v}$ is defined for all complex z with $|z|$ sufficiently large. It satisfies the equation

$$\beta(F(z)) = \beta(z)^3,$$

with $|\beta(z)| > 1$, and with $\beta(z)/z$ converging to $+1$ as $|z| \to \infty$. In particular the co-critical point $2a$ is just large enough so that $\beta(2a)$ is well defined. Now consider the map

$$\hat{\beta}: \mathcal{E} \to \mathbb{C} \setminus \overline{\mathbb{D}} \quad \text{defined by} \quad \hat{\beta}(F_{a,v}) = \beta_{a,v}(2a).$$

It is not hard to check that $\hat{\beta}$ is holomorphic and locally bijective, and that $|\hat{\beta}(F)|$ converges to $+1$ as F converges towards the connectedness locus. In order to show that it is a covering map, we must describe its behavior near infinity.

As in the proof of Lemma 3.3 Equation (3.11), we can estimate the behavior of $\hat{\beta}$ as $|a|$ or $|v|$ tends to infinity, yielding the asymptotic formula

$$\hat{\beta}(F_{a,v}) \sim \sqrt[3]{4}\, a \quad \text{as} \quad |a| \to \infty.$$

Thus $\hat{\beta}: \mathcal{E} \to \mathbb{C} \setminus \overline{\mathbb{D}}$ is proper and locally bijective. Hence it is a covering map of some degree $\mu \geq 1$, as required. $\square$

In particular, it follows that we can choose a conformal isomorphism $\zeta: \mathcal{E} \to \mathbb{D} \setminus \{0\}$ satisfying $\zeta(F)^\mu = 1/\hat{\beta}(F)$. In fact ζ is uniquely defined up to multiplication by μ-th roots of unity. If $\mathcal{E}^+$ denotes the Riemann surface which is obtained from $\mathcal{E}$ by adjoining a single ideal point at infinity, then ζ extends to a conformal isomorphism from $\mathcal{E}^+$ onto the open unit disk $\mathbb{D}$. Now a can be expressed as a meromorphic function on $\mathcal{E}^+$ with a pole of order μ. Writing this as $a = \phi(\zeta)/\zeta^\mu$ where $\phi: \mathcal{E} \to \mathbb{D}$ is holomorphic with $\phi(0) \neq 0$, we can choose a smooth μ-th root of $\phi(\zeta)$ near the origin. Hence the formula

$$\xi = 1/\sqrt[\mu]{a} = \zeta/\sqrt[\mu]{\phi(\zeta)}$$

provides an alternative parametrization of a neighborhood of the base point $\zeta = 0$ in $\mathcal{E}^+$, with ξ^μ precisely equal to $1/a$.

Remark 5.10. Using Lemma 5.9, we can talk about equipotentials and external rays within any escape region. In particular, we can study the landing points of periodic and preperiodic rays. This provides an important tool for understanding the dynamics associated with nearby points of the connectedness locus.

Here is a more geometric interpretation of the multiplicity. Recall that $\mathcal{S}_p$ can be described as an affine curve in the space $\mathbb{C}^2$ with coordinates (a, v).

Lemma 5.11. *For any constant a_0 with $|a_0|$ large, the number of intersections of the line $a = a_0$ in $\mathbb{C}^2$ with the escape region $\mathcal{E} \subset \mathcal{S}_p$ is equal to the multiplicity $\mu(\mathcal{E})$.*

In fact, using this parameter ξ, the μ intersection points correspond precisely to the μ possible choices for an μ-th root of a. □

Corollary 5.12. *The number of escape regions in $\mathcal{S}_p$, counted with multiplicity, is equal to the degree $\nu_3(p)/3$.*

Proof: This follows immediately, since a generic line intersects $\mathcal{S}_p$ exactly $\nu_3(p)/3$ times. □

We can make a corresponding count of the number of escape regions in the quotient curve $\mathcal{S}_p/\mathcal{I}$. The easiest procedure is just to define the *multiplicity* for an escape region $\mathcal{E}/\mathcal{I} \subset \mathcal{S}_p/\mathcal{I}$ to be the sum of the multiplicities of its preimages in $\mathcal{S}_p$. In other words, for each escape region $\mathcal{E}$ in $\mathcal{S}_p$ we set

$$\mu(\mathcal{E}/\mathcal{I}) \;=\; \begin{cases} \mu(\mathcal{E}) & \text{if} \quad \mathcal{E} = \mathcal{I}(\mathcal{E}), \\ 2\mu(\mathcal{E}) & \text{if} \quad \mathcal{E} \neq \mathcal{I}(\mathcal{E}). \end{cases}$$

With this definition, we clearly get the following statement.

Corollary 5.13. *The number of escape regions in $\mathcal{S}_p/\mathcal{I}$, counted with multiplicity, is also equal to $\nu_3(p)/3$.*

5B   The Kneading Sequence of an Escape Region

Any bounded hyperbolic component of $\mathcal{S}_p$ can be concisely labeled by two complex numbers: the a and v coordinates of its center point. However, it is not so easy to label escape components. This section will describe a preliminary classification based on two invariants: the *kneading sequence*, which is a sequence of zeros and ones with period q dividing p, and the *associated quadratic map*, which is a critically periodic quadratic map with period p/q. For periods $p \leq 3$, these invariants suffice to give a complete classification of escape regions in the moduli space $\mathcal{S}_p/\mathcal{I}$, but for larger periods they provide only a partial classification. (A complete classification, based on the Puiseux expansion at infinity, will be described in Part 2 of this paper. Compare [Ki06].)

Let $F = F_{a,v}$ be any map such that the marked critical point a belongs to the filled Julia set $K(F)$ while the orbit of $-a$ escapes to infinity. Then the orbit of any point $z \in K(F)$ can be described roughly by a symbol sequence $\sigma(z) \in \{0, 1\}^{\mathbb{N}}$, as follows. There is a unique external ray, with angle say t, which lands at the escaping co-critical point $2a$, while two rays

of angles $t \pm 1/3$ land at the escaping critical point $-a$. (Compare Figure 9.)
These two rays cut the complex plane into two regions, with a on one side
and $-2a$ on the other. In fact the equipotential through $-a$ and $2a$ cuts
the plane into two bounded regions U_0 and U_1, numbered so that $a \in U_0$
and $-2a \in U_1$, together with one unbounded region where orbits escape to
infinity more rapidly. Now any orbit $z_0 \mapsto z_1 \mapsto \cdots$ in $K(F)$ determines a
symbol sequence

$$\sigma(z_0) = (\sigma_0, \sigma_1, \ldots) \quad \text{with} \quad \sigma_j \in \{0, 1\} \quad \text{and} \quad z_j \in U_{\sigma_j} \quad \text{for all} \quad j \geq 0 .$$

In particular, any periodic point determines a periodic symbol sequence.
Thus, if F belongs to an escape region in $\mathcal{S}_p$, then the critical point a
determines a periodic sequence $\sigma(a) \in \{0, 1\}^{\mathbb{N}}$, with $\sigma_{j+p}(a) = \sigma_j(a)$, and
with $\sigma_0(a) = 0$.

Definition 5.14. The periodic sequence $\sigma_1(a), \sigma_2(a), \ldots$, starting with the
symbol $\sigma_1(a)$ for the critical value, will be called the *kneading sequence*
for the map. It will be convenient to denote this sequence briefly as
$\overline{\sigma_1 \cdots \sigma_{p-1} 0}$, where the overline indicates infinite repetition. Evidently the
(minimal) period q of this kneading sequence is always a divisor of the
period p of a. In particular, $1 \leq q \leq p$.

For each such $F \in \mathcal{S}_p$, let $K_0(F) \subset K(F)$ be the connected component
of a in the filled Julia set. The following result (stated somewhat differently) is due to [BH92]. It makes use of *hybrid equivalence* in the sense
of [DH85].

Theorem 5.15. *The map F restricted to a neighborhood of $K_0(F)$ is hybrid
equivalent to a unique quadratic polynomial, which has periodic critical orbit
of period equal to the quotient p/q, where q is the period of the kneading
sequence $\sigma(a)$.*

In fact, Branner and Hubbard show that every connected component
of $K(F)$ is either a copy of $K_0(F)$ or a point.

Proof of Theorem 5.15 (Outline): Consider the open set U_0, as illustrated
in Figure 9. Let $V_0 \subset U_0$ be the connected component of $F^{-q}(U_0)$ which
contains a, and let $V_j = F^{\circ j}(V_0)$. Evidently $V_j \subset U_{\sigma_j}$ for $0 \leq j \leq q$, with
$V_q = U_0$. Note that the sets $V_1, V_2, \ldots, V_{q-1}$ cannot contain any critical
point. For if $a \in V_j$ then $V_0 \subset V_j$, and it would follow easily that the
kneading sequence has period dividing j. By definition, this cannot happen
for $0 < j < q$. It then follows that $F^{\circ q}$ mapping V_0 onto $V_q = U_0$ is a proper
map with only one critical point. Thus it is a degree two polynomial-like
map with non-escaping critical orbit. Therefore, according to [DH85], the
filled Julia set K_0 of this polynomial-like mapping is hybrid-equivalent to

$K(z \mapsto z^2 + c)$ for some unique c in the Mandelbrot set. It is not hard to see that K_0 can be identified with the connected component of a in $K(F)$. $\qquad\qquad\square$

Here is another interpretation of the kneading sequence $\{\sigma_j\}$.

Theorem 5.16. *Each point* $a_j = F^{\circ j}(a)$ *in the periodic critical orbit is asymptotic to either* a *or* $-2a$ *as* $|a| \to \infty$, *with*

$$a_j \sim \begin{cases} a & \text{if} \quad \sigma_j = 0, \\ -2a & \text{if} \quad \sigma_j = 1, \end{cases}$$

or briefly $a_j \sim (1 - 3\sigma_j)\, a$. *In fact the difference*

$$a_j - (1 - 3\sigma_j)a \tag{5.15}$$

extends to a bounded holomorphic function from $\mathcal{E}^+ = \mathcal{E} \cup \infty$ *to* $\mathbb{C}$.

Proof: Using the defining equation

$$a_{j+1} = a_j^3 - 3a^2 a_j + 2a^3 + v \qquad \text{with} \quad a_0 = a\,,$$

the difference of Equation (5.15) can clearly be expressed as a meromorphic function on $\mathcal{E} \cup \infty$, holomorphic throughout $\mathcal{E}$. Since it is clearly bounded on the intersection of $\mathcal{E}$ with any compact subset of $\mathcal{S}_p$, the only problem is to prove boundedness as $|a| \to \infty$.

As in the proof of Lemma 5.9, we can parametrize a neighborhood of infinity in $\mathcal{E}$ by a branch of $\sqrt[\mu]{a}$. Hence we can expand each a_j as a Puiseux series of the form

$$\sum_{n \le n_0} k_n a^{n/\mu} = k_{n_0} a^{n_0/\mu} + \cdots + k_1 a^{1/\mu} + k_0 + k_{-1} a^{-1/\mu} + \cdots,$$

with leading coefficient $k_{n_0} \ne 0$. (Compare [Ki06].) First consider this expansion for $a_1 = v$. If the leading term had degree $n_0/\mu > 1$, then the $a^{3n_0/\mu}$ term in the series for $F(a_1) = a_2$ would dominate, and it would follow easily that the series for the successive a_j would have leading terms of degree tending rapidly to infinity, which would contradict periodicity. On the other hand, if the leading term of the series for $v = a_1$ had degree $n_0/\mu < 1$, then the $2a^3$ term in the series for $F(a_1)$ would dominate, and again the successive degree would increase rapidly. Thus we must have $n_0 = \mu$. A completely analogous argument now shows that the series for every a_j has leading coefficient of degree $n_0/\mu = 1$. Thus, for each j, this series has the form $a_j = k_\mu a + $ (lower order terms), so that

$$a_{j+1} = F(a_j) = (k_\mu^3 - 3k_\mu + 2)a^3 + \text{(lower order terms)}\,.$$

But by the previous argument, the coefficient of a^3 must be zero. Therefore

$$k_\mu^3 - 3k_\mu + 2 = (k_\mu - 1)^2(k_\mu + 2) = 0.$$

This proves that the leading coefficient k_μ for the expansion of any a_j must be either $+1$ or -2. That is, each a_j must be asymptotic to either a or $-2a$.

First suppose that $a_j \sim a$. Let $\epsilon = a_j - a$. We must prove that ϵ remains bounded as $|a| \to \infty$. Otherwise the Puiseux expansion for ϵ would start with a term of degree n'/n with $0 < n' < \mu$. Using the identity

$$a_{j+1} = F(z + \epsilon) = v + 3a\epsilon^2 + \epsilon^3,$$

the $3a\epsilon^2$ would dominate, and would have degree > 1 which is impossible. Thus $\epsilon = O(1)$ as required.

In the case $a_j \sim -2a$, a completely analogous argument using the identity

$$F(-2a + \epsilon) = 9a^2\epsilon + 6a\epsilon^2 + \epsilon^3$$

proves the even sharper statement that $\epsilon = O(1/a)$ as $|a| \to \infty$. This completes the proof of Theorem 5.16.   □

We can sharpen the count in Lemma 5.11 as follows. Given a kneading sequence $\{\sigma_j\}$ of period dividing p, let $n = \sum_1^p (1 - \sigma_j)$ be the number of indices $1 \le j \le p$ with $\sigma_j = 0$. Thus $1 \le n \le p$.

Lemma 5.17. *The number of escape regions in $\mathcal{S}_p^{\uplus}$ with kneading sequence $\{\sigma_j\}$, counted with multiplicity, is equal to 2^{n-1}.*

Proof: We can embed $\mathcal{S}_p^{\uplus}$ into $\mathbb{C}^p$ by mapping each point with periodic critical orbit

$$a = a_0 \mapsto a_1 \mapsto \cdots \mapsto a_p = a$$

to the p-tuple $(a_1, a_2, \ldots, a_{p-1}, a) \in \mathbb{C}^p$. Such p-tuples form a 1-dimensional affine variety, characterized by the polynomial equations:

$$F(a_j) = a_{j|1} \qquad \text{for} \qquad 1 < j < p,$$

where $F(z) = z^3 - 3a^2 z + 2a^3 + a_1$ so that

$$a_{j+1} - a_1 = (a_j - a)^2(a_j + 2a). \tag{5.16}$$

Now embed $\mathbb{C}^p$ into the projective space $\mathbb{CP}^p$ by identifying each

$$(a_1, \ldots, a_{p-1} : a) \in \mathbb{C}^p \quad \text{with the point} \quad (1 : a_1 : \ldots : a_{p-1} : a)$$

in projective space. Intersect the resulting 1-dimensional projective variety with the $(p - 1)$-plane at infinity consisting of all points of the form

$$(0 : a_1 : \ldots : a_{p-1} : a).$$

The resulting intersection can be described by deleting all terms of lower degree from Equation (5.16), leaving only the cubic equation

$$(a_j - a)^2(a_j + 2a) \; = \; 0. \tag{5.17}$$

This yields an alternative proof that, as a tends to infinity, the ratio a_j/a must tend to either $+1$ or -2. In fact, according to Equation (5.15), we know that

$$\lim_{a \to \infty} a_j/a \; = \; 1 - 3\sigma_j.$$

In other words, the resulting zero-dimensional variety at infinity consists precisely of the points

$$(0 : 1 - 3\sigma_1 : \cdots : 1 - 3\sigma_{p-1} : 1)$$

associated with different possible kneading sequences. The squared factor in equation (5.17) means that each point with $\sigma_j = +1$, $1 \le j < p$, must be counted double, for a total intersection multiplicity of 2^{n-1}. Now approximating the plane at infinity by a plane $a = $ large constant, the number of intersections (counted with multiplicity) remains unchanged, and the conclusion follows.   □

5C Examples

Here are more explicit descriptions of $\mathcal{S}_p$ and $\mathcal{S}_p/\mathcal{I}$ for the cases with $p \le 4$. (For the cases with $p \le 3$, each end of the curve $\mathcal{S}_p$ has multiplicity $\mu = 1$, so that there are exactly $\nu_3(p)/3$ ends.) In specifying periodic kneading sequences, recall that infinite repetition is indicated by an overline so that, for example, $\overline{010}$ stands for $010010010 \cdots$.

Period 1. The curve $\overline{\mathcal{S}}_1 \cong \mathbb{C}$ has genus zero with one puncture of multiplicity one, namely the point at infinity. (Compare Figure 5.) The projection to $\overline{\mathcal{S}}_1/\mathcal{I}$ is a 2-fold branched covering, branched at the puncture point and at the center of the principal hyperbolic component. (Compare Figure 6.)

Period 2. The curve $\mathcal{S}_2$ has genus zero, and two ends of multiplicity one. In fact a polynomial $F \in \mathcal{S}_2$ can be uniquely specified by the "displacement" $\delta = F(a) - a$, which can take any value except zero and infinity. Given $\delta \in \mathbb{C}\backslash\{0\}$, we can solve for $a = -(\delta + \delta^{-1})/3$ and $v = a + \delta$. The

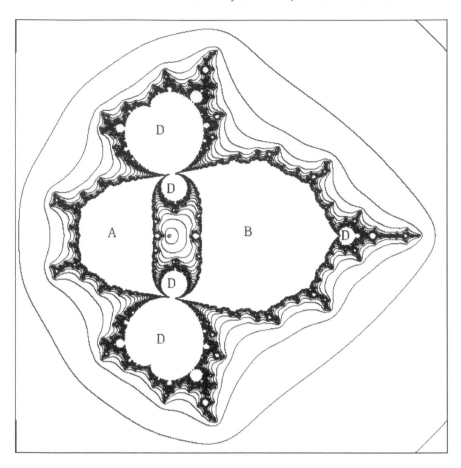

Figure 16. Connectedness locus in $\mathcal{S}_2/\mathcal{I}$, with coordinate $\delta^2 = (F(a) - a)^2$.

projection $\overline{\mathcal{S}}_2 \to \overline{\mathcal{S}}_2/\mathcal{I}$ is branched over the two punctures. Thus $\mathcal{S}_2/\mathcal{I}$ also has genus zero and two ends, with uniformizing parameter $\delta^2 \neq 0$. (See Figures 16–19.) This quotient surface contains one hyperbolic component of type A, one of type B, and infinitely many of types C and D. It contains two "escape regions", consisting of maps for which the orbit of the free critical point $-a$ escapes to infinity. Figure 18 shows a representative Julia set for a point in the inner escape region, centered at the origin, $\delta^2 = 0$. Every connected component of $K(F)$ is either a point or a homeomorphic copy of the filled Julia set for the "basilica" map $Q(z) = z^2 - 1$. Figure 19 shows a representative Julia set for the outer escape region centered at ∞. Here each component of $K(F)$ is either a point or a copy of the closed unit

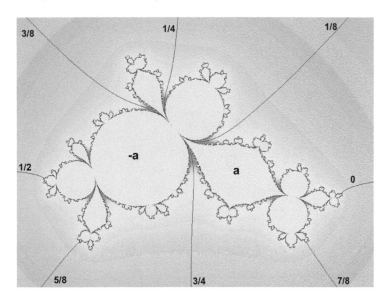

Figure 17. Julia set for a map $F \in \mathcal{S}_2$ which is a limit point of six different hyperbolic components, four bounded and two unbounded, as shown in Figure 16. (This is the higher of the two such maps, with $\delta^2 \approx .0620 + 1.4183\,i$.) Here the rays of angle $1/8$, $1/4$, $3/8$, and $3/4$ land at a common parabolic fixed point of multiplier -1. By arbitrarily small deformations, this parabolic point can split up in eight different ways, yielding a new map F' which either belongs to one of four large bounded hyperbolic components, or one of the two escape regions. (The Julia set for the center of the smallest of these hyperbolic components is shown at the top of Figure 36.) Furthermore, F can be deformed into either escape region in two essentially different ways, depending on whether the rays of angle $\{1/8, 3/8\}$ or those of angle $\{1/4, 3/4\}$ continue to land at a common fixed point. (See Plate III.)

disk (the filled Julia set for $Q(z) = z^2$). In both cases, the filled Julia set is partitioned into two subsets by the two external rays which crash together at the escaping critical point $-a$. These rays are shown in white.

Near either of these two puncture points, the curve $\mathcal{S}_2$ can be described by a Laurent expansion of the form

$$v(a) = \begin{cases} a - 1/(3a) - 1/(3a)^3 - 2/(3a)^5 - 5/(3a)^7 - \cdots & \text{for } \delta \text{ near } 0, \\ -2a + 1/(3a) + 1/(3a)^3 + 2/(3a)^5 + 5/(3a)^7 + \cdots & \text{for } \delta \text{ near } \infty \end{cases}$$

(compare [Ki06]).

However, one cannot expect that all hyperbolic Julia sets will be so easy to understand. The interaction between the two critical orbits can lead to remarkable richness and complexity, even in the hyperbolic case. The

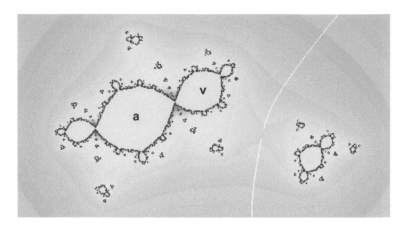

Figure 18. Julia set for a map F in the inner escape region of $\mathcal{S}_2$, with kneading sequence $\overline{0}$ and with associated quadratic map $Q(z) = z^2 - 1$. (Here $\delta = .7 + .3i$.) (See Plate III.)

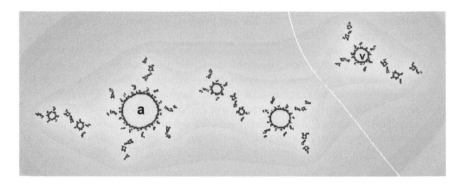

Figure 19. Julia set for a map $F \in \mathcal{S}_2$ in the outer escape region, with kneading sequence $\overline{10}$ and with associated quadratic map $Q(z) = z^2$. (Parameter value $\delta = 2 + .5i$.) In both of these figures, the two external rays which crash together at the escaping critical point $-a$ are shown in white. (See Plate III.)

following two examples both represent points of the curve $\mathcal{S}_2$. Figure 20 shows an example of Type D with parameter $\delta = 2.03614 + 0.05431\,i$. The free critical point $-a$ lies on one of the seven branches emanating from a repelling periodic point (below the figure) with period two and rotation number $2/7$. Figure 21 shows a nearby example of Type C with $\delta = 2.03540 + 0.05316\,i$.

Figure 20. Detail from the Julia set of a more complicated hyperbolic map of type D. The Fatou component of the critical point $-a$ (in the center) has period 42, while the critical point $+a$ (far outside to the left) has period 2. Preimages of $-a$ are surrounded by foliage, while preimages of $+a$ are in the open. (See Plate IV.)

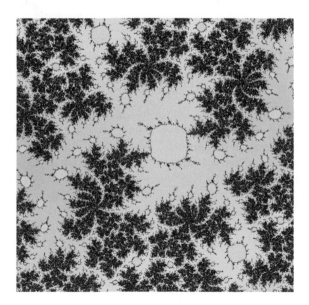

Figure 21. Detail from the Julia set of a hyperbolic map of type C. The critical point $+a$ (far to the left) has period 2, while the Fatou component of $-a$ (in the center) maps to the Fatou component of $+a$ after 85 iterations. Preimages of $-a$ are connected to the foliage on both sides, while the other preimages of $+a$ are connected on one side only. (See Plate IV.)

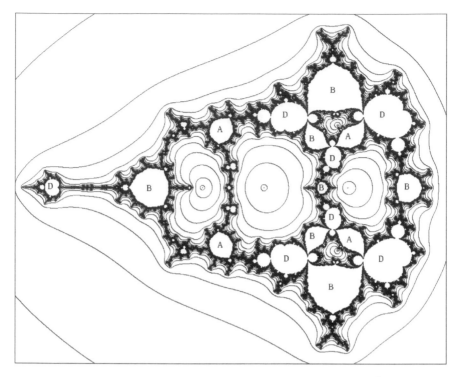

Figure 22. Connectedness locus in $\mathcal{S}_3/\mathcal{I}$, with coordinate $c = \frac{F^{\circ 2}(a) - F(a)}{F(a) - a}$.

Period 3. The curve $\mathcal{S}_3$ has genus one, being diffeomorphic to an eight times punctured torus. However, the quotient $\mathcal{S}_3/\mathcal{I}$ has genus zero, with six ends. (Thus the involution $\mathcal{I}$ fixes four of the ends of $\mathcal{S}_3$ but permutes the other four in pairs.) More explicitly, a map in $\mathcal{S}_3$ is determined up to affine conjugation by the "shape" of its superattracting orbit. After an affine conjugation which moves the critical point a to zero and the critical value $F(a)$ to one, our map F will be replaced by something of the form $w \mapsto \alpha w^3 + \beta w^2 + 1$, where say $0 \mapsto 1 \mapsto 1 + c \mapsto 0$. Here c is a parameter which specifies the "shape" of the critical orbit. It is not difficult to solve for

$$\alpha = -\frac{c^3 + 2c^2 + c + 1}{c(c+1)^2}, \qquad \beta = c - \alpha = \frac{c^4 + 3c^3 + 3c^2 + c + 1}{c(c+1)^2}.$$

Thus the coefficients $\alpha \neq 0$ and β are uniquely determined by the shape parameter c, which can take any finite value with the exception of $c = 0$, -1, where the denominator vanishes, and $c = -0.1225 \pm 0.7448\,i$ or -1.7548 (to four significant figures), where the numerator of the expression for α vanishes. These five points, together with the point at infinity, represent punctures in the Riemann surface $\overline{\mathcal{S}}_3/\mathcal{I} \approx \widehat{\mathbb{C}}$. The associated connectedness locus in the c-plane, or in other words in $\mathcal{S}_3/\mathcal{I}$, is shown in Figure 22. To obtain a corresponding picture for the curve $\mathcal{S}_3$ itself, we would have to form the 2-fold covering of $\widehat{\mathbb{C}}$, ramified at four of the six puncture points. In fact this ramified covering is a (punctured) elliptic curve which can be identified with the Riemann surface of the function

$$c \mapsto \sqrt{c(c^3 + 2c^2 + c + 1)} = c(c+1)\sqrt{\alpha},$$

ramified at $c = 0$ and at the three values of c which represent period three centers in the Mandelbrot set.

Remark 5.18. This period three moduli space $\mathcal{S}_3/\mathcal{I}$ has six escape regions, corresponding to maps for which the orbit of the free critical point escapes to infinity. Each of these regions is canonically isomorphic to a punctured disk. The five finite puncture points can be characterized as the five values of c for which the quadratic map $z \mapsto z^2 + c$ has critical orbit of period one, two or three. In other words, they are the center points of the five hyperbolic components in the Mandelbrot set which have period ≤ 3. For F in three of these escape regions, corresponding to the period 3 centers in the Mandelbrot set, the Julia set $J(F)$ contains infinitely many copies of the associated quadratic Julia set. (This is similar to the situation in Figure 18.) For F in the other three escape regions, the Julia set is made up out of points and circles, as in Figure 19. (Compare [BH92].)

This curve $\mathcal{S}_3/\mathcal{I}$ has four hyperbolic components of Type A. These are the images of the eight Type A centers in $\mathcal{S}_3$, representing maps of the form $F(z) = z^3 + v$ which have critical orbits of period three. The corresponding shape parameters are given by $c = v^2 \approx -1.598 \pm .6666\,i$ or $.0189 \pm .6026\,i$ (the four points where $\beta(c) = 0$). It has eight hyperbolic components of Type B, that is four with $F(a) = -a$, and four with shape parameter $c' = -1 - 1/c$ satisfying $F(-a) = a$. There are infinitely many components of Types C and D.

Period 4. The curve $\mathcal{S}_4$ will be studied in Bonifant and Milnor [BM], the continuation of this paper. Let me simply mention that the situation is more complicated in period 4. In particular, four of the twenty ends of $\mathcal{S}_4$ have multiplicity two, or in other words are ramified over the a-plane.

Proof of Theorem 5.2 (Outline): To prove that the curve $\mathcal{S}_p$ is smooth, we proceed as follows. For any $F \in \widehat{\mathcal{P}}(3)$ which is sufficiently close to $\mathcal{S}_p$, it follows from the implicit function theorem that there is a unique periodic point close to a. Evidently the multiplier $\lambda = \lambda(F)$ of this periodic point depends holomorphically on $F \in \widehat{\mathcal{P}}(3)$. We must prove that the partial derivatives $\partial\lambda/\partial a$ and $\partial\lambda/\partial v$ are not simultaneously zero.

At hyperbolic points, the proof is relatively easy. For each hyperbolic component of $\mathcal{C}(3)$ is canonically biholomorphic to one of four standard models. (See [Mi92b].) Within each of these standard models, the locus of those maps for which a specified one of the critical points is precisely periodic is transparently a smooth complex submanifold.

In a neighborhood of a non-hyperbolic map, we make use of a surgery argument as follows. Evidently the critical point $-a$ cannot belong to the attractive basin of $+a$, hence the immediate basin of a is isomorphic to the open unit disk, parametrized by its Böttcher coordinate. Using quasiconformal surgery, it is not difficult to replace this superattractive basin by a basin with multiplier λ.

This yields a new map F_λ, where λ varies over the open unit disk, which depends smoothly on λ and coincides with the original map when $\lambda = 0$. Since the composition $\lambda \mapsto F_\lambda \mapsto \lambda(F_\lambda)$ is the identity map, this proves Theorem 5.2. $\qquad\square$

Some of the ends in the curve $\mathcal{S}_p$ can be described quite explicitly as follows.

Lemma 5.19. *Each end of the cubic superattracting locus $\mathcal{S}_p$ can be described, for large $|a|$, by a Laurent series having one of the following two forms*

$$\left.\begin{array}{c} v - a \\ v + 2a \end{array}\right\} = k_0 + k_1/a^{1/\mu} + k_2/a^{2/\mu} + \cdots,$$

where μ is the multiplicity. Among these, there are $\nu_2(p)/2$ ends with trivial kneading sequence. These correspond to the $\nu_2(p)/2$ period p centers $z \mapsto z^2 + c_0$ in the Mandelbrot set, and have Laurent series of the form

$$v - a = c_0/3a + k_3/a^3 + k_5/a^5 + \cdots.$$

For these special ends, the non-trivial components of the corresponding filled Julia sets $K(F)$ are homeomorphic to the filled Julia sets of the associated quadratic map $z \mapsto z^2 + c_0$. These ends have the property that, for large $|a|$, the orbit of the periodic critical point a under the associated cubic map F always stays within the $1/|a|$ neighborhood of a. On the other hand, if F belongs to an end of $\mathcal{S}_p$ which is not of this form, with $|a|$ large, then the orbit of a under F must also pass close to the co-critical point $-2a$.

The proof is not difficult, and will be omitted. $\qquad\square$

It follows easily that each of these special ends is carried into itself by the involution $\mathcal{I}$. Using the identity $g(\mathcal{S}_p) = 2g(\mathcal{S}_p/\mathcal{I}) + r/2 - 1$, where g is the genus and r the number of ramification points, this yields the crude inequality $g(\mathcal{S}_p) \geq r/2 - 1 \geq \nu_2(p)/4 - 1$, which implies the following: *The genus of $\mathcal{S}_p$ is non-zero for $p \geq 3$, and tends to infinity as $p \to \infty$.*

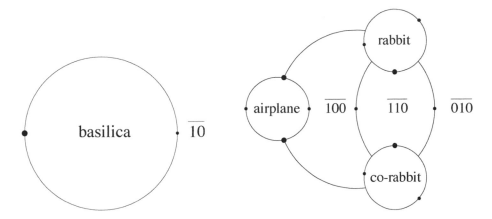

Figure 23. Sketch of the cell subdivision for $\mathcal{S}_2/\mathcal{I}$ and the conjectured cell subdivision for $\mathcal{S}_3/\mathcal{I}$. (Compare Figures 16 and 22.) Here the centers of components of Types A and B have been indicated by heavy and light dots respectively. Each of the complementary 2-cells in $\overline{\mathcal{S}}_p/\mathcal{I}$ is labeled, either with its kneading sequence, or with the nickname for its associated quadratic map if the kneading sequence is $\overline{0}$.

5D   A Conjectured Cell Subdivision of $\mathcal{S}_p$

How can we understand the topology of the curve $\mathcal{S}_p$? Even if we could verify the conjecture that $\mathcal{S}_p$ is connected, this would leave open the problem of computing its genus. This is not a trivial question, since this curve in $\mathbb{C}^2$ has complicated singularities when extended to a variety in the complex projective plane. Here is one approach to understanding the topology of $\mathcal{S}_p$ and of its non-singular compactification $\overline{\mathcal{S}}_p$. We will usually assume that $p \geq 2$, since the structure of $\mathcal{S}_1$ is much simpler.

Since each bounded hyperbolic component in $\mathcal{S}_p$ is conformally isomorphic to the unit disk, with a preferred center point, we can define the concept of a *regulated path* in the connectedness locus $\mathcal{C}(\mathcal{S}_p)$. (Compare the appendix. Of course, since we don't know whether $\mathcal{C}(\mathcal{S}_p)$ is locally connected, there may be real difficulties in proving the existence of such regulated paths.)

Conjecture 5.20. *For each escape region $\mathcal{E}_\iota \subset \mathcal{S}_p$ with $p \geq 2$, there is a unique simple closed regulated curve Γ_ι which separates $\mathcal{E}_\iota$ from the other escape regions. This curve Γ_ι bounds a topological cell $\mathbf{e}_\iota$ within the compactified curve $\overline{\mathcal{S}}_p$, so that $\mathbf{e}_\iota$ contains $\mathcal{E}_\iota$ but is disjoint from the other escape regions. Furthermore, there is a cell subdivision of $\overline{\mathcal{S}}_p$ with the union Γ of the Γ_ι as 1-skeleton, and with the $\mathbf{e}_\iota$ as 2-cells. In particular, the union of the closures $\overline{\mathbf{e}}_\iota = \mathbf{e}_\iota \cup \Gamma_\iota$ of these 2-cells is the entire curve $\overline{\mathcal{S}}_p$.*

We can describe this 1-skeleton Γ as the "*core*" of $\mathcal{S}_p$, since all of the most interesting dynamics seems to be centered around it. In particular, it is conjectured that the center points of all hyperbolic components of Types A and B are contained in Γ.

For the special case $p = 1$, the situation is much simpler. Define the *core* of $\mathcal{S}_1$ to be the center point of its unique hyperbolic component of Type A, so that we obtain a cell subdivision with a single vertex (corresponding to the map $z \mapsto z^3$), with the rest of $\mathcal{S}_1$ as 2-cell.

If this conjecture is true, then it follows easily that the quotient $\overline{\mathcal{S}}_p / \mathcal{I}$ has a corresponding cell structure, with the quotient graph $\Gamma / \mathcal{I}$ as 1-skeleton. See Figure 23, which illustrates the period two and three cases.

For $p \geq 2$, we can divide the bounded hyperbolic components in $\mathcal{S}_p$ into two classes as follows. Call V a *core component* if there are two or more escape regions $\mathcal{E}_\iota$ such that the intersection $\overline{V} \cap \overline{\mathcal{E}}_\iota$ contains more than one point. (In terms of this conjectured cell structure, this means that V is cut into two or more pieces by Γ.) Call V a *peripheral component* otherwise. We can define the *thick core* $\Gamma^+ \subset \mathcal{S}_p$ to be the closure of the union of all core components. Then the complement $\mathcal{S}_p \smallsetminus \Gamma^+$ is a disjoint union of open sets $\mathcal{E}_\iota^+$, one contained in each cell $\mathbf{e}_\iota$. By a *limb* of the connectedness locus within $\mathbf{e}_\iota$ we mean a connected component of $\mathcal{C}(\mathcal{S}_p) \cap \mathcal{E}_\iota^+$. Thus every peripheral hyperbolic component must be contained in some unique limb. *Conjecturally every limb L is attached to Γ^+ at a unique parabolic point $F_0 \in \overline{L} \cap \Gamma^+$, and is separated from the rest of the connectedness locus by two external rays in $\mathcal{E}_\iota$ which land at F_0.*

Again the case $p = 1$ is simpler. The thick core $\Gamma^+ \subset \mathcal{S}_1$ is defined to be the closure of the principal hyperbolic component, and its complement is defined to be $\mathcal{E}^+$.

There are completely analogous descriptions within $\mathcal{S}_p / \mathcal{I}$. The reader should have no difficulty in distinguishing core components and limbs among the reasonably large components in Figures 6, 16, and 22.

6   Quadratic Julia Sets in Cubic Parameter Space

Assuming that the cell subdivision of Section 5D exists, we can get a good idea of the structure of the connectedness locus $\mathcal{C}(\mathcal{S}_p)$ by studying its intersection with each of the conjectured 2-cells. This section will attempt to provide an explicit description for those complementary 2-cells which have trivial kneading sequence, and hence are associated with quadratic polynomials of critical period p. (For an analogous construction, see [BHe01]).

Let Q be a quadratic polynomial with period p critical point, and let $\mathcal{E}_Q \subset \mathcal{S}_p / \mathcal{I}$ be the associated escape region. Thus, for every $F \in \mathcal{E}_Q$, the filled Julia set $K(F)$ contains infinitely many copies of $K(Q)$. Let

$\mathbf{e}_Q \supset \mathcal{E}_Q$ be the 2-cell which contains $\mathcal{E}_Q$ in the conjectured cell subdivision of Section 5D.

Basic Construction. Let $K^\sharp(Q)$ be the compact set which is obtained by cutting open[4] the filled Julia set $K(Q)$ along its minimal Hubbard tree T_Q. Thus the preimage $T_Q^\sharp = \eta^{-1}(T_Q)$, under the natural projection $\eta :$ $K^\sharp(Q) \to K(Q)$ is a topological circle, which can be described as the *inner boundary* of $K^\sharp(Q)$. (Compare Figure 24.) Let $J^\sharp(Q) = \eta^{-1}(J(Q))$ be the preimage of the Julia set in $K^\sharp(Q)$.

Conjecture 6.1. *There exists a dynamically defined canonical embedding ϕ from the cut-open filled Julia set $K^\sharp(Q)$ into the parameter cell $\overline{\mathbf{e}}_Q$ which turns $K^\sharp(Q)$ inside-out so that the inner boundary circle $T_Q^\sharp$ of $K^\sharp(Q)$ maps to the outer boundary circle $\Gamma_Q = \partial \mathbf{e}_Q$.*

Alternatively, for periods ≤ 3, since $\mathcal{S}_p/\mathcal{I}$ has genus zero, we can turn $\mathcal{S}_p/\mathcal{I}$ inside out by mapping it onto the Riemann sphere so that the puncture point in $\mathcal{E}_Q$ goes to the point at infinity. The resulting picture can be compared directly with $K^\sharp(Q)$. As an example, for the case $Q(z) = z^2 - 1$ we can compare Figure 24 with Figure 25. Evidently there is a strong resemblance between the region outside the white circle $T_Q^\sharp$ in Figure 24 and the region outside the black circle Γ_Q in Figure 25. The most striking difference is the presence of many copies of the Mandelbrot set in Figure 25, and also in the right half of Figure 26. (Compare Conjecture 6.2 below.) Figures 27 and 28 provide a similar example for period three.

The embedding $\phi : K^\sharp(Q) \to \mathbf{e}_Q$ can be described intuitively as follows. Each point $\widehat{z} \in K^\sharp(Q) \setminus J^\sharp(Q)$ corresponds to a cubic map $\phi(\widehat{z}) \in \overline{\mathbf{e}}_Q$ which is constructed starting with the quadratic map Q by altering the dynamics so that $\widehat{z}$ will be an additional critical point (or a double critical point in the special case that $\widehat{z}$ is already critical). Thus there are three cases.

Case A. If $\widehat{z}$ belongs to the Fatou component V_0 containing 0 then we will obtain a cubic map of Type A, with both critical points in the same Fatou component.

Case B. If $\widehat{z}$ belongs to some forward image $Q^{\circ j}(V_0)$ with $0 < j < p$, then we will obtain a component of Type B, with both critical points in the same cycle of Fatou components.

Case C. For all other $\widehat{z}$ in the Fatou subset of $K^\sharp(Q) \setminus J^\sharp(Q)$, we will obtain a component of Type C.

This provides a very rough description of the map ϕ within the Fatou region of $K^\sharp(Q)$. It is conjectured that ϕ extends continuously over all of $K^\sharp(Q)$.

---

[4]In the special case $p = 1$, so that $Q(z) = z^2$, no cutting is necessary, and $K^\sharp(Q)$ should be identified with the unit disk $K(Q)$.

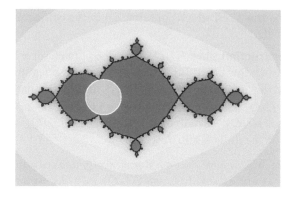

Figure 24. The cut-open filled Julia set $K^\sharp(Q)$ for the "basilica" map $Q(z) = z^2 - 1$. Here the Julia set $J^\sharp(Q)$ is colored black. (See Plate IV.)

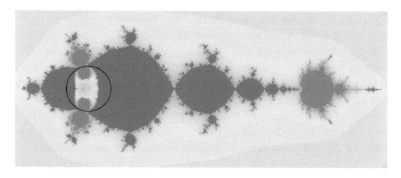

Figure 25. The curve $\mathcal{S}_2$ inverted so that the "basilica" escape region will be on the outside. The black circle approximates the core Γ_Q of $\mathcal{S}_2$. (See Plate IV.)

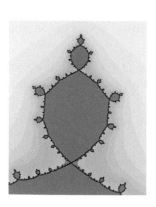
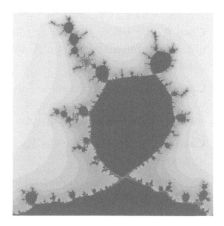

Figure 26. Details near the top of Figures 24 and 25, respectively.

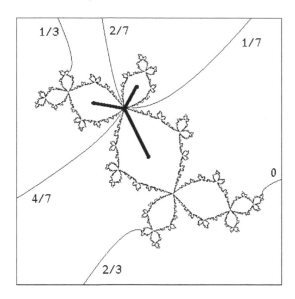

Figure 27. Julia set for the Douady rabbit, with minimal Hubbard tree emphasized.

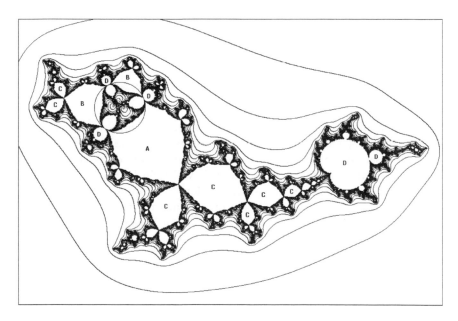

Figure 28. This is Figure 22, inverted in a small circle about the upper right puncture point and then rotated 90°. Our claim is that the region outside of the small circle is homeomorphic to the Figure 27, cut open along its minimal tree, with further decorations added, including Mandelbrot sets or portions of Mandelbrot sets.

To deal with components of Type D, the following supplementary statement is needed. A point in the cut-open Julia set $J^\sharp(Q)$ will be called *periodic* if it is the image of a periodic point of J_Q.

Conjecture 6.2. *For every Julia periodic point $\widehat{z} \in J^\sharp(Q)$, a copy of the Mandelbrot set is attached to $\phi(K^\sharp(Q))$ at the point $\phi(\widehat{z})$ within the curve $\mathcal{S}_p/\mathcal{I}$. More precisely, if $\widehat{z}$ is a periodic point with combinatorial rotation number $m/n \neq 0$, then the (m/n)-limb of the Mandelbrot set is attached within $\mathbf{e}_Q$, with the rest of the Mandelbrot set attached in the complementary parameter region $\mathcal{S}_p/\mathcal{I}\smallsetminus\mathbf{e}_Q$. On the other hand, for periodic points of rotation number zero, the entire Mandelbrot set will be attached within $\mathbf{e}_Q$. In all cases, this attached Mandelbrot set intersects $\phi(K^\sharp(Q))$ only at the point $\phi(\widehat{z})$ itself; and in all cases there are further decorations added to these Mandelbrot sets.*

There is a very close connection between periodic points in the Julia set J_Q and external rays which are periodic under angle doubling. The connection becomes even closer if we pass to the cut-open Julia set $J^\sharp(Q)$. In fact:

Lemma 6.3. *Every periodic external ray lands on a periodic point in $J^\sharp(Q)$, and every periodic point in $J^\sharp(Q)$ is the landing point of exactly one external ray. Hence, if Conjecture 6.2 is correct, there is a one-to-one correspondence between periodic external rays for $K(Q)$ and Mandelbrot copies attached to $\phi(K^\sharp(Q))$.*

Evidently this provides a simple way of labeling those Mandelbrot copies which are attached in regions $\mathbf{e}_Q$.

Outline of the Construction. This conjectured embedding ϕ from $K^\sharp(Q)$ into $\mathbf{e}_Q \subset \mathcal{S}_p$ will be described in the following four subsections. Section 6A will describe $\phi(\widehat{z})$ in the case where $\widehat{z}$ corresponds to the center of a Fatou component in $K(Q)$. The image $\phi(\widehat{z})$ will then be the center of a hyperbolic component in $\mathcal{S}_p/\mathcal{I}$. As in Section 4, a representative cubic map is most easily described by means of its Hubbard tree. In Section 6B, we extend to the case of an arbitrary $\widehat{z}$ in the Fatou subset $K^\sharp(Q)\smallsetminus J^\sharp(Q)$. Section 6C will study the case where $\widehat{z}$ belongs to the Julia set $J^\sharp(Q)$ but is not periodic, and Section 6D will consider the case of a periodic point in the Julia set.

6A   Enramification: The Hubbard Tree

The rather ungainly word "*enramification*" will be used for the operation of constructing a cubic polynomial map F from some given quadratic polynomial map Q by artificially introducing a new critical point (or by replacing the simple critical point by a double critical point). Here three cautionary points should be emphasized:

- Although the construction is known to make sense in many cases, its existence in general is conjectural.

- The construction is not always uniquely defined. More precisely if $\widehat{z}$ belongs to the minimal tree T_Q, and is not an endpoint of this tree, then multiple choices are possible; hence the necessity of cutting-open along T_Q in the discussion above.

- This new map F is well defined only up to affine conjugation, so that we cannot distinguish between F and its image $\mathcal{I}(F) : z \mapsto -F(-z)$. For this reason, we will work with $\mathcal{S}_p/\mathcal{I}$ rather than $\mathcal{S}_p$.

We first discuss enramification as an operation on abstract Hubbard trees. Here is a brief outline to fix notations. (See the appendix for a more detailed presentation.) Let F be a post-critically finite polynomial of degree $d \geq 2$, and let $S \supset F(S)$ be a forward invariant set containing the critical points. The associated *Hubbard tree* $T = T(S)$ is a finite acyclic simplicial complex which has dimensional one, except in the special case where $S = T(S)$ consist of a single point. Its underlying topological space $|T| \subset K(F)$ is the smallest subset of $K(F)$ which contains S and is connected by regulated paths. Define the *valence* $n(z)$ of a point $z \in |T|$ to be the number of connected components of $|T| \smallsetminus \{z\}$. The set V of vertices of T consists of S together with the finitely many points $\mathbf{v}$ of valence $n(\mathbf{v}) \geq 3$. Note that F maps vertices into vertices. Furthermore, if $\mathbf{e}$ is an edge with $\mathbf{v}$ as one of its two boundary points and if U is a small neighborhood of $\mathbf{v}$, then F maps $U \cap \mathbf{e}$ into a unique edge which will be denoted by $F_{\mathbf{v}}(\mathbf{e})$. Similarly, any iterate $F^{\circ k}$ induces a map $F_{\mathbf{v}}^{\circ k}$ from edges at $\mathbf{v}$ to edges at $F^{\circ k}(\mathbf{v})$. As part of the structure of T we include the following:

(1) The map F restricted to the set V of vertices.

(2) The *local degree* function which assigns an integer $d(\mathbf{v}) \geq 1$ to each vertex, with
$$\sum_{\mathbf{v}} (d(\mathbf{v}) - 1) = d - 1 \,.$$
Here $d(\mathbf{v}) > 1$ if and only if $\mathbf{v}$ is a critical point.

(3) The *angle function* $\angle(\mathbf{e}_1, \mathbf{e}_2) \in \mathbb{Q}/\mathbb{Z}$, where $\mathbf{e}_1$ and $\mathbf{e}_2$ are any two edges meeting at a common vertex $\mathbf{v}$. This vanishes if and only if $\mathbf{e}_1 = \mathbf{e}_2$. The map F multiplies all angles at $\mathbf{v}$ by $d(\mathbf{v})$, in the sense that
$$\angle\big(F_{\mathbf{v}}(\mathbf{e}_1), F_{\mathbf{v}}(\mathbf{e}_2)\big) = d(\mathbf{v}) \angle(\mathbf{e}_1, \mathbf{e}_2) \,.$$

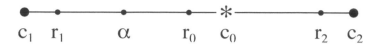

$$c_1 \quad r_1 \qquad\qquad \alpha \qquad\qquad r_0 \quad c_0 \qquad\qquad\qquad r_2 \quad c_2$$

Figure 29. A Hubbard tree for the "airplane" map $Q(z) = z^2 - 1.754866\cdots$, with critical orbit $\{c_j\}$ of period three. The associated root points $\{r_j\}$ form a period three orbit in the Julia set with rotation number zero, while the fixed point α has rotation number $1/2$.

Recall that any periodic point z of period $p \geq 1$ in the Julia set has a well-defined *rotation number* in $\mathbb{Q}/\mathbb{Z}$ which describes the way that the external rays landing at z are rotated by $F^{\circ p}$. (See for example [Mi00].) If we choose a Hubbard tree having z as a vertex, then this rotation number also describes the way in which the edges incident to z are rotated by the correspondence $F_z^{\circ p}$.

For any critically finite polynomial F, there is a unique *minimal tree* $T_{\min}(F)$ which is obtained by taking the union of critical orbits as the generating set S. Our preliminary goal can be described roughly as follows:

> *Given a post-critically finite quadratic polynomial $Q(z) = z^2 + c$, to study cubic polynomials F such that the minimal Hubbard tree for F can be constructed by minor modifications of some Hubbard tree for Q.*

To this end, we start with the minimal Hubbard tree $T_{\min} = T_{\min}(Q)$ for the quadratic map. Given any periodic or pre-periodic point $\mathbf{w}_0 \in K(F)$, let $T(\mathbf{w}_0)$ be the Hubbard tree for Q which has as its generating set S the union of the critical orbit and the orbit of $\mathbf{w}_0$.

Theorem 6.4. *In most cases the extended tree $T = T(\mathbf{w}_0)$, as described above, can be made into the Hubbard tree $\widehat{T}$ of a cubic polynomial simply by replacing the local degree function $d(\mathbf{v})$ by*

$$\widehat{d}(\mathbf{v}) = \begin{cases} d(\mathbf{v}) & \text{if} \quad \mathbf{v} \neq \mathbf{w}_0 \\ d(\mathbf{v}) + 1 & \text{if} \quad \mathbf{v} = \mathbf{w}_0, \end{cases}$$

and by carefully modifying the angle function at $\mathbf{w}_0$ and at all vertices which are iterated pre-images of $\mathbf{w}_0$. In fact, if $\mathbf{w}_0$ has valence $n(\mathbf{w}_0) \geq 1$, then there are precisely $n(\mathbf{w}_0)$ distinct ways of carrying out this angle modification. However, there is one special case where a different construction is necessary, namely the case when $\mathbf{w}_0$ is a periodic point in the Julia set with rotation number zero but with valence $n(\mathbf{w}_0) = 2$.

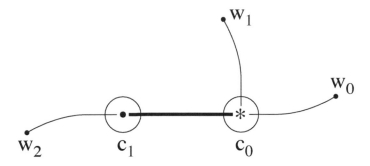

Figure 30. The extended tree $T(\mathbf{w}_0)$ for the "basilica" map $Q(z) = z^2 - 1$, together with the period three orbit $\{\mathbf{w}_0, \mathbf{w}_1, \mathbf{w}_2\}$. The minimal tree $T_{\min} \subset T(\mathbf{w}_0)$ has been emphasized. Note that the edge from $\mathbf{c}_0$ to $\mathbf{w}_0$ maps to the path from $\mathbf{c}_1$ through $\mathbf{c}_0$ to $\mathbf{w}_1$.

For an example of this special case, see Figure 29, and for its resolution see Remark 6.7 and Figure 31.

Note. If $n(\mathbf{w}_0) \le 1$, then the construction is uniquely defined and no angle modification is necessary. In particular, this will be the case whenever $\mathbf{w}_0$ lies strictly outside of the minimal tree. (Compare Figure 30.)

The proof of Theorem 6.4 will be based on two lemmas. Let $Q(z) = z^2 + c$ be a critically finite quadratic polynomial, and let

$$0 = \mathbf{c}_0 \mapsto \mathbf{c}_1 \mapsto \cdots \mapsto \mathbf{c}_k$$

be the distinct points in the critical orbit, with $\mathbf{c}_1 = c$. (We are mainly interested in the critically periodic case where $k = p - 1$ and $Q(\mathbf{c}_{p-1}) = \mathbf{c}_0$; but it is no more difficult to consider the preperiodic case at the same time.)

Lemma 6.5. *Let* $Q(z) = z^2 + c$ *be a critically finite quadratic polynomial, as above. If $k > 0$ then, in the minimal tree $T_{\min}$, we have $n(\mathbf{c}_1) = 1$ and $n(\mathbf{c}_0) \le 2$. In fact, in the critically periodic case of period p we have*

$$1 = n(\mathbf{c}_1) \le n(\mathbf{c}_2) \le \cdots \le n(\mathbf{c}_{p-1}) \le n(\mathbf{c}_0) \le 2.$$

If $\mathbf{w}_0$ is a periodic or preperiodic point which lies strictly outside of $T_{\min}$, then a similar argument applied to the extended tree $T(\mathbf{w}_0)$ shows that $n(\mathbf{w}_0) = 1$.

Proof: Every non-degenerate tree must have at least two vertices $\mathbf{v}$ with $n(\mathbf{v}) = 1$. For otherwise, starting with any edge we could keep finding successive edges on one end or the other until we generate a closed loop, which is impossible. Since every vertex outside the critical orbit has valence $n(\mathbf{v}) \geq 3$ by the definition of the minimal tree, it follows that there are at least two of the $\mathbf{c}_j$ with $n(\mathbf{c}_j) = 1$.

Therefore, since F is injective in the neighborhood of any non-critical point, we have

$$1 = n(\mathbf{c}_1) = n(\mathbf{c}_2) \leq \cdots \leq n(\mathbf{c}_k) \leq n(Q(\mathbf{c}_k)).$$

Furthermore, since F is at most 2-to-1 in a neighborhood of the critical point $\mathbf{c}_0$, it follows that $n(\mathbf{c}_0) \leq 2$. In the critically periodic case, since $Q(\mathbf{c}_{p-1}) = \mathbf{c}_0$, it follows that $n(\mathbf{c}_i) \leq 2$ for all points of the critical orbit.

Now consider the extended tree $T(\mathbf{w}_0)$ in the case that $\mathbf{w}_0 \notin |T_{\min}|$. If $\mathbf{w}_0 \mapsto \mathbf{w}_1 \mapsto \cdots$, then a similar argument shows that $n(\mathbf{w}_h) \leq n(\mathbf{w}_{h+1})$ provided that $\mathbf{w}_h \neq 0$. On the other hand, this extended tree must have at least one endpoint outside of $|T_{\min}|$. For otherwise, starting with any edge $\mathbf{e}$ outside of $|T_{\min}|$ and extending in both directions, we could either construct a closed loop outside of $T_{\min}$, or else construct a path from $|T_{\min}|$ to itself which passes through $\mathbf{e}$. Since both cases are impossible, it follows easily that $\mathbf{w}_0$ is an endpoint of $T(\mathbf{w}_0)$, as asserted. (Note that this last argument works even in the degenerate case $Q(z) = z^2$.) $\qquad\square$

Lemma 6.6. *Let* $\mathbf{w}_0$ *be periodic of period* $p \geq 1$ *in the quadratic Julia set* $J(Q)$. *If* $n(\mathbf{w}_0) \geq 3$, *then the map* $Q_{\mathbf{w}_0}^{\circ p}$ *permutes the* $n(\mathbf{w}_0)$ *incident edges cyclically.*

Proof: An analogous statement for external rays landing at $\mathbf{w}_0$ is proved in [Mi00, Lemma 2.7]; and the corresponding statement for the tree $T(\mathbf{w}_0)$ follows easily. $\qquad\square$

Proof of Theorem 6.4: We distinguish six cases.

Case 1. Suppose that $\mathbf{w}_0$ is an *endpoint* of the tree $T(\mathbf{w}_0)$, in the sense that $n(\mathbf{w}_0) \leq 1$. Then we need only replace the local degree $d(\mathbf{w}_0)$ by

$$\widehat{d}(\mathbf{w}_0) = d(\mathbf{w}_0) + 1.$$

The given angle function is consistent without any change. Evidently this construction is uniquely defined. Using Poirier's Theorem (stated as Theorem 7.3 in the appendix), it yields a cubic map which is uniquely defined up to affine conjugation.

By Lemma 6.5 this includes all cases where the point $\mathbf{w}_0$ does not belong to $|T_{\min}|$. *For the rest of the proof, it will be tacitly understood that* $\mathbf{w}_0 \in |T_{\min}|$, *and that* $n(\mathbf{w}_0) \geq 2$.

Case 2. Suppose Q is critically periodic, and that $\mathbf{w}_0$ is one of the points $\mathbf{c}_h$ in the critical orbit, but is not the critical point $\mathbf{c}_0 = 0$. Since we have assumed that $n(\mathbf{w}_0) \geq 2$, it follows by Lemma 6.5 that $n(\mathbf{w}_0) = 2$, and it is not hard to check that the angle $\angle(\mathbf{e}_1, \mathbf{e}_2)$ between the two edges which meet at $\mathbf{w}_0$ must be equal to $1/2$. To make a cubic tree, we must first redefine the local degree at $\mathbf{w}_0$ to be $\widehat{d}(\mathbf{w}_0) = 2$. Let $\mathbf{e}_1'$ and $\mathbf{e}_2'$ be the two edges which meet at $Q(\mathbf{w}_0)$. Since the cubic map $\widehat{F}_{\mathbf{w}_0}$ must double angles, the angles for the cubic map must satisfy

$$2 \angle(\mathbf{e}_1, \mathbf{e}_2) \;=\; \angle(\mathbf{e}_1', \mathbf{e}_2') \;=\; 1/2 \,.$$

In other words, we must change the angle at $\mathbf{w}_0$ to either $\angle(\mathbf{e}_1, \mathbf{e}_2) = \frac{1}{4}$ or $\angle(\mathbf{e}_1, \mathbf{e}_2) = \frac{3}{4}$. Furthermore, we must make exactly the same change at every iterated pre-image $\mathbf{c}_j$, $0 < j \leq h$, for which $n(\mathbf{c}_j) = 2$. Thus, in this case there are just two essentially distinct ways of carrying out the construction.

As an example, if we start with a *real* quadratic map in Case 2, then the construction will yield two different cubic maps which are complex conjugate to each other, but are not affinely conjugate to any real map.

Case 3. Now suppose that $\mathbf{w}_0$ the critical point 0. Then we must replace $d(\mathbf{w}_0) = 2$ by $\widehat{d}(\mathbf{w}_0) = 3$. Whether $\mathbf{w}_0$ is periodic or pre-periodic, there are just two incident edges. Again there are two possible choices for the modified angles, but in this case the possible choices are $\frac{1}{3}$ and $\frac{2}{3}$.

Case 4. Next suppose that $\mathbf{w}_0$ is strictly preperiodic. (Compare [Bi89].) Then the local map from a neighborhood of $\mathbf{w}_0$ to a neighborhood of $Q(\mathbf{w}_0)$ preserves the angles between neighboring edges, say $\theta_1, \ldots, \theta_n$. The angles at $Q(\mathbf{w}_0)$ will not change; but we must choose new angles $\widehat{\theta}_j$ at $\mathbf{w}_0$ satisfying $2\widehat{\theta}_j \equiv \theta_j \pmod{\mathbb{Z}}$. In order to satisfy the condition that $\sum \widehat{\theta}_j = \sum \theta_j = 1$, we must choose $\widehat{\theta}_j = (\theta_j + 1)/2$ for one of the n angles, and $\widehat{\theta}_j = \theta_j/2$ for the $n - 1$ remaining angles. Again, any choice of angles at $\mathbf{w}_0$ must be propagated backwards to any vertices which eventually map to $\mathbf{w}_0$; and hence again there are exactly n allowable choices. If $\mathbf{w}_0 = 0$ hence $n(\mathbf{w}_0) = 2$, the argument is similar, but now the allowable angles are $\frac{1}{3}$ and $\frac{2}{3}$.

Case 5. Finally, suppose that $\mathbf{w}_0 \in |T_{\min}|$ is periodic of period $q \geq 1$ and belongs to the Julia set. If $n(\mathbf{w}_0) \geq 3$, then by Lemma 6.6 the first return map $Q^{\circ q}$ permutes the $n = n(\mathbf{w}_0)$ edges which meet at $\mathbf{w}_0$ cyclically. The same will be true for $n(\mathbf{w}_0) = 2$, proved that the rotation number at $\mathbf{w}_0$ is equal to $1/2$. Number these edges as $\{\mathbf{e}_j\}$ with $j \in \mathbb{Z}/q$ so that $Q^{\circ q \mathbf{w}_0}(\mathbf{e}_j) = \mathbf{e}_{j+1}$. Since $\mathbf{w}_0$ is in the Julia set, the angle θ_j between $\mathbf{e}_j$ and the next edge in positive cyclic order is equal to $1/n$ by definition. However, in order to make $T(\mathbf{w}_0)$ into a cubic tree, we must choose new angles $\widehat{\theta}_j$. Since $Q^{\circ n}$ has degree two at $\mathbf{w}_0$, the required condition on these new angles is that

$$\widehat{\theta}_{j+1} \equiv 2\widehat{\theta}_j \qquad (\mathrm{mod}\ \mathbb{Z}).$$

Iterating q times, this yields $\widehat{\theta}_j \equiv 2^q\widehat{\theta}_j$ (mod $\mathbb{Z}$), so that each $\widehat{\theta}_j$ must have the form $k/(2^q - 1)$. The only possible solution is that these angles form some cyclic permutation of the sequence

$$1/(2^q - 1),\ \ 2/(2^q - 1),\ \ 2^2/(2^q - 1),\ \ \ldots,\ \ 2^{q-1}/(2^q - 1).$$

In fact, using the requirement that $0 < \widehat{\theta}_j < 1$ with $\sum \widehat{\theta}_j = 1$, we can write $2\widehat{\theta}_j = \widehat{\theta}_{j+1} + \epsilon_j$ with $\epsilon_j \in \{0, 1\}$. Summing over j, it follows that $\sum \epsilon_j = 1$; and the conclusion follows easily. Thus there are exactly n distinct solutions. Evidently, any choice of angles for $\mathbf{w}_0$ can easily be propagated backward to any vertices of $T(\mathbf{w}_0)$ which eventually map to $\mathbf{w}_0$ (including all vertices on its periodic orbit).

Since this covers all possibilities (except the case of rotation number zero and valence two, which has been excluded), it completes the proof of Theorem 6.4. □

Remark 6.7. In the exceptional case of rotation number zero and valence two, this construction cannot work, since we would have to find a non-zero solution to the equation $\theta \equiv 2\theta$ (mod $\mathbb{Z}$), which is impossible. However this does not cause any real difficulty. It merely requires a somewhat different modification in which extra branches are added to the tree. As a typical example, if we start with the airplane tree of Figure 29 and want to enramify the point $\mathbf{r}_1$ of period three, then we must add three short branches to the tree, as shown in Figure 31. (Here, following the conventions of Section 7, the angles between consecutive edges at the periodic Julia vertices of valence three must all be $120°$.) In such examples, there are always two possible choices since we can put the new critical point to either side of the original tree.

It is interesting to ask which trees contain such a vertex of valence two and rotation number zero. Conjecturally, the Hubbard tree for $Q(z) = z^2 + c$ contains such an exceptional periodic orbit if and only if the induced mapping from T to itself has positive topological entropy, or if and only if c belongs to the central "cactus" in the Mandelbrot set. (For a study of Hubbard tree entropy, see [Li07]. Following [CM89], the central *cactus* is the smallest compact subset of the Mandelbrot set which contains the central cardioid component and all of its iterated satellites.)

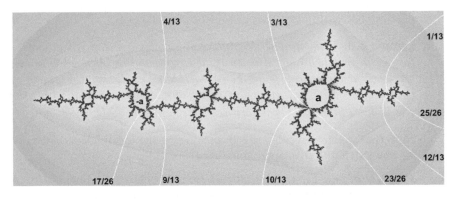

Figure 31. Above: the airplane tree of Figure 29, with a new period 3 critical orbit added. Below: the corresponding Julia set. (Here $a = c_0 \cong .828 + .019 i$ and $v = c_1 \cong -.758 - .152 i$, with $\mathbf{w}_0 = -a$.) Note that the Fatou components containing c_j and $\mathbf{w}_{j-1}$ have the repelling point r_j as a common boundary point. See Plate V.)

6B Components of Type A, B, C

Now consider an arbitrary point $\hat{z} \in K^\sharp(Q) \setminus J^\sharp(Q)$. The corresponding Fatou component in $K(Q)$ contains a unique precritical point. According to Theorem 6.4 we can construct an associated cubic Hubbard tree, and hence an associated cubic map which is postcritically finite and hyperbolic. We want to find a map in the same hyperbolic component W which corresponds to $\hat{z}$.

Consider the case of an interior point $w \in K(Q)$. Within the component U of w, there will always be one and only one pre-critical point $\mathbf{w}_0$. As in Section 4A, we can make the extended Hubbard tree $T'(\mathbf{w}_0)$ into a cubic Hubbard tree. (In the special case where $\mathbf{w}_0 \in T$, there are $n(\mathbf{w}_0)$

different ways of carrying out this construction.) Each such cubic Hubbard tree determines a post-critically finite hyperbolic cubic polynomial F_0, and we can now obtain the required polynomial F by "tuning" F_0.

To understand this tuning construction, first look at an arbitrary component U of the interior of $K(Q)$. Then some iterate $Q^{\circ m}$ maps U diffeomorphically onto the component $\mathbf{e}_0$ which contains the critical point. Since the first return map from U_0 to itself has degree two, there is a canonical Böttcher diffeomorphism β from U_0 onto the open unit disk. Hence U itself is canonically diffeomorphic to the open unit disk under $\beta \circ Q^{\circ m}$.

To study the corresponding open set in cubic parameter space, let W be the component of the polynomial F_0 within the open set of hyperbolic polynomials within $\mathcal{S}_p$. We will prove the following.

Lemma 6.8. *This open set W in the parameter curve $\mathcal{S}_p$ is either a one-, two- or three-fold branched cover of the corresponding open set U in $K(Q)$, branched at the central point $F_0 \mapsto \mathbf{w}_0$. More explicitly, the degree of this covering is two if $\mathbf{w}_0$ coincides with the quadratic critical point, three if $\mathbf{w}_0$ is one of the $p-1$ post-critical points, and $W \overset{\approx}{\to} U$ in all other cases.*

Note that we have avoided this branching, in the formulation of Conjecture 6.1, by cutting $K(Q)$ open along its Hubbard tree.

Proof (Outline): First note that each bounded Fatou component $U \subset K(Q)$ of the quadratic filled Julia set is canonically biholomorphic to the open unit disk $\mathbb{D}$. To see this, let $U_0 = Q^{\circ n}(U)$, $n \geq 0$, be the first forward image of U which contains the critical point. Then we can first map U biholomorphically onto U_0 by $Q^{\circ n}$, and then map U_0 biholomorphically onto $\mathbb{D}$, using the Böttcher coordinate associated with the degree two self-map $Q^{\circ p} : U_0 \to U_0$.

Similarly, for any bounded hyperbolic component $W \subset \mathcal{S}_p$, there is a canonical holomorphic map from W to $\mathbb{D}$. The composition $W \to \mathbb{D} \leftrightarrow U$ then yield the required holomorphic covering map from W onto $U \subset K(Q)$.

For components of Type A, B or C one proceeds as follows.[5]

For $F \in W$, let $n \geq 0$ be the smallest integer such that $F^{\circ n}(-a)$ belongs to the Fatou component U_a of the periodic point $+a$. If W is of Type C, then the Böttcher coordinate for the degree two map $F^{\circ p} : U_a \to U_a$ is well defined, and we can simply define $\beta(F)$ to be the Böttcher coordinate of $F^{\circ n}(-a)$. Evidently β maps W biholomorphically onto $\mathbb{D}$.

For Type A (with $n = 0$) or Type B (with $0 < n < p$), we have to work a little harder. For the central point $F_0 \in W$, the image $F_0^{\circ n}(-a)$ is precisely equal to $+a$, and we set $\beta(F_0) = 0$. For $F \neq F_0$, we will see

---

[5]For Type D one would use a quite different construction, mapping $F \in W$ to the multiplier of the periodic orbit associated with $-a$; but that will not concern us here.

that the Böttcher coordinate for the map $F^{op} : U_a \rightarrow U_a$ can be defined in a neighborhood of $+a$ which is large enough to contain $F^{on}(-a)$. The Böttcher coordinate of $F^{on}(-a)$ will then be the required invariant $\beta(F)$; and it is not hard to check that the correspondence $F \mapsto \beta(F)$ from the open set $W \subset S_p$ to $\mathbb{D}$ is proper and holomorphic, and is locally bijective except at F_0. We will prove that it has degree two (for Type A) or degree three (for Type B) by studying local behavior near the central point F_0. (Compare the proof of Lemma 3.6.)

First consider a component $W \subset S_p$ of Type A. It will be convenient to set $z = a + w$, and $v = F(a) = a + \delta$. Using w as independent variable, we obtain a polynomial map

$$\Psi(w) = \delta + 3aw^2 + w^3,$$

which is affinely conjugate to F, with critical points $w = 0$ and $-2a$. A brief computation shows that the n-th iterate of Ψ has the form

$$\Psi^{on}(w) = \delta_n + 3ac_nw^2 + c_nw^3 + O(w^4).$$

where δ_n and c_n are polynomial functions of δ and a, with $\delta_n = \Psi^{on}(0)$. In particular, if the critical point 0 has period p, then $\delta_p = 0$ so that

$$\Psi^{op}(w) = 3ac_pw^2 + c_pw^3 + O(w^4).$$

Here the coefficient c_p must be non-zero, since the orbit of zero has period exactly p.

The Böttcher coordinate associated with Ψ^{op} has a power series expansion of the form

$$\beta(w) = 3ac_pw + \text{(higher order terms)},$$

and converges for $|w| < |2a|$. Hence we have the asymptotic estimate

$$\beta(w) \sim 3ac_pw \quad \text{as} \quad w/a \rightarrow 0.$$

We can apply this estimate to the critical value

$$\Psi^{op}(-2a) = 3ac_p(-2a)^2 + c_p(-2a)^3 + O(a^4) = 4c_pa^3 + O(a^4).$$

Using the identity $\beta(\Psi^{op}(w)) = \beta(w)^2$, since

$\beta(\Psi^{op}(-2a)) \sim 3ac_p(4c_pa^3)$ we obtain $\beta(-2a) \sim \sqrt{12c_p^2a^4} = \pm 2\sqrt{3}\, c_p\, a^2$

as $a \rightarrow 0$. This is the required asymptotic estimate, proving that $\beta : W \rightarrow \mathbb{D}$ has degree two.

For Type B components, the construction is as follows. Suppose that

$$F^{\circ m}(U_a) = U_{-a} \text{ and } F^{\circ n}(U_{-a}) = U_a \,, \text{ where } m + n = p.$$

Let $F^{\circ m}(a) = -a + \epsilon$, where $|\epsilon|$ is small. It will be convenient to say that two variables have the same *order* as $\epsilon \to 0$ if each one is asymptotic to a constant multiple of the other.

Since $-a$ is a simple critical point, it follows that $F(-a + \epsilon) - F(-a)$ has the order of ϵ^2 as $\epsilon \to 0$, and that the first derivative $F'(-a + \epsilon)$ has the order of ϵ. It follows easily that $a - F^{\circ n}(-a)$ also has the order of ϵ^2 as $\epsilon \to 0$. Furthermore, the second derivative of $F^{\circ p}$ at a has the order of ϵ. Hence the Böttcher coordinate of $a + w$ has the order of ϵw as both ϵ and w/ϵ tend to zero. Taking $a + w$ equal to $F^{\circ n}(-a)$, so that w has the order of ϵ^2, it follows that the corresponding Böttcher coordinate has the order of ϵ^3, as required. $\square$

As an immediate consequence of Lemma 6.8, it follows that the Poincaré geodesic joining the critical point of Q to its root point, lifts to a pair of curves in a component of type A, or a tripod of components in a component of Type B. One of the sectors which is cut by these Poincaré geodesics will correspond to the intersection of this component with the conjectured cell e_Q.

6C   The Non-Periodic Julia Case

This is perhaps the easiest case to understand. (Compare Figure 2.) If $\widehat{z} \in J^\sharp(Q)$ is not periodic, then there is a canonical embedding $\iota : K(Q) \hookrightarrow K(F)$ which satisfies $\iota \circ Q = f \circ \iota$. The set $K(F)$ can be obtained from the embedded image $\iota(K(Q))$ by adjoining a limb L_z homeomorphic to $K(F)$ at $\iota(z)$ for every $z \in K(Q)$ which is either equal to $\widehat{z}$ or to some iterated preimage of $\widehat{z}$. This limb L_z maps homeomorphically onto $L(Q(z))$ for $z \neq \widehat{z}$, while $L_{\widehat{z}}$ maps homeomorphically onto all of $K(F)$. There is a canonical retraction $K(F) \to \iota(K(Q))$ which maps each L_z to its attaching point $\iota(z)$. Assuming local connectivity (as in Remark 2.10), we can construct a simple topological model for $K(F)$ as follows: An orbit $z_0 \mapsto z_1 \mapsto \cdots$ in $K(F)$ is uniquely determined by its image

$$(r(z_0), \ r(z_1), \ \ldots) \ \in \ \iota(K(Q))^{\mathbb{N}}.$$

Furthermore, a given sequence can occur if and only if $r(z_j) \mapsto r(z_{j+1})$ for all j such that $r(z_j) \neq \iota(\widehat{z})$.

Thus $K(F)$ can be uniquely described as a topological dynamical system. However, this analysis does not specify just how $K(F)$ is embedded in $\mathbb{C}$. In fact, if the image of the point $\widehat{z}$ under the projection $K^\sharp(Q) \rightarrow K(Q)$ cuts $K(Q)$ into n distinct components, or equivalently if there are n distinct external rays landing at this image point in $K(Q)$, then there are n essentially distinct ways of embedding the dynamical system $K(F)$ into $\mathbb{C}$. In fact if we cut $\mathbb{C}$ open along any one of these n rays which land on $K(Q)$, then we can paste the entire limb $L_{\widehat{z}}$ into the resulting gap. These n choices correspond precisely to the n ways of lifting the image in $K(Q)$ up to the cut-open set $K^\sharp(Q)$.

6D   The Periodic Julia Case

If $\widehat{z} \in J^\sharp(Q)$ corresponds to a periodic point in $J(Q)$, then the situation is similar but more complicated, as illustrated in Figure 3. The repelling periodic point $\widehat{z}$ will be replaced by a parabolic periodic point in $K(F)$, on the boundary of a new parabolic Fatou component. Perhaps the easiest construction is to use the Hubbard tree argument of Theorem 6.4 or Remark 6.7 to construct an associated hyperbolic map. The required $\phi(\widehat{z})$ will then be the root point of the corresponding hyperbolic component.

7   Appendix on Hubbard Trees

This will be an exposition of Hubbard trees, as originally described in [DH84, Section IV], with more precise statements due to Alfredo Poirier. It also describes slightly modified "puffed-out" Hubbard trees.

The *Hubbard tree* T associated with a post-critically finite polynomial F can be defined as follows. Each component of the Fatou set $\widehat{\mathbb{C}} \smallsetminus J(F)$ contains a unique periodic or pre-periodic point which will be called its *center*. A path in the filled Julia set $K(F)$ is *regulated* if its intersection with each Fatou component consists of at most two Poincaré geodesics, each joining the center to a boundary point. Now let $S \subset K(F)$ be a finite set which contains all critical points and satisfies $F(S) \subset S$. The *associated tree* $T = T_S$ is the regulated convex closure: that is the smallest set containing a regulated path between any two points of S. This is a topological tree, and can be triangulated so that the set of vertices consists of the given set S, together with a finite number of points where three or more edges come together. The given mapping F carries each edge of T homeomorphically onto some union of edges, namely the unique regulated path joining the images of its two endpoints within T.

For a polynomial map with a superattracting cycle, there is a modified version of this definition which is sometimes helpful, since it more closely resembles the Julia set. Let U_T be the union of all Fatou components in

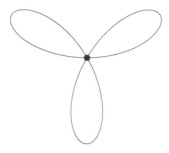

$K(F)$ which contain vertices of T (or which contain points of the finite set S). Then the boundary $\mathcal{P}(T)$ of the union $T \cup U_T$ will be called the *puffed-out Hubbard tree*. It consists of ∂U_T (the union of the boundary circles of these Fatou components), together with $T \setminus U_T$ (that part of the tree which lies outside of U_T). As an example, the minimal tree T for the Douady rabbit is the dark tripod shown in Figure 27, while the puffed-out tree $\mathcal{P}(T)$ is homeomorphic to the above sketch. Other examples of puffed-out trees are shown in Figures 34 and 35. We will see that T and $\mathcal{P}(T)$ contain the same information; so that we can use whichever one seems more convenient.

In this paper we will usually concentrate on the *minimal* Hubbard tree, taking S to be the union of the orbits of the critical points. However, the construction works equally well taking a larger finite set, for example by also including one or more periodic orbits. If $T \subset K(F)$ is an arbitrary Hubbard tree, then each iterated preimage is also a Hubbard tree, so that we form an ascending sequence

$$T \subset F^{-1}(T) \subset F^{-2}(T) \subset \cdots \subset K(F).$$

If we exclude the case where T is a single point (the minimal tree for $F(z) = z^d$), then the union of the $F^{-n}(T)$ is everywhere dense in the filled Julia set $K(F)$. Similarly, the puffed-out trees $\mathcal{P}(F^{-n}(T)) \subset K(F)$ provide better and better approximations to the Julia set as $n \to \infty$.

We will ignore complications in the geometry of T and think of it simply as a one-dimensional acyclic simplicial complex. Three additional elements of structure are immediately apparent:

(1) There is a prescribed map (the restriction of F) from the set V of vertices to itself. This carries the two endpoints of any edge to distinct points, so that it can be extended to a map from T to itself which is one-to-one on each edge.

(2) We must specify which vertices are critical points, and with what multiplicity. It will be convenient to describe this by the *local degree*

function $d : V \rightarrow \{1, 2, 3, \ldots\}$, where we set $d(\mathbf{v}) = 1$ if the vertex $\mathbf{v}$ is non-critical and $d(\mathbf{v}) = m + 1 \geq 2$ if $\mathbf{v}$ is a critical point of multiplicity m.

(3_0) If three or more edges meet at a vertex, then the cyclic order of these edges in the positive direction around this vertex must be specified. Note that this cyclic order determines, up to isotopy, how the tree is to be embedded into $\mathbb{C}$.

However, this data is not sufficient to uniquely determine the affine conjugacy class of F. For example Figures 32 and 33 illustrate Julia sets which are not affinely conjugate to their mirror images, although this fact cannot be deduced if we are given only the data above. Similarly, the three puffed-out trees at the top of Figure 35 cannot be distinguished without further information. For this reason we introduce the *angles* between edges of the tree as an essential part of the structure.

Let $\mathbf{e}_1, \mathbf{e}_2, \ldots, \mathbf{e}_n$ be the edges incident to a single vertex, listed in positive cyclic order, where the subscripts are interpreted as integers modulo n so that $\mathbf{e}_0 = \mathbf{e}_n$. Then the angles between successive edges $\mathbf{e}_j$ and $\mathbf{e}_{j+1}$ are to be positive rational numbers with sum

$$\angle(\mathbf{e}_0, \mathbf{e}_1) + \angle(\mathbf{e}_1, \mathbf{e}_2) + \cdots + \angle(\mathbf{e}_{n-1}, \mathbf{e}_n) = 1.$$

More generally, the angle (in the positive direction) between any two edges meeting at a common vertex is well defined. We will think of this angle as an element of the circle $\mathbb{Q}/\mathbb{Z}$, and replace the hypothesis (3_0) by the following sharper hypothesis. It will be convenient to use the notation $F_{\mathbf{v}}(\mathbf{e})$ for the unique edge which contains $F(U \cap \mathbf{e})$, where U is a small neighborhood of $\mathbf{v}$.

(3) This angle $\angle(\mathbf{e}, \mathbf{e}') \in \mathbb{Q}/\mathbb{Z}$ is well defined for any two edges meeting at a common vertex $\mathbf{v}$, and satisfies

$$\angle(\mathbf{e}, \mathbf{e}') + \angle(\mathbf{e}', \mathbf{e}'') \equiv \angle(\mathbf{e}, \mathbf{e}'') \pmod{\mathbb{Z}},$$

with $\angle(\mathbf{e}, \mathbf{e}') \equiv 0$ only if $\mathbf{e} = \mathbf{e}'$. Furthermore, for the image of two such edges near $\mathbf{v}$, we have

$$\angle\Big(F_{\mathbf{v}}(\mathbf{e}), F_{\mathbf{v}}(\mathbf{e}')\Big) \equiv d(\mathbf{v}) \angle(\mathbf{e}, \mathbf{e}') \pmod{\mathbb{Z}}. \tag{7.18}$$

For the definition of the angle between two edges, we must distinguish two cases, as follows.

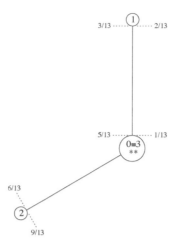

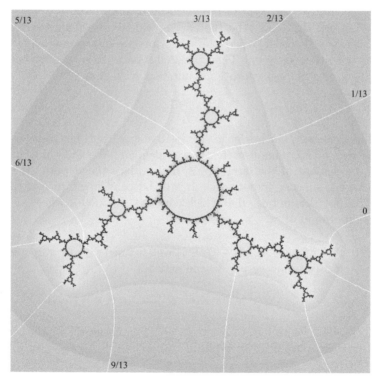

Figure 32. The top sketch shows a cubic Hubbard tree with a double critical point at which the two adjacent branches form an angle of $1/3$ ($= 120°$). External angles along the root orbit have been indicated. The lower figure shows the corresponding Julia set. All external rays with angles of the form $k/13$ are shown in white. (See Plate V.)

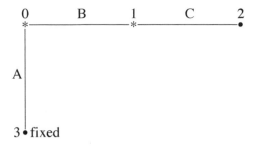

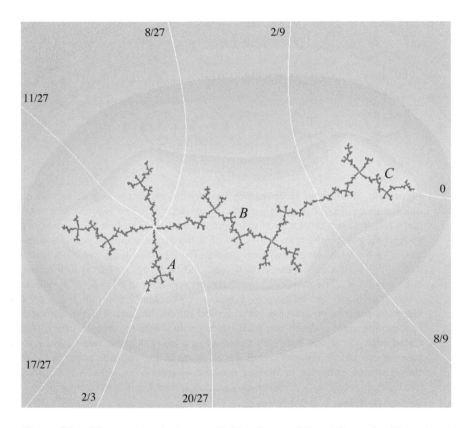

Figure 33. The top sketch shows a Hubbard tree with a right angle. The critical orbit maps as $0 \mapsto 1 \mapsto 2 \mapsto 3 \circlearrowleft$, where 0 and 1 are simple critical points, while the edges of the tree map as $A \mapsto C \mapsto A \cup B \cup C$, $B \mapsto \overline{B} \cup \overline{A}$, where the overline stands for reversal of orientation. The corresponding cubic Julia set is shown below. Here the critical points 0 and 1 are the landing points of the 8/27 and 2/9 rays. (See Plate V.)

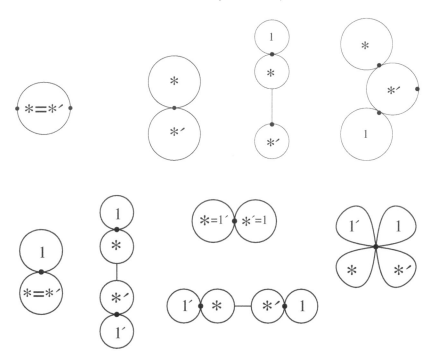

Figure 34. Puffed-out Hubbard tree representations for the nine essentially distinct maps with both critical orbits periodic of period ≤ 2. The top row illustrates four maps which have a critical fixed point, denoted by $*$. The first two represent points in $\mathcal{S}_1 \cap \mathcal{S}_1'$, namely the map $z \mapsto z^3$ with a fixed double critical point, and the map $z \mapsto z^3 + \frac{3}{2}z$ with two critical fixed points. The next two maps correspond to points in $\mathcal{S}_2 \cap \mathcal{S}_1'$, with a period two critical orbit, indicated by $* \leftrightarrow 1$, as well as the fixed critical point $*'$. The remaining five diagrams represent in $\mathcal{S}_2 \cap \mathcal{S}_2'$ with both critical orbits of period 2, labeled by $* \leftrightarrow 1$ and $*' \leftrightarrow 1'$. The dots on the boundary circles represent periodic points of minimal period on the boundary of the corresponding Fatou component. (When two such Fatou components touch each other, the associated edge of the tree has been collapsed to a point in these two figures.) In each case we can obtain representatives for all maps in the corresponding $\mathcal{S}_p \cap \mathcal{S}_q'$ from the illustrated examples by making use of conjugation by $z \leftrightarrow -z$ (180° degree rotation), together with complex conjugation (reflection in a horizontal line).

Definition 7.1. Call a vertex $\mathbf{v} \in T \subset K(F)$ either a *Fatou vertex* or a *Julia vertex* according as it belongs to the Fatou set or the Julia set of F. In fact we can make this distinction just from the structures (1) and (2) described above: *It is easy to check that $\mathbf{v}$ is a Fatou vertex if and only if some forward image of $\mathbf{v}$ is a periodic critical point.* In the figures, we will usually emphasize this distinction by replacing each Fatou vertex by a small circle.

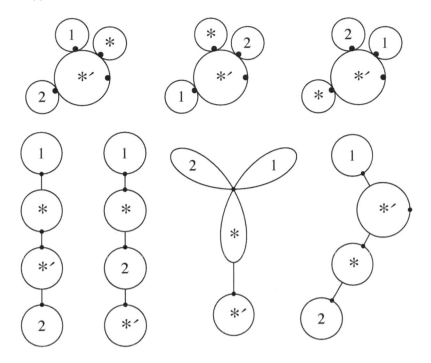

Figure 35. Trees for the seven essentially different maps in $\mathcal{S}_3 \cap \mathcal{S}'_1$. Here the period three critical orbit is labeled by the symbols $* \mapsto 1 \mapsto 2 \mapsto *$, while the fixed critical point is labeled by $*'$. Note the $1/7$, $2/7$, $4/7$ internal angles for the top three examples, and the $1/3$, $2/3$ internal angles for the last example.

In the case of two edges of T which meet at the center of a Fatou component U of F, the definition of this angle is straightforward. There is an essentially unique conformal diffeomorphism taking U to the unit disk and taking each edge intersected with U to a radius. We can then use the usual angle, as measured within the unit disk. Since the map F from U to $F(U)$ corresponds to the map $w \mapsto w^{d(\mathbf{v})}$ between disks, the identity (7.18) follows easily.

In the case of edges meeting at a Julia vertex $\mathbf{v}$, Poirier defines this angle as follows. If $\mathbf{v}$ is periodic under F, and if n edges meet at $\mathbf{v}$, then we simply define the angle between two edges which are consecutive in cyclic order to be $1/n$. In the more general case where $\mathbf{v}$ is not periodic, we must choose these angles so that Equation (7.18) is satisfied. However, it may happen that there is more than one possible choice. In that case, we can resolve the difficulty as follows. Suppose that $F^{\circ m}(\mathbf{v})$ is periodic. Then $F^{-m}(T)$ will have a full complement of edges meeting at $\mathbf{v}$. If there are n such edges, then we again define the angle between two which are consecutive in cyclic order to be $1/n$. As an example, in Figure 33, the angle $\angle(A, B)$ could a priori be either $1/4$ or $3/4$. In this case, it suffices to pass to $F^{-1}(T)$ which has four edges meeting at this point, in order to determine that the correct angle is $1/4$.

Definition 7.2. By an *abstract tree* we will mean a topological tree which has been provided with a mapping from vertices to vertices, a local degree function, and an angle function satisfying the conditions (1), (2) and (3) above. Note that the cyclic order (3_0) is uniquely determined by the angular structure (3).

The problem is now to characterize which abstract trees can actually be realized as the Hubbard trees of polynomials. Poirier provides a very simple answer as follows.

Theorem 7.3 (Poirier). *An abstract tree can be realized as the Hubbard tree of a polynomial if an only if two conditions are satisfied:*

1. *(Expansiveness) For every edge $\mathbf{e}$, either at least one of its two boundary points is a Fatou vertex, or else some forward image $F^{\circ k}(\mathbf{e})$ covers two or more edges.[6]*

2. *(Normalization) The consecutive angles around any periodic Julia vertex are all equal.*

When these conditions are satisfied, the resulting polynomial is unique up to affine conjugation, and has degree d satisfying $d-1 = \sum_{\mathbf{v}} \Big(d(\mathbf{v}) - 1 \Big)$.

For the proof, the reader is referred to [Po93].   □

We can also use these ideas to construct the puffed-out Hubbard tree, starting only with the abstract tree. Simply replace each Fatou vertex by a small circle. Now for any edge $\mathbf{e}$ which does not contain a pre-critical point in its interior (or in other words, any edge such that no forward image crosses through a critical point), collapse that portion $\mathbf{e}_0$ of $\mathbf{e}$ which is outside of the small circles to a single point. It is not difficult to check that the result is homeomorphic to the puffed-out tree as described above. As an example, in Figure 12, neither edge has an interior pre-critical point, hence both edges must be collapsed, yielding the top right diagram of Figure 34.

---

[6] An equivalent expansivity condition, used by Bruin, Kaffl, and Schleicher [BS01, BS08, BKS], is the following: *For every edge $\mathbf{e}$ there must be some forward image $F^{\circ k}(\mathbf{e})$ which contains a critical point (perhaps on its boundary).* It is not difficult to show that this Bruin-Schleicher condition is completely equivalent to Poirier's condition.

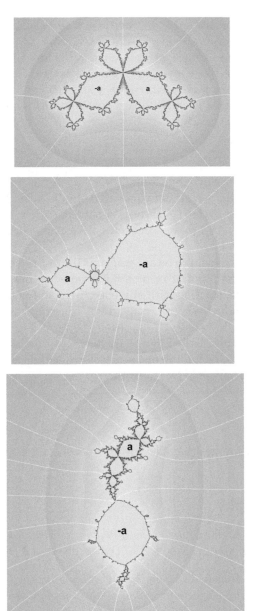

Figure 36. Julia sets for three representative examples. The "bow tie" Julia set above, with polynomial $z^3 - \frac{3}{4}z + i\sqrt{7}/4$, corresponds to the last tree in Figure 34. All rays with denominator $3^2 - 1 = 8$ are shown. The next, corresponds to the top right tree in Figure 35, with $f(a)$ in the limb enclosed by the $1/26$ and $2/26$ rays. The last "dancing rabbit" example corresponds to the next to last tree in Figure 35. All rays of denominator $3^3 - 1 = 26$ are shown in these two cases. (See Plate VI.)

To conclude this appendix, here are a few elementary remarks.

As in Section 6A, define the *valence* $n(\mathbf{v})$ to be the number of edges of T which are incident at the vertex $\mathbf{v}$. Thus $n(\mathbf{v}) \geq 1$, except in the trivial case of a tree consisting of a single vertex. The tree T is said to be *minimal* if every vertex with $n(\mathbf{v}) \leq 2$ is critical or post-critical. It is not hard to check that every Hubbard tree contains a unique minimal one.

Note the inequality

$$n(F(\mathbf{v})) \geq n(\mathbf{v})/d(\mathbf{v}), \tag{7.19}$$

which follows easily from formula (7.18). As an example, in the case of a periodic Julia vertex it follows that $n(\mathbf{v})$ takes the same value at all vertices of the cycle. (This was assumed earlier when we assigned the angle $\angle(\mathbf{e}, \mathbf{e}') = 1/n(\mathbf{v})$ between consecutive edges at a periodic Julia vertex $\mathbf{v}$.)

A vertex $\mathbf{v}$ is said to be an *endpoint* (or a *free vertex*) of T if $n(\mathbf{v}) \leq 1$. As noted in the proof of Lemma 6.5, *every tree has at least one endpoint*. (For otherwise, starting with any edge, we could pass to an adjacent edge. Continuing inductively, we could then construct a closed loop, which is impossible.)

Closely related is the Euler characteristic formula $2\chi(T) = \sum_{\mathbf{v} \in V} (2 - n(\mathbf{v}))$, which holds for any one dimensional simplicial complex. In the case of a tree, we have $\chi(T) = 1$, so that

$$\sum_{\mathbf{v}} \left(2 - n(\mathbf{v}) \right) = 2. \tag{7.20}$$

This clearly implies that there is at least one vertex with $n(\mathbf{v}) \leq 1$.

Bibliography

[At92]   P. Atela, *Bifurcations of dynamic rays in complex polynomials of degree two*, Ergod. Th. Dynam. Syst. **12** (1992), 401–423.

[Bi89]   B. Bielefeld, *Changing the order of critical points of polynomials using quasiconformal surgery*, Thesis, Cornell University, 1989.

[Bl84]   P. Blanchard, *Complex analytic dynamics on the Riemann sphere*, Bull. Amer. Math. Soc. **11** (1984), 85–141.

[BDK91] P. Blanchard, R. Devaney, and L. Keen, *The dynamics of complex polynomials and automorphisms of the shift*, Invent. Math. **104** (1991), 545–580.

[BM]   A. Bonifant and J. Milnor, *Cubic polynomial maps with periodic critical orbit, part 2*, in preparation.

[Br86]   B. Branner, *The parameter space for cubic polynomials*, in: "Chaotic Dynamics and Fractals", Barnsley and Demko Eds., Acad. Press, 1986, 169–179.

[Br93]   B. Branner, *Cubic polynomials, turning around the connectedness locus,* in: Topological Methods in Mathematics, Goldberg and Phillips Eds., Publish or Perish, Houston, 1993, 391–427.

[BH88]   B. Branner and J. H. Hubbard, *The iteration of cubic polynomials I: the global topology of parameter space*, Acta Math. **160** (1988), 143–206.

[BH92]   B. Branner and J. H. Hubbard, *The iteration of cubic polynomials II: patterns and parapatterns,* Acta Math. **169** (1992), 229–325.

[Brn60] M. Brown, *A proof of the generalized Schoenflies theorem,* Bull. Amer. Math. Soc. **66** (1960), 74–76. See also: *The monotone union of open n-cells is an open n-cell,* Proc. Amer. Math. Soc. **12** (1961), 812–814.

[BKS]   H. Bruin, A. Kaffl, and D. Schleicher, *Symbolic dynamics of quadratic polynomials,* monograph, in preparation.

[BS01]   H. Bruin and D. Schleicher, *Symbolic dynamics of quadratic polynomials,* Preprint 7, Institut Mittag-Leffler (2001/02). To appear as [BKS].

[BS08]   H. Bruin and D. Schleicher, *Admissibility of kneading sequences and structure of Hubbard trees for quadratic polynomials,* Journal London Math. Soc. **78** 2 (2008), 502–522.

[BHe01] X. Buff and C. Henriksen, *Julia sets in parameter spaces,* Comm. Math. Phys. **220** (2001), 333–375.

[CM89]   P. Cvitanović and J. Myrheim, *Complex universality,* Comm. Math. Phys. **121** (1989), 225–254.

[DH82]   A. Douady and J. H. Hubbard, *Itération des polynômes quadratiques complexes,* C. R. Acad. Sci. Paris **294** (1982), 123–126.

[DH84]   A. Douady and J. H. Hubbard, *Étude dynamique des polynômes complexes, Parts I and II*, Publ. Math. d'Orsay, 1984.

[DH85]   A. Douady and J. H. Hubbard, *On the dynamics of polynomial-like mappings*, Ann. Sci. Ec. Norm. Sup. (Paris) **18** (1985), 287–343.

[Fa92]   D. Faught, *Local connectivity in a family of cubic polynomials*, Thesis, Cornell University, 1992.

[GM93]   L. Goldberg and J. Milnor, *Fixed points of polynomial maps II: fixed point portraits,* Ann. Sci. École Norm. Sup. **26** (1993), 51–98.

[H93]   J. H. Hubbard, *Local connectivity of Julia sets and bifurcation loci: three theorems of J.-C. Yoccoz*, in: Topological Methods in Mathematics, L. Goldberg and A. Phillips Eds., Publish or Perish, Houston, 1993, 467–511.

[Ka06]   A. Kaffl, *Hubbard trees and kneading sequences for unicritical and cubic polynomials*, Thesis, International University Bremen, 2006.

[Ki06]   J. Kiwi, *Puiseux series, polynomial dynamics and iteration of complex cubic polynomials*, Ann. Inst. Fourier (Grenoble) **56** (2006), 1337–1404.

[La89]   P. Lavaurs, *Systèmes dynamiques holomorphes: explosion de points périodiques paraboliques,* Thesis, Université Paris-Sud, 1989.

[Li07]   T. Li, *A monotonicity conjecture for the entropy of Hubbard trees*, Thesis, Stony Brook University, 2007.

[Lu95]   J. Luo, *Combinatorics and holomorphic dynamics: captures, matings, Newton's method,* Thesis, Cornell University, 1995.

[Mi89]   J. Milnor, *Self-similarity and hairiness in the Mandelbrot set*, in: Computers in Geometry, M. Tangora Ed., Lect. Notes Pure Appl. Math. **114**, Dekker, 1989, 211–257.

[Mi92a] J. Milnor, *Remarks on itcrated cubic maps*, Experiment. Math. **1** (1992), 5–24.

[Mi92b] J. Milnor, *Hyperbolic components in spaces of polynomial maps*, with an appendix by A. Poirier, Stony Brook IMS preprint 3, 1992.

[Mi93]   J. Milnor, *Geometry and dynamics of quadratic rational maps*, Experiment. Math. **2** (1993), 37–83.

[Mi00] J. Milnor, *Periodic orbits, external rays, and the Mandelbrot set*, in: Géometrie Complexe et Systèmes Dynamiques, Astérisque **261** (2000), 277–333.

[Mi06] J. Milnor, *Dynamics in One Complex Variable*, 3rd Edition, Princeton University Press, 2006.

[PT04] C. L. Petersen and Tan L., *The central hyperbolic component of cubic polynomials*, manuscript, see http://www.u-cergy.fr/rech/pages/tan/ (2004).

[Po93] A. Poirier, *On postcritically finite polynomials, part 2: Hubbard trees*, Stony Brook IMS preprint 7, 1993.

[Re90] M. Rees, *Components of degree two hyperbolic rational maps*, Invent. Math. **100** (1990), 357–382.

[Re92] M. Rees, *A partial description of parameter space of rational maps of degree two*, Part I: Acta Math. **168** (1992), 11–87.

[Re95] M. Rees, *A partial description of parameter space of rational maps of degree two*, Part II: Proc. London Math. Soc. **70** (1995), 644–690.

[Ro99] P. Roesch, *Puzzles de Yoccoz pour les applications à allure rationnelle*, L'Enseign. Math. **45** (1999), 133–168.

[Ro06] P. Roesch, *Hyperbolic components of polynomials with a fixed critical point of maximal order*, arXiv:math/0612172 (2006), to appear.

[Sch04] D. Schleicher, *On fibers and local connectivity of Mandelbrot and Multibrot sets*, Proc. Symp. Pure Math. **72 I**, Fractal Geometry and Applications, Amer. Math. Soc., 2004, 477–517.

[Ta97] Tan L., *Branched coverings and cubic Newton maps*, Fund. Math. **154** (1997), 207–260.

[TY96] Tan L. and Yin Y., *Local connectivity of the Julia set for geometrically finite rational maps*, (English summary) Sci. China Ser. A **39** (1996), 39–47.

[Wi88] B. Wittner, *On the bifurcation loci of rational maps of degree two*, Thesis, Cornell University (1988).

10 Analytic Coordinates Recording Cubic Dynamics

Carsten Lunde Petersen and Tan Lei

Let $\mathcal{H}$ be the central hyperbolic component of cubic polynomials (i.e., the one containing $z \mapsto z^3$). Works of Milnor on the classification of hyperbolic components show that $\mathcal{H}$ enjoys a kind of universality property. We construct an analytic coordinate on $\mathcal{H}$ that records dynamical invariants. We use quadratic dynamics as parameter models. This coordinate has good extension properties to at least a large part of the boundary of $\mathcal{H}$. We illustrate this by proving some of these extension properties. At the end, we outline applications/further developments and review the Milnor classification of hyperbolic components.

1 Introduction

This paper concerns parameter spaces of rational maps viewed as dynamical systems through iteration. The parameter space of a family of rational maps often has a natural decomposition into the hyperbolic locus and the non-hyperbolic locus: a rational map is *hyperbolic* if and only if all its critical points are attracted to attracting periodic orbits. The hyperbolic locus is open (and conjecturally dense, in any reasonable family). A *hyperbolic component* is a connected component of the hyperbolic locus. Maps within the same hyperbolic component have essentially the same macroscopic dynamics.

However, this quantitative change of dynamics becomes qualitative and often drastic when one moves to the boundary of the hyperbolic component. Therefore a major research interest in this field is to understand the boundary structure of a hyperbolic component, including its topology, geometry and the bifurcations of the dynamics that occurs.

Given such a component H, in order to study its boundary, the first step is actually to study in detail the inner structure of H, to put our hand on an effective measuring of the (infinitesimal) changes of the dynamics within H.

One way to address this problem is to find some suitable model space $\mathcal{X}$ together with a map $\Phi : H \to \mathcal{X}$ that satisfies the following properties:

1. (Coordinate) Φ is injective in H and $\Phi(f)$ depends analytically on f.

2. (Dynamics) $\Phi(f)$ records a complete set of dynamical invariants of f. In other words, from these invariants (together with some combinatorial data) one can reconstruct f up to conformal conjugacy.

Once we have such a coordinate, we may use $\Phi(H)$ to represent H and use $\partial\Phi(H)$ to study ∂H. This will depend on boundary extension properties of Φ. Thus the next step will be to study how well Φ or Φ^{-1} extends to the boundary and to which extent it still reflects the dynamical invariants.

We will call Φ a *dynamical-analytic* coordinate on H.

Douady and Hubbard have already constructed such coordinates for the hyperbolic components of the family of quadratic polynomials. Furthermore their construction generalizes easily to other families with only one critical point or singular value.

Concerning hyperbolic components with two or more critical points, remarkable pioneering work has been done by Mary Rees on the family of quadratic rational maps with marked critical points [Re90]. Rees provides topological-dynamical (non-analytic) coordinates for the hyperbolic components of this family, together with some radial extensions to the boundary.

The main purpose of this paper is to construct a dynamical-analytic coordinate on the principal hyperbolic component of the family of cubic polynomials, and then prove that this coordinate extends continuously to a large part of the boundary and records the bifurcations of the dynamics. Once this is done, we would automatically obtain a similar coordinate for many other hyperbolic components via previous work of Milnor (however the boundary extensions may vary from one another, and may require extra studies).

Our coordinate is similar in spirit to that of Mary Rees. However instead of using Blaschke products, we will use quadratic polynomial dynamics as part of dynamical invariants (similar ideas can be found in [GK90], [Mi]). There are several advantages of doing so. On the one hand, quadratic polynomials are well understood and are more convenient to deal with. On the other hand, they lead to the analyticity of our coordinate, and thus better reflect the analytic structure of the parameter space. This analyticity is also absolutely essential in our study of the boundary, as it will allow us to do holomorphic motions between parameter slices. In particular, it will allow us to extend a powerful result of Faught-Roesch on a single slice to all the other slices. Moreover we use parts of dynamical planes of quadratics to model cubic parameter slices. In the case of quadratic polynomials with an indifferent fixed point, there are still many unsolved questions. Any future progress on the study of such quadratic polynomials will contribute to our further understanding of cubic polynomials.

Acknowledgments. We wish to thank J. Milnor for communicating his ideas about the subject with us, A. Douady for inspiring discussions and the anonymous referee for a careful reading. This research has been partially supported by British EPSRC Grant GR/L60999. The first author also wishes to thank the CNRS-UMR 8088 at Cergy-Pontoise for several invitations and the Danish Natural Sciences Research Council for travel support.

1.1 Cubic Parameter Space and Quadratic Models

Let $\mathcal{M}^3$ denote the space of affine conjugacy classes $[P]$ of cubic polynomials, and let $\mathcal{H}$ denote the central hyperbolic component of $\mathcal{M}^3$, i.e., the component containing the class $[z \mapsto z^3]$. This $\mathcal{H}$ can also be characterized as the set of conjugacy classes $[P]$ of cubic polynomials such that P has an attracting fixed point in $\mathbb{C}$ whose immediate basin contains two critical points, or equivalently as the set of conjugacy classes $[P]$ of hyperbolic cubic polynomials with a Jordan curve Julia set.

For $(\lambda, a) \in \mathbb{C}^2$, define $P_{\lambda,a}(z) = \lambda z + \sqrt{a}z^2 + z^3$. Then the mapping $(\lambda, a) \mapsto [P_{\lambda,a}]$ is well defined from $\mathbb{C}^2$ onto $\mathcal{M}^3$ (as the two choices of $\sqrt{a}$ will lead to conjugate polynomials), but it is not globally injective (as one may normalize different fixed points to be at 0). However, if we assume λ belongs to $\mathbb{D}$, the unit disk in $\mathbb{C}$, and write $\mathbf{a} = (\lambda, a)$, then $P_{\mathbf{a}}$ has an attracting fixed point with multiplier λ and immediate basin $B'_{\mathbf{a}}$. It then easily follows that we may parametrize $\mathcal{H}$ as

$$\mathcal{H} = \{\mathbf{a} \mid \lambda \in \mathbb{D}, B'_{\mathbf{a}} \text{ contains both finite critical points of } P_{\mathbf{a}}\} \,,$$

and this makes $\mathcal{H}$ a connected and bounded open subset of $\mathbb{C}^2$. We define also for each $\lambda \in \mathbb{D}$ the slice $\mathcal{H}_\lambda := \{(\lambda, a) \in \mathcal{H}\}$.

Our first main result in this paper will be:

Theorem A (Coordinatization). *There exists a complex manifold $\mathcal{X}$ that is isomorphic to $\mathbb{D} \times \overline{\mathbb{C}}$ and a dynamical-analytic coordinate Φ of $\mathcal{H}$ that maps $\mathcal{H}$ onto an open subset $\mathcal{Y}$ of $\mathcal{X}$.*

A more precise statement will be given in Theorem A' below.

The definition of Φ is in fact quite easy to describe. It is based on the idea that a cubic polynomial $P_{\mathbf{a}} = P_{\lambda,a} : z \mapsto \lambda z + \sqrt{a}z^2 + z^3$ should behave like the quadratic polynomial $Q_\lambda : z \mapsto \lambda z + z^2$ together with an extra critical point. More precisely, we will produce a semi-conjugacy $\eta_{\mathbf{a}}$

from $P_{\mathbf{a}}$ to Q_λ whose domain of definition contains both critical points, and which maps the "first" attracted critical point to the critical point of Q_λ. Now the "second" critical point $c_{\mathbf{a}}^1$ will have some position under $\eta_{\mathbf{a}}$. Our map $\Phi(\mathbf{a})$ will be roughly $(\lambda, \eta_{\mathbf{a}}(c_{\mathbf{a}}^1)) \in \mathbb{D} \times \overline{\mathbb{C}}$.

But the notions of "first" and "second" attracted critical points are not always well defined. To get rid of this ambiguity, we perform a kind of cut and zip surgery on $\mathbb{D} \times \overline{\mathbb{C}}$, and thereby also obtain a quotient space $\mathcal{X}$ for $\Phi(\mathcal{H})$ to live in.

There is a close relation between our coordinate Φ and Douady-Hubbard theory on the quadratic family $\{f_c : z \mapsto z^2 + c\}$.

The unique unbounded hyperbolic component H_∞ of this family is characterized as the set of c such that the critical value escapes to ∞ under the iterations of f_c. Douady and Hubbard defined a conformal isomorphism $\varphi : H_\infty \to \mathbb{C} \setminus \overline{\mathbb{D}}$, where $\varphi(c)$ records the Böttcher position of the escaping critical value. It turns out that this φ is a dynamical-analytic coordinate of H_∞.

Now a parameter $c \notin H_\infty$ is in the hyperbolic locus if f_c has an attracting periodic orbit $z_0, \ldots, z_{n-1}$. For each hyperbolic component $H \neq H_\infty$, Douady and Hubbard proved that the multiplier map

$$\lambda_H : H \to \mathbb{D}, \quad c \mapsto (f_c^n)'(z_0) = \prod_{i=0}^{n-1} f_c'(z_i)$$

is a dynamical-analytic coordinate on H. (Following the pioneering work of Douady and Hubbard, this kind of coordinate has been used in many settings and even generalized to the anti-holomorphic case; see, e.g., [NS])

In a way, the first coordinate in our map $\Phi : \mathcal{H} \to \mathcal{X}$ is the Douady-Hubbard multiplier map and the second is an adapted version of the Douady-Hubbard φ-map.

1.2　Boundary Extensions

The Douady-Hubbard multiplier map λ_H has a homeomorphic extension to the closure, and, conjecturally, the inverse of the φ-map has a continuous extension to the closure. Many partial results have been obtained in this direction.

Denote by $\overline{\mathcal{Y}}$ the closure of $\mathcal{Y}$ in $\mathcal{X}$. The boundary $\partial\mathcal{H} \subset \mathbb{C}^2$ naturally splits into two parts: The *tame* part, denoted $\partial\mathcal{H}_{\mathbb{D}} := \{(\lambda, a) \in \partial\mathcal{H} | \lambda \in \mathbb{D}\}$, and the *wild* part $\partial\mathcal{H}_{\mathbb{S}^1} := \{(\lambda, a) \in \partial\mathcal{H} | |\lambda| = 1\}$. Set $\overline{\mathcal{H}}_{\mathbb{D}} := \mathcal{H} \cup \partial\mathcal{H}_{\mathbb{D}}$.

Our second main result in this paper is:

Theorem B (Boundary Extension). *The map* $\Phi : \mathcal{H} \to \mathcal{Y}$ *given by Theorem A extends as a homeomorphism* $\Phi : \overline{\mathcal{H}}_{\mathbb{D}} \longrightarrow \overline{\mathcal{Y}}$.

In forthcoming papers, we will study the wild part of the boundary of $\mathcal{H}$, using (often highly non-trivial) results about its quadratic counterparts. See Section 8 for a more detailed description.

1.3 Outline of the Paper

In Section 1, we will give precise definitions and a restatement of Theorem A. The theorem is proved in the following three sections. In Section 6, we prove Theorem B.

In the appendix, we will translate Theorem A to the settings of monic centered polynomials (Theorem C). We then recall results of Milnor and use them to carry our coordinate to other hyperbolic components of the same type (Corollary 9.2).

We provide a table of notation at the end of the paper.

2   Basic Definitions

Let $f : W \longrightarrow \mathbb{C}$ be holomorphic and suppose $f(\alpha) = \alpha \in W$ is an attracting fixed point with multiplier $f'(\alpha) = \lambda \in \mathbb{D}$. Define the attracting basin to be $B(\alpha) = \{z \in W \,|\, f^n(z) \underset{n \to \infty}{\longrightarrow} \alpha\}$. We say that $B(\alpha)$ is a proper basin, if the restriction $f : B(\alpha) \longrightarrow B(\alpha)$ is a proper map. Similar definitions apply when α belongs to an attracting cycle, i.e., when $f^k(\alpha) = \alpha \in W$ for some $k \geq 1$.

Attracting basins and attracting cycles come in two flavors: $\lambda \in \mathbb{D}^*$ are called *attracting* and $\lambda = 0$ is called *super attracting*. If α is super attracting then,

$$f(z) = \alpha + a_k (z - \alpha)^k + \mathcal{O}(z - \alpha)^{k+1},$$

with $a_k \neq 0$, where $k > 0$ is the local degree at α. Moreover there exists a Böttcher coordinate around α, that is, a univalent map $\phi : V \longrightarrow \mathbb{C}$, $V \subseteq B(\alpha)$, with $\phi(\alpha) = 0$ and $\phi \circ f = (\phi)^k$. The Böttcher coordinate is unique up to multiplication by a $(k-1)$th root of unity. In the particular case $k = 2$ (the case we are considering in this paper), ϕ is unique. If α is attracting, there exists a Schröder/linearizing coordinate or, in short, a linearizer for f. This is a univalent map $\phi : V \longrightarrow \mathbb{C}$, $V \subseteq B(\alpha)$ such that $\phi \circ f = \lambda \phi$. A linearizer extends to a holomorphic, locally finite branched covering $\phi : B'(\alpha) \longrightarrow \mathbb{C}$ (where $B'(\alpha)$ is the connected component of $B(\alpha)$ containing α). A linearizer on $B'(\alpha)$ is unique up to multiplication by a non-zero complex number.

Refer to the upper part of Figure 1, as well as Figure 2, for the following definitions.

Definition of U^0 and L^0 When $\lambda \in \mathbb{D}^*$. There is a unique domain $U^0 \subset B(\alpha)$ containing α and characterized by being the largest set which is mapped univalently by a linearizer ϕ onto a round disk centered at 0. Define L^0 to be the connected component of $f^{-1}(\overline{f(U^0)})$ containing α. It is a pinched closed disk. The sets U^0, L^0 are evidently independent of the choice of ϕ. For a choice of ϕ, denote by ψ the local inverse of ϕ, defined on $\phi(U^0)$ and mapping 0 to α.

Definition of c^0, v^0 and Normalization of the Linearizer. In case $\lambda = 0$, α is also a critical point. Set $c^0 = \alpha$. In case $\lambda \in \mathbb{D}^*$, ∂U^0 contains at least one critical point. We choose one of them and name it c^0; it is a *first attracted critical point*. Let $v^0 = f(c^0)$ be the corresponding critical value. It is contained in U^0. Unless explicitly stated otherwise, we shall assume ϕ is normalized so that $\phi(c^0) = 1$ and hence $U^0 = \psi(\mathbb{D})$ and $v^0 = \psi(\lambda)$.

Definition of the (Filled) Potential Function κ When $\lambda \in \mathbb{D}^*$. The *Schröder potential function* is the sub-harmonic function

$$\widehat{\kappa} : B'(\alpha) \longrightarrow [-\infty, \infty[\, , \qquad\qquad \widehat{\kappa}(z) = \frac{\log |\phi(z)|}{\log \frac{1}{|\lambda|}} \, .$$

It satisfies $\widehat{\kappa}(f(z)) = \widehat{\kappa}(z) - 1$ and $\widehat{\kappa}(\alpha) = -\infty$, as well as $\widehat{\kappa}(z) = -\infty$ for any iterated preimage z of α. For $t \in \mathbb{R}$, we let $U(t)$ denote the connected component of $\widehat{\kappa}^{-1}([-\infty, t[)$ containing α and, similarly, we let $L(t)$ denote the connected component of $\widehat{\kappa}^{-1}([-\infty, t])$ containing α or, equivalently, $L(t) = \bigcap_{s>t} U(s)$. Each $U(t)$ is a Jordan domain contained in the compact subset $L(t) \subset B'(\alpha)$. We have $\overline{U(t)} = L(t)$ if and only if $\partial U(t)$ does not contain a (pre)critical point of f. Note that $U(0)$ (respectively $L(0)$) coincides with U^0 (respectively L^0), previously defined, due to our normalization of ϕ. The *Milnor filled potential function* $\kappa : B'(\alpha) \longrightarrow [-\infty, \infty[$ is defined by

$$\kappa(z) = \inf\{s | z \in U(s)\}.$$

It is sub-harmonic with α as its sole pole, and it has the properties

$$\kappa^{-1}([-\infty, t[) = U(t) = f(U(t+1)),$$
$$\kappa^{-1}([-\infty, t]) - L(t) = f(L(t + 1)),$$
$$\kappa^{-1}(t) = L(t) \setminus U(t).$$

Consider the family of quadratic polynomials $Q_\lambda(z) = \lambda z + z^2$ and the family of cubic polynomials $P_{\mathbf{a}}(z) = P_{\lambda,a}(z) = \lambda z + \sqrt{a}z^2 + z^3$. For $\lambda \in \mathbb{D}$, the point $\alpha = 0$ is a (super)attracting fixed point for both Q_λ and $P_{\mathbf{a}}$. Denote by B_λ (respectively $B_{\mathbf{a}}$) the attracting basin of 0 for Q_λ (respectively for $P_{\mathbf{a}}$). Objects (such as ϕ, ψ, U^0, c^0, v^0, etc.) with a subscript λ will be those related to Q_λ, and those with a subscript $\mathbf{a}$ will be related to $P_\mathbf{a}$. However we omit the superscript 0 on c_λ^0 and v_λ^0 since they are unique; in particular, $c_\lambda = -\frac{\lambda}{2}$ and $\phi_\lambda(c_\lambda) = 1$. For the special parameter $\lambda = 0$, we define $\phi_0 = \psi_0 = \mathrm{id}$. Recall that

$$\mathcal{H} = \{\mathbf{a} \mid B_\mathbf{a}' = B_\mathbf{a} \text{ contains both finite critical points of } P_\mathbf{a}\} \ .$$

Set $\mathcal{H}^* = \mathcal{H} \setminus \{\mathbf{0}\}$ and $\mathcal{H}_0^* = \mathcal{H}_0 \setminus \{\mathbf{0}\}$.

Definition of $co_\mathbf{a}^0$ and $c_\mathbf{a}^1$. For $\mathbf{a} \in \mathcal{H}^*$, the critical point labelled $c_\mathbf{a}^0$ is the choice (generically the unique choice) of a first attracted critical point. The other critical point is labelled $c_\mathbf{a}^1$. We denote by $co_\mathbf{a}^0$ the *co-critical point* of $c_\mathbf{a}^0$, that is, the unique point with $P_\mathbf{a}(co_\mathbf{a}^0) = P_\mathbf{a}(c_\mathbf{a}^0)$ for which $co_\mathbf{a}^0 \neq c_\mathbf{a}^0$ if and only if $c_\mathbf{a}^0 \neq c_\mathbf{a}^1$ and $co_\mathbf{a}^0 = c_\mathbf{a}^0$ if and only if $c_\mathbf{a}^0 = c_\mathbf{a}^1$ is a double critical point.

Definition of $t_\mathbf{a}^1$, $U_\mathbf{a}$ and $L_\mathbf{a}$. Assume at first that $\lambda \neq 0$. For $\mathbf{a} \in \mathcal{H} \setminus \mathcal{H}_0$, define $t_\mathbf{a}^1 = \kappa_\mathbf{a}(c_\mathbf{a}^1)$, $U_\mathbf{a} = U_\mathbf{a}(t_\mathbf{a}^1)$ and $L_\mathbf{a} = L_\mathbf{a}(t_\mathbf{a}^1)$. Note that $c_\mathbf{a}^1 \in L_\mathbf{a}^0$ iff $t_\mathbf{a}^1 = 0$. Assume now $\lambda = 0$. For $\mathbf{a} \in \mathcal{H}_0^*$ and $\phi_\mathbf{a}$ the unique Böttcher coordinate, define $t_\mathbf{a}^1 = \log|\phi_\mathbf{a}(c_\mathbf{a}^1)|$, $U_\mathbf{a} \subset B_\mathbf{a}$ to be the largest set mapped univalently by $\phi_\mathbf{a}$ to a round disk, and $L_\mathbf{a} = \overline{U_\mathbf{a}}$. We necessarily have $\phi_\mathbf{a}(U_\mathbf{a}) = D(0, e^{t_\mathbf{a}^1})$. For $\mathbf{a} = \mathbf{0}$, set $t_\mathbf{a}^1 = -\infty$, $U_\mathbf{a} = \emptyset$ and $L_\mathbf{a} = \{0\}$. Define $\psi_0 = \mathrm{id}$ and $L_0(t) = \overline{\mathbb{D}(0, e^t)}$.

For $\mathbf{a} \in \mathcal{H}_0^*$, define $\Omega_\mathbf{a} = \overline{U_\mathbf{a}}$. Then $\eta_\mathbf{a} = \psi_0 \circ \phi_\mathbf{a} = \phi_\mathbf{a} : \Omega_\mathbf{a} \longrightarrow L_0(t_\mathbf{a}^1)$ is a homeomorphic conjugacy which extends to a holomorphic semi-conjugacy on a neighborhood of $\Omega_\mathbf{a}$. The following proposition extends this idea to the case when $\lambda \in \mathbb{D}^*$.

Proposition 2.1. *For $\mathbf{a} \in \mathcal{H} \setminus \mathcal{H}_0$, the local conjugacy $\eta_\mathbf{a} := \psi_\lambda \circ \phi_\mathbf{a}$ from $P_\mathbf{a}$ to Q_λ extends to a continuous surjective semi-conjugacy $\eta_\mathbf{a} : \Omega_\mathbf{a} \longrightarrow L_\lambda(t_\mathbf{a}^1)$, where $\Omega_\mathbf{a}$ is a closed, connected and simply connected set with $U_\mathbf{a} \cup \{c_\mathbf{a}^1\} \subset \Omega_\mathbf{a} \subset L_\mathbf{a}$. Moreover $\eta_\mathbf{a}$ extends holomorphically to a neighborhood of $\Omega_\mathbf{a} \setminus \{co_\mathbf{a}^0\}$, and $\kappa_\lambda(\eta_\mathbf{a}(c_\mathbf{a}^1)) = t_\mathbf{a}^1 = \kappa_\mathbf{a}(c_\mathbf{a}^1)$.*

The map $\mathbf{a} \mapsto \eta_\mathbf{a}(c_\mathbf{a}^1)$ takes values in $B_\lambda \setminus U_\lambda^0$. It is single-valued if $c_\mathbf{a}^1 \notin \partial U_\mathbf{a}^0$, and double-valued if $c_\mathbf{a}^1 \in \partial U_\mathbf{a}^0$ and $\phi_\mathbf{a}(c_\mathbf{a}^1) \neq \pm 1$ (due to the two ways of labeling $c_\mathbf{a}^0$ and $c_\mathbf{a}^1$). In the latter case, its two values z_0, z_1 belong to ∂U_λ^0 and satisfy $\phi_\lambda(z_0) \cdot \phi_\lambda(z_1) = 1$.

This proposition will be proved in the next section, where more details on $\Omega_\mathbf{a}$ are also provided.

We want to use $\mathbf{a} \mapsto \eta_\mathbf{a}(c_\mathbf{a}^1)$ to parametrize $\mathcal{H}$. In order to turn this multi-valued function into a single-valued one, we define an equivalence relation $\sim_\lambda$ on $\overline{\mathbb{C}} \setminus U_\lambda^0$ (see Figure 1) by

$$\lambda \in \mathbb{D}^*, \quad z_1 \sim_\lambda z_2 \Leftrightarrow [z_1 = z_2] \text{ or } [(z_1, z_2 \in \partial U_\lambda^0) \text{ and } (\phi_\lambda(z_1) = \overline{\phi_\lambda(z_2)})] \ ;$$
$$\lambda = 0, \quad z_1 \sim_\lambda z_2 \Leftrightarrow [z_1 = z_2] \ .$$

Note that $\sim_\lambda$ can be given an intrinsic meaning depending only on ∂U_λ^0 (but not on the normalization of ϕ_λ).

(Incidentally, $\sim_\lambda$ can be also thought of as an equivalence relation on $\overline{\mathbb{C}}$, where $z_1 \sim_\lambda z_2 \Leftrightarrow [z_1 = z_2]$ or $[(z_1, z_2 \in \overline{U_\lambda^0})$ and $Re(\phi_\lambda(z_1)) = Re(\phi_\lambda(z_2))]$, thus eliminating the need for the excision of U_λ^0.)

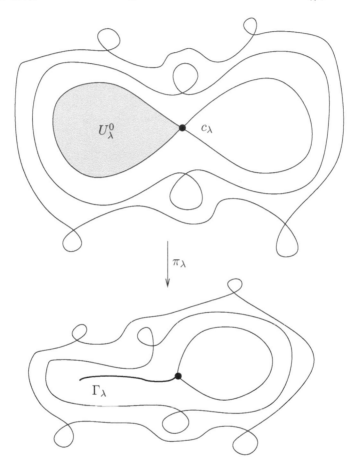

Figure 1. The sets U_λ^0, L_λ^0 (the closed set bounded by the figure eight) and the cut-glue operation.

Define $\mathcal{U}^0 = \{(\lambda, z) | \lambda \in \mathbb{D}^*$ and $z \in U_\lambda^0\}$ and an equivalence relation on $\mathbb{D} \times \overline{\mathbb{C}} \setminus \mathcal{U}^0$ as follows

$$(\lambda_1, z_1) \sim (\lambda_2, z_2) \Leftrightarrow (\lambda_1 = \lambda_2) \text{ and } (z_1 \sim_{\lambda_1} z_2) .$$

Let $\Pi : \mathbb{D} \times \overline{\mathbb{C}} \setminus \mathcal{U}^0 \longrightarrow (\mathbb{D} \times \overline{\mathbb{C}} \setminus \mathcal{U}^0)/ \sim$ denote the natural projection, and let π_λ denote its restriction to the λ-slice.

Definition 2.2. We set

$$\begin{aligned} \mathcal{X} &= \Pi(\mathbb{D} \times \overline{\mathbb{C}} \setminus \mathcal{U}^0), \quad \text{the cut-glued space,} \\ \mathcal{Y} &= \Pi(\mathcal{B} \setminus \mathcal{U}^0), \quad \text{the cut-glued basin,} \\ \Gamma &= \Pi(\partial \mathcal{U}^0), \quad \text{the scar.} \end{aligned}$$

Denote by $\mathcal{X}_\lambda := \Pi(\{\lambda\} \times \overline{\mathbb{C}} \setminus U_\lambda)$, $\mathcal{Y}_\lambda := \Pi(\{\lambda\} \times B_\lambda \setminus U_\lambda)$ and $\Gamma_\lambda := \Pi(\{\lambda\} \times \partial U_\lambda)$, $\lambda \neq 0$ and $\Gamma_0 := \Pi(0, 0)$ the corresponding slices related to $\{\lambda\} \times \overline{\mathbb{C}}$. Set $\mathcal{X}^* := \mathcal{X} \setminus \Pi(\{0\})$, $\mathcal{Y}^* = \mathcal{Y} \setminus \Pi(\{0\})$.

These spaces naturally come with the (Hausdorff) quotient topology. There is a canonical complex structure on $\mathcal{X} \setminus \Gamma$, for which Π is holomorphic.

We may now restate Theorem A:

Theorem A' (Modelization).

(a) *The space $\mathcal{X}$ equipped with the quotient topology is Hausdorff, $\Pi(\mathbf{0}) \in \mathcal{Y}$ and $\mathcal{Y}$ is open in $\mathcal{X}$.*

(b1) *$\mathcal{X}$ is a 2-dimensional complex manifold.*

(b2) *$\mathcal{X}$ is isomorphic to $\mathbb{D} \times \overline{\mathbb{C}}$ by an isomorphism that commutes with the projection to the λ coordinate.*

(c) *The map* $\Phi : \mathcal{H} \longrightarrow \mathcal{Y}$, $\begin{cases} \mathbf{0} & \mapsto \Pi(\mathbf{0}) \\ \mathbf{a} & \mapsto \Pi(\lambda, \eta_a(c_a^1)) \text{ for } \mathbf{a} \neq \mathbf{0} \end{cases}$ *is a well-defined biholomorphic homeomorphism.*

Part (a) is trivial. Parts (b1) and (c) are proved in Sections 2–5 in the following way: We first prove that there is a natural complex structure on $\mathcal{X}^*$ (and therefore on the open set $\mathcal{Y}^*$). We then show that Φ is a homeomorphism that maps $\mathbf{0}$ to $\Pi(\mathbf{0})$ and is analytic from $\mathcal{H}^*$ onto $\mathcal{Y}^*$. This completes the atlas on $\mathcal{X}$ and gives both (b1) and (c). We then prove Theorem B using a result of Faught-Roesch together with Słodkovski extensions of holomorphic motions. This is the content of Section 6. Finally, we establish part (b2) of Theorem A' as a consequence of Theorem B.

3   Definition of Ω_a and Extension of η_a

The main purpose of this section is to prove Proposition 2.1. We shall also, however, introduce some auxiliary notions and some preliminary results which will be useful in subsequent sections.

The following properties of the sets $L_\lambda(s)$ are easily verified. For future reference, we state them as a lemma the proof of which we leave to the reader.

Lemma 3.1 (Structure of $L_\lambda(s)$). *Assume $\lambda \in \mathbb{D}^*$.*

- *For $s \in \mathbb{R}$, the set $\overline{U_\lambda(s)}$ is a closed Jordan domain.*

- *For $s \notin \mathbb{N}$, $L_\lambda(s) = \overline{U_\lambda(s)}$.*

- *For $s \in \mathbb{N}$ (starting from $s = 0$), $\overline{U_\lambda(n)} \subseteq L_\lambda(n)$, and $L_\lambda(n) \setminus \overline{U_\lambda(n)}$ consists of 2^n components each attached to $\overline{U_\lambda(n)}$ by a point in $Q_\lambda^{-n}(c_\lambda)$.*

- *In particular, $L_\lambda(0)$ is bounded by a figure eight whose branching point is the critical point c_λ.*

- *The restrictions $Q_\lambda : \overline{U_\lambda(s)} \to \overline{U_\lambda(s-1)}$ are homeomorphisms for $s \leq 0$ and coverings of degree two branched over v_λ for $s > 0$; the restrictions $Q_\lambda : L_\lambda(s) \to L_\lambda(s-1)$ are homeomorphisms for $s < 0$ and coverings of degree two branched over v_λ for $s \geq 0$.*

The following lemmas show that the structure of $L_{\mathbf{a}}(s)$ is biholomorphically the same as the structure of $L_\lambda(s)$ when $s < t_{\mathbf{a}}^1 := \kappa_{\mathbf{a}}(c_{\mathbf{a}}^1)$.

Lemma 3.2 (Local Biholomorphic Conjugacy). *For $\mathbf{a} = (\lambda, a) \in \mathcal{H} \setminus \mathcal{H}_0$, the conformal isomorphism*

$$\eta_{\mathbf{a}} := \psi_\lambda \circ \phi_{\mathbf{a}} : U_{\mathbf{a}}^0 \longrightarrow U_\lambda^0$$

satisfies $\eta_{\mathbf{a}}(v_{\mathbf{a}}^0) = v_\lambda$ and $\eta_{\mathbf{a}} \circ P_{\mathbf{a}} = Q_\lambda \circ \eta_{\mathbf{a}}$ on $U_{\mathbf{a}}^0$. It continuously extends to a semi-conjugacy $\eta_{\mathbf{a}} : L_{\mathbf{a}}^0 \longrightarrow L_\lambda^0$, is analytic on a neighborhood of $L_{\mathbf{a}}^0 \setminus \{co_{\mathbf{a}}^0\}$ and is unique if $c_{\mathbf{a}}^0 \neq co_{\mathbf{a}}^0$. We have $c_{\mathbf{a}}^0 = co_{\mathbf{a}}^0$ iff $c_{\mathbf{a}}^1 = c_{\mathbf{a}}^0$, and $co_{\mathbf{a}}^0 \in L_{\mathbf{a}}^0$ iff $c_{\mathbf{a}}^1 \in L_{\mathbf{a}}^0$. It is a conjugacy if $t_{\mathbf{a}}^1 > 0$.

The proof is easy and is left to the reader. See Figure 2 for a schematic illustration.

We want to extend $\eta_{\mathbf{a}}$ further if $c_{\mathbf{a}}^1 \notin L_{\mathbf{a}}^0$ (i.e., $t_{\mathbf{a}}^1 > 0$). For the rest of this section we shall omit $\mathbf{a}$ as a subscript.

Lemma 3.3. *For $\mathbf{a} \in \mathcal{H} \setminus \mathcal{H}_0$, the local conjugacy η has a unique biholomorphic extension $\eta : U = U(t^1) \longrightarrow U_\lambda(t^1)$, where $t^1 = \max\{0, \kappa(v^1)+1\}$.*

Proof: By construction, the local conjugacy $\eta : U(0) \longrightarrow U_\lambda(0)$ is biholomorphic and preserves critical values: $\eta(P(c^0)) = Q_\lambda(c_\lambda)$. Write $\hat{t} = \kappa(v^1)$. If $\hat{t} \leq -1$, then $c^1 \in L(0) \setminus U(0)$; hence $t^1 = 0$, and the conclusion of the lemma holds.

Suppose $\hat{t} > -1$, and let n be the unique integer such that $n - 1 \leq \hat{t} < n$. Then the restriction $P : U(n) \longrightarrow U(n-1)$ has degree 2, and hence we can extend η recursively to unique biholomorphic conjugacies

$\eta : U(k) \longrightarrow U_\lambda(k)$ by lifting $\eta \circ P$ on $U(k)$ to Q_λ on $U_\lambda(k)$ for $k = 0 \cdots n$. For any $s > \hat{t}$, the degree of $P : U(\hat{t}+1) \longrightarrow U(\hat{t})$ is 2 and the degree of $P : U(s+1) \longrightarrow U(s)$ is 3 because $v^1 \in U(s) \setminus U(\hat{t})$ for any such s. It follows from the latter case that $t^1 \le 1+\hat{t}$, and it follows from the former case that we can extend η to a uniquely determined biholomorphic conjugacy $\eta : U(\hat{t}+1) \longrightarrow U_\lambda(\hat{t}+1)$ by lifting as above. Hence $0 < \hat{t}+1 \le t^1$. □

The above biholomorphic conjugacy extends uniquely as a homeomorphic conjugacy of the closures $\eta : \overline{U(t^1)} \longrightarrow \overline{U_\lambda(t^1)}$, because both open sets $U(t^1)$ and $U_\lambda(t^1)$ are Jordan domains. Moreover this extension is (except when $co^0 \in \partial U(t^1)$, or, equivalently, when $c^1 \in \partial U(t^1)$) locally the restriction of a holomorphic function. We have

$$
\begin{array}{ccc}
(c^0,0),\ \overline{U(t)} & \xrightarrow{\ \eta\ } & \overline{U_\lambda(t)},\ (c_\lambda,0) \\
{\scriptstyle P}\big\downarrow & & \big\downarrow{\scriptstyle Q_\lambda} \\
(v^0,0),\ \overline{U(t-1)} & \xrightarrow[\text{for } t \le t^1]{\ \eta\ } & \overline{U_\lambda(t-1)},\ (v_\lambda,0)
\end{array}
$$

and

$$
\begin{array}{ccc}
L(t) & \xrightarrow{\ \eta\ } & L_\lambda(t) \\
{\scriptstyle P}\big\downarrow & & \big\downarrow{\scriptstyle Q_\lambda} \\
L(t-1) & \xrightarrow[\text{for } t < t^1]{\ \eta\ } & L_\lambda(t-1)\ .
\end{array}
$$

3.1 Proof of Proposition 2.1

Proof: We shall construct Ω and η.

Case 0. $\lambda = 0$. We take $\Omega = \overline{U_a}$. Then $\eta = \phi : \Omega \longrightarrow \overline{\mathbb{D}(0, e^{t^1})}$ is a homeomorphism which is the restriction of an analytic map.

Assume from now on that $\lambda \ne 0$.

Recall that we are looking for a closed, connected, simply connected set Ω such that $c^1 \in \Omega$, $\overline{U(t^1)} \subseteq \Omega \subseteq L(t^1)$ and such that the following diagram commutes:

$$
\begin{array}{ccc}
\Omega & \xrightarrow{\ \eta\ } & L_\lambda(t^1) \\
{\scriptstyle P}\downarrow & & \downarrow{\scriptstyle Q_\lambda} \\
L(t^1-1) & \xrightarrow{\ \eta\ } & L_\lambda(t^1-1)
\end{array}
\ .
$$

Recall also that $U = U(t^1)$, $U(0) = U^0$ and $L(0) = L^0$.

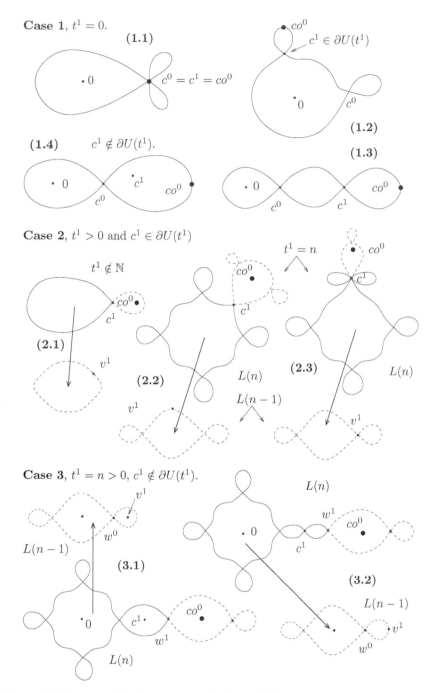

Figure 2. The set Ω is the closed region bounded by the solid curves. We have $co\Omega = L(n) \setminus \Omega$ and $co^0 \notin \Omega$ in Cases 2 and 3, and $co\Omega = \emptyset$ and $co^0 \in \partial\Omega$ in Case 1.

Case 1. $t^1 = 0$. We take $\Omega = L(0) = L^0$. Then η has a continuous extension $\eta : \Omega \longrightarrow L_\lambda(t^1)$, which is analytic except at the boundary point co^0 of Ω. Its value at c^1 is uniquely determined.

Case 2. $t^1 > 0$ and $c^1 \in \overline{U(t^1)} = \overline{U}$. Then c^1 is a separating point for $L(t^1)$, and the critical value v^1 is a point in $\partial L(t^1 - 1)$. Let $co\Omega$ denote the connected component of $P^{-1}(L(t^1 - 1) \setminus \{v^1\})$ containing the co-critical point co^0. Define $\Omega = L(t^1) \setminus co\Omega$. Then $\eta \circ P : \Omega \longrightarrow L_\lambda(t^1 - 1) \subset U(t^1)$ has a unique lift $\eta : \Omega \longrightarrow L_\lambda(t^1)$ to Q_λ that extends η to $U(t^1)$.

Case 3. $t^1 > 0$ and $c^1 \notin \overline{U(t^1)}$. In this case the truncation of $L(t^1)$ is slightly more delicate: We have $t^1 = n > 0$, where n is an integer, $c^1 \in L(n) \setminus \overline{U(n)}$ and $v^1 \in L(n - 1) \setminus \overline{U(n - 1)}$ by Lemma 3.3. The first attracted critical point c^0 is a separation point for $L(0)$, i.e., $L(0) \setminus \{c^0\}$ has two components. It follows by induction that for $m \le n$ there are 2^m points of $P^{-m}(c^0)$ on the boundary of $U(m)$, and that for $m < n$ they are precisely the separation points of $L(m)$. Let w^0 be the separation point of $L(n - 1)$ which separates 0 from the critical value v^1, and let $L'(n - 1)$ denote the connected component of $L(n - 1) \setminus w^0$ containing 0. Moreover let $co\Omega$ denote the connected component of $P^{-1}(L'(n - 1))$ containing co^0. We set $\Omega = L(n) \setminus co\Omega$. Then $\eta \circ P : \Omega \longrightarrow L_\lambda(n - 1) \subset U(n)$ has a unique lift $\eta : \Omega \longrightarrow L_\lambda(n)$ to Q_λ that extends η on $U(n)$. This extension is locally the restriction of a holomorphic function.

This ends the construction of Ω and η.

Clearly $c^1 \notin \partial U^0$ if and only if c^0 is the unique first attracted critical point and $\mathbf{a} \mapsto \eta_a(c_\mathbf{a}^1)$ is single valued. Suppose $\mathbf{a}$ is such that both critical points belong to ∂U^0, and let z_0 and z_1 denote the two values of $\eta(c^1) = \psi_\lambda \circ \phi(c^1) \in \partial U_\lambda^0$. That is, fix a labeling c^0, c^1 of the critical points and let ϕ_0, ϕ_1 denote the two normalizations of ϕ with $\phi_i(c^i) = 1$ for $i = 0, 1$; then $z_i = \eta_i(c^{1-i}) = \psi_\lambda \circ \phi_i(c^{1-i})$ and $\phi_\lambda(z_i) = \phi_i(c^{1-i})$, $i = 0, 1$. Moreover $\phi_1 = \rho \cdot \phi_0$ for some ρ of modulus 1. In particular, $\rho = \rho \cdot 1 = \phi_\lambda(z_1)$ and $\rho \cdot \phi_\lambda(z_0) = 1$, so $\phi_\lambda(z_1) \cdot \phi_\lambda(z_0) = 1$.

Finally, that $\kappa_\lambda(\eta_a(c_\mathbf{a}^1)) = t_\mathbf{a}^1 = \kappa_a(c_\mathbf{a}^1)$ follows from the constructions above. This ends the proof of Proposition 2.1. $\qquad\square$

Corollary 3.4. *The map* $\Phi : \mathcal{H} \longrightarrow \mathcal{Y}$, $\mathbf{0} \mapsto \Pi(0)$ *and* $\mathbf{a} \mapsto \Pi(\lambda, \eta_a(c_\mathbf{a}^1))$ *is well defined. Moreover if* $\mathbf{a} \ne \mathbf{0}$, *then* $\Phi(\mathbf{a}) \ne \Pi(0)$.

Proof: The only ambiguity in the definition of Φ occurs when $\partial U_\mathbf{a}^0$ contains two distinct critical points. In this case different choices of the labeling $c_\mathbf{a}^0, c_\mathbf{a}^1$ lead to two different maps η_a and to two values z_0, z_1 for $\eta_a(c_\mathbf{a}^1)$, with $\phi_\lambda(z_1) \cdot \phi_\lambda(z_0) = 1$. Thus $\Pi(\lambda, z_0) = \Pi(\lambda, z_1)$ and is independent of the choices involved. Hence Φ is well defined.

Clearly $\Phi(\mathbf{a}) \ne \Pi(0)$ if $\mathbf{a} \ne \mathbf{0}$, and $\Phi(\mathcal{H}) \subset \mathcal{Y}$. $\qquad\square$

4 Analyticity of the Map $(\mathbf{a}, z) \mapsto \eta_{\mathbf{a}}(z)$

The purpose of this section is to prove Proposition 4.5 below. For this we will rely on a result of Petersen (Proposition 4.1 below).

For $\mathbf{a} = (0, a) \in \mathcal{H}_0^*$, define the potential function $\widehat{g}_{\mathbf{a}} : B_{\mathbf{a}} \to [-\infty, 0[$ by $\widehat{g}_{\mathbf{a}}(z) = \log |\phi_{\mathbf{a}}(z)|$ near 0, and extend it to the entire basin by the recursive relation

$$\widehat{g}_{\mathbf{a}}(z) = \widehat{g}_{\mathbf{a}}(P_{\mathbf{a}}(z))/2. \tag{4.1}$$

For $t \in \mathbb{R}_-$, let $U_{\mathbf{a}}(t)$ denote the connected component of $\widehat{g}_{\mathbf{a}}^{-1}([-\infty, t[)$ containing 0. Define also the filled potential

$$g_{\mathbf{a}}(z) = \inf\{s | z \in U_{\mathbf{a}}(s)\}.$$

Then $t_{\mathbf{a}}^1 = g_{\mathbf{a}}(c_{\mathbf{a}}^1)$, $U_{\mathbf{a}} = U_{\mathbf{a}}(t_{\mathbf{a}}^1)$ and $\Omega_{\mathbf{a}} = \overline{U_{\mathbf{a}}(t_{\mathbf{a}}^1)}$.
  Define

$$\mathcal{F} := \Phi^{-1}(\Gamma) = \{\mathbf{0}\} \cup \{\mathbf{a} \in \mathcal{H} \setminus \mathcal{H}_0 \mid \text{both critical points are on } \partial U_{\mathbf{a}}^0\}$$

and $\mathcal{F}_\lambda = \mathcal{F} \cap \mathcal{H}_\lambda$. Set

$$\mathcal{U} = \{(\lambda, a, z) | \mathbf{a} = (\lambda, a) \in \mathcal{H} \setminus \mathcal{F} \text{ and } z \in U_{\mathbf{a}}\} .$$

Proposition 4.1. *The set $\mathcal{U}$ is open in $\mathbb{C}^3$, and the map $(\mathbf{a}, z) \mapsto \eta_{\mathbf{a}}(z)$ is complex analytic on $\mathcal{U}$.*

Proof: See Petersen [Pe04], where the proposition is proved for the family $\widehat{P}_{\lambda,b}(z) = \lambda z + bz^2 + z^3$, which double covers the family $P_{\mathbf{a}}$. Moreover there is no obstruction in passing the result to the quotient, because we are avoiding the parameters in $\mathcal{F}$.   □

Definition 4.2. In case that $\lambda \neq 0$ and $t_{\mathbf{a}}^1 > 0$, set $pC_\lambda = \overline{\{Q_\lambda^n(c_\lambda) | n \geq 1\}}$, $pC_{\mathbf{a}} = \overline{\{P_{\mathbf{a}}^n(c_{\mathbf{a}}^0) | n \geq 1\}}$ and $copC_{\mathbf{a}} = P_{\mathbf{a}}^{-1}(pC_{\mathbf{a}}) \setminus L_{\mathbf{a}}^0$. The last two sets depend analytically on $\mathbf{a}$.

Lemma 4.3. *Fix $\mathbf{a} \in \mathcal{H}$ with $\lambda \neq 0$ and $t_{\mathbf{a}}^1 > 0$. Let $X \subset B_{\mathbf{a}}(0) \setminus copC_{\mathbf{a}}$ be any simply connected forward invariant $(P_{\mathbf{a}}(X) \subset X)$ domain, with $U_{\mathbf{a}}(T) \subset X$ for some $0 < T \leq t_{\mathbf{a}}^1$.*
  Then the restriction of $\eta_{\mathbf{a}}$ to $U_{\mathbf{a}}^0$ has an analytic extension $\widehat{\eta}$ to X satisfying $\widehat{\eta} \circ P_{\mathbf{a}} = Q_\lambda \circ \widehat{\eta}$. This extension is uniquely determined by X, but in general depends on X. However if $\widehat{\eta}(X) \subset U_\lambda(t_{\mathbf{a}}^1)$, then $X \subset U_{\mathbf{a}}(t_{\mathbf{a}}^1)$ and $\widehat{\eta} = \eta_{\mathbf{a}}$ on X. Inversely if $U_{\mathbf{a}}(t) \subset X$ for some $0 < t \leq t_{\mathbf{a}}^1$, then $\eta_{\mathbf{a}} = \widehat{\eta}$ on $U_{\mathbf{a}}(t)$. Furthermore if $\Omega_{\mathbf{a}} \subset X$, then $\eta_{\mathbf{a}} = \widehat{\eta}$ on $\Omega_{\mathbf{a}}$.

Let us remark that η has no continuous extension to a neighborhood of $co_{\mathbf{a}}^0$ as a semi-conjugacy: Assume at first $co_{\mathbf{a}}^0 \neq c_{\mathbf{a}}^0$. Then $P_{\mathbf{a}}$ is locally univalent around $co_{\mathbf{a}}^0$. But $\eta_{\mathbf{a}}(P_{\mathbf{a}}(co_{\mathbf{a}}^0)) = v_\lambda$ and $\eta_{\mathbf{a}}$ is locally univalent at $P(co_{\mathbf{a}}^0)$, and v_λ has a unique preimage by Q_λ which is c_λ and around which Q_λ is non-univalent. The other case, when $co_{\mathbf{a}}^0 = c_{\mathbf{a}}^0$, is similar ($P_{\mathbf{a}}$ is locally of degree 3 this time).

Proof: Uniqueness comes from the uniqueness of analytic continuations on simply connected domains. In what follows, we omit the subscript $\mathbf{a}$ for simplicity.

For $t \in \mathbb{R}$, define X_t to be the connected component of $U(t) \cap X$ containing 0. Then each X_t is a simply connected forward invariant domain (in fact $P(X_t) \subseteq X_{t-1} \subseteq X_t$), and $X_t \cap copC = \emptyset$. We define $\widehat{\eta} = \eta$ on $X_0 = U(0)$ and will define $\widehat{\eta}$ recursively on X_n, $n \geq 0$, as the unique lift of $\widehat{\eta} \circ P$ on X_n to Q_λ that fixes 0.

In order to get the existence of the lifting, we make use of the following classical result:

Lemma 4.4. *Let Y, Z and W be path-connected and locally path-connected Hausdorff spaces with base points $y \in Y, z \in Z$ and $w \in W$. Suppose $p : W \to Y$ is an unbranched covering and $f : Z \to Y$ is a continuous map such that $f(z) = y = p(w)$.*

$$Z, z \quad \overset{\tilde{f}}{\longrightarrow} \quad W, w$$
$$f \searrow \quad \downarrow p$$
$$Y, y \; .$$

Then there exists a continuous lift $\tilde{f}$ of f to p with $\tilde{f}(z) = w$ if and only if

$$f_*(\pi_1(Z, z)) \subset p_*(\pi_1(W, w)) \; , \qquad (4.2)$$

where π_1 denotes the fundamental group. This lift is unique.

Proof: By assumption, $U(T) \subset X$ for some $0 < T \leq t^1$, and by Lemma 3.3, the conjugacy η has a biholomorphic extension $\widehat{\eta} : U(T) \longrightarrow U_\lambda(T)$. We shall prove by induction on $n \geq T$ that $\widehat{\eta}$ has a unique analytic extension to X_n, with $\widehat{\eta}^{-1}(pC_\lambda) = pC \subset U^0$.

Suppose that $\widehat{\eta}$ is analytically extended to X_{n-1} with $\widehat{\eta}^{-1}(\widehat{\eta}(pC)) = pC \subset U^0$, and suppose $n \geq T$. We want to prove that there exists a lift $\widetilde{\eta}$ of the analytic and hence continuous map $\widehat{\eta} \circ P$ on X_n to Q_λ that extends $\widehat{\eta}$. Moreover we want to show that $\widetilde{\eta}$ is injective above pC_λ, i.e., that $\widetilde{\eta}^{-1}(pC_\lambda) = pC \subset U^0$.

$$
\begin{array}{ccc}
X_n \setminus \{c^0\} & \xrightarrow{\;\widetilde{\eta}\;} & B_\lambda \setminus \{c_\lambda\} \\[4pt]
P \downarrow & & \downarrow Q_\lambda \\[4pt]
X_{n-1} \setminus \{v^0\} & \xrightarrow{\;\widehat{\eta}\;} & B_\lambda \setminus \{v_\lambda\} \, .
\end{array}
$$

We set all base points to be 0. Any closed loop γ in $X_n \setminus \{c^0\}$ is homotopic to a closed loop γ' in $U(T) \setminus \{c^0\}$, because X_n is simply connected and contains $U(T)$, for which c^0 is an interior point. Thus $(\widehat{\eta} \circ P)_*(\pi_1(X_n \setminus \{c^0\}, 0)) = (\widehat{\eta} \circ P)_*(\pi_1(U(T) \setminus \{c^0\}, 0)) \subset (Q_\lambda)_*(\pi_1(B_\lambda \setminus \{c_\lambda\}, 0))$ is satisfied, by the existence of $\widehat{\eta}$. Hence the lift $\widetilde{\eta}$ fixing 0 exists and is unique. In particular, $\widetilde{\eta} = \widehat{\eta}$ on the connected subset $X_{n-1} \setminus \{c^0\} \subset X_n \setminus \{c^0\}$, and hence extends to a continuous map on X_n. To see that $\widetilde{\eta}^{-1}$ is injective above pC_λ, we note that since $\widehat{\eta}$ is injective above pC_λ, we have $\widetilde{\eta}(z) \in pC_\lambda$ if and only if $z \in P^{-1}(pC)$. However since $\widetilde{\eta} = \widehat{\eta}$ is injective on $L^0 \subset U(T)$, we have either $z \in pC$ or $z \in copC$. By assumption, $copC \cap X_n = \emptyset$. Thus $\widetilde{\eta}$ is injective above pC_λ. In particular, $\widetilde{\eta}(X_n \setminus \{v^0\}) \subset B_\lambda \setminus \{v_\lambda\}$, and we can continue the induction. $\qquad\square$

Proposition 4.5. *The set $\Omega_{\mathbf{a}}$ varies upper semi-continuously for $\mathbf{a} \in \mathcal{H}^*$ for the Hausdorff topology. The map $\mathbf{a} \mapsto \eta_{\mathbf{a}}(c_{\mathbf{a}}^1)$ is analytic on $\mathcal{H} \setminus \mathcal{F}$.*

Proof: The semi-continuity of $\Omega_{\mathbf{a}}$ is shown by inspecting each case in Figure 2 (and using Proposition 4.1 for the case $\lambda = 0$).

Assume $\lambda \neq 0$, and suppose at first that $t_{\mathbf{a}}^1 > 0$. Choose an open connected and simply connected neighborhood X of $\Omega_{\mathbf{a}}$ with $P_{\mathbf{a}}(\overline{X}) \subset U_{\mathbf{a}}(t_{\mathbf{a}}^1 - \frac{1}{2})$ and with $copC_{\mathbf{a}} \cap \overline{X} = \emptyset$. There exists a neighborhood Λ of $\mathbf{a}$ such that for all $\mathbf{a}' \in \Lambda$ we have $P_{\mathbf{a}'}(X) \subset U_{\mathbf{a}'}(t_{\mathbf{a}'}^1) = U_{\mathbf{a}'}$, $\Omega_{\mathbf{a}'} \subset X$ and $copC_{\mathbf{a}'} \cap \overline{X} = \emptyset$, because $\Omega_{\mathbf{a}'}$ varies upper semi-continuously and $copC_{\mathbf{a}'}$ varies continuously with $\mathbf{a}'$ for the Hausdorff metric on compact subsets. Lemma 4.3 implies that for every $\mathbf{a}' \in \Lambda$, the map $\eta_{\mathbf{a}'}$ has a unique analytic extension $\widehat{\eta}_{\mathbf{a}'}$ to X which agrees with $\eta_{\mathbf{a}'}$ on $\Omega_{\mathbf{a}'}$. We need to prove that these lifts define a continuous and hence analytic map.

Denote by p_1 the projection $\mathbf{a}' = (\lambda', a') \mapsto \lambda'$.

$$
\begin{array}{ccc}
\bigcup_{\mathbf{a}' \in \Lambda}(\mathbf{a}', X \setminus \{c_{\mathbf{a}'}^0\}) & \xrightarrow{(p_1, \widetilde{\eta}_{\mathbf{a}'})} & \bigcup_{\mathbf{a}' \in \mathbb{D}^*}(\lambda', B_{\lambda'} \setminus \{c_{\lambda'}\}) \\[6pt]
(id, P_{\mathbf{a}'}) \downarrow & & \downarrow (id, Q_{\lambda'}) \\[6pt]
\bigcup_{\mathbf{a}' \in \Lambda}(\mathbf{a}', U_{\mathbf{a}'} \setminus \{v_{\mathbf{a}'}^0\}) & \xrightarrow{(p_1, \eta_{\mathbf{a}'})} & \bigcup_{\mathbf{a}' \in \mathbb{D}^*}(\lambda', B_{\lambda'} \setminus \{v_{\lambda'}\}) \, .
\end{array}
$$

Here we set the base points to be $(\mathbf{a}, 0)$ for the left hand sets and $(\lambda, 0)$ for the right hand sets. By Proposition 4.1, the map $(p_1, \eta_{\mathbf{a}'})$ is analytic and therefore continuous. The right hand map is a covering. Now any loop in $\bigcup_{\mathbf{a}' \in \Lambda}(\mathbf{a}', X \setminus \{c_{\mathbf{a}'}^0\})$ based at $(\mathbf{a}, 0)$ is homotopic to a loop on the slice $\mathbf{a}' = \mathbf{a}$. Thus condition (4.2) is satisfied, due to the existence of $\widehat{\eta}_{\mathbf{a}}$, and we can apply Lemma 4.4 to obtain a continuous lift in the above diagram, obviously of the form $(p_1, \widetilde{\eta}_{\mathbf{a}'})$. Using the uniqueness of lifts on each slice, we get $\widetilde{\eta}_{\mathbf{a}'} = \widehat{\eta}_{\mathbf{a}'}$.

This implies that $(\mathbf{a}', z) \to \widehat{\eta}_{\mathbf{a}'}(z)$ on $\Lambda \times X$ is continuous and analytic on each variable, and therefore analytic on the joint variable $(\mathbf{a}', z)$. In particular, $\mathbf{a}' \mapsto \widehat{\eta}_{\mathbf{a}'}(c_{\mathbf{a}'}^1) = \eta_{\mathbf{a}'}(c_{\mathbf{a}'}^1)$ is analytic.

Suppose next that $t_{\mathbf{a}}^1 = 0$ and $c_{\mathbf{a}}^1 \in L_{\mathbf{a}}^0 \setminus \overline{U_{\mathbf{a}}^0}$. Then for $\epsilon > 0$ sufficiently small, the set $\Omega_{\mathbf{a}} \setminus \mathbb{D}(co_{\mathbf{a}'}^0, \epsilon)$ is a closed connected set containing $\overline{U_{\mathbf{a}}^0}$ and $c_{\mathbf{a}}^1$. Using an argument similar to the one above, with a sufficiently small neighborhood X of this set, we may draw a similar conclusion.

In case $\lambda = 0$, choose a connected and simply connected neighborhood X of $\Omega_{\mathbf{a}}$ with $co_{\mathbf{a}}^0 \notin \overline{X}$. Then the same argument as above works. $\qquad \square$

5 The Natural Complex Structure of $\mathcal{X} \setminus \{0\}$

With $\widetilde{\Gamma} = \{(\lambda, z), z \in \overline{U_\lambda^0}\} \cup \{0\}$, the restriction $\Pi : \mathbb{D} \times (\overline{\mathbb{C}} \setminus \widetilde{\Gamma}) \longrightarrow \mathcal{X} \setminus \Gamma$ is a homeomorphism of open sets. It thus defines a complex analytic structure $\widehat{\mathcal{A}}$ on $\mathcal{X} \setminus \Gamma$. We will show that this complex structure is in fact the restriction of a complex analytic structure on $\mathcal{X}^* = \mathcal{X} \setminus \Pi(\mathbf{0})$:

Proposition 5.1. *The space $\mathcal{X}^*$ has a complex analytic structure $\mathcal{A}^*$ containing $\widehat{\mathcal{A}}$. Moreover this structure gives each fiber $\mathcal{X}_\lambda := \{\Pi(\lambda, z) | z \in \overline{\mathbb{C}}\}$ a unique complex structure, and the corresponding Riemann surface is isomorphic to $\overline{\mathbb{C}}$.*

Proof: We prove first that each fiber $\mathcal{X}_\lambda$, $\lambda \in \mathbb{D}$ has a unique complex structure containing the chart $\pi_\lambda^{-1} : \mathcal{X}_\lambda \setminus \Gamma_\lambda \longrightarrow \overline{\mathbb{C}}$, which is complex analytic for the structure $\widehat{\mathcal{A}}$.

For $\lambda = 0$, the map $\pi_0^{-1} : \mathcal{X}_0 \longrightarrow \overline{\mathbb{C}}$ is a homeomorphism, and thus a global chart.

Now fix $\lambda \in \mathbb{D}^*$. We define a chart k_λ on $W_\lambda := \pi_\lambda(U_\lambda(1))$, a neighborhood of Γ_λ, as follows:

$$
\begin{array}{ccccc}
U_\lambda(1) \setminus U_\lambda^0 & \xrightarrow{\theta_\lambda} & \mathbb{C} \setminus U & \xrightarrow{f} & \mathbb{C} \setminus \mathbb{D} \\
\phi_\lambda \searrow & & \nearrow g & & \downarrow h \\
\pi_\lambda \downarrow \qquad \mathbb{C} \supset D(\tfrac{1}{|\lambda|}) \supset \mathbb{D} & & & & \\
W_\lambda & \xrightarrow{k_\lambda} & & & \mathbb{C} \, .
\end{array}
$$

We will use a *model double cover* $g : \overline{\mathbb{C}} \longrightarrow \overline{\mathbb{C}}$, the quadratic polynomial $g(z) = -z^2 + 1$. Note that g has a critical point at 0, with critical value 1, and $g(-1) = 0$.

The restriction $\phi_\lambda : U_\lambda(1) \longrightarrow \mathbb{D}(1/|\lambda|)$ is proper, of degree 2 and has the critical point c_λ with critical value 1. Thus ϕ_λ has a univalent lift to g, $\theta_\lambda : U_\lambda(1) \longrightarrow \mathbb{C}$, with $\theta_\lambda(c_\lambda) = 0$, and $\theta_\lambda(0) = -1$.

Denote by U the connected component of $g^{-1}(\mathbb{D})$ containing -1. Then $\theta_\lambda(U_\lambda^0) = U$. Let $f : \overline{\mathbb{C}} \setminus U \longrightarrow \overline{\mathbb{C}} \setminus \mathbb{D}$, with $f(\infty) = \infty$, be the homeomorphically extended Riemann map normalized by $f(0) = 1$. Then

$$f(z_1) = \overline{f(z_2)} \in \mathbb{S}^1 \quad \text{for} \quad z_1, z_2 \in \partial U, \quad \text{with} \quad g(z_1) = \overline{g(z_2)},$$

because both g and f have only real coefficients. Denote by h the branched double covering $h(z) = z + 1/z : \overline{\mathbb{C}} \longrightarrow \overline{\mathbb{C}}$. It satisfies $h(z) = h(1/z)$ for all z, and in particular $h(z) = h(\overline{z})$ whenever $z \in \mathbb{S}^1$. The map

$$k_\lambda := h \circ f \circ \theta_\lambda \circ \pi_\lambda^{-1} : W_\lambda \longrightarrow \mathbb{C} \tag{5.1}$$

is a well-defined homeomorphism onto its image. Moreover the composition $k_\lambda \circ \pi_\lambda = h \circ f \circ \theta_\lambda : U_\lambda(1) \setminus \overline{U_\lambda^0} \longrightarrow \overline{\mathbb{C}} \setminus [-2, 2]$ is holomorphic. Hence the two charts k_λ and π_λ^{-1} together define a complex analytic atlas on $\mathcal{X}_\lambda$.

Finally, let $\chi : \omega \longrightarrow \mathbb{C}$ be a chart of a complex structure on $\mathcal{X}_\lambda$ and suppose $\chi \circ \pi_\lambda$ is holomorphic on $\pi_\lambda^{-1}(\omega) \setminus \partial U_\lambda^0$. Then $\chi \circ k_\lambda^{-1}$ is continuous where defined and holomorphic on $k_\lambda(\omega) \setminus [-2, 2]$. But then it is holomorphic on all of $k_\lambda(\omega)$ by Morera's theorem. Thus χ is in the atlas defined by k_λ. This proves the uniqueness of the complex structure on $\mathcal{X}_\lambda$ containing the chart π_λ^{-1} on $\mathcal{X}_\lambda \setminus \Gamma_\lambda$.

It remains to be proved that the complex structures on the fibers line up to form a complex structure on $\mathcal{X}^*$. The set $W := \bigcup_{\lambda \in \mathbb{D}^*} \{\lambda\} \times W_\lambda$ is an open neighborhood of $\Gamma^* = \Gamma \setminus \Pi(\mathbf{0})$. Moreover $K : W \longrightarrow \mathbb{C}^2$, given by $K(\lambda, z) = (\lambda, k_\lambda(z))$, is a homeomorphism onto its image, because the map $(\lambda, z) \mapsto k_\lambda(z)$ is continuous on W and a homeomorphism of W_λ onto its image for each fixed $\lambda \in \mathbb{D}^*$.

Finally, the composition

$$K \circ \Pi : \bigcup_{\lambda \in \mathbb{D}^*} \{\lambda\} \times \left(U_\lambda(1) \setminus \overline{U_\lambda(0)} \right) \longrightarrow \mathbb{C}^2$$

is holomorphic. Thus K and Π^{-1} together define a complex analytic atlas on $\mathcal{X}^*$. $\qquad\square$

6 Proof of Theorem A'

6.1 Injectivity of Φ (Rigidity), and a Lifting Process

$$
\begin{array}{ccccccccc}
\cdots & \overline{\mathbb{C}} & \xrightarrow{P_1} & \overline{\mathbb{C}} & \xrightarrow{P_1} & \overline{\mathbb{C}} \supset \Omega_1 & & \xrightarrow{P_1} & P_1(\Omega_1) \\
& h_2 \downarrow & & h_1 \downarrow & & h_0 \downarrow \quad h_0 \downarrow & L_\lambda(t) & & \downarrow h_0 \\
\cdots & \overline{\mathbb{C}} & \xrightarrow{P_2} & \overline{\mathbb{C}} & \xrightarrow{P_2} & \overline{\mathbb{C}} \supset \Omega_2 & & \xrightarrow{P_2} & P_2(\Omega_2)
\end{array}
$$

Given any two (fixed) parameters $\mathbf{a_1} = (\lambda_1, a_1)$ and $\mathbf{a_2} = (\lambda_2, a_2)$ with $\Phi(\mathbf{a_1}) = \Phi(\mathbf{a_2})$, we shall prove that $\mathbf{a_1} = \mathbf{a_2}$. To simplify notation, we use the indices $1, 2$ for all of the involved quantities. In particular, $P_1 = P_{\mathbf{a_1}}$ and $P_2 = P_{\mathbf{a_2}}$. It follows immediately that $\lambda_1 = \lambda_2 = \lambda$, and by Proposition 2.1 that $t := t^1_{\mathbf{a_1}} = \kappa_\lambda(\eta_1(c_1^1)) = \kappa_\lambda(z) = \kappa_\lambda(\eta_2(c_2^1)) = t^1_{\mathbf{a_2}}$. Hence we also have $\eta_1(\Omega_1) = \eta_2(\Omega_2) = L_\lambda(t)$. By possibly exchanging the labels of c_2^0 and c_2^1, we can assume that $\eta_1(c_1^1) = z = \eta_2(c_2^1)$, and $\Phi(\mathbf{a_1}) = \Phi(\mathbf{a_2}) - \Pi(\lambda, z)$.

The injectivity of Φ on $\mathcal{H}_0$ was proved by Milnor. The proof in the case that $\lambda \neq 0$ is similar; the core is the so called pull-back argument. We shall go through the details, however. We shall recursively define quasiconformal maps h_n, which are conformal on larger and larger domains. Extracting a convergent subsequence will provide a conformal conjugacy between P_1 and P_2.

Definition of $h|_{\Omega_1}$. We claim that there exists a homeomorphic conjugacy $h : \Omega_1 \longrightarrow \Omega_2$ of P_1 to P_2 such that h is the restriction of a biholomorphic conjugacy between neighborhoods of the two sets.

Assume at first that $t = t^1_{\mathbf{a_1}} = t^1_{\mathbf{a_2}}$ is not an integer. Then both Ω_1 and Ω_2 are closed Jordan domains. Set $h = \eta_2^{-1} \circ \eta_1$; this is a well-defined homeomorphism and maps $c_1^1 \in \partial U_1(t)$ to $c_2^1 \in \partial U_2(t)$.

Assume next that $t = n \geq 0$. Note that $L_\lambda(n) \setminus \overline{U_\lambda(n)}$ consists of 2^n components, each one attached to $U_\lambda(n)$ by a point in $Q_\lambda^{-n}(c_\lambda)$. Each component, except the one containing z, is homeomorphically covered under each of the maps η_1 and η_2 by similar components of $\Omega_i \setminus \overline{U_i(n)}$, $i = 1, 2$. The component containing z, however, is double covered under both maps, with z as a critical value and c_i as the respective critical points. Thus the existence of h also follows in this case. This proves the claim.

Extension of h. We shall extend h to a homeomorphism $h : L_1(t) \longrightarrow L_2(t)$ that conjugates P_1 to P_2. We remark first that $L_i(t) = \Omega_i \cup co\Omega_i$, where the sets $co\Omega_i$, $i = 1, 2$ are as defined in the proof of Proposition 2.1. Secondly, $\eta_i \circ P_i$ maps $co\Omega_i$ homeomorphically onto a common image. The existence of the homeomorphic extension easily follows.

Denote by A_i the annuli $U_i(t+1) \setminus L_i(t)$ and by B_i the annuli $U_i(t) \setminus L_i(t-1)$, $i = 1, 2$. Then $P_i : A_i \longrightarrow B_i$ are covering maps of degree 3. Moreover by the above argument the restriction $h|_{\overline{B}_1} : \overline{B}_1 \longrightarrow \overline{B}_2$ is a homeomorphism and a conformal isomorphism of the annuli. Also, the restriction $h|_{\partial L_1} : \partial L_1 \longrightarrow \partial L_2$ is a homeomorphism. It easily follows that $h \circ P_1$ on $\overline{A_1}$ has a unique homeomorphic (biholomorphic) lift to P_2 that extends h on $L_1(t)$. One way to see this is to uniformize the annuli as round annuli with concentric boundaries and transport the maps to the uniformized annuli, trivially solve the problem and transport the solution back. Since all of the involved maps are continuous on the closures and all boundaries are locally connected, the continuity and hence analyticity of h on ∂L_1 is assured.

Definition of h_0. Fix T, a non-integer real with $t < T < t + 1$, so that $\kappa_1^{-1}(T)$ and $\kappa_2^{-1}(T)$ are smooth Jordan curves bounding domains $U_1(T)$ and $U_2(T)$. Then $h_{0|} := h : U_1(T) \longrightarrow U_2(T)$ is a biholomorphic conjugacy.

For $i = 1, 2$, let $\phi_{i,\infty}$ denote the Böttcher coordinates for P_i tangent to the identity at infinity, and fix $R > 1$. Define neighborhoods of infinity $V_i = \phi_{i,\infty}^{-1}(\overline{\mathbb{C}} \backslash \overline{\mathbb{D}}(R))$ and first extend h_0 by $h_{0|V_1} := \phi_{2,\infty}^{-1} \circ \phi_{1,\infty} : V_1 \longrightarrow V_2$. Finally, extend h_0 to a globally quasiconformal homeomorphism of $\overline{\mathbb{C}}$ by a, say, C^1 homeomorphism between the two remaining annuli.

Our map h_0 is now globally defined and is conformal on the forward invariant domains V_1 and $U_1(T)$.

The Lifting h_1. We claim that there is a homeomorphism $h_1 : \overline{\mathbb{C}} \to \overline{\mathbb{C}}$ such that

1. $P_2 \circ h_1 = h_0 \circ P_1$, i.e.,

$$
\begin{array}{ccc}
\overline{\mathbb{C}} & \xrightarrow{\ h_1\ } & \overline{\mathbb{C}} \\
P_1 \downarrow & & \downarrow P_2 \\
\overline{\mathbb{C}} & \xrightarrow{\ h_0\ } & \overline{\mathbb{C}}
\end{array}
$$

2. h_1 restricted to $P_1^{-1}(V_1 \cup U_1(T))$ is holomorphic;

3. $h_1|_{U_1(T)} = h_0|_{U_1(T)}$;

4. h_1 is isotopic to h_0 relative to the postcritical set of P_1.

(This is to say that P_1 and P_2 are *c-equivalent*, a notion that generalizes Thurston's combinatorial equivalence to postcritically infinite hyperbolic maps. See [CJS04] and [CT07] for a detailed study of the notion.)

Proof: We want to apply Lemma 4.4 to $h_0 \circ P_1$. Notice that for $i = 1, 2$, the restrictions $P_i : \mathbb{C} \setminus P_i^{-1}(\{v_i^0, v_i^1\}) \longrightarrow \mathbb{C} \setminus \{v_i^0, v_i^1\}$ are covering maps. Set $Z := U_1(T) \setminus P_1^{-1}(\{v_1^0, v_1^1\})$. Now any loop based at 0 in $\mathbb{C} \setminus P_1^{-1}(\{v_1^0, v_1^1\})$ is homotopic to a loop based at 0 in Z. Therefore, $\pi_1(\mathbb{C} \setminus P_1^{-1}(\{v_1^0, v_1^1\}), 0)$ is canonically identified with $\pi_1(Z, 0)$.

The map h_0 restricted to Z satisfies $P_2 \circ h_0 = h_0 \circ P_1$ and $h_0(0) = 0$. Therefore

$$(h_0 \circ P_1)_* \pi_1(\mathbb{C} \setminus P_1^{-1}(\{v_1^0, v_1^1\}), 0) = (h_0 \circ P_1)_* \pi_1(Z, 0)$$

$$= (P_2 \circ h_0)_* \pi_1(Z, 0) = P_{2*} \pi_1(\mathbb{C} \setminus P_2^{-1}(\{v_2^0, v_2^1\}), 0) \ .$$

Thus we can apply Lemma 4.4 to $h_0 \circ P_1$ to guarantee the existence of a continuous h_1 satisfying 1. together with the base point condition $h_1(0) = 0$.

By the uniqueness of the lifting, we have also 3. Property 2. follows.

Now we may exchange the role of P_1 and P_2 to show that h_1 has a continuous inverse map. In other words, h_1 is in fact a homeomorphism.

The postcritical set of P_1 is contained in $U_1(T) \cup \{\infty\}$. We will use the fact that a homeomorphism on the closed unit disk $\overline{\mathbb{D}}$, which is the identity on $\{0\} \cup \partial \mathbb{D}$, is isotopic to the identity relative to $\{0\} \cup \partial \mathbb{D}$. This implies that h_0 and h_1 are isotopic relative to $\overline{U}_1(T) \cup \{\infty\}$ and, in particular, relative to the postcritical set of P_1. This proves 4. and ends the proof of the claim. $\qquad\Box$

Notice that we cannot guarantee that $h_1|_{V_1} = h_0|_{V_1}$; this fact is usually not true.

The Subsequent Liftings h_n. From now on, the argument follows from a general rigidity principle; see for example [Mc98]. We may either apply a general theorem or redo the argument in our setting. We choose the latter option.

Recursively define $h_n : \overline{\mathbb{C}} \longrightarrow \overline{\mathbb{C}}$ for each n by $P_2 \circ h_n = h_{n-1} \circ P_1$ and $h_n(0) = 0$.

Each h_n is quasiconformal with the same maximal dilatation K as h_0 and satisfies $h_n = h_0$ on $U_1(T)$, and h_n is conformal on $P_1^{-n}(U_1(T) \cup V_1)$. The family (h_n) is thus normal. Let (h_{n_k}) be a subsequence converging uniformly to some K-quasi conformal map $H : \overline{\mathbb{C}} \longrightarrow \overline{\mathbb{C}}$. Notice that $\overline{\mathbb{C}} \setminus \bigcup_n P_1^{-n}(U_1(T) \cup V_1)$ coincides with the Julia set J_1 of P_1. Therefore H is conformal except possibly on J_1. However J_1 has Lebesgue measure zero, because P_1 is hyperbolic (in fact J_1 is a quasi-circle). Thus H is globally conformal. But since $H = h_0$ on $U_1(T)$, we have $H \circ P_1 = P_2 \circ H$ on $U_1(T)$. By the isolated zero theorem of holomorphic maps, we actually have $H \circ P_1 = P_2 \circ H$ on $\overline{\mathbb{C}}$.

This shows that P_1 and P_2 are conformally conjugate.

Finally, we have parametrized our family so that each conformal conjugacy class is represented only once. Thus $\mathbf{a_1} = \mathbf{a_2}$.

6.2 Analyticity of Φ

Analyticity of Φ on $\mathcal{H} \setminus \{(\lambda, 3\lambda), (\lambda, 0) | \lambda \in \mathbb{D}\}$. From Proposition 4.5, we immediately obtain that Φ is complex analytic on $\mathcal{H} \setminus \mathcal{F}$, as it is a composition of complex analytic maps. Since

$$\mathcal{H} \setminus \{(\lambda, 3\lambda), (\lambda, 0) | \lambda \in \mathbb{D}\} = (\mathcal{H} \setminus \mathcal{F}) \sqcup (\mathcal{F} \setminus \{(\lambda, 0), (\lambda, 3\lambda) | \lambda \in \mathbb{D}\}),$$

it remains to show the analyticity of Φ at

$$\mathbf{a}_0 \in \mathcal{F} \setminus \{(\lambda, 0), (\lambda, 3\lambda) | \lambda \in \mathbb{D}\}.$$

(The parameters $(\lambda, 3\lambda)$ and $(\lambda, 0)$ in $\mathcal{F}$ correspond to maps with a double critical point, respectively maps with a symmetry (i.e., that commute with $z \mapsto -z$).)

Denote by $\hat{\varphi}_{\mathbf{a}}$ the linearizer of $P_{\mathbf{a}}$ at 0, normalized so that $\hat{\varphi}'_{\mathbf{a}}(0) = 1$. Note that this normalization does not depend on which critical point is the "closest" and is analytic at the parameter $\mathbf{a}$.

Recall that $P_{\mathbf{a}}(z) = z^3 + \sqrt{a}z^2 + \lambda z$. The two critical points are: $c_{\mathbf{a}}^{\pm} = (-\sqrt{a} \pm \sqrt{a - 3\lambda})/3$. Assume that $\mathbf{a}_0 \in \mathcal{H}$ and $\mathbf{a}_0 \neq (\lambda, 0), (\lambda, 3\lambda)$. Then the maps

$$\mathbf{a} \mapsto c_{\mathbf{a}}^-, \quad \mathbf{a} \mapsto c_{\mathbf{a}}^+, \quad \Sigma^+ : \mathbf{a} \mapsto \frac{\hat{\varphi}_{\mathbf{a}}(c_{\mathbf{a}}^-)}{\hat{\varphi}_{\mathbf{a}}(c_{\mathbf{a}}^+)}, \quad \Sigma^- : \mathbf{a} \mapsto \frac{\hat{\varphi}_{\mathbf{a}}(c_{\mathbf{a}}^+)}{\hat{\varphi}_{\mathbf{a}}(c_{\mathbf{a}}^-)} = \frac{1}{\Sigma^+(\mathbf{a})}$$

are locally holomorphic in a neighborhood of $\mathbf{a}_0$ as long as the denominators are non-zero at $\mathbf{a}_0$. This is the case when $\mathbf{a}_0 \in \mathcal{F}$, $\mathbf{a}_0 \neq (\lambda, 3\lambda)$, in which case $\Sigma^{\pm}(\mathbf{a}_0) \in S^1 \setminus \{1\}$.

Assume now that $\mathbf{a}_0 \in \mathcal{F} \setminus \{(\lambda, 0), (\lambda, 3\lambda) | \lambda \in \mathbb{D}\}$. Then

$$\hat{\varphi}_{\mathbf{a}_0}(c_{\mathbf{a}_0}^+) = re^{i\theta^+}, \quad \hat{\varphi}_{\mathbf{a}_0}(c_{\mathbf{a}_0}^-) = re^{i\theta^-}, \quad \theta^+ \neq \theta^-,$$

and the component of 0 in $\hat{\varphi}_{\mathbf{a}_0}^{-1}(\{se^{i\theta^{\pm}}, 0 \leq s < r\})$ reaches $c_{\mathbf{a}_0}^{\pm}$ at one end.

Claim 1. *There is a small polydisk neighborhood Δ of $\mathbf{a}_0$ so that $\hat{\varphi}_{\mathbf{a}}(c_{\mathbf{a}}^{\pm}) \in \{r\sqrt{|\lambda|} < |z| < r/\sqrt{|\lambda|}\}$, the two points $\hat{\varphi}_{\mathbf{a}}(c_{\mathbf{a}}^{\pm})$ are on distinct radial lines, and for the half closed radial segment $K_{\mathbf{a}}^{\pm}$ from 0 to $\hat{\varphi}_{\mathbf{a}}(c_{\mathbf{a}}^{\pm})$ (excluding $\hat{\varphi}_{\mathbf{a}}(c_{\mathbf{a}}^{\pm})$ but including 0), the component of 0 in $\hat{\varphi}_{\mathbf{a}}^{-1}(K_{\mathbf{a}}^{\pm})$ reaches $c_{\mathbf{a}}^{\pm}$ at one end.*

Proof: Only the last part is non-trivial. Fix $k = \pm$. Note at first that the $c_{\mathbf{a}}^k$ are simple critical points of $\hat{\varphi}_{\mathbf{a}}$. Take U^k to be a small disk neighborhood of $c_{\mathbf{a}_0}^k$ such that $\hat{\varphi}_{\mathbf{a}_0}$ has no other critical points in $\overline{U^k}$. This remains true for $\hat{\varphi}_{\mathbf{a}}$ with $\mathbf{a} \in \Delta$ and Δ small. By Rouché's theorem, $\hat{\varphi}_{\mathbf{a}}(c_{\mathbf{a}}^k)$ has two preimages (counting multiplicity) in U^k. But $c_{\mathbf{a}}^k$ is one of them, and it has multiplicity two. So it is in fact the only preimage. Similarly, one can show that the component of 0 in $\hat{\varphi}_{\mathbf{a}}^{-1}(K_{\mathbf{a}}^k)$ ends in U^k and ends at a preimage of $\hat{\varphi}_{\mathbf{a}}(c_{\mathbf{a}}^k)$. By uniqueness, the end is $c_{\mathbf{a}}^k$. $\qquad\square$

Claim 2. *Let $h(z) = z + 1/z$. For any $\mathbf{a} = (\lambda, a) \in \Delta$, we have*

$$\Phi_\lambda(\mathbf{a}) = (h \circ \phi_\lambda \circ \pi_\lambda^{-1})^{-1}(\Sigma^+(\mathbf{a}) + \Sigma^-(\mathbf{a}))$$

$$= (h \circ \phi_\lambda \circ \pi_\lambda^{-1})^{-1} \circ h(\Sigma^+(\mathbf{a})) = (h \circ \phi_\lambda \circ \pi_\lambda^{-1})^{-1} \circ h(\Sigma^-(\mathbf{a})) \ .$$

Proof: For every $\lambda \in \mathbb{D}^*$, one can find a $R_\lambda = 1/|\lambda|$ such that ψ_λ, the inverse of the linearizer ϕ_λ for Q_λ (normalized so that $\phi_\lambda(c_\lambda) = 1$), has a univalent extension to $D(R_\lambda) \setminus [1, R_\lambda]$. Denote its image by U'_λ. Set $W'_\lambda = \pi_\lambda(U'_\lambda)$. It is a neighborhood of $\Gamma_\lambda \setminus \pi_\lambda(0)$. Then for the complex structure of $\mathcal{X}_\lambda$, the map $h \circ \phi_\lambda \circ \pi_\lambda^{-1}$ maps W'_λ univalently onto $h(A')$, where $A' = \{z | 1 \leq |z| < R_\lambda, \quad z \notin \mathbb{R}_+\}$. Therefore $\pi_\lambda \circ \psi_\lambda \circ h^{-1} = (h \circ \phi_\lambda \circ \pi_\lambda^{-1})^{-1}$ maps $h(A')$ holomorphically onto W'_λ, where $h^{-1} : h(A') \to A'$ is the multi-valued function.

Fix $\mathbf{a} = (\lambda, a) \in \Delta$. Denote by $\phi_\mathbf{a}^k$ the linearizer that maps $c_\mathbf{a}^k$ to 1. We have $\phi_\mathbf{a}^{l_0}(z) = \dot{\varphi}_\mathbf{a}(z)/\dot{\varphi}_\mathbf{a}(c_\mathbf{a}^k)$.

Now at least one of the critical points $c_\mathbf{a}^\pm$ is the closest. Assume it is $c_\mathbf{a}^+$. Denote by $U'_\mathbf{a}$ the connected component of 0 in $(\phi_\mathbf{a}^+)^{-1}(D(R_\lambda) \setminus [1, R_\lambda[)$. Then $\eta_\mathbf{a} = \psi_\lambda \circ \phi_\mathbf{a}^+$ on $U'_\mathbf{a}$. By Claim 1, $c_\mathbf{a}^- \in U'_\mathbf{a} \setminus U_\mathbf{a}$. Therefore $h \circ \phi_\mathbf{a}^+(c_\mathbf{a}^-) \in h(A')$, and

$$\Phi_\lambda(\mathbf{a}) = \pi_\lambda \eta_\mathbf{a}(c_\mathbf{a}^-) = (\pi_\lambda \psi_\lambda h^{-1}) h(\phi_\mathbf{a}^+(c_\mathbf{a}^-))$$

$$= (h \circ \phi_\lambda \circ \pi_\lambda^{-1})^{-1} h(\phi_\mathbf{a}^+(c_\mathbf{a}^-)) \in W'_\lambda \ .$$

But $h(\phi_\mathbf{a}^+(c_\mathbf{a}^-)) = h(\frac{\dot{\varphi}_\mathbf{a}(c_\mathbf{a}^-)}{\dot{\varphi}_\mathbf{a}(c_\mathbf{a}^+)}) = \Sigma^+(\mathbf{a}) + \Sigma^-(\mathbf{a})$. We are done with the proof of Claim 2.   □

From Claim 2 it is clear that $\mathbf{a} \mapsto \Phi_\lambda(\mathbf{a})$ is analytic in Δ, as it is the composition of analytic functions. This proves the analyticity of $\Phi(\mathbf{a}) = (\lambda, \Phi_\lambda(\mathbf{a}))$ on

$$\mathcal{H} \setminus \{(\lambda, 0), (\lambda, 3\lambda) | \lambda \in \mathbb{D}\} \ .$$

Analyticity of Φ on $\mathcal{H}^* := \mathcal{H} \setminus \{0\}$. Assume now that $\lambda \in \mathbb{D}^*$. At first, Φ_λ is holomorphic on $\mathcal{H}_\lambda \setminus \{(\lambda, 0), (\lambda, 3\lambda)\}$. The two singularities are clearly removable (by boundedness). Therefore $\Phi_\lambda : \mathcal{H}_\lambda \to \mathcal{Y}_\lambda$ is holomorphic. Next, $\Phi(\mathbf{a})$ is also holomorphic in λ on $\{(\lambda, 3\lambda_0), \lambda \in D$, a small disk about $\lambda_0\}$, since it is holomorphic on the punctured disk $D \setminus \{\lambda_0\}$ and the puncture is again removable. To get analyticity at $(\lambda, 0)$, we just need to make the change of coordinates $(\lambda, a) \to (\lambda + a, a)$ and proceed as above. Therefore $\Phi|_{\mathcal{H}^*}$ is holomorphic in each variable, and hence holomorphic as a function of the two variables (λ, a).

Surjectivity of Φ. It suffices to prove that for every $\lambda \in \mathbb{D}$, the map $\Phi_\lambda : \mathcal{H}_\lambda \longrightarrow \mathcal{Y}_\lambda$ is surjective. For $\lambda = 0$, this was proved by Milnor.

Assume $\lambda \in \mathbb{D}^*$. By the above, $\Phi_\lambda : \mathcal{H}_\lambda \to \mathcal{Y}_\lambda$ is holomorphic. We just need to show that Φ_λ is proper (as proper holomorphic maps over a connected domain are always surjective). It is easily seen that $\mathcal{H}_\lambda$ is bounded in $\mathbb{C}$ and therefore has compact closure. Let $\mathbf{a_n} = (\lambda, a_n) \in \mathcal{H}_\lambda$ be a sequence diverging to the boundary of $\mathcal{H}_\lambda$. We need to prove that $\Phi_\lambda(\mathbf{a_n}) \in \mathcal{Y}_\lambda$ diverges to the boundary of $\mathcal{Y}_\lambda$ or, equivalently, that $\kappa_\lambda(\Phi_\lambda(\mathbf{a_n})) \to +\infty$. Assume not, and note that $\kappa_\lambda(\Phi_\lambda(\mathbf{a_n})) = \kappa_{\mathbf{a_n}}(c^1_{\mathbf{a_n}})$ by definition. Hence there exists a subsequence $\mathbf{a_{n_k}}$ such that $\kappa_{\mathbf{a_{n_k}}}(c^1_{\mathbf{a_{n_k}}})$ is uniformly bounded, say by some $K > 0$ and by relative compactness with $\mathbf{a_{n_k}} \to \mathbf{a}'$. We contend that $\kappa_{\mathbf{a}'}(c^1_{\mathbf{a}'}) \leq K < \infty$ and thus $\mathbf{a}' \in \mathcal{H}_\lambda$, which contradicts the fact that $\mathbf{a_n}$ diverges to the boundary of $\mathcal{H}_\lambda$. If $\kappa_{\mathbf{a}'}(c^1_{\mathbf{a}'}) = 0$, we are through, so assume this is not the case. Then $c^0_{\mathbf{a_{n_k}}}$ converges to $c^0_{\mathbf{a}'}$ and $c^1_{\mathbf{a_{n_k}}}$ converges to $c^1_{\mathbf{a}'}$. Hence the linearizers $\phi_{\mathbf{a_{n_k}}}$ converge uniformly on compact subsets of $B_{\mathbf{a}'}$ to $\phi_{\mathbf{a}'}$. But then $\kappa_{\mathbf{a_{n_k}}}$ also converges uniformly to $\kappa_{\mathbf{a}'}$ on compact subsets of $B_{\mathbf{a}'}$. Thus $\kappa_{\mathbf{a_{n_k}}}(c^1_{\mathbf{a_{n_k}}}) \to \kappa_{\mathbf{a}'}(c^1_{\mathbf{a}'}) \leq K$.

Biholomorphicity of Φ on $\mathcal{H}^*$. This follows from a general result about analytic and bijective maps (cf. [Ch90], page 53). However the fiber structures give a direct proof: On $\mathcal{H} \setminus \mathcal{F}$, the map $\Pi^{-1} \circ \Phi$ has the form $(\lambda, a) \mapsto (\lambda, \eta_\mathbf{a}(c^1_\mathbf{a}))$. Thus

$$Jac(\Pi^{-1} \circ \Phi) = \begin{pmatrix} 1 & 0 \\ * & \frac{d}{da}\eta_\mathbf{a}(c^1_\mathbf{a}) \end{pmatrix}.$$

As Φ_λ is univalent, $\frac{d}{da}\eta_\mathbf{a}(c^1_\mathbf{a}) \neq 0$. Therefore the determinant of the Jacobian is non-zero, i.e., Φ is locally biholomorphic.

On $\mathcal{F}$, the argument is similar, by locally uniformizing $\mathcal{Y}$ to a polydisk (preserving the fibers) via a biholomorphic map F and by studying the Jacobian of $F \circ \Phi$.

A Consequence. The set $\mathcal{F}_\lambda$ coincides with $\Phi^{-1}(\Gamma_\lambda)$; it is a real-analytic arc with endpoints $(\lambda, 3\lambda)$ and $(\lambda, 0)$.

Continuity of Φ and Φ^{-1} at 0. We first show the continuity of Φ, giving a proof by contradiction. Assume that there is a sequence $\mathbf{a_n} = (\lambda_n, a_n) \to \mathbf{0}$ such that $\Phi(\mathbf{a_n}) \to (0, b)$, $b \neq 0$. Then $\lambda_n \neq 0$, since Φ_0 is continuous with $\Phi_0 = 0$, and $(0, b) \in \overline{\mathcal{Y}_0}$.

Claim. *We may assume that $(0, b) \in \mathcal{Y}_0$.*

Proof: We may check by hand that $\Phi(\lambda_n, 3\lambda_n) \to \mathbf{0}$. Choose $K_n \subset \mathcal{H}_{\lambda_n}$ to be a continuum containing both $\mathbf{a_n}$ and $(\lambda_n, 3\lambda_n)$ such that $K_n \to \{\mathbf{0}\}$ in the Hausdorff topology. Then a subsequence of $\Phi(K_n)$ converges to a continuum K in $\overline{\mathcal{Y}_0}$. The set K contains both $\mathbf{0}$ and $(0, b)$. Thus K contains a point in $\mathcal{Y}_0 \setminus \{\mathbf{0}\}$; this concludes the proof. $\square$

On the other hand, $\Phi : \mathcal{H}^* \to \mathcal{Y}^*$ is bijective, so there is a $(0, a)$ with $a \neq 0$ such that $\Phi(0, a) = (0, b)$. But Φ is also open near $(0, a)$ and thus maps a small neighborhood Δ of $(0, a)$ onto a neighborhood of $(0, b)$. We may choose Δ to avoid $\mathbf{a_n}$ for large n. These facts together contradict the fact that Φ is injective. Therefore Φ is continuous at $\mathbf{0}$, with $\Phi(\mathbf{0}) = \mathbf{0}$. A similar argument shows that Φ^{-1} is continuous at $\mathbf{0}$. We conclude that $\Phi : \mathcal{H} \to \mathcal{Y}$ is a homeomorphism and biholomorphic on $\mathcal{H}^*$.

An Atlas for $\mathcal{X}$. On $\mathcal{X} \setminus \{\mathbf{0}\}$, we use the natural complex structure given in the previous section. On $\mathcal{Y}$, we use $\Phi^{-1} : \mathcal{Y} \to \mathcal{H} \subset \mathbb{C}^2$. Due to the analyticity of Φ on $\mathcal{H} \setminus \{\mathbf{0}\}$, we get a collection of charts on $\mathcal{X}$, making it a 2-dimensional complex variety.

This ends the proof of parts a), b1) and c) of Theorem A'.

Part b2) follows from b1) by a general result about analytic varieties fibered over $\mathbb{D}$ whose fibers are isomorphic to $\overline{\mathbb{C}}$. At the end of the next section, we will also provide a direct proof of b2).

7   The Extension of Φ to $\overline{\mathcal{Y}}$; the Proof of Theorem B

The objective here is to prove Theorem B, which claims that the biholomorphism $\Phi : \mathcal{H} \to \mathcal{Y}$ extends to a homeomorphism $\Phi : \overline{\mathcal{H}}_{\mathbb{D}} \to \overline{\mathcal{Y}}$, where the subscript $\mathbb{D}$ indicates the intersection with the cylinder above $\mathbb{D}$.

Note that $\partial \mathcal{Y}_\lambda$ is a quasi-circle and $\partial \mathcal{H}_\lambda$ is not (it contains many cusps pointing outwards). We may thus conclude the following.

Corollary 7.1. Φ_λ *is not a quasiconformal map for any* $\lambda \in \mathbb{D}$.

In order to prove the theorem, we will need information from outside of $\mathcal{H}$ and $\mathcal{Y}$. The essential tool we shall use is that of holomorphic motions, a notion developed by Sullivan and Thurston. A *(classical) holomorphic motion* of a set $\mathbb{X} \subset \overline{\mathbb{C}}$ over a complex analytic manifold Λ is a mapping $H : \Lambda \times \mathbb{X} \to \Lambda \times \overline{\mathbb{C}}$, $(\lambda, z) \mapsto (\lambda, h(\lambda, z))$, such that

1. for all $z \in \mathbb{X}$, the map $\lambda \mapsto h(\lambda, z) =: h_z(\lambda)$ is analytic on Λ;

2. for all $\lambda \in \Lambda$, the map $z \mapsto h(\lambda, z) =: h^\lambda(z)$ is injective on X;

3. the time 0 map h^0 is the identity map.

Being a holomorphic motion is a strong property, and holomorphic motions have some elementary properties which can be easily proved using normal family arguments, for example the facts that H has a unique extension as a holomorphic motion over Λ of the closure $\overline{\mathbb{X}}$ of $\mathbb{X}$, and that H is continuous with respect to (λ, z). The next important property is that for each time λ, the map h^λ is (the restriction of) a quasiconformal homeomorphism. Moreover if $\Lambda = \mathbb{D}$, then h^λ is $\log(1+|\lambda|)/(1-|\lambda|)$ quasiconformal. A far deeper property, however, is the celebrated Słodkovski theorem, which states that a holomorphic motion of some set $\mathbb{X}$ over $\mathbb{D}$ extends to a holomorphic motion of all of $\overline{\mathbb{C}}$ over the same disk $\mathbb{D}$. However the extension is by no means unique in the interior of the complement of $\mathbb{X}$. For details, see for example [BR86], [Do92].

We shall use a slightly generalized definition of holomorphic motions. Let $\mathcal{X}$ be a complex analytic manifold and $p : \mathcal{X} \rightarrow \Lambda$ a holomorphic projection. Denote by $\mathcal{X}_\lambda$ the fiber $p^{-1}(\lambda)$. Let $E \subset \mathcal{X}_0$. A holomorphic motion of E over Λ into $\mathcal{X}$ is a map $H : \Lambda \times E \rightarrow \mathcal{X}$, $(\lambda, z) \mapsto H(\lambda, z)$ such that

1. for any fixed λ, $H(\lambda, \cdot)$ is injective on E and maps E into $\mathcal{X}_\lambda$;

2. for any fixed $z \in E$, $H(\cdot, z)$ is analytic;

3. $H(0, \cdot)$ is the identity.

Fix $P_{\lambda,b}(z) = z^3 + bz^2 + \lambda z$, and set $a = b^2$. Denote by $\mathcal{C}_0$ the connectedness locus in $\{(0, a), a \in \overline{\mathbb{C}}\}$, by $\mathcal{C}$ the connectedness locus in $\{(\lambda, a), (\lambda, a) \in \mathbb{D} \times \overline{\mathbb{C}}\}$, and by $\mathcal{C}^c$ the complement $(\mathbb{D} \times \overline{\mathbb{C}}) \setminus \mathcal{C}$. We shall use the ad hoc notation $\sqrt{\mathcal{A}} = \{(\lambda, b) \mid (\lambda, b^2) \in \mathcal{A}\}$ whenever $\mathcal{A} \subset \mathbb{D} \times \overline{\mathbb{C}}$.

Set $\mathbf{b} = (\lambda, b)$. For any $P_{\mathbf{b}}$, define $\varphi_{\mathbf{b}}(z)$ to be its normalized Böttcher coordinates at ∞; more precisely,

$$
\varphi_{\mathbf{b}}(z) := \lim_{n \to \infty} (P_{\mathbf{b}}^n(z))^{\frac{1}{3^n}}
$$

$$
= \lim_{n \to \infty} z \left(\frac{P_{\mathbf{b}}(z)}{z^3} \right)^{\frac{1}{3}} \left(\frac{P_{\mathbf{b}}^2(z)}{P_{\mathbf{b}}(z)^3} \right)^{\frac{1}{3^2}} \cdots \left(\frac{P_{\mathbf{b}}^{n+1}(z)}{P_{\mathbf{b}}^n(z)^3} \right)^{\frac{1}{3^{n+1}}}
$$

$$
= z \prod_{k=0}^{\infty} \left(\frac{P_{\mathbf{b}}^{k+1}(z)}{P_{\mathbf{b}}^k(z)^3} \right)^{\frac{1}{3^{k+1}}} = z \prod_{k=0}^{\infty} \left(1 + \frac{b}{P_{\mathbf{b}}^k(z)} + \frac{\lambda}{P_{\mathbf{b}}^k(z)^2} \right)^{\frac{1}{3^{k+1}}} .
$$

For $Q_\lambda(z) = z^2 + \lambda z$, we similarly define

$$
\varphi_\lambda(z) := \lim_{n \to \infty} (Q_\lambda^n(z))^{\frac{1}{2^n}}
$$

$$
= \lim_{n \to \infty} z \left(\frac{Q_\lambda(z)}{z^2} \right)^{\frac{1}{2}} \left(\frac{Q_\lambda^2(z)}{Q_\lambda(z)^2} \right)^{\frac{1}{2^2}} \cdots \left(\frac{Q_\lambda^{n+1}(z)}{Q_\lambda^n(z)^2} \right)^{\frac{1}{2^{n+1}}}
$$

$$
= z \prod_{k=0}^{\infty} \left(\frac{Q_\lambda^{k+1}(z)}{Q_\lambda^k(z)^2} \right)^{\frac{1}{2^{k+1}}} = z \prod_{k=0}^{\infty} \left(1 + \frac{\lambda}{Q_\lambda^k(z)} \right)^{\frac{1}{2^{k+1}}} .
$$

Set $\mathcal{V} = \{(\lambda, z), |\lambda| < 1, z \in B_\lambda^c\}$. Then

Lemma 7.2. *The map* $(\lambda, z) \mapsto (\lambda, \varphi_\lambda(z))$ *is a homeomorphism from* $\mathcal{V}$ *onto* $\mathbb{D} \times \mathbb{D}^c$ *and is biholomorphic in the interior. Its inverse map is denoted by* Ψ. *We have* $B_0^c = \overline{\mathbb{C}} \setminus \mathbb{D}$ *and* $\Psi(0, z) \equiv (0, z)$.

Now $\Psi : \mathbb{D} \times B_0^c \to \mathcal{V}$ is a biholomorphic map; in particular, it is a holomorphic motion of $B_0^c = \mathbb{D}^c$.

For any (λ, b) with $(\lambda, b^2) \in \mathcal{C}^c$, there is a unique critical point c of $P_{\lambda, b}$ in the basin of ∞. Denote by $c'(\lambda, b)$ the co-critical point. Then $\varphi_{\lambda, b}$ is well defined at $c'(\lambda, b)$.

Proposition 7.3. *The map* $\sqrt{\mathcal{L}} : (\lambda, b) \mapsto (\lambda, \varphi_{\lambda, b}(c'(\lambda, b)))$ *is biholomorphic from* $\sqrt{\mathcal{C}^c}$ *onto* $\mathbb{D} \times \overline{\mathbb{D}}^c$, *odd with respect to* b, *and induces a biholomorphic map* $\mathcal{L} : \mathcal{C}^c \to \mathbb{D} \times \overline{\mathbb{D}}^c$. *In particular,* $\sqrt{\mathcal{L}_0} : (0, b) \mapsto (0, \varphi_{0, b}(c'(0, b)))$ *is biholomorphic from* $\sqrt{\mathcal{C}_0^c}$ *onto* $\overline{\mathbb{D}}^c$, *odd with respect to* b, *and induces a biholomorphic map* $: \mathcal{C}_0^c \to \overline{\mathbb{D}}^c$.

For a proof, see [BH88]. Define a biholomorphic map $\mathcal{M} = \mathcal{L}^{-1} \circ (Id \times \mathcal{L}_0) : \mathbb{D} \times \mathcal{C}_0^c \to \mathcal{C}^c$. It is a holomorphic motion.

Proof of Theorem B: We write also $\mathcal{M} : \mathbb{D} \times \overline{\mathbb{C}} \longrightarrow \mathbb{D} \times \overline{\mathbb{C}}$ for the extended holomorphic motion, whose existence is assured by the Słodkovski theorem [BR86, Do92]. This extension is uniquely determined on the boundary of $\mathcal{C}_0^c$, which contains the boundary of $\mathcal{H}_0$. Hence it easily follows that for each fixed $\lambda \in \mathbb{D}$, we have $\mathcal{M}(\lambda, \partial \mathcal{H}_0) = \partial \mathcal{H}_\lambda$ and $\mathcal{M}(\lambda, \mathcal{H}_0) = \mathcal{H}_\lambda$. Thus $\mathcal{M}$ restricted to $\mathbb{D} \times \overline{\mathcal{H}_0}$ is a homeomorphism from $\mathbb{D} \times \overline{\mathcal{H}_0}$ onto $\overline{\mathcal{H}_{\mathbb{D}}}$.

We define a new map $\mathcal{N} : \mathbb{D} \times \mathcal{X}_0 \to \mathcal{X}$ by

$$\mathcal{N}(\lambda, z) = \begin{cases} \Pi \circ \Psi(\lambda, z) & \text{if } z \in \mathcal{Y}_0^c = B_0^c = \overline{\mathbb{C}} \setminus \mathbb{D} \\ \Phi \circ \mathcal{M} \circ (Id \times \Phi_0^{-1})(\lambda, z) & \text{if } z \in \mathcal{Y}_0 = \mathbb{D} \end{cases}.$$

It is easy to check that $\mathcal{N}$ is a generalized holomorphic motion.

Claim. *The map* $\mathcal{N}$ *is a homeomorphism.*

Assuming for a moment that this claim is true, we have $\Phi = \mathcal{N} \circ (Id \times \Phi_0) \circ \mathcal{M}^{-1} : \mathcal{H} \to \mathcal{Y}$. However, a deep result of Faught-Roesch [Fa92, Ro06] shows that Φ_0 extends to a homeomorphism (also denoted by Φ_0) $: \overline{\mathcal{H}_0} \to \overline{\mathcal{Y}_0}$. Let us call it the *FR-extension*. Therefore the composition above defines a homeomorphic extension $\Phi : \overline{\mathcal{H}_{\mathbb{D}}} \to \overline{\mathcal{Y}}$. The big diagram below illustrates the various maps and their relationships, where the diagram $*$ is the final step.

Proof of the claim: Clearly, $\mathcal{N}$ is injective and a local homeomorphism at an (λ, x) with $|x| \neq 1$. Fix (λ_0, x_0), with $x_0 \in S^1$ and $|\lambda_0| = r_0 < 1$. Fix $r_0 < r < 1$. Now the set of $z \in \mathcal{Y}_0$ whose leaf intersects $\bigcup_{|\lambda| \leq r} \{\lambda\} \times \Gamma_\lambda = \Gamma_{|\lambda| \leq r}$ is compact in $\mathcal{Y}_0 = \mathbb{D}$, as is the case in $\mathcal{H}_0$. Hence there is a neighborhood $V \subset \mathbb{C}$ of x_0 such that $\mathcal{N}(D_r \times V) \cap \Gamma = \emptyset$. Therefore $\Pi^{-1} \circ \mathcal{N} : D_r \times V \to D_r \times \overline{\mathbb{C}}$ is a holomorphic motion in the classical sense. It is therefore a homeomorphism from $D_r \times V$ onto its image [BR86, Do92]. Thus $\mathcal{N} = \Pi \circ (\Pi^{-1} \circ \mathcal{N})$ is a homeomorphism from $D_r \times V$ onto its image and in particular a local homeomorphism at (λ_0, x_0).   □

Corollary 7.4 (Theorem A', Part b2). *The variety $\mathcal{X}$ is isomorphic to* $\mathbb{D} \times \overline{\mathbb{C}}$.

Proof: The map $\mathcal{N} : \mathbb{D} \times \overline{\mathbb{C}} = \mathbb{D} \times \mathcal{X}_0 \to \mathcal{X}$ is a holomorphic motion, and it is analytic on $\mathbb{D} \times \{|z| \geq R\}$ for some large R. Now for each $\lambda \in \mathbb{D}$, the map $\mathcal{N}_\lambda$ pulls back the complex structure of $\mathcal{X}_\lambda$ to a complex structure σ_λ on $\overline{\mathbb{C}}$, which has compact support. It follows that $\lambda \mapsto \sigma_\lambda$ is holomorphic. By the measurable Riemann mapping theorem, there is a holomorphic motion

$$\mathbb{D} \times \overline{\mathbb{C}} \xleftarrow{\mathcal{O}} \mathbb{D} \times \overline{\mathbb{C}},$$

normalized so that $\mathcal{O}_\lambda$ is tangent to the identity at ∞ and fixes 0, so that $\mathcal{O}_\lambda$ integrates σ_λ. Now

$$\mathcal{N} \circ \mathcal{O}^{-1} : \mathbb{D} \times \overline{\mathbb{C}} \xleftarrow{\mathcal{O}} \mathbb{D} \times \overline{\mathbb{C}} \xrightarrow{\mathcal{N}} \mathcal{X}$$

is biholomorphic and fiber-preserving.   □

8   Boundary Extensions as $|\lambda| \to 1$

Fix $\lambda_0 \in \mathbb{S}^1$. In order to extend our coordinate Φ to the λ_0-slice, we must first define a suitable $\mathcal{Y}_{\lambda_0}$ from the dynamical plane of Q_{λ_0} and then study the limit of Φ as $\lambda \in \mathbb{D}$ and $\lambda \to \lambda_0$ radially.

In the parabolic case $\lambda_0 = e^{i2\pi p/q}$, $(p, q) = 1$, the sets U_λ^0, their boundaries and the equivalence relations $\sim_\lambda$ have describable limits. This will lead to the definition of $\mathcal{Y}_{\lambda_0}$, together with a map Φ_{λ_0} from the corresponding part of $\partial \mathcal{H}$ to $\mathcal{Y}_{\lambda_0}$. We will prove that Φ_{λ_0} is a radially continuous extension of Φ. This will be the content of [PRT].

The case $\lambda_0 = e^{i2\pi\theta}$ with θ irrational is a lot more subtle. Let us denote by $\omega(c_0)$ the ω-limit of the critical orbit.

Yoccoz [Yo95] has proved that, depending on whether θ satisfies the Brjuno condition or not, either Q_{λ_0} has a Siegel disk Δ_{λ_0} and the pointed domains $(U_\lambda^0, 0)$ converge in the sense of Carathéodory to $(\Delta_{\lambda_0}, 0)$, or Q_{λ_0} has a Cremer point and the $(U_\lambda^0, 0)$ diverge in the sense of Carathéodory to the singleton $\{0\}$.

The easiest case to study is the case when Δ_{λ_0} exists and has a Jordan curve boundary that passes through the critical point c_{λ_0}. In this case,

$$
\begin{array}{c}
\mathbb{D} \times \overline{\mathbb{D}}^c \xleftarrow{(\lambda, b^2)} \mathbb{D} \times \overline{\mathbb{D}}^c \quad \text{Böttcher position} \\[4pt]
\nearrow_{Id \times \mathcal{L}_0} \quad \uparrow_{\mathcal{L}} \qquad \uparrow_{\sqrt{\mathcal{L}}} \qquad \text{of the escaping} \\[4pt]
\mathbb{D} \times \mathcal{C}_0^c \xrightarrow{\mathcal{M}} \mathcal{C}^c \xleftarrow{(\lambda, b^2)} \sqrt{\mathcal{C}^c} \quad \text{co-critical point} \\[4pt]
\cap \qquad\qquad \cap \\[4pt]
\mathbb{D} \times \overline{\mathbb{C}} \xrightarrow{\mathcal{M}} \mathbb{D} \times \overline{\mathbb{C}} \quad \boxed{\text{Słodkovski extension}}
\end{array}
$$

$$
\begin{array}{c}
\cup \qquad\qquad \cup \\[6pt]
\mathbb{D} \times \mathcal{H}_0 \subset \mathbb{D} \times \overline{\mathcal{H}_0} \xrightarrow{\mathcal{M}} \overline{\mathcal{H}}_{\mathbb{D}} \supset \mathcal{H} \supset \{\lambda\} \times \mathcal{H}_\lambda \ni \mathbf{a} \\[6pt]
Id \times \Phi_0 \downarrow \quad \overset{\text{FR-extension}}{Id \times \Phi_0} \downarrow \quad * \downarrow \Phi \quad \downarrow \Phi \quad \downarrow \Phi_\lambda \quad \eta_\mathbf{a}(c_\mathbf{a}^1) \downarrow \\[6pt]
\mathbb{D} \times \mathcal{Y}_0 \subset \mathbb{D} \times \overline{\mathcal{Y}_0} \xrightarrow{\mathcal{N}} \overline{\mathcal{Y}} \supset \mathcal{Y} \supset \mathcal{Y}_\lambda \xleftarrow{\pi_\lambda} B_\lambda \\[6pt]
\cup \qquad\qquad \cup \qquad \cap \\[6pt]
\mathbb{D} \times \partial \mathcal{Y}_0 \xrightarrow{\mathcal{N}} \partial \mathcal{Y} \qquad \mathcal{X} \\[6pt]
\cap \qquad\qquad \cap \\[6pt]
\mathbb{D} \times \mathcal{Y}_0^c \xrightarrow{\mathcal{N}} \mathcal{Y}^c \qquad \boxed{\mathcal{Y}_0 = B_0 = \mathbb{D}} \\[6pt]
Id \uparrow \qquad\qquad \uparrow_\Pi \\[6pt]
\mathbb{D} \times B_0^c \xrightarrow{\Psi} \mathcal{V} \\[6pt]
(\lambda, z) \quad \mapsto (\lambda, h(\lambda, z)) \qquad \boxed{\text{Böttcher coordinates at } \infty}
\end{array}
$$

$\omega(c_{\lambda_0}) = \partial \Delta_{\lambda_0}$, and the linearizer $\phi_{\lambda_0} : \Delta_{\lambda_0} \longrightarrow \mathbb{D}$ normalized by $\phi_{\lambda_0}(c_{\lambda_0}) = 1$ induces an equivalence relation $\sim_{\lambda_0}$ on $\overline{\mathbb{C}} \setminus \Delta_{\lambda_0}$ similar to the one in the case $\lambda \in \mathbb{D}$. Therefore $(K_{\lambda_0} \setminus \Delta_{\lambda_0}) / \sim_{\lambda_0}$ would be a good candidate for our $\mathcal{Y}_{\lambda_0}$. For θ of bounded type, S. Zakeri [Za99] has proved that the parameter slice $\{(\lambda_0, a) | a \in \mathbb{C}\}$ contains an embedded copy of $\mathcal{Y}_{\lambda_0}$. But it remains to prove that this copy sits on the boundary of $\mathcal{H}$ and that Φ extends radially continuously on it.

In the remaining cases, the compact sets ∂U_λ^0 conjecturally converge to $\omega(c_{\lambda_0})$ in the Hausdorff topology, and $\sim_\lambda$ converges to some equivalence relation $\sim_{\lambda_0}$. Then we can define Y_{λ_0} accordingly. The difficulty here is to prove that the Hausdorff limit of ∂U_λ^0 does exist and does not exceed

$\omega(c_{\lambda_0})$. We don't yet have definite results in these cases; however new results by Inou-Shishikura on parabolic renormalization will surely contribute to future progress on this problem.

9   Application to Other Hyperbolic Components

In order to apply Milnor's result [Mi92], we will need to work on a different parametrization of cubic polynomials.

For $d \geq 3$, let $\mathcal{P}^d = \{P | P(z) = z^d + a_{d-2}z^{d-2} + \ldots + a_0\} \approx \mathbb{C}^{d-1}$ denote the space of monic centered polynomials of degree d.

We want to carry our result on $\mathcal{H}$ to the central hyperbolic component $\mathcal{H}^3$ of $\mathcal{P}^3$. We show here that $\mathcal{H}^3$ is naturally a double cover of $\mathcal{H}$. We then define an appropriate square root, denoted by $\mathcal{Y}^3$, of $\mathcal{Y}$, and provide a dynamical-analytic coordinate from $\mathcal{H}^3$ onto $\mathcal{Y}^3$.

To be more precise, define the moduli space $\mathcal{M}^d$ as the space of affine conjugacy classes $[P]$ of degree d polynomials P and let $proj^d : \mathcal{P}^d \longrightarrow \mathcal{M}^d$ denote the natural projection. If $\rho^{d-1} = 1$ and $P \in \mathcal{P}^d$, then $\rho P(z/\rho) \in \mathcal{P}^d$, and thus conjugating by any of the rotations $z \mapsto \rho z$, $\rho^{d-1} = 1$ leaves $\mathcal{P}^d$ invariant. On the other hand, any other affine map conjugates any element of $\mathcal{P}^d$ out of $\mathcal{P}^d$. It follows that $proj^d : \mathcal{P}^d \longrightarrow \mathcal{M}^d$ is a degree $d - 1$ branched covering. Let $\mathcal{H}^d_{\mathcal{M}}$ denote the central hyperbolic component of $\mathcal{M}^d$; then the restriction $proj^d : \mathcal{H}^d \longrightarrow \mathcal{H}^d_{\mathcal{M}}$ also has degree $d - 1$. For any $d > 1$, the complex line $\{P \in \mathcal{P}^d | a_1 \in \mathbb{C}, \, a_0 = a_2 = a_3 = \ldots = a_{d-2} = 0\}$ is contained in the branching locus, and for $d = 3, 4$, it equals the branching locus for $proj^d$. For any $d > 2$, the polynomials on this line are characterized by having 0 as a fixed point with multiplier a_1 and a $d - 1$ fold rotational symmetry. Since this line equals the branching locus when $d = 3$ (and 4), the projection onto moduli space is a simple squaring, respectively cubing, around this line.

For $\lambda \in \mathbb{D}^* := \mathbb{D} \setminus \{0\}$, recall that $\psi_\lambda : \overline{\mathbb{D}} \to \mathbb{C}$ denotes the inverse of the linearizer ϕ_λ of Q_λ, and that $\psi_\lambda(1)$ is the critical point of Q_λ. The map $\beta : \mathbb{D} \longrightarrow \mathbb{C}$ given by $\beta(0) = 0$ and $\beta(\lambda) = \psi_\lambda(-1)$ for $\lambda \in \mathbb{D}^*$ is holomorphic by the removable singularities theorem. Define a holomorphic branched double covering $\chi : \mathbb{D} \times \overline{\mathbb{C}} \longrightarrow \mathbb{D} \times \overline{\mathbb{C}}$ by $\chi(\lambda, z) = (\lambda, \chi_\lambda(z)) = (\lambda, z^2 + \beta(\lambda))$ and define $\mathcal{S}^3 = \chi^{-1}(\mathbb{D} \times \overline{\mathbb{C}} \setminus \mathcal{U})$. On the space $\mathcal{S}^3$, we define an equivalence relation $\sim^3$ by declaring that $(\lambda_1, z_1) \sim^3 (\lambda_2, z_2)$ iff $\lambda_1 = \lambda_2 = \lambda$ and either $z_1 = z_2$ or both z_1 and z_2 belong to the same connected component of $\chi_\lambda{}^{-1}(\psi_\lambda(\mathbb{S}^1 \setminus \{-1\}))$ and $\chi_\lambda(z_1) \sim_\lambda \chi_\lambda(z_2)$. Let $\Pi^3 : \mathcal{S}^3 \longrightarrow \mathcal{S}^3 / \sim^3 := \mathcal{X}^3$ denote the natural projection, and equip $\mathcal{X}^3$ with the quotient topology. This topology is easily seen to be Hausdorff. Define also $\mathcal{Y}^3 = \Pi^3(\chi^{-1}(\mathcal{B} \setminus \mathcal{U})) \subset \mathcal{X}^3$, and an involution $\kappa : \mathcal{X}^3 \longrightarrow \mathcal{X}^3$ by $\kappa(\Pi^3(\lambda, z)) = \Pi^3(\lambda, -z)$. The fixed point set of this involution is the set

$\mathcal{C}^3 = \Pi^3(\mathbb{D}, 0)$. Define $\Sigma : \mathcal{Y}^3 \longrightarrow \mathcal{Y}$ by $\Sigma(\Pi^3(\lambda, z)) = \Pi(\chi(\lambda, z))$, so that Σ is a degree 2 branched covering with covering transformation κ, branching locus equal to $\mathcal{C}^3$ and branch value set equal to the complex disk $\mathcal{V} = \Pi(\mathbb{D}, 0) = \Pi(\{(\lambda, \beta(\lambda)) | \lambda \in \mathbb{D}\})$. Then $\mathcal{Y}^3$ and hence $\mathcal{X}^3$ has a (unique) complex structure for which the map Σ is holomorphic. This is the unique structure for which the projection $\Pi^3 : \mathbb{D} \times \overline{\mathbb{C}} \setminus \overline{\mathcal{U}} \longrightarrow \mathcal{X}^3$ is holomorphic.

Theorem C. *There exists a biholomorphic map* $\Phi^3 : \mathcal{H}^3 \longrightarrow \mathcal{Y}^3$ *such that* $\Sigma \circ \Phi^3 = \Phi \circ proj^3$.

Proof: The restriction $proj^3 : \mathcal{H}^3 \longrightarrow \mathcal{H}$ is also a holomorphic degree 2 branched covering whose set of branch values equals $\Phi^{-1}(\mathcal{V})$. It follows that there exists a biholomorphic map $\Phi^3 : \mathcal{H}^3 \longrightarrow \mathcal{Y}^3$ with $\Sigma \circ \Phi^3 = \Phi \circ proj^3$. This map satisfies $\Phi^3(-a_0, a_1) = \kappa(\Phi^3(a_0, a_1))$ and is unique up to post composition by κ or, equivalently, by precomposition with the map $(a_0, a_1) \mapsto (-a_0, a_1) : \mathcal{H}^3 \to \mathcal{H}^3$. $\qquad\square$

Note that it follows from Theorem A' that there exists a biholomorphic map from $\mathcal{X}^3$ onto $\mathbb{D} \times \overline{\mathbb{C}}$ that commutes with the projection to the first coordinates.

9.1 Universality of $\mathcal{H}$

Our coordinate on $\mathcal{H}$ and $\mathcal{H}^3$ can be carried to a similar coordinate to many other hyperbolic components, through a result of Milnor [Mi92]. He has classified the hyperbolic components of $\mathcal{P}^d$ into a finite number of types. He then constructed one abstract complex analytic model for each type and proved that any hyperbolic component is canonically biholomorphic to its abstract model.

In order to be precise, let us recall Milnor's classification [Mi92, M2, pages 3–8] of hyperbolic components in terms of mapping schemes. By definition, a (finite) (critical) *mapping schema* is a weighted finite directed graph $\mathbf{S} = (S, w)$ such that each vertex $v \in S$ is the origin of precisely one directed edge e_v, which is allowed to be a closed loop, and $w(v)$ is a non-negative integer, called the critical weight of v. The vertex v is critical iff $w(v) > 0$. The positive integer $d(e_v) = 1 + w(v)$ is called the degree of the edge e_v. The mapping schema $\mathbf{S}$ is connected iff the underlying non-oriented graph is path connected. Clearly, each connected component of $\mathbf{S}$ contains a unique cycle $v_0, \ldots, v_{p-1}$, i.e., the terminal point of e_{v_i} is the vertex $v_{(i+1) \bmod p}$. It is further assumed that $\mathbf{S}$ is *critical*, i.e., each cycle contains at least one critical vertex, and any entry vertex, i.e., one which is not the terminal point of any edge, is critical. Two mapping schemes are isomorphic if there is a weight preserving bijection between the two underlying oriented graphs.

A mapping schema $\mathbf{S}$ is *reduced* iff every vertex is critical. There is a natural projection map from the space of mapping schemes to the space of reduced mapping schemes given by contracting each chain of degree 1 edges as well as their endpoints onto the terminal vertex of the chain.

(Reduced) Mapping schemes encode the dynamical types of hyperbolic components in a canonical way: Firstly, associated with any hyperbolic polynomial P is a mapping schema $\mathbf{S} = (S, w)$, where the set of vertices of S equals the set of Fatou components U of P, with U containing at least one critical or postcritical point. For a vertex $U \in S$, the edge e_U terminates on the vertex $P(U)$, and the critical weight equals the number of P-critical points in U counted with multiplicity, so that the degree of the edge e_U equals the degree of the restriction $P_{|U}$, by the Riemann-Hurwitz theorem. Secondly, if two polynomials P_1 and P_2 belong to the same hyperbolic component, then they have isomorphic mapping schemes, because they are quasiconformally conjugate in a neighborhood of their Julia sets, or J-equivalent, in the language of Mañé-Sad-Sullivan [MSS83]. Thus any hyperbolic component for which all the Fatou components are simply connected has a unique (isomorphic class of) associated mapping schema.

For cubic polynomials with connected Julia sets, there are four types of hyperbolic components, according to their reduced mapping schema (the names are not completely identical to those of Milnor; there are also two types with disconnected Julia sets that we do not mention — see [BH88, BH92] for related results):

- **Type I, (Truly) Cubics.** Corresponds to polynomials with an attracting periodic orbit for which one component of the immediate basin contains both critical points and hence is mapped properly by degree 3 onto its image.

- **Type II, Bi-transitive.** Corresponds to polynomials with an attracting periodic orbit for which both critical points are in the immediate basin of a single attracting cycle, but in two different connected components. The two components are thus mapped quadratically onto their images, and the first return map on a component of the immediate basin has degree 4.

- **Type III, Capture.** Corresponds to polynomials with an attracting periodic orbit which attracts both critical points, only one of which is in the immediate basin; the other is contained in a strictly preperiodic preimage. Again, the map is quadratic on the critical components, but the first return map on a connected component of the immediate basin is quadratic.

- **Type IV, Bi-quadratic.** Corresponds to maps with two distinct attracting cycles, each with a critical point in the immediate basin.

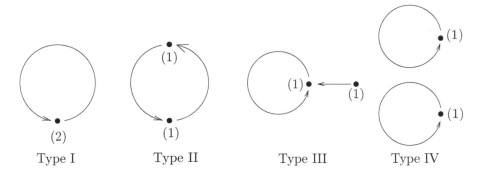

Figure 3. The reduced mapping schemes corresponding to the four types of cubic hyperbolic components. The vertex weight is indicated by the adjacent number in parentheses. This figure is equivalent to [Mi92, Figure 2] but with different notations.

The corresponding reduced mapping schemes are illustrated in Figure 3.

A type IV (bi-quadratic) hyperbolic component H is in some sense not cubic, since the dynamics of the two critical points are completely decoupled. Such a component is simply the product of two one dimensional hyperbolic components, and the direct product of the multiplier maps of the two attracting periodic orbits defines a biholomorphic map onto the bi-disk $\mathbb{D} \times \mathbb{D}$. Bi-quadratics are also bi-renormalizable with two quadratic-like renormalizations; see [EY99] for extended results in this direction.

Let us return to polynomials of general degree. Milnor [Mi92, Definition 2.4] defines a *universal polynomial model space* $\widehat{\mathcal{P}}^{\mathbf{S}}$ associated with every mapping schema $\mathbf{S}$ as follows: The elements of $\widehat{\mathcal{P}}^{\mathbf{S}}$ are the self maps f of $|S| \times \mathbb{C}$ (where $|S|$ denotes the set of vertices of S) such that f maps $v \times \mathbb{C}$ onto $v' \times \mathbb{C}$ as a monic centered polynomial of degree $d(e_v)$, where v' is the terminal vertex of e_v. The hyperbolic locus $\widehat{\mathcal{H}}^{\mathbf{S}} \subset \widehat{\mathcal{P}}^{\mathbf{S}}$ is the subset of maps for which every critical point converges to an attracting periodic cycle. The central hyperbolic component $\widehat{\mathcal{H}}_0^{\mathbf{S}}$ of $\widehat{\mathcal{H}}^{\mathbf{S}}$ is the connected component of $\widehat{\mathcal{H}}^{\mathbf{S}}$ containing the map f which has the form $z^{d(e_v)}$ on each complex plane $v \times \mathbb{C}$. Note that if $\mathbf{S}$ is the weighted graph with only one vertex, with weight $d-1$ and hence with exactly one edge of degree d, we have $\widehat{\mathcal{P}}^{\mathbf{S}} = \mathcal{P}^d$, and $\widehat{\mathcal{H}}^{\mathbf{S}} = \mathcal{H}^d$, where $\mathcal{H}^d$ denotes *the central hyperbolic component* of $\mathcal{P}^d$, namely the one containing z^d. Note also that the dynamics of any $\widehat{\mathcal{P}}^{\mathbf{S}}$ factors into subfactors corresponding to each connected component $\mathbf{S}'$ of $\mathbf{S}$ and that each such factor is universal in the sense that it depends on $\mathbf{S}'$ only and, in particular, not on the other factors corresponding to the other components of $\mathbf{S}$. Evidently, the attracting periodic cycles are in $1 : 1$ correspondence with the connected components of $\mathbf{S}$, e.g., such as with type IV above.

Here is a schematic picture:

{hyperbolic polynomials} $\subset \mathcal{P}^d$ $\longrightarrow$ {H, hyperbolic comp.} $\longrightarrow$

{Julia equivalent classes} $\longrightarrow$ {mapping schemes} $\longrightarrow$

{ **S**, reduced schemes} $\longleftrightarrow$ $\{\widehat{\mathcal{P}}^{\mathbf{S}}\}$ $\longleftrightarrow$ $\{\widehat{\mathcal{H}}_0^{\mathbf{S}}\}$ $\longleftrightarrow$

$\{\bigsqcup \mathbf{S_i} \mid \mathbf{S_i} \text{ con. red. schema}\}$ $\longleftrightarrow$ $\left\{\prod \widehat{\mathcal{H}}_0^{\mathbf{S_i}}\right\}$

Though not explicitly stated, Milnor establishes the following factorization theorem for hyperbolic components:

Theorem 9.1 (Milnor). *Let H be a hyperbolic component in $\mathcal{P}^d$, for some $d \geq 2$, with reduced mapping schema $\mathbf{S}$, and let $\mathbf{S_1}, \ldots, \mathbf{S_n}$ be the connected components of $\mathbf{S}$. Then H is biholomorphic to the direct product $\widehat{\mathcal{H}}_0^{\mathbf{S_1}} \times \ldots \times \widehat{\mathcal{H}}_0^{\mathbf{S_n}}$. In particular, if one $\mathbf{S_i}$ is the type I reduced schema above (i.e., with only one vertex of weight 2), then the central hyperbolic component $\mathcal{H}^3$ of $\mathcal{P}^3$ appears as a factor in H. Consequently, every type I hyperbolic component in $\mathcal{P}^3$ is canonically biholomorphic to $\mathcal{H}^3$.*

(Note that this statement is similar to the discussion of the Teichmüller space of a rational map by McMullen and Sullivan in [MS98].)

Combining this with our result, we obtain the following.

Corollary 9.2. *The dynamical-analytic coordinate for $\mathcal{H}^3$ constructed in Theorem C automatically yields a similar coordinate for any truly cubic (type I) factor of any hyperbolic component in $\mathcal{P}^d$, $d \geq 3$.*

Note that the tools and ideas we have developed in this paper can easily be adapted to the study of type III components.

Milnor's result on isomorphisms of hyperbolic components of the same type is in fact *only valid* for the spaces $\mathcal{P}^d$ of monic centered polynomials and the universal polynomial model spaces $\widehat{\mathcal{P}}^{\mathbf{S}}$. The reason for this is that, for an arbitrary parameter space, a hyperbolic component may be a branched covering of a corresponding component in $\mathcal{P}^d$. In this case, it is not in general biholomorphic to the covered component, or if some component in $\mathcal{P}^d$ has a symmetry, the branching may go the other way. In fact some of the hyperbolic components in $\mathcal{P}^d$ do have symmetries. The most prominent example is the central hyperbolic component $\mathcal{H}^d \subset \mathcal{P}^d$ consisting of maps for which some attracting fixed point attracts all critical points in the immediate basin. Using terminology similar to the above, these polynomials are truly degree d.

10 Table of Notation

$\lambda \in \mathbb{D}$	$P_{\mathbf{a}}(z) = P_{\lambda,a}(z) = \lambda z + \sqrt{a}z^2 + z^3$	$Q_\lambda(z) = \lambda z + z^2$
attracting fixed point	0	0
attracting basin	$B_{\mathbf{a}}$	B_λ
case $\lambda \neq 0$		
univalence domain with max. round co-domain for linearizer	$U_{\mathbf{a}}^0$	U_λ^0
critical points critical values	$c_{\mathbf{a}}^0 \in \partial U_{\mathbf{a}}^0,\ c_{\mathbf{a}}^1$ $v_{\mathbf{a}}^0 \in U_{\mathbf{a}}^0,\ v_{\mathbf{a}}^1$	$c_\lambda^0 \in \partial U_\lambda^0$ $v_\lambda^0 \in U_\lambda^0$
normalized linearizer	$\phi_{\mathbf{a}},\ \phi_{\mathbf{a}}(c_{\mathbf{a}}^0) = 1$ $\hat{\phi}_{\mathbf{a}},\ \hat{\phi}_{\mathbf{a}}'(0) = 1$	$\phi_\lambda,\ \phi_\lambda(c_\lambda^0) = 1$
filled potential funct.	$\kappa_{\mathbf{a}} : B_{\mathbf{a}} \to \mathbb{R} \cup \{-\infty\}$	$\kappa_\lambda : B_\lambda \to \mathbb{R} \cup \{-\infty\}$
open potential domains	$U_{\mathbf{a}}^0(t) = $ c.c. of $\kappa_{\mathbf{a}}^{-1}[-\infty, t[$	$U_\lambda^0(t) = $ c.c. of $\kappa_\lambda^{-1}[-\infty, t[$
closed connected potential sets	$L_{\mathbf{a}}(t) = $ c.c. of $\kappa_{\mathbf{a}}^{-1}[-\infty, t]$	$L_\lambda(t) = $ c.c. of $\kappa_\lambda^{-1}[-\infty, t]$
case $\lambda = 0$, $a \neq 0$		
Böttcher coordinates	$\phi_{\mathbf{a}}$	$\phi_0 = id$
univalence domain with max. round co-domain for ϕ	$U_{\mathbf{a}}^0$	$B_0 = \mathbb{D}$
critical points potential function	$c_{\mathbf{a}}^0 = 0,\ c_{\mathbf{a}}^1 \in \partial U_{\mathbf{a}}^0$ $g_{\mathbf{a}} : B_{\mathbf{a}} \to [-\infty, 0[$	$c_0^0 = 0$ $g_\lambda : B_\lambda \to [-\infty, 0[$
quotient		$\Gamma_\lambda = \partial U_\lambda^0/\sim,$ $\mathcal{Y}_\lambda = B_\lambda \setminus U_\lambda^0/\sim$ $\mathcal{X}_\lambda = \mathbb{C} \setminus U_\lambda^0/\sim$ where $z_1 \sim z_2 \iff$ $z_1 = z_2$ or $z_1, z_2 \in \partial U_\lambda^0$ and $\phi_\lambda(z_1) = \phi_\lambda(z_2)$
parameter space		$\mathcal{H}$ the hyperbolic component of z^3 $\eta_{\mathbf{a}} = $ extension of $\phi_\lambda^{-1} \circ \phi_{\mathbf{a}}: \Omega_{\mathbf{a}} \subseteq B_{\mathbf{a}} \longrightarrow B_\lambda$ $\mathcal{X} := \bigsqcup_\lambda \mathcal{X}_\lambda,\ \mathcal{Y} := \bigsqcup_\lambda \mathcal{Y}_\lambda,\ \Gamma = \bigsqcup_\lambda \Gamma_\lambda$ $\Phi : \mathbf{a} \mapsto [\eta_{\mathbf{a}}(c_{\mathbf{a}}^1)],\ \mathcal{H} \to \mathcal{Y} \subset \mathcal{X},\ \mathcal{F} \to \Gamma$

Bibliography

[BR86]   L. Bers and H. L. Royden, *Holomorphic families of injections*, Acta Math. **157** (1986), 259–286.

[BH88]   B. Branner and J. H. Hubbard, *The iteration of cubic polynomials I: the global topology of parameter space*, Acta Math. **160** (1988), 143–206.

[BH92]   B. Branner and J. H. Hubbard, *The iteration of cubic polynomials II: patterns and parapatterns*, Acta Math. **169** (1992), 229–325.

[Ch90]   B. Chabat, *Introduction à l'analyse complexe, functions de plusieurs variables*, vol. 2, Mir, 1990.

[CJS04]   G. Z. Cui, Y. Jiang, and D. Sullivan, *On geometrically finite branched coverings, realization of rational maps*, in: Complex dynamics and related topics (Y. Jiang and Y. Wang, eds.), The International Press, 2004, 15–29.

[CT07]   G. Z. Cui and L. Tan, *A characterization of hyperbolic rational maps*, preprint, arXiv:math/0703380, 2007.

[Do92]   A. Douady, *Prolongement de mouvements holomorphes [d'après Słodkovski et autres]*, Seminaire Bourbaki No. 755 (N. Bourbaki, ed.), Societé Mathématique de France, 1992.

[EY99]   A. Epstein and M. Yampolsky, *Geography of the cubic connectedness locus: intertwining surgery*, Ann. Sci. École Norm. Sup. (4) 2 **32** (1999), 151–185.

[Fa92]   D. Faught, *Local connectivity in a family of cubic polynomials*, Ph.D. thesis, Cornell University, 1992.

[GK90]   L. Goldberg and L. Keen, *The mapping class group of a generic quadratic rational map and automorphisms of the 2-shift*, Invent. Math. **101** (1990), 335–372.

[MSS83]   R. Mañé, P. Sad, and D. Sullivan, *On the dynamics of rational maps*, Ann. Sci. École Norm. Sup. **16** (1983), 193–217.

[Mc98]   C. T. McMullen, *Self-similarity of Siegel disks and Hausdorff dimension of Julia sets*, Acta Math. **180** 2 (1998), 247–292, Appendix A.

[MS98]   C. T. McMullen and D. Sullivan, *Quasiconformal homeomorphisms and dynamics: the Teichmüller space of a holomorphic dynamical system*, Adv. Math. **135** 2 (1998), 351–395.

[Mi92]   J. Milnor, *Hyperbolic components in spaces of polynomial maps*, Stony Brook IMS Preprint 3, 1992.

[Mi]   J. Milnor, *From quadratic to cubic by 'ramification' of a point (outline)*, Private communication.

[NS]   S. Nakane and D. Schleicher, *On multicorns and unicorns: bifurcations in spaces of antiholomorphic polynomials*, Manuscript in preparation.

[Pe04]   C. L. Petersen, *When Schröder meets Böttcher, convergence of level sets*, C. R. Math. Acad. Sci. Paris, Ser. I **339** 3 (2004), 219–222.

[PRT]   C.L. Petersen, P. Roesch, and L. Tan, *Parabolic slices on the boundary of* $\mathcal{H}$, Manuscript in preparation.

[Re90]   M. Rees, *Components of degree two hyperbolic rational maps*, Invent. Math. **100** (1990), 357–382.

[Ro06]   P. Roesch, *Hyperbolic components of polynomials with a fixed critical point of maximal order*, Prépublication 306, Laboratoire Emile Picard, Université Paul Sabatier, Toulouse, 2006.

[Yo95]   J. C. Yoccoz, *Petits diviseurs en dimension* 1, Astérisque **231**, Société Mathématique de France, 1995.

[Za99]   S. Zakeri, *Dynamics of cubic Siegel polynomials*, Comm. Math. Phys. **206** 1 (1999), 185–233.

11 Cubic Polynomials: A Measurable View of Parameter Space

Romain Dujardin

1 Introduction

The study of closed positive currents associated to bifurcations of holomorphic dynamical systems is a topic of rapidly growing interest [De01, De03, BB07a, BB07b, Ph05, DF08]. The underlying thesis is that bifurcation currents should contain a lot of information on the geography of the parameter spaces of rational maps of the Riemann sphere.

The simplest case in which the currents are not just measures is when the parameter space under consideration has complex dimension 2: a classical such example is the space Poly_3 of cubic polynomials. The purpose of the present article is to give a detailed account of the fine structure of bifurcation currents and the bifurcation measure (to be defined shortly) in this setting. We chose to restrict ourselves to cubics because, thanks to the existing literature, the picture is much more complete in this case; nevertheless, it is clear that many results go through in a wider context. We note in particular that some laminarity properties of bifurcation currents were recently extended to the space of quadratic rational maps by Bassanelli and Berteloot in [BB07b].

The work of Branner and Hubbard [BH88, BH92] is a standard reference on the parameter space of cubic polynomials. A crude way of comparing our approach to theirs would be to say that in [BH88, BH92] the focus is more on the topological properties of objects, whereas here we concentrate on their measurable and complex analytic properties — this is exactly what positive closed currents do.

Let us describe the setting more precisely (see Section 2 for details). Consider the cubic polynomial $f_{c,v}(z) = z^3 - 3c^2z + 2c^3 + v$, with critical points at $\pm c$. Since the critical points play symmetric roles, we can focus on $+c$. We say that c is *passive* near the parameter (c_0, v_0) if the family of holomorphic functions $(c, v) \mapsto f_{c,v}^n(c)$ is locally normal. Otherwise, c is said to be *active*. It is a classical result that bifurcations always occur at parameters with active critical points.

Let $\mathcal{C}^{\pm}$ be the set of parameters for which $\pm c$ has bounded orbit. It is a closed, unbounded subset in parameter space, and it can easily be seen that the locus where c is active is $\partial \mathcal{C}^{+}$. We also denote by $\mathcal{C} = \mathcal{C}^{+} \cap \mathcal{C}^{-}$ the connectedness locus.

The Green function $z \mapsto G_{f_{c,v}}(z)$ depends in a plurisubharmonic way on (c, v), so the formulas $(c, v) \mapsto G_{f_{c,v}}(\pm c)$ define plurisubharmonic functions in the space $\mathbb{C}^{2}_{c,v}$ of cubic polynomials. We define $T^{\pm} = dd^{c} G_{f_{c,v}}(\pm c)$ to be the *bifurcation currents* associated to $\pm c$. Notice that $T_{\mathrm{bif}} = T^{+} + T^{-}$ is the standard bifurcation current associated to the variation of the Lyapunov exponent of the maximal entropy measure, as considered in [De01, De03, BB07a, Ph05]. It is easy to see that $\mathrm{Supp}(T^{+})$ is the activity locus of c, that is, $\mathrm{Supp}(T^{+}) = \partial \mathcal{C}^{+}$.

The geometric intuition underlying positive closed currents is that of analytic subvarieties. Nevertheless, the "geometry" of general positive closed currents can be quite poor. *Laminar currents* form a class of currents with rich geometric structure, well suited for applications in complex dynamics. We say that a positive closed current is *laminar* if it can be written as an integral of compatible disks $T = \int [D_{\alpha}] d\nu(\alpha)$. Here compatible means that two disks do not have isolated intersection points. The local geometry of a laminar current can still be very complicated. On the other hand, we say that T is *uniformly laminar* if the disks D_{α} are organized as laminations. The local geometry is tame in this case: a uniformly laminar current is locally the "product" of the leafwise integration along the leaves by a *transverse measure* on transversals.

The main point in our study is to look for some laminar structure for the bifurcation currents in some regions of parameter space. To understand why some laminarity should be expected, it is useful to mention some basics on *deformations*. We say that two rational maps are deformations of each other if there is a J-stable (in the sense of [MSS83]) family connecting them. There is a natural stratification of the space of cubic polynomials according to the dimension of the space of deformations. The mappings in the bifurcation locus correspond to dimensions 0 and 1. The dimension 1 case occurs, for instance, when one critical point is active and the other is attracted by an attracting cycle. The bifurcation locus contains a lot of holomorphic disks near such a parameter, and it is not surprising that the structure of the bifurcation current reflects this fact. On the other hand, we will see that the bifurcation measure $T^{+} \wedge T^{-} = T^{2}_{\mathrm{bif}}$ concentrates on the "most bifurcating" part of parameter space, that is, on the closure of the set of parameters with zero dimensional deformation space.

Let us be more specific. Assume first that we are outside the connectedness locus; say $-c$ escapes to infinity (and hence is passive). The structure of $\mathcal{C}^{+} \setminus \mathcal{C}$ has been described by Branner-Hubbard [BH88, BH92] — and also Kiwi [Ki06] from a different point of view. There is a canonical deformation of the maps outside $\mathcal{C}$, called the *wringing* operation, and $\mathcal{C}^+$ locally looks like a holomorphic disk times a closed transversal set which is the union of countably many copies of the Mandelbrot set and uncountably many points. Our first result is as follows (Proposition 3.4 and Theorem 3.9).

Theorem A. *The current T^+ is uniformly laminar outside the connectedness locus, and the transverse measure gives full mass to the point components.*

The delicate part of the theorem is the statement about the transverse measure. It follows from an argument of *similarity between the dynamical and parameter spaces* at the measurable level, together with a symbolic dynamics construction in the spirit of [DDS05]. We also give a new and natural proof of the continuity of the wringing operation, which is a central result in [BH88].

Assume now that $-c$ is passive but does not escape to infinity (this corresponds to parameters in $\operatorname{Int}(\mathcal{C}^-)$). This happens when $-c$ is attracted towards an attracting cycle, and, if the hyperbolicity conjecture holds, this is the only possible case. We have the following result (Theorem 4.1 and Proposition 4.2).

Theorem B. *The current T^+ is laminar in $\operatorname{Int}(\mathcal{C}^-)$, and uniformly laminar in components where $-c$ is attracted to an attracting cycle.*

In particular, laminarity holds regardless of the hyperbolic nature of components. The proof is based on a general laminarity criterion due to De Thélin [DT04]. The proof of uniform laminarity in the hyperbolic case is more classical and follows from quasiconformal surgery.

As a corollary of the two previous results, we thus get the following.

Corollary C. *The current T^+ is laminar outside $\partial \mathcal{C}^+ \cap \partial \mathcal{C}^-$.*

In the last part of the paper, we concentrate on the remaining part of parameter space. More precisely, we study the structure of the *bifurcation measure* $\mu_{\mathrm{bif}} = T^+ \wedge T^-$, which is supported in $\partial \mathcal{C}^+ \cap \partial \mathcal{C}^-$. Understanding this measure was already one of the main goals in [DF08], where it was proved that $\operatorname{Supp}(\mu_{\mathrm{bif}})$ is the closure of Misiurewicz points (i.e., strictly critically preperiodic parameters), and where the dynamical properties of μ_{bif}-a.e. parameters were investigated. It is an easy observation that a μ_{bif}-a.e. polynomial has zero dimensional deformation space (Proposition 5.1).

It follows from [DF08] that the currents $T^\pm$ are limits, in the sense of currents, of the curves

$$\operatorname{Per}^\pm(n, k) = \left\{ (c, v) \in \mathbb{C}^2, \ f^n(\pm c) = f^k(\pm c) \right\}.$$

These curves intersect at Misiurewicz points — at least, the components where $\pm c$ are strictly preperiodic do. The laminarity of $T^\pm$ and the density of Misiurewicz points could lead to the belief that the local structure of μ_{bif} is that of the "geometric intersection" of the underlying "laminations" of T^+ and T^-. The next theorem asserts that the situation is in fact more subtle (Theorem 5.6).

Theorem D. *The measure μ_{bif} does not have local product structure on any set of positive measure.*

Observe that this can also be interpreted as saying that generic wringing curves do not admit continuations through $\partial \mathcal{C}$. As a perhaps surprising consequence of this result and previous work of ours [Du03, Du04], we obtain an asymptotic lower bound on the (geometric) genus of the curves $\mathrm{Per}^\pm(n, k(n))$ as $n \to \infty$, where $0 \leq k(n) < n$ is any sequence. This gives some indications on a question raised by Milnor in [Mi09]. Recall that the geometric genus of a curve is the genus of its desingularization; also, we define the genus of a reducible curve as the sum of the genera of its components. The degree of the curve $\mathrm{Per}^+(n, k(n))$ (respectively $\mathrm{Per}^-(n, k(n))$) is equal to 3^{n-1} (respectively 3^n). Since these curves are possibly very singular (for instance at infinity in the projective plane), the genus cannot be directly read from the degree.

Theorem E. *For any sequence $k(n)$ with $0 \leq k(n) < n$,*

$$\frac{1}{3^n} \mathrm{genus}\left(\mathrm{Per}^\pm(n, k(n))\right) \to \infty.$$

Finally, there is a striking analogy between the parameter space of cubic polynomials with marked critical points and dynamical spaces of polynomial automorphisms of $\mathbb{C}^2$. This was noted earlier: for instance, as far as the global topology of the space is concerned, the reader should compare the papers [BH88] and [HO94]. We give in Table 1 a list of similarities and dissimilarities between the two. Of course there is no rigorous link between the two columns, and this table has to be understood more as an intuitive guide. For notation and basic concepts on polynomial automorphisms of $\mathbb{C}^2$, we refer to [BS91a, BLS93, FS92, Si99]. On the other hand, Milnor's article [Mi92] exhibits similar structures between *parameter spaces* of cubic polynomials and quadratic Hénon maps.

The paper is structured as follows. In Section 2, we recall some basics on the dynamics on cubic polynomials and laminar currents. In Section 3, we explain and reprove some results of Branner-Hubbard and Kiwi on the escape locus, and prove Theorem A. Theorem B is proved in Section 4. In Section 5, we discuss a notion of *higher bifurcation* based on the dimension of the space of deformations, and prove Theorems D and E.

Parameter space of cubics	Polynomial automorphisms of $\mathbb{C}^2$
$\mathcal{C}^{\pm}, \partial\mathcal{C}^{\pm}, \mathcal{C}, \partial\mathcal{C}^+ \cap \partial\mathcal{C}^-$ $T^+, T^-, \mu_{\text{bif}}$	$K^{\pm}, J^{\pm}, K, J$ T^+, T^-, μ
$\frac{1}{d^n}\left[\text{Per}^{\pm}(n,k)\right] \to T^{\pm}$ [DF08]	$\begin{cases} \frac{1}{d^n}\left[(f^{\pm n})^*(L)\right] \to T^{\pm} \\ W^{s/u} \text{ are equidistributed} \end{cases}$ [FS92]
Supp(μ_{bif}) is the Shilov boundary of $\mathcal{C}$ [DF08]	Supp(μ) is the Shilov boundary of K [BS91a]
Supp(μ_{bif}) $=$ $\overline{\{\text{Misiurewicz pts}\}}$ [DF08]	Supp(μ) $= \overline{\{\text{saddle pts}\}}$ [BLS93]
$T^{\pm}$ are laminar outside $\partial\mathcal{C}^+ \cap \partial\mathcal{C}^-$ (Th.B and D)	T^+ are laminar [BLS93]
Point components have full transverse measure (Th. A)	The same when it makes sense [BS98, Th. 7.1]
T^+ intersects all algebraic subvarieties but the Per$^+(n)$ (Prop. 2.7)	T^+ intersects all algebraic subvarieties [BS91b]
$T^{\pm}$ are not extremal near infinity (Cor. 3.11)	$T^{\pm}$ are extremal near infinity [FS92, BS98]
$T^+ \wedge T^-$ is not a geometric intersection (Th.D)	$T^+ \wedge T^-$ is a geometric intersection [BLS93]
Supp(μ_{bif}) $\subsetneq \partial\mathcal{C}^+ \cap \partial\mathcal{C}^-$	??
Uniform laminarity outside $\mathcal{C}$	??

Table 1.1. Correspondences between Poly$_3$ and polynomial automorphisms of $\mathbb{C}^2$.

Acknowledgments. This work is a sequel to [DF08], and would never have existed without Charles Favre's input and ideas. I also thank Mattias Jonsson for pointing out to me the problem of understanding bifurcation currents in the space of cubic polynomials, and for several interesting discussions on the topic.

2   Preliminaries

2.1   Stability; Active and Passive Critical Points

The notion of stability considered in this article will be that of J-stability, in the sense of [MSS83].

Definition 2.1. Let $(f_\lambda)_{\lambda \in \Lambda}$ be a family of rational maps, parameterized by a complex manifold Λ. We say that f_λ is J-stable (or simply stable) if there is a holomorphic motion of the Julia sets $J(f_\lambda)$ compatible with the dynamics.

We say that f_λ is a deformation of $f_{\lambda'}$ if there exists a J-stable family connecting them.

It is well known that if f_λ is a deformation of $f_{\lambda'}$, then f_λ and $f_{\lambda'}$ are quasiconformally conjugate in neighborhoods of their respective Julia sets. Notice that there is a stronger notion of stability, where the conjugacy is required on the whole Riemann sphere. This is the notion considered, for instance, in [MS98] for the definition of the Teichmüller space of a rational map. The stronger notion introduces some distinctions which are not relevant from our point of view, like distinguishing the center from the other parameters in a hyperbolic component.

It is a central theme in our study to consider the bifurcations of critical points one at a time. This is formalized in the next classical definition.

Definition 2.2. Let $(f_\lambda, c(\lambda))_{\lambda \in \Lambda}$ be a holomorphic family of rational maps with a marked (i.e., holomorphically varying) critical point. The marked critical point c is *passive* at $\lambda_0 \in \Lambda$ if $\{\lambda \mapsto f_\lambda^n c(\lambda)\}_{n \in \mathbb{N}}$ forms a normal family of holomorphic functions in a neighborhood of λ_0. Otherwise, c is said to be *active* at λ_0.

Notice that the passivity locus is by definition open, whereas the activity locus is closed. This notion is closely related to the bifurcation theory of rational maps, as the following classical proposition shows.

Proposition 2.3 (Lyubich, [Ly83, Mc87]). *Let (f_λ) be a family of rational maps with all its critical points marked (which is always the case, by possibly replacing Λ with some branched cover). Then the family is J-stable if and only if all critical points are passive.*

2.2   The Space of Cubic Polynomials

It is well known that the parameter space of cubic polynomials modulo affine conjugacy has complex dimension 2. It has several possible presentations (see Milnor [Mi92] for a more complete discussion).

The most commonly used parametrization is the following: for $(a, b) \in \mathbb{C}^2$, put, as in [BH88, Mi92],

$$f_{a,b}(z) = z^3 - 3a^2 z + b.$$

The term a^2 allows the critical points ($+a$ and $-a$) to depend holomorphically on f.

Two natural involutions in parameter space are of particular interest. The involution $(a, b) \mapsto (-a, b)$ preserves $f_{a,b}$ but exchanges the marking of critical points. In particular, both critical points play the same role.

The other interesting involution is $(a, b) \mapsto (-a, -b)$. It sends $f_{a,b}$ to $f_{a,-b}$, which is conjugate to f (via $-\mathrm{id}$). It can further be shown that the moduli space of cubic polynomials modulo affine conjugacy is

$$\mathbb{C}^2_{a,b} / \left((a, b) \sim (-a, -b) \right),$$

which is not a smooth manifold.

In the present paper, we will use the following parametrization (cf. Milnor [Mi09] and Kiwi [Ki06]): for $(c, v) \in \mathbb{C}^2$, we put

$$f_{c,v}(z) = z^3 - 3c^2 z + 2c^3 + v = (z - c)^2 (z + 2c) + v.$$

The critical points here are $\pm c$, so they are both marked, and the critical values are v and $v + 4c^3$. The involution that exchanges the marking of critical points then takes the form $(c, v) \mapsto (-c, v + 4c^3)$.

Notice also that the two preimages of v (respectively $v + 4c^3$) are c and $-2c$ (respectively $-c$ and $2c$). The points $-2c$ and $2c$ are called *cocritical*.

The advantage of this presentation is that it is well behaved with respect to the compactification of $\mathbb{C}^2$ into the projective plane. More precisely, it separates the sets $\mathcal{C}^+$ and $\mathcal{C}^-$ at infinity (see Remark 3 below). The motivation for introducing these coordinates in [Mi09] was to reduce the degree of the $\mathrm{Per}^+(n)$ curves.

Notice finally that in [DF08], we used a still different parameterization,

$$f_{c,\alpha}(z) = \frac{1}{3} z^3 - \frac{1}{2} c z^2 + \alpha^3,$$

where the critical points are 0 and c. In these coordinates, the two currents associated to 0 and c have the same mass (compare with Proposition 2.5 below).

2.3 Loci of Interest in Parameter Space

We give a list of notation for the subsets in parameter space which will be of interest to us:

- $\mathcal{C}^\pm$ is the set of parameters for which $\pm c$ has bounded orbit.

- $\text{Per}^+(n)$ is the set of parameters for which the critical point $+c$ has period n, and $\text{Per}^+(n,k)$ is the set of cubic polynomials f for which $f^k(c) = f^n(c)$. $\text{Per}^-(n)$ and $\text{Per}^-(n,k)$ are defined analogously.

- $\mathcal{C} = \mathcal{C}^+ \cap \mathcal{C}^-$ is the *connectedness locus*.

- $\mathcal{E} = \mathbb{C}^2 \setminus \mathcal{C}$ is the *escape locus*.

- $\mathbb{C}^2 \setminus (\mathcal{C}^+ \cup \mathcal{C}^-)$ is the *shift locus*.

The sets $\mathcal{C}^\pm$ and $\mathcal{C}$ are closed. Branner and Hubbard proved in [BH88] that the connectedness locus is compact and connected[1].

The $\text{Per}^\pm(n,k)$ are algebraic curves. Since $\text{Per}^+(n) \subset \mathcal{C}^+$, we see that $\mathcal{C}^+$ is unbounded. It is easy to prove that the activity locus associated to $\pm c$ is $\partial \mathcal{C}^\pm$. Consequently, the bifurcation locus is $\partial \mathcal{C}^+ \cup \partial \mathcal{C}^-$.

Remark 1. Let Ω be a component of the passivity locus associated to, say, $+c$. We say that Ω is hyperbolic if c converges to an attracting cycle throughout Ω. It is an obvious consequence of the density of stability [MSS83] that if the hyperbolicity conjecture holds, then all passivity components are of this type. It would be interesting to describe their geometry, which should be related to the geometry of the $\text{Per}^+(n)$ curves.

2.4　Bifurcation Currents

Here we apply the results of [DF08] in order to define various plurisubharmonic functions and currents in parameter space and list their first properties. We refer to Demailly's survey article [De93] for basics on positive closed currents. The support of a current or measure is denoted by $\text{Supp}(\cdot)$, and the trace measure is denoted by $\sigma_.$. Also, $[V]$ denotes the integration current over the subvariety V.

We identify the parameter $(c,v) \in \mathbb{C}^2$ with the corresponding cubic polynomial f. For $(f,z) = (c,v,z) \in \mathbb{C}^2 \times \mathbb{C}$, the function $(f,z) \mapsto G_f(z) = G_{f_{(c,v)}}(z)$ is continuous and plurisubharmonic. We define

$$G^+(c,v) = G_f(c) \text{ and } G^-(c,v) - G_f(-c). \tag{2.1}$$

The functions G^+ and G^- are continuous, non-negative, and plurisubharmonic. They have the additional property of being pluriharmonic when positive.

We may thus define (1,1) closed positive currents $T^\pm = dd^c G^\pm$, which will be referred to as the *bifurcation currents associated to* $\pm c$. From [DF08,

---

[1]Our notation $\mathcal{C}^+$ corresponds to $\mathcal{B}^-$ in [BH92], and $\mathcal{E}^-$ in [Ki06].

Section 3], we get that $\mathrm{Supp}(T^{\pm})$ is the activity locus associated to $\pm c$, that is, $\mathrm{Supp}(T^{\pm}) = \partial \mathcal{C}^{\pm}$.

It is also classical to consider the Lyapunov exponent $\chi = \log 3 + G^+ + G^-$ of the equilibrium measure, and the corresponding current $T_{\mathrm{bif}} = dd^c \chi = T^+ + T^-$, known as the *bifurcation current* [De01, BB07a]. Its support is the bifurcation locus.

The following equidistribution theorem was proved in [DF08].

Theorem 2.4. *Let* $(k(n))_{n \geq 0}$ *be any sequence of integers such that* $0 \leq k(n) < n$. *Then*

$$\lim_{n \to \infty} \frac{1}{3^n} \left[\mathrm{Per}^{\pm}(n, k(n)) \right] = T^{\pm}.$$

We can compute the mass of the currents $T^{\pm}$. The lack of symmetry is not a surprise since c and $-c$ do not play the same role with respect to v.

Proposition 2.5. *The mass of* T^+ *with respect to the Fubini-Study metric on* $\mathbb{C}^2$ *is* $1/3$. *On the other hand, the mass of* T^- *is* 1.

Proof: We use the equidistribution theorem. Indeed, a direct computation shows that the degree of $\mathrm{Per}^+(n)$ is 3^{n-1}, whereas the degree of $\mathrm{Per}^-(n)$ is 3^n. □

We will mainly be interested in the fine geometric properties, in particular the *laminarity*, of the currents $T^{\pm}$. A basic motivation for this is the following observation.

Proposition 2.6. *Let* Δ *be a holomorphic disk in* $\mathbb{C}^2$, *where* G^+ *and* G^- *are harmonic. Then the family* $\{f \in \Delta\}$ *is* J-*stable.*
This holds in particular when $\Delta \subset \mathcal{C}^+ \setminus \partial \mathcal{C}^-$.

Proof: Recall that G^+ is non-negative, and pluriharmonic where it is positive. Hence if $G^+|_{\Delta}$ is harmonic, either $G^+ \equiv 0$ (i.e., c does not escape) on Δ, or $G^+ > 0$ (i.e., c always escapes) on Δ. In any case, c is passive. Doing the same with $-c$ and applying Proposition 2.3 then finishes the proof. □

The laminarity results will tell us that there are plenty of such disks, and how these disks are organized in space: they are organized as (pieces of) *laminations*.

We can describe the relative positions of T^+ and of algebraic curves, thus filling in a line of Table 1. Notice that G^+ is continuous, so $T^+ \wedge [V] = dd^c(G^+|_V)$ is always well defined. Also, if $V \subset \mathrm{Supp}(T^+)$, then $G^+|_V = 0$, so $T^+ \wedge [V] = 0$.

Proposition 2.7. *Let* V *be an algebraic curve such that* $T^+ \wedge [V] = 0$. *Then* $V \subset \mathrm{Per}^+(n, k)$ *for some* (n, k). *Moreover, if* $\mathrm{Supp}(T^+) \cap V = \emptyset$, *then* $V \subset \mathrm{Per}^+(n)$ *for some* n.

Proof: If $T^+ \wedge [V] = 0$, then c is passive on V, which is an affine algebraic curve, so by [DF08, Theorem 2.5], $V \subset \mathrm{Per}(n, k)$ for some (n, k).

Assume now that $\mathrm{Supp}(T^+) \cap V = \emptyset$, and that V is a component of $\mathrm{Per}^+(n, k)$, where c is strictly preperiodic at generic parameters (i.e., excluding possibly finitely many exceptions). Then there exists a persistent cycle of length $n - k$ along V, on which c falls after at most k iterations. We prove that this leads to a contradiction.

The multiplier of this cycle defines a holomorphic function on V. Since $V \cap \mathrm{Supp}(T^+) = \emptyset$, if $f_0 \in V$, c is passive in a neighborhood of f_0 in $\mathbb{C}^2$, so the multiplier cannot be greater than 1. In particular, the multiplier is constant along V, and this constant has modulus ≤ 1. We now claim that the other critical point must have a bounded orbit: indeed, if the cycle is attracting, there is a critical point in the immediate basin of the cycle, which cannot be c since c is generically strictly preperiodic. Now, if the multiplier is a root of unity or is of Cremer type, then the cycle must be accumulated by an infinite critical orbit — necessarily that of $-c$ (see, e.g., [Mi99]) — and if there is a Siegel disk, its boundary must lie in the closure of an infinite critical orbit. In any case, both critical points have bounded orbits, so K is connected. Since V is unbounded, this contradicts the compactness of the connectedness locus [BH88]. □

2.5 Background on Laminar Currents

Here we briefly introduce the notions of laminarity that will be considered in the paper. It should be mentioned that the definitions of flow boxes, laminations, laminar currents, etc. are tailored to the specific needs of this paper, and hence are not as general as they could be. For more details on these notions, see [BLS93, Du03, Du04]; see also [Gh99] for general facts on laminations and transverse measures.

We first recall the notion of the direct integral of a family of positive closed currents. Assume that $(T_\alpha)_{\alpha \in \mathcal{A}}$ is a measurable family of positive closed currents in some open subset $\Omega \subset \mathbb{C}^2$, and that ν is a positive measure on $\mathcal{A}$ such that (reducing Ω is necessary) $\alpha \mapsto \mathrm{Mass}(T_\alpha)$ is ν-integrable. Then we can define a positive closed current $T = \int T_\alpha d\nu(\alpha)$ by the obvious pairing with test forms:

$$\langle T, \varphi \rangle = \int \langle T_\alpha, \varphi \rangle d\nu(\alpha).$$

Now let T and S be positive closed currents in a ball $\Omega \subset \mathbb{C}^2$. We say that the wedge product $T \wedge S$ is *admissible* if for some (and hence for every) potential u_T of T, u_T is locally integrable with respect to the trace measure σ_S of S. In this case, we may classically define $T \wedge S = dd^c(u_T S)$, which is a positive measure. The next lemma shows that integrating families of positive closed currents is well behaved with respect to taking wedge products.

Lemma 2.8. *Let $T = \int T_\alpha d\nu(\alpha)$ be as above, and assume that S is a positive closed current such that the wedge product $T \wedge S$ is admissible. Then for ν-a.e. α, $T_\alpha \wedge S$ is admissible, and*

$$T \wedge S = \int (T_\alpha \wedge S) d\nu(\alpha).$$

Proof: The result is local, so we may assume that Ω is the unit ball. Assume for the moment that there exists a measurable family of non-positive plurisubharmonic functions u_α, with $dd^c u_\alpha = T_\alpha$ and $\|u_\alpha\|_{L^1(\Omega')} \leq C\mathrm{Mass}(T_\alpha)$, for every $\Omega' \Subset \Omega$. In particular, we get that $\alpha \mapsto \|u_\alpha\|_{L^1(\Omega')}$ is ν-integrable, and the formula $u_T = \int u_\alpha d\nu(\alpha)$ defines a non-positive potential of T. Then since $u_T \in L^1(\sigma_S)$ and all potentials are non-positive, we get that for ν-a.e. α, $u_\alpha \in L^1(\sigma_S)$, and $u_T S = \int (u_\alpha S) d\nu(\alpha)$ by Fubini's Theorem.

It remains to prove our claim. Observe first that if we are able to find a potential u_α of T_α, with $\|u_\alpha\|_{L^1(\Omega')} \leq C\mathrm{Mass}(T_\alpha)$, then, by slightly reducing Ω', we can get a non-positive potential with controlled norm by subtracting a constant, since by the submean inequality, $\sup_{\Omega''} u_\alpha \leq C(\Omega'') \|u_\alpha\|_{L^1(\Omega')}$. Another observation is that it is enough to prove it when T_α is smooth and use regularization.

The classical way of finding a potential for a positive closed current of bidegree (1,1) in a ball is to first use the usual Poincaré lemma for d, and then solve a $\bar{\partial}$ equation for a (0,1) form (see [De93]). The Poincaré lemma certainly preserves the L^1 norm (up to constants), because it boils down to integrating forms along paths. Solving $\bar{\partial}$ with L^1 control in a ball is much more delicate but still possible (see for instance [Ra98, p.300]). This concludes the proof. $\square$

By a *flow box*, we mean a closed family of disjoint holomorphic graphs in the bidisk. In other words, it is the total space of a holomorphic motion of a closed subset in the unit disk, parameterized by the unit disk. It is an obvious consequence of Hurwitz' Theorem that the closure of any family of disjoint graphs in $\mathbb{D}^2$ is a flow box. Moreover, the holonomy map along this family of graphs is automatically continuous. This follows, for instance, from the Λ-lemma of [MSS83].

Let $\mathcal{L}$ be a flow box, written as the union of a family of disjoint graphs as $\mathcal{L} = \bigcup_{\alpha \in \tau} \Gamma_\alpha$, where $\tau = \mathcal{L} \cap (\{0\} \times \mathbb{D})$ is the central transversal. To every positive measure ν on τ, there corresponds a positive closed current in $\mathbb{D}^2$, defined by the formula

$$T = \int_\tau [\Gamma_\alpha] d\nu(\alpha). \tag{2.2}$$

A *lamination* in $\Omega \subset \mathbb{C}^2$ is a closed subset of Ω which is locally biholomorphic to an open subset of a flow box $\mathcal{L} = \bigcup \Gamma_\alpha$. A positive closed current supported on a lamination is said to be *uniformly laminar*, and *subordinate* to the lamination if it is locally of the form (2.2). Not every lamination carries a uniformly laminar current. This is the case if and only if there exists an *invariant transverse measure*, that is, a family of positive measures on transversals, invariant under holonomy. Conversely, a uniformly laminar current induces a natural measure on every transversal to the lamination.

We say that two holomorphic disks D and D' are *compatible* if $D \cap D'$ is either empty or open in the disk topology. A positive current T in $\Omega \subset \mathbb{C}^2$ is *laminar* if there exists a measured family $(\mathcal{A}, \nu)$ of holomorphic disks $D_\alpha \subset \Omega$, such that for every pair (α, β), D_α and D_β are compatible and

$$T = \int_\mathcal{A} [D_\alpha] d\nu(\alpha). \tag{2.3}$$

The difference between this and uniform laminarity is that there is no (even locally) uniform lower bound on the size of the disks in (2.3). In particular, there is no associated lamination, and the notion is strictly weaker. Notice also that the integral representation (2.3) does not prevent T from being closed because of boundary cancellation.

Equivalently, a current is laminar if it is the limit of an increasing sequence of uniformly laminar currents. More precisely, T is laminar in Ω if there exists a sequence of open subsets $\Omega^i \subset \Omega$ such that $\|T\| (\partial \Omega^i) = 0$, together with an increasing sequence of currents $(T^i)_{i \geq 0}$, where each T^i is uniformly laminar in Ω^i, converging to T.

In Section 5, we will consider *woven* currents. The corresponding definitions will be given at that time.

3   Laminarity Outside the Connectedness Locus

In this section, we give a precise description of T^+ outside the connectedness locus. Subsections 3.1 and 3.2 are rather expository. We first recall the *wringing* construction of Branner-Hubbard, and how it leads to uniform laminarity. Then, we explain some results of Kiwi [Ki06] on the geometry of $\mathcal{C}^+$ at infinity. In Section 3.3, based on an argument of similarity between dynamical and parameter spaces and a theorem of [DDS05], we prove that the transverse measure induced by T^+ gives full mass to the point components. The presentation is as self-contained as possible; only the results of [BH92], which depend on the combinatorics of tableaux, are not reproved.

3.1 Wringing and Uniform Laminarity

We start by defining some analytic functions in a part of parameter space, in analogy with the definition of the functions $G^{\pm}$. Recall that for a cubic polynomial f, the Böttcher coordinate φ_f is a holomorphic function defined in an open neighborhood of infinity

$$U_f = \left\{ z \in \mathbb{C}, G_f(z) > \max(G^+, G^-) \right\},$$

and which semiconjugates f to z^3 there, i.e., $\varphi_f \circ f = (\varphi_f)^3$. Also, $\varphi_f = z + O(1)$ at infinity.

Assume that $-c$ is the fastest escaping critical point, i.e., assume that $G_f(c) < G_f(-c)$. The corresponding critical value $v + 4c^3$ has two distinct preimages, $-c$ and $2c$ (because $c \neq 0$), and it turns out that

$$\varphi_f(2c) := \lim_{U_f \ni z \to 2c} \varphi_f(z)$$

is always well defined (whereas there would be an ambiguity in defining $\varphi_f(-c)$). We put $\varphi^-(f) = \varphi_f(2c)$, so that φ^- is a holomorphic function in the open subset $\{G^+ < G^-\}$ satisfying $G^-(f) = \log |\varphi^-(f)|$.

Remark 2. We note that it is possible to continue $(\varphi^-)^3$ to the larger subset $\{G^+ < 3G^-\}$ by evaluating φ_f at the critical value, and similarly for $(\varphi^-)^9$, etc. So locally, it is possible to define branches of φ^- in larger subsets of parameter space by considering inverse powers.

Following Branner and Hubbard [BH88] (we reproduce Branner's exposition [Br91]), we now define the basic operation of *wringing the complex structure*. This provides a holomorphic 1-parameter deformation of a map in the escape locus, which has the advantage of being independent of the number and relative position of escaping critical points.

Let $\mathbb{H}$ be the right half plane, endowed with the following group structure (with 1 as unit element):

$$u_1 * u_2 = (s_1 + it_1) * (s_2 + it_2) = (s_1 + it_1)s_2 + it_2.$$

For $u \in \mathbb{H}$, we define the diffeomorphism $g_u : \mathbb{C} \setminus \mathbb{D} \to \mathbb{C} \setminus \mathbb{D} \to$ by $g_u(z) = z |z|^{u-1}$. Notice that g_u commutes with z^3, and $g_{u_1 * u_2} = g_{u_1} \circ g_{u_2}$, i.e., $u \mapsto g_u$ is a left action on $\mathbb{C} \setminus \mathbb{D}$. Since g_u commutes with z^3, we can define a new invariant almost complex structure on $U_f = \{G_f(z) > \max(G^+, G^-)\}$ by replacing σ_0 with $g_u^*(\sigma_0)$ in the Böttcher coordinate. Here σ_0 is the standard complex structure on $\mathbb{C}$. More precisely, we define σ_u as the unique almost complex structure on $\mathbb{C}$, invariant under f, such that

$$\sigma_u = \begin{cases} \varphi_f^* g_u^* \sigma_0 & \text{on } U_f \\ \sigma_0 & \text{on } K_f \end{cases}.$$

Of course, viewed as a Beltrami coefficient, σ_u depends holomorphically on u, and it can be straightened by a quasiconformal map ϕ_u such that

- $\phi_u \circ f \circ \phi_u^{-1}$ is a monic centered polynomial,

- $g_u \circ \varphi_f \circ \phi_u^{-1}$ is tangent to the identity at infinity.

Under these assumptions, ϕ_u is unique and depends holomorphically on u [BH88, Section 6].

We put $w(f, u) = f_u = \phi_u \circ f \circ \phi_u^{-1}$. This is a holomorphic disk through $f = f_1$ in parameter space. Using the terminology of [DH85b], f_u is *hybrid equivalent* to f, so by the uniqueness of the renormalization for maps with connected Julia sets, if $f \in \mathcal{C}$, then $f_u \equiv f$. On the other hand, the construction is non-trivial in the escape locus, as the next proposition shows.

Proposition 3.1. *Wringing yields a lamination of $\mathbb{C}^2 \setminus \mathcal{C}$ by Riemann surfaces whose leaves are isomorphic to the disk or to the punctured disk.*

This provides, in particular, a new proof of the following core result of [BH88].

Corollary 3.2. *The wring mapping $w : \mathbb{C}^2 \setminus \mathcal{C} \to \mathbb{C}^2 \setminus \mathcal{C}$ is continuous.*

The proof of the proposition is based on the following simple but useful lemma.

Lemma 3.3. *Let $\Omega \subset \mathbb{C}^2$ be a bounded open set, and let $\pi : \Omega \to \mathbb{C}$ be a holomorphic function. Assume that there exists a closed subset Γ and a positive constant c such that*

- *for every $p \in \Gamma$, there exists a holomorphic map $\gamma_p : \mathbb{D} \to \Gamma$, with $\gamma_p(0) = p$ and $|(\pi \circ \gamma_p)'(0)| \geq c$,*

- *the disks are compatible, in the sense that $\gamma_p(\mathbb{D}) \cap \gamma_q(\mathbb{D})$ is either empty or open in each of the two disks.*

Then $\overline{\bigcup_{p \in \Gamma} \gamma_p(\mathbb{D})}$ is a lamination of Γ by holomorphic curves.

Proof: We may assume that $c = 1$. First, since $(\pi \circ \gamma_p)'(0)$ never vanishes, π is a submersion in a neighborhood of Γ. Hence, locally, near $p \in \Gamma$, we may suppose that π is the first projection by choosing adapted coordinates (z, w), i.e., that $\pi(z, w) = z$.

The second claim is that if f is holomorphic and bounded by R on $\mathbb{D}$, with $f(0) = 0$ and $|f'(0)| \geq 1$, then there exists a constant $\varepsilon(R)$ depending only on R such that f is injective on $D(0, \varepsilon(R))$. Then by the Koebe Theorem, the image univalently covers a disk of radius $\varepsilon(R)/4$.

Now observe that since Ω is bounded, if $\Gamma_1 \subset \Gamma$ is relatively compact, then the family of mappings $\{\gamma_p, \ p \in \Gamma_1\}$ is locally equicontinuous. Hence in the local coordinates (z, w) defined above, we can apply the previous observation to the family of functions $\pi \circ \gamma_p$. The claim asserts that, given any z_0 and any point p in the fiber $\Gamma \cap \pi^{-1}(z_0)$, the disk $\gamma_p(\mathbb{D})$ contains a graph for the projection π over the disk $D(z_0, \varepsilon/4)$ for some uniform ε.

Thus all components of $\bigcup_{p \in \Gamma} \gamma_p(\mathbb{D})$ over the disk $D(z_0, \varepsilon/8)$ are graphs over that disk, and they are disjoint. Moreover they are uniformly bounded in the vertical direction by equicontinuity. As observed in Section 2.5, from Hurwitz' Theorem and equicontinuity, we conclude that $\overline{\bigcup_{p \in \Gamma} \gamma_p(\mathbb{D})}$ is a lamination. $\qquad\square$

Proof of the proposition: The idea is to use the previous lemma with $\pi = \varphi^{\pm}$. We need to understand how the $\varphi^{\pm}$ evolve under wringing. We prove that the wringing disks form a lamination in the open set $\{G^+ < G^-\}$ where φ^- is defined — what we need in the sequel is just a neighborhood of $\mathcal{C}^+$. The case of $\{G^- < G^+\}$ is of course symmetric, and Remark 2 allows us to extend our reasoning to a neighborhood of $\{G^+ = G^-\}$.

Recall that $w(f, u) = f_u = \phi_u \circ f \circ \phi_u^{-1}$, so if $f = f_{(c,v)}$, the critical points of f_u are $\phi_u(\pm c)$. Also, an explicit computation shows that $g_u \circ \varphi_f \circ \phi_u^{-1}$ semiconjugates f_u and z^3 near infinity, so, by the normalization we have done, this is the Böttcher function φ_{f_u}. In particular, $\varphi_{f_u}(\phi_u(z)) = g_u(\varphi_f(z))$. At the level of the Green function, we thus get the identity

$$G_{f_u}(\phi_u(z)) = sG_f(z) \text{ where } u = s + it. \qquad (3.4)$$

Assume that $G^+ < G^-$, so that the fastest escaping critical point is $-c$. By (3.4), the fastest critical point of f_u is $\phi_u(-c)$, so

$$\varphi^-(f_u) = g_u \circ \phi_f \circ \phi_u^{-1}(\phi_u(-c)) = g_u(\varphi^-(f)) = \varphi^-(f) \left|\varphi^-(f)\right|^{u-1}. \qquad (3.5)$$

This equation tells us two things: first, that $(f, u) \mapsto w(f, u)$ is locally uniformly bounded, and second, that $f \mapsto \frac{\partial \varphi^-(w(f,u))}{\partial u}\big|_{u=0}$ is locally uniformly bounded from below in parameter space.

Moreover the wringing disks are compatible because of the group action $w(w(f,v),u) = w(f,u*v)$. By using Lemma 3.3, we conclude that they fit together in a lamination.

Now if L is a leaf of the lamination, and if $f \in L$ is any point, then $w(f,\cdot) : \mathbb{H} \to L$ is a universal cover. Indeed it is clearly onto because of the group action, and since both the leaf and the map $w(f,\cdot)$ are uniformly transverse to the fibers of φ^-, it is locally injective.

There are two possibilities: either it is globally injective, or it is not. In the first case, $L \simeq \mathbb{D}$. In the second case, if $w(f,s_1 + it_1) = w(f,s_2 + it_2)$, then by (3.4), $s_1 = s_2$. Since the leaf is a graph with uniform size for the projection φ^- near $w(f,s_1 + it_1)$, there exists a minimal $T > 0$ such that $w(f,s_1 + it_1 + iT) = w(f,s_1 + it_1)$. The group structure on $\mathbb{H}$ has the property that $u*(v+iT) = (u*v)+iT$, from which we deduce that for every $u \in \mathbb{H}$, $w(f,u+iT) = w(f,u)$. In particular, L is conformally isomorphic to the punctured disk. □

Proof of Corollary 3.2: By definition, the holonomy of a lamination is continuous. We only have to worry about the compatibility between the natural parameterization of the leaves and the lamination structure. Here, by (3.5), the correspondence between the parameter u on the leaves and the parameter induced by the transversal fibration $\varphi^- = \mathrm{cst}$ is clearly uniformly bicontinuous — by uniform we mean here locally uniform with respect to the leaf. This proves the corollary. □

Proposition 3.4. *The current T^+ is uniformly laminar in the escape locus, and subordinate to the lamination by wringing curves.*

Proof: The limit of a converging sequence of uniformly laminar currents, all subordinate to the same lamination, is itself subordinate to this lamination: indeed, locally, in a flow box, this is obvious. So by Theorem 2.4, to get the uniform laminarity of T^+, it is enough to prove that the curves $[\mathrm{Per}^+(n)]$ are subordinate to the wringing lamination in $\mathbb{C}^2 \setminus \mathcal{E}$. But this is again obvious: $w(f,u)$ is holomorphically conjugate to f on $\mathrm{Int}(K_f)$, so if f has a superattracting cycle of some period, then so does $w(f,u)$. □

3.2 The Geometry of $\mathcal{C}^+$ at Infinity

In this section, we follow Kiwi [Ki06]. For the sake of convenience, we include most proofs.

The following lemma is [Ki06, Lemma 7.2], whose proof relies on easy explicit estimates. Besides notation, the only difference is the slightly more general hypothesis $|v| < 3\,|c|$. We leave it to the reader to check that this more general assumption is enough to get the conclusions of the lemma.

Lemma 3.5. *If $|v| < 3\,|c|$ and c is large enough, then the following hold:*

- $|f^n(-c)| \geq |c|^{3^n} (\sqrt{2})^{-3^{n-1}-1}$,

- $G^-(c, v) = \log |c| + O(1)$,

- $G^+(c, v) \leq \frac{1}{3} \log |c| + O(1)$.

As a consequence, we get an asymptotic expansion of φ^- (compare [Ki06, Lemma 7.10]). Notice that under the assumptions of the previous lemma, $G^+(c, v) < G^-(c, v)$, so φ^- is well defined.

Lemma 3.6. *When $|v| < 3\,|c|$ and $(c, v) \to \infty$, then*

$$\varphi^-(c, v) = 2^{2/3} c + o(c).$$

Proof: Recall that

$$\varphi^-(c, v) = \varphi_{f_{(c,v)}}(2c) = \lim_{n \to \infty} 2c \left(\frac{f(2c)}{(2c)^3} \right)^{\frac{1}{3}} \cdots \left(\frac{f^n(2c)}{(f^{n-1}(2c))^3} \right)^{\frac{1}{3^n}}.$$

For large z, $f(z)/z^3 = 1 + O(1/z)$, and since $f(2c) = f(-c)$, by Lemma 3.5 we get that for $n > 1$,

$$\frac{f^n(2c)}{(f^{n-1}(2c))^3} = 1 + O\left(\frac{2^{3^n/2}}{|c|^{3^n}} \right),$$

and the product converges uniformly for large c. Moreover, $f(2c)/(2c)^3$ converges to $1/2$ when $(c, v) \to \infty$ and $|v| \leq 3\,|c|$, whereas, for $n \geq 2$, $f^n(2c)/(f^{n-1}(2c))^3$ converges to 1. This gives the desired estimate. $\qquad\square$

We now translate these results into more geometric terms. See Figure 1 for a synthetic picture. Compactify $\mathbb{C}^2$ as the projective plane $\mathbb{P}^2$, and choose homogeneous coordinates $[C : V : T]$ so that $\{[c : v : 1], (c, v) \in \mathbb{C}^2\}$ is our parameter space. Consider the coordinates $x = \frac{1}{c} = \frac{T}{C}$ and $y = \frac{v}{c} = \frac{V}{C}$ near infinity, and the bidisk $B = \{|x| \leq x_0, |y| \leq 1/3\}$. The line at infinity becomes the vertical line $x = 0$ in the new coordinates.

Proposition 3.7 (Kiwi [Ki06]).

(i) *For k_0 large enough, the level sets $\{\varphi^- = k, |k| \geq k_0\}$ are vertical holomorphic graphs in B. They fit together with the line at infinity, $x = 0$, as a holomorphic foliation.*

(ii) *For small enough x_0, the leaves of the wringing lamination are graphs over the first coordinate in $B \setminus \{x = 0\}$.*

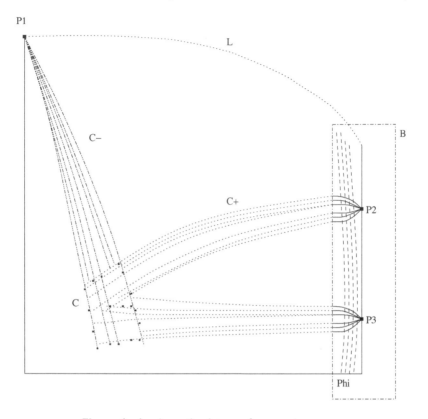

Figure 1. A schematic picture of parameter space.

Proof: In the new coordinates, we have

$$\varphi^-(x,y) = \frac{2^{2/3}}{x}(1 + \delta(x,y)),$$

where $\delta(x,y)$ is a holomorphic function outside $\{x = 0\}$. Since $\delta \to 0$ as $x \to 0$, δ extends holomorphically to B. Let $k' = 1/k$; the equation $\varphi^- = k$ is rewritten as

$$x = 2^{2/3}(1 + \delta(x,y))k'.$$

For small enough k' and fixed y, this equation has exactly one solution in x, which depends holomorphically on y; this means that $\{\varphi^- = k\}$ is a vertical graph close to $x = 0$. Clearly,

$$\{x = 0\} \cup \bigcup_{|k| \geq k_0} \{\varphi^- = k\}$$

is a lamination near $\{x = 0\}$. If we fix two small holomorphic transversals to $\{x = 0\}$, the holonomy map is holomorphic outside the origin, so it extends holomorphically. We conclude that this lamination by vertical graphs extends as a holomorphic foliation across $\{x = 0\}$.

On the other hand, we know that the wringing curves are transverse to the fibers of φ^-, i.e., they are graphs for the projection φ^-. So by using the estimate of Lemma 3.6 again, and by Rouché's Theorem, we find that the wringing leaves are graphs over the c coordinate for large c. This gives the second part of the proposition. $\square$

The next proposition asserts that the wringing curves contained in $\mathcal{C}^+$ cluster only at two points at infinity. In particular, these points are singular points of the wringing lamination. The results of [Ki06, Section 7] may be thought of as the construction of an abstract "desingularization" of this lamination.

Proposition 3.8 (Kiwi [Ki06]). *The closure of $\mathcal{C}^+$ in $\mathbb{P}^2$ is $\overline{\mathcal{C}^+} = \mathcal{C}^+ \cup \{[1:1:0], \ [1:-2:0]\}$. Likewise, $\overline{\mathcal{C}^-} = \mathcal{C}^- \cup \{[0:1:0]\}$.*

Remark 3. The choice of the (c, v) parameterization is due precisely to this proposition. By an easy computation, the reader may check that, in the (a, b) coordinates of Section 2.2, $\mathcal{C}^+$ and $\mathcal{C}^-$ both cluster at the same point $[0:1:0]$ at infinity, an unwelcome feature.

Proof: The starting point is the fact that $G^+ + G^-$ is proper in $\mathbb{C}^2$: see [BH88, Section 3]. Consider now the following open neighborhood of $\mathcal{C}^+$:

$$\mathcal{V} = \left\{ (c, v) \in \mathbb{C}^2, \ G_f(v) = 3G_f(c) < \frac{1}{3}G_f(-c) \right\} = \left\{ 3G^+ < \frac{1}{3}G^- \right\}.$$

We will show that $\overline{\mathcal{V}}$ intersects the line at infinity in $[1:1:0]$ and $[1:-2:0]$ by proving that when $G_f(-c)$ tends to infinity in $\mathcal{V}$, v/c converges to either 1 or -2. Showing that both points are actually reached is easy: this is already the case for the $\mathrm{Per}^+(2)$ curve, which has equation

$$(v - c)^2(v + 2c) + (v - c) = 0.$$

Let us introduce some standard concepts. If $(c, v) \in \mathcal{V}$, in the dynamical plane, then the level curves $\{G_f(z) = r\}$ are smooth Jordan curves for $r > G_f(-c)$, and $\{G_f(z) = G_f(-c)\}$ is a figure eight curve with a self intersection at $-c$. By a *disk at level n*, we mean a connected component of $\{z, \ G_f(z) < 3^{-n+1}G_f(-c)\}$. If z is a point deeper than level n, we define $D_f^n(z)$ to be the connected component of $\{G_f(z) < 3^{-n+1}G_f(-c)\}$ containing z. Also,

$$A_f^0 = \{z, \ G_f(-c) < G_f(z) < 3G_f(-c)\}$$

is an annulus of modulus

$$\frac{1}{2\pi} \log \frac{e^{3G_f(-c)}}{e^{G_f(-c)}} = \frac{1}{\pi} G_f(-c).$$

If $(c, v) \in \mathcal{V}$, then v, c, and $-2c$ (the cocritical point) are points of level ≥ 2. There are two disks of level 1 corresponding to the two inner components of the figure eight: $D_f^1(c)$ and $D_f^1(-2c)$. In particular, $D_f^1(v) \setminus \overline{D_f^2(v)}$ is an annulus of modulus $\geq \frac{1}{2}\text{modulus}(A_f^0)$ that either separates c and v from $-2c$ or separates $-2c$ and v from c, depending on which of the two disks contains v.

When (c, v) tends to infinity in $\mathcal{V}$, $\text{modulus}(A_f^0) = G_f(-c) \to \infty$, and standard estimates in conformal geometry imply that $|v - c| = o(c)$ in the first case and that $|v + 2c| = o(c)$ in the second, that is, v/c respectively converges to 1 or -2. To prove this, we may for instance use the fact that an annulus with large modulus contains an essential round annulus with almost the same modulus [Mc94].

From this we easily get the corresponding statement for $\mathcal{C}^-$, by noting that the involution which exchanges the markings $(c, v) \mapsto (-c, v + 4c^3)$ contracts the line at infinity with $[1 : 0 : 0]$ deleted onto the point $[0 : 1 : 0]$. □

3.3 A Transverse Description of T^+

Branner and Hubbard gave in [BH92] a very detailed picture of $\mathcal{C}^+ \setminus \mathcal{C}$, both from the point of view of the dynamics of an individual mapping $f \in \mathcal{C}^+ \setminus \mathcal{C}$ and from the point of view of describing $\mathcal{C}^+ \setminus \mathcal{C}$ as a subset in parameter space. Roughly speaking, a map in $\mathcal{C}^+ \setminus \mathcal{C}$ can be either quadratically renormalizable or not. If $f \in \mathcal{C}^+ \setminus \mathcal{C}$, K_f is disconnected, so it has infinitely many components. We denote by $C(+c)$ the connected component of the non-escaping critical point. Then:

- **Renormalizable Case.** If $C(+c)$ is periodic, then f admits a quadratic renormalization, and $C(+c)$ is qc-homeomorphic to a quadratic Julia set. Moreover a component of $K(f)$ is not a point if and only if it is a preimage of $C(+c)$.

- **Non-renormalizable Case.** If $C(+c)$ is not periodic, then K_f is a Cantor set.

There is a very similar dichotomy in parameter space. Notice that near infinity, $\mathcal{C}^+ \subset B$, where B is the bidisk of Proposition 3.7. For large k, consider a disk Δ of the form $\Delta = \{\varphi^- = k\} \cap B$. Then $\mathcal{C}^+ \cap \Delta$ is a disconnected compact subset of Δ. The connected components are of two different types:

- **Point Components.** These correspond to non-quadratically renormalizable maps.

- **Copies of M.** The quadratically renormalizable parameters are organized into Mandelbrot-like families [DH85b], giving rise to countably many homeomorphic copies of the Mandelbrot set in Δ.

Remark 4. These results imply that the components of $\text{Int}(\mathcal{C}^+) \setminus \mathcal{C}$ are exactly the open subsets obtained by holomorphically moving (under wringing) the components of the Mandelbrot copies. In particular, if the *quadratic* hyperbolicity conjecture holds, then all these components are hyperbolic.

Since T^+ is uniformly laminar and subordinate to the lamination by wringing curves, in order to have a good understanding of T^+, it is enough to describe the transverse measure $T^+ \wedge [\Delta]$.

Theorem 3.9. *The transverse measure induced by T^+ on a transversal gives full mass to the point components.*

In particular, point components are dense in $\partial \mathcal{C}^+ \cap \Delta$. Notice that the Mandelbrot copies are also dense, since they contain the points $\text{Per}^+(n) \cap \Delta$.

Proof: The proof of the theorem will follow from the similarity between the dynamical and parameter spaces, and a symbolic dynamics argument in the style of [DDS05].

In the domain of parameter space under consideration, $G^+ < G^-$, so in the dynamical plane, the open set $\{G_f(z) < G^-(f)\}$ is bounded by a figure eight curve (see Figure 2). Let U_1 and U_2 be its two connected components, and assume that $c \in U_2$, so that $f|_{U_i} : U_i \to \{G_f < 3G^-\}$ is proper with topological degree i. We denote by f_i the restriction $f|_{U_i}$, where $d_i = \deg(f_i)$, and by U the topological disk $\{G_f < 3G^-\}$.

Thus, classically, we get a decomposition of $K(f)$ in terms of itineraries in the symbol space $\Sigma := \{1, 2\}^{\mathbb{N}}$. More precisely, for a sequence $\alpha \in \Sigma$, let

$$K_\alpha = \left\{ z \in \mathbb{C}, \ f^i(z) \in U_{\alpha(i)} \right\}.$$

For every α, K_α is a non-empty closed subset, and the K_α form a partition of K.

For example, for $\alpha = \overline{2} = (2222\cdots)$, $K_{\overline{2}}$ is the filled Julia set of the quadratic-like map $f|_{U_2} : U_2 \to U$. On the other hand, $K_{\overline{1}}$ is a single repelling fixed point. Notice also that $K_{\overline{12}}$ is not a single point, even if the sequence $\overline{12}$ contains infinitely many 1's.

We now explain how the decomposition $K = \bigcup_{\alpha \in \Sigma} K_\alpha$ is reflected in the Brolin measure μ. Let p be any point outside the filled Julia set, and

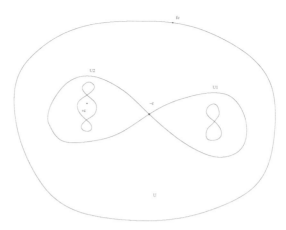

Figure 2. Level curves of the Green function.

write

$$\frac{1}{3^n}(f^n)^*\delta_p = \frac{1}{3^n} \sum_{(\alpha_0,\dots,\alpha_{n-1})\in\{1,2\}^n} f^*_{\alpha_0}\cdots f^*_{\alpha_{n-1}}\delta_p. \tag{3.6}$$

In [DDS05], we proved that for *any* sequence $\alpha \in \Sigma$, and for any p outside the Julia set, the sequence of measures

$$\frac{1}{d_{\alpha_0}\cdots d_{\alpha_{n-1}}} f^*_{\alpha_0}\cdots f^*_{\alpha_{n-1}}\delta_p$$

converges to a probability measure μ_α independently of p and is supported on ∂K_α. The measure μ_α is the analogue of the Brolin measure for the sequence (f_{α_k}). This was stated as Theorem 4.1 and Corollary 4.6 in [DDS05] for horizontal-like sequences in the unit bidisk, but it is easy to translate in the setting of polynomial-like maps in $\mathbb{C}$.

Let ν be the shift invariant measure on Σ gives mass $(d_{\alpha_0}\cdots d_{\alpha_{n-1}})/3^n$ to the cylinder of sequences starting with $\alpha_0,\dots,\alpha_{n-1}$ — the $(\frac{1}{3},\frac{2}{3})$ measure on Σ. Then (3.6) may be rewritten as

$$\frac{1}{3^n}(f^n)^*\delta_p = \sum_{(\alpha_0,\dots,\alpha_{n-1})\in\{1,2\}^n} \frac{d_{\alpha_0}\cdots d_{\alpha_{n-1}}}{3^n}\left[\frac{1}{d_{\alpha_0}\cdots d_{\alpha_{n-1}}} f^*_{\alpha_0}\cdots f^*_{\alpha_{n-1}}\delta_p\right],$$

so at the limit we get the decomposition $\mu = \int_\Sigma \mu_\alpha d\nu(\alpha)$.

We fix a transversal disk Δ as in Proposition 3.7. If Δ close enough to the line at infinity, then $G^+ > 0$ on $\partial\Delta$. Observe that the measure $T^+\wedge\Delta$ induced by T^+ on Δ is the "bifurcation current" of the family of cubic polynomials parameterized by Δ. We denote it by $T^+|_\Delta$.

Following [DF08, Section 3], consider $\widehat{\Delta} = \Delta \times \mathbb{C}$ and the natural map $\hat{f} : \widehat{\Delta} \to \widehat{\Delta}$, defined by $\hat{f}(\lambda, z) = (\lambda, f_\lambda(z))$. Consider any graph $\Gamma_\gamma = \{ z = \gamma(\lambda) \}$ over the first coordinate such that $\gamma(\lambda)$ escapes under iteration by f_λ for $\lambda \in \Delta$ (for instance $\gamma(\lambda) = -c(\lambda)$). Then the sequence of currents $\frac{1}{3^n}(\hat{f}^*)^n[\Gamma]$ converges to a current $\widehat{T}$ such that $T^+|_\Delta = (\pi_1)_*(\widehat{T}|_{\Gamma_{+c}})$. Here $\pi_1 : \widehat{\Delta} \to \Delta$ is the natural projection.

Now there is no choice involved in the labelling of U_1 and U_2, so since $G^+ < G^-$ on Δ, the decomposition $f_\lambda^{-1}U = U_1 \cup U_2$ can be followed continuously throughout Δ. So using self explanatory notation, we get, for $i = 1, 2$, a map $\hat{f}_i : \widehat{U}_i \to \widehat{U}$. This gives rise to a coding of $\hat{f}$ orbits, in the same way as before. We thus get a decomposition of $\frac{1}{3^n}(\hat{f}^*)^n[\Gamma]$ as

$$\frac{1}{3^n}(\hat{f}^*)^n[\Gamma] = \frac{1}{3^n} \sum_{(\alpha_0,\dots,\alpha_{n-1}) \in \{1,2\}^n} \hat{f}^*_{\alpha_0} \cdots \hat{f}^*_{\alpha_{n-1}}[\Gamma].$$

The convergence theorem of [DDS05] implies that for any $\alpha \in \Sigma$, the sequence of currents

$$\widehat{T}_{\alpha,n} = \frac{1}{d_{\alpha_0} \cdots d_{\alpha_{n-1}}} \hat{f}^*_{\alpha_0} \cdots \hat{f}^*_{\alpha_{n-1}}[\Gamma]$$

converges to a current $\widehat{T}_\alpha$ in $\Delta \times \mathbb{C}$. Indeed, for every $\lambda \in \Delta$, the sequence of measures $\frac{1}{d_{\alpha_0} \cdots d_{\alpha_{n-1}}}(f_\lambda)^*_{\alpha_0} \cdots (f_\lambda)^*_{\alpha_{n-1}} \delta_{\gamma(\lambda)}$ converges, by [DDS05]. On the other hand, the currents $\widehat{T}_{\alpha,n}$ are contained in a fixed vertically compact subset of $\Delta \times \mathbb{C}$, and all cluster values of this sequence have the same slice measures on the vertical slices $\{\lambda\} \times \mathbb{C}$. Currents with horizontal support being determined by their vertical slices, we conclude that the sequence converges (see [DDS05, Section 2] for basics on horizontal currents).

At the limit, we again get a decomposition of $\widehat{T}$ as an integral of positive closed currents, $\widehat{T} = \int_\Sigma \widehat{T}_\alpha d\nu(\alpha)$, where ν is the $(\frac{1}{3}, \frac{2}{3})$ measure on Σ. Furthermore, for every α, the wedge product $\widehat{T}_\alpha \wedge [\Gamma_c]$ is well defined. This follows for instance from a classical transversality argument: recall that on $\partial\Delta$, c escapes, so $\operatorname{Supp}\widehat{T}_\alpha \cap \Gamma_c \Subset \Gamma_c$. In particular, any potential $\widehat{G}_\alpha$ of $\widehat{T}_\alpha$ is locally integrable[2] on Γ_c. Hence by taking the wedge product with the graph Γ_c and projecting down to Δ, we get a decomposition

$$T^+|_\Delta = (\pi_1)_*(\widehat{T} \wedge [\Gamma_c]) = \int_\Sigma (\pi_1)_*(\widehat{T}_\alpha \wedge [\Gamma_c]) d\nu(\alpha) = \int_\Sigma T^+_\alpha|_\Delta d\nu(\alpha). \quad (3.7)$$

The second equality is justified by Lemma 2.8.

Here $T^+_\alpha|_\Delta$ should be understood as the contribution of the combinatorics of α to the measure $T^+|_\Delta$ in parameter space; it is supported on the

---

[2] The results in [DDS05] imply that $\widehat{T}_\alpha$ has a continuous potential for ν-a.e. α.

set of parameters for which $c \in K_\alpha$. Notice also that, since by definition $c \in U_2$, for $T_\alpha^+|_\Delta$ to be non-zero, it is necessary that α starts with the symbol 2.

Furthermore, since there are no choices involved in the labelling of the U_i, the decompositions $K = \bigcup K_\alpha$ and $\mu = \int \mu_\alpha d\nu(\alpha)$ are invariant under wringing. Hence the decomposition $T^+ = \int T_\alpha^+ d\nu(\alpha)$, which was defined locally, makes sense as a global decomposition of T^+ in $\mathbb{C}^2 \setminus \mathcal{C}$.

We summarize this discussion in the following proposition.

Proposition 3.10. *There exists a decomposition of T^+ as an integral of positive closed currents in $\mathbb{C}^2 \setminus \mathcal{C}$,*

$$T^+ = \int_\Sigma T_\alpha^+ d\nu(\alpha),$$

where $\Sigma = \{1,2\}^\mathbb{N}$ and ν is the $(\frac{1}{3}, \frac{2}{3})$-measure on Σ. Moreover T_α^+ is supported on the set of parameters for which c has itinerary α with respect to the natural decomposition $\{G_f(z) < G^-\} = U_1 \cup U_2$, with $c \in U_2$.

We now conclude the proof of Theorem 3.9. We combine the previous discussion with the results of [BH92]. In dynamical space, if a component of the Julia set is not a point, then it is preperiodic. Hence its associated sequence $\alpha \in \Sigma$ is preperiodic. Since the measure ν does not charge points, the set of preperiodic symbolic sequences has measure zero. Consequently, by the decomposition $\mu = \int \mu_\alpha d\nu(\alpha)$, the set of preperiodic components has zero μ-measure.

In parameter space, if the connected component of λ in $\mathcal{C}^+ \cap \Delta$ is not a point, then $C(+c)$ is periodic. Hence the itinerary sequence of c is periodic, and this once again corresponds to a set of sequences of zero ν-measure. In view of the decomposition $T^+|_\Delta = \int T_\alpha^+|_\Delta d\nu(\alpha)$, we conclude that the set of such parameters has zero $T^+|_\Delta$-measure. $\square$

We say that a current is *extremal* if it is an extremal point in the cone of closed positive current. The extremality properties of invariant currents play an important role in higher dimensional holomorphic dynamics. It follows from Proposition 3.8 that the current T^+ is not extremal in a small neighborhood of the line at infinity. Notice that the two branches of T^+ at $[1 : 1 : 0]$ (respectively $[1 : -2 : 0]$) correspond to sequences starting with the symbol 21 (respectively 22). More generally, we have the following corollary, which fills in a line of Table 1.

Corollary 3.11. *The current T^+ is not extremal in any open subset of the escape locus $\mathbb{C}^2 \setminus \mathcal{C}$.*

Proof: Since the measure ν on Σ does not charge points, and the currents T_α^+ have disjoint supports, the decomposition of Proposition 3.10 is non-trivial in any open subset of $\mathbb{C}^2 \setminus \mathcal{C}$. $\square$

4 Laminarity at the Boundary of the Connectedness Locus

We start by a general laminarity statement in $\text{Int}(\mathcal{C}^-)$ which, together with Proposition 3.4, completes the proof of the laminarity of T^+ outside $\partial\mathcal{C}^-$. Notice that since $\text{Supp}(T^+) = \partial\mathcal{C}^+$, this is really a statement about the boundary of the connectedness locus.

Theorem 4.1. *The current T^+ is laminar in $\text{Int}(\mathcal{C}^-)$.*

Proof: Consider a parameter $f_0 \in \text{Int}(\mathcal{C}^-) \cap \text{Supp}(T^+) = \text{Int}(\mathcal{C}^-) \cap \partial\mathcal{C}^+ \subset \partial\mathcal{C}$. The critical point $+c$ is active at f_0, whereas $-c$ is passive. Let $U \ni f_0$ be a small ball for which $-c$ remains passive. There are nearby parameters for which $+c$ escapes, and we find that parameters in $U \setminus \mathcal{C}$ have a stable quadratic renormalization (see Remark 4).

Because T^+ has a continuous potential $G^+ \geq 0$, we can write

$$T^+ = dd^c G^+ = \lim_{\varepsilon \to 0} dd^c \max(G^+, \varepsilon) = \lim_{\varepsilon \to 0} T_\varepsilon^+.$$

Moreover G^+ is pluriharmonic where it is positive, so $dd^c \max(G^+, \varepsilon)$ has laminar structure. We will study the topology of the leaves in U and conclude by using a result of De Thélin [DT04].

In the open set U, $-c$ never escapes, so the Böttcher function $\varphi^+(f) = \varphi_f(-2c)$ is well defined, and $G^+ = \log|\varphi^+|$. The real hypersurface $\{G^+ = \varepsilon\}$ is foliated by the holomorphic curves $\{\varphi^+ = \exp(\varepsilon + i\theta)\}$, and it is well known (see, e.g., [De93]) that $dd^c \max(G^+, \varepsilon)$ is uniformly laminar and subordinate to this foliation — the fact that both $\{G^+ = \varepsilon\}$ and the leaves are non-singular will be a consequence of the wringing argument below.

The point is that in U the leaves of this foliation are closed and biholomorphic to the disk. Indeed, recall that the wring deformation $f \mapsto w(f, u)$ is a homeomorphism in $\mathbb{C}^2 \setminus \mathcal{C}$ for fixed u: it can be inverted by using the group action. In particular, $w(\cdot, s)$ maps $U \cap \{G^+ = \varepsilon\}$ homeomorphically into a neighborhood of infinity in $\text{Int}(\mathcal{C}^-)$, where the foliation $\varphi^+ = \text{cst}$ is well understood. By using Proposition 3.7 i, with $+$ and $-$ swapped, we get that the leaves are planar Riemann surfaces. We conclude that the leaves are disks. Indeed if $L = \{\varphi^+ = \exp(\varepsilon + i\theta)\}$ is such a leaf in U, $w(\cdot, s)(L)$ is an open subset in a vertical disk Δ (see Proposition 3.7 i), and hence $w(\cdot, s)^{-1}(\Delta)$ is a disk traced on $\{\varphi^+ = \exp(\varepsilon + i\theta)\}$ containing L. Thus its intersection with the ball B is a disk, by the maximum principle.

De Thélin's Theorem [DT04] asserts that if Δ_n is a sequence of (possibly disconnected) submanifolds of zero genus in the unit ball, then any cluster value of the sequence of currents $[\Delta_n]/\text{Area}(\Delta_n)$ is a laminar current. Pick a sequence $\varepsilon_n \to 0$. The approximating currents $T_{\varepsilon_n}^+$ are *integrals* of families of holomorphic disks. Thus, for every n, we first approximate $T_{\varepsilon_n}^+$ by a finite combination of leaves and denote the resulting approximating sequence of currents by $(S_{n,j})_{j \geq 0}$. For every j, $S_{n,j}$ is a normalized finite sum of currents of integration along disks. Hence, by choosing an appropriate subsequence, we can ensure that $S_{n,j(n)} \to T^+$, and applying De Thélin's result finishes the proof. □

Using quasiconformal deformations, we obtain a much more precise result in hyperbolic (and hence conjecturally all) components. Notice that, due to Remark 4, the quadratic hyperbolicity conjecture is enough to ensure the uniform laminarity of T^+ outside $\partial \mathcal{C}^+ \cap \partial \mathcal{C}^-$.

Proposition 4.2. *The current T^+ is uniformly laminar in hyperbolic components of* $\mathrm{Int}(\mathcal{C}^-)$.

Proof: The proof will use a very standard quasiconformal surgical argument (cf. e.g., [CG93, VIII.2]). The major step is to prove that the foliation by disks of the form $\{\varphi^+ = \varepsilon\}$ considered in the previous proposition does not degenerate in a neighborhood of $\partial \mathcal{C}^+$.

Before going into the details, we sketch the argument. Assume $f_0 \in \partial \mathcal{C}^+ \cap \Omega$, where Ω is a hyperbolic component of $\mathrm{Int}(\mathcal{C}^-)$. The multiplier of the attracting cycle defines a natural holomorphic function on Ω. Using quasiconformal surgery, we construct a transverse section of this function through f_0, which is a limit of sections of the form $\{\varphi^+ = \varepsilon\}$. Using Lemma 3.3, we conclude that the foliation $\{\varphi^+ = \varepsilon\}$ extends as a lamination to $\partial \mathcal{C}^+$. Then, approximating T^+ by T_ε^+ as before implies that T^+ is uniformly laminar.

Let Ω be a hyperbolic component of $\mathrm{Int}(\mathcal{C}^-)$, and let $f_0 \in \Omega \cap \overline{\mathcal{E}}$. The critical point $-c$ is attracted by an attracting periodic point of period m that persists through Ω. Let ρ be the multiplier of the cycle; ρ is a holomorphic function on Ω. Let $U_0, \ldots, U_m = U_0$ be the cycle of Fatou components containing the attracting cycle. Since f_0 is in the escape locus or at its boundary, $+c$ is not attracted by the attracting cycle, so its orbit does not enter $U_0, \ldots, U_m$.

The map $f_0^m : U_0 \to U_0$ is conjugate to a Blaschke product of degree 2: there exists a conformal map $\varphi : U_0 \to \mathbb{D}$ such that

$$\varphi \circ f_0^m \circ \varphi^{-1} = z \frac{z + \lambda_0}{1 + \overline{\lambda_0} z} = B_{\lambda_0}(z),$$

where $\lambda_0 = \rho(f_0)$ is the multiplier of the cycle. For simplicity, we assume that $m = 1$; see [CG93] for an adaptation to the general case and for more details.

Choose a small ε and then an r so that $|\lambda_0| < 1 - \varepsilon < r < 1$ and for every $\lambda \in D(0, 1 - \varepsilon)$, we have

$$\overline{D(0, r)} \Subset B_\lambda^{-1}(\overline{D(0, r)}).$$

For $\lambda \in D(0, 1 - \varepsilon)$, choose a smoothly varying family of diffeomorphisms

$$\psi_\lambda : f^{-1}\varphi^{-1}(\overline{D(0,r)}) \longrightarrow B_\lambda^{-1}(\overline{D(0,r)})$$

with $\psi_{\lambda_0} = \varphi$ and such that

- $B_\lambda \circ \psi_\lambda = \varphi \circ f$ on $\partial f^{-1}\varphi^{-1}(\overline{D(0,r)})$,
- $\psi_\lambda = \varphi$ in $\varphi^{-1}(\overline{D(0,r)})$.

We can thus define a map $g_\lambda : \mathbb{C} \to \mathbb{C}$ by

$$\begin{cases} g_\lambda = \varphi^{-1} \circ B_\lambda \circ \psi_\lambda \text{ in } f^{-1}\varphi^{-1}(\overline{D(0,r)}) \\ g_\lambda = f \text{ outside } f^{-1}\varphi^{-1}(\overline{D(0,r)}). \end{cases}$$

By the definition of ψ_λ, the two definitions agree on $\partial f^{-1}\varphi^{-1}(\overline{D(0,r)})$.

Now define an almost complex structure σ_λ by setting $\sigma_\lambda = \psi_\lambda^* \sigma_0$ on $f^{-1}\varphi^{-1}(\overline{D(0,r)})$, extending it by invariance under g_λ, and setting $\sigma_\lambda = \sigma_0$ outside the grand orbit (under f_0) of the attracting Fatou component. By construction, any orbit under g_λ hits the region where g_λ is not holomorphic at most once. Thus we can straighten σ_λ, and we find that g_λ is quasiconformally conjugate on $\mathbb{C}$ to a cubic polynomial f_λ, which varies continuously with $\lambda \in D(0, 1 - \varepsilon)$. By construction, $f_{\lambda_0} = f_0$ and $\rho(f_\lambda) = \lambda$ (because the conjugacy is holomorphic in a neighborhood of the attracting fixed point).

The family of maps f_λ that we have constructed defines a continuous section of ρ through f_0. Now, if $f_0 \notin \mathcal{C}$, $\varphi^+(f_\lambda)$ is constant because f_0 and f_λ are holomorphically conjugate outside the Julia set. So f_λ is forced to stay in the one dimensional local leaf $L = \{\varphi^+ = \varphi^+(f_0)\}$. Since ρ is holomorphic on L and f_λ is a section of ρ, we find that f_λ depends holomorphically on λ, and that ρ is a local biholomorphism on L.

Now assume f_0 is at the boundary of the escape locus (that is, at the boundary of $\mathcal{C}^+$). Let (f_n) be a sequence of maps in $\mathcal{E}$ converging to f_0, say with $\rho(f_n) = \lambda_0$. There is a holomorphic disk $\Delta_n = \bigcup_{\lambda \in D(0, 1 - \varepsilon)} f_n(\lambda)$ attached to each f_n, with $\rho(f_n(\lambda)) = \lambda$. Notice that any two such disks are equal or disjoint. Let Δ_0 be any cluster value of the sequence of disks (Δ_n), in the topology of uniform convergence on compact subsets. This disk defines a section of ρ through f_0, so, in particular, we deduce that ρ is a submersion at f_0.

We claim that Δ_0 does not depend on the sequence (Δ_n). Indeed, ρ is a submersion at f_0, and hence a fibration in a neighborhood of f_0, so the conclusion follows from Hurwitz' Theorem.

By construction, this holomorphic disk is contained in $\partial \mathcal{C}^+$, because G^+ is constant along Δ_n and tends to zero as $n \to \infty$. Thus both critical points are passive on Δ_0, and the dynamics along Δ_0 is J-stable.

In conclusion, we have constructed a family of holomorphic sections of ρ in $\Omega \cap (\mathcal{E} \cup \partial\mathcal{C}^+)$. In $\mathcal{E}$, the sections are compatible in the sense of Lemma 3.3, because they are subordinate to the foliation by $\{\varphi^+ = \text{cst}\}$. As seen before, since the sections of $\partial\mathcal{C}^+$ are obtained by taking limits, they are also compatible by Hurwitz' Theorem. Hence Lemma 3.3 applies and tells us that these disks laminate $\Omega \cap (\mathcal{E} \cup \partial\mathcal{C}^+)$. Hence we have proved that the natural foliation of $\Omega \cap \mathcal{E}$ admits a continuation, as a lamination, to the boundary of $\mathcal{E}$.

It is now clear that T^+ is uniformly laminar in Ω, because T^+ can be written as $T^+ = \lim T_\varepsilon^+ = \lim dd^c \max(G^+, \varepsilon)$, and, as we have seen in the proof of the previous theorem, T_ε^+ is subordinate to the natural foliation of $\Omega \cap \mathcal{E}$. $\qquad\square$

5   Rigidity at the Boundary of the Connectedness Locus

5.1   Higher-Order Bifurcations and the Bifurcation Measure

So far, we have seen that many parameters in the bifurcation locus admit a one parameter family of deformations. Due to laminarity, this holds for T_{bif}-a.e. parameters outside $\partial\mathcal{C}^+ \cap \partial\mathcal{C}^-$. In this section, we will concentrate on rigid parameters, that is, on cubic polynomials with no deformations.

Let μ_{bif} be the positive measure defined by $\mu_{\text{bif}} = T^+ \wedge T^-$. An easy calculation shows that $T_{\text{bif}}^2 = (T^+ + T^-)^2 = 2\mu_{\text{bif}}$. One major topic in [DF08] was the study of the properties of μ_{bif}, which is in many respects the right analogue for cubics of the harmonic measure of the Mandelbrot set. In particular, it was proved that $\text{Supp}(\mu_{\text{bif}})$ is the closure of Misiurewicz points and a proper subset of $\partial\mathcal{C}^+ \cap \partial\mathcal{C}^-$.

We define the *second bifurcation locus* Bif_2 to be the closure of the set of rigid parameters, that is, parameters that do not admit deformations[3]. By the density of Misiurewicz points, it is clear that $\text{Supp}(\mu_{\text{bif}}) \subset \text{Bif}_2$. In the next proposition, we get a considerably stronger statement. Notice that the result is valid for polynomials of all degrees.

Proposition 5.1. *The set of parameters f for which there exists a holomorphic disk with $f \in \Delta \subset \mathcal{C}$ has zero μ_{bif}-measure.*

In view of Proposition 2.6, this can be understood as a generic rigidity result.

Corollary 5.2. *A μ_{bif} generic polynomial admits no deformations.*

---

[3]It is unclear whether taking the closure is necessary here, i.e., whether the set of rigid polynomials is closed.

Proof of Proposition 5.1: Recall from [DF08, Section 6] that $\mu_{\text{bif}} = (dd^cG)^2$, where $G = \max(G^+, G^-)$. Now if Δ is a holomorphic disk contained in $\mathcal{C}$, $G = 0$ on Δ. It is known that if E is a subset of $\text{Supp}(dd^cG)^2$ such that, through every point in E, there is a holomorphic disk on which G is harmonic, then E has zero $(dd^cG)^2$ measure (see [Si99, Corollary A.10.3]).

Another argument goes as follows: by [DF08, Theorem 10], a μ_{bif}-a.e. polynomial f satisfies the Topological Collet Eckmann (TCE) property. Also, both critical points are on the Julia set, so the only possible deformations come from invariant line fields [MS98]. On the other hand, the Julia set of a TCE polynomial has Hausdorff dimension strictly less than 2 [PR98], so it has no invariant line fields. □

The next result gives some rough information on Bif_2. We do not know whether the first inclusion is proper.

Proposition 5.3. *The second bifurcation locus satisfies*

$$\text{Supp}(\mu_{\text{bif}}) \subset \text{Bif}_2 \subsetneq (\partial\mathcal{C}^+ \cap \partial\mathcal{C}^-) \cup \mathcal{N} \subsetneq \partial\mathcal{C},$$

where the exceptional set $\mathcal{N}$ is empty if the hyperbolicity conjecture holds.

Proof: The relationship of Bif_2 with active and passive critical points is as follows:

- If one critical point is passive while the other is active, it is expected that the passive critical point will give rise to a modulus of deformation. This is the case when it is attracted by a (super)attracting cycle. Otherwise, we do not know how to prove it. Nevertheless, it is clear that there are nearby parameters with deformations, and moreover that these will be generic in the measure theoretic sense (see Theorem 4.1). The existence of such parameters contradicts the hyperbolicity conjecture anyway.

- When the two critical points are active, there may exist deformations. The list of possibilities is as follows (see [MS98]):

  1. There is a parabolic cycle attracting both critical points, and their grand orbits differ.
  2. There is a Siegel disk containing a postcritical point.
  3. There is an invariant line field — of course it is conjectured that this is not the case.

In [DF08], we gave an example (due to Douady) of a cubic polynomial $f_0 \in (\partial\mathcal{C}^+ \cap \partial\mathcal{C}^-) \setminus \text{Supp}(\mu_{\text{bif}})$. This map has the property of having a parabolic fixed point attracting both critical points, as in Case 1 above.

Furthermore, every nearby parameter either has a parabolic point, and is in fact conjugate to f_0, or has an attracting point. In particular, there is, locally, a disk of such parameters, conjugate to f_0. It is then clear that such a parameter is not approximated by rigid ones, so $f_0 \notin \mathrm{Bif}_2$.   □

We have no precise information on the location of parameters in Case 2 above. Here is a specific question: let $\mathrm{Per}_1(\theta)$ be the subvariety of parameters with a fixed point of multiplier $e^{2i\pi\theta}$. Assume that θ satisfies a Diophantine condition, so that every $f \in \mathrm{Per}_1(\theta)$ has a Siegel disk. For some parameters (see Zakeri [Za99]), it happens that one of the critical points falls into the Siegel disk after finitely many iterations (the other one is necessarily in the Julia set). In parameter space, this gives rise to disks contained in $\partial\mathcal{C}^+ \cap \partial\mathcal{C}^-$. Are these disks contained in $\mathrm{Supp}(\mu_{\mathrm{bif}})$, or in Bif_2?

5.2   Woven Currents

Just as foliations can be generalized to webs, there is a class of woven currents that extends laminar currents. They were introduced by T.C. Dinh [Di05].

Definition 5.4. Let $\Omega \subset \mathbb{C}^2$ be open. A web in Ω is any family of analytic subsets of Ω whose volume is bounded by some constant c.

A positive closed current T in Ω is uniformly woven if and only if there exists a constant c and a web $\mathcal{W}$ as above, whose leaves are of volume bounded by c, endowed with a positive measure ν such that

$$T = \int_{\mathcal{W}} [V]d\nu(V).$$

Notice that, by Bishop's Theorem, any family of subvarieties with uniformly bounded volume is relatively compact in the Hausdorff topology on compact subsets of Ω. By analogy with the laminar case, we also define general woven currents.

Definition 5.5. A current T in Ω is said to be woven if there exists a sequence of open subsets $\Omega_i \subset \Omega$ such that T gives zero mass to $\partial\Omega_i$, and an increasing sequence of currents (T_i), converging to T, such that T_i is uniformly woven in Ω_i.

5.3   Non-geometric Structure of μ_{bif}

In this section, we investigate the structure of the bifurcation measure μ_{bif} and prove that it is not the "geometric intersection" of the laminar currents

$T^{\pm}$. As a consequence, we will derive an asymptotic estimate on the genera of the Per(n, k) curves.

It is natural to wonder whether generic wringing curves in $\partial \mathcal{C}^{\pm}$ admit continuations across $\partial \mathcal{C}$ (compare with the work of Willumsen [Wi97]). Many wringing curves are subordinate to algebraic subsets, so the continuation indeed exists: this is the case for the Per$^{\pm}(n, k)$ curves as well as the subsets in parameter space defined by the condition that a periodic cycle of given length has a given indifferent multiplier. Similarly, we may wonder whether the disks constituting the laminar structure of $T^{\pm}$ on $\partial \mathcal{C} \setminus (\partial \mathcal{C}^{+} \cap \partial \mathcal{C}^{-})$ admit continuations across $\partial \mathcal{C}^{+} \cap \partial \mathcal{C}^{-}$.

We will look at the continuation property from the point of view of the structure of the bifurcation measure. If E is a closed subset in $\mathrm{Supp}(\mu)$, we say that μ_{bif} has *local product structure on* E if there exist uniformly laminar currents $S^{\pm} \leq T^{\pm}$ defined in a neighborhood of E such that $\mu_{\mathrm{bif}}|_E = S^{+} \wedge S^{-}$. Abusing conventions, we will extend this definition to the case where S^{+} and S^{-} are merely uniformly woven currents and $0 < S^{+} \wedge S^{-} \leq \mu_{\mathrm{bif}}|_E$.

This terminology is justified by the fact that the wedge product of uniformly laminar currents coincides with the natural geometric intersection. More specifically, if S^{+} and S^{-} are uniformly laminar currents in Ω, written as integrals of disks of the form

$$S^{\pm} = \int [\Delta^{\pm}] d\mu^{\pm},$$

then if the wedge product $S^{+} \wedge S^{-}$ is well defined,

$$S^{+} \wedge S^{-} = \int [\Delta^{+} \cap \Delta^{-}] d\mu^{+} d\mu^{-},$$

where $[\Delta^{+} \cap \Delta^{-}]$ is the sum of point masses at isolated intersections of Δ^{+} and Δ^{-}. We refer the reader to [BLS93, Du04] for more details on these notions. Analogous results hold for uniformly woven currents.

The next theorem shows that the structure of the bifurcation measure is somewhat more complicated. Loosely speaking, this means that, in the measure theoretic sense, neither wringing curves nor the disks of Section 4 admit continuations across $\mathrm{Supp}(\mu_{\mathrm{bif}})$.

Theorem 5.6. *The measure μ_{bif} does not have local product structure on any set of positive measure.*

Proof: Assume μ_{bif} has local product structure on a set E of positive measure. Let $S^{\pm} = \int [\Delta^{\pm}] d\mu^{\pm}$ be uniformly woven currents as in the previous discussion, with $0 < S^{+} \wedge S^{-} \leq \mu_{\mathrm{bif}}|_E$. It is a classical result (see, e.g., [BLS93, Lemma 8.2]) that the currents $S^{\pm}$ have continuous potentials.

Here the $\Delta^{\pm}$ are possibly singular curves of bounded area. By Lemma 2.8, $S^+ \wedge S^-$ admits a decomposition as $\int [\Delta^+] \wedge S^- d\mu^+$. Since each Δ^+ has finitely many singular points, the induced measure $[\Delta^+] \wedge S^-$ gives full mass to the regular part of Δ^+, so we find that the variety Δ^+ is smooth around $S^+ \wedge S^-$-a.e. point.

In particular, by Proposition 5.1 above, and by discarding a set of varieties of zero μ^+ measure if necessary, we may assume that for every Δ^+, $\Delta^+ \setminus \mathcal{C} \neq \emptyset$. Outside $\mathcal{C}$, T^+ is uniformly laminar and subordinate to the lamination by wringing curves, so S^+ itself is subordinate to this lamination and the transverse measure of S^+ is dominated by that of T^+ (see [BLS93, Section 6], [Du06]).

From this discussion and Theorem 3.9, we conclude that for μ^+-almost every Δ^+, $\Delta^+ \setminus \mathcal{C}$ is contained in at most countably many wringing curves and the polynomials in $\Delta^+ \setminus \mathcal{C}$ have Cantor Julia sets. Removing a set of zero μ^+-measure once again, we may assume that this property holds for all Δ^+. Lastly, we will similarly assume that Δ^+ is not contained in a $\mathrm{Per}^+(n, k)$ curve.

By [DF08, Corollary 11], there exists a set of parameters of full μ_{bif}-measure $E' \subset E$ for which the orbits of both critical points are dense in the Julia set (this is a consequence of the TCE property). By Fubini's Theorem, we may further assume that every Δ^+ intersects E'.

Now since any variety Δ^+ is contained in $\mathcal{C}^+$, the point $+c$ is passive on Δ^+. By [DF08, Theorem 4], if there exists a parameter $f \in \Delta^+$ for which $+c$ is preperiodic, then

- either it is persistently preperiodic,

- or it is persistently attracted by an attracting periodic point or lies in a persistent Siegel disk.

These two properties contradict the genericity assumptions made on Δ^+. Thus we infer that $+c$ is never preperiodic on Δ^+, in which case its orbit can be followed by a holomorphic motion. But on E', the orbit of $+c$ is dense on the Julia set, which is connected, whereas on $\Delta \setminus \mathcal{C}$ it is contained in a Cantor set. We have reached a contradiction. □

Not much seems to be known about the geometry of the $\mathrm{Per}^{\pm}(n)$ and $\mathrm{Per}^{\pm}(n, k)$ curves. It has been proved by Milnor [Mi09] that the $\mathrm{Per}^{\pm}(n)$ curves are smooth in $\mathbb{C}^2$. For every $0 \leq k < n$, $\mathrm{Per}(n, k)$ is an algebraic curve of degree 3^n. The geometry of such curves is unknown. In particular, these curves might be very singular (this already happens for the $\mathrm{Per}^{\pm}(n)$ at infinity) and have many irreducible components, so their genus cannot be directly estimated.

As a consequence of Theorem 5.6 and previous work of ours [Du03, Du04], we get a rough asymptotic estimate on the genus of these curves.

Recall that the geometric genus of a compact singular Riemann surface is the genus of its desingularization. In the next theorem, "genus" means "geometric genus" and, as usual, Per^+ may be replaced by Per^-.

Theorem 5.7. *Let $k(n)$ be any sequence satisfying $0 \leq k(n) < n$, and write the decomposition of $\mathrm{Per}^+(k(n), n)$ into irreducible components as $\mathrm{Per}^+(k(n), n) = \sum m_{i,n} C_{i,n}$. Then*

$$\frac{1}{3^n} \sum_i m_{i,n} \cdot \mathrm{genus}(C_{i,n}) \to \infty. \tag{5.8}$$

Consequently, $\max_i \mathrm{genus}(C_{i,n}) \to \infty$.

Before embarking on the proof, let us give a more precise statement in the case $k(n) = 0$. The curve $\mathrm{Per}^+(n)$ is reducible because $\mathrm{Per}^+(n) = \bigcup_{d|n} \mathrm{Per}^+(d)$. Following Milnor's notation [Mi09, Section 5], we denote by $\mathcal{S}_n$ the curve of polynomials such that $+c$ has period exactly n.

Corollary 5.8. *Write the decomposition of $\mathcal{S}_n$ into irreducible components as $\mathcal{S}_n = \sum m_{i,n} D_{i,n}$. Then $m_{i,n} = 1$, and*

$$\frac{1}{3^n} \mathrm{genus}(\mathcal{S}_n) = \frac{1}{3^n} \sum_i \mathrm{genus}(D_{i,n}) \to \infty. \tag{5.9}$$

Notice that it is conjectured in [Mi09] that for all n, $\mathcal{S}_n$ is connected, and hence irreducible because $\mathcal{S}_n$ is smooth. Of course the estimate (5.9) becomes more transparent in that case.

Proof (Sketch): We first prove that the multiplicities $m_{i,n}$ equal 1. To achieve this, consider a point f_0 of intersection between $\mathcal{S}_n$ and a disk of the form $\Delta = \{\varphi^- = k\} \cap B$ (we use notation as in Section 3.3). This point will be the center of a hyperbolic component of a copy of the Mandelbrot set. Furthermore, f_0 is embedded in a *Mandelbrot-like family* along Δ, that is, there is a one to one correspondence between cubic polynomials $f \in \Delta$ near f_0 and their quadratic renormalizations (see [BH92] for details). On the other hand, it is known (see [DH85a]) that the centers of hyperbolic components of the Mandelbrot set are simple roots of the equation $g_c^k(0) = 0$ (where $g_c(z) = z^2 + c$). From this we conclude that the multiplicity of f_0 as a solution of $f_{c,v}^n(c) = c$ along Δ is 1, and hence the multiplicity of the component of $\mathrm{Per}^+(n)$ containing f_0 is 1.

To conclude the proof, it remains to control the contribution of the curves $\mathrm{Per}^+(d)$ to equation (5.8), for d strictly dividing n. Of course if $d|n$ and $d < n$, then $d \leq n/2$. Decompose $\mathrm{Per}^+(d)$ into its irreducible components $\sum C_i$. We have that $\sum \deg(C_i) \leq \sum_{i \leq d} 3^i = O(3^d)$. Now recall that an algebraic curve of degree d has genus $\leq (d-1)(d-2)/2$, and hence

$$\sum \mathrm{genus}(C_i) \leq \frac{1}{2} \sum \deg(C_i)^2 \leq \frac{1}{2} \left(\sum \deg(C_i) \right)^2 = O(3^{2d}) = O(3^n),$$

which finishes the proof.   □

Proof of Theorem 5.7: In [Du03], we proved that if a sequence of algebraic curves C_n in $\mathbb{P}^2$ satisfies certain geometric estimates, then the cluster values of the sequence of currents $[C_n]/\deg(C_n)$ are laminar. In [Du04], we proved that if T_1 and T_2 are two such laminar currents with continuous potentials, then the wedge product measure $T_1 \wedge T_2$ has local product structure. By Theorem 5.6, we know that μ_{bif} does not have product structure on any set of positive measure. Therefore, the geometric estimates of [Du03] are not satisfied. By inspecting the geometry of the $\mathrm{Per}^+(n, k(n))$ curves, we will see that this leads to (5.8). A delicate issue is that [Du03] requires an assumption on the singularities which may not be satisfied when $k(n) \neq 0$; therefore, we need to generalize it slightly. This is where woven currents are needed.

We start with the simpler case, where $k(n) = 0$. By Milnor's result, the curves $\mathrm{Per}^+(n)$ are smooth in $\mathbb{C}^2$, so in $\mathbb{P}^2$ they can have singular points only at $[1 : 1 : 0]$ and $[1 : -2 : 0]$. Let $\mathrm{Per}^+(n) = \sum m_{i,n} C_{i,n}$ be the decomposition into irreducible components. We have to count the number of "bad components" when projecting $C_{i,n}$ from a generic point in $\mathbb{P}^2$. They are defined as follows: let p be a generic point in $\mathbb{P}^2$, and consider the central projection $\pi_p : \mathbb{P}^2 \setminus \{p\} \to \mathbb{P}^1$. For a subdivision $\mathcal{Q}$ of $\mathbb{P}^1$ by squares, and for $Q \in \mathcal{Q}$, we say that a *connected* component of $C_{i,n} \cap \pi_p^{-1}(Q)$ is *good* if it is a graph over Q, and *bad* otherwise.

We remark that the curves $C_{i,n}$ do not intersect in $\mathbb{C}^2$, so when considering the union of the $C_{i,n}$, no additional bad components arise from intersecting branches of different irreducible components of $\mathrm{Per}^+(n)$. Also, the number of local irreducible components at every singular point of $C_{i,n}$ is bounded by $\deg(C_{i,n})$, since a nearby line cannot have more than $\deg(C_{i,n})$ intersection points with $C_{i,n}$. By [Du03, Proposition 3.3], the number of bad components for each $C_{i,n}$ is thus bounded by $4\mathrm{genus}(C_{i,n}) + 6\deg(C_{i,n})$. By summing this estimate for all irreducible components, we find that the number (with multiplicity) of bad components is bounded by $4\sum_i m_{i,n} \cdot \mathrm{genus}(C_{i,n}) + O(3^n)$, and by [Du03, Proposition 3.4], if (5.8) does not hold for a certain subsequence $n_i \to \infty$, we find that T^+ is laminar everywhere, and strongly approximable in the sense of [Du04, Def. 4.1]. The same holds for T^-, since $+c$ and $-c$ are exchanged by a birational involution. In particular, by [Du04], $T^+ \wedge T^-$ must have local product structure, which contradicts Theorem 5.6.

The proof in the general case involves a modification of [Du03, Du04], which was already considered in [DT06] (see also [Di05]). The required modification is contained in the following proposition. We define the genus of a reducible Riemann surface as the sum of the genera of its components.

Proposition 5.9. *Let C_n be a sequence of (singular) algebraic curves in $\mathbb{P}^2$, with $\deg(C_n) \to \infty$. Assume that $\mathrm{genus}(C_n) = O(\deg(C_n))$. Then the cluster values of the sequence $[C_n]/\deg(C_n)$ are woven currents.*

Proof: This is just a careful inspection of [Du03]. As before, we fix a generic point in $\mathbb{P}^2$, and consider the central projection $\pi_p : \mathbb{P}^2 \setminus \{p\} \to \mathbb{P}^1$. For a subdivision $\mathcal{Q}$ of $\mathbb{P}^1$ by squares, and for $Q \in \mathcal{Q}$, we look for *good components*, that is, *irreducible* components of $C_n \cap \pi_p^{-1}(Q)$ which are graphs over Q. The difference with the laminar case is that good components are now allowed to intersect (hence the word "irreducible" rather than "connected").

For each irreducible component $C_{i,n}$ of C_n, the number of bad components is controlled by using the Riemann-Hurwitz formula for the natural map $\widehat{C}_{i,n} \to \mathbb{P}^1$, where $\widehat{C}_{i,n}$ is the normalization of $C_{i,n}$. A straightforward adaptation of [Du03, Proposition 3.3] shows that the number of bad components is bounded by $O(\mathrm{genus}(C_{i,n})) + O(\deg(C_{i,n}))$. By summing for all $C_{i,n}$ and noting that any union of good components is good, we find that, under the assumption of the lemma, the total number of bad components is $O(\deg(C_n))$.

We can now proceed with the proof of [Du03], by replacing "laminar" with "woven" everywhere, and obtain that the cluster values are woven currents. $\qquad\square$

The woven currents we obtain in this way satisfy explicit estimates which allow us to adapt [Du04] and study their intersection. This was already observed in [DT06, Section 3] — note that woven currents are called geometric there.

Proposition 5.10. *Let T_1, T_2, be woven currents obtained as cluster values of curves satisfying the assumption of Proposition 5.9. Furthermore, assume that T_1 and T_2 have continuous potentials in some open set Ω. Then the wedge product $T_1 \wedge T_2$ has local product structure in Ω.*

Proof: The scheme is as follows: consider two distinct linear projections π_1 and π_2 in $\mathbb{C}^2$, and subdivisions by squares of size r of the projection bases. This gives rise to a subdivision by cubes of size $O(r)$ in $\mathbb{C}^2$, which we denote by $\mathcal{Q}$. If T is a woven current, as given by Proposition 5.9, then for a generic such subdivision $\mathcal{Q}$, and for every $Q \in \mathcal{Q}$, there exists a uniformly woven current T_Q in Q such that the mass of $T - \sum_{Q \in \mathcal{Q}} T_Q$ is $O(r^2)$. If T_1

and T_2 are two such currents, with continuous potentials, it is then easy to adapt [Du04, Theorem 4.1] and find that the wedge product $T_1 \wedge T_2$ is approximated by $\sum_{Q \in \mathcal{Q}} T_{1,Q} \wedge T_{2,Q}$, which has a geometric interpretation, by Lemma 2.8. We conclude that the measure $T_1 \wedge T_2$ has local product structure. □

The conclusion of the proof of the theorem is now clear. If for some sequence $\mathrm{Per}^+(n, k(n))$, the estimate (5.8) is violated, then by the previous proposition, the measure $\mu = T^+ \wedge T^-$ would have product structure, which is not the case, by Theorem 5.6. □

Bibliography

[BB07a]   G. Bassanelli and F. Berteloot, *Bifurcation currents in holomorphic dynamics on* $\mathbf{P}^k$, J. Reine Angew. Math. **608** (2007), 201–235.

[BB07b]   G. Bassanelli and F. Berteloot, *Bifurcation currents and holomorphic motions in bifurcation loci,* Preprint, 2007.

[BLS93]   E. Bedford, M. Lyubich, and J. Smillie, *Polynomial diffeomorphisms of* $\mathbb{C}^2$ *IV: The measure of maximal entropy and laminar currents,* Invent. Math. **112** (1993), 77–125.

[BS91a]   E. Bedford and J. Smillie, *Polynomial diffeomorphisms of* $\mathbb{C}^2$*: Currents, equilibrium measure, and hyperbolicity,* Invent. Math. **103** (1991), 69–99.

[BS91b]   E. Bedford and J. Smillie, *Fatou-Bieberbach domains arising from polynomial automorphisms,* Indiana Univ. Math. J. **40** (1991), 789–792.

[BS98]   E. Bedford and J. Smillie, *Polynomial diffeomorphisms of* $\mathbb{C}^2$ *VI: Connectivity of J,* Ann. of Math. (2) **148** (1998), 695–735.

[Br91]   B. Branner, *Cubic polynomials: turning around the connectedness locus,* in: Topological Methods in Modern Mathematics, Publish or Perish, Houston, TX, 1993, 391–427.

[BH88]   B. Branner and J. H. Hubbard, *The iteration of cubic polynomials I: the global topology of parameter space,* Acta Math. **160** 3–4 (1988), 143–206.

[BH92]   B. Branner and J. H. Hubbard, *The iteration of cubic polynomials II: patterns and parapatterns,* Acta Math. **169** 3–4 (1992), 229–325.

[CG93]   L. Carleson and T. W. Gamelin, *Complex dynamics,* Universitext: Tracts in Mathematics, Springer, New York, 1993.

[De93]   J. P. Demailly, *Monge-Ampère operators, Lelong numbers and intersection theory,* in: Complex Analysis and Geometry, Univ. Ser. Math., Plenum, NY, 1993, 115–193.

[De01]   L. DeMarco, *Dynamics of rational maps: a current on the bifurcation locus,* Math. Res. Lett. **8** 1–2 (2001), 57–66.

[De03]   L. DeMarco, *Dynamics of rational maps: Lyapunov exponents, bifurcations, and capacity,* Math. Ann. **326** 1 (2003), 43–73.

[DT04]   H. de Thélin, *Sur la laminarité de certains courants,* Ann. Sci. École Norm. Sup. (4) **37** (2004), 304–311.

[DT06]   H. de Thélin, *Sur la construction de mesures selles,* Ann. Inst. Fourier (Grenoble) **56** 2 (2006), 337–372.

[Di05]   T. C. Dinh, *Suites d'applications méromorphes multivaluées et courants laminaires,* J. Geom. Anal. **15** (2005), 207–227.

[DDS05]   T. C. Dinh, R. Dujardin, and N. Sibony, *On the dynamics near infinity of some polynomial mappings in* $\mathbb{C}^2$, Math. Ann. **333** 4 (2005), 703–739.

[DH85a]   A. Douady and J. H. Hubbard, *Etude dynamique des polynômes complexes, partie II,* Publications Mathématiques d'Orsay, Université de Paris-Sud, Orsay, 1985.

[DH85b]   A. Douady and J. H. Hubbard, *On the dynamics of polynomial-like mappings,* Ann. Scient. Éc. Norm. Sup. **18** (1985), 287–343.

[Du03]   R. Dujardin, *Laminar currents in* $\mathbb{P}^2$, Math. Ann. **325** (2003), 745–765.

[Du04]   R. Dujardin, *Sur l'intersection des courants laminaires,* Pub. Mat. **48** (2004), 107–125.

[Du06]   R. Dujardin, *Approximation des fonctions lisses sur certaines laminations,* Indiana Univ. Math. J. **55** 2 (2006), 579–592.

[DF08]   R. Dujardin and C. Favre, *Distribution of rational maps with a preperiodic critical point,* Amer. Journ. Math. **130** (2008), 979–1032.

[FS92]   J. E. Fornæss and N. Sibony, *Complex Hénon mappings in* $\mathbb{C}^2$ *and Fatou-Bieberbach domains,* Duke Math. J. **65** (1992), 345–380.

[Gh99]   E. Ghys, *Laminations par surfaces de Riemann,* in: Dynamique et Géométrie Complexes, Panoramas et Synthèses **8**, Soc. Math. France, Paris, 1999, 49–95.

[HO94]   J. H. Hubbard and R. W. Oberste-Vorth, *Hénon mappings in the complex domain I: the global topology of dynamical space,* Publ. Math. IHÉS, **79** (1994), 5–46.

[Ki06]   J. Kiwi, *Puiseux series, polynomial dynamics, and iteration of complex cubic polynomials,* Ann. Inst. Fourier (Grenoble) **56** 5 (2006), 1337–1404.

[Ly83]   M. Lyubich, *Some typical properties of the dynamics of rational mappings,* Russian Math. Surveys **38:5** (1983), 154–155.

[MSS83]   R. Mañé, P. Sad, and D. Sullivan, *On the dynamics of rational maps,* Ann. Sci. École Norm. Sup. **16** (1983), 193–217.

[Mc87]   C. T. McMullen, *Families of rational maps and iterative root-finding algorithms,* Ann. of Math. **125** (1987), 467–493.

[Mc94]   C. T. McMullen, *Complex dynamics and renormalization,* Ann. Math. Stud. **135**, Princeton Univ. Press, 1994.

[MS98]   C. T. McMullen and D. Sullivan, *Quasiconformal homeomorphisms and dynamics III: The Teichmüller space of a holomorphic dynamical system,* Adv. Math. **135** 2 (1998), 351–395.

[Mi92]   J. Milnor, *Remarks on iterated cubic maps,* Experiment. Math. **1** (1992), 5–24.

[Mi99]   J. Milnor, *Dynamics in one complex variable: introductory lectures,* Friedr. Vieweg & Sohn, Braunschweig, 1999.

[Mi09]   J. Milnor, "Cubic polynomial maps with periodic critical orbit, part I", in: *Complex Dynamics: Families and Friends* (this volume), Chapter 9. A K Peters, Wellesley, MA, 2009, 333–411. (Preprint issued in 1991 under the title "On cubic polynomials with periodic critical points").

[Ph05]   N. Pham, *Lyapunov exponents and bifurcation currents for polynomial-like maps,* Preprint, 2005. Available at `arxiv.org` as `math.DS/0512557`.

[PR98] F. Przytycki and S. Rohde, *Porosity of Collet-Eckmann Julia sets,* Fund. Math. **155** 2 (1998), 189–199.

[Ra98] R. M. Range, *Holomorphic functions and integral representations in several complex variables,* Graduate Texts in Math. **108**, Springer-Verlag, 1998.

[Si99] N. Sibony, *Dynamique des applications rationnelles de $\mathbb{P}^k$,* in: Dynamique et Géométrie Complexes, Panoramas et Synthèses **8**, Soc. Math. France, Paris, 1999, 97–185.

[Wi97] P. Willumsen, *On accumulation of stretching rays,* PhD thesis, Danmarks Tekniske Univ, 1997.

[Za99] S. Zakeri, *Dynamics of cubic Siegel polynomials,* Comm. Math. Phys. **206** (1999), 185–233.

12 Bifurcation Measure and Postcritically Finite Rational Maps

Xavier Buff and Adam Epstein

Bassanelli and Berteloot [BB07] have defined a bifurcation measure μ_{bif} on the moduli space $\mathcal{M}_d$ of rational maps of degree $d \geq 2$. They have proved that it is a positive measure of finite mass and that its support is contained in the closure of the set $\mathcal{Z}_d$ of conjugacy classes of rational maps of degree d having $2d - 2$ indifferent cycles.

Denote by $\mathcal{X}_d$ the set of conjugacy classes of strictly postcritically finite rational maps of degree d which are not flexible Lattès maps. We prove that $\text{Supp}(\mu_{\text{bif}}) = \overline{\mathcal{X}_d} = \overline{\mathcal{Z}_d}$. Our proof is based on a transversality result due to the second author.

A similar result was obtained with different techniques by Dujardin and Favre [DF08] for the bifurcation measure on moduli spaces of polynomials of degree $d \geq 2$.

1 Introduction

A result of Lyubich [Ly83] asserts that for each rational map $f : \mathbb{P}^1 \to \mathbb{P}^1$ of degree $d \geq 2$, there is a unique probability measure μ_f of maximal entropy $\log d$. It is ergodic, satisfies $f^* \mu_f = d \cdot \mu_f$ and is carried outside the exceptional set of f. This measure is the *equilibrium measure* of f.

The *Lyapunov exponent* of f with respect to the measure μ_f may be defined by

$$\mathfrak{L}(f) := \int_{\mathbb{P}^1} \log \|Df\| \, \mathrm{d}\mu_f,$$

where $\| \cdot \|$ is any smooth metric on $\mathbb{P}^1$. Since μ_f is ergodic, the quantity $e^{\mathfrak{L}(f)}$ records the average rate of expansion of f along a typical orbit with respect to μ_f.

The space Rat_d of rational maps of degree d is a smooth complex manifold of dimension $2d + 1$. The function $\mathcal{L} : \mathrm{Rat}_d \to \mathbb{R}$ is continuous [Ma88] and plurisubharmonic [De03] on Rat_d. The positive $(1, 1)$-current

$$T_{\mathrm{bif}} := dd^c \mathcal{L}$$

is called the *bifurcation current* in Rat_d.

The *bifurcation locus* in Rat_d is the closure of the set of discontinuity of the map $\mathrm{Rat}_d \ni f \mapsto J_f$, where J_f stands for the Julia set of f and the continuity is for the Hausdorff topology for compact subsets of $\mathbb{P}^1$. DeMarco [De03] proved that the support of the bifurcation current T_{bif} is equal to the bifurcation locus.

For $f \in \mathrm{Rat}_d$, denote by $\mathcal{O}(f)$ the set of rational maps which are conjugate to f by a Möbius transformation. This set is a 3 dimensional complex analytic submanifold of Rat_d. In fact, $\mathcal{O}(f)$ is biholomorphic to $\mathrm{Aut}(\mathbb{P}^1)/\Gamma$, where $\mathrm{Aut}(\mathbb{P}^1) \simeq \mathrm{PSL}(2, \mathbb{C})$ is the group of Möbius transformations, and where $\Gamma \subset \mathrm{Aut}(\mathbb{P}^1)$ is the finite subgroup of Möbius transformations which commute with f.

The *moduli space* $\mathcal{M}_d$ is the quotient $\mathrm{Rat}_d/\mathrm{Aut}(\mathbb{P}^1)$, where $\mathrm{Aut}(\mathbb{P}^1)$ acts on Rat_d by conjugation. It is an orbifold of complex dimension $2d - 2$. It is a normal, quasiprojective variety. We denote by $\mathfrak{p} : \mathrm{Rat}_d \to \mathcal{M}_d$ the canonical projection.

The Lyapunov exponent is invariant under holomorphic conjugacy; thus it is constant on the orbits $\mathcal{O}(f)$. The map $\mathcal{L} : \mathrm{Rat}_d \to \mathbb{R}$ descends to a map $\hat{\mathcal{L}} : \mathcal{M}_d \to \mathbb{R}$ which is continuous, bounded from below and plurisubharmonic on $\mathcal{M}_d$ (see [BB07] Proposition 6.2). The measure

$$\mu_{\mathrm{bif}} := (dd^c \hat{\mathcal{L}})^{\wedge(2d-2)}$$

is called the *bifurcation measure* on $\mathcal{M}_d$. By construction, we have

$$\mathfrak{p}^* \mu_{\mathrm{bif}} = T_{\mathrm{bif}}^{\wedge(2d-2)}.$$

We denote by $\mathcal{C}(f)$ the set of critical points of f and by $\mathcal{V}(f) := f(\mathcal{C}(f))$ the set of critical values of f. The *postcritical set* is

$$\mathcal{P}(f) := \bigcup_{c \in \mathcal{C}(f)} \bigcup_{n \geq 1} f^{\circ n}(c).$$

A rational map f is *postcritically finite* if $\mathcal{P}(f)$ is finite. It is *strictly postcritically finite* if $\mathcal{P}(f)$ is finite and if f does not have any superattracting cycle. The map f is a *Lattès map* if it is obtained as the quotient of an affine map $A : z \mapsto az + b$ on a complex torus $\mathbb{C}/\Lambda$: there is a finite-to-one holomorphic map $\Theta : \mathbb{C}/\Lambda \to \mathbb{P}^1$ such that the following diagram commutes:

$$
\begin{array}{ccc}
\mathbb{C}/\Lambda & \xrightarrow{\;A\;} & \mathbb{C}/\Lambda \\
\Theta \downarrow & & \downarrow \Theta \\
\mathbb{P}^1 & \xrightarrow[\;f\;]{} & \mathbb{P}^1.
\end{array}
$$

A Lattès map is strictly postcritically finite (see [Mi06]). It is a *flexible Lattès map* if we can choose Θ with degree 2 and $A(z) = az + b$ with $a > 1$ an integer.

Let us introduce the following notation:

- $\mathcal{X}_d \subset \mathcal{M}_d$ for the set of conjugacy classes of strictly postcritically finite rational maps of degree d which are not flexible Lattès maps;

- $\mathcal{X}_d^* \subset \mathcal{X}_d$ for the subset of conjugacy classes of maps which have only simple critical points and satisfy $\mathcal{C}(f) \cap \mathcal{P}(f) = \emptyset$;

- $\mathcal{Z}_d \subset \mathcal{M}_d$ for the set of conjugacy classes of maps which have $2d - 2$ indifferent cycles (we do not count multiplicities).

Bassanelli and Berteloot [BB07] proved that

- the bifurcation measure does not vanish identically on $\mathcal{M}_d$ and has finite mass,

- the conjugacy class of any non-flexible Lattès map lies in the support of μ_{bif},

- the support of μ_{bif} is contained in the closure of $\mathcal{Z}_d$.

Our main result is the following.

Main Theorem. *The support of the bifurcation measure in $\mathcal{M}_d$ is:*

$$
\mathrm{Supp}(\mu_{\mathrm{bif}}) = \overline{\mathcal{X}_d^*} = \overline{\mathcal{X}_d} = \overline{\mathcal{Z}_d}.
$$

Remark. We shall prove in Section 2 that

$$
\overline{\mathcal{X}_d^*} = \overline{\mathcal{X}_d} = \overline{\mathcal{Z}_d}. \tag{1}
$$

Then, due to the results in [BB07], we will only have to prove the inclusion $\mathcal{X}_d^* \subseteq \mathrm{Supp}(\mu_{\mathrm{bif}})$.

Remark. We do not know how to prove that the conjugacy classes of flexible Lattès maps lie in the support of μ_{bif}. It would be enough to prove that every flexible Lattès map can be approximated by strictly postcritically finite rational maps which are not Lattès maps.

Remark. The underlying idea of the proof is a potential-theoretic interpretation of a result of Tan Lei [Ta90] concerning the similarities between the Mandelbrot set and Julia sets.

Our proof relies on a transversality result that we will now present. Elements of $(\mathbb{P}^1)^{2d-2}$ are denoted $\underline{z} = (z_1, \ldots, z_{2d-2})$.

Let $f \in \mathrm{Rat}_d$ be a postcritically finite rational map with $2d - 2$ distinct critical points $c_1, \ldots, c_{2d-2}$, satisfying $\mathcal{C}(f) \cap \mathcal{P}(f) = \emptyset$. There are integers $\ell_j \geq 1$ such that $\alpha_j := f^{\circ \ell_j}(c_j)$ are periodic points of f. Set $\underline{c} := (c_1, \ldots, c_{2d-2})$ and $\underline{\alpha} := (\alpha_1, \ldots, \alpha_{2d-2})$. The critical points are simple and the periodic points are repelling. By the Implicit Function Theorem, there are

- an analytic germ $\underline{c} : (\mathrm{Rat}_d, f) \rightarrow ((\mathbb{P}^1)^{2d-2}, \underline{c})$ such that for g near f, $\mathfrak{c}_j(g)$ is a critical point of g and

- an analytic germ $\underline{a} : (\mathrm{Rat}_d, f) \rightarrow ((\mathbb{P}^1)^{2d-2}, \underline{\alpha})$ such that for g near f, $\mathfrak{a}_j(g)$ is a periodic point of g.

Let $\underline{v} : (\mathrm{Rat}_d, f) \rightarrow ((\mathbb{P}^1)^{2d-2}, \underline{\alpha})$ be defined by

$$\underline{v} := (v_1, \ldots, v_{2d-2}) \quad \text{with} \quad v_j(g) := g^{\circ \ell_j}\big(\mathfrak{c}_j(g)\big).$$

Denote by $D_f \underline{v}$ and $D_f \underline{a}$ the differentials of $\underline{v}$ and $\underline{a}$ at f. The transversality result we are interested in is the following.

Theorem 1.1. *If f is not a flexible Lattès map, then the linear map*

$$D_f \underline{v} - D_f \underline{a} : T_f \mathrm{Rat}_d \rightarrow \bigoplus_{j=1}^{2d-2} T_{\alpha_j} \mathbb{P}^1$$

is surjective. The kernel of $D_f \underline{v} - D_f \underline{a}$ is the tangent space to $\mathcal{O}(f)$ at f.

There are several proofs of this result. Here, we include a proof due to the second author. A different proof was obtained by van Strien [vS00]. His proof covers a more general setting where the critical orbits are allowed to be preperiodic to a hyperbolic set. Our proof covers the case of maps commuting with a non-trivial group of Möbius transformations (a slight modification of van Strien's proof probably also covers this case).

Remark. A similar transversality theorem holds in the space of polynomials of degree d and the arguments developed in this article lead to an alternative proof of the result of Dujardin and Favre.

Acknowledgments. We would like to thank our colleagues in Toulouse who brought this problem to our attention, in particular François Berteloot, Julien Duval and Vincent Guedj. We would like to thank John Hubbard for his constant support.

2 The Sets $\mathcal{X}_d^*$, $\mathcal{X}_d$ and $\mathcal{Z}_d$

In this section, we prove (1). Since $\mathcal{X}_d^* \subseteq \mathcal{X}_d$, it is enough to prove that

$$\mathcal{X}_d \subseteq \overline{\mathcal{Z}_d} \quad \text{and} \quad \mathcal{Z}_d \subseteq \overline{\mathcal{X}_d^*}.$$

2.1 Tools

Our proof relies on the following three results.

The first result is an immediate consequence of the Fatou-Shishikura inequality on the number of non-repelling cycles of a rational map (see [Sh87] and/or [Ep99]). For $f \in \mathrm{Rat}_d$, denote by

- $N_{\mathrm{att}}(f)$ the number of attracting cycles of f,

- $N_{\mathrm{ind}}(f)$ the number of distinct indifferent cycles of f and

- $N_{\mathrm{crit}}(f)$ the number of critical points of f, counting multiplicities, whose orbits are strictly preperiodic to repelling cycles.

Theorem 2.1. *For any rational map $f \in \mathrm{Rat}_d$, we have:*

$$N_{\mathrm{crit}}(f) + N_{\mathrm{ind}}(f) + N_{\mathrm{att}}(f) \leq 2d - 2.$$

The second result is a characterization of *stability* due to Mañé, Sad and Sullivan [MSS83] (compare with [Mc94] Section 4.1). Let Λ be a complex manifold. A family of rational maps $\Lambda \ni \lambda \mapsto f_\lambda$ is an *analytic family* if the map $\Lambda \times \mathbb{P}^1 \ni (\lambda, z) \mapsto f_\lambda(z) \in \mathbb{P}^1$ is analytic. The family is *stable at* $\lambda_0 \in \Lambda$ if the number of attracting cycles of f_λ is locally constant at λ_0.

Theorem 2.2. *Let $\Lambda \ni \lambda \mapsto f_\lambda$ be an analytic family of rational maps parametrized by a complex manifold Λ. The following assertions are equivalent.*

- *The family is stable at λ_0.*

- *For all $m \in \mathbb{S}^1$ and $p \geq 1$, the number of cycles of f_λ having multiplier m and period p is locally constant at λ_0.*

- *For all $\ell \geq 1$ and $p \geq 1$, the number of critical points c of f_λ such that $f^{\circ \ell}(c)$ is a repelling periodic point of period p is locally constant at λ_0.*

- *The maximum period of an indifferent cycle of f_λ is locally bounded at λ_0.*

- *The maximum period of a repelling cycle of f_λ contained in the post-critical set $\mathcal{P}(f_\lambda)$ is locally bounded at λ_0.*

The third result is due to McMullen [Mc87]. Let Λ be an irreducible quasiprojective complex variety. A family of rational maps $\Lambda \ni \lambda \mapsto f_\lambda$ is an *algebraic family* if the map $\Lambda \times \mathbb{P}^1 \ni (\lambda, z) \mapsto f_\lambda(z) \in \mathbb{P}^1$ is a rational mapping. The family is *trivial* if all its members are conjugate by Möbius transformations. The family is *stable* if it is stable at every $\lambda \in \Lambda$.

Theorem 2.3. *A stable algebraic family of rational maps is either trivial or it is a family of flexible Lattès maps.*

2.2 Spaces of Rational Maps with Marked Critical Points and Marked Periodic Points

Here and henceforth, it will be convenient to consider the set $\mathrm{Rat}_d^{\mathrm{crit},\mathrm{per}_n}$ of rational maps of degree d with marked critical points and marked periodic points of period dividing n. This set may be defined as follows.

First, the unordered sets of m points in $\mathbb{P}^1$ may be identified with $\mathbb{P}^m$. This yields a rational map $\sigma_m : (\mathbb{P}^1)^m \to \mathbb{P}^m$ which may be defined as follows:

$$\sigma_m\big([x_1 : y_1], \ldots, [x_m : y_m]\big) = [a_0 : \cdots : a_m]$$

with

$$\sum_{j=0}^m a_j x^j y^{m-j} = \prod_{i=1}^m (xy_i - yx_i).$$

Second, to a rational map $f \in \mathrm{Rat}_d$, we associate the unordered set $\{c_1, \ldots, c_{2d-2}\}$ of its $2d - 2$ critical points, listed with repetitions according to their multiplicities. This induces a rational map $\mathrm{crit} : \mathrm{Rat}_d \to \mathbb{P}^{2d-2}$ which may be defined as follows: if $f\big([x : y]\big) = \big[P(x,y) : Q(x,y)\big]$ with P and Q homogeneous polynomials of degree d, then

$$\mathrm{crit}(f) = [a_0 : \cdots : a_{2d-2}] \quad \text{with} \quad \sum_{j=0}^{2d-2} a_j x^j y^{2d-2-j} = \frac{\partial P}{\partial x}\frac{\partial Q}{\partial y} - \frac{\partial P}{\partial y}\frac{\partial Q}{\partial x}.$$

Third, given an integer $n \geq 1$, to a rational map $f \in \mathrm{Rat}_d$, we associate the unordered set $\{\alpha_1, \ldots, \alpha_{d^n+1}\}$ of the fixed points of $f^{\circ n}$, listed with repetitions according to their multiplicities. This induces a rational map $\mathrm{per}_n : \mathrm{Rat}_d \to \mathbb{P}^{d^n+1}$ which may be defined as follows: if $f^{\circ n}\big([x : y]\big) = \big[P_n(x,y) : Q_n(x,y)\big]$ with P_n and Q_n homogeneous polynomials of degree d^n, then

$$\mathrm{per}_n(f) = [a_0 : \cdots : a_{d^n+1}]$$

with

$$\sum_{j=0}^{d^n+1} a_j x^j y^{d^n+1-j} = yP_n(x,y) - xQ_n(x,y).$$

The space of rational maps of degree d with marked critical points and marked periodic points of period dividing n is the set:

$$\mathrm{Rat}_d^{\mathrm{crit,per}_n} := \left\{ \begin{array}{l} (f, \underline{c}, \underline{\alpha}) \in \mathrm{Rat}_d \times (\mathbb{P}^1)^{2d-2} \times (\mathbb{P}^1)^{d^n+1} \text{ such that} \\ \mathfrak{crit}(f) = \sigma_{2d-2}(\underline{c}) \text{ and } \mathfrak{per}_n(f) = \sigma_{d^n+1}(\underline{\alpha}) \end{array} \right\}.$$

Then, $\mathrm{Rat}_d^{\mathrm{crit,per}_n}$ is an algebraic subset[1] of $\mathrm{Rat}_d \times (\mathbb{P}^1)^{2d-2} \times (\mathbb{P}^1)^{d^n+1}$. Since there are $2d - 2 + d^n + 1$ equations, the dimension of any irreducible component of $\mathrm{Rat}_d^{\mathrm{crit,per}_n}$ is at least that of Rat_d, i.e., $2d + 1$. Since the fibers of the projection $\mathrm{Rat}_d^{\mathrm{crit,per}_n} \to \mathrm{Rat}_d$ are finite (a rational map has finitely many critical points and finitely many periodic points of period dividing n), the dimension of any component is exactly $2d + 1$.

2.3 The Inclusion $\mathcal{X}_d \subseteq \overline{\mathcal{Z}_d}$

Let $f_0 \in \mathrm{Rat}_d$ be a strictly postcritically finite rational map but not a flexible Lattès map. We must show that any neighborhood of f_0 in Rat_d contains a rational map with $2d - 2$ indifferent cycles. This is an immediate consequence (by induction) of the following lemma.

Lemma 2.4. *Let $r \geq 1$ and $s \geq 0$ be integers such that $r + s = 2d - 2$. Assume $f_0 \in \mathrm{Rat}_d$ is not a flexible Lattès map, $N_{\mathrm{crit}}(f_0) = r$ and $N_{\mathrm{ind}}(f_0) = s$. Then, arbitrarily close to f_0, we may find a rational map f_1 such that $N_{\mathrm{crit}}(f_1) = r - 1$ and $N_{\mathrm{ind}}(f_1) = s + 1$.*

Proof: Let $p_1, \ldots, p_s$ be the periods of the indifferent cycles of f_0. Denote by p their least common multiple. A point $\lambda \in \mathrm{Rat}_d^{\mathrm{crit,per}_p}$ is of the form

$$(g_\lambda, c_1(\lambda), \ldots, c_{2d-2}(\lambda), \alpha_1(\lambda), \ldots \alpha_{d^p+1}(\lambda)).$$

We choose $\lambda_0 \in \mathrm{Rat}_d^{\mathrm{crit,per}_p}$ so that

- $g_{\lambda_0} = f_0$,

- $c_1(\lambda_0), \ldots c_r(\lambda_0)$ are preperiodic to repelling cycles of f_0 and

- $\alpha_1(\lambda_0), \ldots, \alpha_s(\lambda_0)$ are indifferent periodic points of f_0 belonging to distinct cycles.

For $i \in [1, r]$, let $k_i \geq 1$ be the least integer such that

$$g_{\lambda_0}^{\circ k_i}(c_i(\lambda_0)) = g_{\lambda_0}^{\circ(k_i+p_i)}(c_i(\lambda_0)).$$

---

[1] By algebraic we mean a quasiprojective variety in some projective space. It is classical that any product of projective spaces embeds in some $\mathbb{P}^N$. Thus, it makes sense to speak of algebraic subsets of products of projective spaces

For $j \in [1, s]$, let m_j be the multiplier of $g_{\lambda_0}^{\circ p}$ at $\alpha_j(\lambda_0)$. In particular, $m_j \in \mathbb{S}^1$.

Consider the algebraic subset

$$\left\{ \lambda \in \mathrm{Rat}_d^{\mathrm{crit,per}_p} \; ; \; \begin{array}{l} \forall i \in [1, r-1], \; g_\lambda^{\circ k_i}\big(c_i(\lambda)\big) = g_\lambda^{\circ(k_i+p_i)}\big(c_i(\lambda)\big) \text{ and} \\ \forall j \in [1, s], \text{ the multiplier of } g_\lambda^{\circ p} \text{ at } \alpha_j(\lambda) \text{ is } m_j \end{array} \right\}$$

and let Λ be an irreducible component containing λ_0. Note that there are $r - 1 + s = 2d - 3$ equations. It follows that the dimension of Λ is at least $(2d + 1) - (2d - 3) = 4$. In particular, the image of Λ by the projection $\mathrm{Rat}_d^{\mathrm{crit,per}_p} \to \mathrm{Rat}_d$ cannot be contained in $\mathcal{O}(f_0)$.

Embedding Λ in some $\mathbb{P}^N$ (with N sufficiently large) and slicing with an appropriate projective subspace, we deduce that there is an algebraic curve $\Gamma \subseteq \Lambda$ containing λ_0, whose image by the projection $\mathrm{Rat}_d^{\mathrm{crit,per}_p} \to \mathrm{Rat}_d$ is not contained in $\mathcal{O}(f_0)$. Desingularizing the algebraic curve, we obtain a smooth quasiprojective curve Σ (i.e., a Riemann surface of finite type) and an algebraic map $\Sigma \to \Gamma$ which is surjective and generically one-to-one. We let $\sigma_0 \in \Sigma$ be a point which is mapped to $\lambda_0 \in \Gamma$. With an abuse of notation, we write g_σ, $c_i(\sigma)$ and $\alpha_j(\sigma)$ in place of $g_{\lambda(\sigma)}$, $c_i\big(\lambda(\sigma)\big)$ and $\alpha_j\big(\lambda(\sigma)\big)$. Then, we have an algebraic family of rational maps $\Sigma \ni \sigma \mapsto g_\sigma$ parametrized by a smooth quasiprojective curve Σ, coming with marked critical points $c_i(\sigma)$ and marked periodic points $\alpha_j(\sigma)$.

The family is not trivial and g_{σ_0} is not a flexible Lattès map. Assume that the family were stable at σ_0. Then, according to Theorem 2.2, the number of critical points which are preperiodic to repelling cycles would be locally constant at σ_0. Thus, the critical orbit relation

$$g_\sigma^{\circ k_r}\big(c_r(\sigma)\big) = g_\sigma^{\circ(k_r+p_r)}\big(c_r(\sigma)\big)$$

would hold in a neighborhood of σ_0 in Σ, thus for all $\sigma \in \Sigma$ by analytic continuation. As a consequence, for all $\sigma \in \Sigma$, we would have

$$N_{\mathrm{crit}}(g_\sigma) + N_{\mathrm{ind}}(g_\sigma) = 2d - 2.$$

According to Theorem 2.1, this would imply that $N_{\mathrm{att}}(g_\sigma) = 0$ for all $\sigma \in \Sigma$. The family would be stable which would contradict Theorem 2.3.

Thus, the family $\Sigma \ni \sigma \mapsto g_\sigma$ is not stable at σ_0. According to Theorem 2.2, the maximum period of an indifferent cycle of g_σ is not locally bounded at σ_0. Thus, we may find a parameter $\sigma_1 \in \Sigma$ arbitrarily close to σ_0 such that g_{σ_1} has at least $s + 1$ indifferent cycles. $\qquad\square$

2.4   The Inclusion $\mathcal{Z}_d \subseteq \overline{\mathcal{X}_d^*}$

The proof follows essentially the same lines as the previous one. We begin with a map $f_0 \in \mathrm{Rat}_d$ having $2d - 2$ indifferent cycles. We must show that arbitrarily close to f_0, we may find a postcritically finite map $f_1 \in \mathrm{Rat}_d$ which has only simple critical points, satisfies $\mathcal{C}(f_1) \cap \mathcal{P}(f_1) = \emptyset$ and is not a flexible Lattès map.

The following lemma implies (by induction) that arbitrarily close to f_0, we may find a rational map $f_1 \in \mathrm{Rat}_d$ whose postcritical set contains $2d-2$ distinct repelling cycles. Automatically, the $2d-2$ critical points of f_1 have to be simple, strictly preperiodic, with disjoint orbits. In particular, $\mathcal{C}(f_1) \cap \mathcal{P}(f_1) = \emptyset$. In addition, Lattès maps lie in a Zariski closed subset of Rat_d. Since f_0 has indifferent cycles, it cannot be a Lattès map. So, if f_1 is sufficiently close to f_0, then f_1 is not a Lattès map.

Lemma 2.5. *Let $r \geq 0$ and $s \geq 1$ be integers such that $r + s = 2d - 2$. Let $f_0 \in \mathrm{Rat}_d$ satisfy $N_{\mathrm{crit}}(f_0) = r$, $N_{\mathrm{ind}}(f_0) = s$, the postcritical set of f_0 containing r distinct repelling cycles. Then, arbitrarily close to f_0, there is a rational map f_1 which satisfies $N_{\mathrm{crit}}(f_1) = r + 1$ and $N_{\mathrm{ind}}(f_1) = s - 1$, the postcritical set of f_1 containing $r + 1$ distinct repelling cycles.*

Proof: The proof is similar to the proof of Lemma 2.4; we do not give all the details. First, note that since $s \geq 1$, f_0 has an indifferent cycle, and thus, is not a Lattès map. Second, we may find a non-trivial algebraic family of rational maps $\Sigma \ni \sigma \mapsto g_\sigma$ parametrized by a smooth quasiprojective curve Σ with marked critical points $c_1(\sigma)$, ..., $c_{2d-2}(\sigma)$ and marked periodic points $\alpha_1(\sigma)$, ..., $\alpha_{d^p+1}(\sigma)$ such that

- $g_{\sigma_0} = f_0$,

- for $j \in [1, s]$, the periodic points $\alpha_j(\sigma_0)$ belong to distinct indifferent cycles of f_0,

- for all $\sigma \in \Sigma$ and all $i \in [1, r]$, the critical point $c_i(\sigma)$ is preperiodic to a periodic cycle of g_σ and

- for all $\sigma \in \Sigma$ and all $j \in [1, s - 1]$, the periodic points $\alpha_j(\sigma)$ are indifferent periodic points of g_σ.

If the family were stable at σ_0, the indifferent periodic point $\alpha_s(\sigma_0)$ would be persistently indifferent in a neighborhood of σ_0 in Σ, thus in all Σ by analytic continuation. The relation $N_{\mathrm{crit}}(g_\sigma) + N_{\mathrm{ind}}(g_\sigma) = 2d - 2$ would hold throughout Σ. Thus, for all $\sigma \in \Sigma$, we would have $N_{\mathrm{att}}(g_\sigma) = 0$ and the family would be stable. This is not possible since g_{σ_0} is not a Lattès map and since the family $\Sigma \ni \sigma \mapsto g_\sigma$ is not trivial.

It follows that the period of a repelling cycle contained in the postcritical set of g_σ is not locally bounded at σ_0. In addition, for $i \in [1, r]$, the critical points $c_i(\sigma_0)$ of g_{σ_0} are preperiodic to distinct repelling cycles. Thus, we may find a rational map g_{σ_1}, with σ_1 arbitrarily close to σ_0, such that g_{σ_1} has $r + 1$ critical points preperiodic to distinct repelling cycles. $\quad\square$

3   Transversality

Here and henceforth, we will consider various holomorphic families $t \mapsto \gamma_t$ defined near 0 in $\mathbb{C}$. We will employ the notation

$$\gamma := \gamma_0 \quad \text{and} \quad \dot{\gamma} := \frac{d\gamma_t}{dt}\Big|_{t=0}.$$

3.1   The Tangent Space to $\mathcal{O}(f)$

Here we characterize the vectors $\xi \in T_f \mathrm{Rat}_d$ which are tangent to $\mathcal{O}(f)$ for some rational map $f \in \mathrm{Rat}_d$.

Note that if $t \mapsto f_t$ is a holomorphic family of rational maps, then for every $z \in \mathbb{P}^1$, the vector $\dot{f}(z)$ belongs to the tangent space $T_{f(z)}\mathbb{P}^1$. Thus, if $\xi \in T_f \mathrm{Rat}_d$, then for every $z \in \mathbb{P}^1$, we have $\xi(z) \in T_{f(z)}\mathbb{P}^1$. If $\xi \in T_f \mathrm{Rat}_d$ there is a unique vector field η_ξ, meromorphic on $\mathbb{P}^1$ with poles in $\mathcal{C}(f)$, such that

$$Df \circ \eta_\xi = -\xi.$$

Indeed, if z is not a critical point of f, then $D_z f : T_z\mathbb{P}^1 \to T_{f(z)}\mathbb{P}^1$ is an isomorphism, whence we may define $\eta_\xi(z)$ by

$$\eta_\xi(z) := -\left(D_z f\right)^{-1}\left(\xi(z)\right).$$

Moreover, in this situation, it follows from the Implicit Function Theorem that there is a unique holomorphic germ $t \mapsto z_t$ with $z_0 = z$ such that $f_t(z_t) = f(z)$, and furthermore $\eta_{\dot{f}}(z) = \dot{z} \in T_z\mathbb{P}^1$.

Remark. If f has simple critical points, then the vector field η_ξ has simple poles or removable singularities along $\mathcal{C}(f)$. There is a removable singularity at $c \in \mathcal{C}(f)$ if and only if $\xi(c) = 0$. This can be seen by working in coordinates, using the fact that f' has simple zeroes at points of $\mathcal{C}(f)$.

Recalling that $\mathrm{Aut}(\mathbb{P}^1)$ is a Lie group, we denote by $\mathrm{aut}(\mathbb{P}^1)$ the corresponding Lie algebra: that is, the tangent space to $\mathrm{Aut}(\mathbb{P}^1)$ at the identity map. Thus, $\mathrm{aut}(\mathbb{P}^1)$ is canonically isomorphic to the space of globally holomorphic vector fields.

If $X \subseteq \mathbb{P}^1 - \mathcal{C}(f)$ and if θ is a vector field defined on $f(X)$, then the vector field $f^*\theta$ is defined on X by

$$f^*\theta(z) := (D_z f)^{-1}\left(\theta \circ f(z)\right).$$

If $\theta \in \mathrm{aut}(\mathbb{P}^1)$, the vector field $f^*\theta$ is the unique meromorphic vector field on $\mathbb{P}^1$ such that $Df \circ f^*\theta = \theta \circ f$.

Proposition 3.1. *A vector* $\xi \in T_f\text{Rat}_d$ *is tangent to* $\mathcal{O}(f)$ *if and only if*

$$\eta_\xi = \theta - f^*\theta$$

for some $\theta \in \text{aut}(\mathbb{P}^1)$.

Proof: The derivative at the identity of

$$\text{Aut}(\mathbb{P}^1) \ni \phi \mapsto \phi \circ f \circ \phi^{-1} \in \text{Rat}_d$$

is the linear map

$$\text{aut}(\mathbb{P}^1) \ni \theta \mapsto \theta \circ f - Df \circ \theta \in T_f\text{Rat}_d.$$

Thus, $\xi \in T_f\text{Rat}_d$ is tangent to $\mathcal{O}(f)$ if and only if $\xi = \theta \circ f - Df \circ \theta$ for some $\theta \in \text{aut}(\mathbb{P}^1)$. Since $\theta \circ f - Df \circ \theta = Df \circ (f^*\theta - \theta)$, it follows that $\xi \in T_f\text{Rat}_d$ is tangent to $\mathcal{O}(f)$ if and only if $\eta_\xi = \theta - f^*\theta$ for some $\theta \in \text{aut}(\mathbb{P}^1)$. $\quad\square$

3.2   Guided Vector Fields

Let θ be a vector field, defined and holomorphic on a neighborhood of the critical value set of some rational map $f \in \text{Rat}_d$. Given $\xi \in T_f\text{Rat}_d$, we want to understand under which conditions the vector field $f^*\theta + \eta_\xi$ is holomorphic on a neighborhood of the critical point set of f.

Lemma 3.2. *Let* c *be a simple critical point of* $f \in \text{Rat}_d$ *and let* θ *be a vector field, holomorphic near* $v = f(c)$. *For any* $\xi \in T_f\text{Rat}_d$, *the vector field* $f^*\theta + \eta_\xi$ *is holomorphic near* c *if and only if* $\theta(v) = \xi(c)$.

Proof: Since c is a simple critical points of f, it follows from the Implicit Function Theorem that there is a unique holomorphic germ $t \mapsto c_t$ with $c_0 = c$ such that c_t is a critical point of f_t. Let $v_t := f_t(c_t)$ be the corresponding critical values. Note that $\dot{v} = \dot{f}(c) + D_c f(\dot{c}) = \xi(c)$, since $\dot{f} = \xi$ and $D_c f = 0$.

Let $t \mapsto \phi_t$ be a holomorphic family of Möbius transformations sending v to v_t, with $\phi_0 = \text{Id}$. Note that there is a holomorphic family of local biholomorphisms ψ_t sending c to c_t, with $\psi_0 = \text{Id}$ and

$$\phi_t \circ f = f_t \circ \psi_t.$$

Differentiating this identity with respect to t and evaluating at $t = 0$ yields $\dot{\phi} \circ f = \xi + Df \circ \dot{\psi}$, whence $Df \circ (\dot{\psi} - f^*\dot{\phi}) = -\xi$. Consequently, $\dot{\psi} - f^*\dot{\phi} = \eta_\xi$ whence $f^*\dot{\phi} + \eta_\xi$ is holomorphic in a neighborhood of c.

It follows that $f^*\theta + \eta_\xi$ is holomorphic near c if and only if $f^*(\theta - \dot{\phi})$ is holomorphic near c. Since $\theta - \dot{\phi}$ is holomorphic near v, this is the case if and only if $\theta - \dot{\phi}$ vanishes at v, i.e., if and only if $\theta(v) = \dot{\phi}(v) = \dot{v} = \xi(c)$. $\quad\square$

For a finite set $X \subset \mathbb{P}^1$, we denote by $\mathcal{T}(X)$ the linear space of vector fields on X; note that $\mathcal{T}(X)$ is canonically isomorphic to $\bigoplus_{x \in X} T_x \mathbb{P}^1$.

Let $f \in \mathrm{Rat}_d$ have $2d - 2$ simple critical points, let $A \subset \mathbb{P}^1 - \mathcal{C}(f)$ be finite, and set $B = f(A) \cup \mathcal{V}(f)$. We shall say that a vector field $\tau \in \mathcal{T}(B)$ is *guided* by $\xi \in T_f \mathrm{Rat}_d$ if

$$\tau = f^* \tau + \eta_\xi \text{ on } A \quad \text{and} \quad \tau \circ f = \xi \text{ on } \mathcal{C}(f).$$

Note that a priori, there might be distinct critical points c_1 and c_2 with $f(c_1) = f(c_2)$ but $\xi(c_1) \neq \xi(c_2)$. In this case, no vector field can be guided by ξ.

3.3   Quadratic Differentials

Recall that a quadratic differential is a section of the complex line bundle obtained as $\otimes$-square of the holomorphic cotangent bundle. For a finite set $X \subset \mathbb{P}^1$, we denote by $\mathcal{Q}(\mathbb{P}^1, X)$ the set of all meromorphic quadratic differentials whose poles are all simple and lie in X. This is a vector space of dimension $\max(|X| - 3, 0)$.

Given $q \in \mathcal{Q}(\mathbb{P}^1, X)$ and a vector field τ, defined and holomorphic near $x \in X$, we regard the product $q \otimes \tau$ as a meromorphic 1-form defined in a neighborhood of x, whence there is a residue $\mathrm{Res}_x(q \otimes \tau)$ at x. If τ_1 and τ_2 agree at x, then $\mathrm{Res}_x(q \otimes \tau_1) = \mathrm{Res}_x(q \otimes \tau_2)$, since q has at worst a simple pole at x. Thus it makes sense to talk about $\mathrm{Res}_x(q \otimes \tau)$ even when τ is only defined at x.

Given $q \in \mathcal{Q}(\mathbb{P}^1, X)$ and $\tau \in \mathcal{T}(X)$, we define

$$\langle q, \tau \rangle := 2i\pi \sum_{x \in X} \mathrm{Res}_x(q \otimes \tau).$$

Lemma 3.3. *Given $\tau \in \mathcal{T}(X)$, let θ be a $\mathcal{C}^\infty$ vector field on $\mathbb{P}^1$ which agrees with τ on X and is holomorphic in a neighborhood of X. Then for every $q \in \mathcal{Q}(\mathbb{P}^1, X)$,*

$$\langle q, \tau \rangle = - \int_{\mathbb{P}^1} q \otimes \bar{\partial}\theta.$$

Proof: Let U be a finite union of smoothly bounded disks, with pairwise disjoint closures, each enclosing a unique point of X, and such that θ is holomorphic in a neighborhood of $\overline{U}$. Then for any $q \in \mathcal{Q}(\mathbb{P}^1, X)$, we have

$$\langle q, \tau \rangle = \int_{\partial U} q \otimes \theta = - \int_{\mathbb{P}^1 - \overline{U}} q \otimes \bar{\partial}\theta = - \int_{\mathbb{P}^1} q \otimes \bar{\partial}\theta,$$

where the first equality is due to the Residue Theorem, the second to Stokes' Theorem, and the last from $\bar{\partial}\theta = 0$ in a neighborhood of $\overline{U}$.   □

Given a rational map $f : \mathbb{P}^1 \to \mathbb{P}^1$ and a meromorphic quadratic differential q on $\mathbb{P}^1$, we define the push-forward f_*q as follows. At a point $w \in \mathbb{P}^1$ which is neither a critical value nor the image of a pole, we set

$$f_*q(w)(\tau_1, \tau_2) = \sum_{z \in f^{-1}(w)} q(z)\big((D_z f)^{-1}\tau_1, (D_z f)^{-1}\tau_2\big).$$

The resulting quadratic differential f_*q is in fact globally meromorphic. Moreover, if $q \in \mathcal{Q}(\mathbb{P}^1, A)$ then $f_*q \in \mathcal{Q}(\mathbb{P}^1, B)$ with $B = f(A) \cup \mathcal{V}(f)$. Checking that f_*q belongs to $\mathcal{Q}(\mathbb{P}^1, B)$ requires some justifications which can be found in [DH93] for example.

We denote by $\nabla : \mathcal{Q}(\mathbb{P}^1, A) \to \mathcal{Q}(\mathbb{P}^1, B)$ the linear map defined by

$$\nabla q := q - f_*q.$$

Lemma 3.4. *Let $f \in \mathrm{Rat}_d$ be a rational map with all critical points simple and let $A \subset \mathbb{P}^1 - \mathcal{C}(f)$ be finite. Set $B = f(A) \cup \mathcal{V}(f)$ and let $\tau \in \mathcal{T}(B)$ be a vector field guided by $\xi \in T_f\mathrm{Rat}_d$. Then, for all $q \in \mathcal{Q}(\mathbb{P}^1, A)$, we have*

$$\langle \nabla q, \tau \rangle = 0.$$

Proof: Let θ be a C^∞ vector field on $\mathbb{P}^1$ which agrees with τ on B and is holomorphic in a neighborhood of B. Since $\tau \circ f = \xi$ on $\mathcal{C}(f)$, Lemma 3.2 implies that the vector field $f^*\theta + \eta_\xi$ is holomorphic on a neighborhood of $\mathcal{C}(f)$. It follows that $f^*\theta + \eta_\xi$ is C^∞ on $\mathbb{P}^1$, holomorphic in a neighborhood of A and agrees with $f^*\tau + \eta_\xi = \tau$ on A. Consequently, for any $q \in \mathcal{Q}(\mathbb{P}^1, A)$

$$\langle f_*q, \tau \rangle = -\int_{\mathbb{P}^1} f_*q \otimes \bar{\partial}\theta = -\int_{\mathbb{P}^1} q \otimes f^*\bar{\partial}\theta = -\int_{\mathbb{P}^1} q \otimes \bar{\partial}(f^*\theta + \eta_\xi) = \langle q, \tau \rangle$$

where the first and last equalities follow from Lemma 3.3, the second from a change of variable and the third from $\bar{\partial}\eta_\xi = 0$ on $\mathbb{P}^1 - A$. □

Lemma 3.5. *Let $f \in \mathrm{Rat}_d$ be postcritically finite. If f is not a flexible Lattès map, then the linear endomorphism $\nabla : \mathcal{Q}(\mathbb{P}^1, \mathcal{P}(f)) \to \mathcal{Q}(\mathbb{P}^1, \mathcal{P}(f))$ is injective.*

Proof: See [DH93]. □

Proposition 3.6. *Let $f \in \mathrm{Rat}_d$ be postcritically finite with all critical points simple and $\mathcal{C}(f) \cap \mathcal{P}(f) = \emptyset$. Assume further that f is not a Lattès map. If $\xi \in T_f\mathrm{Rat}_d$ guides $\tau \in \mathcal{T}(\mathcal{P}(f))$ then $\xi \in T_f\mathcal{O}(f)$.*

Proof: Since, the vector space $\mathcal{Q}(\mathbb{P}^1, \mathcal{P}(f))$ is finite dimensional, the injectivity of $\nabla : \mathcal{Q}(\mathbb{P}^1, \mathcal{P}(f)) \to \mathcal{Q}(\mathbb{P}^1, \mathcal{P}(f))$ implies its surjectivity. Thus, it follows from lemma 3.4 that $\langle q, \tau \rangle = 0$ for every $q \in \mathcal{Q}(\mathbb{P}^1, \mathcal{P}(f))$.

Lemma 3.7. *A vector field $\tau \in T(X)$ extends holomorphically to $\mathbb{P}^1$ if and only if $\langle q, \tau \rangle = 0$ for every $q \in \mathcal{Q}(\mathbb{P}^1, X)$.*

Proof: We may clearly assume without loss of generality that X contains at least three distinct points x_1, x_2, x_3. Let θ be the unique holomorphic vector field on $\mathbb{P}^1$ which coincides with τ at x_1, x_2 and x_3. We must show that $\theta(x) = \tau(x)$ for any $x \in X - \{x_1, x_2, x_3\}$. Up to scale, there is a unique meromorphic quadratic differential q with simple poles at x_1, x_2, x_3 and x. The globally meromorphic 1-form $q \otimes \theta$ has only simple poles, and these must lie in $\{x_1, x_2, x_3, x\}$. The sum of residues of a meromorphic 1-form on $\mathbb{P}^1$ is 0. It follows that τ and θ coincide at x if and only if $\mathrm{Res}_x(q \otimes \theta) = \mathrm{Res}_x(q \otimes \theta)$, whence

$$\sum_{y \in \{x_1, x_2, x_3, x\}} \mathrm{Res}_y(q \otimes \tau) = \sum_{y \in \{x_1, x_2, x_3, x\}} \mathrm{Res}_y(q \otimes \theta)$$
$$= \sum_{y \in \mathbb{P}^1} \mathrm{Res}_y(q \otimes \theta) = 0. \qquad \square$$

Consequently, τ admits a globally holomorphic extension $\theta \in \mathrm{aut}(\mathbb{P}^1)$. Since τ is guided by ξ, it follows from Lemma 3.2 that the vector field $f^*\theta + \eta_\xi$ is holomorphic on $\mathbb{P}^1$. Moreover, $f^*\theta + \eta_\xi$ agrees with θ on $\mathcal{P}(f)$. Since a rational map whose postcritical set contains only two points is conjugate to $z \mapsto z^{\pm d}$, the set $\mathcal{P}(f)$ contains at least three points, whence the globally holomorphic vector fields are equal. That is to say $\eta_\xi = \theta - f^*\theta$ with $\theta \in \mathrm{aut}(\mathbb{P}^1)$. Since $\xi \in T_f\mathcal{O}(f)$ in view of Proposition 3.1, this completes the proof of Proposition 3.6. $\qquad \square$

3.4   Proof of Theorem 1.1

In this section, we prove Theorem 1.1. By assumption, $f \in \mathrm{Rat}_d$ is postcritically finite with $2d - 2$ distinct critical points, $\mathcal{C}(f) \cap \mathcal{P}(f) = \emptyset$ and f is not a Lattès map. Let the analytic germs

$$\underline{v} : (\mathrm{Rat}_d, f) \to (\mathbb{P}^1)^{2d-2} \quad \text{and} \quad \underline{a} : (\mathrm{Rat}_d, f) \to (\mathbb{P}^1)^{2d-2}$$

be defined as in the Introduction. We shall show that the linear map $D_f\underline{v} - D_f\underline{a}$ is surjective and that its kernel is $T_f\mathcal{O}(f)$.

Note that $T_f\mathrm{Rat}_d$ has complex dimension $2d + 1$. The map $D_f\underline{v} - D_f\underline{a}$ has maximal rank $2d - 2$ if and only if the kernel has dimension 3. Now on $\mathcal{O}(f)$, we have $\underline{v} \equiv \underline{a}$, whence $T_f\mathcal{O}(f) \subseteq \mathrm{Ker}(D_f\underline{v} - D_f\underline{a})$. Since $\mathcal{O}(f)$ has complex dimension 3, it suffices to show

$$\mathrm{Ker}(D_f\underline{v} - D_f\underline{a}) \subseteq T_f\mathcal{O}(f).$$

Henceforth, we assume that ξ belongs to $\mathrm{Ker}(D_f\mathfrak{v} - D_f\mathfrak{a})$. In view of Proposition 3.6, it suffices to show that ξ guides a vector field $\tau \in \mathcal{T}\big(\mathcal{P}(f)\big)$.

We begin by specifying τ on $\mathcal{P}(f)$. Let $t \mapsto f_t$ be a family of rational maps of degree d such that $f_0 = f$ and $\dot{f} = \xi$. If α is a repelling periodic point of f then, by the Implicit Function Theorem, there is a unique germ $t \mapsto \alpha_t$ with $\alpha_0 = \alpha$ such that α_t is a periodic point of f_t. For periodic $\alpha \in \mathcal{P}(f)$, we set

$$\tau(\alpha):=\dot{\alpha} \in T_\alpha \mathbb{P}^1.$$

Note that if $\beta = f(\alpha)$, then $\beta_t = f_t(\alpha_t)$ is a periodic point of f_t. Evaluating derivatives at $t = 0$ yields

$$\tau(\beta) = \dot{\beta} = \dot{f}(\alpha) + D_\alpha f(\dot{\alpha}) = \xi(\alpha) + D_\alpha f\big(\tau(\alpha)\big).$$

Since $D_\alpha f$ is invertible, we deduce that

$$(D_\alpha f)^{-1}\big(\tau(\beta)\big) = (D_\alpha f)^{-1}\big(\xi(\alpha)\big) + \tau(\alpha) \,,$$

whence

$$f^*\tau(\alpha) = -\eta_\xi(\alpha) + \tau(\alpha).$$

Thus, on the set of repelling periodic points contained in $\mathcal{P}(f)$, we have

$$\tau = f^*\tau + \eta_\xi. \tag{2}$$

Since there are no critical points in $\mathcal{P}(f)$, there is a unique extension of τ to the whole postcritical set such that (2) remains valid. Note that since no $z \in \mathcal{P}(f)$ is precritical, there is a unique analytic germ $t \mapsto z_t$ such that z_t is preperiodic under f_t, and we have $\tau(z) = \dot{z}$.

To complete the proof of Theorem 1.1, it suffices to show that $\tau \circ f = \xi$ on $\mathcal{C}(f)$. So, let $c \in \mathcal{C}$ and for $k \geq 1$, define $v_t^k:=f_t^{\circ k}(c_t)$. The critical point c is the j-th critical point of f as listed in the Introduction. We have $\mathfrak{v}_j(f_t) = v_t^\ell$ for some integer $\ell \geq 1$ and $\alpha_t:=\mathfrak{a}_j(f_t)$ is a periodic point of f_t with $\alpha = v^\ell$. By assumption, $\xi \in \mathrm{Ker}(D_f\mathfrak{v} - D_f\mathfrak{a})$, whence

$$\dot{v}^\ell = D_f\mathfrak{v}_j(\xi) = D_f\mathfrak{a}_j(\xi) = \dot{\alpha} = \tau(\alpha).$$

Differentiating $f_t(v_t^k) = v_t^{k+1}$ with respect to t yields

$$\xi(v^k) + D_{v^k} f(\dot{v}^k) = \dot{v}^{k+1}.$$

Applying $(D_{v^k} f)^{-1}$ gives $-\eta_\xi(v^k) + \dot{v}^k = (D_{v_k} f)^{-1}(\dot{v}^{k+1})$, whence

$$\dot{v}^k = (D_{v^k} f)^{-1}(\dot{v}^{k+1}) + \eta_\xi(v^k).$$

We now proceed by decreasing induction on $k \geq 1$. Since (2) holds on $\mathcal{P}(f)$, if $\dot{v}^{k+1} = \tau(v^{k+1})$, then

$$\dot{v}^k = (D_{v^k} f)^{-1}\big(\tau \circ f(v^k)\big) + \eta_\xi(v^k) = (f^*\tau + \eta_\xi)(v^k) = \tau(v^k).$$

The desired result is obtained by taking $k = 1$: $\xi(c) = \dot{v}^1 = \tau(v^1) = \tau \circ f(c)$.

4   The Bifurcation Measure

We will now prove that $\mathcal{X}_d^* \subseteq \mathrm{Supp}(\mu_{\mathrm{bif}})$. This will complete the proof of our main theorem. Here and henceforth, we let $f \in \mathrm{Rat}_d$ be a postcritically finite map with only simple critical points, which satisfies $\mathcal{C}(f) \cap \mathcal{P}(f) = \emptyset$ and is not a flexible Lattès map. It suffices to exhibit a $2d-2$-dimensional complex manifold $\Sigma \subset \mathrm{Rat}_d$ containing f and a basis of neighborhoods Σ_n of f in Σ, such that

$$\int_{\Sigma_n} (T_{\mathrm{bif}})^{\wedge(2d-2)} > 0.$$

4.1   Another Definition of the Bifurcation Current

We will use a second definition of the bifurcation current T_{bif} due to De-Marco [De01] (see [De03] or [BB07] for the equivalence of the two definitions). The current T_{bif} may be defined by considering the behavior of the critical orbits as follows.

Set $\mathcal{J}:=\{1,\ldots,2d-2\}$. Let $\pi : \mathbb{C}^2 - \{0\} \to \mathbb{P}^1$ be the canonical projection. Denote by $\tilde{x}:=(x_1, x_2)$ the points in $\mathbb{C}^2$. Let $U \subset \mathrm{Rat}_d$ be a sufficiently small neighborhood of f so that there are:

- holomorphic functions $\{\mathfrak{c}_j : U \to \mathbb{P}^1\}_{j \in \mathcal{J}}$ following the critical points of g as g ranges in U,

- holomorphic functions $\{\tilde{\mathfrak{c}}_j : U \to \mathbb{C}^2 - \{0\}\}_{j \in \mathcal{J}}$ such that $\pi \circ \tilde{\mathfrak{c}}_j = \mathfrak{c}_j$, and

- an analytic family $U \ni g \mapsto \tilde{g}$ of non-degenerate homogeneous polynomials of degree d such that $\pi \circ \tilde{g} = g \circ \pi$.

The map $\mathcal{G} : U \times \mathbb{C}^2 \to \mathbb{R}$ defined by

$$\mathcal{G}(g, \tilde{x}):= \lim_{n \to +\infty} \frac{1}{d^n} \log \|\tilde{g}^{\circ n}(\tilde{x})\|$$

is plurisubharmonic on $U \times \mathbb{C}^2$. DeMarco [De03] proved that

$$T_{\mathrm{bif}}\big|_U = \sum_{j=1}^{2d-2} \mathrm{dd}^c \mathcal{G}_j ,$$

where $\mathcal{G}_j : U \to \mathbb{R}$ is defined by $\mathcal{G}_j(g):=\mathcal{G}\big(g, \tilde{\mathfrak{c}}_j(g)\big)$.

4.2 Definition of Σ

Let the analytic germs $\mathfrak{c}, \mathfrak{a}, \mathfrak{v} : (\mathrm{Rat}_d, f) \to (\mathbb{P}^1)^{2d-2}$ be defined as in the Introduction. Set $\underline{c}:=\underline{c}(f)$ and $\underline{\alpha}:=\underline{a}(f) = \underline{v}(f)$. Given a rational map $g \in \mathrm{Rat}_d$, let $\underline{g} : (\mathbb{P}^1)^{2d-2} \to (\mathbb{P}^1)^{2d-2}$ be the map defined by

$$\underline{g}(\underline{z}):=\big(g(z_1), \ldots, g(z_{2d-2})\big).$$

Recall that for g near f, we have $\mathfrak{v}_j(g) = g^{\circ \ell_j} \circ \mathfrak{c}_j(g)$ for some integer ℓ_j. Let p_j be the period of α_j. Let p be the least common multiple of the periods p_j. For g near f, let $\mathfrak{m}_j(g)$ be the multiplier of α_j as a fixed point of $g^{\circ p}$. Denote by $\vec{x} = (x_1, \ldots, x_{2d-2})$ the elements of $\mathbb{C}^{2d-2}$ and let $\vec{M}_g : \mathbb{C}^{2d-2} \to \mathbb{C}^{2d-2}$ be the linear map defined by

$$\vec{M}_g(\vec{x}):=\big(\mathfrak{m}_1(g) \cdot x_1, \ldots, \mathfrak{m}_{2d-2}(g) \cdot x_{2d-2}\big).$$

For every $j \in \mathcal{J}$, α_j is repelling. It follows that for g near f, there is a local biholomorphism $\underline{\mathrm{Lin}}_g : (\mathbb{C}^{2d-2}, \vec{0}) \to \big((\mathbb{P}^1)^{2d-2}, \underline{a}(g)\big)$ linearizing $\underline{g}^{\circ p}$, that is

$$\underline{\mathrm{Lin}}_g \circ \vec{M}_g = \underline{g}^{\circ p} \circ \underline{\mathrm{Lin}}_g.$$

In addition, we may choose $\underline{\mathrm{Lin}}_g$ such that the germ $(g, \vec{x}) \mapsto \underline{\mathrm{Lin}}_g(\vec{x})$ is analytic near $(f, \vec{0})$ and the germ $(g, \underline{z}) \mapsto \underline{\mathrm{Lin}}_g^{-1}(\underline{z})$ is analytic near $(f, \underline{\alpha})$.

Lemma 4.1. *There exists an analytic germ* $S : (\mathbb{C}^{2d-2}, \vec{0}) \to (\mathrm{Rat}_d, f)$ *such that for $\vec{x}$ near $\vec{0}$*

$$\mathfrak{v} \circ S(\vec{x}) = \underline{\mathrm{Lin}}_{S(\vec{x})}.$$

Proof: Let $\vec{\mathfrak{h}} : (\mathrm{Rat}_d, f) \to (\mathbb{C}^{2d-2}, \vec{0})$ be the analytic germ defined by $\vec{\mathfrak{h}}(g):=\underline{\mathrm{Lin}}_g^{-1} \circ \underline{v}(g)$. Then $\underline{v}(g) = \underline{\mathrm{Lin}}_g \circ \vec{\mathfrak{h}}(g)$ and $\underline{a}(g) = \underline{\mathrm{Lin}}_g(\vec{0})$. Differentiating with respect to g and evaluating at $g = f$ yields

$$D_f \underline{v} = D_f \underline{a} + D_{\vec{0}}\underline{\mathrm{Lin}}_f \circ D_f \vec{\mathfrak{h}}.$$

According to Theorem 1.1, the linear map

$$D_f \underline{v} - D_f \underline{a} : T_f \mathrm{Rat}_d \to \bigoplus_{j \in \mathcal{J}} T_{\alpha_j} \mathbb{P}^1$$

is surjective. Since $D_{\vec{0}}\underline{\mathrm{Lin}}_f : \mathbb{C}^{2d-2} \to \bigoplus_{j \in \mathcal{J}} T_{\alpha_j} \mathbb{P}^1$ is invertible. Thus $D_f \vec{\mathfrak{h}} : T_f \mathrm{Rat}_d \to \mathbb{C}^{2d-2}$ has maximal rank. It follows from the Implicit Function Theorem that there is a section $S : (\mathbb{C}^{2d-2}, \vec{0}) \to (\mathrm{Rat}_d, f)$ with $\vec{\mathfrak{h}} \circ S = \mathrm{Id}$. This may be rewritten as $\underline{v} \circ S(\vec{x}) = \underline{\mathrm{Lin}}_{S(\vec{x})}$. $\square$

For each $j \in \mathcal{J}$, choose a neighborhood V_j of α_j in $\mathbb{P}^1$ such that there is a holomorphic section $\sigma_j : V_j \to \mathbb{C}^2 - \{0\}$ of $\pi : \mathbb{C}^2 - \{0\} \to \mathbb{P}^1$. Fix $r > 0$ small enough that the germ S given by Lemma 4.1 and the germ $\underline{\mathrm{Lin}}_f$ are both defined and analytic on a neighborhood of $\overline{\Delta}^{2d-2}$ where $\Delta := D(0, r)$, and such that

$$\underline{\mathrm{Lin}}_f(\overline{\Delta}^{2d-2}) \subset \underline{V} := \prod_{j \in \mathcal{J}} V_j.$$

We set

$$\Sigma := S(\Delta^{2d-2}).$$

By definition, this subset of Rat_d is a submanifold of complex dimension $2d - 2$.

4.3 Definition of Σ_n

For $n \geq 1$, set

$$S_n := S \circ \overrightarrow{M}_f^{-n} : \Delta^{2d-2} \to \Sigma \quad \text{and} \quad \Sigma_n := S_n(\Delta^{2d-2}) \subset \Sigma.$$

Clearly, the sets Σ_n form a basis of neighborhoods of f in Σ. For $n \geq 1$, let $\underline{\mathfrak{v}}^n : \Sigma \to \mathbb{P}^1$ be the map defined by

$$\underline{\mathfrak{v}}^n(g) := \underline{g}^{\circ(np)} \circ \underline{\mathfrak{v}}(g).$$

Note that for $j \in \mathcal{J}$, we have $\mathfrak{v}_j^n(g) = g^{\circ(\ell_j + np)} \circ \mathfrak{c}_j(g)$.

Lemma 4.2. The sequence $(\underline{\mathfrak{v}}^n \circ S_n)$ converges uniformly to $\underline{\mathrm{Lin}}_f$ on Δ^{2d-2}.

Proof: Note that the sequence (S_n) converges to f uniformly and exponentially on $\overline{\Delta}^{2d-2}$ as n tends to ∞. It follows that the sequence $(\vec{x} \mapsto \overrightarrow{M}_{S_n(\vec{x})})$ converges uniformly and exponentially to $\overrightarrow{M}_f$. Consequently the sequence $(\vec{x} \mapsto \overrightarrow{M}_{S_n(\vec{x})}^n \circ M_f^{-n}(\vec{x}))$ converges uniformly to the identity. Thus, for n large enough and for any $\vec{x} \in \Delta^{2d-2}$, setting $g_n := S_n(\vec{x})$, we have

$$\underline{\mathrm{Lin}}_{g_n} \circ \overrightarrow{M}_{g_n}^n \circ \overrightarrow{M}_f^{-n}(\vec{x}) = \underline{g_n}^{\circ(np)} \circ \underline{\mathrm{Lin}}_{g_n} \circ \overrightarrow{M}_f^{-n}(\vec{x})$$

$$= \underline{g_n}^{\circ(np)} \circ \underline{\mathfrak{v}} \circ S \circ M_f^{-n}(\vec{x})$$

$$= \underline{\mathfrak{v}}^n \circ S_n(\vec{x}).$$

The result follows easily. $\square$

4.4 The Proof

Here, we shall use the notation of Section 4.1. We assume that n is large enough that Σ_n is contained in U, whence every map $g \in \Sigma_n$ has a lift $\tilde{g}$ to homogeneous coordinates and there is a potential function $\mathcal{G}$ defined on $\Sigma_n \times \mathbb{C}^2$ such that

$$\mathcal{G}\big(g, \tilde{g}(\tilde{x})\big) = d \cdot \mathcal{G}(g, \tilde{x}) \quad \text{and} \quad \forall \lambda \in \mathbb{C}^*, \quad \mathcal{G}(g, \lambda \tilde{x}) = \mathcal{G}(g, \tilde{x}) + \log|\lambda|.$$

Recall that by Lemma 4.2, the sequence $(\underline{\mathfrak{v}}^n \circ S_n)$ converges uniformly to $\underline{\mathrm{Lin}}_f$ on Δ^{2d-2} and by assumption, $\underline{\mathrm{Lin}}_f(\overline{\Delta}^{2d-2}) \subset \underline{V}$. From now on, let n be sufficiently large so that $\underline{\mathfrak{v}}^n(\Sigma_n) \subseteq \underline{V}$. In that case, for each $j \in \mathcal{J}$, the map

$$\tilde{\mathfrak{v}}^n_j := \sigma_j \circ \mathfrak{v}^n_j : \Sigma_n \to \mathbb{C}^2 - \{0\}$$

is well defined. In this case, we may define plurisubharmonic functions $\mathcal{G}^n_j : \Delta^{2d-2} \dashrightarrow \mathbb{R}$ by

$$\mathcal{G}^n_j(\underline{x}) := \mathcal{G}\big(S_n(\underline{x}), \tilde{\mathfrak{v}}^n_j \circ S_n(\underline{x})\big).$$

Lemma 4.3. *If n is sufficiently large, then*

$$S^*_n(T_{\mathrm{bif}}) = d^{-np} \sum_{j \in \mathcal{J}} d^{-\ell_j} \mathrm{dd}^c \mathcal{G}^n_j.$$

Proof: Note that for $g \in \Sigma_n$ and $j \in \mathcal{J}$, we have

$$\pi \circ \tilde{g}^{\circ(\ell_j + np)}\big(\tilde{\mathfrak{c}}_j(g)\big) = g^{\circ(\ell_j + np)}\big(\mathfrak{c}_j(g)\big) = \mathfrak{v}^n_j(g) = \pi \circ \tilde{\mathfrak{v}}^n_j(g),$$

whence

$$\tilde{g}^{\circ n}\big(\tilde{\mathfrak{c}}_j(g)\big) = \lambda^n_j(g) \cdot \tilde{\mathfrak{v}}^n_j(g)$$

for some holomorphic functions $\lambda^n_j : \Sigma_n \to \mathbb{C}^*$. Thus, if $\vec{x} \in \Delta^{2d-2}$ and $g_n := S_n(\vec{x})$, then

$$\begin{aligned}
\mathcal{G}^n_j(\vec{x}) &= \mathcal{G}\big(g_n, \tilde{g}_n^{\circ(\ell_j + np)}\big(\tilde{\mathfrak{c}}_j(g_n)\big)\big) - \log\big|\lambda^n_j(g_n)\big| \\
&= d^{\ell_j + np} \mathcal{G}\big(g_n, \tilde{\mathfrak{c}}_j(g_n)\big) - \log\big|\lambda^n_j(g_n)\big| \\
&= d^{\ell_j + np} \mathcal{G}_j(g_n) - \log\big|\lambda^n_j(g_n)\big| \\
&= d^{\ell_j + np} \mathcal{G}_j \circ S_n(\vec{x}) - \log\big|\lambda^n_j \circ S_n(\vec{x})\big|.
\end{aligned}$$

Since $\log|\lambda^n_j \circ S_n|$ is pluriharmonic on Δ^{2d-2}, we have

$$\mathrm{dd}^c \mathcal{G}^n_j = d^{\ell_j + np} \cdot \mathrm{dd}^c(\mathcal{G}_j \circ S_n) = d^{\ell_j + np} \cdot S^*_n(\mathrm{dd}^c \mathcal{G}_j).$$

Consequently, in view of DeMarco's formula, we have

$$S^*_n(T_{\mathrm{bif}}) = \sum_{j \in \mathcal{J}} S^*_n(\mathrm{dd}^c \mathcal{G}_j) = d^{-np} \sum_{j \in \mathcal{J}} d^{-\ell_j} \mathrm{dd}^c \mathcal{G}^n_j.$$

$\square$

Set

$$M_n := \int_{\Sigma_n} T_{\mathrm{bif}}{}^{\wedge(2d-2)}.$$

To conclude the proof of the main theorem, we will now show that $M_n > 0$ for n large enough. In fact, we will show that there is a constant $m > 0$ such that

$$M_n \underset{n \to +\infty}{\sim} \frac{m}{d^{(2d-2)np}}.$$

Set $|\ell| := \sum_{j \in \mathcal{J}} \ell_j$. For $j \in \mathcal{J}$, let $\mathrm{lin}_j : \Delta \to \mathbb{P}^1$ be the map defined by

$$\underline{\mathrm{Lin}}_f(\vec{x}) = \big(\mathrm{lin}_1(x_1), \dots, \mathrm{lin}_{2d-2}(x_{2d-2})\big)$$

and set

$$W_j := \mathrm{lin}_j(\Delta).$$

Recall that μ_f is the equilibrium measure of f.

Lemma 4.4. *We have*

$$\lim_{n \to +\infty} d^{(2d-2)np} \cdot M_n = (2d-2)! \cdot d^{-|\ell|} \cdot \prod_{j \in \mathcal{J}} \mu_f(W_j).$$

Proof: By Lemma 4.2, the sequences of functions $\mathfrak{v}_j^n \circ S_n$ converge uniformly on Δ^{2d-2} to $\underline{x} \mapsto \mathrm{lin}_j(x_j)$. So, the sequences of functions $\mathcal{G}_j^n : \Delta^{2d-2} \to \mathbb{R}$ converge uniformly to

$$\mathcal{G}_j^\infty : \underline{x} \mapsto \mathcal{G}\big(f, \sigma_j \circ \mathrm{lin}_j(x_j)\big).$$

Due to the uniform convergence of the potentials, we may write

$$\lim_{n \to +\infty} \int_{\Delta^{2d-2}} \left(\sum_{j \in \mathcal{J}} d^{-\ell_j} \mathrm{dd}^c \mathcal{G}_j^n \right)^{\wedge(2d-2)} = \int_{\Delta^{2d-2}} \left(\sum_{j \in \mathcal{J}} d^{-\ell_j} \mathrm{dd}^c \mathcal{G}_j^\infty \right)^{\wedge(2d-2)}.$$

Note that $\mathcal{G}_j^\infty$ only depends on the j-th coordinate. It follows that

$$\left(\sum_{j=1}^{2d-2} d^{-\ell_j} \mathrm{dd}^c \mathcal{G}_j^\infty \right)^{\wedge(2d-2)} = (2d-2)! \, d^{-|\ell|} \cdot \bigwedge_{j \in \mathcal{J}} \mathrm{dd}^c \mathcal{G}_j^\infty.$$

In addition, $\mathcal{G}_j^\infty(\underline{x}) = G_j \circ \mathrm{lin}_j(x_j)$ with $G_j : W_j \to \mathbb{R}$ the subharmonic function defined by $G_j(z) = \mathcal{G}\big(f, \sigma_j(z)\big)$. We have $\mathrm{dd}^c G_j = \mu_f|_{W_j}$. Therefore, according to Fubini's theorem,

$$\int_{\Delta^{2d-2}} \left(\bigwedge_{j \in \mathcal{J}} \mathrm{dd}^c \mathcal{G}_j^\infty \right) = \prod_{j \in \mathcal{J}} \left(\int_\Delta \mathrm{dd}^c \left(G_j \circ \mathrm{lin}_j \right) \right) = \prod_{j \in \mathcal{J}} \mu_f(W_j). \qquad \square$$

We now complete the proof. Since the periodic points α_j are repelling, they are in the support of the equilibrium measure μ_f. Thus, for every $j \in \mathcal{J}$, $\mu_f(W_j) > 0$. As a consequence,

$$M_n \underset{n \to +\infty}{\sim} \frac{m}{d^{(2d-2)np}} \quad \text{with} \quad m := (2d-2)! \cdot d^{-|\ell|} \cdot \prod_{j=1}^{2d-2} \mu_f(W_j) > 0,$$

f is in the support of $T_{\mathrm{bif}}{}^{\wedge(2d-2)}$ and the conjugacy class of f is in the support of μ_{bif}.

Bibliography

[BB07] G. Bassanelli and F. Berteloot, *Bifurcation currents in holomorphic dynamics on* $\mathbb{P}^k$, J. Reine Angew. Math. **608** (2007), 201–235.

[De01] L. DeMarco, *Dynamics of rational maps: a current on the bifurcation locus*, Math. Res. Lett. **8** 1–2 (2001), 57–66.

[De03] L. DeMarco, *Dynamics of rational maps: Lyapunov exponents, bifurcations, and capacity*, Math. Ann. **326** 1 (2003), 43–73.

[DH93] A. Douady and J. H. Hubbard, *A proof of Thurston's topological characterization of rational functions*, Acta Math. **171** 2 (1993), 263–297.

[DF08] R. Dujardin and C. Favre, *Distribution of rational maps with a preperiodic critical point*, Amer. Journ. Math. **130** (2008), 979–1032.

[Ep99] A. L. Epstein, *Infinitesimal Thurston rigidity and the Fatou-Shishikura inequality*, Stony Brook IMS Preprint 1, 1999.

[HS94] J. H. Hubbard and D. Schleicher, *The spider algorithm*, in: Complex Dynamical Systems, Proc. Sympos. Appl. Math. **49**, Amer. Math. Soc., Providence, RI, 1994, 155–180.

[Ly83] M. J. Lyubich, *Entropy properties of rational endomorphisms of the Riemann sphere*, Ergod. Th. Dynam. Sys. **3** 3 (1983), 351–385.

[Ma88] R. Mañé, *The Hausdorff dimension of invariant probabilities of rational maps*, Lecture Notes in Math. **1331**, Springer, Berlin, 1988, 86–117.

[MSS83] R. Mañé, P. Sad, and D. Sullivan, *On the dynamics of rational maps*, Ann. Sci. École Norm. Sup. (4) **16** 2 (1983), 193–217.

[Mc87]   C. T. McMullen, *Families of rational maps and iterative root-finding algorithms*, Ann. of Math. (2) **125** 3 (1987), 467–493.

[Mc94]   C. T. McMullen, *Complex dynamics and renormalization*, Annals of Mathematics Studies, **135**, Princeton University Press, Princeton, NJ, 1994.

[Mi06]   J. Milnor, *On Lattès maps*, Dynamics on the Riemann sphere, Eur. Math. Soc., Zürich, 2006, 9–43.

[Sh87]   M. Shishikura, *On the quasiconformal surgery of rational functions*, Ann. Sci. École Norm. Sup. (4) **20** 1 (1987), 1–29.

[Ta90]   L. Tan, *Similarity between the Mandelbrot set and Julia sets*, Comm. Math. Phys. **134** 3 (1990), 587–617.

[vS00]   S. van Strien, *Misiurewicz maps unfold generically (even if they are critically non-finite)*, Fund. Math. **163** 1 (2000), 39–54.

13 Real Dynamics of a Family of Plane Birational Maps: Trapping Regions and Entropy Zero

Eric Bedford and Jeffrey Diller

1 Introduction

We consider dynamics of the one parameter family of birational maps

$$f = f_a : (x, y) \mapsto \left(y\frac{x + a}{x - 1}, x + a - 1 \right). \tag{1.1}$$

This family was introduced and studied by Abarenkova, Anglès d'Auriac, Boukraa, Hassani, and Maillard, with results published in [A1–7]. We consider here real (as opposed to complex) dynamics, treating f_a as a self-map of $\mathbf{R}^2$, and we restrict our attention to parameters $a > 1$. In order to discuss our main results, we let $\mathcal{B}^+, \mathcal{B}^- \subset \mathbf{R}^2$ be the sets of points with orbits diverging locally uniformly to infinity in forward/backward time, and we take $K \subset \mathbf{R}^2$ to be the set of points $p \in \mathbf{R}^2$ whose full orbits $(f^n(p))_{n \in \mathbf{Z}}$ are bounded.

In [BD05] we studied the dynamics of f_a for the parameter region $a < 0$, $a \neq -1$. In this case, $\mathcal{B}^+$ and $\mathcal{B}^-$ are dense in $\mathbf{R}^2$; and the complement $\mathbf{R}^2 - \mathcal{B}^+ \cup \mathcal{B}^-$ is a non-compact set on which the action of f_a is very nearly hyperbolic and essentially conjugate to the golden mean subshift. In particular, f_a is topologically mixing on $\mathbf{R}^2 - \mathcal{B}^+ \cup \mathcal{B}^-$, and most points therein have unbounded orbits. The situation is quite different when $a > 1$.

Theorem 1.1. *If $a > 1$, then $\mathbf{R}^2 - \mathcal{B}^+ \cup \mathcal{B}^- = K$. Moreover, the set K is compact in $\mathbf{R}^2$, contained in the square $[-a, 1] \times [-1, a]$.*

The sets $\mathcal{B}^+$ and K are illustrated for typical parameter values $a > 1$ in Figure 1. Note that all figures in this paper are drawn with $\mathbf{R}^2$ compactified as a torus $S^1 \times S^1 \cong (\mathbf{R} \cup \{\infty\}) \times (\mathbf{R} \cup \{\infty\})$. The plane is parametrized so that infinity is visible, with top/bottom and left/right sides identified

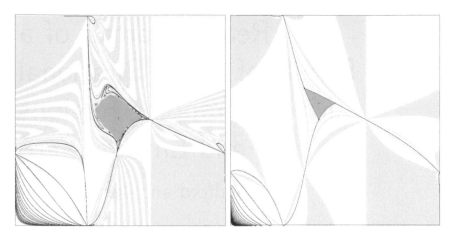

Figure 1. Dynamics of f for parameter values $a = 1.1$ (left) and $a = 2$ (right). Forward orbits of white points escape to infinity traveling up and to the right. Yellow points escape by alternating between the bottom right and upper left corners. The black curves are stable manifolds of a saddle three cycle (the vertices of the "triangle"). The green regions consist of points whose orbits are bounded in forward and backward time. Reflection about the line $y = -x$ corresponds to replacing f by f^{-1}. It leaves the green region invariant and exchanges stable and unstable manifolds. (See Plate VI.)

with the two circles at infinity. All four corners of the square correspond to the point (∞, ∞), which is a parabolic fixed point for f_a.

For each $a \neq -1$ there is a unique fixed point $p_{fix} = ((1 - a)/2, (a - 1)/2) \in \mathbf{R}^2$. For $a < 0$, p_{fix} is a saddle point, and for $a > 0$, p_{fix} is indifferent with $Df_a(p_{fix})$ conjugate to a rotation. For generic $a > 0$, f_a acts on a neighborhood of p_{fix} as an area-preserving twist map with non-zero twist parameter (see Proposition 5.1). As is shown in Figure 2, f_a exhibits KAM behavior for typical $a > 0$. In particular K has non-empty interior, and the restriction $f_a : K \circlearrowleft$ is neither topologically mixing nor hyperbolic.

It is known (see [DF01] Section 9) that f_a is integrable for the parameter values $a = -1, 0, \frac{1}{3}, \frac{1}{2}, 1$. In particular f_a has topological entropy zero for these parameters. On the basis of computer experiments, [AABM00, ABM99] conjectured that $a = 3$ is the (unique) other parameter where entropy vanishes. The map f_3 is not integrable, and in fact the complexified map $f_3 : \mathbf{C}^2 \circlearrowleft$ has entropy $\log 1 + \frac{\sqrt{5}}{2} > 0$ (see [BD05] and [Du06]). Nevertheless, we prove in Section 6 that not only does $f_3 : \mathbf{R}^2 \circlearrowleft$ have zero entropy; all points except p_{fix} in $\mathbf{R}^2$ are transient:

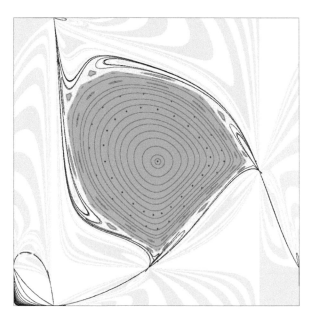

Figure 2. This is the left side of Figure 1 redrawn with coordinates changed to magnify the behavior near p_{fix}. Shown are several invariant circles about p_{fix}, as well as some intervening cycles and elliptic islands of large period. (See Plate VII.)

Theorem 1.2. *For $a = 3$, $K = \{p_{fix}\}$. The stable (and unstable) set of p_{fix} consists of three analytic curves. The three curves meet transversely at p_{fix} and have no other pairwise intersections in $\mathbf{R}^2$.*

Thus, p_{fix} is the only periodic and the only non-wandering point in $\mathbf{R}^2$. Figure 3 illustrates this theorem. As the figure makes evident, the stable arcs for p_{fix} intersect pairwise on a countable set at infinity. This reflects the fact that f is not a homeomorphism.

Often in this paper we will bound the number of intersections between two real curves by computing the intersection number of their complexifications. We will also use the special structure (see Figure 4) of f_a to help track the forward and backward images of curves. The versatility of these two techniques is seen from the fact that they apply in cases where maps have maximal entropy [BD05, BD] and in the present situation where we will show that f_3 has zero entropy.

Acknowledgments. The first and second authors are supported by NSF grants DMS-0601965 and DMS-0653678, respectively.

Figure 3. Dynamics of f when $a = 3$. The three cycle has now disappeared, and with it the twist map dynamics. The indifferent fixed point is now parabolic, and its stable and unstable sets are shown in red and green, respectively. (See Plate VII.)

2 Background

Here we recall some basic facts about the family of maps (1.1). Most of these are discussed at greater length in [BD05]. The maps (1.1) preserve the singular two form

$$\eta := \frac{dx \wedge dy}{y - x + 1}.$$

Each map $f = f_a$ is also reversible, which is to say equal to a composition of two involutions [BHM01, BHM98]. Specifically, $f = \tau \circ \sigma$

$$\tau(x, y) := \left(x\frac{a - y}{1 + y}, a - 1 - y \right), \quad \sigma(x, y) := (-y, -x).$$

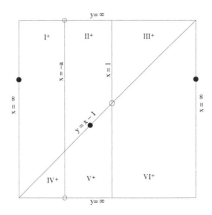
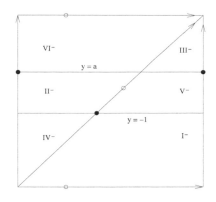

Figure 4. Partition of $\mathbf{R}^2$ by invariant curves and critical sets of f (left side) and f^{-1} (right side). Points of indeterminacy for f appear as hollow circles, whereas those for f^{-1} are shown as solid circles. Arrows on the right side indicate the direction in which f^2 translates points along $\operatorname{supp}(\eta)$.

In particular, $f^{-1} = \sigma \circ \tau$ is conjugate to f by either involution, a property that will allow us to infer much about f^{-1} directly from facts about f.

Though our goal is to understand the dynamics of f acting on $\mathbf{R}^2$, it will be convenient to extend the domain of f first by complexification to all of $\mathbf{C}^2$ and then by compactifying each coordinate separately to $\mathbf{P}^1 \times \mathbf{P}^1$. This gives us a convenient way of keeping track of the complexity of algebraic curves in $\mathbf{P}^1 \times \mathbf{P}^1$. Any such curve V is given as the zero set of a rational function $R : \mathbf{P}^1 \times \mathbf{P}^1 \to \mathbf{P}^1$. The *bidegree* $(j, k) \in \mathbf{N}^2$ of R, obtained by taking the degrees of R as a function of the first and second variables in turn, encodes the second homology class of V. In particular, if V and V' are curves with bidegrees (j, k) and (j', k') having no irreducible components in common, then the number of (complex) intersections, counted with multiplicity, between V and V', is

$$V \cdot V' = jk' + j'k. \tag{2.2}$$

We will use this fact at key points as a convenient upper bound for the number of *real* intersections between two algebraic curves.

Bidegrees transform linearly under our maps. Taking f^*V to be the zero set of $R \circ f$, we have

$$\operatorname{bideg} f^*V = \begin{pmatrix} 1 & 1 \\ 1 & 0 \end{pmatrix} \operatorname{bideg} V. \tag{2.3}$$

Similarly, f_*V (the zero set of $R \circ f^{-1}$) has bidegree given by

$$\text{bideg}\, f_*V = \begin{pmatrix} 0 & 1 \\ 1 & 1 \end{pmatrix} \text{bideg}\, V. \tag{2.4}$$

The divisor (η) of η, regarded as a meromorphic two form on $\mathbf{P}^1 \times \mathbf{P}^1$, is supported on the three lines $\{x = \infty\}$, $\{y = \infty\}$, $\{y = x - 1\}$ where η has simple poles. It follows from $f^*\eta = \eta$ that $\text{supp}\,(\eta)$ is invariant under f. Specifically, f interchanges the lines at infinity according to

$$(\infty, y) \mapsto (y, \infty) \mapsto (\infty, y + a - 1)$$

and maps $\{y = x-1\}$ to itself by $(x, x-1) \mapsto (x+a, x+a-1)$. Thus f^2 acts by translation on each of the three lines separately. The directions of the translations divide parameter space into three intervals: $(-\infty, 0)$, $(0, 1)$, and $(1, \infty)$. In particular, the directions are the same for all $a \in (1, \infty)$, which is the range that concerns us here.

It should be stressed that $f : \mathbf{P}^1 \times \mathbf{P}^1 \circlearrowright$ is *not* a diffeomorphism, nor indeed even continuous at all points. In particular, the critical set $C(f)$ consists of two lines $\{x = -a\}$ and $\{x = 1\}$, which are mapped by f to points $(0, 1)$ and (∞, a), respectively. Applying the involution σ, one finds that $\{y = -1\}$ and $\{y = a\}$ are critical for f^{-1}, mapping backward to $(-a, \infty)$ and $(-1, 0)$, respectively. Clearly, f cannot be defined continuously at the latter two points. Hence we call each a *point of indeterminacy* and refer to $I(f) = \{(-a, \infty), (-1, 0)\}$ as the *indeterminacy set* of f. Likewise, $I(f^{-1}) = \{(0, 1), (\infty, a)\}$. For convenience, we let

$$I^\infty(f) := \bigcup_{n \in \mathbf{N}} I(f^n) = \bigcup_{n \in \mathbf{N}} f^{-n}I(f)$$

$$C^\infty(f) := \bigcup_{n \in \mathbf{N}} C(f^n) = \bigcup_{n \in \mathbf{N}} f^{-n}C(f)$$

denote the set of all points which are indeterminate/critical for high enough forward iterates of f. We point out that both $I^\infty(f)$ and $I^\infty(f^{-1})$ are contained in $\text{supp}\,(\eta)$. Since $I(f)$ is contained in $\text{supp}\,(\eta)$, it follows that $I^\infty(f)$ is a discrete subset of $\text{supp}\,(\eta)$ that accumulates only at (∞, ∞). We also observe that for the parameter range $a > 1$, we have $I^\infty(f) \cap I^\infty(f^{-1}) = \emptyset$, a fact which will be useful to us below.

The closure $\overline{\mathbf{R}^2}$ of the real points in $\mathbf{P}^1 \times \mathbf{P}^1$ is just the torus $S^1 \times S^1$. The left side of Figure 4 shows how the critical set of f together with $\text{supp}\,(\eta)$ partition $\overline{\mathbf{R}^2}$ into six open sets, which we have labeled I^+ to VI^+. The right side shows the partition by $\text{supp}\,(\eta)$ and the critical set of f^{-1}. As f is birational, each piece of the left partition maps diffeomorphically onto a piece of the right partition. We have chosen labels on the right side so that I^+ maps to I^-, etc.

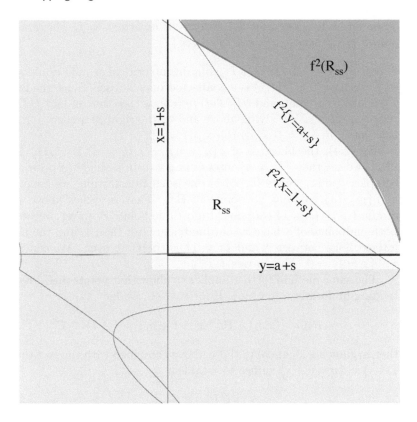

Figure 5. First trapping region (shaded yellow/green) for f. Also shown are the subregion R_{ss} (the square with lower left corner $(1+s, a+s)$ bounded by the thick black lines) and $f^2(R_{ss})$ (green region) for $s = .3$. The parameter value is $a = 1.2$. (See Plate VII.)

3 Trapping Regions

For the rest of this paper we assume $a > 1$. Figure 4 is useful for determining images and preimages of real curves by f. In this section, we combine the information presented in the figure with intersection data gleaned from bidegrees to help identify two "trapping regions" through which orbits are forced to wander to infinity. Our first trapping region is the set

$$T_0^+ := \{(x, y) \in \mathbf{R}^2 : x > 1, y > a\}$$

of points lying to the right of the critical set of f and above the critical set of f^{-1}.

Theorem 3.1. *The region T_0^+ is forward invariant by f. Forward orbits of points in T_0^+ tend uniformly to (∞, ∞).*

Proof: Since $T_0^+ \subset III^+ \cup VI^+$ contains no critical or indeterminacy points of f, we have that $f(T_0^+)$ is a connected open subset. Since the left side of T_0^+ maps to a point, and $\partial T_0^+ \cap I(f) = \emptyset$, we see that in fact $f(T_0^+)$ is the region in $III^- \cup VI^-$ lying above and to the right of an arc $\gamma \subset f\{y = a\}$ that joins $(a, \infty) = f(\infty, a)$ to $(\infty, a) = f\{x = 1\}$.

By (2.4), the bidegree of $f\{y = a\} = f_*(y = a)$ is $(1, 1)$. Hence by (2.2), we see that $f\{y = a\}$ intersects $y = a$ in a single (a priori, possibly complex) point. Since (a, ∞) is one such intersection, we conclude that $\gamma \cap \{y = a\}$ contains no points in $\mathbf{R}^2$. That is, γ lies above $\{y = a\}$. Similarly, $\gamma \cap \{x = 1\}$ contains at most one point. However, because $a > 1$, both endpoints of γ have x coordinates greater than 1, and the number of intersections between γ and $\{x = 1\}$ is therefore even. We conclude that $\gamma \cap \{x = 1\} = \emptyset$. This proves $f(T_0^+) \subset T_0^+$.

The same method further applies to show that points in T_0^+ have orbits tending uniformly to (∞, ∞). If for $s, t > 0$, we let

$$R_{st} = \{(x, y) \in \mathbf{R}^2 : x > 1 + s, y > 1 + t\} \subset T_0^+,$$

then arguments identical to those above combined with the fact that $f\{x = 1 + s\} = \{y = a + s\}$ suffice to establish

$$f(R_{st}) \subset R_{t+a, s}.$$

Hence $f^2(R_{ss}) \subset R_{s+a-1, s+a-1}$. Since $a > 1$ and the closures $\overline{R_{ss}} \subset \overline{\mathbf{R}^2}$ decrease to the point (∞, ∞) as $s \to \infty$, the proof is finished. □

Our second trapping region is more subtle. In particular, it has two connected components which are interchanged by f. The first component is

$$A := \{(x, y) \in \mathbf{R}^2 : x > 1, y < -x\}.$$

Lemma 3.2. *We have $A \cap f^{-1}(A) = \emptyset$ and $f(A) \subset f^{-1}(A)$.*

Proof: The set A lies entirely in region I^- shown on the right side of Figure 4, and therefore $f^{-1}(A)$ lies in region I^+ on the left side. In particular $A \cap f^{-1}(A) = \emptyset$.

To see that $f(A) \subset f^{-1}(A)$, we note that A is bounded in $\overline{\mathbf{R}^2}$ by portions of three lines, one of which $\{x = 1\}$ is critical for f. Moreover, $\overline{A}$ contains no point in $I(f)$ or $I(f^{-1})$ and A itself avoids the critical sets of f and f^{-1}. Therefore, $f(A)$ is a connected open subset of $\mathbf{R}^2$, bounded on the left by the segment $\{(-\infty, y) : y \geq a\}$ and the right by an arc γ in

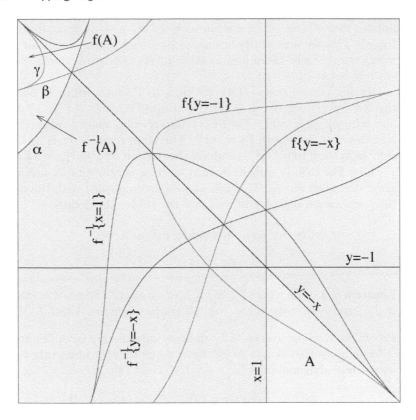

Figure 6. The second trapping region for (∞, ∞), shown here in yellow together with the curves used in the proof of Lemma 6. Lines used in the proof are black, images of lines are red, and preimages are blue. The parameter value is $a = 3$. (See Plate VII.)

$f\{y = -x\}$ that joins $(-\infty, a)$ to $(-\infty, \infty)$. We claim that $f(A)$ lies below $\{y = -x\}$ and above the arc β in $f\{y = -1\} \cap VI^-$ that joins $(-\infty, a)$ to $(-a, \infty)$. Both these claims can be verified in the same fashion; we give the details only for $\{y = -x\}$.

Clearly $(-\infty, a) \in \gamma$ lies below $\{y = -x\}$, and $(-\infty, \infty) \in \gamma$ lies exactly on $\{y = -x\}$. Hence to establish that $f(A)$ is below $\{y = -x\}$ it suffices to show that γ does not intersect $\{y = -x\}$ at any other point. But the curve $f_*\{y = -x\}$ containing γ has bidegree $(1, 2)$ and $\{y = -x\}$ has bidegree $(1, 1)$, so there are a total of three complex intersections between the two curves. The intersection at $(-\infty, \infty)$ is actually a tangency, which accounts for two of the three intersections. The fact that $\{y = -x\}$ joins the critical line $\{x = -a\}$ to the critical line $\{x = 1\}$ through regions II^+ and V^+

implies that $f\{y = -x\}$ joins through an arc passing right from $(1,0)$ to (∞, a). This arc necessarily intersects $\{y = -x\}$ and accounts for the third intersection. Since there are no other intersections we conclude that γ is entirely below $\{y = -x\}$.

Having pinned down $f(A)$, we turn to $f^{-1}(A)$, which is bounded on the left by $\{(-\infty, y) : y \geq 1\}$, on the right by $f^{-1}\{x = 1\}$ and from above by $f^{-1}\{y = -x\}$. The top boundary component is in fact just $\sigma(\gamma)$ and therefore lies above $\{y = -x\}$. Therefore to show $f(A) \subset f^{-1}(A)$, it suffices to show that $f(A)$ lies above the portion α of $f^{-1}\{x = 1\}$ bounding $f^{-1}(A)$. For this, it suffices in turn to show simply that α lies below the arc β described above. This can also be accomplished with the method of the previous paragraph and we spare the reader the details. □

We now define our second trapping region to be

$$T_1^+ := A \cup f^{-1}(A).$$

Theorem 3.3. *The region T_1^+ is forward invariant. Forward orbits of points in T_1^+ tend uniformly to (∞, ∞), alternating between A and $f^{-1}(A)$.*

Proof: Forward invariance of T_1^+ follows immediately from Lemma 3.2. In order to show that points in T_1^+ have orbits tending uniformly to infinity, we consider diagonal lines

$$L_t := \{y = t(x-1)\}, \quad L_t' := \{y = tx - 1\}$$

passing through the points $(1,0) \in I(f)$ and $(0, -1) \in I(f^{-1})$, respectively.

Lemma 3.4. *The curve $f^2(L_t) \cap A$ (non-empty only for $t < -1$) lies strictly below L_t'.*

Proof: One computes easily that $f(L_t) = L_{1/t}'$ (we only include the *strict* transform of L_t in the image $f(L_t)$; the line $\{y = a\} = f(1,0)$ appearing in $f_* L_t$ is omitted). Because $Df_{(\infty,\infty)}^2 = \mathrm{id}$, it follows that $f^2(L_t) = f(L_{1/t}')$ is tangent to L_t' at (∞, ∞). Moreover, $f^2(L_t) \cap II^- = f(L_{1/t}' \cap II^+)$ joins $(0, -1)$ to $(a, 0)$ and therefore intersects L_t' at $(0, -1)$. This gives us a total of three points (counting multiplicity) in $f^2(L_t) \cap L_t' - A$. On the other hand both L_t' and L_t have bidegree $(1,1)$ regardless of t, so from (2.2) and (2.4), we find $L_t' \cdot f_* L_{1/t}' = 3$. It follows that

$$f^2(L_t) \cap L_t' \cap A = \emptyset.$$

Finally, since $f^2(L_t) \cap I^- = f(L_{1/t}' \cap I^+)$ joins $(0, -1)$ to (∞, ∞), we conclude that it lies strictly below L_t'. □

To complete the proof of Theorem 3.3, let $p \in T_1^+$ be given. We may assume in fact that $p \in A$. Then each point $p_n := (x_n, y_n) := f^{2n}(p)$, $n \geq 0$ lies in A. Therefore $p_n \in L_{t_n}$ for some $t_n < -1$. Because L_t' is below L_t, the previous lemma implies that (t_n) is a decreasing sequence. More precisely, $t_{n+1} < r t_n$, where $r = r(x_n) < 1$ increases to 1 as $x_n \to \infty$.

Now if (p_n) does not converge to (∞, ∞), we have a subsequence (p_{n_j}) such that $x_{n_j} < M < \infty$. The previous paragraph implies that by further refining this subsequence, we may assume that $p_{n_j} \to (x, \infty)$ for some $x < \infty$. This, however, contradicts the facts that f is continuous on $\overline{A}$ and that $f^n(x, \infty) \to (\infty, \infty)$. $\qquad\square$

Let us define the forward basin $\mathcal{B}^+$ of (∞, ∞) to be the set of points $p \in \overline{\mathbf{R}^2}$ for which there exists a neighborhood $U \ni p$ such that $f^n | U$ is well defined for all $n \in \mathbf{N}$ and converges uniformly to (∞, ∞) on U. Our definition of $\mathcal{B}^+$ here differs slightly from the one given in the introduction in that we now allow points at infinity. Note that $(\infty, \infty) \notin \mathcal{B}^+$, because $I^\infty(f)$ accumulates at (∞, ∞).

Theorem 3.5. *The basin $\mathcal{B}^+$ is a forward and backward invariant, connected and open set that contains all points in* $\mathrm{supp}\,(\eta) \cup C^\infty(f) - I^\infty(f) - \{(\infty, \infty)\}$. *Moreover, the following are equivalent for a point $p \in \mathbf{R}^2$.*

- *The forward orbit of p is well defined and unbounded.*

- $p \in \mathcal{B}^+$.

- $f^n(p)$ *is in the interior of* $\overline{T_0^+ \cup T_1^+}$ *for $n \in \mathbf{N}$ large enough.*

We remark that in this context, it might be more relevant to consider orbits that accumulate on $\mathrm{supp}\,(\eta)$ rather than orbits which are unbounded. Theorem 3.5 holds regardless.

Proof: The basin $\mathcal{B}^+$ is open and invariant by definition. Since $I^\infty(f) \subset \mathrm{supp}\,(\eta)$, we have that $\overline{I^\infty(f)}$ is discrete and accumulates only at (∞, ∞). Therefore connectedness of $\mathcal{B}^+$ follows from invertibility of f.

Observe now that the $\overline{T_0^+ \cup T_1^+}$ contains a neighborhood of any point $(t, \infty), (\infty, t), (t, t-1) \in \mathrm{supp}\,(\eta)$ for which t is large enough. From this it follows easily that $\mathrm{supp}\,(\eta) - \overline{I^\infty(f)} \subset \mathcal{B}^+$. Since $C(f) - I(f)$ maps to $I(f^{-1})$ and $I^\infty(f^{-1}) \cap I^\infty(f) = \emptyset$, it further follows that $C^\infty(f) - I^\infty(f) \subset \mathcal{B}^+$.

The statements in the final assertion are listed from weakest to strongest, so it suffices to prove the first implies the last. So suppose that $p \in \mathbf{R}^2$ is a point whose forward orbit is well defined and unbounded. If $(f^n(p))_{n \geq 0}$ accumulates at $q \in \mathrm{supp}\,(\eta) - I^\infty(f)$, then it also accumulates at every point in the forward orbit of q. We observed in the previous paragraph that $f^n(q)$ lies in the interior of $\overline{T_0^+} \cup \overline{T_1^+}$. Therefore since f^n is continuous on a neighborhood of q, it follows that $f^m(p) \in \mathcal{B}^+$ for some large m. By invariance, we conclude $p \in \mathcal{B}^+$, too.

The remaining possibility is that the forward orbit of p accumulates at infinity only at points in $I^\infty(f)$. This is impossible for the following reason. Since $I^\infty(f)$ is disjoint from $I^\infty(f^{-1})$, and since f and f^{-1} are conjugate via the automorphism σ, we have that the backward iterates (f^{-n}) converge uniformly to (∞, ∞) on a neighborhood $U \ni I^\infty(f)$. Thus points in U cannot recur to points in $I^\infty(f)$, and in particular, the forward orbit of p cannot accumulate on $I^\infty(f)$. $\square$

Because f and f^{-1} are conjugate via $(x, y) \mapsto (-y, -x)$, we immediately obtain trapping regions

$$T_0^- = \sigma(T_0^+), \quad T_1^- = \sigma(T_1^+)$$

for f^{-1}, for which the exact analogues of Theorems 3.1, 3.3, and 3.5 hold. In particular, the backward basin $\mathcal{B}^- = \sigma(\mathcal{B}^+)$ of (∞, ∞) includes all points in $C^\infty(f)$ and all points in $\mathrm{supp}\,(\eta) - I^\infty(f^{-1}) - \{(\infty, \infty)\}$. Using once again the fact that $I^\infty(f)$ is disjoint from $I^\infty(f^{-1})$, we have the following.

Corollary 3.6. *All points except (∞, ∞) with unbounded forward or backward orbits are wandering. Such points include all of $\mathrm{supp}\,(\eta)$, $I^\infty(f)$, $I^\infty(f^{-1})$, $C^\infty(f)$ and $C^\infty(f^{-1})$*

4   Points with Bounded Orbits

In this section we turn our attention to the set

$$K = \{p \in \mathbf{R}^2 : (f^n(p))_{n \in \mathbf{Z}} \text{ is bounded}\}$$

of points with bounded forward and backward orbits. We begin by emphasizing an immediate implication of Theorem 3.5.

Corollary 4.1. *K is a compact, totally invariant subset of $\mathbf{R}^2 - \mathrm{supp}\,(\eta)$ containing no critical or indeterminate point for any iterate of f. Any non-wandering point in $\overline{\mathbf{R}^2}$ except (∞, ∞) belongs to K.*

To study K we define several auxiliary subsets of $\mathbf{R}^2$. Letting $(x_0, y_0) := (\frac{1-a}{2}, \frac{a-1}{2})$ denote the coordinates of the unique finite fixed point for f, we set

$S_0 := [-a, 1] \times [-1, a]$, $S_1 := [x_0, 1] \times [a, \infty]$,
$S_2 := \{(x, y) : x \le -1, \frac{1}{2}(a - 1) \le y \le -x\}$, $S_3 := [x_0, 1] \times [-\infty, -1]$,
$S_4 := [1, \infty] \times [y_0, a]$.

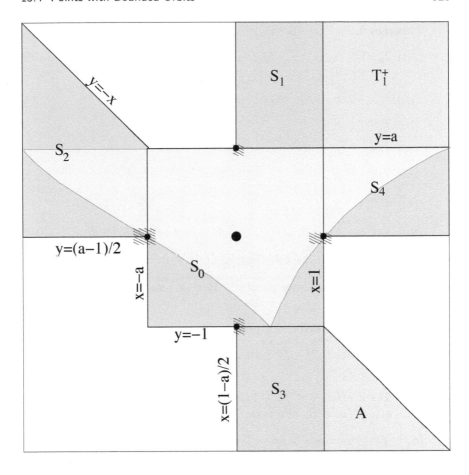

Figure 7. The regions S_0 through S_4 are quadrilaterals. The dot in the center is the fixed point, and the pink shaded region is $f(S_0)$. The regions T_0^+ and A are also shown. Observe that if all shaded regions are reflected about the line $y = -x$ (i.e., by the involution σ), then S_0 is sent to itself and the images of the remaining regions exactly fill the complementary white regions. (See Plate VII.)

These sets are shown in Figure 7 with the image $f(S_0)$ superimposed. Observe that S_0 is just the square in $\mathbf{R}^2$ cut out by the critical sets of f and f^{-1}, and that the other four regions, three rectangles and a trapezoid (S_2), are arranged around S somewhat like blades on a fan. Theorem 4.2 shows that f "rotates" the blades counterclockwise. It also shows that S_0 acts as a kind of reverse trapping region for K since a point in S_0 whose orbit leaves S_0 cannot return again.

Theorem 4.2. *The sets $S_0, \ldots, S_4$ satisfy the following:*

(i) $T_0^+, T_1^+, S_0, \ldots, S_4$, *together with their images under the involution σ, cover $\mathbf{R}^2$;*

(ii) $f(S_0) \subset S_0 \cup S_2 \cup S_4$;

(iii) $f(S_1) \subset S_2$;

(iv) $f(S_2) \subset S_3 \cup A$;

(v) $f(S_3) \subset S_4$;

(vi) $f(S_4) \subset S_1 \cup T_0^+$;

(vii) *any point $p \in S_1 \cup \cdots \cup S_4$ has a forward orbit tending to (∞, ∞).*

From this theorem, we have immediately that

Corollary 4.3. *For every point $p \in \mathbf{R}^2$, we exactly one of the following is true.*

- $f^n(p) \notin S_0$ *for all $n \in \mathbf{N}$.*

- *There exists $N \in \mathbf{N}$ such that $f^n(p) \notin S_0$ for $n < N$ and $f^n(p) \in K$ for all $n \geq N$.*

- *There exist $N \leq M \in \mathbf{N}$ such that $f^n(p) \in S_0$ for all $N \leq n \leq M$ and $f^n(p) \notin S_0$ otherwise.*

In the first two cases, p has a forward orbit tending to (∞, ∞) through T_0^+ or T_1^+.

Since f and f^{-1} are conjugate by σ, we further have

Corollary 4.4. $K \subset S_0$.

Theorem 3.5 and Corollary 4.4 combine to yield Theorem 1.1 in the introduction. We spend the rest of this section proving Theorem 4.2. The first assertion is obvious from Figure 7.

Proof of Assertion (ii).   Since S_0 lies between the critical lines of f in regions II^+ and V^+ (see Figure 4), $f(S_0)$ is the closure of a connected open set in regions II^- and V^-. Moreover, ∂S_0 is contains the point $(1, 0) \in I(f)$, which maps to $y = a$. Hence $y = a$ constitutes the upper boundary of S_0.

The vertical sides of S_0 are critical for f and map to points, so S_0 is bounded below by the intersections α and β of $f\{y = a\}$ and $f\{y = -1\}$, respectively, with $II^- \cup V^-$. From Figure 4 one finds that α stretches right from $(-\infty, a)$ to $(0, -1)$ and that β stretches left from $(0, -1)$ to (∞, a). Moreover, one computes from 2.4 and 2.2 that each of these curves intersects each horizontal line no more than once. Hence α has non-positive slope at all points and β has non-negative slope. Since $(-a, y_0) = f(x_0, a) \in \alpha$ and $(1, y_0) = f(x_0, -1) \in \beta$, it follows that $f(S_0) - S_0$ is contained in S_2 and S_4. □

To verify the remaining assertions, it is useful to choose non-negative "adapted" coordinates on S_1 through S_4 as follows. Each boundary ∂S_j, $1 \leq j \leq 4$, contains a segment ℓ of either the horizontal or vertical line passing through the fixed point (x_0, y_0). We choose coordinates $(x_j, y_j) = \psi_j(x, y)$ on S_j so that $\ell \cap S_0$ becomes the origin (marked in Figure 7, distance to ℓ becomes the x_j coordinate and distance along ℓ to $\ell \cap S_0$ becomes the y_j coordinate. Thus, for example, we obtain coordinates $(x_1, y_1) = \psi_1(x, y) := (x - x_0, y - a)$ on S_1 and coordinates $(x_2, y_2) = \psi_2(x, y) := (y - y_0, 1 - x)$ on S_2, etc.

Proof of Assertions (iii) to (vi). To establish assertion (iii), we let $(x_1, y_1) = \psi_1(x, y)$ be the adapted coordinates of a point $(x, y) \in S_1$ and $(x_2, y_2) = \psi_2 \circ f(x, y)$ the coordinates of its image. Then $0 \leq x_1 \leq \frac{a+1}{2}$ and $0 \leq y_1$, and direct computation gives that

$$(x_2, y_2) := \psi_2 \circ f \circ \psi_1^{-1}(x_1, y_1) = \left(x_1, \frac{4ax_1 + y_1 + ay_1 + 2x_1 y_1}{1 + a - 2x_1} \right).$$

Hence (x_2, y_2) satisfy the same inequalities as (x_1, y_1), and it follows that $f(x, y) \in S_2$.

Verifying assertion (iii) is more or less the same, though messier because of S_2 is not a rectangle. If $(x, y) \in S_2$, then $(x_2, y_2) := \psi(x, y)$ satisfies $0 \leq x_2 \leq y_2 + \frac{a+1}{2}$ and $0 \leq y_2$. One computes

$$(x_3, y_3) := \psi_3 \circ f(x, y) = \left(\frac{-1 + a^2 + 2y_2(x_2 + a - 1)}{2(y_2 + a + 1)}, y_2 \right).$$

Since $a > 1$, we see that $x_3, y_3 \geq 0$. So to complete the proof, it suffices to show that $x_3 \leq y_3 + \frac{a+1}{2}$:

$$y_3 + \frac{a+1}{2} - x_3 = \frac{2 + 2a + y_2(5 + a + 2(y_2 - x_2))}{2(y_2 + a + 1)} > \frac{y_2(5 + a + (a+1))}{2(y_2 + a + 1)} > 0.$$

We leave it to the reader to verify assertions (v) and (vi). □

Lemma 4.5. *Let $p \in S_1 \cup S_2 \cup S_3 \cup S_4$ be any point. Then the adapted x-coordinate of $f^2(p)$ is positive. If, moreover, the adapted y-coordinate of p is positive and $f^2(p) \in S_1 \cup S_2 \cup S_3 \cup S_4$, then the adapted y-coordinate of $f(p)$ is larger than that of p.*

Proof: In the same way we proved assertions (iii) through (vi) above, one may verify that the restriction of f to $S_1 \cup S_3$ preserves the adapted x-coordinate of a point, and the images of S_2 and S_4 do not contain points with adapted x-coordinate equal to zero. Hence $f^2(p)$ cannot have adapted x-coordinate equal to zero.

Similarly, one finds that the restriction of f to S_2 and S_4 preserves adapted y-coordinates, whereas f increases the adapted y-coordinates of points in S_1 and S_3. For example, if $p = (x_1, y_1) \in S_1$, we computed previously that $f(p)$ has adapted y-coordinate

$$y_2 = \frac{4ax_1 + y_1 + ay_1 + 2x_1y_1}{1 + a - 2x_1} \geq y_1 + \frac{4ax_1}{1 + a - 2x_1} \geq y_1$$

with equality throughout if and only if $x_1 = 0$. $\qquad\square$

Proof of Assertion (vii). Suppose the assertion is false. Then by assertions (iii) through (vi) it follows that there is a point p whose forward orbit is entirely contained $S_1 \cup S_2 \cup S_3 \cup S_4$. Moreover, by Theorem 3.5 the forward orbit of p must be bounded. Therefore, we can choose $q \in S_1 \cup S_2 \cup S_3 \cup S_4$ the be an accumulation point of $(f^n(p))_{n \in \mathbf{N}}$ whose adapted y-coordinate is as large as possible.

So on the one hand $f^2(q)$ is an accumulation point of the forward orbit of p and cannot, by definition of q, have adapted y-coordinate larger than that of q. But on the other hand, the first assertion of Lemma 4.5 implies that q cannot have adapted x-coordinate equal to zero, and therefore the second assertion tells us that $f^2(q)$ must have larger adapted y-coordinate than q. This contradiction completes the proof. $\qquad\square$

5   Behavior Near the Fixed Point

When $a \geq 0$, the eigenvalues of $Df(p_{fix})$ are a complex conjugate pair $\lambda, \bar{\lambda}$ of modulus one, with

$$\lambda = e^{i\gamma_0} = \frac{i + \sqrt{a}}{i - \sqrt{a}}.$$

Hence $Df(p_{fix})$ is conjugate to rotation by an angle γ_0 that decreases from 0 to $-\pi$ as a increases from 0 to ∞. When $\gamma_0 \notin \mathbf{Q}\pi$ is an irrational angle, it is classical that f can be put formally into Birkhoff normal form:

Proposition 5.1. *If λ is not a root of unity, there is a formal change of coordinate $z = x + iy$ in which $p_{fix} = 0$, and f becomes*

$$z \mapsto z \cdot e^{i(\gamma_0 + \gamma_2 |z|^2 + \cdots)}$$

where γ_0 is as above, and

$$\gamma_2(a) = \frac{4(3a-1)}{\sqrt{a}(a-3)(1+a)^2}.$$

Hence the "twist" parameter γ_2 is well defined and non-zero everywhere except $a = \frac{1}{3}$ and $a = 3$, changing signs as it passes through these two parameters.

For the proof of the proposition, we briefly explain how the Birkhoff normal form and attendant change of coordinate are computed (see [SM95, Section 23] for a more complete explanation). We start with a matrix C whose columns are complex conjugates of each other and which satisfies $C^{-1}Df(p_{fix})C = \operatorname{diag}(\lambda, \bar{\lambda})$. Conjugating $C^{-1} \circ f_a \circ C$, we "complexify" the map f_a, using new variables $(s, t) = (z, \bar{z})$, in which the map becomes

$$f : \quad s \mapsto \lambda s + p(s, t), \quad t \mapsto \bar{\lambda}t + q(s, t),$$

and the power series coefficients of q are the complex conjugates of the coefficients of p. We look for a coordinate change $s = \phi(\xi, \eta) = \xi + \cdots$, $t = \psi(\xi, \eta) = \eta + \cdots$, such that the coefficients of ϕ and ψ are complex conjugates, and which satisfies $u \cdot \phi = p(\phi, \psi)$, for some $u(\xi, \eta) = \alpha_0 + \alpha_2 \xi \eta + \alpha_4 \xi^2 \eta^2 + \cdots$. Solving for the coefficients, we find $u = \exp i(\gamma_0 + \gamma_2 \xi \eta + \cdots)$, with γ_0 and γ_2 as above. Returning to the variables $s = z$ and $t = \bar{z}$ gives the normal form.

For the rest of this section, we restrict our attention to the case $a = 3$. As was noted in [BHM01], this is the parameter value for which a 3-cycle of saddle points coalesces with p_{fix}. When $a = 3$, we have $\gamma_0 = -2\pi/3$. Translating coordinates so that p_{fix} becomes the origin, we have

$$f^3(x, y) = (x, y) + Q + O(|(x, y)|^3)$$
$$Q = (Q_1, Q_2) = (x^2/2 + xy + y^2, x^2 + xy + y^2/2).$$

Let us recall a general result of Hakim [Ha98] on the local structure of a holomorphic map which is tangent to the identity at a fixed point. A vector v is said to be characteristic if Qv is a multiple of v. The characteristic vectors v for the map f_a are $(1, -\frac{1}{2})$, $(1, 1)$, and $(1, 2)$. Hakim [Ha98] shows that for each characteristic v with $Qv \neq 0$, there is a holomorphic embedding $\varphi_v : \Delta \to \mathbf{C}^2$ which extends continuously to $\bar{\Delta}$ and such that $\varphi_v(1) = (0, 0)$, and the disk $\varphi_v(\Delta)$ is tangent to v at $(0, 0)$. Further, $f(\varphi_v(\Delta)) \subset \varphi_v(\Delta)$, and for every $z \in \varphi_v(\Delta)$, $\lim_{n \to \infty} f^n z = (0, 0)$. The disk $\varphi_v(\Delta)$ is a $\mathbf{C}^2$ analogue of an "attracting petal" at the origin. Applied to our real function f_a, this means that for each of the three vectors v, there is a real analytic "stable arc" $\gamma_v^s \subset \mathbf{R}^2$, ending at $(0, 0)$ and tangent to v. Considering f^{-1}, we have an "unstable arc" γ_v^u approaching $(0, 0)$ tangent to $-v$. These two arcs fit together to make a C^1 curve, but they are not in general analytic continuations of each other.

Let us set $r(1, y) = Q_2(1, y)/Q_1(1, y)$ and write $v = (1, \eta)$. Then $a(v) := r'(\eta)/Q_1(1, \eta)$ is an invariant of the map at the fixed point. If we make a linear change of coordinates so that the characteristic vector $v = (1, 0)$ points in the direction of the x-axis, then we may rewrite f in local coordinates as

$$(x, u) \mapsto (x - x^2 + O(|u|x^2, x^3), u(1 - ax) + O(|u|^2 x, |u|x^2)), \qquad (6)$$

(see [Ha98]). For the function (5), we find that $a(v) = -3$ for all three characteristic vectors. We conclude that the stable arcs γ_v^s are weakly repelling in the normal direction.

Another approach, which was carried out in [ABM99], is to find the formal power series expansion of a uniformization of γ_v^s at p_{fix}.

6   The Case $a = 3$

In this section we continue with the assumption $a = 3$, our aim being to give a global treatment of the stable arcs for p_{fix}. The basic idea here is that the behavior of $f|_{S_0}$ is controlled by invariant cone fields on a small punctured neighborhood of p_{fix}.

In order to proceed, we fix some notation. Let $\phi(x, y) = \phi_0(x, y) = x + 1$ and for each $j \in \mathbf{Z}$, let $\phi_j(x, y) = \phi \circ f^{-j}(x, y)$. Then $\phi_{j+k} = \phi_j \circ f^{-k}$ and in particular, $\phi_j(p_{fix}) = \phi_0 \circ f^{-j}(p_{fix}) = \phi_0(p_{fix}) = 0$ for every $j \in \mathbf{Z}$. We will be particularly concerned with the cases $j = -1, 0, 1, 2$ and observe for now that the level set $\{\phi_j = s\}$ is

- a horizontal line $\{x = -1 + s\}$ when $j = 0$;

- a vertical line $\{y = 1 + s\}$ when $j = 1$;

- a hyperbola with asymptotes $\{x = -3\}$, $\{y = -1 + s\}$ when $j = -1$; and

- a hyperbola with asymptotes $\{x = 1 - s\}$, $\{y = 3\}$ when $j = 2$.

Most of our analysis will turn on the interaction between level sets of ϕ_{-1} and ϕ_3.

Proposition 6.1. $\{\phi_2 = 0\}$ *is tangent to* $\{\phi_{-1} = 0\}$ *at* p_{fix}. *Moreover* $S_0 \cap \{\phi_2 > 0\} \subset S_0 \cap \{\phi_{-1} > 0\}$.

Proof: The first assertion is a consequence of the facts that $f^3\{\phi_{-1} = 0\} = \{\phi_2 = 0\}$ and that Df^3 is the identity at p_{fix}. Since both curves in question are hyperbolas with horizontal and vertical asymptotes, it follows that p_{fix} is the *only* point where the curves meet.

The asymptotes of $\{\phi_{-1} = 0\}$ are the bottom and left sides of S_0, whereas those of $\{\phi_2 = 0\}$ are the top and right sides. Therefore the zero level set of ϕ_{-1} intersects S_0 in a connected, concave up arc; and the zero level set of ϕ_2 meets S_0 in a connected concave down arc. It follows that the first level set is above and to the right of the second. Finally, direct computation shows that ϕ_{-1} and ϕ_2 are positive at the lower left corner of S_0. This proves the second assertion in the proposition. $\square$

Using the functions ϕ_j, $j = -1, 0, 1$, we define "unstable wedges" $W_j \subset S_0$ emanating from p_{fix}:

$$\begin{aligned}
W_0 &= \{(x, y) \in S_0 : 0 \leq \phi_{-1}, \phi_0\} \\
W_1 &= \{(x, y) \in S_0 : 0 \leq \phi_0, \phi_1\} \\
W_2 &= \{(x, y) \in S_0 : 0 \leq \phi_{-1}, \phi_1\}
\end{aligned}$$

Proposition 6.2. *We have* $f(W_0) \cap S_0 \subset W_1$, $f(W_1) \cap K \subset W_2$, *and* $f(W_2 \cap S_0) \subset W_0$.

Proof: It is immediate from definitions that $f(W_0) \cap S_0 \subset \{\phi_0, \phi_1 \geq 0\} \cap S_0 = W_1$.

Likewise,

$$f(W_1) \cap S_0 \subset \{\phi_1, \phi_2 \geq 0\} \cap S_0 \subset \{\phi_1, \phi_{-1} \geq 0\} \cap S_0 = W_2.$$

The second inclusion follows from the Proposition 6.1. Similar reasoning shows that $f(W_2) \cap S_0 \subset W_0$. $\square$

Lemma 6.3. *For each neighborhood U of p_{fix}, there exists $m = m(U) > 0$ such that*

$$\phi_{-1}(p) + m \leq \phi_2(p) \leq -m.$$

for every $p \in W_0 - U$.

Proof: Given $p \in W_1$, we write $p = (-1 + s, 1 + t)$ where $0 \leq s, t \leq 2$ and compute

$$\phi_2(p) = \frac{-2s - 2t - st}{t + 2} = \frac{s(-2 + t/2) + t(-2 + s/2)}{t + 2} < -\frac{s + t}{t + 2} \leq -\frac{s + t}{4};$$

and

$$\phi_2(p) - \phi_{-1}(p) = \frac{2(s^2 + t^2) + st(8 + s - t)}{(2 - s)(t + 2)} \geq \frac{2(s^2 + t^2)}{4}$$

Since the quantities $s^2 + t^2$ and $s + t$ are both bounded below by positive constants on $W_0 - U$, the lemma follows. $\square$

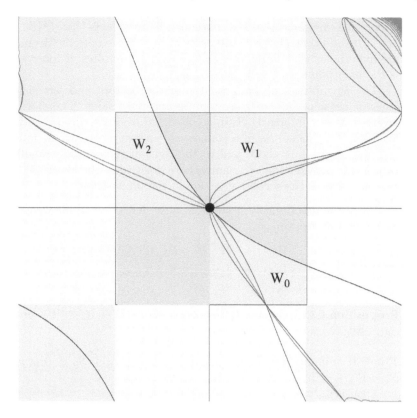

Figure 8. Unstable wedges. The square in the center is S_0, with the fixed point in the middle. Boundaries of the wedges are the black curves. The images of the wedges under f^3 are bounded by the blue curves. The unstable curves for the fixed point are red. (See Plate VII.)

Proposition 6.4. *Let U be any neighborhood of p_{fix}. Then there exists $N > 0$ such that $f^n(p) \notin S_0$ for every $p \in W_0 \cup W_1 \cup W_2 - U$ and every $n \geq N$.*

Proof: By Proposition 6.2, we can assume that the point p in the statement of the lemma lies in $W_0 - U$. Since ϕ_2 is continuous on W_0 with $\phi_2(p_{fix}) = 0$ and $\phi_2 < 0$ elsewhere on W_0, we may assume that U is of the form $\{\phi_2 > -\epsilon\}$ for some $\epsilon > 0$.

Suppose for the moment that $f^3(p) \in S_0$. Then Proposition 6.2 implies that $f^3(p) \in W_0$. By Lemma 6.3, we then have

$$\phi_2 \circ f^3(p) - \phi_2(p) = \phi_{-1}(p) - \phi_2(p) < -m(U).$$

In particular, $f^3(p) \in W_0 - U$. Repeating this reasoning, we find that if

$p \in W_0 - U$ and $f^{3j}(p) \in S_0$ for $j = 1, \ldots, J$, then

$$\phi_2 \circ f^{3J}(p) < -Jm(U).$$

On the other hand ϕ_2 is bounded below on W_0 (e.g., by -4). So if $J > 4/m(U)$, it follows that $f^{3j}(p) \notin S_0$ for some $j \leq J$. By Corollary 4.3, we conclude that $f^n(p) \notin S_0$ for all $n \geq 3j$.   □

Let us define the local stable set of p_{fix} to be

$$W^s_{loc}(p_{fix}) := \{p \in \mathbf{R}^2 : f^n(p) \in S_0 \text{ for all } n \in \mathbf{N} \text{ and } f^n(p) \to p_{fix}\},$$

with the local unstable set $W^u_{loc}(p_{fix})$ defined analogously.

Corollary 6.5. *The wedges W_0, W_1, W_2 meet $W^s_{loc}(p_{fix})$ only at p_{fix}, whereas they entirely contain $W^u_{loc}(p_{fix})$. Thus*

$$W^s_{loc}(p_{fix}) = W^s(p_{fix}) \cap S_0 = \{p \in \mathbf{R}^2 : f^n(p) \in S_0 \text{ for all } n \in \mathbf{N}\}, \quad (6.5)$$

and similarly for W^u_{loc}.

Proof: The first assertion about $W^s_{loc}(p_{fix})$ follows from Propositions 6.2 and 6.4. Reversibility of f then implies the first assertion about $W^u_{loc}(p_{fix})$ is disjoint from the sets $\sigma(W_0), \sigma(W_1), \sigma(W_2)$.

We will complete the proof of the first assertion about W_0

$$S_0 \subset \bigcup_{j=0}^{2} W_j \cup \sigma(W_j).$$

Observe that $\phi_0 \circ \sigma = -\phi_1$. Hence,

$$\phi_{-1} \circ \sigma = \phi_0 \circ f \circ \sigma = -\phi_1 \circ f^{-1} = -\phi_2.$$

Thus $\sigma(W_0) = \{\phi_1, \phi_2 \leq 0\}$, etc. The desired inclusion is therefore an immediate consequence of Proposition 6.1.

The first equality in (6.5) follows from Corollary 4.3, and the second from the analogue of Proposition 6.4 for f^{-1}.   □

Each of the wedges W_j admits an "obvious" coordinate system identifying a neighborhood of p_{fix} in W_j with a neighborhood of $(0,0)$ in $\mathbf{R}^2_+ := \{x, y \geq 0\}$. Namely, we let $\Psi_0 : W_0 \to \mathbf{R}^2$ be given by $\Psi_0 = (\phi_{-1}, \phi_0)$, $\Psi_1 : W_1 \to \mathbf{R}^2$ by $\Psi_1 = (\phi_0, \phi_1)$, and $\Psi_1 : W_1 \to \mathbf{R}^2$ by $\Psi_2 = (\phi_1, \phi_{-1})$. If $U \subset \mathbf{R}^2_+$ is a sufficiently small neighborhood of $(0,0)$, then in light of Proposition 6.2, the maps $f_{01}, f_{12}, f_{20} : U \to \mathbf{R}^2_+$ given by $f_{ij} = \Psi_j \circ f \circ \Psi_i^{-1}$ are all well defined. In fact, f_{01} is just the identity map. The other two are more complicated, but in any case a straightforward computation shows

Proposition 6.6. *If $x, y \geq 0$ are sufficiently small, the entries of the matrix $Df_{ij}(x, y)$ are non-negative.*

The proposition tells us that there are forward invariant cone fields defined near p_{fix} on each of the W_j; i.e., the cones consisting of vectors with non-negative entries in the coordinate system Ψ_j. In this spirit, we will call a connected arc $\gamma : [0, 1] \in W_j \cap \overline{U}$ admissible if $\gamma(0) = p_{fix}$, $\gamma(1) \in \partial U$ and both coordinates of the function $\Psi_j \circ \gamma : [0, 1] \to \mathbf{R}^2$ are non-decreasing. We observe in particular, that each of the two pieces of $\partial W_j \cap \overline{U}$ are admissible arcs.

By Proposition 6.4, we can choose the neighborhood U of p_{fix} above so that its intersections with the wedges W_j are "pushed out" by f. That is, $f(W_j \cap bU) \cap U = \emptyset$. This assumption and Propositions 6.2 and 6.6 imply immediately that

Corollary 6.7. *If $\gamma : [0, 1] \to W_i$ is an admissible arc, then (after reparametrizing) so is $f_{ij} \circ \gamma$.*

Theorem 6.8. *For each $j = 0, 1, 2$, the set $\bigcap_{n \in \mathbf{N}} f^{3n}(W_j) \cap U$ is an admissible arc.*

Proof: For any closed $E \subset \overline{U}$, we let Area (E) denote the area with respect to the f-invariant two form η. Since U avoids the poles of η, we have Area $(U) < \infty$.

By Proposition 6.2 and our choice of U, the sets $f^{3n}(W_j) \cap \overline{U}$ decrease as n increases. In particular, Proposition 6.4 tells us that Area $(f^{3n}(W_j) \cap \overline{U}) \to 0$. Finally, for every $n \in \mathbf{N}$, $U \cap \partial f^{3n}(W_j) = f^{3n}(\partial W_j) \cap U$ consists of two admissible curves meeting at p.

Admissibility implies that in the coordinates Ψ_j, the two curves bounding $f^{3n}(W_j)$ in U are both graphs over the line $y = x$ of functions with Lipschitz constant no larger than 1. The fact that the region between these two curves decreases as n increases translates into the statement that the graphing functions are monotone in n. The uniformly bounded Lipschitz constant together with the fact that the area between the curves is tending to zero, implies further that the graphing functions converge uniformly to the same limiting function and that this limiting function also has Lipschitz constant no larger than 1. $\qquad\square$

Observe that by Corollary 6.5, we have just characterized the intersection of the unstable set of p_{fix} with U. Since f and f^{-1} are conjugate via the involution σ, the theorem also characterizes the stable set of p_{fix}. Our final result, stated for the stable rather than the unstable set of p_{fix}, summarizes what we have established. It is an immediate consequence of Proposition 6.4, and Corollary 4.3 and Theorem 6.8.

Corollary 6.9. *When* $a = 3$ *the set* $W^s(p_{fix}) \cap S_0$ *consists of three arcs meeting transversely at* p_{fix}, *and it coincides with the set of points whose forward orbits are entirely contained in* S_0. *All points not in* $W^s(p_{fix})$ *have forward orbits tending to infinity. In particular,* $K = \{p_{fix}\}$

Theorem 1.2 is an immediate consequence.

Bibliography

[AAB99a] N. Abarenkova, J. C. Anglès d'Auriac, S. Boukraa, S. Hassani, and J. M. Maillard, *From Yang-Baxter equations to dynamical zeta functions for birational transformations*, Statistical physics on the eve of the 21st century, Ser. Adv. Statist. Mech., vol. 14, World Sci. Publishing, River Edge, NJ, 1999, 436–490.

[AAB99b] N. Abarenkova, J. C. Anglès d'Auriac, S. Boukraa, S. Hassani, and J. M. Maillard, *Topological entropy and Arnold complexity for two-dimensional mappings*, Phys. Lett. A **262** 1 (1999), 44–49.

[AAB99c] N. Abarenkova, J. C. Anglès d'Auriac, S. Boukraa, S. Hassani, and J. M. Maillard, *Rational dynamical zeta functions for birational transformations*, Phys. A **264** 1–2 (1999), 264–293.

[AAB00] N. Abarenkova, J. C. Anglès d'Auriac, S. Boukraa, S. Hassani, and J. M. Maillard, *Real Arnold complexity versus real topological entropy for birational transformations*, J. Phys. A **33** 8 (2000), 1465–1501.

[AAdBM99] N. Abarenkova, J. C. Anglès d'Auriac, S. Boukraa, and J.-M. Maillard, *Growth-complexity spectrum of some discrete dynamical systems*, Phys. D **130** 1–2 (1999), 27–42.

[AABM00] N. Abarenkova, J. C. Anglès d'Auriac, S. Boukraa, and J.-M. Maillard, *Real topological entropy versus metric entropy for birational measure-preserving transformations*, Phys. D **144** 3–4 (2000), 387–433.

[ABM99] J. C. Anglès d'Auriac, S. Boukraa, and J.-M. Maillard, *Functional relations in lattice statistical mechanics, enumerative combinatorics, and discrete dynamical systems*, Ann. Comb. **3** 2–4 (1999), 131–158.

[BD05]   E. Bedford and J. Diller, *Real and complex dynamics of a family of birational maps of the plane: the golden mean subshift*, Amer. J. Math. **127** 3 (2005), 595–646.

[BD]   E. Bedford and J. Diller, *Dynamics of a family of plane birational maps: maximal entropy*, J. Geom. Analysis, to appear.

[BHM98]   S. Boukraa, S. Hassani, and J.-M. Maillard, *Product of involutions and fixed points*, Alg. Rev. Nucl. Sci. **2** 1 (1998), 1–16.

[BHM01]   S. Boukraa, S. Hassani, and J.-M. Maillard, *Properties of fixed points of a two-dimensional birational transformation*, Alg. Rev. Nucl. Sci. **3** 1–2 (2001).

[DF01]   J. Diller and C. Favre, *Dynamics of bimeromorphic maps of surfaces*, Amer. J. Math. **123** 6 (2001), 1135–1169.

[Du06]   R. Dujardin, *Laminar currents and birational dynamics*, Duke Math. J. **131** 2 (2006), 219–247.

[GH78]   P. Griffiths and J. Harris, *Principles of algebraic geometry*, Wiley Classics Library, John Wiley & Sons Inc., New York, 1994. Reprint of the 1978 original.

[Gu05]   V. Guedj, *Entropie topologique des applications méromorphes*, Ergod. Th. Dynam. Syst. **25** 6 (2005), 1847–1855.

[Ha98]   M. Hakim, *Analytic transformations of* $(\mathbf{C}^p, 0)$ *tangent to the identity*, Duke Math. J. **92** 2 (1998), 403–428.

[KH95]   A. Katok and B. Hasselblatt, *Introduction to the modern theory of dynamical systems*, Encyclopedia of Mathematics and its Applications, vol. 54, Cambridge University Press, Cambridge, 1995, with a supplementary chapter by A. Katok and L. Mendoza.

[SM95]   C. L. Siegel and J. K. Moser, *Lectures on celestial mechanics*, Classics in Mathematics, Springer-Verlag, Berlin, 1995, Translated from the German by C. I. Kalme. Reprint of the 1971 translation.

IV. Making New Friends

14   The Hunt for Julia Sets with Positive Measure

Arnaud Chéritat

This article is based on the talk given by the author for the conference. We will explain the status that had reached, during Hubbard's 60th birthday conference, Douady's plan to produce Julia sets with positive Lebesgue measure. We will also quickly mention the progress done since then.

1   Introduction

The title of the talk I gave at Hubbard's 60th birthday conference was "Are there Julia sets with positive measure?" Four months later I gave a talk in Denmark where the first two words were interchanged and the question mark became a period.

I do not know when Adrien Douady set up his plan for finding a Julia set with positive measure, but in 1998 he offered me to study it for my PhD thesis. He added that his opinion about whether or not such Julia sets existed depended on his mood. If the plan worked it would solve an open conjecture. But in case of failure, it would not say anything on the measure of the objects he was looking at (quadratic Julia sets with a Cremer point).[1]

When I finished my thesis in 2001, I had convinced Adrien that his plan would work, by proving it would follow from a pair of reasonable conjectures, supported by analogy with known results for Julia sets with a bounded type Siegel disk, also supported by computer pictures.

Adrien told me cylinder renormalization could help in proving these conjectures and had me and Xavier Buff meet and begin a very productive collaboration on quadratic Julia sets. I also went to Toronto for six months where I met Yampolsky and had many discussions with him. However cylinder renormalization turned out to be difficult. We lacked first an invariant class, second an argument to turn this into a proof of my conjectures. An invariant class was found by Inou and Shishikura in 2003.

---

[1] We now have examples with positive measure, but it is still unknown whether or not some of them have measure 0.

To turn it into a proof still required some time and effort. I remember three key periods. First, I was asked to write a review for the French Mathematical Society on the book *The Mandelbrot Set, Theme and Variations* edited by Tan Lei [Ta00]. In the process, I learnt important methods in holomorphic dynamics, and this knowledge proved very useful. Second, this is in the favorable atmosphere of Hubbard's 60th birthday conference that I saw the end of the tunnel, i.e., how things may fit together to provide a proof. The third period was when Xavier and I worked hard during the following summer holiday to make it work.

2 Crash-Course

This section introduces some terminology for the non-experts or readers in the distant future. See for instance [Mi00] for a more thorough introduction.

In this article, it is understood that we work with complex polynomials, i.e., with some $P \in \mathbb{C}[X]$. A *quadratic polynomial* is just any degree 2 polynomial. We are interested in the *iteration* of P, i.e., in the behavior of sequences z_n defined by $z_{n+1} = P(z_n)$, and how this behavior depends on z_0 and on P. Here P^n will refer to the composition $P \circ \cdots \circ P$, and not to the product.

The *orbit* of a point is the set $\{P^n(z) \mid n \in \mathbb{N}\}$. A *periodic point* is a point z_0 such that $z_k = z_0$ for some $k > 0$. Its *period* is the least value of k such that this hold. A *cycle* is the orbit of a periodic point. A period one periodic point is also called a *fixed* point. Its *multiplier* is the derivative $\lambda = (P^k)'(z_0)$. The periodic point is *repelling* if $|\lambda| > 1$, *indifferent* if $|\lambda| = 1$, *attracting* if $|\lambda| < 1$ and *superattracting* if $\lambda = 0$. Superattracting is considered as a subcase of attracting. For instance, ∞ is a superattracting fixed point of any polynomial of degree > 1.

The periodic point z_0 is *linearizable* when there exists a homeomorphism ϕ from a neighborhood of z_0 to a neighborhood of 0 such that $\phi \circ P^k \circ \phi^{-1}(z) = \lambda z$ holds in a neighborhood of z_0. For dimension one analytic maps, the existence of a continuous conjugacy to the linear part is known to imply the existence of an analytic one.

If $|\lambda| \neq 0$ and $|\lambda| \neq 1$, then the periodic point is automatically linearizable. If $|\lambda| = 0$ then it can be conjugated to $z \mapsto z^d$ in a neighborhood of 0, for some integer $d \geq 2$. If $|\lambda| = 1$, then write $\lambda = e^{2i\pi\theta}$ for some $\theta \in \mathbb{R}$. This θ is called the *rotation number*. If $\theta \in \mathbb{Q}$ ($\iff$ λ is a root of unity), then z_0 is a *parabolic* point. It is most of times non-linearizable. If $\theta \notin \mathbb{Q}$, then z_0 is called an *irrationally indifferent* periodic point. Several subcases occur. In the linearizable case, the *Siegel disk* refers to the maximal domain on which a conjugacy to a rotation on a disk is possible.

In the non-linearizable case, z_0 is called a *Cremer point*, and the dynamics becomes very complicated.

The *basin* of an attracting cycle is the set of points whose orbit tends to the cycle (i.e., end up staying in every neighborhood of the cycle). The *filled-in Julia set* $K(P)$ of a polynomial P of degree ≥ 2 is the complement of the basin of ∞. It is a non-empty compact subset of $\mathbb{C}$.

The *Julia set* $J(P)$ has three equivalent and standard definitions. It is the boundary of $K(P)$ (which is the same as the boundary of the basin of ∞). It is the closure of the set of repelling periodic points. It is the complement of the Fatou set $\mathcal{F}$, where $z \in \mathcal{F} \iff$ there exists a neighborhood V of z on which the family $\{P^n\}_{n \in \mathbb{N}}$ is normal.

A degree 2 polynomial has a connected Julia set $J(P) \iff K(P)$ is connected $\iff$ the (unique) critical point of P belongs to $K(P)$.

The *Mandelbrot set* M is the set of $c \in \mathbb{C}$ such that the quadratic polynomial $P_c(z) = z^2 + c$ has a connected Julia set. It is thus also the set of c such that the critical point $z = 0$ of P_c has a bounded orbit. The latter definition is used for making computer pictures of M.

A *hyperbolic* polynomial P is one which has an *iterate* P^n, $n > 0$, whose derivative is expanding (for the spherical metric) on all its Julia set.

3 Status of the Hunt at the Time of the Conference

This section chiefly presents the content of the talk I gave for Hamal's 60th birthday conference, in a more detailed way and completed by references.

3.1 Motivation

The Julia set is the locus of chaos. To have a Julia set of positive measure means that a randomly chosen point in the Riemann sphere has a non zero probability of being in the locus of chaos. It is interesting in itself to know whether or not this may happen for polynomials.

Yet, the question has strong links with a set of still open conjectures:

Conjecture (Fatou). *In the set of degree 2 polynomials, hyperbolic ones are dense.*[2]

Conjecture (MLC). *The Mandelbrot set is locally connected.*

Conjecture (NILF). *The Julia set of a quadratic polynomial carries no invariant line field.*[3]

---

[2]Fatou also conjectured that this holds in the set of degree d polynomials and in the set of degree d rational maps.

[3]See [Mc94] for the definitions of a line field and the invariance thereof.

Let us also label this article's main question:

Conjecture (Zero). *The Lebesgue measure of Julia sets of quadratic polynomials is always equal to 0.*

It was also conjectured for higher degree polynomials, and also for rational maps whose Julia set is not the whole Riemann sphere (there exist rational maps whose Julia set is the whole Riemann sphere).

We have the following known relations between them (as of 2005):

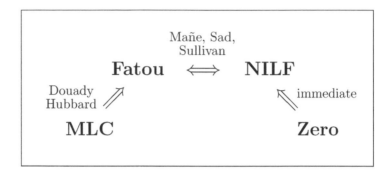

For MLC $\implies$ Fatou, see [DH84] and [Sch04].
For Fatou $\iff$ NILF, see [MSS83] and [Mc94].

The Zero conjecture thus fitted well in the picture, all the more so as its analog for Kleinian group, Ahlfors' conjecture, was proved in 2004.
Yet, the Zero conjecture fails [BC].

3.2 Known Results (Where Not to Look at) for Quadratic Polynomials

In the rest of this article P denotes a polynomial of degree ≥ 2 (and most of times $= 2$), $J(P)$ its Julia set, Leb the Lebesgue measure, $\dim_H$ the Hausdorff dimension, M the Mandelbrot set, and Pol_2 the set of polynomials of degree 2, also called *quadratic* polynomials. All degree 2 polynomials are $\mathbb{C}$-affine conjugated to a unique polynomial of the form $Q_c = z^2 + c$.

Recall that a hyperbolic polynomial or rational fraction is one which has an iterate whose derivative is expanding (for the spherical metric) on all its Julia set. This is equivalent to all the critical points belonging to attracting or superattracting basins. The Lebesgue measure of such a Julia set is then equal to 0 (see [DH84]), and even better (see [Su83]):

Theorem 3.1 (Sullivan). *A hyperbolic rational fraction has a Julia set J of Hausdorff dimension < 2 (thus zero Lebesgue measure):*

$$\dim_H J < 2.$$

Let's recall a basic fact (yet not obvious):

Theorem 3.2 (Fatou, Julia, Douady, Hubbard). *A degree 2 polynomial has at most one non repelling cycle.*

The proof of this theorem can be found in [Do83] and [Bl84]. Next, by [Ly91]:

Theorem 3.3 (Lyubich, Shishikura). *If P has no indifferent periodic points and is not infinitely renormalizable, then*

$$\text{Leb } J(P) = 0.$$

So, to find positive measure in Pol_2, we must look either

- at infinitely renormalizable polynomials

- at polynomials with an indifferent cycle

These two conditions are mutually exclusive, and prevent P from being hyperbolic.

The three theorems mentioned above show that there are plenty of "small" Julia sets, with different notions of what is "small". Let us cite a theorem by [Sh98] providing plenty, in some sense, of Julia sets that are "big", in some sense:

Theorem 3.4 (Shishikura).

(S1) *For a Baire generic set of values of $c \in \partial M$, $P = z^2 + c$ has a Hausdorff dimension 2 Julia set.*

(S2) *For a Baire generic set of values of $\theta \in \mathbb{R}$, $P = e^{2i\pi\theta}z + z^2$ has a Hausdorff dimension 2 Julia set.*

Remark 1. This may sound like an encouragement, but in case (S1), for a Baire generic set of values of $c \in \partial M$, P has no indifferent cycle, and is not renormalizable, thus by a previously mentioned theorem: $\text{Leb } J(P) = 0$.

Let us focus on quadratic polynomials with an indifferent cycle. According to the Douady-Hubbard theory of tuning (quadratic-like renormalization of quadratic maps, [DH85]) it is enough to study the case where the cycle has period 1. Then, P is $\mathbb{C}$-affine conjugated to

$$P_\theta : \quad z \mapsto e^{2i\pi\theta}z + z^2$$

for a unique $\theta \in \mathbb{R}$. In the rational case we have [ADU93]:

Figure 1. The Julia set of P_θ for θ = the golden mean = $\frac{1+\sqrt{5}}{2}$. The basin of infinity is in white, the Julia set in black and the rest in gray. The latter consists in the Siegel disk and its iterated preimages. We also drew a few invariant curves within the Siegel disk and marked its center $z = 0$ with a cross.

Theorem 3.5 (Aaronson, Denker, Urbanski). *If $\theta \in \mathbb{Q}$ then*

$$\dim_H J(P_\theta) < 2.$$

Recall that the set of bounded type irrationals, i.e., of numbers $\theta = a_0 + \cfrac{1}{a_1 + \cfrac{}{\ddots}}$ with a_n bounded, has measure equal to 0. From [Mc98]:

Theorem 3.6 (McMullen). *If θ is a bounded type irrational then*

$$\dim_H J(P_\theta) < 2.$$

Let PZ be the set of irrational $\theta = a_0 + \cfrac{1}{a_1 + \cfrac{}{\ddots}}$ such that $\ln a_n = \mathcal{O}(\sqrt{n})$. This is a set of full measure in $\mathbb{R}$. By [PZ04]:

Theorem 3.7 (Petersen, Zakeri). *If $\theta \in \mathrm{PZ}$, then $z = 0$ is linearizable and*

$$\mathrm{Leb}\, J(P) = 0.$$

3.3 Douady's Plan

It takes place in the family

$$P_\theta(z) = e^{2i\pi\theta} z + z^2$$

with $\theta \in \mathbb{R}$. We recall that $z = 0$ is an indifferent fixed point and that all other periodic points are repelling (by Theorem 3.2).

Conjecture (Douady). *There exists a value of $\theta \in \mathbb{R}$ such that $z = 0$ is a Cremer point of P_θ and such that*

$$\mathrm{Leb}\, J(P_\theta) > 0.$$

In short, the idea is to look at the filled-in Julia set $K(P_\theta)$ when its interior is not empty and perturb θ so as to lose much interior but not too much measure. Let us detail this.

Let us list a few properties concerning $J = J(P)$ and $K = K(P)$ where P is a polynomial of degree at least 2:

- (Fatou, Julia) $J = \partial K$,

- (Fatou, Julia, Sullivan) $\mathrm{int}(K) \neq \varnothing \iff K$ has a cycle which is either attracting, parabolic or Siegel,

- (Douady [Do94]) the map $P \mapsto \mathrm{Leb}(K)$ is upper semi-continuous.

The idea is then to construct a sequence of $\theta_n \in \mathbb{R}$ such that

(D1) $\theta_n \longrightarrow \theta \in \mathbb{R}$,

(D2) each P_{θ_n} is Siegel or parabolic at 0 (then $\mathrm{int}\, K(P_{\theta_n}) \neq \varnothing$),

(D3) P_θ is Cremer at $z = 0$,

(D4) $\mathrm{Leb}\,\mathrm{int}(K(P_{\theta_n}))$ is bounded below,

i.e., we do not lose too much measure between two successive θ_n. Then, by (D1), (D4) and the semi-continuity mentioned above, $\mathrm{Leb}\, K(P_\theta) > 0$. By (D3), $\mathrm{int}(K(P_\theta)) = \varnothing$, i.e., $J(P_\theta) = K(P_\theta)$:

$$\mathrm{Leb}\, J(P_\theta) > 0.$$

To build such a sequence, θ_{n+1} is defined as a *perturbation* of θ_n. The main problem is to ensure that the measure loss for the interior of the filled-in Julia set is small, so as to have (D4) hold. Adrien proposed to do this in two steps:

- Start from a $\theta_n \in \mathbb{R} \setminus \mathbb{Q}$ with P_{θ_n} having a Siegel disk, and take $\theta_{n+1} \in \mathbb{Q}$ to be one of its convergents.

- Start from $\theta_{n+1} \in \mathbb{Q}$ and take $\theta_{n+2} \in \mathbb{R} \setminus \mathbb{Q}$ close to θ_{n+1} to be one of its special irrational perturbations, as described below.

  Such a $P_{\theta_{n+2}}$ has a Siegel disk.

It is easy to get (D3) if the perturbations can be taken arbitrarily small.

The special sequence of irrational perturbations $(\alpha_k)_{k \in \mathbb{N}}$ of a rational p/q is defined as follows: write one of the two (necessarily finite) continued fraction expansions

$$p/q = [a_0, \ldots, a_m] = a_0 + \cfrac{1}{a_1 + \cfrac{1}{\ddots + \cfrac{1}{a_m}}}$$

then let[4]

$$\alpha_k = [a_0, a_1, \ldots, a_m, k, 1, 1, \ldots].$$

So here is a wish list he asked me to check as a PhD thesis subject:

Wish 1. *Let α be such that P_α has a Siegel disk (with possible additional conditions, like "α has bounded type"). Let p_n/q_n be the continued fraction convergents of α. Then*

$$\liminf_{n \to +\infty} \mathrm{Leb}\, \overset{\circ}{K}(P_{p_n/q_n}) \geq \mathrm{Leb}\, \overset{\circ}{K}(P_\alpha),$$

in other words, the loss of measure of interiors is tame.

This allows to do Adrien's first step: replacing some θ_n for which P_{θ_n} has a Siegel disk by one of its convergent without losing too much measure, and calling this convergent θ_{n+1}.

Yet, when one perturbs a rational, the theory of parabolic implosion [Do94] easily implies that there is a definite loss of measure, which necessarily complicates the second step:

---

[4]It is chosen so that in the cylinder renormalization associated to the parabolic implosion at θ_{n+1}, the virtual multiplier = the golden mean.

$$\limsup_{\alpha \longrightarrow p/q} \operatorname{Leb} K(P_\alpha) < \operatorname{Leb} K(P_{p/q}) = \operatorname{Leb} \overset{\circ}{K}(P_{p/q}).$$

To miminize this loss, Adrien proposed to take the special sequence $\alpha_k \longrightarrow p/q$ described above.

Let $\operatorname{loss}(p/q)$ be defined by:

$$\frac{\liminf\limits_{k\to+\infty} \operatorname{Leb} \overset{\circ}{K}(P_{\alpha_k})}{\operatorname{Leb} K(P_{p/q})} = 1 - \operatorname{loss}(p/q),$$

which is sort of a measure of the area loss for this special sequence. One has $0 < \operatorname{loss}(p/q) < 1$ (the lower bound follows from the definite loss aforementioned, the upper bound from the fact that the Siegel disk of P_{α_k} cannot shrink too much in terms of its area, which is proved using the the theory of parabolic implosion and a lower bound on the size of Siegel disks with a fixed rotation number (diophantine in our case), due to Siegel. See also my thesis [Ch01] completed with [Ch]).

Wish 2. *With possible additional conditions, for big values of q, $\operatorname{loss}(p/q)$ is arbitrarily close to 0.*

Then in the second step of Adrien's plan (defining the irrational θ_{n+2} once the rational θ_{n+1} has been fixed), to keep the loss of measure small so that (D4) can hold, it is necessary that in the previous step (deciding which approximant of the irrational θ_n the rational θ_{n+1} shall be), we took the approximant deep enough, so that its denominator is big, so that the quantity $\operatorname{loss}(\theta_{n+1})$ is small.

3.4 Analysis of the Plan and Reduction

Based on the preparatory work of Jellouli [Je94], I gave a positive answer to Wish 1 in my thesis [Ch01]:

Theorem 3.8 (Chéritat). *Wish 1 is granted, without any additional condition on α.*

To prove this theorem, it was enough to prove that the set U_n of points which under iteration of P_{p_n/q_n} never leave the Siegel disk Δ of the unperturbed map, takes most of the area of Δ, in the sense that $\operatorname{Leb}(\Delta \backslash U_n) \longrightarrow 0$ as $n \longrightarrow +\infty$. This is done by comparing the dynamics to a vector field. For the expert reader, let us mention that the comparison is made possible thanks to a control on parabolic explosion, allowed by Yoccoz's inequality on the size of the limbs of the Mandelbrot set. See [Ch01] for details.

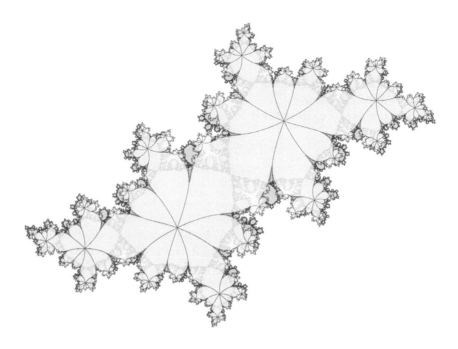

Figure 2. On this picture, we superposed (semi-transparent) the Julia set of P_α for the approximant $\alpha = 13/8$ of the golden mean over the Julia set for $\alpha = $ the golden mean. Adrien's Wish 1 was based on that kind of pictures. In particular, notice how thin is the intersection between the unperturbed Siegel disk and the basin of infinity for the perturbed Julia set.

Towards Wish 2, I gave in my thesis an upper bound on $\mathrm{loss}(p/q)$ in terms of the image of the basin of ∞ in Fatou coordinates (see Figure 5 for an illustration): let us put the natural Euclidean metrics on $\mathbb{C}/\mathbb{Z}$ inherited from $\mathbb{C}$, and do the same with the Lebesgue measure. For $p/q \in \mathbb{Q}$, the basin of ∞ for $P_{p/q}$ is an open set, and its image in a repelling cylinder of the parabolic point is a bounded non empty open set $U = U(p/q)$. Let

- $A(p/q)$ be its Lebesgue measure, and

- $h(p/q)$ its height: $h(p/q) = \sup\limits_{z \in U} \mathrm{Im}\, z - \inf\limits_{z \in U} \mathrm{Im}\, z.$

Figure 5 shows an example.

Figure 3. The fat Douady rabbit ($\alpha = 1/3 = [0,3] = [0,2,1]$), and one of its perturbations ($\alpha = [0,2,1,k,\frac{1+\sqrt{5}}{2}]$, for $k = 150$).

Theorem 3.9 (Chéritat). *There exists a function $m(A',h')$ of $A' > 0$ and $h' > 0$ such that*

1. *$\forall p/q \in \mathbb{Q}$, if $A(p/q) \le A'$ and $h(p/q) \le h'$ then*

$$\mathrm{loss}(p/q) \le m(A',h'),$$

2. *for all fixed $h' > 0$, $m(A',h') \longrightarrow 0$ as $A' \longrightarrow 0$.*

See [Ch01]. Hence if $U(p/q)$ has small area and not too big height, then the loss is small, allowing Adrien's plan to work:

Corollary 3.10. *If the next conjecture holds, then there exists a θ such that* Leb $J(P_\theta) > 0$.

To state this conjecture, let us say that an irrational α is gilded if its continued fraction expansion has its entries eventually all equal to 1: $\alpha = [a_0, \dots, a_k, 1, 1, \dots]$.

Conjecture (Chéritat). *For all gilded irrational α, denoting p_n/q_n its convergents, we have*

(C1) $A(p_n/q_n) \longrightarrow 0$,

(C2) $h(p_n/q_n)$ *is bounded.*

Figure 4. The Julia set for $\alpha = 89/55$ (9-th approximant to the golden mean). Compare with Figure 1.

Backing this conjecture, Figure 6 is the kind of experiment that allowed me to convince Adrien that his method worked.

Proof of of Corollary 3.10: The above conjecture with Theorem 3.9 imply that for all gilded irrational α, $\mathrm{loss}(p_n/q_n) \longrightarrow 0$ where p_n/q_n denote the convergents of α.

Now we want to construct a sequence θ_n satisfying conditions (D1) to (D4) of Douady's plan. Let θ_0 be any gilded irrational, for instance the golden mean. Let θ_n be defined by induction according to the two steps of Douady's plan, which are reminded here: θ_{2p} is irrational, θ_{2p+1} is one of its convergents and is thus rational, θ_{2p+2} is a special gilded irrational perturbation of the rational number θ_{2p+1}. We claim we can choose the sequence (θ_n) so that for all $n > 0$:

$$\mathrm{Leb}\,\overset{\circ}{K}(P_{\theta_n}) > \left(1 - 2^{-n}\right)\,\mathrm{Leb}\,\overset{\circ}{K}(P_{\theta_{n-1}}). \tag{3.1}$$

Given θ_{2p}, the existence of a convergent θ_{2p+1} satisfying (3.1) with $n = 2p + 1$ is granted by Theorem 3.8. We moreover choose θ_{2p+1} so that $\mathrm{loss}(\theta_{2p+1}) < 2^{-2p-2}$. Therefore, there exists a special perturbation θ_{2p+2} of θ_{2p+1} so that (3.1) holds for $n = 2p + 2$.

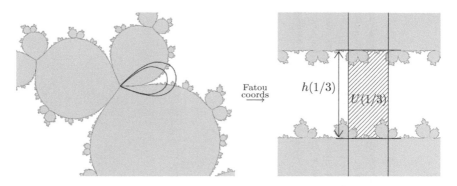

Figure 5. Left: close-up at the parabolic fixed point on the Julia set of P_α for $\alpha = 1/3$. We sketched a fundamental domain. By $P_{1/3}^3$, its inner boundary is mapped to the outer, and the quotient is a Riemann surface isomorphic to $\mathbb{C}/\mathbb{Z}$, called the Ecalle Voronin cylinder. The isomorphism is called the Fatou coordinate. Right: image of the Julia set by the Fatou coordinate. The cylinder $\mathbb{C}/\mathbb{Z}$ has been unwrapped: the picture is invariant by $z \mapsto z+1$ and vertical lines mark a fundamental domain of the cylinder $\mathbb{C}/\mathbb{Z}$. We hatched its intersection with the basin of ∞. The quantities h and A are the euclidean height and area of the hatched set.

The sequence $(\theta_n)_{n\in\mathbb{N}}$ thus satisfies (D2) and (D4). To have it also satisfy (D1) it is enough to require at each step that $|\theta_{n+1} - \theta_n| < 2^{-n}$, which is possible since in the construction above, there is a whole sequence of valid choices of θ_{n+1}, and this sequence tends to θ_n. So now the sequence θ_n tends to some θ. To also satisfy (D3), it is enough to have a θ that satisfies Cremer's Liouvillian sufficient condition [Mi00] for non-linearizability: if q is the denominator of θ_{2p-1}, then this can be achieved for instance by requiring that $\max\left(|\theta_{2p+1} - \theta_{2p}|, |\theta_{2p} - \theta_{2p-1}|\right) < \exp(-p \cdot 2^q)$. □

3.5 Analogy with McMullen's Results

This subsection is rather informal, because first the details are a bit complicated, and second at that time we did not know how to make it work. The results we are referring to will be found in [Mc98].

McMullen proved that the critical point is a Lebesgue density point of the interior of the filled-in Julia set of P_α for any bounded type α. Now, the pictures of the parabolic filled-in Julia sets $K(P_{p_n/q_n})$ in Fatou coordinates bear some kind of resemblance to that of zooms on the critical point of the Siegel disk bearing filled-in Julia set $K(P_\alpha)$. This is illustrated in Figure 7.

Could McMullen's density results be adapted to our setting, i.e., to the parabolic filled-in Julia sets seen in Fatou coordinates, and yield my conjecture, henceforth positive measure?

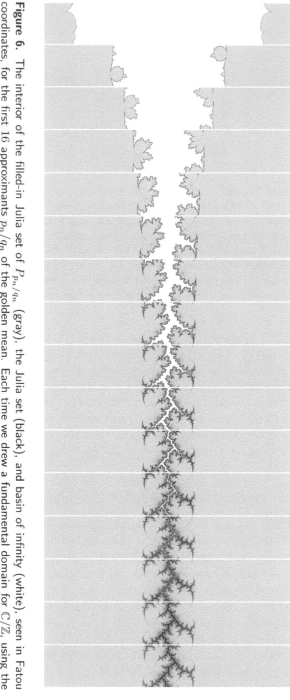

Figure 6. The interior of the filled-in Julia set of P_{p_n/q_n} (gray), the Julia set (black), and basin of infinity (white), seen in Fatou coordinates, for the first 16 approximants p_n/q_n of the golden mean. Each time we drew a fundamental domain for $\mathbb{C}/\mathbb{Z}$, using the same scale, and flipped a strip upside down every two strips, so that a convergence is clearly visible. Of course we put only a finite part of each infinite strip but what is missed is just plain grayness. It really looks like the white set has bounded height and measure tending to 0. But keep in mind that the black set is in the closure of the white, so that some measure may hide there even if it looks unlikely (for instance the fjords are extremely thin).

McMullen proved more: for instance, when the entries of the continued fraction of α are eventually periodic of period p, then there exist a map $\lambda : z \mapsto \tau z$ if p is even and $\tau\bar{z}$ if p is odd, with $\tau \in \mathbb{C}$, $0 < |\tau| < 1$ and analytic functions g_k such that the sequence

$$\lambda^{-n} \circ P_{\alpha}^{q_{n+kp}} \circ \lambda^{n} \xrightarrow[n \to +\infty]{} g_k.$$

We conjecture[5] the existence of a similar tower of limits $\tilde{g}_k$

$$\lambda^{-n} \circ P_{p_{n+kp}/q_{n+kp}}^{q_{n+kp}} \circ \lambda^{n} \xrightarrow[n \to +\infty]{} \tilde{g}_k.$$

Figure 8 shows close-ups on the Julia sets for the convergents p_n/q_n of the golden mean (whose period $p = 1$), with n ranging from 5 to 13. In each frame, the scale has been set so that the distance between the critical point (at the center of the frame) and its q_n-th iterate, is constant (about a 16th of the frame width). The convergence of these Julia sets echoes the aforementioned conjectured convergence of the rescaled q_n-th iterate.

3.6   What Remained to Be Done

Together with Xavier Buff, we were addressing my conjecture (on the areas $A(p_n/q_n)$, and heights $h(p_n/q_n)$) via the study of *cylinder renormalization* (Buff, Epstein, Shishikura, Yampolsky, ...).[6] Let $\mathcal{R}$ denote this renormalization.

H. Inou and M. Shishikura had recently proved the existence of a stable class for $\mathcal{R}$. This was, and still is, done by perturbation of the parabolic case thus requires the rotation number to be close enough to 0. In particular, there is an integer $N > 0$ such that this class contains the quadratic polynomials P_α for which α has all its continued fraction entries $\geq N$.

In the discussion above, we used *gilded* numbers, i.e., numbers whose continued fraction entries are eventually constant equal to 1. The whole discussion is valid if instead we look at numbers whose continued fraction entries are eventually constant equal to N. With this modification, Inou

---

[5]This conjecture is still unproven today. We found another way to transfer McMullen's Lebesgue density result to parabolic filled-in Julia sets seen in cylinders.

[6]Roughly, it consists in the following: draw a well-chosen curve C between the fixed point 0 and the other fixed point of P_α. Among other conditions, we want that P_α be injective on C and send it to a curve disjoint from C apart from its endpoints. Consider the bounded region D bounded by C and $P_\alpha(C)$. Consider the first return map from D to D. Now glue the boundary of D by identifying $z \in C$ with $P_\alpha(z)$. Remove the two endpoints from this. One gets a Riemann surface isomorphic to a cylinder, or if one prefers to $\mathbb{C}^*$. The first return map is conjugated by this isomorphism to a map which is the cylinder renormalization of P_α. The fixed point 0 of P_α corresponds to one end of the cylinder, and drawing another curve from this end to a fixed point of the first return map allows to renormalize the renormalized map, and so on...

$$\frac{233}{144}$$

$$\frac{1+\sqrt{5}}{2}$$

Figure 7. The image on the left shows the Julia set of P_{p_n/q_n} in Fatou coordinate for some approximant p_n/q_n of the golden mean. The image in the middle shows a close-up on the same Julia set, but in the initial coordinates, centered on the critical point. The image on the right shows a close-up at the same scale, also centered on the critical point, on the Julia set of P_α for $\alpha =$ the golden mean. There is some kind of resemblance, especially when one looks at the thick part of the basin of infinity.

Figure 8. Close-ups of the Julia sets for the convergents p_n/q_n of the golden mean, with n ranging from 5 to 13. In each frame, the scale has been set so as to have a constant distance between the critical point, at the center of the frame, and its q_n-th iterate.

and Shishikura's result proved part (C2) of my conjecture (i.e., the height $h(p_n/q_n)$ is bounded).

We then had a framework to attack part (C1) (which claims that the area $A(p_n/q_n)$ should tend to 0).

4 After the Conference

4.1 Present Results

Xavier Buff and I finished the proof during the summer following the conference (2005). To bound the area $A(p_n/q_n)$, we first used Inou-Shishikura's results to prove that the post-critical set of P_{p_n/q_n} stays arbitrarily close to the Siegel disk Δ of P_α:

Theorem 4.1. *Let α be any irrational with every continued fraction entry $\geq N$ and with moreover bounded type, i.e., the entries form a bounded sequence. Then $\forall \varepsilon > 0$, $\exists n_0$ such that $\forall n \geq n_0$, every point in the post-critical set of P_{p_n/q_n} is at distance $< \varepsilon$ from Δ (or in Δ).*

Then, using Wish 1 (which I proved earlier in my thesis), McMullen's theorem of Lebesgue density of int $K(P_\alpha)$ near Δ, a pull-back argument, a lemma à la Vitali, and the bound on $h(p_n/q_n)$, it is easy to conclude and prove part (C1) of my conjecture, thus finishing the proof of the existence of a Julia set with positive Lebesgue measure.

We also extended the method to yield examples of Julia sets with positive measure for quadratic polynomials with a Siegel disk, and also for infinitely renormalizable quadratic polynomials.

The preprint [IS] presenting the near-parabolic renormalization of Inou and Shishikura has been just been submitted; our preprint [BC] giving the full details of our positive measure result too.

4.2 Prospects

There are still open questions:[7]

- Does there exist a polynomial or a rational map, of degree ≥ 2, with a Cremer point, but whose Julia set has Lebesgue measure $= 0$?

- Quadratic Julia sets bearing a Cremer point, or a fixed Siegel disk not containing the critical point in its boundary, are non-locally connected. There is also a construction by Dan Sørensen of some infinitely renormalizable quadratic polynomials with a non-locally connected Julia set [Sø00].

---

[7]Thanks to the people who formulated some of them.

Could it happen that for quadratic polynomials, J is non-locally connected $\Longleftrightarrow$ J has positive Lebesgue measure?

- Similarly, is it equivalent for a quadratic Julia set with a fixed Siegel disk Δ, to have a positive measure Julia set and to have Δ not containing the critical point?

- Is it possible to characterize the set of irrational numbers α such that Leb $J(P_\alpha) > 0$?

- Can we produce accurate computer pictures of quadratic Julia sets with positive measure?

- Would one be able to see anything on an accurate picture? The problem is that around the Cremer fixed point, the Julia set could be so dense that it would not be possible to see the difference between J and a neighborhood of the Cremer point, unless one draws an astronomical number of pixels.

- Is P ergodic on J in our examples?

Moreover, let us mention the following open conjecture, which if it were true, would give another proof of the existence of Cremer point bearing quadratic Julia sets with positive measure:

Conjecture (Hubbard).

$$\inf_{|\lambda|<1} \text{Leb}\, K(\lambda z + z^2) > 0 \, .$$

Bibliography

[ADU93] J. Aaronson, M. Denker, and M. Urbanski, *Ergodic theory for Markov fibered systems and parabolic rational maps*, Trans. Amer. Math. Soc. **337** 2, 495–548 (1993).

[Bl84] P. Blanchard, *Complex analytic dynamics on the Riemann sphere*, Bull. Amer. Math. Soc., **11** 2 (1984), 85–141.

[BC] X. Buff and A. Chéritat, *Quadratic Julia sets with positive area*, arXiv:math/0605514.

[Ch01] A. Chéritat, *Recherche d'ensembles de Julia de mesure de Lebesgue positive*, Thesis, Université Paris Sud, Orsay, 2001.

[Ch] A. Chéritat, *Semi-continuity of Siegel disks under parabolic implosion*, arXiv:math/0507250.

[Do83]   A. Douady, *Systèmes dynamique holomorphes*, Séminaire Bour-baki, 35ᵉ année, 1982/83, no. 599. Astérisque 105–106, Soc. Math. France, 1983.

[Do94]   A. Douady, *Does a Julia set depend continuously on the polyno-mial?*, in: Complex dynamical systems: the mathematics behind the Mandelbrot set and Julia sets, R. L. Devaney Ed., Proc. Symp. in Appl. Math. **49**, Amer. Math. Soc. (1994), 91–138.

[DH84]   A. Douady and J. H. Hubbard, *Étude dynamique des polynômes complexes I/II*, Publications mathématiques d'Orsay (1984/85).

[DH85]   A. Douady and J. H. Hubbard, *On the dynamics of polynomial-like mappings*, Ann. Scient. Éc. Norm. Sup., Sér. 4 **18** 2 (1985), 287–343.

[IS]   H. Inou and M. Shishikura, *The renormalization for parabolic fixed points and their perturbation*, preprint.

[Je94]   H. Jellouli, *Sur la densité intrinsèque pour la mesure de Lebesgue et quelques problèmes de dynamique holomorphe*, Thesis, Univer-sité Paris Sud, Orsay, 1994.

[Ly91]   M. Lyubich, *On the Lebesgue measure of a quadratic polynomial*, arXiv:math/9201285, Stony Brook IMS Preprint 10, 1991.

[MSS83] R. Mañé, P. Sad, and D. Sullivan, *On the dynamics of rational maps*, Ann. Scient. Éc. Norm. Sup., Sér. 4 **16** 2 (1983), 193–217.

[Mc94]   C. T. McMullen, *Complex dynamics and renormalization*, Ann. Math. Studies **135** (1994).

[Mc98]   C. T. McMullen, *Self-similarity of Siegel disks and Hausdorff di-mension of Julia sets*, Acta Math. **180** 2 (1998), 247–292.

[Mi00]   J. Milnor, *Dynamics in one complex variable: Introductory lec-tures*, 2nd Edition, Vieweg Verlag, Wiesbaden, 2000.

[PZ04]   C. Petersen and S. Zakeri, *On the Julia set of a typical quadratic polynomial with a Siegel disk*, Ann. Math. (2) **159** 1 (2004), 1–52.

[Sch04] D. Schleicher, *On fibers and local connectivity of Mandelbrot and multibrot sets*, in: Fractal Geometry and Applications: A Jubilee of Benoît Mandelbrot, Proc. Symp. Pure Math. **72** I, Amer. Math. Soc., 2004, 477–517.

[Sh98]   M. Shishikura, *The Hausdorff dimension of the boundary of the Mandelbrot set and Julia sets*, Ann. Math. (2) **147** 2 (1998), 225–267.

[Sø00]   D. Sørensen, *Infinitely renormalizable quadratic polynomials with non-locally connected Julia set*, J. Geom. Anal. **10** 1 (2000), 169–206.

[Su83]   D. Sullivan, *Conformal dynamical systems*, in: Geometric Dynamics, Springer Verlag Lecture Notes **1007**, 1983, 725–752.

[Ta00]   Tan Lei (ed.), *The Mandelbrot Set: Theme and Variations*, London Math. Soc. Lecture Note Series **274**, Cambridge University Press, 2000.

15 On Thurston's Pullback Map

Xavier Buff, Adam Epstein, Sarah Koch, and Kevin Pilgrim

Let $f : \mathbb{P}^1 \to \mathbb{P}^1$ be a rational map with finite postcritical set P_f. Thurston showed that f induces a holomorphic map $\sigma_f : \mathrm{Teich}(\mathbb{P}^1, P_f) \to \mathrm{Teich}(\mathbb{P}^1, P_f)$ of the Teichmüller space to itself. The map σ_f fixes the basepoint corresponding to the identity map $\mathrm{id} : (\mathbb{P}^1, P_f) \to (\mathbb{P}^1, P_f)$. We give examples of such maps f showing that the following cases may occur:

1. the basepoint is an attracting fixed point, the image of σ_f is open and dense in $\mathrm{Teich}(\mathbb{P}^1, P_f)$ and the pullback map $\sigma_f : \mathrm{Teich}(\mathbb{P}^1, P_f) \to \sigma_f\big(\mathrm{Teich}(\mathbb{P}^1, P_f)\big)$ is a covering map,

2. the basepoint is a superattracting fixed point, the image of σ_f is $\mathrm{Teich}(\mathbb{P}^1, P_f)$ and $\sigma_f : \mathrm{Teich}(\mathbb{P}^1, P_f) \to \mathrm{Teich}(\mathbb{P}^1, P_f)$ is a ramified Galois covering map, or

3. the map σ_f is constant.

1 Introduction

In this article, Σ is an oriented 2-sphere. All maps $\Sigma \to \Sigma$ are assumed to be orientation-preserving. The map $f : \Sigma \to \Sigma$ is a branched covering of degree $d \geqslant 2$. A particular case of interest is when Σ can be equipped with an invariant complex structure for f. In that case, $f : \Sigma \to \Sigma$ is conjugate to a rational map $F : \mathbb{P}^1 \to \mathbb{P}^1$.

According to the Riemann-Hurwitz formula, the map f has $2d - 2$ critical points, counting multiplicities. We denote Ω_f the set of critical points and $V_f := f(\Omega_f)$ the set of critical values of f. The *postcritical set* of f is the set

$$P_f := \bigcup_{n > 0} f^{\circ n}(\Omega_f).$$

The map f is *postcritically finite* if P_f is finite. Following the literature, we refer to such maps simply as *Thurston maps*.

Two Thurston maps $f : \Sigma \to \Sigma$ and $g : \Sigma \to \Sigma$ are *equivalent* if there are homeomorphisms $h_0 : (\Sigma, P_f) \to (\Sigma, P_g)$ and $h_1 : (\Sigma, P_f) \to (\Sigma, P_g)$ for which $h_0 \circ f = g \circ h_1$ and h_0 is isotopic to h_1 through homeomorphisms agreeing on P_f. In particular, we have the following commutative diagram:

$$
\begin{array}{ccc}
(\Sigma, P_f) & \xrightarrow{h_1} & (\Sigma, P_g) \\
\downarrow{\scriptstyle f} & & \downarrow{\scriptstyle g} \\
(\Sigma, P_f) & \xrightarrow{h_0} & (\Sigma, P_g).
\end{array}
$$

In [DH93], Douady and Hubbard, following Thurston, give a complete characterization of equivalence classes of rational maps among those of Thurston maps. The characterization takes the following form.

A branched covering $f : (\Sigma, P_f) \to (\Sigma, P_f)$ induces a holomorphic self-map

$$
\sigma_f : \mathrm{Teich}(\Sigma, P_f) \to \mathrm{Teich}(\Sigma, P_f)
$$

of Teichmüller space (see Section 2 for the definition). Since it is obtained by lifting complex structures under f, we will refer to σ_f as the *pullback map* induced by f. The map f is equivalent to a rational map if and only if the pullback map σ_f has a fixed point. By a generalization of the Schwarz lemma, σ_f does not increase Teichmüller distances. For most maps f, the pullback map σ_f is a contraction, and so a fixed point, if it exists, is unique.

In this note, we give examples showing that the contracting behavior of σ_f near this fixed point can be rather varied.

Theorem 1.1. *There exist Thurston maps f for which σ_f is contracting, has a fixed point τ and:*

1. *the derivative of σ_f is invertible at τ, the image of σ_f is open and dense in $\mathrm{Teich}(\mathbb{P}^1, P_f)$ and $\sigma_f : \mathrm{Teich}(\mathbb{P}^1, P_f) \to \sigma_f\big(\mathrm{Teich}(\mathbb{P}^1, P_f)\big)$ is a covering map,*

2. *the derivative of σ_f is not invertible at τ, the image of σ_f is equal to $\mathrm{Teich}(\mathbb{P}^1, P_f)$ and $\sigma_f : \mathrm{Teich}(\mathbb{P}^1, P_f) \to \mathrm{Teich}(\mathbb{P}^1, P_f)$ is a ramified Galois covering map,[1] or*

3. *the map σ_f is constant.*

In Section 2, we establish notation, define Teichmüller space and the pullback map σ_f precisely, and develop some preliminary results used in our subsequent analysis. In Sections 3, 4, and 5.1, respectively, we give concrete

---

[1] A ramified covering is Galois if the group of deck transformations acts transitively on the fibers.

examples which together provide the proof of Theorem 1.1. We supplement these examples with some partial general results. In Section 3, we state a fairly general sufficient condition on f under which σ_f evenly covers it image. This condition, which can sometimes be checked in practice, is excerpted from [Ko07] and [Ko08]. Our example illustrating (2) is highly symmetric and atypical; we are not aware of any reasonable generalization. In Section 5.2, we state three conditions on f equivalent to the condition that σ_f is constant. Unfortunately, each is extremely difficult to verify in concrete examples.

Acknowledgments. We would like to thank Curt McMullen who provided the example showing (3).

2 Preliminaries

Recall that a Riemann surface is a connected oriented topological surface together with a *complex structure*: a maximal atlas of charts $\phi : U \to \mathbb{C}$ with holomorphic overlap maps. For a given oriented, compact topological surface X, we denote the set of all complex structures on X by $\mathcal{C}(X)$. It is easily verified that an orientation-preserving branched covering map $f : X \to Y$ induces a map $f^* : \mathcal{C}(Y) \to \mathcal{C}(X)$; in particular, for any orientation-preserving homeomorphism $\psi : X \to X$, there is an induced map $\psi^* : \mathcal{C}(X) \to \mathcal{C}(X)$.

Let $A \subset X$ be finite. The Teichmüller space of (X, A) is

$$\text{Teich}(X, A) := \mathcal{C}(X)/\!\sim_A$$

where $c_1 \sim_A c_2$ if and only if $c_1 = \psi^*(c_2)$ for some orientation-preserving homeomorphism $\psi : X \to X$ which is isotopic to the identity relative to A. In view of the homotopy-lifting property, if

- $B \subset Y$ is finite and contains the critical value set V_f of f, and

- $A \subseteq f^{-1}(B)$,

then $f^* : \mathcal{C}(Y) \to \mathcal{C}(X)$ descends to a well-defined map σ_f between the corresponding Teichmüller spaces:

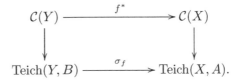

This map is known as the *pullback map* induced by f.

In addition if $f : X \to Y$ and $g : Y \to Z$ are orientation-preserving branched covering maps and if $A \subset X$, $B \subset Y$ and $C \subset Z$ are such that

- B contains V_f and C contains V_g,

- $A \subseteq f^{-1}(B)$ and $B \subseteq g^{-1}(C)$,

then C contains the critical values of $g \circ f$ and $A \subseteq (g \circ f)^{-1}(C)$. Thus

$$\sigma_{g \circ f} : \mathrm{Teich}(Z, C) \to \mathrm{Teich}(X, A)$$

can be decomposed as $\sigma_{g \circ f} = \sigma_f \circ \sigma_g$:

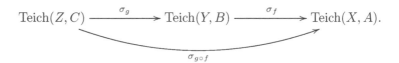

For the special case of $\mathrm{Teich}(\mathbb{P}^1, A)$, we may use the Uniformization Theorem to obtain the following description. Given a finite set $A \subset \mathbb{P}^1$ we may regard $\mathrm{Teich}(\mathbb{P}^1, A)$ as the quotient of the space of all orientation-preserving homeomorphisms $\phi : \mathbb{P}^1 \to \mathbb{P}^1$ by the equivalence relation $\sim$ whereby $\phi_1 \sim \phi_2$ if there exists a Möbius transformation μ such that $\mu \circ \phi_1 = \phi_2$ on A, and $\mu \circ \phi_1$ is isotopic to ϕ_2 relative to A. Note that there is a natural basepoint $\circledast$ given by the class of the identity map on $\mathbb{P}^1$. It is well known that $\mathrm{Teich}(\mathbb{P}^1, A)$ has a natural topology and complex manifold structure (see [H06]).

The *moduli space* is the space of all injections $\psi : A \hookrightarrow \mathbb{P}^1$ modulo post-composition with Möbius transformations. The moduli space will be denoted as $\mathrm{Mod}(\mathbb{P}^1, A)$. If ϕ represents an element of $\mathrm{Teich}(\mathbb{P}^1, A)$, the restriction $[\phi] \mapsto \phi|_A$ induces a universal covering $\pi : \mathrm{Teich}(\mathbb{P}^1, A) \to \mathrm{Mod}(\mathbb{P}^1, A)$ which is a local biholomorphism with respect to the complex structures on $\mathrm{Teich}(\mathbb{P}^1, A)$ and $\mathrm{Mod}(\mathbb{P}^1, A)$.

Let $f : \mathbb{P}^1 \to \mathbb{P}^1$ be a Thurston map with $|P_f| \geqslant 3$. For any $\Theta \subseteq P_f$ with $|\Theta| = 3$, there is an obvious identification of $\mathrm{Mod}(\mathbb{P}^1, P_f)$ with an open subset of $(\mathbb{P}^1)^{P_f - \Theta}$. Assume $\tau \in \mathrm{Teich}(\mathbb{P}^1, P_f)$ and let $\phi : \mathbb{P}^1 \to \mathbb{P}^1$ be a homeomorphism representing τ with $\phi|_\Theta = \mathrm{id}|_\Theta$. By the Uniformization Theorem, there exist

- a unique homeomorphism $\psi : \mathbb{P}^1 \to \mathbb{P}^1$ representing $\tau' := \sigma_f(\tau)$ with $\psi|_\Theta = \mathrm{id}|_\Theta$, and

- a unique rational map $F : \mathbb{P}^1 \to \mathbb{P}^1$, such that the following diagram commutes:

$$
\begin{array}{ccc}
(\mathbb{P}^1, P_f) & \xrightarrow{\ \ \psi\ \ } & (\mathbb{P}^1, \psi(P_f)) \\
\big\downarrow{f} & & \big\downarrow{F} \\
(\mathbb{P}^1, P_f) & \xrightarrow{\ \ \phi\ \ } & (\mathbb{P}^1, \phi(P_f)).
\end{array}
$$

Conversely, if we have such a commutative diagram with F holomorphic, then

$$\sigma_f(\tau) = \tau',$$

where $\tau \in \mathrm{Teich}(\mathbb{P}^1, P_f)$ and $\tau' \in \mathrm{Teich}(\mathbb{P}^1, P_f)$ are the equivalence classes of ϕ and ψ, respectively. In particular, if $f : \mathbb{P}^1 \to \mathbb{P}^1$ is rational, then $\sigma_f : \mathrm{Teich}(\mathbb{P}^1, P_f) \to \mathrm{Teich}(\mathbb{P}^1, P_f)$ fixes the basepoint $\circledast$.

3   Proof of (1)

In this section, we prove that there are Thurston maps $f : \Sigma \to \Sigma$ such that σ_f

- is contracting,

- has a fixed point $\tau \in \mathrm{Teich}(\Sigma, P_f)$, and

- is a covering map over its image that is open and dense in $\mathrm{Teich}(\Sigma, P_f)$.

In fact, we show that this is the case when $\Sigma = \mathbb{P}^1$ and $f : \mathbb{P}^1 \to \mathbb{P}^1$ is a polynomial whose critical points are all periodic. The following is adapted from [Ko08].

Proposition 3.1. *If $f : \mathbb{P}^1 \to \mathbb{P}^1$ is a polynomial of degree $d \geqslant 2$ whose critical points are all periodic, then*

- $\sigma_f\big(\mathrm{Teich}(\mathbb{P}^1, P_f)\big)$ *is open and dense in* $\mathrm{Teich}(\mathbb{P}^1, P_f)$, *and*

- $\sigma_f : \mathrm{Teich}(\mathbb{P}^1, P_f) \to \sigma_f\big(\mathrm{Teich}(\mathbb{P}^1, P_f)\big)$ *is a covering map.*

In particular, the derivative $D\sigma_f$ is invertible at the fixed point $\circledast$.

This section is devoted to the proof of this proposition.

Let $n = |P_f| - 3$. We will identify $\mathrm{Mod}(\mathbb{P}^1, P_f)$ with an open subset of $\mathbb{P}^n$ as follows. First enumerate the finite postcritical points as $p_0, \ldots, p_{n+1}$. Any point of $\mathrm{Mod}(\mathbb{P}^1, P_f)$ has a representative $\psi : P_f \hookrightarrow \mathbb{P}^1$ such that

$$\psi(\infty) = \infty \quad \text{and} \quad \psi(p_0) = 0.$$

Two such representatives are equal up to multiplication by a non-zero complex number. We identify the point in $\mathrm{Mod}(\mathbb{P}^1, P_f)$ with the point

$$[x_1 : \ldots : x_{n+1}] \in \mathbb{P}^n \quad \text{where} \quad x_1 := \psi(p_1) \in \mathbb{C}, \ldots, x_{n+1} := \psi(p_{n+1}) \in \mathbb{C}.$$

In this way, the moduli space $\mathrm{Mod}(\mathbb{P}^1, P_f)$ is identified with $\mathbb{P}^n - \Delta$, where Δ is the *forbidden locus*:

$$\Delta := \big\{ [x_1 : \ldots : x_{n+1}] \in \mathbb{P}^n \ ; \ (\exists i, \ x_i = 0) \text{ or } (\exists i \neq j, \ x_i = x_j) \big\}.$$

The universal cover $\pi : \mathrm{Teich}(\mathbb{P}^1, P_f) \to \mathrm{Mod}(\mathbb{P}^1, P_f)$ is then identified with a universal cover $\pi : \mathrm{Teich}(\mathbb{P}^1, P_f) \to \mathbb{P}^n - \Delta$.

Generalizing a result of Bartholdi and Nekrashevych [BN06], the thesis [Ko07] showed that when $f : \mathbb{P}^1 \to \mathbb{P}^1$ is a unicritical polynomial there is an analytic endomorphism $g_f : \mathbb{P}^n \to \mathbb{P}^n$ for which the following diagram commutes:

$$
\begin{array}{ccc}
\mathrm{Teich}(\mathbb{P}^1, P_f) & \xrightarrow{\ \sigma_f\ } & \mathrm{Teich}(\mathbb{P}^1, P_f) \\
\Big\downarrow{\scriptstyle \pi} & & \Big\downarrow{\scriptstyle \pi} \\
\mathbb{P}^n & \xleftarrow{\ g_f\ } & \mathbb{P}^n .
\end{array}
$$

We show that the same result holds when $f : \mathbb{P}^1 \to \mathbb{P}^1$ is a polynomial whose critical points are all periodic.

Proposition 3.2. *Let $f : \mathbb{P}^1 \to \mathbb{P}^1$ be a polynomial of degree $d \geqslant 2$ whose critical points are all periodic. Set $n := |P_f| - 3$. Then,*

1. *there is an analytic endomorphism $g_f : \mathbb{P}^n \to \mathbb{P}^n$ for which the following diagram commutes:*

$$
\begin{array}{ccc}
\mathrm{Teich}(\mathbb{P}^1, P_f) & \xrightarrow{\ \sigma_f\ } & \mathrm{Teich}(\mathbb{P}^1, P_f) \\
\Big\downarrow{\scriptstyle \pi} & & \Big\downarrow{\scriptstyle \pi} \\
\mathbb{P}^n & \xleftarrow{\ g_f\ } & \mathbb{P}^n
\end{array}
$$

2. *σ_f takes its values in $\mathrm{Teich}(\mathbb{P}^1, P_f) - \pi^{-1}(\mathcal{L})$ with $\mathcal{L} := g_f^{-1}(\Delta)$,*

3. *$g_f(\Delta) \subseteq \Delta$, and*

4. *the critical point locus and the critical value locus of g_f are contained in Δ.*

Proof of Proposition 3.1 assuming Proposition 3.2: Note that $\mathcal{L}$ is a codimension 1 analytic subset of $\mathbb{P}^n$, whence $\pi^{-1}(\mathcal{L})$ is a codimension 1 analytic subset of $\text{Teich}(\mathbb{P}^1, P_f)$. Thus, the complementary open sets are dense and connected. Since $g_f : \mathbb{P}^n - \mathcal{L} \to \mathbb{P}^n - \Delta$ is a covering map, the composition

$$g_f \circ \pi : \text{Teich}(\mathbb{P}^1, P_f) - \pi^{-1}(\mathcal{L}) \to \mathbb{P}^n - \Delta$$

is a covering map. Moreover,

$$\pi(\circledast) = g_f \circ \pi \circ \sigma_f(\circledast) = g_f \circ \pi(\circledast).$$

By universality of the covering map $\pi : \text{Teich}(\mathbb{P}^1, P_f) \to \mathbb{P}^n - \Delta$, there is a unique map $\sigma : \text{Teich}(\mathbb{P}^1, P_f) \to \text{Teich}(\mathbb{P}^1, P_f) - \pi^{-1}(\mathcal{L})$ such that

- $\sigma(\circledast) = \circledast$ and

- the following diagram commutes:

$$\begin{array}{ccc} \text{Teich}(\mathbb{P}^1, P_f) & \xrightarrow{\quad \sigma \quad} & \text{Teich}(\mathbb{P}^1, P_f) - \pi^{-1}(\mathcal{L}) \\ {\scriptstyle \pi} \downarrow & \swarrow {\scriptstyle g_f \circ \pi} & \\ \mathbb{P}^n - \Delta. & & \end{array}$$

Furthermore, $\sigma : \text{Teich}(\mathbb{P}^1, P_f) \to \text{Teich}(\mathbb{P}^1, P_f) - \pi^{-1}(\mathcal{L})$ is a covering map. Finally, by uniqueness we have $\sigma_f = \sigma$. □

Proof of Proposition 3.2. The proof consists of several steps.

(1) We first show the existence of the endomorphism $g_f : \mathbb{P}^n \to \mathbb{P}^n$. We start with the definition of g_f.

The restriction of f to P_f is a permutation which fixes ∞. Denote by $\mu : [0, n+1] \to [0, n+1]$ the permutation defined by:

$$p_{\mu(k)} = f(p_k)$$

and denote by ν the inverse of μ.

For $k \in [0, n+1]$, let m_k be the multiplicity of p_k as a critical point of f (if p_k is not a critical point of f, then $m_k := 0$).

Set $a_0 := 0$ and let $Q \in \mathbb{C}[a_1, \ldots, a_{n+1}, z]$ be the homogeneous polynomial of degree d defined by

$$Q(a_1, \ldots, a_{n+1}, z) := \int_{a_{\nu(0)}}^{z} \left(d \prod_{k=0}^{n+1} (w - a_k)^{m_k} \right) dw.$$

Given $\mathbf{a} \in \mathbb{C}^{n+1}$, let $F_{\mathbf{a}}$ be the unique monic polynomial of degree d which vanishes at $a_{\nu(0)}$, and whose critical points are exactly those points a_k for which $m_k > 0$, counted with multiplicity m_k:

$$F_{\mathbf{a}}(z) := Q(a_1, \ldots, a_{n+1}, z).$$

Let $G_f : \mathbb{C}^{n+1} \to \mathbb{C}^{n+1}$ be the homogeneous map of degree d defined by

$$G_f \begin{pmatrix} a_1 \\ \vdots \\ a_{n+1} \end{pmatrix} := \begin{pmatrix} F_{\mathbf{a}}(a_{\nu(1)}) \\ \vdots \\ F_{\mathbf{a}}(a_{\nu(n+1)}) \end{pmatrix} = \begin{pmatrix} Q(a_1, \ldots, a_{n+1}, a_{\nu(1)}) \\ \vdots \\ Q(a_1, \ldots, a_{n+1}, a_{\nu(n+1)}) \end{pmatrix}.$$

We claim that $G_f^{-1}(\mathbf{0}) = \{\mathbf{0}\}$ and thus, $G_f : \mathbb{C}^{n+1} \to \mathbb{C}^{n+1}$ induces an endomorphism $g_f : \mathbb{P}^n \to \mathbb{P}^n$. Indeed, let us consider a point $\mathbf{a} \in \mathbb{C}^{n+1}$. By definition of G_f, if $G_f(\mathbf{a}) = \mathbf{0}$, then the monic polynomial $F_{\mathbf{a}}$ vanishes at $a_0, a_1, \ldots, a_{n+1}$. The critical points of $F_{\mathbf{a}}$ are those points a_k for which $m_k > 0$. They are all mapped to 0 and thus, $F_{\mathbf{a}}$ has only one critical value in $\mathbb{C}$. A calculation using the Riemann-Hurwitz formula implies that all of the preimages of this critical value must therefore coincide, and since $a_0 = 0$, they all coincide at 0. Thus, for all $k \in [0, n+1]$, $a_k = 0$.

Let us now prove that for all $\tau \in \text{Teich}(\mathbb{P}^1, P_f)$, we have

$$\pi(\tau) = g_f \circ \pi \circ \sigma_f(\tau).$$

Let τ be a point in $\text{Teich}(\mathbb{P}^1, P_f)$ and set $\tau' := \sigma_f(\tau)$.

We will show that there is a representative ϕ of τ and a representative ψ of τ' such that $\phi(\infty) = \psi(\infty) = \infty$, $\phi(p_0) = \psi(p_0) = 0$ and

$$G_f\big(\psi(p_1), \ldots, \psi(p_{n+1})\big) = \big(\phi(p_1), \ldots, \phi(p_{n+1})\big). \tag{3.1}$$

It then follows that

$$g_f\big([\psi(p_1) : \ldots : \psi(p_{n+1})]\big) = [\phi(p_1) : \ldots : \phi(p_{n+1})]$$

which concludes the proof since

$$\pi(\tau') = [\psi(p_1) : \ldots : \psi(p_{n+1})] \quad \text{and} \quad \pi(\tau) = [\phi(p_1) : \ldots : \phi(p_{n+1})].$$

To show the existence of ϕ and ψ, we may proceed as follows. Let ϕ be any representative of τ such that $\phi(\infty) = \infty$ and $\phi(p_0) = 0$. Then, there is a representative $\psi : \mathbb{P}^1 \to \mathbb{P}^1$ of τ' and a rational map $F : \mathbb{P}^1 \to \mathbb{P}^1$ such that the following diagram commutes:

$$
\begin{array}{ccc}
\mathbb{P}^1 & \xrightarrow{\psi} & \mathbb{P}^1 \\
{\scriptstyle f} \downarrow & & \downarrow {\scriptstyle F} \\
\mathbb{P}^1 & \xrightarrow{\phi} & \mathbb{P}^1.
\end{array}
$$

We may normalize ψ so that $\psi(\infty) = \infty$ and $\psi(p_0) = 0$. Then, F is a polynomial of degree d. Multiplying ψ by a non-zero complex number, we may assume that F is a monic polynomial.

We now check that these homeomorphisms ϕ and ψ satisfy the required Property (3.1). For $k \in [0, n+1]$, set

$$x_k := \psi(p_k) \quad \text{and} \quad y_k := \phi(p_k).$$

We must show that

$$G_f(x_1, \ldots, x_{n+1}) = (y_1, \ldots, y_{n+1}).$$

Note that for $k \in [0, n+1]$, we have the following commutative diagram:

$$
\begin{array}{ccc}
p_{\nu(k)} & \xmapsto{\;\psi\;} & x_{\nu(k)} \\
{\scriptstyle f}\downarrow & & \downarrow{\scriptstyle F} \\
p_k & \xmapsto{\;\phi\;} & y_k.
\end{array}
$$

Consequently, $F(x_{\nu(k)}) = y_k$. In particular $F(x_{\nu(0)}) = 0$. In addition, the critical points of F are exactly those points x_k for which $m_k > 0$, counted with multiplicity m_k. As a consequence, $F = F_{\mathbf{x}}$ and

$$
G_f \begin{pmatrix} x_1 \\ \vdots \\ x_{n+1} \end{pmatrix} = \begin{pmatrix} F_{\mathbf{x}}(x_{\nu(1)}) \\ \vdots \\ F_{\mathbf{x}}(x_{\nu(n+1)}) \end{pmatrix} = \begin{pmatrix} F(x_{\nu(1)}) \\ \vdots \\ F(x_{\nu(n+1)}) \end{pmatrix} = \begin{pmatrix} y_1 \\ \vdots \\ y_{n+1} \end{pmatrix}.
$$

(2) To see that σ_f takes its values in $\mathrm{Teich}(\mathbb{P}^1, P_f) - \pi^{-1}(\mathcal{L})$, we may proceed by contradiction. Assume

$$\tau \in \mathrm{Teich}(\mathbb{P}^1, P_f) \quad \text{and} \quad \tau' := \sigma_f(\tau) \in \pi^{-1}(\mathcal{L}).$$

Then, since $\pi = g_f \circ \pi \circ \sigma_f$, we obtain

$$\pi(\tau) = g_f \circ \pi(\tau') \in \Delta.$$

But if $\tau \in \mathrm{Teich}(\mathbb{P}^1, P_f)$, then $\pi(\tau)$ cannot be in Δ, and we have a contradiction.

(3) To see that $g_f(\Delta) \subseteq \Delta$, assume

$$\mathbf{a} := (a_1, \ldots, a_{n+1}) \in \mathbb{C}^{n+1}$$

and set $a_0 := 0$. Set

$$(b_0, b_1, \ldots, b_{n+1}) := \big(0, F_{\mathbf{a}}(a_{\nu(1)}), \ldots, F_{\mathbf{a}}(a_{\nu(n+1)})\big).$$

Then,

$$G_f(a_1, \ldots, a_{n+1}) = (b_1, \ldots, b_{n+1}).$$

Note that
$$a_i = a_j \quad \Longrightarrow \quad b_{\mu(i)} = b_{\mu(j)}.$$

In addition $[a_1 : \ldots : a_{n+1}]$ belongs to Δ precisely when there are integers $i \neq j$ in $[0, n+1]$ such that $a_i = a_j$. As a consequence,

$$[a_1 : \ldots : a_{n+1}] \in \Delta \quad \Longrightarrow \quad [b_1 : \ldots : b_{n+1}] \in \Delta.$$

This proves that $g_f(\Delta) \subseteq \Delta$.

(4) To see that the critical point locus of g_f is contained in Δ, we must show that $\mathrm{Jac}\, G_f : \mathbb{C}^{n+1} \to \mathbb{C}$ does not vanish outside Δ. Since $g_f(\Delta) \subseteq \Delta$, we then automatically obtain that the critical value locus of g_f is contained in Δ.

Note that $\mathrm{Jac}\, G_f(a_1, \ldots, a_{n+1})$ is a homogeneous polynomial of degree $(n+1) \cdot (d-1)$ in the variables $a_1, \ldots, a_{n+1}$. Consider the polynomial $J \in \mathbb{C}[a_1, \ldots, a_{n+1}]$ defined by

$$J(a_1, \ldots, a_{n+1}) := \prod_{0 \leqslant i < j \leqslant n+1} (a_i - a_j)^{m_i + m_j} \quad \text{with} \quad a_0 := 0.$$

Lemma 3.3. *The Jacobian* $\mathrm{Jac}\, G_f$ *is divisible by* J.

Proof: Set $a_0 := 0$ and $G_0 := 0$. For $j \in [1, n+1]$, let G_j be the j-th coordinate of $G_f(a_1, \ldots, a_{n+1})$, i.e.,

$$G_j := d \int_{a_{\nu(0)}}^{a_{\nu(j)}} \prod_{k=0}^{n+1} (w - a_k)^{m_k} dw.$$

For $0 \leqslant i < j \leqslant n+1$, note that setting $w = a_i + t(a_j - a_i)$, we have

$$G_{\mu(j)} - G_{\mu(i)} = d \int_{a_i}^{a_j} \prod_{k=0}^{n+1} (w - a_k)^{m_k} dw$$

$$= d \int_0^1 \prod_{k=0}^{n+1} (a_i + t(a_j - a_i) - a_k)^{m_k} \cdot (a_j - a_i) dt$$

$$= (a_j - a_i)^{m_i + m_j + 1} \cdot H_{i,j},$$

with

$$H_{i,j} := d \int_0^1 t^{m_i} (t-1)^{m_j} \prod_{\substack{k \in [0, n+1] \\ k \neq i, j}} \left(a_i - a_k + t(a_j - a_i) \right)^{m_k} dt.$$

In particular, $G_{\mu(j)} - G_{\mu(i)}$ is divisible by $(a_j - a_i)^{m_i + m_j + 1}$.

For $k \in [0, n+1]$, let L_k be the row defined as:

$$L_k := \begin{bmatrix} \dfrac{\partial G_k}{\partial a_1} & \cdots & \dfrac{\partial G_k}{\partial a_{n+1}} \end{bmatrix}.$$

Note that L_0 is the zero row, and for $k \in [1, n+1]$, L_k is the k-th row of the Jacobian matrix of G_f. According to the previous computations, the entries of $L_{\mu(j)} - L_{\mu(i)}$ are the partial derivatives of $(a_j - a_i)^{m_i + m_j + 1} \cdot H_{i,j}$. It follows that $L_{\mu(j)} - L_{\mu(i)}$ is divisible by $(a_j - a_i)^{m_i + m_j}$. Indeed, $L_{\mu(j)} - L_{\mu(i)}$ is either the difference of two rows of the Jacobian matrix of G_f, or such a row up to sign, when $\mu(i) = 0$ or $\mu(j) = 0$. As a consequence, Jac G_f is divisible by J. $\qquad\square$

Since $\sum m_j = d - 1$, an easy computation shows that the degree of J is $(n+1) \cdot (d-1)$. Since J and Jac G_f are homogeneous polynomials of the same degree and since J divides Jac G_f, they are equal up to multiplication by a non-zero complex number. This shows that Jac G_f vanishes exactly when J vanishes, i.e., on a subset of Δ. This completes the proof of Proposition 3.2. $\qquad\square$

4   Proof of (2)

In this section we present an example of a Thurston map f such that the pullback map $\sigma_f : \mathrm{Teich}(\mathbb{P}^1, P_f) \to \mathrm{Teich}(\mathbb{P}^1, P_f)$ is a ramified Galois covering and has a fixed critical point.

Let $f : \mathbb{P}^1 \to \mathbb{P}^1$ be the rational map defined by:

$$f(z) = \frac{3z^2}{2z^3 + 1}.$$

Note that f has critical points at $\Omega_f = \{0, 1, \omega, \bar{\omega}\}$, where

$$\omega := -1/2 + i\sqrt{3}/2 \quad \text{and} \quad \bar{\omega} := -1/2 - i\sqrt{3}/2$$

are cube roots of unity. Notice that

$$f(0) = 0, \ f(1) = 1, \ f(\omega) = \bar{\omega} \text{ and } f(\bar{\omega}) = \omega.$$

So, $P_f = \{0, 1, \omega, \bar{\omega}\}$ and f is a Thurston map. We illustrate the critical dynamics of f with the following *ramification portrait*:

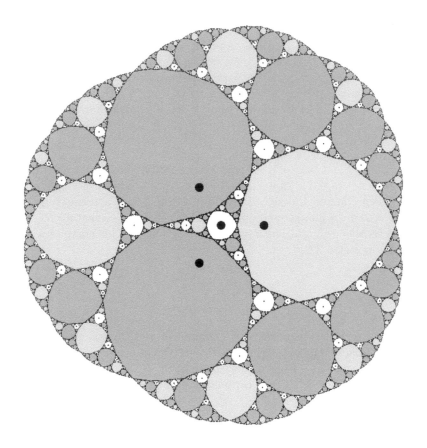

Figure 1. The Julia set of the rational map $f : z \mapsto 3z^2/(2z^3 + 1)$. The basin of 0 is white. The basin of 1 is light grey. The basin of $\{\omega, \bar{\omega}\}$ is dark grey.

Since $|P_f| = 4$, the Teichmüller space $\text{Teich}(\mathbb{P}^1, P_f)$ has complex dimension 1.

Set $\Theta := \{1, \omega, \bar{\omega}\} \subset P_f$. We identify the moduli space $\text{Mod}(\mathbb{P}^1, P_f)$ with $\mathbb{P}^1 - \Theta$. More precisely, if $\phi : P_f \hookrightarrow \mathbb{P}^1$ represents a point in $\text{Mod}(\mathbb{P}^1, P_f)$ with $\phi|_\Theta = \text{id}|_\Theta$, then we identify the class of ϕ in $\text{Mod}(\mathbb{P}^1, P_f)$ with the point $\phi(0)$ in $\mathbb{P}^1 - \Theta$. Consequently, the universal covering $\pi : \text{Teich}(\mathbb{P}^1, P_f) \to \text{Mod}(\mathbb{P}^1, P_f)$ is identified with a universal covering $\pi : \text{Teich}(\mathbb{P}^1, P_f) \to \mathbb{P}^1 - \Theta$ and $\pi(\circledast)$ is identified with 0.

Assume $\tau \in \mathrm{Teich}(\mathbb{P}^1, P_f)$ and let $\phi : \mathbb{P}^1 \to \mathbb{P}^1$ be a homeomorphism representing τ with $\phi|_\Theta = \mathrm{id}|_\Theta$. There exists a unique homeomorphism $\psi : \mathbb{P}^1 \to \mathbb{P}^1$ representing $\tau' := \sigma_f(\tau)$ and a unique cubic rational map $F : \mathbb{P}^1 \to \mathbb{P}^1$ such that

- $\psi|_\Theta = \mathrm{id}|_\Theta$ and

- the following diagram commutes

$$
\begin{array}{ccc}
\mathbb{P}^1 & \xrightarrow{\;\psi\;} & \mathbb{P}^1 \\
{\scriptstyle f}\downarrow & & \downarrow{\scriptstyle F} \\
\mathbb{P}^1 & \xrightarrow{\;\phi\;} & \mathbb{P}^1 .
\end{array}
$$

We set

$$y := \phi(0) = \pi(\tau) \quad \text{and} \quad x := \psi(0) = \pi(\tau').$$

The rational map F has the following properties:

(P1) 1, ω and $\bar{\omega}$ are critical points of F, $F(1) = 1$, $F(\omega) = \bar{\omega}$, $F(\bar{\omega}) = \omega$ and

(P2) $x \in \mathbb{P}^1 - \Theta$ is a critical point of F and $y = F(x) \in \mathbb{P}^1 - \Theta$ is the corresponding critical value.

For $\alpha = [a : b] \in \mathbb{P}^1$, let F_α be the rational map defined by

$$F_\alpha(z) := \frac{az^3 + 3bz^2 + 2a}{2bz^3 + 3az + b}.$$

Note that $f = F_0$.

We first show that $F = F_\alpha$ for some $\alpha \in \mathbb{P}^1$. For this purpose, we may write $F = P/Q$ with P and Q polynomials of degree $\leqslant 3$. Note that if $\widehat{F} = \widehat{P}/\widehat{Q}$ is another rational map of degree 3 satisfying Property (P1), then $F - \widehat{F}$ and $(F - \widehat{F})'$ vanish at 1, ω and $\bar{\omega}$. Since

$$F - \widehat{F} = \frac{P\widehat{Q} - Q\widehat{P}}{Q\widehat{Q}}$$

and since $P\widehat{Q} - Q\widehat{P}$ has degree $\leqslant 6$, we see that $P\widehat{Q} - Q\widehat{P}$ is equal to $(z^3 - 1)^2$ up to multiplication by a complex number.

A computation shows that F_0 and F_∞ satisfy Property (P1). We may write $F_0 = P_0/Q_0$ and $F_\infty = P_\infty/Q_\infty$ with

$$P_0(z) = 3z^2, \quad Q_0(z) = 2z^3 + 1, \quad P_\infty(z) = z^3 + 2 \quad \text{and} \quad Q_\infty(z) = 3z.$$

The previous observation shows that $PQ_0 - QP_0$ and $PQ_\infty - QP_\infty$ are both scalar multiples of $(z^3 - 1)^2$, and thus, we can find complex numbers a and b such that

$$a \cdot (PQ_\infty - QP_\infty) + b \cdot (PQ_0 - QP_0) = 0,$$

whence

$$P \cdot (aQ_\infty + bQ_0) = Q \cdot (aP_\infty + bP_0).$$

This implies that

$$F = \frac{P}{Q} = \frac{aP_\infty + bP_0}{aQ_\infty + bQ_0} = F_\alpha \quad \text{with} \quad \alpha = [a : b] \in \mathbb{P}^1.$$

We now study how $\alpha \in \mathbb{P}^1$ depends on $\tau \in \mathrm{Teich}(\mathbb{P}^1, P_f)$. The critical points of F_α are 1, ω, $\bar{\omega}$ and α^2. We therefore have

$$x = \alpha^2 \quad \text{and} \quad y = F_\alpha(\alpha^2) = \frac{\alpha(\alpha^3 + 2)}{2\alpha^3 + 1} = \frac{x^2 + 2\alpha}{2x\alpha + 1}.$$

In particular,

$$\alpha = \frac{x^2 - y}{2xy - 2}.$$

Consider now the holomorphic maps $X : \mathbb{P}^1 \to \mathbb{P}^1$, $Y : \mathbb{P}^1 \to \mathbb{P}^1$ and $A : \mathrm{Teich}(\mathbb{P}^1, P_f) \to \mathbb{P}^1$ defined by

$$X(\alpha) := \alpha^2, \quad Y(\alpha) := \frac{\alpha(\alpha^3 + 2)}{2\alpha^3 + 1}$$

and

$$A(\tau) := \frac{x^2 - y}{2xy - 2} \quad \text{with} \quad y = \pi(\tau) \quad \text{and} \quad x = \pi \circ \sigma_f(\tau).$$

Observe that

$$X^{-1}(\{1, \omega, \bar{\omega}\}) = Y^{-1}(\{1, \omega, \bar{\omega}\}) = \Theta' := \{1, \omega, \bar{\omega}, -1, -\omega, -\bar{\omega}\}.$$

Thus, we have the following commutative diagram,

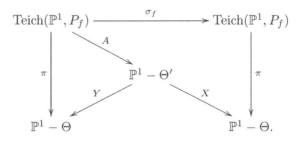

In this paragraph, we show that σ_f has local degree two at the fixed basepoint. Since $f = F_0$, we have $A(\circledast) = 0$. Also, $\pi(\circledast) = \pi \circ \sigma_f(\circledast) = 0$. Since $Y(\alpha) = 2\alpha + \mathcal{O}(\alpha^2)$, the germ $Y : (\mathbb{P}^1, 0) \to (\mathbb{P}^1, 0)$ is locally invertible at 0. Since $\pi : \mathrm{Teich}(\mathbb{P}^1, P_f) \to \mathrm{Mod}(\mathbb{P}^1, P_f)$ is a universal covering, the germ $\pi : \big(\mathrm{Teich}(\mathbb{P}^1, P_f), \circledast\big) \to \big(\mathrm{Mod}(\mathbb{P}^1, P_f), \circledast\big)$ is also locally invertible at 0. Since $X(\alpha) = \alpha^2$, the germ $X : (\mathbb{P}^1, 0) \to (\mathbb{P}^1, 0)$ has degree 2 at 0. It follows that σ_f has degree 2 at $\circledast$ as required.

Finally, we prove that σ_f is a surjective Galois orbifold covering. First, note that the critical value set of Y is Θ whence $Y : \mathbb{P}^1 - \Theta' \to \mathbb{P}^1 - \Theta$ is a covering map. Since $\pi = Y \circ A$ and since $\pi : \mathrm{Teich}(\mathbb{P}^1, P_f) \to \mathbb{P}^1 - \Theta$ is a universal covering map, we see that $A : \mathrm{Teich}(\mathbb{P}^1, P_f) \to \mathbb{P}^1 - \Theta'$ is a covering map (hence a universal covering map).

Second, note that $X : \mathbb{P}^1 - \Theta' \to \mathbb{P}^1 - \Theta$ is a ramified Galois covering of degree 2, ramified above 0 and ∞ with local degree 2. Let M be the orbifold whose underlying surface is $\mathbb{P}^1 - \Theta$ and whose weight function takes the value 1 everywhere except at 0 and ∞ where it takes the value 2. Then, $X : \mathbb{P}^1 - \Theta' \to M$ is a covering of orbifolds and $X \circ A : \mathrm{Teich}(\mathbb{P}^1, P_f) \to M$ is a universal covering of orbifolds.

Third, let T be the orbifold whose underlying surface is $\mathrm{Teich}(\mathbb{P}^1, P_f)$ and whose weight function takes the value 1 everywhere except at points in $\pi^{-1}(\{0, \infty\})$ where it takes the value 2. Then $\pi : T \to M$ is a covering of orbifolds. We have the following commutative diagram:

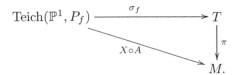

It follows that $\sigma_f : \mathrm{Teich}(\mathbb{P}^1, P_f) \to T$ is a covering of orbifolds (thus a universal covering). Equivalently, the map $\sigma_f : \mathrm{Teich}(\mathbb{P}^1, P_f) \to \mathrm{Teich}(\mathbb{P}^1, P_f)$ is a ramified Galois covering, ramified above points in $\pi^{-1}(\{0, \infty\})$ with local degree 2. Figure 2 illustrates the behavior of the map σ_f.

5 Proof of (3)

5.1 Examples

Here, we give examples of Thurston maps f such that

- P_f contains at least 4 points, so $\mathrm{Teich}(\mathbb{P}^1, P_f)$ is not reduced to a point, and

- $\sigma_f : \mathrm{Teich}(\mathbb{P}^1, P_f) \to \mathrm{Teich}(\mathbb{P}^1, P_f)$ is constant.

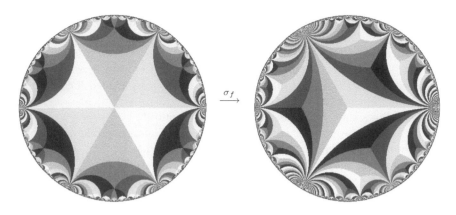

$$\xrightarrow{\sigma_f}$$

Figure 2. For $f(z) = 3z^2/(2z^3 + 1)$, the pullback map σ_f fixes $0 = \circledast$. It sends hexagons to triangles. There is a critical point with local degree 2 at the center of each hexagon and a corresponding critical value at the center of the image triangle. On the left, the map $X \circ A$ sends lighter grey hexagons to the unit disk in $\mathbb{P}^1 - \Theta$ and darker grey hexagons to the complement of the unit disk in $\mathbb{P}^1 - \Theta$. On the right, the map π sends lighter grey triangles to the unit disk in $\mathbb{P}^1 - \Theta$ and darker grey triangles to the complement of the unit disk in $\mathbb{P}^1 - \Theta$.

The main result, essentially due to McMullen, is the following.

Proposition 5.1. *Let* $s : \mathbb{P}^1 \to \mathbb{P}^1$ *and* $g : \mathbb{P}^1 \to \mathbb{P}^1$ *be rational maps with critical value sets* V_s *and* V_g. *Let* $A \subset \mathbb{P}^1$ *be finite. Assume* $V_s \subseteq A$ *and* $V_g \cup g(A) \subseteq s^{-1}(A)$. *Then*

- $f := g \circ s$ *is a Thurston map,*

- $V_g \cup g(V_s) \subseteq P_f \subseteq V_g \cup g(A)$, *and*

- *the dimension of the image of* $\sigma_f : \mathrm{Teich}(\mathbb{P}^1, P_f) \to \mathrm{Teich}(\mathbb{P}^1, P_f)$ *is at most* $|A| - 3$.

Remark. If $|A| = 3$ the pullback map σ_f is constant.

Proof: Set $B := V_g \cup g(A)$. The set of critical values of f is the set

$$V_f = V_g \cup g(V_s) \subseteq B.$$

By assumption,
$$f(B) = g \circ s(B) \subseteq g(A) \subseteq B.$$

So, the map f is a Thurston map and $V_g \cup g(V_s) \subseteq P_f \subseteq B$.

Note that $B \subseteq s^{-1}(A)$ and $A \subseteq g^{-1}(B)$. According to the discussion at the beginning of Section 2, the rational maps s and g induce pullback maps

$$\sigma_s : \text{Teich}(\mathbb{P}^1, A) \to \text{Teich}(\mathbb{P}^1, B) \quad \text{and} \quad \sigma_g : \text{Teich}(\mathbb{P}^1, B) \to \text{Teich}(\mathbb{P}^1, A).$$

In addition,

$$\sigma_f = \sigma_s \circ \sigma_g.$$

The dimension of the Teichmüller space $\text{Teich}(\mathbb{P}^1, A)$ is $|A| - 3$. Thus, the rank of $D\sigma_g$, and so that of $D\sigma_f$, at any point in $\text{Teich}(\mathbb{P}^1, A)$ is at most $|A| - 3$. This completes the proof of the proposition. □

Let us now illustrate this proposition with some examples.

Example 1. We are not aware of any rational map $f : \mathbb{P}^1 \to \mathbb{P}^1$ of degree 2 or 3 for which $|P_f| \geqslant 4$ and $\sigma_f : \text{Teich}(\mathbb{P}^1, P_f) \to \text{Teich}(\mathbb{P}^1, P_f)$ is constant. We have an example in degree 4: the polynomial f defined by

$$f(z) = 2i \left(z^2 - \frac{1+i}{2} \right)^2.$$

This polynomial can be decomposed as $f = g \circ s$ with

$$s(z) = z^2 \quad \text{and} \quad g(z) = 2i \left(z - \frac{1+i}{2} \right)^2.$$

The critical value set of s is

$$V_s = \{0, \infty\} \subset A := \{0, 1, \infty\}.$$

The critical value set of g is

$$V_g = \{0, \infty\} \subset \{0, \infty, -1, 1\} = s^{-1}(A).$$

In addition, $g(0) = -1$, $g(1) = 1$ and $g(\infty) = \infty$, so

$$g(A) = \{-1, 1, \infty\} \subset s^{-1}(A).$$

According to the previous proposition, $f = g \circ s$ is a Thurston map and since $|A| = 3$, the map $\sigma_f : \text{Teich}(\mathbb{P}^1, P_f) \to \text{Teich}(\mathbb{P}^1, P_f)$ is constant.

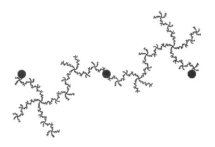

Figure 3. The Julia set of the degree 4 polynomial $f : z \mapsto 2i\left(z^2 - \frac{1+i}{2}\right)^2$ is a dendrite. There is a fixed critical point at ∞. Its basin is white. The point $z = 1$ is a repelling fixed point. All critical points are in the backward orbit of 1.

Note that $V_f = \{0, -1, \infty\}$ and $P_f = \{0, 1, -1, \infty\}$. The ramification portrait for f is as follows.

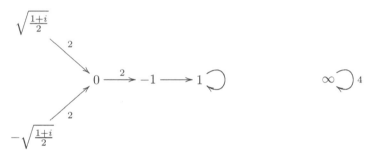

Example 2. We also have examples of rational maps $f : \mathbb{P}^1 \to \mathbb{P}^1$ for which $\sigma_f : \mathrm{Teich}(\mathbb{P}^1, P_f) \to \mathrm{Teich}(\mathbb{P}^1, P_f)$ is constant and $|P_f| \geqslant 4$ is an arbitrary integer. Assume $n \geqslant 2$ and consider $s : \mathbb{P}^1 \to \mathbb{P}^1$ and $g : \mathbb{P}^1 \to \mathbb{P}^1$ the polynomials defined by

$$s(z) = z^n \quad \text{and} \quad g(z) = \frac{(n+1)z - z^{n+1}}{n}.$$

Set $A := \{0, 1, \infty\}$. The critical value set of s is $V_s = \{0, \infty\} \subset A$.

The critical points of g are the n-th roots of unity and g fixes those points; the critical values of g are the n-th roots of unity. In addition, $g(0) = 0$. Thus

$$V_g \cup g(V_s) = V_g \cup g(A) = s^{-1}(A).$$

According to Proposition 5.1, $P_f = s^{-1}(A)$ and the pullback map σ_f is constant. In particular, $|P_f| = n + 2$.

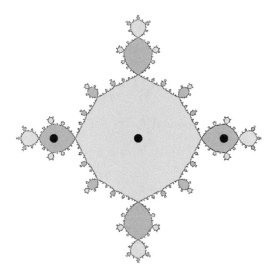

Figure 4. The Julia set of the degree 6 polynomial $f : z \mapsto z^2(3 - z^4)/2$. There are superattracting fixed points at $z = 0$, $z = 1$ and $z = \infty$. All other critical points are in the backward orbit of 1. The basin of ∞ is white. The basin of 0 is light grey. The basin of 1 is dark grey.

For $n = 2$, f has the following ramification portrait.

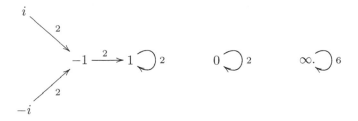

Example 3. Proposition 5.1 can be further exploited to produce examples of Thurston maps f where σ_f has a *skinny image*, which is not just a point.

For $n \geqslant 2$, let A_n be the union of $\{0, \infty\}$ and the set of n-th roots of unity. Let $s_n : \mathbb{P}^1 \to \mathbb{P}^1$ and $g_n : \mathbb{P}^1 \to \mathbb{P}^1$ be the polynomials defined by

$$s_n(z) = z^n \quad \text{and} \quad g_n(z) = \frac{(n+1)z - z^{n+1}}{n}.$$

The critical points of g_n are the n-th roots of unity and g_n fixes those points; the critical values of g_n are the n-th roots of unity. In particular,

$V_{g_n} \subset A_n$. In addition, $g_n(0) = 0$, and so,

$$g_n(A_n) = A_n.$$

Assume $n \geqslant 2$ and $m \geqslant 1$ are integers with m dividing n, let's say $n = km$. Note that

$$V_{s_k} \subset A_m \quad \text{and} \quad V_{g_n} \cup g_n(A_n) = A_n = s_k^{-1}(A_m).$$

It follows that the polynomial $f : \mathbb{P}^1 \to \mathbb{P}^1$ defined by

$$f := g_n \circ s_k$$

is a Thurston map and

$$A_n = V_{g_n} \cup g_n(V_{s_k}) \subseteq P_f \subseteq V_{g_n} \cup g_n(A_n) = A_n \quad \text{so,} \quad P_f = A_n.$$

In particular, the dimension of the Teichmüller space $\mathrm{Teich}(\mathbb{P}^1, P_f)$ is $n-1$.

Claim 5.2. *The image of $\sigma_f : \mathrm{Teich}(\mathbb{P}^1, P_f) \to \mathrm{Teich}(\mathbb{P}^1, P_f)$ has dimension $m - 1$. Thus, its codimension is $(k-1)m$.*

Proof: On the one hand, since g_n is a polynomial whose critical points are all fixed, Proposition 3.1 implies that $\sigma_{g_n} : \mathrm{Teich}(\mathbb{P}^1, A_n) \to \mathrm{Teich}(\mathbb{P}^1, A_n)$ has open image. Composing with the forgetful projection

$$\mathrm{Teich}(\mathbb{P}^1, A_n) \to \mathrm{Teich}(\mathbb{P}^1, A_m),$$

we deduce that $\sigma_{g_n} : \mathrm{Teich}(\mathbb{P}^1, A_n) \to \mathrm{Teich}(\mathbb{P}^1, A_m)$ has open image.

On the other hand, since $s_k : \mathbb{P}^1 - A_n \to \mathbb{P}^1 - A_m$ is a covering map, it follows from general principles that $\sigma_{s_k} : \mathrm{Teich}(\mathbb{P}^1, A_m) \to \mathrm{Teich}(\mathbb{P}^1, A_n)$ is a holomorphic embedding with everywhere injective derivative. □

Question. If $f : \mathbb{P}^1 \to \mathbb{P}^1$ is a Thurston map such that the pullback map $\sigma_f : \mathrm{Teich}(\mathbb{P}^1, P_f) \to \mathrm{Teich}(\mathbb{P}^1, P_f)$ is constant, then is it necessarily of the form described above? In particular, is there a Thurston map $f : \mathbb{P}^1 \to \mathbb{P}^1$ with constant $\sigma_f : \mathrm{Teich}(\mathbb{P}^1, P_f) \to \mathrm{Teich}(\mathbb{P}^1, P_f)$, such that $\deg(f)$ is prime?

5.2   Characterizing When σ_f is Constant

Suppose $f : \mathbb{P}^1 \to \mathbb{P}^1$ is a Thurston map with $|P_f| \geqslant 4$.

Let $\mathcal{S}$ denote the set of free homotopy classes $[\gamma]$ of simple, closed, unoriented curves γ in $\mathbb{P}^1 - P_f$ such that each component of $\mathbb{P}^1 - \gamma$ contains at least two elements of P_f. Let $\mathbb{R}[\mathcal{S}]$ denote the free $\mathbb{R}$-module generated by $\mathcal{S}$. Given $[\gamma]$ and $[\widetilde{\gamma}]$ in $\mathcal{S}$, define the *pullback relation* on $\mathcal{S}$, denoted $\underset{f}{\leftarrow}$, by defining $[\gamma]\underset{f}{\leftarrow}[\tilde{\gamma}]$ if and only if there is a component δ of $f^{-1}(\gamma)$ which, as a curve in $\mathbb{P}^1 - P_f$, is homotopic to $\tilde{\gamma}$.

The *Thurston linear map*

$$\lambda_f : \mathbb{R}[\mathcal{S}] \to \mathbb{R}[\mathcal{S}]$$

is defined by specifying the image of basis elements $[\gamma] \in \mathcal{S}$ as follows:

$$\lambda_f([\gamma]) = \sum_{[\gamma]\underset{f}{\leftarrow}[\gamma_i]} d_i[\gamma_i].$$

Here, the sum ranges over all $[\gamma_i]$ for which $[\gamma]\underset{f}{\leftarrow}[\gamma_i]$, and

$$d_i = \sum_{f^{-1}(\gamma)\supset\delta\simeq\gamma_i} \frac{1}{|\deg(\delta \to \gamma)|},$$

where the sum ranges over components δ of $f^{-1}(\gamma)$ homotopic to γ_i.

Let $\mathrm{PMCG}(\mathbb{P}^1, P_f)$ denote the pure mapping class group of $(\mathbb{P}^1, P_f)$ — that is, the quotient of the group of orientation-preserving homeomorphisms fixing P_f pointwise by the subgroup of such maps isotopic to the identity relative to P_f. Thus,

$$\mathrm{Mod}(\mathbb{P}^1, P_f) = \mathrm{Teich}(\mathbb{P}^1, P_f)/\mathrm{PMCG}(\mathbb{P}^1, P_f).$$

Elementary covering space theory and homotopy-lifting facts imply that there is a finite-index subgroup $H_f \subset \mathrm{PMCG}(\mathbb{P}^1, P_f)$ consisting of those classes represented by homeomorphisms h lifting under f to a homeomorphism $\tilde{h}$ which fixes P_f pointwise. This yields a homomorphism

$$\phi_f : H_f \to \mathrm{PMCG}(\mathbb{P}^1, P_f)$$

defined by

$$\phi_f([h]) = [\tilde{h}] \quad \text{with} \quad h \circ f = f \circ \tilde{h}.$$

Following [BN06] we refer to the homomorphism ϕ_f as the *virtual endomorphism of* $\mathrm{PMCG}(\mathbb{P}^1, P_f)$ associated to f.

Theorem 5.1. *The following are equivalent:*

1. $\underset{f}{\leftarrow}$ *is empty;*

2. λ_f *is constant;*

3. ϕ_f *is constant;*

4. σ_f *is constant.*

Proof: We will show that (1) $\implies$ (2) $\implies$ (3) $\implies$ (4), and that failure of (1) implies failure of (4). That (1) $\iff$ (2) is an immediate consequence of the definitions.

Suppose (2) holds. Since $\mathrm{PMCG}(\mathbb{P}^1, P_f)$ is generated by Dehn twists, it suffices to show $\phi_f(g) = 1$ for every Dehn twist g. We may represent g by a homeomorphism h which is the identity off an annulus A. By (2), the set $f^{-1}(A)$ is contained in a disjoint union of disks D_i, each of which contains at most one point of P_f. There is a lift $\tilde{h}$ of h under f which is the identity outside $\mathbb{P}^1 - \cup_i D_i$. Since the disks D_i are peripheral, $\tilde{h}$ represents the trivial element of $\mathrm{PMCG}(\mathbb{P}^1, P_f)$.

Suppose (3) holds. We first recall the action of the mapping class group on Teichmüller space. Suppose $g : \mathbb{P}^1 \to \mathbb{P}^1$ represents an element $[g]$ of $\mathrm{PMCG}(\mathbb{P}^1, P_f)$ and $\psi : \mathbb{P}^1 \to \mathbb{P}^1$ represents an element τ of $\mathrm{Teich}(\mathbb{P}^1, P_f)$. Then $\psi \circ g$ represents another element of $\mathrm{Teich}(\mathbb{P}^1, P_f)$. The transformation $\psi \mapsto \psi \circ g$ descends to a map on Teichmüller space which is independent of the representative g of $[g]$, and defines a properly discontinuous right action of $\mathrm{PMCG}(\mathbb{P}^1, P_f)$ on $\mathrm{Teich}(\mathbb{P}^1, P_f)$ by biholomorphisms. The image of τ under the action of $[g]$ is denoted $\tau.[g]$. The quotient $\mathrm{Teich}(\mathbb{P}^1, P_f)/\mathrm{PMCG}(\mathbb{P}^1, P_f)$ is the moduli space $\mathrm{Mod}(\mathbb{P}^1, P_f)$.

If $h : \mathbb{P}^1 \to \mathbb{P}^1$ represents an element of H_f, then by definition $h \circ f = f \circ \tilde{h}$ for some homeomorphism $\tilde{h} : \mathbb{P}^1 \to \mathbb{P}^1$ fixing P_f pointwise. If $\psi : \mathbb{P}^1 \to \mathbb{P}^1$ represents $\tau \in \mathrm{Teich}(\mathbb{P}^1, P_f)$ then $\sigma_f(\tau)$ is represented by a homeomorphism $\tilde{\psi} : \mathbb{P}^1 \to \mathbb{P}^1$ for which there exists a holomorphic map $f_\tau : \mathbb{P}^1 \to \mathbb{P}^1$ satisfying $\psi \circ f = f_\tau \circ \tilde{\psi}$. Combining these observations implies that the diagram below commutes:

$$
\begin{array}{ccccc}
\mathbb{P}^1 & \xrightarrow{\ \tilde{h}\ } & \mathbb{P}^1 & \xrightarrow{\ \tilde{\psi}\ } & \mathbb{P}^1 \\
\downarrow{\scriptstyle f} & & \downarrow{\scriptstyle f} & & \downarrow{\scriptstyle f_\tau} \\
\mathbb{P}^1 & \xrightarrow{\ h\ } & \mathbb{P}^1 & \xrightarrow{\ \psi\ } & \mathbb{P}^1.
\end{array}
$$

Hence $\sigma_f(\tau.[h]) = \sigma_f(\tau).\phi_f([h])$ for all $\tau \in \mathrm{Teich}(\mathbb{P}^1, P_f)$ and all $[h] \in H_f$.

If now $\phi_f \equiv 1$, the identity element of $\mathrm{PMCG}(\mathbb{P}^1, P_f)$, then σ_f assumes the same value at all points lying in the same H_f-orbit, and so σ_f descends to a holomorphic map

$$
\overline{\sigma}_f : \mathrm{Teich}(\mathbb{P}^1, P_f)/H_f \to \mathrm{Teich}(\mathbb{P}^1, P_f).
$$

We now use a classical fact pointed out to us by M. Bainbridge: the domain of $\overline{\sigma}_f$ is a finite cover of $\mathrm{Mod}(\mathbb{P}^1, P_f)$, hence is a quasiprojective variety. Since $\mathrm{Teich}(\mathbb{P}^1, P_f)$ is biholomorphic to a bounded domain in $\mathbb{C}^n$, the induced map $\overline{\sigma}_f$ is a bounded analytic function on a quasiprojective variety, and is therefore constant, by a generalization of the classical Liouville theorem.

Suppose (1) fails. For $\tau \in \mathrm{Teich}(\mathbb{P}^1, P_f)$ and $[\delta] \in \mathcal{S}$ let $\ell_\tau([\delta])$ denote the length of the unique simple closed geodesic in $[\delta]$ on $\mathbb{P}^1 - P_f$, equipped with a hyperbolic metric corresponding to a complex structure representing τ. It is known (see [DH93]) that if $\ell_\tau([\gamma])$ is sufficiently small, then whenever $[\gamma] \underset{f}{\leftarrow} [\widetilde{\gamma}]$, one has $\ell_{\sigma_f(\tau)}([\widetilde{\gamma}]) \leqslant c \cdot \ell_\tau([\gamma])$ where c depends only on $|P_f|$ and on $\deg f$. In particular, $\ell_{\sigma_f(\tau)}([\widetilde{\gamma}]) \to 0$ as $\ell_\tau([\gamma]) \to 0$. Hence σ_f cannot be constant. $\qquad\square$

Bibliography

[BN06]   L. Bartholdi and V. Nekrashevych, *Thurston equivalence of topological polynomials*, Acta. Math. **197** (2006), 1–51.

[DH93]   A. Douady and J. H. Hubbard, *A proof of Thurston's characterization of rational functions*, Acta Math. (2) **171** (1993), 263–297.

[H06]   J. H. Hubbard, *Teichmüller theory and applications to geometry, topology, and dynamics, volume 1: Teichmüller theory*, Matrix Editions, Ithaca, 2006.

[Ko07]   S. Koch, *Teichmüller theory and endomorphisms of $\mathbb{P}^n$*, PhD thesis, Université de Provence, 2007.

[Ko08]   S. Koch, *A new link between Teichmüller theory and complex dynamics*, PhD thesis, Cornell University, 2008.

16   On the Boundary Behavior of Thurston's Pullback Map

Nikita Selinger

1   Introduction

In the early 1980s, Thurston came up with one of the most important theorems in the field of complex dynamics. Thurston's characterization theorem gives a criterion for whether a combinatorics (prescribed by a branched cover) can be realized by a rational map, and also provides a rigidity statement: two rational maps are equivalent if and only if they are conjugate by a Möbius transformation.

The theorem has proved to be useful in a number of applications. The combinatorial classification of postcritically finite polynomials in terms of Hubbard trees in [Po] as well as the results of Tan Lei [Ta92] on the combinatorics of certain rational maps viewed as "matings" of two polynomials of equal degrees (see also [Mi04]) are built on the foundation of Thurston's theorem. The first proof of Thurston's theorem was published by Douady and Hubbard in [DH93]. Most parts of this article have become classical results over the last 20 years. We will need most of them in the sequel.

The proof of Thurston's theorem relates the original question to whether a holomorphic function on Teichmüller space has a fixed point. In this paper, we introduce a new approach to the latter question which involves studying the behavior of the map at the boundary of Teichmüller space. The question of extending the Thurston pullback operation to the boundary of Teichmüller space was independently raised in [Pi01]. This paper provides a modified proof of Thurston's theorem and the main result of [Pi01]. This is a preliminary research announcement.

We denote by $\mathbb{P}$ the Riemann sphere and by $\mathbb{S}^2$ the topological 2-sphere. All of the maps considered in this paper are orientation-preserving. We first briefly recall the setup of [DH93].

Acknowledgments. I am grateful to Adam Epstein and Kevin Pilgrim for many useful discussions and help with making this text better. I also thank Jan Cannizzo for editing the article.

1.1   Thurston's Theorem

A continuous map $f\colon \mathbb{S}^2 \to \mathbb{S}^2$ is a *branched cover* if each point of $\mathbb{S}^2$ has a neighborhood U such that on each connected component of $f^{-1}(U)$ the map f is either a homeomorphism or topologically conjugate to $z \mapsto z^d$ near 0. *Critical points* of f are those points for which f has local degree greater than 1. Note that every branched cover has a well-defined mapping degree. In the rest of the paper, d always stands for the degree of f; we always assume that $d > 1$. A branched cover is *postcritically finite* if all critical points have finite orbits, i.e., they are periodic or pre-periodic. Postcritically finite branched covers are also called *Thurston maps*. Let P_f denote the postcritical set of f. Two postcritically finite maps $f, g\colon \mathbb{S}^2 \to \mathbb{S}^2$ are Thurston equivalent if there are homeomorphisms $h_1, h_2\colon \mathbb{S}^2 \to \mathbb{S}^2$ such that

$$
\begin{array}{ccc}
(\mathbb{S}^2, P_f) & \xrightarrow{\ h_1\ } & (\mathbb{S}^2, P_g) \\
\Big\downarrow{\scriptstyle f} & & \Big\downarrow{\scriptstyle g} \\
(\mathbb{S}^2, P_f) & \xrightarrow{\ h_2\ } & (\mathbb{S}^2, P_g)
\end{array}
$$

commutes, $h_1(P_f) = h_2(P_f) = P_g$, $h_1|_{P_f} = h_2|_{P_f}$, and h_1 and h_2 are homotopic relative to P_f. We omit the precise definition of *hyperbolic orbifolds* (see [DH93]): all postcritically finite branched covers have hyperbolic orbifolds, except for a few explicitly understood cases. In what follows, we always assume that all maps have hyperbolic orbifolds.

A simple closed curve $\gamma \subset \mathbb{S}^2 \setminus P_f$ is *essential* if every component of $\mathbb{S}^2 \setminus \gamma_i$ intersects P_f in at least two points. A *multicurve* is a finite collection $\Gamma = \{\gamma_1, \ldots, \gamma_k\}$ of disjoint essential pairwise non-homotopic simple closed curves in $\mathbb{S}^2 \setminus P_f$. A multicurve is *$f$-invariant* (or just invariant when it is clear which map is considered) if each component of $f^{-1}(\gamma_i)$ is either non-essential or homotopic (in $\mathbb{S}^2 \setminus P_f$) to a curve in Γ. Every f-invariant multicurve Γ has its associated *Thurston matrix* $M_\Gamma = (m_{i,j})$, given by

$$
m_{i,j} = \sum_{\gamma_{i,j,k}} (\deg f|_{\gamma_{i,j,k}} \colon \gamma_{i,j,k} \to \gamma_j)^{-1}
$$

where $\gamma_{i,j,k}$ ranges through all preimages of γ_j that are homotopic to γ_i. Since all entries of M_Γ are positive real, the leading eigenvalue λ_Γ of M_Γ is positive and real. The multicurve Γ is a *Thurston obstruction* if $\lambda_\Gamma \geq 1$.

With these definitions at hand, we can formulate Thurston's theorem:

Theorem 1.1 (Thurston's Theorem [DH93]). *A postcritically finite branched cover $f\colon \mathbb{S}^2 \to \mathbb{S}^2$ with hyperbolic orbifold is either Thurston-equivalent to a rational map g (which is then necessarily unique up to conjugation by a Möbius transformation), or it has a Thurston obstruction.*

1.2   The Setup

Teichmüller space $\mathcal{T}_f$ is the space of all homeomorphisms $\phi\colon (\mathbb{S}^2, P_f) \to \mathbb{P}$ quotiented by an equivalence relation generated by $Aut(\mathbb{P})$ and by isotopies of ϕ relative to P_f. Equivalently, one can view $\mathcal{T}_f$ as the quotient of the space of smooth almost-complex structures on $\mathbb{S}^2$ so that two almost-complex structures μ_1, μ_2 are identified if there exists a diffeomorphism $h\colon \mathbb{S}^2 \to \mathbb{S}^2$, with $h|_{P_f} = $ id and h homotopic to the identity rel P_f, such that $\mu_1 = h^*\mu_2$. The moduli space $\mathcal{M}_f$ can be viewed as the space of all injections of P_f into $\mathbb{P}$ modulo $Aut(\mathbb{P})$, with the obvious forgetful projection map $\pi\colon \mathcal{T}_f \to \mathcal{M}_f$.

The mapping $\mu \mapsto f^*\mu$ on almost-complex structures induces an analytic mapping $\sigma_f : \mathcal{T}_f \to \mathcal{T}_f$ which is called the *Thurston pullback map*. The proof of Theorem 1.1 is based on the fact that f is equivalent to a rational map if and only if σ_f has a fixed point in $\mathcal{T}_f$ (see Proposition 2.3 in [DH93]).

Let μ and μ' be almost-complex structures corresponding to points τ and $\sigma_f(\tau)$ of the Teichmüller space $\mathcal{T}_f$; let ϕ and ϕ' be isomorphisms of $(\mathbb{S}^2, \mu)$ and $(\mathbb{S}^2, \mu')$ with $\mathbb{P}$, respectively. The cotangent space to the Teichmüller space $\mathcal{T}_f$ at a point τ can be canonically identified with the space $Q(\mathbb{P}, P)$ of all meromorphic quadratic differentials on $\mathbb{P}$ with at most simple poles, all in P (or, in other words, the space of all integrable holomorphic differentials in $\mathbb{P} \setminus P$), where $P = \phi(P_f)$ is the set of coordinates of the postcritical set P_f in the complex structure on $\mathbb{S}^2$ corresponding to τ. The coderivative of $d_\tau^*\sigma_f$ is equal to the push-forward operator $(f_\tau)_*\colon Q(\mathbb{P}, P') \to Q(\mathbb{P}, P)$, given by the formula

$$((f_\tau)_*q)(v) = \sum_{y \in (f_\tau)^{-1}(x)} q\left([Df_\tau(y)]^{-1}(v)\right), \tag{1.1}$$

where $P' = \phi'(P_f)$ and f_τ is the rational map $f_\tau = \phi \circ f \circ (\phi')^{-1}$ (see Proposition 3.2 in [DH93]).

Each point in $\mathcal{T}_f$ defines hyperbolic structure on $\mathbb{S}^2 \setminus P_f$. We denote by $l(\tau, \gamma)$ the length of the hyperbolic geodesic in the hyperbolic structure defined by $\tau \in \mathcal{T}_f$, which is homotopic to an essential simple closed curve γ (i.e., the length is measured in the Poincaré metric of $\mathbb{P} \setminus \tau(P_f)$; we assume that the Poincaré metric has constant curvature -1). Note that $\log l(\tau, \gamma)$ is 1-Lipschitz in τ with respect to the Teichmüller metric of $\mathcal{T}_f$ (see Proposition 7.3 in [DH93]; the constant is different due to different definitions of the Teichmüller and Poincaré metrics).

1.3   Teichmüller Geometry and Augmented Teichmüller Space

There are two well-studied metrics on Teichmüller spaces: the Teichmüller metric and the Weil-Petersson (WP) metric. Note that if q is a quadratic differential, then $|q|$ is an area form. For $q \in T^*_\tau \mathcal{T}_f \cong Q(\mathbb{P}, P)$, define the Teichmüller norm by $\|q\|_T = \int |q|$ and the WP norm by

$$\|q\|_{WP} = \left(\int \frac{|q|^2}{\rho^2} \right)^{\frac{1}{2}},$$

where ρ^2 is the hyperbolic area element of $\mathbb{P} \setminus P$. Both norms give rise to the corresponding norms on $T_\tau \mathcal{T}_f$ and, by integrating the dual norm along paths in $\mathcal{T}_f$, to the corresponding metrics on $\mathcal{T}_f$. The Thurston pullback map σ_f is contracting in the Teichmüller metric (see Proposition 3.3 in [DH93]) and is $\sqrt{d}$-Lipschitz (see the proof of Theorem 2.1) in the WP metric. The Teichmüller metric is "finer" than the WP metric: there exists a constant V such that $\|v\|_{WP} \leq V\|v\|_T$ for any vector v in the tangent bundle $T\mathcal{T}$.

We will use the following property of the WP metric: it is not complete, and the completion of the Teichmüller space $\mathcal{T}_f$ with respect to the WP metric is canonically homeomorphic to the augmented Teichmüller space $\overline{\mathcal{T}}_f$. The mapping class group acts on the Teichmüller space by WP-isometries; the action of the mapping class group extends continuously to the boundary of $\overline{\mathcal{T}}_f$, and the quotient of $\overline{\mathcal{T}}_f$ by the mapping class group is (topologically) the compactified moduli space of stable curves $\overline{\mathcal{M}}_f$ (see [Wo02] for a more detailed description and further references).

While the moduli space $\mathcal{M}$ of a finite type surface is the space of all Riemann surfaces of given type, the compactified moduli space of stable curves $\overline{\mathcal{M}}$ is the space of all noded Riemann surfaces of that type. The augmented Teichmüller space $\overline{\mathcal{T}}$ is the space of all marked noded Riemann surfaces of given type, and the projection $\pi \colon \overline{\mathcal{T}} \to \overline{\mathcal{M}}$ simply forgets the marking. Note that $l(\tau, \gamma)$ extends continuously to $\overline{\mathcal{T}}$ for every γ (as a map from $\overline{\mathcal{T}}$ to $[0, +\infty]$). The boundary of $\overline{\mathcal{T}}$ is a stratified space where each boundary stratum corresponds to a collection of pinched curves that are non-homotopic to each other and boundary components (in our case it must be a multicurve Γ), the multicurve for which $l(\tau, \gamma) = 0$ if and only if $\gamma \in \Gamma$ for all points in the stratum. We denote by $\mathcal{S}_\Gamma$ the stratum of $\overline{\mathcal{T}}_f$ corresponding to the multicurve Γ, where $\mathcal{S}_\emptyset = \mathcal{T}_f$. We will also make use of the fact that the length of a geodesic is intimately related to the moduli of annuli homotopic to this geodesic. It follows from the Collaring Lemma that for a geodesic γ of length l, there exists a homotopic annulus of modulus at least $\frac{\pi}{l} - 1$. On the other hand, the core curve of an annulus of modulus m has length at least $\frac{\pi}{m}$. Thus, there is no annulus homotopic to γ with modulus greater than $\frac{\pi}{l}$, since γ is the shortest curve in its homotopy class.

2 The Proof

As was shown in [DH93], if σ_f has no fixed points, then the orbit under the action of σ_f of any point in $\mathcal{T}_f$ tends to infinity (i.e., eventually leaves every compact subset). Therefore it is very important to understand the boundary behavior of σ_f. We start by extending the action of σ_f to augmented Teichmüller space.

Theorem 2.1 (Thurston Pullback on Augmented Teichmüller Space).
The Thurston pullback map σ_f of a postcritically finite branched cover $f : \mathbb{S}^2 \to \mathbb{S}^2$ extends continuously to the augmented Teichmüller space.

Proof: We will prove the theorem by estimating the norm of the coderivative of σ_f. We keep the notation introduced in Section 1.2 and put $g = f_\tau$ for simplicity. If we take some small domain $U \subset (\mathbb{P} \setminus P)$ with local coordinate ζ such that it has exactly d disjoint preimages $U_i, i = 1, \dots, d$, with $g_i : I = U_i \to U$ the local branches of g^{-1}, then we can write formula 1.1 in the form

$$g_* q|_U = \sum_i g_i^* q.$$

Let ρ^2 and ρ_1^2 stand for the hyperbolic area elements on $\mathbb{P} \setminus P$ and $\mathbb{P} \setminus P'$. The hyperbolic area element on $\mathbb{P} \setminus g^{-1}(P)$ is given by $g^* \rho^2$. The inclusion map $I : \mathbb{P} \setminus g^{-1}(P) \to \mathbb{P} \setminus P'$ is length-decreasing; therefore $\rho_1^2 \leq g^* \rho^2$.

Now we can locally estimate:

$$\int_U \frac{|g_* q|^2}{\rho^2} = \int_U \frac{|\sum_i g_i^* q|^2}{\rho^2} \leq d \sum_i \int_U \frac{|g_i^* q|^2}{\rho^2} = d \sum_i \int_{U_i} \frac{|q|^2}{g^* \rho}$$
$$\leq d \int_{g^{-1}(U)} \frac{|q|^2}{\rho_1^2},$$

where the second inequality follows from the fact that

$$\left| \sum_{i=1}^d a_i \right|^2 \leq d \sum_{i=1}^d |a_i|^2.$$

Combining local estimates, we get

$$\|(f_\tau)_* q\|_{WP} \leq \sqrt{d} \|q\|_{WP},$$

where d is the degree of f. This means that $\|d_\tau \sigma_f\|_{WP} = \|d_\tau^* \sigma_f\|_{WP} \leq \sqrt{d}$. Hence σ_f is a Lipschitz function with respect to the Weil-Petersson metric. It is evident that every Lipschitz function on a metric space can be continuously extended to its completion. $\square$

Denote by $f^{-1}(\Gamma)$ the multicurve of all essential mutually non-homotopic preimages of curves in Γ.

Proposition 2.2. *For any multicurve Γ, we have*

$$\sigma_f(\mathcal{S}_\Gamma) \subset \mathcal{S}_{f^{-1}(\Gamma)}.$$

Proof: Suppose $\tau \in \mathcal{S}_\Gamma$. Pick a sequence $\tau_n \in \mathcal{T}_f$ so that $\tau_n \to \tau$. Then $l(\tau_n, \gamma) \to 0$ for every $\gamma \in \Gamma$, and there exist annuli A_n of moduli M_n tending to infinity in the homotopy class of γ in $\mathbb{P} \setminus \tau_n(P_f)$. Each preimage of A_n is mapped to A_n by a map of degree at most d, so all these preimages have moduli at least $M_n/d \to \infty$, and the lengths of their core curves tend to 0. Hence, for any essential preimage α of a curve in Γ, we have $l(\sigma_f(\tau), \alpha) = \lim l(\sigma_f(\tau_n), \alpha) = 0$.

On the other hand, if $l(\sigma_f(\tau_n), \alpha) = \varepsilon$ is small enough, then (see [DH93, Theorem 7.1]) there exists a simple closed geodesic $\beta \in \mathbb{P} \setminus \sigma_f(\tau_n)(f^{-1}(P_f))$ of length $L \leq \varepsilon(dp+1)$ (where p is the cardinality of P_f) that is homotopic to $\sigma_f(\tau_n)(\alpha)$ in $\mathbb{P} \setminus \sigma_f(\tau_n)(P_f)$. Then $\delta = f_{\tau_n}(\beta)$ is a very short geodesic (of length at most L) in $\mathbb{P} \setminus \tau_n(P_f)$; in particular, from Proposition 6.7 in [DH93], it follows that δ_n is a simple closed curve. Since $\tau_n \to \tau$ and $\inf l(\tau, \delta) > 0$, where the infimum is taken over all essential simple closed curves not in Γ (in fact, there are only finitely many short non-homotopic curves) we get that $\tau_n^{-1}(\delta)$ must be homotopic to some $\gamma \in \Gamma$ for n large and ε small enough. But then α is homotopic to a component of $f^{-1}(\gamma)$ and therefore $\alpha \in f^{-1}(\Gamma)$. $\square$

Note that Γ is f-invariant if and only if $f^{-1}(\Gamma) \subset \Gamma$. We say that Γ is *completely invariant* if $f^{-1}(\Gamma) = \Gamma$. A boundary stratum $\mathcal{S}$ will be called *invariant* if $\sigma_f(\mathcal{S}) \subset \mathcal{S}$.

Corollary 2.3. *There is a natural bijective correspondence between σ_f-invariant strata and completely invariant multicurves. In particular, if τ and $\sigma_f(\tau)$ both lie in the same boundary stratum $\mathcal{S}_\Gamma$, then Γ is completely invariant.*

2.1   Necessity of the Criterion

In this section, we would like to mention a simpler argument for the necessity of the criterion than the one in [DH93]. The argument has been in the air for a while; it was used to prove the necessity part of an analogous statement in [HSS09] but in a slightly less general setting, when every obstruction is a Levy cycle.

We want to classify invariant boundary strata $\mathcal{S}_\Gamma$ using λ_Γ. An invariant stratum $\mathcal{S}$ of $\overline{\mathcal{T}}_f$ will be called *weakly attracting* if there exists a nested decreasing sequence of neighborhoods U_n such that $\sigma_f(U_n) \subset U_n$

and $\bigcap U_n = \overline{\mathcal{S}}$. An invariant stratum $\mathcal{S}$ of $\overline{\mathcal{T}}_f$ will be called *weakly repelling* if for any compact set $K \subset \mathcal{S}$ there exists a neighborhood $K \subset U$ such that every point of $U \cap \mathcal{T}_f$ escapes from U after finitely many iterations (for every $\tau \in U \cap \mathcal{T}_f$, there exists an $n \in \mathbb{N}$ such that $\sigma_f^{\circ n}(\tau) \notin U$).

We call Γ *positive* if there exists a leading eigenvector of M_Γ with positive entries. Note that each multicurve has a positive sub-multicurve with the same leading eigenvalue (the curves corresponding to non-zero components of any leading eigenvector form a multicurve).

Proposition 2.4. *If Γ is an invariant positive multicurve and $\lambda_\Gamma \geq 1$, then $\mathcal{S}_\Gamma$ is weakly attracting.*

Proof: Take a leading eigenvector v of M_Γ. All of its entries are non-zero because Γ is positive. Consider $U_n \subset \mathcal{T}_f$, the set of all points of the augmented Teichmüller space for which there exist non-intersecting annuli A_i homotopic to γ_i such that $\mod A_i \geq nv_i$. Then it follows from the Grötzsch inequality that $\sigma_f(U_n) \subset U_n$. From the relation between the moduli of annuli and the lengths of simple closed curves discussed in Section 1.3, it follows that $\bigcap \overline{U_n} = \overline{\mathcal{S}_\Gamma}$. $\square$

Proof of necessity: Let f be equivalent to a rational function. Then σ_f has a fixed point τ_0 in $\mathcal{T}_f$, and the orbit of every point converges to τ_0 (see [DH93]). Suppose f has a Thurston obstruction. Then f also has a positive obstruction Γ. Take U from the previous proposition, so that $\mathcal{S}_\Gamma \subset U$ and $\tau_0 \notin U$ (τ_0 is in the Teichmüller space $\mathcal{T}_f$, and $\mathcal{S}_\Gamma$ is on the boundary of $\overline{\mathcal{T}}_f$; hence $\tau_0 \notin \overline{\mathcal{S}_\Gamma}$). But then the orbit of every point of U stays in U and cannot converge to τ_0. This is a contradiction. $\square$

2.2 Sufficiency of the Criterion

Lemma 2.5. *Let $M \in \mathbb{R}_{n \times n}$ be a matrix such that all eigenvalues of M have absolute value strictly less than 1. Then there exists a norm $\| \cdot \|_M$ on $\mathbb{R}_n$ such that the $\|M\|_M < 1$. Moreover, if all entries of M are non-negative, then one can choose a norm given by $\|v\|_M = \max_i\{a_i|v_i|\}$, where a_i are some positive constants.*

The proof of the lemma is an exercise in linear algebra.

Proposition 2.6. *If $\Gamma = \{\gamma_1, \ldots, \gamma_n\}$ is a completely invariant multicurve and $\lambda_\Gamma < 1$, then $\mathcal{S}_\Gamma$ is weakly repelling.*

Proof: Let K be any compact subset of $\mathcal{S}_\Gamma$; assume that $D > 0$. Then

$$\inf_{\tau \in K, \gamma \notin \Gamma} l(\tau, \gamma) = k > 0,$$

where the infimum is taken over all points $\tau \in K$ and essential simple closed curves $\gamma \notin \Gamma$. Put $v_\tau = (1/l(\tau, \gamma_1), \ldots, 1/l(\tau, \gamma_n))^T$.

Set

$$U(\Gamma, t, k) = \overline{\{\tau \in \mathcal{T}_f | \inf_{\tau \in K, \gamma \notin \Gamma} l(\tau, \gamma) \geq k, \|v_\tau\|_{M_\Gamma} \geq t\}},$$

where the norm $\|\cdot\|_{M_\Gamma}$ is chosen as in the previous lemma, so that $\|M_\Gamma\|_{M_\Gamma} = \lambda < 1$.

Clearly, $K \subset U(\Gamma, t, k/2)$ for any t.

The same argument as in the proof of Proposition 8.2 in [DH93] shows that for all points τ in $U(\Gamma, t, k/2) \cap \mathcal{T}_f$, we have $v_{\sigma_f(\tau)} \leq M_\Gamma v_\tau + v'$, where v' is a constant positive vector (v' depends only on the bound of short curves that are not in Γ, which is uniformly bounded by $k/2$ in our case). Hence, if t was chosen large enough (namely, if $t > 2\|v'\|_{M_\Gamma}/(1-\lambda)$), then for any $\tau \in V$,

$$\begin{aligned} \|v_{\sigma_f(\tau)}\|_{M_\Gamma} &\leq \|M_\Gamma\|_{M_\Gamma}\|v_\tau\|_{M_\Gamma} + \|v'\|_{M_\Gamma} \\ &< \lambda\|v_\tau\|_{M_\Gamma} + \frac{1-\lambda}{2}\|v_\tau\|_{M_\Gamma} = (1-\lambda/2)\|v_\tau\|_{M_\Gamma}. \end{aligned}$$

This shows that the forward orbit of a point in $U(\Gamma, t, k/2) \cap \mathcal{T}_f$ cannot stay in $U(\Gamma, t, k/2)$ forever. $\qquad\square$

The following lemma [DH93, Lemma 5.2] plays a significant role in understanding the behavior of σ_f.

Lemma 2.7. *There exists an intermediate cover $\mathcal{M}_f'$ of $\mathcal{M}_f$ (so that the maps $\pi_1 \colon \mathcal{T}_f \to \mathcal{M}_f'$ and $\pi_2 \colon \mathcal{M}_f' \to \mathcal{M}_f$ are covers and $\pi_2 \circ \pi_1 = \pi$) such that π_2 is finite and the diagram*

$$\begin{array}{ccc} \mathcal{T}_f & \xrightarrow{\sigma_f} & \mathcal{T}_f \\ \pi_1 \downarrow & & \downarrow \pi \\ \mathcal{M}_f' & \xrightarrow{\sigma_f^1} & \mathcal{M}_f \end{array}$$

commutes.

The space $\mathcal{M}_f'$ is the quotient of $\mathcal{T}_f$ by a subgroup G of the mapping class group. We can also form a compactification $\overline{\mathcal{M}_f'} = \overline{\mathcal{T}}_f/G$. It is indeed compact, since $\pi_2 \colon \mathcal{M}_f' \to \mathcal{M}_f$ has finite degree.

Note that, since the mapping class group acts by isometries on $\mathcal{T}_f$, one can define the Teichmüller and WP metrics on $\mathcal{M}_f'$ so that π_1 is a local isometry.

The boundary of $\overline{\mathcal{M}}'_f$ also has stratified structure, so that every stratum $\mathcal{S}_{[\Gamma]} \subset \overline{\mathcal{M}}'_f$ corresponds to an equivalence class $[\Gamma]$ of multicurves under the action of G and has infinitely many preimages in $\overline{\mathcal{T}}_f$. We will use the fact that the full preimage $\pi_1^{-1}(U)$ is "discrete" for all U; for instance, the WP-distance between any two disjoint connected components of U is greater than some $c(U) > 0$. The previous lemma says, in particular, that σ_f acts in the same way on every preimage of a neighborhood of $\mathcal{S}_{[\Gamma]}$ which is invariant (it commutes with the corresponding deck transformation).

Proof of sufficiency: Suppose that σ_f has no fixed points and f has no Thurston obstructions.

Fix a point $\tau_1 \in \mathcal{T}_f$. Put $\tau_n = \sigma_f^{\circ n}(\tau_1), m'_n = \pi_1(\tau_n), m_n = \pi(\tau_n)$ and $\delta = d_T(\tau_1, \tau_2)$. If (m'_n) has an accumulation point in $\mathcal{M}'_f$, then (m_n) has an accumulation point in $\mathcal{M}_f$, and it follows (see the proof of Proposition 5.1 in [DH93]) that τ_n converges to a fixed point of σ_f. Thus, all accumulation points (the set is obviously non-empty, since $\overline{\mathcal{M}}'_f$ is compact) of (m_n) lie on the boundary of $\overline{\mathcal{M}}'_f$.

Take an accumulation point $m'_{n_i} \to l'_0$ on a stratum $\mathcal{S}_{[\Gamma]}$ of minimal possible dimension. Then m_{n_i+1} converges (this part of the statement follows from Lemma 2.7) to a point l_1 in $\pi_2(\mathcal{S}_{[\Gamma]})$. Indeed, when m'_{n_i} is close enough to l'_0, all curves of some $\Gamma_1 \in [\Gamma]$ are short in τ_{n_i}, and since $d_T(\tau_{n_i}, \tau_{n_i+1}) \le d_T(\tau_1, \tau_2) = \delta$, they are also short (at most e^δ times longer) in τ_{n_i+1}. Hence $l_1 \in \pi_2(\overline{\mathcal{S}_{[\Gamma]}})$. On the other hand, $\pi_2^{-1}(l_1)$ is finite and therefore contains an accumulation point of m'_n which by assumption cannot lie in a stratum of lower dimension. Thus $l_1 \in \pi_2(\mathcal{S}_{[\Gamma]})$.

Choose a small enough neighborhood U' of $\mathcal{S}_{[\Gamma]} \subset \overline{\mathcal{M}}'_f$. Put $U'' = \pi_2(U')$. Take U' so that $\pi^{-1}(U'')$ consists of disjoint neighborhoods of preimage strata and the Teichmüller distance between any two preimages is bounded below by a constant $C > \delta$. It will also be the case that the WP distance between any two connected components of the preimages $\pi^{-1}(U'')$ is bounded below by some positive number c.

There exists an N such that for all $i > N$, both $m'_{n_i} \in U'$ and $m_{n_i+1} \in U''$. Since the Teichmüller distance between τ_{n_i} and τ_{n_i+1} is less than δ, it follows that they lie in the same preimage component U of $\pi^{-1}(U'')$. There exists a multicurve Γ_1 in $[\Gamma]$ such that U contains $\mathcal{S}_{\Gamma_1}$. We may as well assume that $\Gamma = \Gamma_1$. Define the point $\tau_0 = \mathcal{S}_\Gamma \cap \pi_1^{-1}(l_0)$ — the closest (in the WP metric) to τ_{n_i} point in the fiber $\pi_1^{-1}(l_0)$. Since $d_{WP}(\tau_{n_i}, \tau_0) = d_{WP}(m_{n_i}, l_0) \to 0$, we can assume that N was chosen so that $d_{WP}(\tau_{n_i}, \tau_0) < c/\sqrt{d}$. Then $d_{WP}(\tau_{n_i+1}, \sigma_f(\tau_0)) < c$, and hence $\sigma_f(\tau_0) \in U$ (from the commutative diagram in Lemma 2.7, it follows that $\pi(\sigma_f(\tau_0)) = l_1 \in U''$). Therefore both $\sigma_f(\tau_0)$ and τ_0 lie in the same boundary stratum $\mathcal{S}_\Gamma$.

From Corollary 2.3, it follows that Γ is completely invariant. We assumed that Γ is not an obstruction; hence $\mathcal{S}_\Gamma$ is weakly repelling.

Denote by K' the set of all accumulation points of (τ_n) in $S_{[\Gamma]}$, and let K be the preimage of K' that lies within U. Since K' is closed (there are no accumulation points on $\partial S_{[\Gamma]}$), it is compact. Then

$$\inf_{\tau \in K, \gamma \notin \Gamma} l(\tau, \gamma) = 2k > 0.$$

Moreover, it follows from a compactness argument that there exists a T such that for all $t > T$, the neighborhood $\pi_1(U(\Gamma, t, 0)) \subset U'$ of $\mathcal{S}_{[\Gamma]}$ (where $U(\Gamma, t, k)$ is defined as in Proposition 2.6) satisfies

$$(m_n) \cap \pi_1(U(\Gamma, t, 0)) \subset \pi_1(U(\Gamma, t, k)).$$

In other words, all points τ_n that are close to K are far from the boundary of $\mathcal{S}_\Gamma$.

Take k, T, N as defined the previous paragraphs and $t > T$ such that $m_i \notin \pi_1(U(\Gamma, t, k))$ for all $i < N$. Take the first n such that $m_n \in \pi_1(U(\Gamma, e^\delta t, k))$. Then m_{n-1} is in $\pi_1(U(\Gamma, t, 0))$ and hence in $\pi_1(U(\Gamma, t, k))$. We know that τ_{n-1} and τ_n lie in the same preimage of $\pi_1(U(\Gamma, t, k))$; we may as well assume that it actually is $U(\Gamma, t, k)$. But it then follows from the same estimates as in the proof of Proposition 2.6 that τ_{n-1} must lie in $U(\Gamma, e^\delta t, k)$ if t was chosen large enough. We have a contradiction, because then m_{n-1} also lies in $U(\Gamma, e^\delta t, k)$. This contradiction shows that Γ should have been an obstruction. $\qquad\square$

2.3 Canonical Obstructions

Using our understanding of the boundary behavior, we can also prove the existence of canonical obstructions as stated in [Pi01]. Take an orbit $\tau_n = \sigma_f^{\circ n}(\tau_1)$ in $\mathcal{T}_f$.

Theorem 2.8 (Theorem 1.1 in [Pi01]). *Let f be a Thurston map with hyperbolic orbifold, and let Γ_c denote the set of all homotopy classes of essential simple closed curves γ in $\mathbb{S}^2 \setminus P_f$ such that $l(\tau_n, \gamma) \to 0$. Then either Γ_c is empty and f is equivalent to a rational map, or Γ_c is a Thurston obstruction.*

Proof: If f is equivalent to a rational function, then (τ_n) converges in $\mathcal{T}_f$, and hence Γ_c is empty.

If f is not equivalent to a rational function, then, as we have shown in the proof of the previous theorem, (τ_n) has an accumulation point on a stratum of lowest possible dimension $\mathcal{S}_{[\Gamma]}$, so that Γ is an obstruction. But then $\mathcal{S}_\Gamma$ is weakly attracting, as are all fixed strata in $\pi_1^{-1}(\mathcal{S}_{[\Gamma]})$. This means that after (τ_n) enters a small neighborhood of some stratum $\mathcal{S}_{\Gamma_c}$, it never leaves it again (and the size of this neighborhood is the same for all preimages). Thus, (τ_n) actually tends to $\mathcal{S}_{\Gamma_c}$. Then Γ_c is indeed the set of all homotopy classes of essential simple closed curves γ in $\mathbb{S}^2 \setminus P_f$ such that $l(\tau_n, \gamma) \to 0$ and is completely invariant. $\qquad\square$

Bibliography

[DH93] A. Douady and J. H. Hubbard, *A proof of Thurston's topological characterization of rational functions*, Acta Math. **171** (1993), 263–297.

[HSS09] J. H. Hubbard, D. Schleicher, and M. Shishikura, *Exponential Thurston maps and limits of quadratic differentials*, J. Amer. Math. Soc. **22** (2009), 77–117.

[Mi04] J. Milnor, *Pasting together Julia sets: a worked out example of mating*. Experiment. Math. **13** 1 (2004), 55–92.

[Pi01] K. Pilgrim, *Canonical Thurston obstructions*, Adv. Math. **158** (2001), 154–168.

[Po] A. Poirier, *On postcritically finite polynomials, part 2: Hubbard trees*, Stony Brook Preprint **7** (1993).

[Ta92] Tan Lei, *Matings of quadratic polynomials*, Ergod. Thy. Dynam. Syst. **12** 3 (1992), 589–620.

[Wo02] S. A. Wolpert, *Geometry of the Weil-Petersson completion of Teichmüller space*, Surveys in Differential Geometry, Vol. VIII, Boston, MA, 2002, 357–393.

17 Computing Arithmetic Invariants for Hyperbolic Reflection Groups

Omar Antolín-Camarena, Gregory R. Maloney, and Roland K. W. Roeder

1 Introduction

Suppose that P is a finite-volume polyhedron in $\mathbb{H}^3$ each of whose dihedral angles is an integer submultiple of π. Then the group $\Lambda(P)$ generated by reflections in the faces of P is a discrete subgroup of $\mathrm{Isom}(\mathbb{H}^3)$. If one restricts attention to the subgroup $\Gamma(P)$ consisting of orientation-preserving elements of $\Lambda(P)$, one naturally obtains a discrete subgroup of $PSL(2,\mathbb{C}) \cong \mathrm{Isom}^+(\mathbb{H}^3)$. This very classical family of finite-covolume Kleinian groups is known as the family of *polyhedral reflection groups*.

There is a complete classification of hyperbolic polyhedra with non-obtuse dihedral angles, and hence of hyperbolic reflection groups, given by Andreev's Theorem [An70,RHD07] (see also [RH93,Ho92,BS96] for alternatives to the classical proof); however, many more detailed questions about the resulting reflection group remain mysterious. We will refer to finite-volume hyperbolic polyhedra with non-obtuse dihedral angles as *Andreev Polyhedra* and finite-volume hyperbolic polyhedra with integer submultiple of π dihedral angles *Coxeter Polyhedra*.

A fundamental question for general Kleinian groups is: given Γ_1 and Γ_2 does there exist an appropriate conjugating element $g \in PSL(2,\mathbb{C})$ so that Γ_1 and $g\Gamma_2 g^{-1}$ both have a finite-index subgroup in common? In this case, Γ_1 and Γ_2 are called *commensurable*. Commensurable Kleinian groups have many properties in common, including coincidences in the lengths of closed geodesics (in the corresponding orbifolds) and a rational relationship (a commensurability) between their covolumes, if the groups are of finite-covolume. See [NR92] for many more interesting aspects of commensurability in the context of Kleinian groups.

If Γ_1 and Γ_2 are fundamental groups of hyperbolic manifolds M_1 and M_2, commensurability is the same as the existence of a common finite-sheeted cover $\widetilde{M}$ of M_1 and of M_2. Similarly, if $\Gamma(P_1)$ and $\Gamma(P_2)$ are

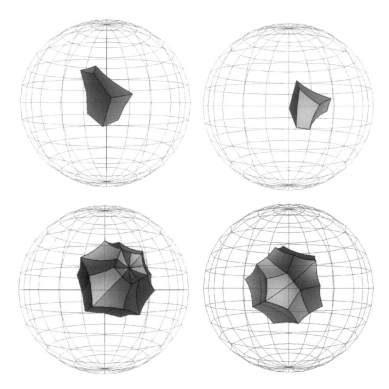

Figure 1. Two commensurable polyhedra P_1 (left top) and P_2 (right top) which tile a common larger polyhedron Q, here the right-angled dodecahedron.

polyhedral reflection groups, they are commensurable if and only if there is a larger polyhedron Q that is tiled both by P_1 under reflections in the faces (of P_1) and by P_2 under reflections in the faces (of P_2.) The existence of such a polyhedron Q is clearly a fundamental and delicate question from hyperbolic geometry. See Figure 1 for an example of two commensurable polyhedra. These coordinates for these polyhedra were computed using [Ro07] and displayed in the conformal ball model of $\mathbb{H}^3$ using Geomview [Ge].

A pair of sophisticated number-theoretic invariants has been developed by Reid and others to distinguish between commensurability classes of general finite-covolume Kleinian groups. See the recent textbook [MR03] and the many references therein. Given Γ, these invariants are a number field $k(\Gamma)$ known as the *invariant trace field* and a quaternion algebra $A(\Gamma)$ over $k(\Gamma)$ known as the *invariant quaternion algebra*. In fact, the invariant trace field is obtained by intersecting all of the fields generated by traces of elements of the finite-index subgroups of Γ. It is no surprise that such a field is related to commensurability because the trace of a loxodromic element a of Γ is related to the translation distance d along the axis of a by $2\cosh(d) = \mathrm{Re}(tr(A))$.

The pair $(k(\Gamma), A(\Gamma))$ does a pretty good job to distinguish commensurability classes, but there are examples of incommensurable Kleinian groups with the same $(k(\Gamma), A(\Gamma))$. For arithmetic groups, however, the pair $(k(\Gamma), A(\Gamma))$ is a complete commensurability invariant. Thus, one can find unexpected commensurable pairs of groups by computing these two invariants and by verifying that each group is arithmetic. See Subsection 7.5 for examples of such pairs that were discovered in this way. The precise definitions of the invariant trace field, the invariant quaternion algebra, and an arithmetic group will be given in Section 3.

It can be rather difficult to compute the invariant trace field and invariant quaternion algebra of a given Kleinian group "by hand." However there is a beautiful computer program called SNAP [GHN98] written by Coulson, Goodman, Hodgson, and Neumann, as described in [CGHN00]. They have computed the invariant trace field and invariant quaternion algebra, as well as many other interesting invariants, for many of the manifolds in the Hildebrand-Weeks census [HW89] and in the Hodgson-Weeks census [HW]. The basic idea used in SNAP is to compute a high-precision decimal approximation for an ideal triangulation of the desired manifold M using Newton's Method and then to use the LLL algorithm [LLL82] to guess exact algebraic numbers from the approximate values. These guessed values can be checked for correctness using the gluing equations describing M, and if the values are correct, the invariant trace field and invariant quaternion algebra can be computed from the exact triangulation.

SNAP provides a vast source of examples, also seen in the appendix of the book [MR03], and adds enormous flavor to the field. The fundamental techniques used in SNAP provide inspiration for our current work with polyhedral reflection groups.

In the case of polyhedral reflection groups there is a simplified description of the invariant trace field and the invariant quaternion algebra in terms of the Gram matrix of the polyhedron [MR98]. This theorem avoids the rather tedious trace calculations and manipulation of explicit generators of the group. Following the general technique used in the program SNAP we compute a set of outward unit normals to the faces of the polyhedron P to a high decimal precision and then use the LLL algorithm to guess the exact normals as algebraic numbers. From these normals the Gram matrix is readily computed, both allowing us to check whether the guessed algebraic numbers are in fact correct, and providing the exact data needed to use the theorem from [MR98] in order to compute the invariant trace field and invariant quaternion algebra for $\Gamma(P)$.

Our technique is illustrated for a simple example in Section 5 and a description of our program (available to download, see [ACMR]) is given in Section 6. Section 4 provides details on how to interpret the quaternion algebra.

Finally in Section 7 we provide many results of our computations. Among these, the highlights are (1) the completion of Vesnin's classification of arithmetic Löbell polyhedra [Ve91], showing that the n-th one is arithmetic if and only if $n = 5, 6$, or 8, (2) discovery of previously unknown commensurable pairs of polyhedra, and (3) an example of a non-arithmetic pair for which the invariants we consider do not distinguish commensurability class.

Acknowledgments. We thank Colin Maclachlan and Alan Reid for their beautiful work and exposition on the subject in [MR03]. We also thank Alan Reid for his many helpful comments.

The anonymous referees have provided helpful comments and suggestions that help to focus the direction of this paper. We thank them for their time.

We thank the authors of SNAP [GHN98] and the corresponding paper [CGHN00], which inspired this project (including the choice of name for our collection of scripts). Among them Craig Hodgson has provided helpful comments. We thank Andrei Vesnin who informed us about his result about arithmeticity of Löbell polyhedra. We effusively thank the writers of PARI/GP [PARI05], the system in which we have written our entire program and which is also used in SNAP [GHN98].

The third author thanks Mikhail Lyubich and Ilia Binder for their financial support and interest in the project. He also thanks John Hubbard, to whom this volume is dedicated, for introducing him to hyperbolic geometry and for his enthusiasm for mathematics in general and experimental mathematics in particular.

2   Hyperbolic Polyhedra and the Gram Matrix

We briefly recall some fundamental hyperbolic geometry, including the definition of a hyperbolic polyhedron and of the Gram matrix of a polyhedron.

Let $E^{3,1}$ be the four-dimensional Euclidean space with the indefinite metric $\|\mathbf{x}\|^2 = -x_0^2 + x_1^2 + x_2^2 + x_3^2$. Then hyperbolic space $\mathbb{H}^3$ is the component having $x_0 > 0$ of the subset of $E^{3,1}$ given by

$$\|\mathbf{x}\|^2 = -x_0^2 + x_1^2 + x_2^2 + x_3^2 = -1$$

with the Riemannian metric induced by the indefinite metric

$$-dx_0^2 + dx_1^2 + dx_2^2 + dx_3^2.$$

The hyperplane orthogonal to a vector $\mathbf{v} \in E^{3,1}$ intersects $\mathbb{H}^3$ if and only if $\langle \mathbf{v}, \mathbf{v} \rangle > 0$. Let $\mathbf{v} \in E^{3,1}$ be a vector with $\langle \mathbf{v}, \mathbf{v} \rangle > 0$, and define

$$P_{\mathbf{v}} = \{\mathbf{w} \in \mathbb{H}^3 | \langle \mathbf{w}, \mathbf{v} \rangle = 0\}$$

to be the hyperbolic plane orthogonal to $\mathbf{v}$; and the corresponding closed half space:

$$H_{\mathbf{v}}^+ = \{\mathbf{w} \in \mathbb{H}^3 | \langle \mathbf{w}, \mathbf{v} \rangle \geq 0\}.$$

It is a well-known fact that given two planes $P_{\mathbf{v}}$ and $P_{\mathbf{w}}$ in $\mathbb{H}^3$ with $\langle \mathbf{v}, \mathbf{v} \rangle = 1$ and $\langle \mathbf{w}, \mathbf{w} \rangle = 1$, they:

- intersect in a line if and only if $\langle \mathbf{v}, \mathbf{w} \rangle^2 < 1$, in which case their dihedral angle is $\arccos(-\langle \mathbf{v}, \mathbf{w} \rangle)$;

- intersect in a single point at infinity if and only if $\langle \mathbf{v}, \mathbf{w} \rangle^2 = 1$, in this case their dihedral angle is 0;

- are disjoint if and only if $\langle \mathbf{v}, \mathbf{w} \rangle^2 > 1$, in which case the distance between them is $\operatorname{arccosh}(-\langle \mathbf{v}, \mathbf{w} \rangle))$.

Suppose that $\mathbf{e}_1, \ldots, \mathbf{e}_n$ satisfy $\langle \mathbf{e}_i, \mathbf{e}_i \rangle > 0$ for each i. Then, a *hyperbolic polyhedron* is an intersection

$$P = \bigcap_{i=0}^{n} H_{\mathbf{e}_i}^+$$

having non-empty interior.

If we normalize the vectors $\mathbf{e}_i$ that are orthogonal to the faces of a polyhedron P, the *Gram Matrix* of P is given by $M_{ij}(P) = 2\langle \mathbf{e}_i, \mathbf{e}_j \rangle$. It is also common to define the Gram matrix without this factor of 2, but our definition is more convenient for arithmetic reasons. By construction, a Gram matrix is symmetric and has 2s on the diagonal. Notice that the Gram matrix encodes information about both the dihedral angles between adjacent faces of P and the hyperbolic distances between non-adjacent faces.

3 Invariant Trace Field, Invariant Quaternion Algebra, and Arithmeticity

The *trace field* of a subgroup Γ of $PSL(2,\mathbb{C})$ is the field generated by the traces of its elements; that is, $\mathbb{Q}(\operatorname{tr}\Gamma) := \mathbb{Q}(\operatorname{tr}\gamma : \gamma \in \Gamma)$.[1] This field is not a commensurability invariant as shown by the following example found in [MR03].

Consider the group Γ generated by

$$A = \begin{pmatrix} 1 & 1 \\ 1 & 0 \end{pmatrix}, \quad B = \begin{pmatrix} 1 & 0 \\ -\omega & 1 \end{pmatrix},$$

where $\omega = (-1 + i\sqrt{3})/2$. The trace field of Γ is $\mathbb{Q}(\sqrt{-3})$. Now let $X = \begin{pmatrix} i & 0 \\ 0 & -i \end{pmatrix}$. It is easy to see that X normalizes Γ and its square is the identity (in $PSL(2,\mathbb{C})$), so that $\Lambda = \langle \Gamma, X \rangle$ contains Γ as a subgroup of index 2 and is therefore commensurable with Γ. But Λ also contains $XBA = \begin{pmatrix} i & i \\ i\omega & -i + i\omega \end{pmatrix}$, so the trace field of Λ contains i in addition to ω.

The easiest way to fix this, that is, to get a commensurability invariant related to the trace field, is to associate to Γ the intersection of the trace fields of all finite index subgroups of Γ; this is the *invariant trace field* denoted $k\Gamma$.

While this definition clearly shows commensurability invariance, it does not lend itself to practical calculation. The proof of Theorem 3.3.4 in [MR03] brings us closer: it shows that instead of intersecting many trace fields, one can look at a single one, namely, the invariant trace field of Γ equals the trace field of its subgroup $\Gamma^{(2)} := \langle \gamma^2 : \gamma \in \Gamma \rangle$. (This also shows that the invariant trace field is non-trivial which is not entirely clear from the definition as an intersection.) When a finite set of generators for the group is known, this is actually enough to compute the invariant trace field. Indeed, Lemma 3.5.3 in [MR03] establishes that if $\Gamma = \langle \gamma_1, \gamma_2, \ldots, \gamma_n \rangle$, the trace field of Γ is generated by $\{\operatorname{tr}(\gamma_i) : 1 \le i \le n\} \cup \{\operatorname{tr}(\gamma_i\gamma_j) : 1 \le i < j \le n\} \cup \{\operatorname{tr}(\gamma_i\gamma_j\gamma_k) : 1 \le i < j < k \le n\}$. Applying this result to $\Gamma^{(2)}$ we get a finite procedure for computing the invariant trace field.

For reflection groups a more efficient description can be given in terms of the Gram matrix. The description given above uses roughly $n^3/6$ generators for a polyhedron with n faces; the following description will use only around $n^2/2$. But before we state it we need to define a certain quadratic space over the field

$$k(P) := \mathbb{Q}(a_{i_1 i_2} a_{i_2 i_3} \cdots a_{i_r i_1} : \{i_1, i_2, \ldots, i_r\} \subset \{1, 2, \ldots, n\})$$

---

[1]Note that for $\gamma \in PSL(2,\mathbb{C})$, the trace $\operatorname{tr}\gamma$ is only defined up to sign.

associated to a polyhedron; this space will also appear in the next section in the theorem used to calculate the invariant quaternion algebra.

As in the previous section, given a polyhedron P we will denote the outward-pointing normals to the faces by $\mathbf{e}_1, \ldots, \mathbf{e}_n$ and the Gram matrix by (a_{ij}). Define $M(P)$ as the vector space over $k(P)$ spanned by of all the vectors of the form $a_{1i_1} a_{i_1 i_2} \cdots a_{i_{r-1} i_r} \mathbf{e}_{i_r}$ where $\{i_1, i_2, \ldots, i_r\}$ ranges over the subsets of $\{1, 2, \ldots, n\}$ and n is the number of faces of P. This space $M(P)$ will be equipped with the restriction of the quadratic form with signature $(3, 1)$ used in $\mathbb{H}^3$. We recall that the discriminant of a non-degenerate symmetric bilinear form $\langle \cdot, \cdot \rangle$ is defined as $\det\left(\langle v_i, v_j \rangle \right)_{ij}$ where $\{v_i\}_i$ is a basis for the vector space on which the form is defined. The discriminant does depend on the choice of basis, but for different bases the discriminants differ by multiplication by a square in the ground field: indeed, if $u_i = \sum_j \alpha_{ij} v_j$, the discriminant for the basis $\{u_i\}_i$ is that of the basis $\{v_i\}_i$ multiplied by $\det(\alpha_{ij})^2$.

Now we can state the theorem we use to calculate the invariant trace field, Theorem 10.4.1 in [MR03]:

Theorem 3.1. *Let P be a Coxeter polyhedron and let Γ be the reflection group it determines. Let (a_{ij}) be the Gram matrix of P. The invariant trace field of Γ is $k(P)(\sqrt{d})$, where d is the discriminant of the quadratic space $M(P)$ and $k(P) = \mathbb{Q}(a_{i_1 i_2} a_{i_2 i_3} \cdots a_{i_r i_1} : \{i_1, i_2, \ldots, i_r\} \subset \{1, 2, \ldots, n\})$ is the field defined previously.*

3.1   The Invariant Quaternion Algebra

A *quaternion algebra* over a field F is a four-dimensional associative algebra A with basis $\{1, i, j, k\}$ satisfying $i^2 = a1$, $j^2 = b1$ and $ij = ji = -k$ for some $a, b \in F$. Note that $k^2 = (ij)^2 = ijij = -ijji = -ab1$. The case $F = \mathbb{R}$, $a = b = -1$ gives Hamilton's quaternions.

The quaternion algebra defined by a pair a, b of elements of F is denoted by its *Hilbert symbol* $\left(\frac{a,b}{F} \right)$. A quaternion algebra does not uniquely determine a Hilbert symbol for it, since, for example, $\left(\frac{a,b}{F} \right) = \left(\frac{a,-ab}{F} \right) = \left(\frac{au^2, bv^2}{F} \right)$ for any invertible elements $u, v \in F$. Fortunately, there is a computationally effective way of deciding whether two Hilbert symbols give the same quaternion algebra. This will be discussed in Section 4; for now we will just define the invariant quaternion algebra of a subgroup of $PSL(2, \mathbb{C})$ and state the theorem we use to calculate a Hilbert symbol for it.

Given any non-elementary[2] subgroup Γ of $PSL(2, \mathbb{C})$, we can form the algebra $A_0\Gamma := \{\sum a_i \gamma_i : a_i \in \mathbb{Q}(\operatorname{tr}\Gamma), \gamma_i \in \Gamma\}$. (Abusing notation slightly

---

[2]This means that the action of Γ on $\mathbb{H}^3 \cup \hat{\mathbb{C}}$ has no finite orbits. Reflection groups determined by finite-volume polyhedra are always non-elementary.

we consider the elements of Γ as matrices defined up to sign.) This turns out to be a quaternion algebra over the trace field $\mathbb{Q}(\operatorname{tr}\Gamma)$ (see Theorem 3.2.1 in [MR03]).

Just as with the trace fields, we define the *invariant quaternion algebra* of Γ, denoted by $A\Gamma$, as the intersection of all the quaternion algebras associated to finite-index subgroups of Γ.

When Γ is finitely generated in addition to non-elementary, we are in a situation similar to that of the invariant trace field in that the invariant quaternion algebra is simply the quaternion algebra associated to the subgroup $\Gamma^{(2)}$ of Γ, or in symbols, $A\Gamma = A_0\Gamma^{(2)}$. To see this, note that Theorem 3.3.5 in [MR03] states that for finitely generated non-elementary Γ, the quaternion algebra $A_0\Gamma^{(2)}$ is a commensurability invariant. Now, given an arbitrary finite-index subgroup Λ of Γ we have $A\Gamma \subset A_0\Gamma^{(2)} = A_0\Lambda^{(2)} \subset A_0\Lambda$.

In the case where Γ is the reflection group of a polyhedron P, $A\Gamma$ can be identified as the even-degree subalgebra of a certain Clifford algebra associated with P. Let us briefly recall the basic notions related to Clifford algebras. Given an n-dimensional vector space V over a field F equipped with a non-degenerate symmetric bilinear form $\langle \cdot, \cdot \rangle$ and associated quadratic form $\|\cdot\|^2$, the Clifford algebra it determines is the 2^n-dimensional F-algebra generated by all formal products of vectors in V subject to the condition $v^2 = \langle v, v \rangle 1$ (where 1 is the empty product of vectors). If $\{v_1, v_2, \ldots, v_n\}$ is an orthogonal basis of V, a basis for the Clifford algebra $C(V)$ is $\{v_{i_1} v_{i_2} \cdots v_{i_r} : 1 \le r \le n, 1 \le i_1 < i_2 < \cdots < i_r \le n\}$. There is a $\mathbb{Z}_2$-grading of $C(V)$ given on monomials by the parity of the number of vector factors.

Now we can state the result mentioned above: $A\Gamma$ is the even-degree subalgebra of $C(M)$ where M is the vector space over the invariant trace field of Γ that appears in Theorem 3.1. This relationship between the invariant trace algebra and M allows one to prove a theorem giving an algorithm for computing the Hilbert symbol for the invariant quaternion algebra, part of Theorem 3.1 in [MR98]:

Theorem 3.2. *The invariant quaternion algebra of the reflection group Γ of a polyhedron P is given by*

$$A\Gamma = \left(\frac{-\|u_1\|^2 \|u_2\|^2, -\|u_1\|^2 \|u_3\|^2}{k\Gamma} \right),$$

where $\{u_1, u_2, u_3, u_4\}$ is an orthogonal basis for the quadratic space $M(P)$ defined in the previous section.

In many cases when studying an orbifold $O = \mathbb{H}^3/\Gamma$, a simple observation about the singular locus of O leads to the fact that the invariant

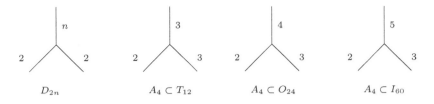

Figure 2. The possible vertices in the singular locus of a 3-dimensional orbifold. The last three have vertex stabilizer containing A_4.

quaternion algebra can represented by the Hilbert symbol $\left(\frac{-1,-1}{k\Gamma}\right)$. This happens particularly often for polyhedral reflection groups.

Any vertex in the singular locus must be trivalent and must have labels from the short list shown in Figure 2. See [BP01] for more information on orbifolds, and in particular page 24 from which Figure 2 is essentially copied.

The last three have vertex stabilizer containing A_4, so if $\mathbb{H}^3/\Gamma$ has singular locus containing such a vertex, Γ must contain A_4 as a subgroup. In this case, the invariant quaternion algebra can be represented by the Hilbert symbol $\left(\frac{-1,-1}{k\Gamma}\right)$, see [MR03], Lemma 5.4.1. (See also Lemma 5.4.2.)

For a polyhedral reflection group generated by a Coxeter polyhedron P, the corresponding orbifold $\mathbb{H}^3/\Gamma$ has underlying space $\mathbb{S}^3$ and the singular set is (an unknotted) copy of the edge graph of P. The label at each edge having dihedral angle $\frac{\pi}{n}$ is merely n. In many cases, this simplification makes it easier to compute the Hilbert symbol of a polyhedral reflection group. We do not automate this check within our program, but it can be useful to the reader.

4   Quaternion Algebras and Their Invariants

As mentioned before, a quaternion algebra is fully determined by its Hilbert symbol $\left(\frac{a,b}{F}\right)$, although this is by no means unique. For example,

$$\left(\frac{b,a}{F}\right), \qquad \left(\frac{a,-ab}{F}\right), \qquad \text{and} \qquad \left(\frac{ax^2,by^2}{F}\right)$$

all determine the same algebra (here x and y are arbitrary invertible elements of F.)

Taking $F = \mathbb{R}$, the Hilbert symbol $\left(\frac{-1,-1}{\mathbb{R}}\right)$ represents the ordinary quaternions (or Hamiltonians,) denoted by $\mathcal{H}$. If F is any field, the Hilbert symbol $\left(\frac{1,1}{F}\right)$ is isomorphic to $M_2(F)$, the two-by-two matrices over F.

A natural question now arises; namely, when do two Hilbert symbols represent the same quaternion algebra? This question is pertinent for us — especially in the case when F is a number field — because the invariant quaternion algebra is an invariant for a reflection group. For these purposes, we will need some way of classifying quaternion algebras over number fields. All of the following material on quaternion algebras appears in the reference [MR03].

The first step in the classification of quaternion algebras is the following theorem (see Theorem 2.1.7 from [MR03]):

Theorem 4.1. *Let A be a quaternion algebra over a field F. Then either A is a division algebra or A is isomorphic to $M_2(F)$.*

In the latter case, we say that A *splits*. There are several different ways of expressing this condition, one of which will be particularly useful for us (see Theorem 2.3.1 from [MR03]):

Theorem 4.2. *The quaternion algebra $A = \left(\frac{a,b}{F}\right)$ splits over F if and only if the equation*

$$ax^2 + by^2 = 1 \qquad (4.1)$$

has a solution in $F \times F$. We call this equation the Hilbert equation *of A.*

When F is a number field, it turns out that in order to classify the quaternion algebras over F completely we need to look at quaternion algebras over the completions of F with respect to its valuations. The first chapter of [MR03] contains a brief introduction to number fields and valuations, and [La94] is a standard text on the subject.

Definition 4.3. Let F be any field. A *valuation* on F is a map $\nu : F \to \mathbb{R}^+$ such that

(i) $\nu(x) \geq 0$ for all $x \in F$ and $\nu(x) = 0$ if and only if $x = 0$,

(ii) $\nu(xy) = \nu(x)\nu(y)$ for all $x, y \in F$, and

(iii) $\nu(x + y) \leq \nu(x) + \nu(y)$ for all $x, y \in F$.

Any field admits a trivial valuation $\nu(x) = 1$ for all $x \neq 0$. When F is a subfield of the real (or complex) numbers, the ordinary absolute value (or modulus) function is a valuation when restricted to F. In general, valuations on a field fall into two different classes.

Definition 4.4. If a valuation ν on a field F also satisfies

(iv) $\nu(x + y) \leq max\{\nu(x), \nu(y)\}$ for all $x, y \in F$, then ν is called a *non-Archimedean* valuation. If the valuation ν does not satisfy (iv), then it is called *Archimedean*.

There is also a notion of equivalence between valuations.

Definition 4.5. Two valuations ν_1 and ν_2 on F are called *equivalent* if there exists some $\alpha \in \mathbb{R}^+$ such that $\nu_2(x) = \left(\nu_1(x) \right)^\alpha$ for all $x \in F$.

When F is a number field, it is possible to classify all valuations on F up to this notion of equivalence. Let σ be a real or complex embedding of F. Then a valuation ν_σ can be defined by $\nu_\sigma(x) = |\sigma(x)|$, where $|\cdot|$ is the absolute value on $\mathbb{R}$ or modulus on $\mathbb{C}$. This is an Archimedean valuation on F, and up to equivalence these are the only Archimedean valuations that F admits.

Denote by R_F the ring of integers of F — i.e., the set of elements of F satisfying some monic polynomial equation with integer coefficients — which is a subring of F. Let $\mathcal{P}$ be a prime ideal in R_F. Define a function $n_\mathcal{P} : R_F \to \mathbb{Z}$ by $n_\mathcal{P}(a) = m$, where m is the largest integer such that $a \in \mathcal{P}^m$. Since F is the field of fractions of R_F, $n_\mathcal{P}$ can be extended to all of F by the rule $n_\mathcal{P}(a/b) = n_\mathcal{P}(a) - n_\mathcal{P}(b)$. Now pick c with $0 < c < 1$. The function $\nu_\mathcal{P} : R_F \to \mathbb{R}^+$ given by $\nu_\mathcal{P}(x) = c^{n_\mathcal{P}(x)}$ is a non-Archimedean valuation on F. Moreover, all non-Archimedean valuations on F are equivalent to a valuation of this form.

We summarize these facts in the following theorem (Theorem 0.6.6 from [MR03]).

Theorem 4.6. *Let F be a number field. Then every Archimedean valuation of F is equivalent to ν_σ for some real or complex embedding σ of F, and every non-Archimedean valuation of F is equivalent to $\nu_\mathcal{P}$ for some prime ideal $\mathcal{P}$ of R_F. The former are sometimes called* infinite places, *while the latter are called* finite places.

A valuation ν on a field F defines a metric on F by $d(x, y) = \nu(x - y)$. The completion of F with respect to this metric is denoted by F_ν. Equivalent valuations give rise to the same completions. If $\nu = \nu_\sigma$ for a real or complex embedding σ of F, then F_ν is isomorphic to $\mathbb{R}$ or $\mathbb{C}$, respectively. If $\nu = \nu_\mathcal{P}$ for some prime ideal $\mathcal{P} \subseteq R_F$, then F_ν is called a $\mathcal{P}$-*adic field*.

Let A be a quaternion algebra over a number field F, and let F_ν be the completion of F with respect to some valuation ν. Then we can construct the tensor product $A \otimes_F F_\nu$, which turns out to be a quaternion algebra over F_ν. Indeed, if $A = \left(\frac{a,b}{F} \right)$, then $A \otimes_F F_\nu = \left(\frac{a,b}{F_\nu} \right)$. Ultimately, the classification of quaternion algebras over F will be reduced to the classification of quaternion algebras over completions of F with respect to valuations ν. The two following theorems will be useful in this regard.

Theorem 4.7. *Let $\mathbb{R}$ be the real number field. Then $\mathcal{H}$ is the unique quaternion division algebra over $\mathbb{R}$. (*Compare to Theorem 2.5.1 from [MR03].)

Theorem 4.8. *Let F_ν be a $\mathcal{P}$-adic field. Then there is a unique quaternion division algebra over F_ν.* (Compare to Theorem 2.6.4 from [MR03].)

Thus when A is a quaternion algebra over a number field F and A_ν is the corresponding quaternion algebra over a real or $\mathcal{P}$-adic completion F_ν of F, by Theorem 4.1 there are two possibilities: A_ν is the unique quaternion division algebra over F_ν, or $A_\nu \cong M_2(F_\nu)$. In the former case, we say that A *ramifies* at ν, while in the latter case we say that A *splits* at ν. When $F_\nu = \mathbb{R}$, there is a simple test to determine which algebra is represented by the Hilbert symbol $\left(\frac{a,b}{\mathbb{R}} \right)$:

$$\text{if } a \text{ and } b \text{ are both negative, then } A \cong \mathcal{H}, \text{ otherwise } A \text{ splits.} \qquad (4.2)$$

Various tests exist for $\mathcal{P}_\nu$, but often the simplest test is to determine if there exists a solution to equation (4.1). (See the appendix of this paper.) Notice that by Theorem 4.2 every quaternion algebra over the complex numbers is isomorphic to $M_2(\mathbb{C})$.

The following theorem provides the necessary criterion for distinguishing between quaternion algebras over number fields.

Theorem 4.9 (Vignéras [Vi80]). *Let F be a number field. For each quaternion algebra A over F, denote by $Ram(A)$ the set of all real or finite places at which A ramifies. Then two quaternion algebras A and A' over F are equal if and only if $Ram(A) = Ram(A')$.* (See also Theorem 2.7.5 from [MR03].)

Thus the complete identification of a quaternion algebra A over a number field F amounts to determining $Ram(A)$. It is easy to check if A ramifies at the real infinite places of F. Let α be a primitive element of F (i.e., an element whose powers form a basis for F over $\mathbb{Q}$.) Every embedding of F in $\mathbb{R}$ can be obtained from a real root α_i of the minimal polynomial of α over $\mathbb{Q}$ by extending the map $\sigma_i : \alpha \to \alpha_i$ linearly to F. If $A = \left(\frac{a,b}{F} \right)$ and a and b are expressed as polynomials in α, then it is straightforward to check if condition (4.2) holds for $\sigma_i(a)$ and $\sigma_i(b)$.

The finite places of F are more difficult to check. The most straightforward method is to check to see if the Hilbert equation (4.1) has a solution; our procedure for doing this is detailed in the appendix.

4.1   Arithmeticity

The notion of an arithmetic group comes from the theory of algebraic groups and is a standard way of producing finite-covolume discrete subgroups of semi-simple Lie groups. To see how this general theory relates to Kleinian groups, see [HLMA92] or [MR03].

In the case of Kleinian groups, the following definition coincides with the most general one, and is naturally related to the quaternion algebras which we have already mentioned.

Let A be a quaternion algebra over a number field F and denote by R_F the ring of integers in F. An *order* $\mathcal{O}$ in A is an R_F-lattice (spanning A over F) that is also a ring with unity. For every complex place ν of F there is an embedding of $A \longrightarrow M_2(\mathbb{C})$ determined by the isomorphism $A \otimes_F F_\nu \cong M_2(\mathbb{C})$. Given a complex place ν and an order $\mathcal{O}$, we can construct a subgroup of $SL(2, \mathbb{C})$, and hence of $PSL(2, \mathbb{C})$, by taking the image $\Gamma_\mathcal{O}^\nu$ of the elements of $\mathcal{O}$ with unit norm under the embedding $A \longrightarrow M_2(\mathbb{C})$ defined above.

In the case that F has a unique complex place ν and that A ramifies over every real place of F, then $\Gamma_\mathcal{O} := \Gamma_\mathcal{O}^\nu$ is a discrete subgroup of $PSL(2, \mathbb{C})$ (see Sections 8.1 and 8.2 of [MR03]).

Definition 4.10. A Kleinian group Γ is called *arithmetic* if it is commensurable with $\Gamma_\mathcal{O}$ for some order $\mathcal{O}$ of a quaternion algebra that ramifies over every real place and is defined over a field with a unique complex place.

Viewing $\mathrm{Isom}^+(\mathbb{H}^3)$ as $SO^+(3, 1)$ furnishes an alternative construction of arithmetic Kleinian groups as follows. Let F be a real number field, and (V, q) a four-dimensional quadratic space over F with signature $(3, 1)$. Any F-linear map $\sigma : V \longrightarrow V$ preserving q can be identified with an element of $SO^+(3, 1)$, and thus of $\mathrm{Isom}^+(\mathbb{H}^3)$, by extension of scalars from F to $\mathbb{R}$. Given an R_F-lattice $L \subset V$ of rank 4 over R_F, the group $SO(L) := \{\sigma \in SO^+(3, 1) \cap GL(4, F) : \sigma(L) = L\}$ is always discrete and arithmetic.

Moreover, the groups of the form $SO(L)$ give representatives for the commensurability classes of all Kleinian groups that possess a non-elementary Fuchsian subgroup. Since all reflection groups determined by finite-volume polyhedra have non-elementary Fuchsian subgroups, for our purposes this can be considered the definition of arithmeticity. See [HLMA92, page 143] for a discussion. The distinction between arithmetic groups arising from quaternion algebras and those arising from quadratic forms is also discussed in [VS93, pages 217–221], whose authors call the latter "arithmetic groups of the simplest kind".

Aside from its relationship to algebraic groups, arithmeticity is interesting for many reasons including the fact that for arithmetic groups Γ the pair $(k\Gamma, A\Gamma)$ is a complete commensurability invariant (see Section 8.4 of [MR03]). This will allow us to identify several unexpected pairs of commensurable reflection groups which are presented in Section 7.5.

To decide whether a given reflection group Γ determined by a polyhedron P is arithmetic there is a classical theorem due to Vinberg [Vi67]:

Theorem 4.11. *Let (a_{ij}) be the Gram matrix of a Coxeter polyhedron P. Then the reflection group determined by P is arithmetic if and only if the following three conditions hold:*

1. $K := \mathbb{Q}(a_{ij})$ *is totally real.*

2. *For every embedding $\sigma : K \longrightarrow \mathbb{C}$ such that $\sigma|_{k(P)} \neq \mathrm{id}$ (where $k(P)$ is the field defined in Theorem 3.1), the matrix $(\sigma(a_{ij}))$ is positive semi-definite.*

3. *The a_{ij} are algebraic integers.*

(The reader should note that this is Theorem 10.4.5 from [MR03], but with a typographical error: k should be K.)

More generally, for any finite-covolume Kleinian group Γ, Maclachlan and Reid have proved a similar result, Theorem 8.3.2 in [MR03]:

Theorem 4.12. *A finite-covolume Kleinian group Γ is arithmetic if and only if the following three conditions hold:*

1. $k\Gamma$ *has exactly one complex place.*

2. $A\Gamma$ *ramifies at every real place of $k\Gamma$.*

3. $\mathrm{tr}\gamma$ *is an algebraic integer for each $\gamma \in \Gamma$.*

5   Worked Example

A Lambert cube is a compact polyhedron realizing the combinatorial type of a cube, with three disjoint non-coplanar edges chosen and assigned dihedral angles $\frac{\pi}{l}$, $\frac{\pi}{m}$, and $\frac{\pi}{n}$, and the remaining edges assigned dihedral angles $\frac{\pi}{2}$. It is easy to verify that if $l, m, n > 2$, then, such an assignment of dihedral angles satisfies the hypotheses of Andreev's Theorem. The resulting polyhedron is called the (l, m, n)-Lambert Cube, which we will denote by $P_{l,m,n}$.

In this section we illustrate our techniques by computing the invariant trace field and invariant quaternion algebra associated to the $(3, 3, 6)$ Lambert cube.

The starting point of our computation is a set of low-precision decimal approximations of outward-pointing normal vectors $\{\mathbf{e}_1, \ldots, \mathbf{e}_6\}$ to the six faces of our cube. For a given compact hyperbolic polyhedron, it is non-trivial to construct such a set of outward-pointing normal vectors. One way is to use the collection of Matlab scripts described in [Ro07]. Throughout this paper we will always assume the following normalization for the location of our polyhedron: the first three faces meet at a vertex, the first face has normal vector $(0, 0, 0, *)$, the second face has normal vector $(0, 0, *, *)$, and the third has form $(0, *, *, *)$, where $*$ indicates that no condition is placed on that number.

We then use Newton's Method with extended-precision decimals to improve this set of approximate normals until they are very precise. (Here, we do this with precision 40 numbers, but we display fewer digits for the reader.) The vectors $\{e_1, \ldots e_6\}$ are displayed as rows in the following matrix:

$$
\begin{bmatrix}
0.0 & 0.0 & 0.0 & -0.999999 \\
0.0 & 0.0 & 0.866025 & 0.500000 \\
0.0 & -1.000000 & 0.0 & 0.0 \\
1.389410 & 0.866025 & -0.738319 & 1.278806 \\
0.797085 & 1.278806 & 0.0 & 0.0 \\
0.627285 & 0.0 & -1.180460 & 0.0
\end{bmatrix}
$$

The normalization we have chosen for the location of our polyhedron assures us that each of these decimals should approximate an algebraic number of some (low) degree. There are commands in many computer algebra packages for guessing the minimal polynomial that is most likely satisfied by a given decimal approximate. Most of these commands are ultimately based on the LLL algorithm [LLL82]. (We have used the command minpoly() in Maple, the command algdep() in Pari/GP, and the command RootApproximant[] in Mathematica 6.) Each of these commands requires a parameter specifying up to what degree of polynomials to search. In this case we specify degree 30. The resulting matrix of guessed minimal polynomials is:

$$
\begin{bmatrix}
X & X & X & 1 + X \\
X & X & -3 + 4X^2 & -1 + 2X \\
X & 1 + X & X & X \\
-9 - 88X^2 + 48X^4 & -3 + 4X^2 & 1 - 28X^2 + 48X^4 & 3 - 28X^2 + 16X^4 \\
-9 + 4X^2 + 16X^4 & 3 - 28X^2 + 16X^4 & X & X \\
-1 - X^2 + 9X^4 & X & -3 + X + 3X^2 & X
\end{bmatrix}
$$

The next step is to specify which root of each given minimal polynomial is closest to the decimal approximate above. In the current case, the solutions of each polynomial are easily expressed by radicals, so we merely pick the appropriate expression. (For more complicated examples, our computer program uses a more sophisticated way of expressing algebraic numbers as described in Section 6.) For the current example, the following matrix contains as rows our guessed exact values for $\{e_1, \ldots e_6\}$.

$$N := \begin{bmatrix} 0 & 0 & 0 & -1 \\ 0 & 0 & \frac{\sqrt{3}}{2} & \frac{1}{2} \\ 0 & -1 & 0 & 0 \\ \frac{\sqrt{33+6\sqrt{37}}}{6} & \frac{\sqrt{3}}{2} & -\frac{\sqrt{42+6\sqrt{37}}}{12} & \frac{\sqrt{14+2\sqrt{37}}}{4} \\ \frac{\sqrt{-2+2\sqrt{37}}}{4} & \frac{\sqrt{14+2\sqrt{37}}}{4} & 0 & 0 \\ \frac{\sqrt{2+2\sqrt{37}}}{6} & 0 & \frac{-1-\sqrt{37}}{6} & 0 \end{bmatrix}$$

Corresponding to this set of guessed normal vectors we have the Gram Matrix $G_{i,j} = 2\langle e_i, e_j \rangle$, in which $a = \frac{-\sqrt{14+2\sqrt{37}}}{2}$ and $b = \frac{-\sqrt{3}(1+\sqrt{37})}{6}$:

$$\begin{bmatrix} 2 & -1 & 0 & a & 0 & 0 \\ -1 & 2 & 0 & 0 & 0 & b \\ 0 & 0 & 2 & -\sqrt{3} & a & 0 \\ a & 0 & -\sqrt{3} & 2 & 0 & 0 \\ 0 & 0 & a & 0 & 2 & -1 \\ 0 & b & 0 & 0 & -1 & 2 \end{bmatrix}$$

By checking that there are 2's down the diagonal of G and that there is $-2\cos(\alpha_{ij})$ in the ij-th entry of G if faces i and j are adjacent, we can see that the guessed matrix N whose rows represent outward-pointing normal vectors was correct. Here there are $4 \cdot 6 - 6 = 18$ equations that we have checked, consistent with the number of guessed values in the matrix N. Consequently G is the exact Gram matrix for the $(3,3,6)$ Lambert cube. We can use this to compute the invariant trace field and the invariant quaternion algebra.

The non-trivial cyclic products from G correspond to non-trivial cycles in the Coxeter symbol for $P_{3,3,6}$, which is depicted in Figure 3 with the appropriate element of the Gram matrix written next to each edge. (For those unfamiliar with Coxeter symbols, see Section 6.)

The non-trivial cyclic products correspond to closed loops in the Coxeter symbol. Always included are the squares of each entry of G, of which the only two irrational ones are: $(g_{14})^2 = (g_{35})^2 = \frac{7+\sqrt{37}}{2}$ and $(g_{26})^2 = \frac{19+\sqrt{37}}{6}$. The other non-trivial cyclic product corresponds to the closed loop in the Coxeter symbol: $g_{12}g_{26}g_{65}g_{53}g_{34}g_{41} = 11 + 2\sqrt{37}$. Thus, $k(P_{3,3,6}) = \mathbb{Q}(\sqrt{37})$.

Notice that the four vectors $v_1 := e_1$, $v_{12} := g_{12}e_2 = -e_2$, $v_{14} := g_{14}e_4 = \frac{-\sqrt{14+2\sqrt{37}}}{2}e_4$, and $v_{143} := g_{14}g_{43}e_3 = \frac{-\sqrt{14+2\sqrt{37}}}{2} \cdot \sqrt{3}e_3$ are

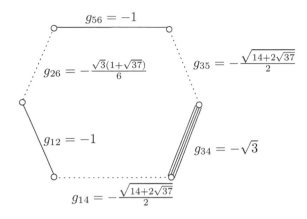

Figure 3. Coxeter symbol for the $(3, 3, 6)$ Lambert cube. Next to each edge we display the corresponding element of the Gram matrix G.

linearly independent, so that they span the quadratic space (M, q). (See the description of Theorems 3.1 and 3.2.) We now compute the matrix representing q with respect to the basis $\{\mathbf{v}_1, \mathbf{v}_{12}, \mathbf{v}_{143}, \mathbf{v}_{14}\}$.

$$
\begin{bmatrix}
2 & 1 & 0 & \frac{7+\sqrt{37}}{2} \\
1 & 2 & 0 & 0 \\
0 & 0 & 21 + 3\sqrt{37} & \frac{21+3\sqrt{37}}{2} \\
\frac{7+\sqrt{37}}{2} & 0 & \frac{21+3\sqrt{37}}{2} & 7 + \sqrt{37}
\end{bmatrix}
$$

The determinant of this matrix, and hence the discriminant of the quadratic form q (a number well-defined up to a square in the field $k(P_{3,3,6})$) is $d = \frac{2973-489\sqrt{37}}{2}$, consequently $k\Gamma_{3,3,6} = k(P_{3,3,6})\left(\sqrt{\frac{2973-489\sqrt{37}}{2}}\right)$. Since d is primitive for $k(P_{3,3,6})$, we have $k\Gamma_{3,3,6} = \mathbb{Q}\left(\sqrt{\frac{2973-489\sqrt{37}}{2}}\right)$. This expression still looks rather cumbersome, and by writing a minimal polynomial for $\sqrt{\frac{2973-489\sqrt{37}}{2}}$ and using the "polredabs()" command in Pari, we can check that $\sqrt{-10 - 2\sqrt{37}}$ also generates this field, hence $k\Gamma_{3,3,6} = \mathbb{Q}\left(\sqrt{-10 - 2\sqrt{37}}\right)$.

In order to use Theorems 3.1 and 3.2 to compute the invariant quaternion algebra we need to express the quadratic form q with respect to an orthogonal basis $\{\mathbf{w}_1, \mathbf{w}_2, \mathbf{w}_3, \mathbf{w}_4\}$. The result is:

$$\begin{bmatrix} 2 & 0 & 0 & 0 \\ 0 & \frac{3}{2} & 0 & 0 \\ 0 & 0 & 21+3\sqrt{37} & 0 \\ 0 & 0 & 0 & \frac{-151-25\sqrt{37}}{12} \end{bmatrix}$$

Thus, a Hilbert symbol describing $A\Gamma_{3,3,6}$ is given by

$$\left(\frac{-q(\mathbf{w}_1)q(\mathbf{w}_2),\, -q(\mathbf{w}_1)q(\mathbf{w}_3)}{k\Gamma_{3,3,6}} \right) = \left(\frac{-3,\, -42-6\sqrt{37}}{\mathbb{Q}\left(\sqrt{-10-2\sqrt{37}} \right)} \right).$$

Without using the machinery described in Section 4 is difficult to interpret this Hilbert symbol. The invariant trace field $k\Gamma_{3,3,6}$ has two real places, and $A\Gamma_{3,3,6}$ is ramified at each of these two places. By attempting to solve the Hilbert Equation (see the appendix of this paper) we also observe that $A\Gamma_{3,3,6}$ is ramified over exactly two finite prime ideals in the ring of integers from $k\Gamma_{3,3,6}$. These prime ideals lie over the rational prime 3 and we denote them by $\mathcal{P}_3$ and $\mathcal{P}'_3$. Thus, according to Vignéras [Vi80], this finite collection of ramification data provides a complete invariant for

the isomorphism class of $A\Gamma_{3,3,6} \cong \left(\frac{-3,\, -42-6\sqrt{37}}{\mathbb{Q}\left(\sqrt{-10-2\sqrt{37}} \right)} \right).$

Using either Theorem 4.11 or Theorem 4.12, one can also deduce that $\Gamma_{3,3,6}$ is not arithmetic because the Gram matrix contains the element $-\sqrt{\frac{14+2\sqrt{37}}{2}}$, which is not an algebraic integer. However the other two hypotheses of each of these theorems are satisfied.

6   Description of the Program

We've written a collection of PARI/GP scripts that automate the procedure from Section 5 for (finite-volume) hyperbolic polyhedra. The scripts take as input two matrices: the matrix of face normals, whose rows are low-precision decimal approximations to the outward-pointing normal vectors (normalized in $\mathbb{H}^3$ as in the previous section); and the matrix of edge labels, a square matrix (n_{ij}) whose diagonal entries must be one, and whose off-diagonal terms n_{ij} describe the relation between the i-th and j-th faces: $n_{ij} = 0$ means the faces are non-adjacent and any other value means that they meet at a dihedral angle of π/n_{ij}. The matrix of edge labels actually determines the polyhedron uniquely (up to hyperbolic isometry); one can use [Ro07] to compute the approximate face normals from the edge labels.

In a typical session, after loading the input matrices, the user runs Newton's Method to obtain a higher-precision approximation for the face normals, and then can have the computer guess and verify exact values for the normal vectors and for the Gram matrix. With the exact Gram matrix, the user can request the invariant trace field (described by a primitive element) and the invariant quaternion algebra (described by a Hilbert symbol, or, after an additional command, by ramification data). There is also a function to test the polyhedron for arithmeticity — this too takes the Gram matrix as input.

The scripts are available at [ACMR]; the package includes a sample session and a user guide. The scripts, as mentioned before, are written in the high-level language GP, which helped in our effort to make the source code as readable as possible. In fact, the reader can consider the source code as executable statements of the theorems in the previous sections.

The worked example in Section 5 closely follows the algorithms used in our scripts, with two exceptions: in that example all of the algebraic numbers were cleanly expressed by radicals and the Coxeter symbol was very simple being a single closed loop, neither of which are typical. Thus, the main aspects of the algorithms not explained in that example are choosing an appropriate representation of general algebraic numbers, and choosing a systematic way of listing the non-zero cyclic products $a_{i_1 i_2} a_{i_2 i_3} \cdots a_{i_r i_1}$ from the Gram matrix.

We describe a given algebraic number α as a pair $(p(z), \tilde{\alpha})$, where $p(z)$ is the minimal polynomial for α over $\mathbb{Q}$ and $\tilde{\alpha}$ is a decimal approximation to α. This representation is rather common — it is described in the textbook on computational number theory by Cohen [Co93]. While this representation is not already available in PARI/GP [PARI05], it is easy to program an algebraic number package working within PARI/GP for this representation.

A disadvantage of this representation for algebraic numbers is that when performing arithmetic on algebraic numbers, it is usually necessary to find composita of the fields generated by each number. This is not only a programming difficulty, it is the slowest part of our program. In some cases, when we know that we will do arithmetic with a given set of numbers $\{\alpha_1, \ldots, \alpha_n\}$, we are able to speed up our calculations by computing a primitive element β for $\mathbb{Q}(\alpha_1, \ldots, \alpha_n)$ and re-expressing each α_i as an element of $\mathbb{Q}[z]/\langle p(z) \rangle$ where $p(z)$ is the minimal polynomial for β over $\mathbb{Q}$. (PARI/GP has a data type called "polmod" for this representation.) Once expressed in terms of a common field, algebraic computations in terms of these polmods are extremely fast in PARI/GP.

Of course we don't typically have *a priori* knowledge of the outward-pointing normal vectors or of the entries in the Gram matrix for our polyhedron P. However, decimal approximations can be obtained using [Ro07]. Just like in Section 5, we use the LLL algorithm [LLL82] to guess minimal polynomials for the algebraic numbers represented by these decimal approximations. Typically a rather high-precision approximation is needed (sometimes 100 digits of precision) to obtain a correct guess. In this case, we use Newton's Method and the high-precision capabilities of PARI/GP to improve the precision of the normal vectors obtained from [Ro07]. Correctness of the guess is verified once we use the guessed algebraic numbers for the outward-pointing normals to compute the Gram matrix and verify that each entry corresponding to a dihedral angle $\frac{\pi}{n}$ has the correct minimal polynomial for $-2\cos\left(\frac{\pi}{n}\right)$. Additionally we check that the diagonal entries are exactly 2. In some cases the guessed polynomials are not correct, but typically, by sufficiently increasing the number of digits of precision, reapplying Newton's Method, and guessing again, we arrive at correct guesses.

Note that the number of checks, i.e., equations, equals the number of variables. If P has n faces, there are $4n - 6$ variables: one for each coordinate of each outward-pointing normal, minus the six coordinates normalized to 0. Since we deal with compact Coxeter polyhedra and these necessarily have three faces meeting at each vertex, the number of edges is $3n - 6$; there is one equation for each of these, and there are n additional equations for the diagonal entries of the Gram matrix, giving a total of $4n - 6$ equations.

The guess and check philosophy is inspired by SNAP [GHN98], which also goes through the process of improving an initial approximation to the hyperbolic structure, guessing minimal polynomials, and verifying. Such techniques are also central to many areas of experimental mathematics in which the LLL algorithm is used to guess linear dependencies that are verified *a posteriori*.

However, there are two major differences between our program and SNAP. Instead of guessing shape parameters for a ideal tetrahedral decomposition of a manifold and checking that the guessed values satisfy the gluing equations, we guess exact outward normal vectors and verify that the guessed values have the right inner products. One could imagine using SNAP's approach for polyhedra, applying it to the orbifold $\mathbb{H}^3/\Gamma$, where Γ is the reflection group generated by P, however it is conceptually simpler to describe the face normals of P than to find a decomposition of $\mathbb{H}^3/\Gamma$ into ideal tetrahedra. On the other hand, our approach is computationally more difficult, since for a typical example like the dodecahedron we guess 48 algebraic numbers, while a typical manifold studied in SNAP is described by around 12 algebraic numbers.

The other difference is that instead of working explicitly with traces of group elements, we use Theorem 3.1 working with elements of the Gram matrix, which, as mentioned earlier in Section 3, is more efficient. This technique requires listing the non-zero cyclic products from the Gram matrix. In doing this, we try to avoid finding more of them than necessary to compute the field they generate. It is easier to discuss these products in terms of the Coxeter symbol for the polyhedron. Recall that the Coxeter symbol is a graph whose vertices are the faces of the polyhedron with edges between pairs of non-adjacent faces and also between pairs of adjacent non-perpendicular faces. Typically edges between non-adjacent faces are drawn dashed, and edges between adjacent ones meeting at an angle of π/n are labeled $n - 2$. A non-zero cyclic product in the Gram matrix corresponds to a closed path in the Coxeter symbol. An example Coxeter symbol appears in Figure 3.

Our method, then, is to list all the squares of the elements of the Gram matrix, one for each edge in the Coxeter symbol, and a set of cycles that form a basis for the $\mathbb{Z}/2\mathbb{Z}$-homology of the Coxeter symbol. These allow one to express any cyclic product in the Gram matrix as a product of a number of basic cycles and squares or inverses of squares of Gram matrix elements corresponding to edges traversed more than once.

To get such a basis one can take any spanning tree for the graph and then, for each non-tree edge, take the cycle formed by that edge and the unique tree-path connecting its endpoints. Finding a spanning tree for a graph is a classical computer science problem for which we use breadth-first search (see [CLR90]). The spanning tree is also used for the task of finding a basis for $M(P)$, as this requires finding paths in the Coxeter symbol from a fixed vertex to four others. Breadth-first search has the advantage of producing a short bushy tree which in turn gives short paths and cycles.

7 Many Computed Examples

In this section we present a number of examples including some commensurable pairs of groups. We also present some borderline cases of groups that are incommensurable, but have some of the invariants in common.

The most obvious way to specify the groups we deal with is to give their Coxeter symbols, both because it is standard in the literature and because our algorithm uses them. However, it would not be practical to do this due to the large number of dashed edges occurring in our examples. For instance, the Coxeter symbol of the right-angled dodecahedron has no solid edges but 36 dashed ones, making the symbol non-planar and hard to read. The reason so many Coxeter symbols appear in the literature is that they are especially useful for tetrahedra in high dimensions and those are planar.

For this reason, instead of Coxeter symbols, we draw the polyhedra and label each edge with dihedral angle π/n by the integer n, omitting the labels on edges with dihedral angle $\pi/2$.

7.1   Lambert Cubes

Recall from Section 5 that a Lambert cube is a compact polyhedron realizing the combinatorial type of a cube, with three disjoint non-coplanar edges chosen and assigned dihedral angles $\frac{\pi}{l}$, $\frac{\pi}{m}$, and $\frac{\pi}{n}$ with $l, m, n > 2$, and the remaining edges assigned dihedral angles $\frac{\pi}{2}$. Any reordering of (l, m, n) can be obtained by applying an appropriate (possibly orientation reversing) isometry, so when studying Lambert cubes it suffices to consider triples with $l \leq m \leq n$. In Table 7.1 we provide the invariant trace fields and the ramification data for the invariant quaternion algebras for Lambert cubes with small l, m, and n.

It is interesting to notice that many of the above computations are done by hand in [HLMA92], where the authors determine which of the "Borromean Orbifolds" are arithmetic by recognizing that they are 8-fold covers of appropriate Lambert cubes. (They call the Lambert cubes "pyritohedra".) Our results are consistent with theirs.

In this table and all that follow, we specify the invariant trace field $k\Gamma$ by a canonical minimal polynomial $p(z)$ for a primitive element of the field and, below this polynomial, a decimal approximation of the root that corresponds to this primitive element.

We specify the real ramification data for the invariant quaternion algebra $A\Gamma$ by a vector $\mathbf{v}$ whose length is the degree of $p(z)$. If the i-th root of $p(z)$ (with respect to Pari's internal root numbering scheme) is real, we place in the i-coordinate of $\mathbf{v}$ a 1 if $A\Gamma$ is ramified over corresponding completion of $k\Gamma$, otherwise we place a 0. If the i-th root of $p(z)$ is not real then we place a -1 in the i-coordinate of $\mathbf{v}$.

Below this vector $\mathbf{v}$ we provide an indication of whether $A\Gamma$ ramifies at a prime ideal $\mathcal{P}_p$ over a rational prime p. In the case that $A\Gamma$ ramifies at multiple prime ideals over the same p, we list multiple symbols $\mathcal{P}_p, \mathcal{P}'_p, \ldots$, etc. More detailed information about the generators of these prime ideals can be obtained in PARI's internal format using our scripts.

The column labeled "arith?" indicates whether the group is arithmetic. In the case that the group satisfies conditions (1) and (2) from Theorems 4.11 and 4.12 but has elements with non-integral traces, we place a star next to the indication that the group is not arithmetic. (Some authors refer to such groups as "pseudo-arithmetic".)

7.2   Truncated Cubes

Another very simple family of hyperbolic reflection groups is obtained by truncating a single vertex of a cube, assigning dihedral angles $\frac{\pi}{l}, \frac{\pi}{m}$, and $\frac{\pi}{n}$ to the edges entering the vertex that was truncated and $\frac{\pi}{2}$ dihedral angles at all of the remaining edges. Since the three edges entering the vertex that was truncated form a *prismatic 3-circuit*, Andreev's Theorem provides the

(l, m, n)	$k\Gamma$	disc	ramification	arith?
$(3, 3, 3)$	$x^4 - x^3 - x^2 - x + 1$ root: $-0.6513 - 0.7587i$	-507	$[1, 1, -1, -1]$ $\emptyset$	Yes
$(3, 3, 4)$	$x^2 + 1$ root: i	-4	$[-1, -1]$ $\emptyset$	No*
$(3, 3, 5)$	$x^8 - x^7 - 3x^6 + x^4 - 3x^2 - x + 1$ root: $0.7252 - 0.6885i$	102378125	$[1, 1, 1, 1,$ $-1, -1, -1, -1]$ $\emptyset$	No
$(3, 3, 6)$	$x^4 + 5x^2 - 3$ root: $-2.3540i$	-4107	$[1, 1, -1, -1]$ $\mathcal{P}_3, \mathcal{P}_3'$	No*
$(3, 4, 4)$	$x^4 - 2x^3 - 2x + 1$ root: $-0.3660 + 0.9306i$	-1728	$[1, 1, -1, -1]$ $\emptyset$	Yes
$(3, 4, 5)$	$x^8 - x^7 - 18x^6 - 18x^5 + 95x^4 + 218x^3 + 182x^2 + 71x + 11$ root: $-2.1422 + 1.1460i$	249761250000	$[1, 1, 1, 1,$ $-1, -1, -1, -1]$ $\emptyset$	No
$(3, 4, 6)$	$x^4 - x^3 - 11x^2 + 33x - 6$ root: $2.3860 - 1.4419i$	-191844	$[1, 1, -1, -1]$ $\mathcal{P}_2, \mathcal{P}_2'$	No*
$(3, 5, 5)$	$x^8 - 3x^7 + x^5 + 3x^4 + x^3 - 3x + 1$ root: $-0.2252 + 0.9743i$	184280625	$[1, 1, 1, 1,$ $-1, -1, -1, -1]$ $\emptyset$	No
$(3, 5, 6)$	$x^8 - x^7 - 19x^6 + 30x^5 + 74x^4 - 72x^3 - 310x^2 + 413x - 71$ root: $-1.6938 - 1.2086i$	876653128125	$[1, 1, 1, 1,$ $-1, -1, -1, -1]$ $\emptyset$	No
$(3, 6, 6)$	$x^4 - x^3 - x^2 - x + 1$ root: $-0.6514 - 0.7587i$	-507	$[1, 1, -1, -1]$ $\emptyset$	Yes
$(4, 4, 4)$	$x^4 - x^2 - 1$ root: $0.7862i$	-400	$[1, 1, -1, -1]$ $\emptyset$	Yes
$(4, 4, 5)$	$x^8 - 4x^7 + 4x^6 - 6x^5 + 19x^4 - 14x^3 + 4x^2 - 6x + 1$ root: $-0.7486 + 1.4344i$	1548800000	$[1, 1, 1, 1,$ $-1, -1, -1, -1]$ $\emptyset$	No
$(4, 4, 6)$	$x^4 - 2x^3 - 6x + 9$ root: $-0.8229 + 1.5241i$	-9408	$[1, 1, -1, -1]$ $\mathcal{P}_3, \mathcal{P}_3'$	No*
$(4, 5, 5)$	$x^8 + 5x^6 - 3x^4 - 20x^2 + 16$ root: $1.6754i$	2714410000	$[1, 1, 1, 1,$ $-1, -1, -1, -1]$ $\mathcal{P}_2, \mathcal{P}_2'$	No
$(4, 5, 6)$	$x^8 - 3x^7 - 25x^6 + 95x^5 + 50x^4 - 3x^3 - 1751x^2 + 2600x - 995$ root: $-1.7137 - 1.9493i$	21059676450000	$[1, 1, 1, 1,$ $-1, -1, -1, -1]$ $\emptyset$	No
$(4, 6, 6)$	$x^4 - x^3 - x^2 + 7x - 2$ root: $1.2808 - 1.3861i$	-10404	$[1, 1, -1, -1]$ $\mathcal{P}_2, \mathcal{P}_2'$	No*
$(5, 5, 5)$	$x^4 - x^3 + x^2 - x + 1$ root: $0.8090 - 0.5878i$	125	$[-1, -1, -1, -1]$ $\emptyset$	No
$(5, 5, 6)$	$x^8 + 9x^6 + 13x^4 - 27x^2 + 9$ root: $2.0277i$	5863730625	$[1, 1, 1, 1,$ $-1, -1, -1, -1]$ $\mathcal{P}_3, \mathcal{P}_3', \mathcal{P}_5, \mathcal{P}_5'$	No
$(5, 6, 6)$	$x^8 - 2x^7 - 2x^6 + 16x^5 - 15x^4 - 19x^3 + 43x^2 - 22x + 1$ root: $1.3532 - 1.5790i$	10637578125	$[1, 1, 1, 1,$ $-1, -1, -1, -1]$ $\emptyset$	No
$(6, 6, 6)$	$x^4 - x^3 - 3x^2 - x + 1$ root: $-0.8956 - 0.4448i$	-1323	$[1, 1, -1, -1]$ $\emptyset$	Yes

Table 7.1. Commensurability invariants for Lambert Cubes.

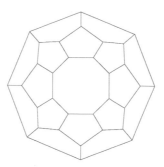

Figure 4. The Löbell polyhedron for $n = 8$.

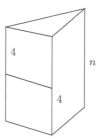

Figure 5. The "hexahedron" H_n. Edges labeled by an integer n are assigned dihedral angle $\frac{\pi}{n}$ and unlabeled edges are assigned $\frac{\pi}{2}$.

necessary and sufficient condition that $\frac{1}{l} + \frac{1}{m} + \frac{1}{n} < 1$ for the existence of such a polyhedron. In Table 7.2 we list the (l, m, n) truncated cubes for $l, m, n \leq 6$.

7.3   Löbell Polyhedra

For each $n \geq 5$, there is a radially-symmetric combinatorial polyhedron having two n-sided faces and having $2n$ faces with 5 sides, which provides a natural generalization of the dodecahedron. This combinatorial polyhedron is depicted in Figure 4 for $n = 8$.

Andreev's Theorem provides the existence of a compact right-angled polyhedron L_n realizing this abstract polyhedron because it contains no prismatic 3-circuits or prismatic 4-circuits.

An alternative construction of L_n is obtained by grouping $2n$ copies of the "hexahedron" shown in Figure 5 around the edge labeled n [Ve87]. This construction is shown for L_{10} in Figure 7. We denote this polyhedron by H_n, and note that, by construction, L_n and H_n are commensurable for each n.

(l, m, n)	$k\Gamma$	disc	ramification	arith?
$(3, 3, 4)$	$x^4 - 2x^3 + x^2 + 2x - 1$ root: $1.2071 - 0.9783i$	-448	$[1, 1, -1, -1]$ $\emptyset$	Yes
$(3, 3, 5)$	$x^4 - x^2 - 1$ root: $0.7862i$	-400	$[1, 1, -1, -1]$ $\emptyset$	Yes
$(3, 3, 6)$	$x^4 + 2x^2 - 11$ root: $2.1128i$	-6336	$[1, 1, -1, -1]$ $\emptyset$	No*
$(3, 4, 4)$	$x^2 + 2$ root: $1.4142i$	-8	$[-1, -1]$ $\mathcal{P}_3, \mathcal{P}_3'$	Yes
$(3, 4, 5)$	$x^8 + 14x^6 + 57x^4 + 86x^2 + $ 41 root: $1.3571i$	26869760000	$[-1, -1, -1, -1,$ $-1, -1, -1, -1]$ $\emptyset$	No
$(3, 4, 6)$	$x^4 + 12x^2 + 81$ root: $1.2247 - 2.7386i$	57600	$[-1, -1, -1, -1]$ $\emptyset$	No
$(3, 5, 5)$	$x^4 + 7x^2 + 11$ root: $-2.1490i$	4400	$[-1, -1, -1, -1]$ $\emptyset$	No
$(3, 5, 6)$	$x^8 - 4x^7 + 30x^6 - 64x^5 + $ $262x^4 - 384x^3 + 978x^2 - $ $684x + 1629$ root: $1.3660 - 1.7150i$	2734871040000	$[-1, -1, -1, -1,$ $-1, -1, -1, -1]$ $\emptyset$	No
$(3, 6, 6)$	$x^2 - x + 4$ root: $0.5000 + 1.9365i$	-15	$[-1, -1]$ $\mathcal{P}_2, \mathcal{P}_2'$	Yes
$(4, 4, 4)$	$x^4 - 2x^2 - 1$ root: $0.6436i$	-1024	$[1, 1, -1, -1]$ $\emptyset$	Yes
$(4, 4, 5)$	$x^4 + 3x^2 + 1$ root: $1.6180i$	400	$[-1, -1, -1, -1]$ $\emptyset$	No
$(4, 4, 6)$	$x^4 + 2x^2 - 2$ root: $1.6529i$	-4608	$[1, 1, -1, -1]$ $\emptyset$	No*
$(4, 5, 5)$	$x^8 - 2x^6 - 8x^5 - 5x^4 + $ $8x^3 + 12x^2 + 4x - 1$ root: $-0.7071 + 0.5442i$	368640000	$[1, 1, 1, 1,$ $-1, -1, -1, -1]$ $\emptyset$	No
$(4, 5, 6)$	$x^8 + 2x^6 - 39x^4 - 130x^2 - $ 95 root: $-1.0532i$	-5042995200000	$[1, 1, -1, -1,$ $-1, -1, -1, -1]$ $\emptyset$	No
$(4, 6, 6)$	$x^4 + 2x^2 + 4$ root: $0.7071 - 1.2247i$	576	$[-1, -1, -1, -1]$ $\emptyset$	No
$(5, 5, 5)$	$x^4 - 5$ root: $1.4953i$	-2000	$[1, 1, -1, -1]$ $\emptyset$	Yes
$(5, 5, 6)$	$x^8 + 6x^6 - 13x^4 - 66x^2 + $ 61 root: $2.4662i$	12648960000	$[1, 1, 1, 1,$ $-1, -1, -1, -1]$ $\emptyset$	No
$(5, 6, 6)$	$x^4 + 21x^2 + 99$ root: $2.6732i$	39600	$[-1, -1, -1, -1]$ $\emptyset$	No
$(6, 6, 6)$	$x^4 - 2x^3 - 2x + 1$ root: $-0.3660 + 0.9306i$	-1728	$[1, 1, -1, -1]$ $\emptyset$	Yes

Table 7.2. Commensurability invariants for some truncated cubes.

Of historical interest is that the first example of a closed hyperbolic manifold was constructed by Löbell [Lö31] in 1931 by an appropriate gluing of 8 copies of L_6. (See also [Ve87] for an exposition in English, and generalizations.) This gluing corresponds to constructing an index 8 subgroup of the reflection group in the faces of L_6, hence the Löbell manifold has the same commensurability invariants as those presented for L_6 in the table above. It is also true that the Löbell manifold is arithmetic, because this underlying reflection group is arithmetic. Arithmeticity of the classical Löbell manifold was previously observed by Andrei Vesnin, but remained unpublished [Ve].

Furthermore, Vesnin observed in [Ve91] that the if the Löbell polyhedron L_n is arithmetic, then $n = 5, 6, 7, 8, 10, 12$, or 18. He shows that the reflection group generated by L_n contains a $(2, 4, n)$ triangle group which must be arithmetic if L_n is arithmetic, and applies the classification of arithmetic triangle groups by Takeuchi [Ta77]. In combination with our computations, Vesnin's observation yields:

Theorem 7.1. *The Löbell polyhedron L_n is arithmetic if and only if $n = 5, 6,$ or 8.*

Table 7.3 contains data for the first few Löbell polyhedra L_n and, consequently, the first few "hexahedra" H_n.

It is interesting to notice that L_7 and L_{14} have isomorphic invariant trace fields, which are actually not the same field (one can check that the specified roots generate different fields). This alone suffices to show that L_7 and L_{14} are incommensurable. Further, albeit unnecessary, justification is provided by the fact that their invariant quaternion algebras are not isomorphic, since $A\Gamma(L_7)$ has no finite ramification, whereas $A\Gamma(L_{14})$ is ramified at the two finite prime ideals $\mathcal{P}_7, \mathcal{P}_7'$ lying over the rational prime 7.

7.4   Doubly-Truncated Prisms

For $q = 4$ and 5, there exist compact polyhedra realizing two doubly-truncated prisms pictured in Figure 6. We will denote by Γ_1 and Γ_2 the reflections groups generated by the two polyhedra in Figure 6, numbered from left to right.

When $q = 4$, the invariant trace fields are equal, both equal to $\mathbb{Q}(a)$, with a is an imaginary fourth root of 2, and $A\Gamma_1 = A\Gamma_2 \cong \left(\frac{-1, -1}{\mathbb{Q}(a)}\right)$. Indeed, one can show this using the fact that Γ_1 and Γ_2 contain subgroups isomorphic to A_4, see the comment at the end of Section 3. However, Γ_1 and Γ_2 are incommensurable since Γ_1 is arithmetic, while Γ_2 has non-integral traces.

If we repeat the calculation with $q = 5$ both groups have invariant trace field generated by an imaginary fourth root of 20 and have isomor-

L_n	$k\Gamma$	disc	ramification	arith?
L_5	$x^4 - x^2 - 1$ root: $0.7862i$	-400	$[1, 1, -1, -1]$ $\emptyset$	Yes
L_6	$x^2 + 2$ root: $1.4142i$	-8	$[-1, -1]$ $\mathcal{P}_3, \mathcal{P}_3'$	Yes
L_7	$x^6 + x^4 - 2x^2 - 1$ root: $-0.6671i$	153664	$[1, 1, -1, -1,$ $-1, -1]$ $\emptyset$	No
L_8	$x^4 - 2x^2 - 1$ root: $0.6436i$	-1024	$[1, 1, -1, -1]$ $\emptyset$	Yes
L_9	$x^6 + 3x^4 - 3$ root: $1.5913i$	1259712	$[1, 1, -1, -1,$ $-1, -1]$ $\emptyset$	No
L_{10}	$x^4 + 3x^2 + 1$ root: $0.6180i$	400	$[-1, -1, -1, -1]$ $\emptyset$	No
L_{11}	$x^{10} + 4x^8 + 2x^6 - 5x^4 -$ $2x^2 + 1$ root: $1.6378i$	-219503494144	$[1, 1, 1, 1, -1, -1,$ $-1, -1, -1, -1]$ $\emptyset$	No
L_{12}	$x^4 + 2x^2 - 2$ root: $1.6529i$	-4608	$[1, 1, -1, -1]$ $\emptyset$	No*
L_{13}	$x^{12} + 5x^{10} + 5x^8 - 6x^6 -$ $7x^4 + 2x^2 + 1$ root: $1.6646i$	564668382613504	not computed (long computation)	No
L_{14}	$x^6 + x^4 - 2x^2 - 1$ root: $1.3424i$	153664	$[1, 1, -1, -1,$ $-1, -1]$ $\mathcal{P}_7, \mathcal{P}_7'$	No
L_{15}	$x^8 + 5x^6 + 5x^4 - 5x^2 - 5$ root: $1.6814i$	-1620000000	$[1, 1, -1, -1,$ $-1, -1, -1, -1]$ $\emptyset$	No
L_{16}	$x^8 + 4x^6 + 2x^4 - 4x^2 - 1$ root: $1.6875i$	-1073741824	$[1, 1, -1, -1,$ $-1, -1, -1, -1]$ $\emptyset$	No
L_{17}	not computed (long computation.)			
L_{18}	$x^6 - 3x^2 - 1$ root: $0.5893i$	419904	$[1, 1, -1, -1,$ $-1, -1]$ $\mathcal{P}_3, \mathcal{P}_3'$	No

Table 7.3. Commensurability invariants for the Löbell polyhedra. Note that L_5 is the right-angled regular dodecahedron.

phic invariant quaternion algebras. However, both groups have non-integral traces, so neither is arithmetic. Thus, we cannot determine whether these two groups are commensurable, or not. (See Subsection 7.6, below, for similar examples of non-arithmetic polyhedra with matching pairs $(k\Gamma, A\Gamma)$.)

There is a good reason why these pairs have the same invariant trace field: The invariant trace field for an amalgamated product of two Kleinian groups is the compositum of the corresponding invariant trace fields. See Theorem 5.6.1 from [MR03]. The group on the left can be expressed as an

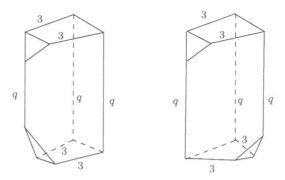

Figure 6. Edges labeled by an integer n are assigned dihedral angle $\frac{\pi}{n}$ and unlabeled edges are assigned dihedral angle $\frac{\pi}{2}$. (In particular, the triangular faces are at right angles to each adjacent face.)

amalgamated product obtained by gluing the "top half," a singly truncated prism to the "bottom half," another singly truncated prism (that is congruent to the top half) along a (q, q, q)-triangle group. The group on the right can be expressed in the same way, just with a different gluing along the (q, q, q) triangle group. This construction is similar to the construction of "mutant knots," for which commensurability questions are also delicate. See page 190 of [MR03].

7.5 Commensurable Pairs

As noted in Section 3, one way to find pairs of commensurable groups is to verify that two groups are arithmetic, have the same invariant trace field, and have isomorphic invariant quaternion algebras. In this section, we describe the commensurability classes of arithmetic reflection groups in which we have found more than one group. It is very interesting to notice that Ian Agol has proven that there are a finite number of commensurability classes of arithmetic reflection groups in dimension 3, [Ag06], however there is no explicit bound.

The reader may want to refer to the Arithmetic Zoo section of [MR03] to find other Kleinian groups within the same commensurability classes.

It is easy to see that the dodecahedral reflection group is an index 120 and 60 subgroup of the reflection groups in the tetrahedra T_2 and T_4, respectively. However, for some of these commensurable pairs the commensurability is difficult to "see" directly by finding a bigger polyhedron Q that is tiled by reflections of each of the polyhedra within the commensurability class.

For example, the authors first noticed the commensurability of the right-angled dodecahedron L_5 with the $(4, 4, 4)$-Lambert cube using our

$k\Gamma$	disc	finite ramification
$x^2 + 2$, root: $1.4142i$	-8	$\mathcal{P}_3, \mathcal{P}_3'$
The $(3, 4, 4)$ truncated cube		vol ≈ 1.0038
The Löbell 6 polyhedron		vol $\approx 6 \cdot 1.0038$
$x^4 - x^2 - 1$, root: $0.7861i$	-400	$\emptyset$
Compact tetrahedron T_2 from p. 416 of [MR03]		vol ≈ 0.03588
Compact tetrahedron T_4 from p. 416 of [MR03]		vol $\approx 2 \cdot 0.03588$
The $(4, 4, 4)$ Lambert cube		vol $\approx 15 \cdot 0.03588$
The $(3, 3, 5)$ truncated cube		vol $\approx 26 \cdot 0.03588$
The Löbell 5 polyhedron (dodecahedron)		$\approx 120 \cdot 0.03588$
$x^4 - x^3 - x^2 - x + 1$, root: $-0.6514 - 0.7587i$	-507	$\emptyset$
The $(3, 3, 3)$ Lambert cube		vol ≈ 0.324423449
The $(3, 6, 6)$ Lambert cube		vol $\approx \frac{5}{3} \cdot 0.324423449$
$x^4 - 2x^2 - 1$, root: $0.6436i$	-1024	$\emptyset$
The $(4, 4, 4)$ truncated cube		vol ≈ 1.1273816
The Löbell 8 polyhedron		vol $\approx 8 \cdot 1.1273816$
$x^4 - 2x^3 - 2x + 1$, root: $-0.3660 + 0.9306i$	-1728	$\emptyset$
The $(3, 4, 4)$ Lambert cube		vol ≈ 0.4506583058
The $(6, 6, 6)$ truncated cube		vol $\approx 6 \cdot 0.4506583058$

Table 7.4. Commensurability classes of arithmetic reflection groups.

computations and only later did the authors find the explicit tiling of the dodecahedron by 8 copies of the $(4, 4, 4)$-Lambert cube shown in Figure 1. (In fact, the commensurability has been observed elsewhere via the connection with the Borromean Rings Orbifold, but to our knowledge we are the first to draw the picture.)

An unexpected (and previously unknown) pair of commensurable polyhedra is the $(3, 3, 5)$-truncated cube and the dodecahedron. We leave it as a challenge to the reader to exhibit this commensurability via tiling. (At least 13 copies of the dodecahedron are required!)

For $n = 3$ and 4 we see the commensurability between the Löbell $2n$ polyhedron and the $(4, 4, n)$ truncated cube. This commensurability is a well-known fact for any n: one can group $2n$ of the $(4, 4, n)$ truncated cubes around the edge with label n forming the right-angled $2n$ Löbell polyhedron. This construction is shown for $n = 5$ in Figure 7.

The approximate volumes were computed using Damian Heard's program Orb [He06]. (Orb provides twice the volume of the polyhedron, which is the volume of the orbifold obtained by gluing the polyhedron to its mirror image along its boundary.)

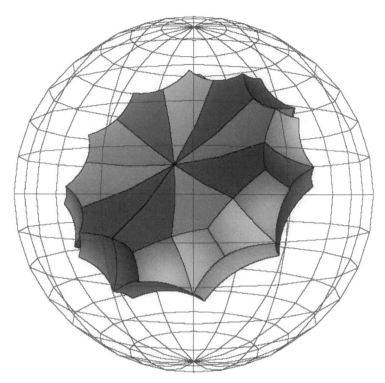

Figure 7. Tiling the Löbell 10 polyhedron with 10 copies of the $(4, 4, 5)$ truncated cube.

7.6   Pairs Not Distinguished by $k\Gamma$ and $A\Gamma$

We found three pairs of non-arithmetic polyhedral reflection groups which are indistinguishable by the invariant trace field and the invariant quaternion algebra. The first of these is the pair of doubly truncated prisms with $q = 5$, from Subsection 7.4. Each of these has the same volume, approximately 2.73952694 since the second can be obtained from the first by cutting in a horizontal plane and applying a $\frac{1}{3}$ twist. We leave it as an open question to the reader whether the two polyhedra in the first pair are commensurable.

The second and third pairs are the $(4, 4, 5)$ truncated cube and the Löbell 10 polyhedron and the $(4, 4, 6)$ truncated cube and the Löbell 12 polyhedron, each of which fit into the general commensurability between the $(4, 4, n)$ truncated cube and the Löbell $2n$ polyhedron, as described in the previous subsection.

8   Appendix: Computing Finite Ramification of Quaternion Algebras

Recall from Section 4 that to determine whether two quaternion algebras over a number field F are isomorphic it suffices to compare their ramification over all real and finite places of F.

To compute the finite ramification of a quaternion algebra $A \cong \left(\frac{a,b}{F}\right)$ we first recall that there are a finite number of candidate primes over which A can ramify: A necessarily splits (is unramified) over any prime ideal not dividing the ideal $\langle 2ab \rangle$. To check those primes $\mathcal{P}$ dividing $\langle 2ab \rangle$, we apply Theorem 4.2 which states that A splits over $\mathcal{P}$ if and only if there is a solution to the Hilbert Equation $aX^2 + bY^2 = 1$ in the completion F_ν. Here ν is the valuation given by $\nu(x) = c^{n_\mathcal{P}(x)}$ on R_F, see Section 4.

Because it is difficult to do computer calculations in the completion F_ν, where elements are described by infinite sequences of elements of F, we ultimately want to reduce our calculations to be entirely within R_F. Hensel's Lemma, see [La94], is the standard machinery for this reduction. Happily, for the problem at hand, the proper use of Hensel's Lemma has previously been worked out, see [CGHN00] (whose authors write $\nu_\mathcal{P}$ instead of our $n_\mathcal{P}$ and $|\cdot|$ instead of our ν). We use the same techniques they do but without many of the optimizations, which we find unnecessary for our program (this is not the bottleneck in our code). The following theorem provides the necessary reduction:

Theorem 8.1. *Let $\mathcal{P}$ be a prime ideal in R_F, let $a, b \in R_F$ be such that $n_\mathcal{P}(a), n_\mathcal{P}(b) \in \{0, 1\}$, and define an integer m as follows: if $\mathcal{P} | 2$, $m = 2n_\mathcal{P}(2) + 3$ and if $\mathcal{P} \nmid 2$ then $m = 1$ if $n_\mathcal{P}(a) = n_\mathcal{P}(b) = 0$ and $m = 3$ otherwise.*

Let S be a finite set of representatives for the ring $R_F / \mathcal{P}^m$. The Hilbert Equation

$$aX^2 + bY^2 = 1$$

has a solution with X and $Y \in F_\nu$ if and only if there exist elements X', Y', and $Z' \in S$ such that

$$\nu(aX'^2 + bY'^2 - Z'^2) \leq c^m$$

and $\max\{\nu(X'), \nu(Y'), \nu(Z')\} = 1$.

Recall that $0 < c < 1$ is the arbitrary constant that appears in the definition of ν. Also note that the condition that $n_\mathcal{P}(a), n_\mathcal{P}(b) \in \{0, 1\}$ is no real restriction since one can always divide the elements of F appearing in the Hilbert symbol by any squares in the field without changing A.

Our Theorem 8.1 is a minor extension of Proposition 4.9 of [CGHN00]. Their proposition only applies to dyadic primes, corresponding to $m = 2n_{\mathcal{P}}(2) + 3$. The other two cases can be proved analogously using, instead of their Lemma 4.8, the following statement:

Lemma 8.2. *Suppose that ν is the valuation corresponding to a non-dyadic prime $\mathcal{P}$. Let X, X' be in F_ν and suppose that $\nu(X) \leq 1$ and $\nu(X - X') \leq c^k$ for some non-negative integer k. Then, $\nu(X^2 - X'^2) \leq c^{2k}$.*

The reason why we make this minor extension of Proposition 4.9 from [CGHN00] is that we do not implement the optimizations for non-dyadic primes that appear in SNAP, instead choosing to use the Hilbert Equation in all cases.

Thus, the problem of determining finite ramification of A reduces to finding solutions in R_F to the Hilbert inequality from Theorem 8.1. Our program closely follows SNAP [GHN98], solving the equation by exploring, in depth-first order [CLR90], a tree whose vertices are quadruples (X, Y, Z, n), where $\nu(aX^2 + bY^2 - Z^2) \leq c^n$. The children of the vertex (X, Y, Z, n) are the vertices $(X_1, Y_1, Z_1, n + 1)$ with $X_1 \equiv X, Y_1 \equiv Y$, and $Z_1 \equiv Z \pmod{\mathcal{P}^n}$. Fortunately, the condition $\max\{\nu(X), \nu(Y), \nu(Z)\} = 1$ means that the search space is considerably reduced because there must be a solution to the inequality with one of X, Y, Z equal to 1. Indeed, if $\nu(Y) = 1$ then Y is invertible modulo any power of $\mathcal{P}$ and $(XY^{-1}, 1, ZY^{-1})$ is also a solution. So, our program only searches the three sub-trees given by fixing X, Y, or Z at 1. In the case that we search all three trees up to level $n = m$ unsuccessfully, then Theorem 8.1 guarantees that there is no solution to the Hilbert Equation, and consequently A ramifies over $\mathcal{P}$.

Bibliography

[Ag06]   I. Agol, *Finiteness of arithmetic Kleinian reflection groups*, International Congress of Mathematicians, vol. II, Eur. Math. Soc., Zürich (2006), 951–960.

[An70]   E. M. Andreev, *On convex polyhedra in Lobacevskii spaces*, English Translation, Math. USSR Sbornik **10** (1970), 413–440.

[ACMR]   O. Antolín-Camarena, G. R. Maloney, and R. K. W. Roeder, *SNAP-HEDRON: a computer program for computing arithmetic invariants of polyhedral reflection groups*, http://www. math.toronto.edu/~rroeder/SNAP-HEDRON.

[BP01] M. Boileau and J. Porti, *Geometrization of 3-orbifolds of cyclic type*, Astérisque **272** (2001), Appendix A by M. Heusener and J. Porti.

[BS96] P. Bowers and K. Stephenson, *A branched Andreev-Thurston theorem for circle packings of the sphere*, Proc. London Math. Soc. (3) **73** 1 (1996), 185–215.

[Co93] H. Cohen, *A course in computational algebraic number theory*, Graduate Texts in Mathematics **138**, Springer-Verlag, Berlin, 1993.

[CLR90] T. H. Cormen, C. E. Leiserson, and R. L. Rivest, *Introduction to algorithms*, The MIT Electrical Engineering and Computer Science Series, MIT Press, Cambridge, MA, 1990.

[CGHN00] D. Coulson, O. A. Goodman, C. D. Hodgson, and W. D. Neumann, *Computing arithmetic invariants of 3-manifolds*, Experiment. Math. **9** 1 (2000), 127–152.

[Ge] Geomview, *www.geomview.org*.

[GHN98] O. A. Goodman, C. D. Hodgson, and W. D. Neumann, *Snap home page*, 1998, http://www.ms.unimelb.edu.au/~snap. Includes source distribution and extensive tables of results of Snap computations.

[He06] D. Heard, *Orb*, 2006, http://www.ms.unimelb.edu.au/~snap/orb. html.

[HW89] M. Hildebrand and J. Weeks, *A computer generated census of cusped hyperbolic 3-manifolds*, in: Computers and Mathematics, Springer, New York, 1989, 53–59.

[HLMA92] H. M. Hilden, M. T. Lozano, and J. M. Montesinos-Amilibia, *On the Borromean orbifolds: Geometry and arithmetic*, in: Topology '90, Ohio State Univ. Math. Res. Inst. Publ., vol. 1, de Gruyter, Berlin, 1992, 133–167.

[Ho92] C. D. Hodgson, *Deduction of Andreev's theorem from Rivin's characterization of convex hyperbolic polyhedra*, in: Topology '90, Ohio State Univ. Math. Res. Inst. Publ., vol. 1, de Gruyter, Berlin, 1992, 185–193.

[HW] C. Hodgson and J. Weeks, *A census of closed hyperbolic 3-manifolds*, ftp://www.geometrygames.org/priv/weeks/SnapPea/ SnapPeaCensus/ClosedCens%us/.

[LLL82]   A. K. Lenstra, H. W. Lenstra, Jr., and L. Lovász, *Factoring poly-nomials with rational coefficients*, Math. Ann. **261** 4 (1982), 515–534.

[Lö31]   F. Löbell, *Beispiele geschlossener dreidimensionaler Clifford-Kleinscher Räume negativer Krümmung*, Ber. Sächs. Akad. Wiss. **83** (1931), 168–174.

[La94]   S. Lang, *Algebraic number theory*, Second Ed., Graduate Texts in Mathematics **110**, Springer-Verlag, New York, 1994.

[MR98]   C. Maclachlan and A. W. Reid, *Invariant trace-fields and quaternion algebras of polyhedral groups*, J. London Math. Soc. (2) **58** 3 (1998), 709–722.

[MR03]   C. Maclachlan and A. W. Reid, *The arithmetic of hyperbolic 3-manifolds*, Graduate Texts in Mathematics **219**, Springer-Verlag, New York, 2003.

[NR92]   W. D. Neumann and A. W. Reid, *Arithmetic of hyperbolic manifolds*, in: Topology '90, Ohio State Univ. Math. Res. Inst. Publ., vol. 1, de Gruyter, Berlin, 1992, 273–310.

[PARI05] The PARI Group, *PARI/GP, version* 2.1.7, Bordeaux, 2005, Available at http://pari.math.u-bordeaux.fr/.

[RH93]   I. Rivin and C. D. Hodgson, *A characterization of compact convex polyhedra in hyperbolic 3-space*, Invent. Math. **111** (1993), 77–111.

[Ro07]   R. K. W. Roeder, *Constructing hyperbolic polyhedra using Newton's method*, Experiment. Math. **16** (2007), 463–492.

[RHD07]   R. K. W. Roeder, J. H. Hubbard, and W. D. Dunbar, *Andreev's theorem on hyperbolic polyhedra*, Ann. Inst. Fourier **57** 3 (2007), 825–882.

[Ta77]   K. Takeuchi, *Arithmetic triangle groups*, J. Math. Soc. Japan **29** 1 (1977), 91–106.

[Ve87]   A. Vesnin, *Three-dimensional hyperbolic manifolds of Löbell type*, Siberian Math. J **28** 5 (1987), 731–733.

[Ve91]   A. Vesnin, *Three-dimensional hyperbolic manifolds with a common fundamental polyhedron*, Translation, Math. Notes **49** 5–6 (1991), 575–577.

[Ve]   A. Vesnin, Personal communication, 2007.

[Vi80]   M. F. Vignéras, *Arithmétique des algèbres de quaternions*, Lecture Notes in Mathematics **800**, Springer, Berlin, 1980.

[Vi67]   È. B. Vinberg, *Discrete groups generated by reflections in Lobačevskiĭ spaces*, Mat. Sb. (N.S.) **72 (114)** (1967), 471–488; Correction, Ibid. **73 (115)** (1967), 303.

[VS93]   È. B. Vinberg and O. V. Shvartsman, *Discrete groups of motions of spaces of constant curvature*, Geometry, II, Encyclopaedia Math. Sci. **29**, Springer, Berlin, 1993, 139–248.

Contributors

Omar Antolín-Camarena, Department of Mathematics, University of Toronto, Toronto, Ontario, Canada M5S 2E4, `oantolin@math.utoronto.ca`

Eric Bedford, Department of Mathematics, Indiana University, Bloomington, IN 47405, USA, `bedford@indiana.edu`

Paul Blanchard, Department of Mathematics, Boston University, 111 Cummington Street, Boston, MA 02215, USA, `paul@bu.edu`

Alexander Blokh, Department of Mathematics, University of Alabama at Birmingham, Birmingham, AL 35294-1170, USA, `ablokh@math.uab.edu`

Xavier Buff, Institut de Mathématiques de Toulouse, Université Paul Sabatier, 118 Route de Narbonne, 31062 Toulouse Cedex 4, France, `buff@picard.ups-tlse.fr`

Arnaud Chéritat, Institut de Mathématiques de Toulouse, Université Paul Sabatier, 118 Route de Narbonne, 31062 Toulouse Cedex 4, France, `cheritat@picard.ups-tlse.fr`

Robert L. Devaney, Department of Mathematics, Boston University, 111 Cummington Street, Boston, MA 02215, USA, `bob@bu.edu`

Jeffrey Diller, Department of Mathematics, University of Notre Dame, Notre Dame, IN 46556, USA, `diller.1@nd.edu`

Romain Dujardin, Institut de Mathématiques de Jussieu et UFR de mathématiques, Université Paris Diderot (Paris 7), 175 rue du Chevaleret, 75013 Paris, France, `dujardin@math.jussieu.fr`

Adam Epstein, Mathematics Institute, University of Warwick, Coventry CV47AL, UK, `A.L.Epstein@warwick.ac.uk`

Núria Fagella, Departament de Matemàtica Aplicada i Anàlisi, Universitat de Barcelona, Gran Via 585, 08007 Barcelona, Spain, `fagella@maia.ub.es`

Antonio Garijo, Dep. d'Eng. Informàtica i Matemàtiques, Universitat Rovira i Virgili, Av. Països Catalans 26, 43007 Tarragona, Spain, antonio.garijo@urv.cat

Christian Henriksen, Department of Mathematics, Technical University of Denmark, Matematiktorvet, Building 303 S, 2800 Kongens Lyngby, Denmark, christian.henriksen@mat.dtu.dk

Jeremy Kahn, Institute for Mathematical Sciences, Stony Brook University, NY 11794, USA, kahn@math.sunysb.edu

Sarah Koch, Department of Mathematics, Malott Hall, Cornell University, Ithaca, NY 14853, USA, kochs@math.cornell.edu

Mikhail Lyubich, Institute for Mathematical Sciences, Stony Brook University, NY 11794, USA, mlyubich@math.sunysb.edu

Gregory Maloney, Department of Mathematics, University of Toronto, Toronto, Ontario, Canada M5S 2E4, maloneyg@math.utoronto.ca

Sebastian Marotta, Department of Mathematics, Boston University, 111 Cummington Street, Boston, MA 02215, USA, smarotta@bu.edu

John Milnor, Institute for Mathematical Sciences, Stony Brook University, NY 11794, USA, jack@math.sunysb.edu

Volodymyr Nekrashevych, Department of Mathematics, Texas A&M University, College Station, TX 77843-3368, USA, nekrash@math.tamu.edu

Lex Oversteegen, Department of Mathematics, University of Alabama at Birmingham, Birmingham, AL 35294-1170, USA, overstee@math.uab.edu

Carsten Lunde Petersen, IMFUFA, NSM, Roskilde University, Postbox 260, DK-4000 Roskilde, Denmark, lunde@ruc.dk

Kevin M. Pilgrim, Department of Mathematics, Indiana University, Bloomington, IN 47405 USA, pilgrim@indiana.edu

Roland Roeder, Department of Mathematics, University of Toronto, Toronto, Ontario, Canada M5S 2E4, rroeder@math.utoronto.ca

Elizabeth Russell, Department of Mathematics, Boston University, 111 Cummington Street, Boston, MA 02215, USA, erussell@bu.edu

Dierk Schleicher, Jacobs University Bremen, Research I, Postfach 750 561, 28725 Bremen, Germany, `dierk@jacobs-university.de`

Nikita Selinger, Jacobs University Bremen, Research I, Postfach 750 561, 28725 Bremen, Germany, `n.selinger@jacobs-university.de`

Mitsuhiro Shishikura, Department of Mathematics, Faculty of Science, Kyoto University, Kyoto 606-8502, Japan, `mitsu@math.kyoto-u.ac.jp`

Tan Lei, Unité CNRS-UMR 8088, Département de Mathématiques, Université de Cergy-Pontoise, 2 Av. A. Chauvin, 95302 Cergy-Pontoise, `tanlei@math.u-cergy.fr`

William Thurston, Department of Mathematics, Malott Hall, Cornell University, Ithaca, NY 14853, USA, `wpl@math.cornell.edu`

Yin Yongcheng, School of Mathematical Sciences, Fudan University, Shanghai, 200433, P. R. China, `yin@zju.edu.cn`

Jean-Christophe Yoccoz, Collège de France, Département de Mathématiques, 3, rue d'Ulm, 75231 Paris Cedex 05, France, `jean-c.yoccoz@college-de-france.fr`

T - #0294 - 071024 - C8 - 229/152/29 - PB - 9780367384944 - Gloss Lamination